**Math Finance Law 18**, *(Math Fin Law 18)*, Public Listed Firm Rule No.56125-58942

# Math Finance Law 18

## Mathematical Financial Law

## Public Listed Firm Rule No.56125-589452

by

# Steve Asikin

O
Steve Asikin ISBN 13: **978-1564912361**, ISBN 10: **1564912363**

# MATH FINANCE LAW 18
Mathematical Financial Laws
Public Listed Firm Rule No. 56125-58942
## By Steve Asikin

Creative Team:Steve Asikin
         & BMC Research (Non Profit). New York

An Amazon Paperback
First Published in United States 2017
By Amazon Createspace
This 1st edition published in 2017
AMZN, PO. Box 81226, Seattle
WA 98108-1226, United States

ISBN-13: 978-1546912361
ISBN-10: 1546912363

`

1

## PREFACE

Math Finance Law Series are truly an Artwork of US (American) Culture for the Benefit of the World (every Individual, depends on each understanding and power to implement the ideas). *They are Not possible to be results of any other culture!*

These Pacioli's Du Pont-Limlingan constructions are accurate maths with very powerful visual Finance Architecture Board that makes corporate meeting much easier and much more useful. Now we could coordinate Sales-Marketing, Production, Human Resources, Operation and Finance altogether, in a productive meeting that thousands of people could understand it quickly.

(1)Each page was set as beautiful Art of <u>American Caligraphy</u>, that now starts competing to common Arabic or Kanji visual art.
(2)More beautiful than that, are its <u>Word Meanings</u> as nice Philosophical Poems for those who knows & dream by Senses.
(3)Even beautier than that, is their <u>Math Formulations</u>, that add precision & accuracy to the great ideas combining Arts & Logic.
(4)Better than Math, are its B/S-I/S <u>Finance Accounting</u> marvel of IFRS-GAAP optimization of Money Matters impacting most.
(5)More <u>Construction Technology</u> benefit can be Engineered for those who are good both in Math and Accountancy as well.
(6)Better than both Math & Accounting, is its <u>Finance Control & Assurance</u>, to realize its optimization of Healthy Corporations.
(7)On top of them all, its <u>Regulating Power</u> could make many companies practicing those all at simultaneous time, for better Finance Security and bring Wealth to all persons in the world!

Let us now start making Math experts help us solving Finance Problems (which are more realistic, urgent & needed by all).

2

Steve Asikin ISBN 13: **978-1564912361**, ISBN 10: **1564912363**

**Math Finance Law 18**, *(Math Fin Law 18)*, Public Listed Firm Rule No.56125-58942

## TABLE OF CONTENTS

`

Steve Asikin ISBN 13: **978-1564912361**, ISBN 10: **1564912363**

# INTRODUCTION:
## The Math Fin Law Terminologies

Income Statement

| S | V | | | | | |
|---|---|---|---|---|---|---|
| | M | F | | | | |
| | | O | I | | | |
| | | | B | T | | |
| | | | | A | D | Y |
| | | | | | | H |
| | | | | R | | |

Balance Sheet

| K | C | W | L | X | G |
|---|---|---|---|---|---|
| J | | | | | Z |
| P | | | | Q | |
| | N | | E | U | |
| | | | | R | |

Please carefully look that all of 26 characters (**A** to **Z**) used once, except the **R**, that in Accountancy must be appeared in both diagrams at same identical value, with glossary (of well known abbreviations) in the next page. Only 2 (two) of them are historical as **U=E'** and **S'** means carried forward legal numbers from previous year.

The 13 (thirteen) common measures (**c**, **d**, **f**, **g**, **i**, **j**, **l**, **p**, **q**, **s**, **t**, **v**, and **y**), are put up front as desired criteria along the way, so then later analyst could be surprised.

In Balance Sheet, we found that **W= C+N= K+J+P+N= L+E= X+Q+U+R= G+Z+Q+U+R** as well, so then we put **W** in the middle of its T-Account balance.

In Income Statement, same applied as **S= V+F+I +T+Y+H+R= V+M= V+F+ O= V+F+I+B= V+F+I+T+A= V+F+I+B= V+F+I+T+D+R**.

Steve Asikin ISBN 13: **978-1564912361**, ISBN 10: **1564912363**

**A**= After Tax Income;

**B**= Before Tax Income,

**C**= Current Assets, **c**= C/X= Current Ratio,

**D**= Dividend Paid, **d**= D/A=Dividend Portion or Payout,

**E**= Equity or Capital, **E'**= Equity of Past Year,

**F**= Fixed Cost, **f**= F/S=Fixed Portion,

**G**= Goods or Trade Payables,

     **g**= 360G/V= Goods Payable Days,

**H**= Home Taken Dividend,

**I**= Interest Expense, **i**= I/S= Interest Portion,

**J**= Jobs or Account Receivable,

     **j**= 360J/S= Jobs or Trade Account ReceivableDays,

**K**= Kind of Cash,

**L**= Liabilities or Debt, **l**= L/E= Leverage or Gearing Ratio,

**M**= Margin of Contribution,

**N**= Non Current Assets,

**O**= Operational Surplus,

**P**= Procured Inventories,

     **p**= 360P/V= Procured Inventory Days,

**Q**= Quoted Longterm Liabilities,

     **q**= [C-P]/X= Quick or Acid Test Ratio,

**R**= Retained Earnings,

**S**= Sales or Revenues, **S'**= Sales of Past Year,

     **s**= [S/S']-1= Sales Growth,

**T**= Tax Paid, **t**= T/B= Tax Rate,

**U**= Utilized or Starting Capital,

**V**= Variable Cost, **v**= V/S= Variable Portion,

**W**= Wealth or Total Assets,

**X**= Xpress or Current Debt,

**Y**= Yielding Tax of Dividend,

     **y**= Y/D= Yielding Tax Rate of Dividend,

**Z**= Zero Trade Current Debt.

Steve Asikin ISBN 13: **978-1564912361**, ISBN 10: **1564912363**

# CHAPTER-25:

## N= Non Current Asset

## Continued from Book 1-10 and 11-17:

Rule-56125:

If both ($I$), ($U$), ($S'$), ($s$), ($V$), ($f$), ($i$), ($t$), ($D$), ($K$), ($J$) and ($P$) are known, then its Non Current Asset planned is:
planned is:

$$N = [1+I(U+\{S'[1+s]-V-S'f[1+s]-S'i[1+s]\}[1-t]-D)-[K+J+P]$$

Rule-56126:

If both ($I$), ($U$), ($S'$), ($s$), ($V$), ($f$), ($i$), ($t$), ($D$), ($K$), ($J$) and ($P$) are known, then its Non Current Asset planned is:

$$I = \{N+K+J+P\}/(U+\{S'[1+s]-V-S'f[1+s]-S'i[1+s]\}[1-t]-D)-1$$

Rule-56127:

If both ($I$), ($N$), ($S'$), ($s$), ($V$), ($f$), ($i$), ($t$), ($D$), ($K$), ($J$) and ($P$) are known, then its Utilized or Starting Capital must be:

$$U = (\{N+K+J+P\}-[1+I]\{S'[1+s]-V-S'f[1+s]-S'i[1+s]\}[1-t]-D)/[1+I]$$

`

Steve Asikin ISBN 13: **978-1564912361**, ISBN 10: **1564912363**

Rule-56128:

> If both (**/**), (**U**), (**S'**), (**s**), (**V**), (**f**), (**i**), (**t**), (**D**), (**K**), (**J**) and (**P**) are known, then its Non Current Asset planned is:
>
> $$S' = (\{[N+K=J+P]/[1+/-U+D\}/[1-t]+V) / \{[1+s][1-f-i]\}$$

Rule-56129:

> If both (**/**), (**U**), (**S'**), (**N**), (**V**), (**f**), (**i**), (**t**), (**D**), (**K**), (**J**) and (**P**) are known, then its Sales Growth planned is:
>
> $$s = (\{[N+K=J+P]/[1+/-U+D\}/[1-t]+V) / \{S'[1-f-i]\} - 1$$

Rule-56130:

> If both (**/**), (**U**), (**S'**), (**s**), (**N**), (**f**), (**i**), (**t**), (**D**), (**K**), (**J**) and (**P**) are known, then its Variable Portion planned is:
>
> $$V = [[1+/(U+\{S'[1+s]-S'f[1+s]-S'i[1+s]\}[1-t]-D) - \{N+K+J+P\}]/\{[1+/[1-t]\}$$

Rule-56131:

> If both (**/**), (**U**), (**S'**), (**s**), (**V**), (**N**), (**i**), (**t**), (**D**), (**K**), (**J**) and (**P**) are known, then its Fixed Portion planned is:
>
> $$f = [[1+/(U+\{S'[1+s]-V-S'i[1+s]\}[1-t]-D) - \{N+K+J+P\}]/\{S'[1+s][1+/[1-t]\}$$

Steve Asikin ISBN 13: **978-1564912361**, ISBN 10: **1564912363**

Rule-56132:

If both (**/**), (**U**), (**S'**), (**s**), (**V**), (**f**), (**N**), (**t**), (**D**), (**K**), (**J**) and (**P**) are known, then its Interest Portion planned is:

$$i= [[1+/(U+\{S'[1+s]-V-S'f[1+s]\}[1-t]-D) -\{N+K+J+P\}]/\{S'[1+s][1+/[1-t]\}$$

Rule-56133:

If both (**/**), (**U**), (**S'**), (**s**), (**V**), (**f**), (**i**), (**N**), (**D**), (**K**), (**J**) and (**P**) are known, then its Non Current Asset planned is:

$$t= 1-\{[N+K+J+P]/[1+/-U+D\} /\{S'[1+s]-V-S'f[1+s]-S'i[1+s]\}$$

Rule-56134:

If both (**/**), (**U**), (**S'**), (**s**), (**V**), (**f**), (**i**), (**t**), (**N**), (**K**), (**J**) and (**P**) are known, then its Dividend planned is:

$$D= U+\{S'[1+s]-V-S'f[1+s]-S'i[1+s]\}[1-t] -[N+K+J+P]/[1+/$$

Rule-56135:

If both (**/**), (**U**), (**S'**), (**s**), (**V**), (**f**), (**i**), (**t**), (**D**), (**N**), (**J**) and (**P**) are known, then its Kind of Cash planned is:

$$K= [1+/(U+\{S'[1+s]-V-S'f[1+s]-S'i[1+s]\} [1-t]-D)-\{N+J+P\}$$

8

Rule-56136:

If both $(\textbf{\textit{I}})$, $(\textbf{U})$, $(\textbf{\$'})$, $(\textbf{s})$, $(\textbf{V})$, $(\textbf{f})$, $(\textbf{i})$, $(\textbf{t})$, $(\textbf{D})$, $(\textbf{K})$, $(\textbf{N})$ and $(\textbf{P})$ are known, then its Job or Trade Account Receivable planned is:

$$J= [1+I](U+\{\$'[1+s]-V-\$'f[1+s]-\$'i[1+s]\}[1-t]-D) -\{N+K+P\}$$

Rule-56137:

If both $(\textbf{\textit{I}})$, $(\textbf{U})$, $(\textbf{\$'})$, $(\textbf{s})$, $(\textbf{V})$, $(\textbf{f})$, $(\textbf{i})$, $(\textbf{t})$, $(\textbf{D})$, $(\textbf{K})$, $(\textbf{J})$ and $(\textbf{N})$ are known, then its Procured Inventory planned is:

$$P= [1+I](U+\{\$'[1+s]-V-\$'f[1+s]-\$'i[1+s]\} [1-t]-D)-\{N+K+J\}$$

Rule-56138:

If both $(\textbf{\textit{I}})$, $(\textbf{U})$, $(\textbf{\$'})$, $(\textbf{s})$, $(\textbf{V})$, $(\textbf{f})$, $(\textbf{i})$, $(\textbf{t})$, $(\textbf{D})$, $(\textbf{K})$, $(\textbf{J})$ and $(\textbf{p})$ are known, then its Non Current Asset planned is:

$$N= [1+I](U+\{\$'[1+s]-V-\$'f[1+s]-\$'i[1+s]/360]\} [1-t]-D)-[K+J+Vp/360]$$

Rule-56139:

If both $(\textbf{\textit{I}})$, $(\textbf{U})$, $(\textbf{\$'})$, $(\textbf{s})$, $(\textbf{V})$, $(\textbf{f})$, $(\textbf{i})$, $(\textbf{t})$, $(\textbf{D})$, $(\textbf{K})$, $(\textbf{J})$ and $(\textbf{p})$ are known, then its Non Current Asset planned is:

$$I= [N+K+J+Vp/360]/(U+\{\$'[1+s]-V-\$'f[1+s] -\$'i[1+s]\}[1-t]-D)-1$$

9

Steve Asikin ISBN 13: **978-1564912361**, ISBN 10: **1564912363**

Rule-56140:

If both (**/**), (**U**), (**S'**), (**s**), (**V**), (**f**), (**i**), (**t**), (**D**), (**K**), (**J**) and (**p**) are known, then its Non Current Asset planned is:

$$U= ([N+K+J+Vp/360]-[1+/]\{S'[1+s]-V-S'f[1+s]$$
$$-S'i[1+s]\}[1-t]-D)/[1+/]$$

Rule-56141:

If both (**/**), (**U**), (**N**), (**s**), (**V**), (**f**), (**i**), (**t**), (**D**), (**K**), (**J**) and (**p**) are known, then its Sales Past must be:

$$S'= (\{[N+K+J+Vp/360]/[1+/]-U+D\}/[1-t]+V)$$
$$/\{[1+s][1-f-i]\}$$

Rule-56142:

If both (**/**), (**U**), (**S'**), (**N**), (**V**), (**f**), (**i**), (**t**), (**D**), (**K**), (**J**) and (**p**) are known, then its Sasles Growth planned is:

$$s= (\{[N+K+J+Vp/360]/[1+/]-U+D\}/[1-t]+V)$$
$$/\{S'[1-f-i]\}-1$$

Rule-56143:

If both (**/**), (**U**), (**S'**), (**s**), (**N**), (**f**), (**i**), (**t**), (**D**), (**K**), (**J**) and (**p**) are known, then its Variable Portion planned is:

$$V= [[1+/](U+\{S'[1+s]-S'f[1+s]-S'i[1+s]\}[1-t]-D)$$
$$-[N+K+J]]/\{p/360+[1+/][1-t]\}$$

10

Rule-56144:

> If both $(I)$, $(U)$, $(S')$, $(s)$, $(V)$, $(N)$, $(i)$, $(t)$, $(D)$, $(K)$, $(J)$ and $(p)$ are known, then its Fixed Portion planned is:
>
> $$f= [[1+I(U+\{S'[1+s]-V-S'i[1+s]\}[1-t]-D) \\ -[N+K+J+Vp/360]]/\{S'[1+s][1+I][1-t]\}$$

Rule-56145:

> If both $(I)$, $(U)$, $(S')$, $(s)$, $(V)$, $(f)$, $(N)$, $(t)$, $(D)$, $(K)$, $(J)$ and $(p)$ are known, then its Interest Portion planned is:
>
> $$i= [[1+I(U+\{S'[1+s]-V-S'f[1+s]\}[1-t]-D) \\ -[N+K+J+Vp/360]]/\{S'[1+s][1+I][1-t]\}$$

Rule-56146:

> If both $(I)$, $(U)$, $(S')$, $(s)$, $(V)$, $(f)$, $(i)$, $(N)$, $(D)$, $(K)$, $(J)$ and $(p)$ are known, then its Tax Rate planned is:
>
> $$t= 1-\{[N+K+J+Vp/360]/[1+I]+D\} \\ /(U+\{S'[1+s]-V-S'f[1+s]-S'i[1+s]\}$$

Rule-56147:

> If both $(I)$, $(U)$, $(S')$, $(s)$, $(V)$, $(f)$, $(i)$, $(t)$, $(N)$, $(K)$, $(J)$ and $(p)$ are known, then its Dividend planned is:
>
> $$D= (U+\{S'[1+s]-V-S'f[1+s]-S'i[1+s]\}[1-t] \\ -[N+K+J+Vp/360]/[1+I]$$

11

Steve Asikin ISBN 13: **978-1564912361**, ISBN 10: **1564912363**

Rule-56148:

If both $(I)$, $(U)$, $(S')$, $(s)$, $(V)$, $(f)$, $(i)$, $(t)$, $(D)$, $(N)$, $(J)$ and $(p)$ are known, then its Kind of Cash planned is:

$$K = [1+I(U+\{S'[1+s]-V-S'f[1+s]-S'i[1+s]\} \\ [1-t]-D)-[N+J+Vp/360]$$

Rule-56149:

If both $(I)$, $(U)$, $(S')$, $(s)$, $(V)$, $(f)$, $(i)$, $(t)$, $(D)$, $(K)$, $(N)$ and $(p)$ are known, then its Job or Trade Receivable Days planned is:

$$J = [1+I(U+\{S'[1+s]-V-S'f[1+s]-S'i[1+s]\}[1-t]-D) \\ -\{N+K+Vp/360\}$$

Rule-56150:

If both $(I)$, $(U)$, $(S')$, $(s)$, $(V)$, $(f)$, $(i)$, $(t)$, $(D)$, $(K)$, $(J)$ and $(N)$ are known, then its Procured Inventory Days planned is:

$$p = 360[[1+I(U+\{S'[1+s]-V-S'f[1+s]-S'i[1+s]\} \\ [1-t]-D)-[N+K+J]]/V$$

Rule-56151:

If both $(I)$, $(U)$, $(S')$, $(s)$, $(V)$, $(f)$, $(i)$, $(t)$, $(D)$, $(K)$, $(J)$, $(S)$, $(v)$ and $(p)$ are known, then its Non Current Asset planned is:

$$N = [1+I(U+\{S'[1+s]-V-S'f[1+s]-S'i[1+s]\} \\ [1-t]-D)-[K+J+Svp/360]$$

12

Steve Asikin ISBN 13: **978-1564912361**, ISBN 10: **1564912363**

Rule-56152:

If both **(N)**, **(U)**, **(S')**, **(s)**, **(V)**, **(f)**, **(i)**, **(t)**, **(D)**, **(K)**, **(J)**, **(S)**, **(v)** and **(p)** are known, then its Leverage or Gearing Ratio planned is:

$$\textit{l} = [N+K+J+Svp/360]/(U+\{S'[1+s]-V-S'f[1+s] \\ -S'i[1+s]\}[1-t]-D)-1$$

Rule-56153:

If both **(/)**, **(N)**, **(S')**, **(s)**, **(V)**, **(f)**, **(i)**, **(t)**, **(D)**, **(K)**, **(J)**, **(S)**, **(v)** and **(p)** are known, then its Utilized or Starting Capital must be:

$$U = ([N+K+Sj/360+Svp/360]-[1+\textit{l}]\{S'[1+s]-V \\ -S'f[1+s]-S'i[1+s]\}[1-t]-D)/[1+\textit{l}]$$

Rule-56154:

If both **(/)**, **(U)**, **(N)**, **(s)**, **(V)**, **(f)**, **(i)**, **(t)**, **(D)**, **(K)**, **(J)**, **(S)**, **(v)** and **(p)** are known, then its Sales Past must be:

$$S' = (\{[N+K+J+Svp/360]/[1+\textit{l}]-U+D\}/[1-t]+V) \\ /\{[1+s][1-f-i]\}$$

$$\text{or}$$

$$S' = S/[1+s]$$

Steve Asikin ISBN 13: **978-1564912361**, ISBN 10: **1564912363**

Rule-56155:

If both (**/**), (**U**), (**S'**), (**N**), (**V**), (**f**), (**i**), (**t**), (**D**), (**K**), (**J**), (**S**), (**v**) and (**p**) are known, then its Sales Growth planned is:

$$s = (\{[N+K+J+Sup/360]/[1+/]-U+D\}/[1-t]+V)$$
$$/\{S'[1-f-i]\}-1$$
or
$$s = S/S'-1$$

Rule-56156:

If both (**/**), (**U**), (**S'**), (**s**), (**N**), (**f**), (**i**), (**t**), (**D**), (**K**), (**J**), (**S**), (**v**) and (**p**) are known, then its Variable Cost planned is:

$$V = [[1+/](U+\{S'[1+s]-S'f[1+s]-S'i[1+s]\}[1-t]-D)$$
$$-[N+K+J+Sup/360]]/\{[1+/][1-t]\}$$
or
$$V = Sv$$

Rule-56157:

If both (**/**), (**U**), (**S'**), (**s**), (**V**), (**N**), (**i**), (**t**), (**D**), (**K**), (**J**), (**S**), (**v**) and (**p**) are known, then its Fixed Portion planned is:

$$f = [[1+/](U+\{S'[1+s]-V-S'i[1+s]\}[1-t]-D)$$
$$-[N+K+J+Sup/360]]/\{S'[1+s][1+/][1-t]\}$$

14

Steve Asikin ISBN 13: **978-1564912361**, ISBN 10: **1564912363**

Rule-56158:

If both $(I)$, $(U)$, $(S')$, $(s)$, $(V)$, $(f)$, $(N)$, $(t)$, $(D)$, $(K)$, $(J)$, $(S)$, $(v)$ and $(p)$ are known, then its Interest Portion planned is:

$$i= [[1+I](U+\{S'[1+s]-V-S'f[1+s]\}[1-t]-D) \\ -[N+K+J+Svp/360]]/\{S'[1+s][1+I][1-t]\}$$

Rule-56159:

If both $(I)$, $(U)$, $(S')$, $(s)$, $(V)$, $(f)$, $(i)$, $(N)$, $(D)$, $(K)$, $(J)$, $(S)$, $(v)$ and $(p)$ are known, then its Tax Rate planned is:

$$t= 1-\{[N+K+J+Svp/360]/[1+I]-U+D\} \\ /\{S'[1+s]-V-S'f[1+s]-S'i[1+s]\}$$

Rule-56160:

If both $(I)$, $(U)$, $(S')$, $(s)$, $(V)$, $(f)$, $(i)$, $(t)$, $(N)$, $(K)$, $(J)$, $(S)$, $(v)$ and $(p)$ are known, then its Dividend planned is:

$$D= (U+\{S'[1+s]-V-S'f[1+s]-S'i[1+s]\}[1-t] \\ -[N+K+J+Svp/360]/[1+I]$$

Rule-56161:

If both $(I)$, $(U)$, $(S')$, $(s)$, $(V)$, $(f)$, $(i)$, $(t)$, $(D)$, $(N)$, $(J)$, $(S)$, $(v)$ and $(p)$ are known, then its Kind of Cash planned is:

$$K= [1+I](U+\{S'[1+s]-V-S'f[1+s]-S'i[1+s]\} \\ [1-t]-D)-[N+J+Svp/360]$$

Steve Asikin ISBN 13: **978-1564912361**, ISBN 10: **1564912363**

Rule-56162:

If both (**/**), (**U**), (**S'**), (**s**), (**V**), (**f**), (**i**), (**t**), (**D**), (**K**), (**N**), (**S**), (**v**) and (**p**) are known, then its Job or Tade Account Receivable planned is:

$$J= [1+\text{/}(U+\{S'[1+s]-V-S'f[1+s]-S'i[1+s]\}[1-t]-D) -\{N+K+Svp/360\}$$

Rule-56163:

If both (**/**), (**U**), (**S'**), (**s**), (**V**), (**f**), (**i**), (**t**), (**D**), (**K**), (**J**), (**N**), (**v**) and (**p**) are known, then its Sales or Revenue planned is:

$$S= 360[[1+\text{/}(U+\{S'[1+s]-V-S'f[1+s]-S'i[1+s]\} [1-t]-D)-[N+K+J]]/[vp]$$

or

$$S= S'[1+s]$$

or

$$S= V/v$$

Rule-56164:

If both (**/**), (**U**), (**S'**), (**s**), (**V**), (**f**), (**i**), (**t**), (**D**), (**K**), (**J**), (**S**), (**N**) and (**p**) are known, then its Variable Portion planned is:

$$v= 360[[1+\text{/}(U+\{S'[1+s]-V-S'f[1+s]-S'i[1+s]\} [1-t]-D)-[N+K+J]]/[Sp]$$

or

$$v= V/S$$

Steve Asikin ISBN 13: **978-1564912361**, ISBN 10: **1564912363**

Rule-56165:

If both (**/**), (**U**), (**S'**), (**s**), (**V**), (**f**), (**i**), (**t**), (**D**), (**K**), (**J**), (**S**), (**v**) and (**n**) are known, then its Procured Inventory Days planned is:

$$\mathbf{p} = 360[[1+\text{/}(\mathbf{U}+\{\mathbf{S'}[1+s]-\mathbf{V}-\mathbf{S'f}[1+s]-\mathbf{S'i}[1+s]\}$$
$$[1-\mathbf{t}]-\mathbf{D})-[\mathbf{N}+\mathbf{K}+\mathbf{J}]]/[\mathbf{Sv}]$$

Rule-56166:

If both (**/**), (**U**), (**S'**), (**s**), (**V**), (**f**), (**i**), (**t**), (**D**), (**K**), (**J**), (**v**) and (**p**) are known, then its Non Current Asset planned is:

$$\mathbf{N} = [1+\text{/}(\mathbf{U}+\{\mathbf{S'}[1+s]-\mathbf{V}-\mathbf{S'f}[1+s]-\mathbf{S'i}[1+s]\}$$
$$[1-\mathbf{t}]-\mathbf{D})-\{\mathbf{K}+\mathbf{J}+\mathbf{S'vp}[1+s]/360\}$$

Rule-56167:

If both (**N**), (**U**), (**S'**), (**s**), (**V**), (**f**), (**i**), (**t**), (**D**), (**K**), (**J**), (**v**) and (**p**) are known, then its Leverage or Gearing Ratio planned is:

$$\mathbf{/} = \{\mathbf{N}+\mathbf{K}+\mathbf{J}+\mathbf{S'vp}[1+s]/360\}/(\mathbf{U}+\{\mathbf{S'}[1+s]-\mathbf{V}$$
$$-\mathbf{S'f}[1+s]-\mathbf{S'i}[1+s]\}[1-\mathbf{t}]-\mathbf{D})-1$$

Rule-56168:

If both (**/**), (**N**), (**S'**), (**s**), (**V**), (**f**), (**i**), (**t**), (**D**), (**K**), (**J**), (**v**) and (**p**) are known, then its Utilized or Starting Capital must be:

$$\mathbf{U} = (\{\mathbf{N}+\mathbf{K}+\mathbf{J}+\mathbf{S'vp}[1+s]/360\}-[1+\text{/}\{\mathbf{S'}[1+s]-\mathbf{V}$$
$$-\mathbf{S'f}[1+s]-\mathbf{S'i}[1+s]\}[1-\mathbf{t}]-\mathbf{D})/[1+\text{/}$$

Steve Asikin ISBN 13: **978-1564912361**, ISBN 10: **1564912363**

Rule-56169:

If both (**I**), (**U**), (**N**), (**s**), (**V**), (**f**), (**i**), (**t**), (**D**), (**K**), (**J**), (**v**) and (**p**) are known, then its Sales Past must be:

$$S' = ([1+I]\{U-V[1-t]-D\}-[N+K+J])$$
$$/([1+s]\{vp/360+[1+I][f+i-1][1-t]\})$$

Rule-56170:

If both (**I**), (**U**), (**S'**), (**N**), (**V**), (**f**), (**i**), (**t**), (**D**), (**K**), (**J**), (**v**) and (**p**) are known, then its Sales Growth planned is:

$$s = ([1+I]\{U-V[1-t]-D\}-[N+K+J])$$
$$/(S'\{vp/360+[1+I][f+i-1][1-t]\})-1$$

or

$$s = S/S'-1$$

Rule-56171:

If both (**I**), (**U**), (**S'**), (**s**), (**N**), (**f**), (**i**), (**t**), (**D**), (**K**), (**J**), (**v**) and (**p**) are known, then its Variable Cost planned is:

$$V = [[1+I](U+\{S'[1+s]-S'f[1+s]-S'i[1+s]\}[1-t]-D)$$
$$-\{N+K+J+S'vp[1+s]/360\}]/\{[1+I][1-t]\}$$

Steve Asikin ISBN 13: **978-1564912361**, ISBN 10: **1564912363**

Rule-56172:

If both (**/**), (**U**), (**S'**), (**s**), (**V**), (**N**), (**i**), (**t**), (**D**), (**K**), (**J**), (**v**) and (**p**) are known, then its Fixed Portion planned is:

$$f= [[1+/(U+\{S'[1+s]-V-S'i[1+s]\}[1-t]-D)$$
$$-\{N+K+J+S'vp[1+s]/360\}]$$
$$/\{S'[1+s][1+/[1-t]\}$$

Rule-56173:

If both (**/**), (**U**), (**S'**), (**s**), (**V**), (**f**), (**N**), (**t**), (**D**), (**K**), (**J**), (**v**) and (**p**) are known, then its Interest Portion planned is:

$$i= [[1+/(U+\{S'[1+s]-V-S'f[1+s]\}[1-t]-D)$$
$$-\{N+K+J+S'vp[1+s]/360\}]$$
$$/\{S'[1+s][1+/[1-t]\}$$

Rule-56174:

If both (**/**), (**U**), (**S'**), (**s**), (**V**), (**f**), (**i**), (**N**), (**D**), (**K**), (**J**), (**v**) and (**p**) are known, then its Tax Rate planned is:

$$t= 1-(\{N+K+J+S'vp[1+s]/360\}/[1+/-U+D)$$
$$/\{S'[1+s]-V-S'f[1+s]-S'i[1+s]\}$$

Rule-56175:

If both (**/**), (**U**), (**S'**), (**s**), (**V**), (**f**), (**i**), (**t**), (**N**), (**K**), (**J**), (**v**) and (**p**) are known, then its Dividend Payout planned is:

$$D= U+\{S'[1+s]-V-S'f[1+s]-S'i[1+s]\}[1-t]$$
$$-\{N+K+J+S'vp[1+s]/360\}/[1+/$$

Steve Asikin ISBN 13: **978-1564912361**, ISBN 10: **1564912363**

Rule-56176:

If both $(I)$, $(U)$, $(S')$, $(s)$, $(V)$, $(f)$, $(i)$, $(t)$, $(D)$, $(N)$, $(J)$, $(v)$ and $(p)$ are known, then its Kind of Cash planned is:

$$K = [1 + I(U + \{S'[1+s] - V - S'f[1+s] - S'i[1+s]\} [1-t] - D) - \{N + J + S'vp[1+s]/360\}$$

Rule-56177:

If both $(I)$, $(U)$, $(S')$, $(s)$, $(V)$, $(f)$, $(i)$, $(t)$, $(D)$, $(K)$, $(N)$, $(v)$ and $(p)$ are known, then its Job or Trade Account Receivable planned is:

$$J = [[1 + I(U + \{S'[1+s] - V - S'f[1+s] - S'i[1+s]\} [1-t] - D) - \{N + K + S'vp[1+s]/360\}$$

Rule-56178:

If both $(I)$, $(U)$, $(S')$, $(s)$, $(V)$, $(f)$, $(i)$, $(t)$, $(D)$, $(K)$, $(J)$, $(N)$ and $(p)$ are known, then its Variable Portion planned is:

$$v = 360[[1 + I(U + \{S'[1+s] - V - S'f[1+s] - S'i[1+s]\} [1-t] - D) - [N + K + J]]/\{S'p[1+s]\}$$

Rule-56179:

If both $(I)$, $(U)$, $(S')$, $(s)$, $(V)$, $(f)$, $(i)$, $(t)$, $(D)$, $(K)$, $(J)$, $(v)$ and $(N)$ are known, then its Procured Inventory Days planned is:

$$p = 360[[1 + I(U + \{S'[1+s] - V - S'f[1+s] - S'i[1+s]\} [1-t] - D) - [N + K + J]]/\{S'v[1+s]\}$$

`

Steve Asikin ISBN 13: **978-1564912361**, ISBN 10: **1564912363**

Rule-56180:

If both (**∫**), (**U**), (**$'**), (**s**), (**V**), (**f**), (**i**), (**t**), (**D**), (**K**), (**$**), (**j**) and (**P**) are known, then its Non Current Asset planned is:

$$N= [1+\smallint](U+\{\$'[1+s]-V-\$'f[1+s]-\$'i[1+s]\}$$
$$[1-t]-D)-[K+\$j/360+P]$$


Rule-56181:

If both (**N**), (**U**), (**$'**), (**s**), (**V**), (**f**), (**i**), (**t**), (**D**), (**K**), (**$**), (**j**) and (**P**) are known, then its Leverage or Gearing Ratio planned is:

$$\smallint= \{N+K+\$j/360+P\}/(U+\{\$'[1+s]-V-\$'f[1+s]$$
$$-\$'i[1+s]\}[1-t]-D)-1$$


Rule-56182:

If both (**∫**), (**N**), (**$'**), (**s**), (**V**), (**f**), (**i**), (**t**), (**D**), (**K**), (**$**), (**j**) and (**P**) are known, then its Utilized or Starting Capital must be:

$$U= (\{N+K+\$j/360+P\}-[1+\smallint]\{\$'[1+s]-V-\$'f[1+s]$$
$$-\$'i[1+s]\}[1-t]-D)/[1+\smallint]$$


Rule-56183:

If both (**∫**), (**U**), (**N**), (**s**), (**V**), (**f**), (**i**), (**t**), (**D**), (**K**), (**$**), (**j**) and (**P**) are known, then its Sales Past must be:

$$\$'= (\{[N+K+\$j/360+P]/[1+\smallint]-U+D\}/[1-t]+V)$$
$$/\{[1+s][1-f-i]\}$$

or

$$\$'= \$/\$'-1$$

Steve Asikin ISBN 13: **978-1564912361**, ISBN 10: **1564912363**

Rule-56184:

If both $(\textbf{\textit{I}})$, $(\textbf{U})$, $(\textbf{S'})$, $(\textbf{N})$, $(\textbf{V})$, $(\textbf{f})$, $(\textbf{i})$, $(\textbf{t})$, $(\textbf{D})$, $(\textbf{K})$, $(\textbf{S})$, $(\textbf{j})$ and $(\textbf{P})$ are known, then its Sales Growth planned is:

$$s = (\{[N+K+Sj/360+P]/[1+I]-U+D\}/[1-t]+V)$$
$$/\{S'[1-f-i]\}-1$$

or

$$s = S/S'-1$$

Rule-56185:

If both $(\textbf{\textit{I}})$, $(\textbf{U})$, $(\textbf{S'})$, $(\textbf{s})$, $(\textbf{N})$, $(\textbf{f})$, $(\textbf{i})$, $(\textbf{t})$, $(\textbf{D})$, $(\textbf{K})$, $(\textbf{S})$, $(\textbf{j})$ and $(\textbf{P})$ are known, then its Variable Cost planned is:

$$V = [[1+I(U+\{S'[1+s]-S'f[1+s]-S'I[1+s]\}[1-t]-D)$$
$$-\{N+K+Sj/360+P\}]/\{[1+I][1-t]\}$$

Rule-56186:

If both $(\textbf{\textit{I}})$, $(\textbf{U})$, $(\textbf{S'})$, $(\textbf{s})$, $(\textbf{V})$, $(\textbf{N})$, $(\textbf{i})$, $(\textbf{t})$, $(\textbf{D})$, $(\textbf{K})$, $(\textbf{S})$, $(\textbf{j})$ and $(\textbf{P})$ are known, then its Fixed Portion planned is:

$$f = [[1+I(U+\{S'[1+s]-V-S'i[1+s]\}[1-t]-D)$$
$$-\{N+K+Sj/360+P\}]/\{S'[1+s][1+I][1-t]\}$$

Rule-56187:

If both $(\textbf{\textit{I}})$, $(\textbf{U})$, $(\textbf{S'})$, $(\textbf{s})$, $(\textbf{V})$, $(\textbf{f})$, $(\textbf{N})$, $(\textbf{t})$, $(\textbf{D})$, $(\textbf{K})$, $(\textbf{S})$, $(\textbf{j})$ and $(\textbf{P})$ are known, then its Interest Portion planned is:

$$i = [[1+I(U+\{S'[1+s]-V-S'f[1+s]\}[1-t]-D)$$
$$-\{N+K+Sj/360+P\}]/\{S'[1+s][1+I][1-t]\}$$

22

Steve Asikin ISBN 13: **978-1564912361**, ISBN 10: **1564912363**

Rule-56188:

If both (**/**), (**U**), (**S'**), (**s**), (**V**), (**f**), (**i**), (**t**), (**D**), (**K**), (**S**), (**j**) and (**P**) are known, then its Non Current Asset planned is:

$$t= 1-\{[N+K+Sj/360+P]/[1+/-U+D\}$$
$$/\{S'[1+s]-V-S'f[1+s]-S'i[1+s]\}$$

Rule-56189:

If both (**/**), (**U**), (**S'**), (**s**), (**V**), (**f**), (**i**), (**t**), (**N**), (**K**), (**S**), (**j**) and (**P**) are known, then its Dividend planned is:

$$D= U+\{S'[1+s]-V-S'f[1+s]-S'i[1+s]\}[1-t]$$
$$-[N+K+Sj/360+P]/[1+/]$$

Rule-56190:

If both (**/**), (**U**), (**S'**), (**s**), (**V**), (**f**), (**i**), (**t**), (**D**), (**N**), (**S**), (**j**) and (**P**) are known, then its Kind of Cash planned is:

$$K= [1+/(U+\{S'[1+s]-V-S'f[1+s]-S'i[1+s]\}$$
$$[1-t]-D)-[N+Sj/360+P]$$

Rule-56191:

If both (**/**), (**U**), (**S'**), (**s**), (**V**), (**f**), (**i**), (**t**), (**D**), (**K**), (**N**), (**j**) and (**P**) are known, then its Sales or Revenue planned is:

$$S= 360[[1+/(U+\{S'[1+s]-V-S'f[1+s]-S'i[1+s]\}$$
$$[1-t]-D)-\{N+K+P\}]/j$$
$$\text{or}$$
$$S= S'[1+s]$$

Steve Asikin ISBN 13: **978-1564912361**, ISBN 10: **1564912363**

Rule-56192:

If both (**/**), (**U**), (**$'**), (**s**), (**V**), (**f**), (**i**), (**t**), (**D**), (**K**), (**$**), (**N**) and (**P**) are known, then its Job or Trade Receivable Days planned is:

$$\mathbf{j}= 360\mathbf{[}[1+\mathbf{/}(\mathbf{U}+\{\mathbf{\$'}[1+\mathbf{s}]-\mathbf{V}-\mathbf{\$'f}[1+\mathbf{s}]-\mathbf{\$'i}[1+\mathbf{s}]\} \\ [1-\mathbf{t}]-\mathbf{D})-\{\mathbf{N}+\mathbf{K}+\mathbf{P}\}\mathbf{]}/\mathbf{\$}$$

Rule-56193:

If both (**/**), (**U**), (**$'**), (**s**), (**V**), (**f**), (**i**), (**t**), (**D**), (**K**), (**$**), (**j**) and (**N**) are known, then its Procured Inventory planned is:

$$\mathbf{P}= [1+\mathbf{/}(\mathbf{U}+\{\mathbf{\$'}[1+\mathbf{s}]-\mathbf{V}-\mathbf{\$'f}[1+\mathbf{s}]-\mathbf{\$'i}[1+\mathbf{s}]\} \\ [1-\mathbf{t}]-\mathbf{D})-\{\mathbf{N}+\mathbf{K}+\mathbf{\$j}/360\}$$

Rule-56194:

If both (**/**), (**U**), (**$'**), (**s**), (**V**), (**f**), (**i**), (**t**), (**D**), (**K**), (**$**), (**j**) and (**p**) are known, then its Non Current Asset planned is:

$$\mathbf{N}= [1+\mathbf{/}(\mathbf{U}+\{\mathbf{\$'}[1+\mathbf{s}]-\mathbf{V}-\mathbf{\$'f}[1+\mathbf{s}]-\mathbf{\$'i}[1+\mathbf{s}]\} \\ [1-\mathbf{t}]-\mathbf{D})-[\mathbf{K}+\mathbf{\$j}/360+\mathbf{Vp}/360]$$

Rule-56195:

If both (**N**), (**U**), (**$'**), (**s**), (**V**), (**f**), (**i**), (**t**), (**D**), (**K**), (**$**), (**j**) and (**p**) are known, then its Leverage or Gearing Ratio planned is:

$$\mathbf{/}= [\mathbf{N}+\mathbf{K}+\mathbf{\$j}/360+\mathbf{Vp}/360]/(\mathbf{U}+\{\mathbf{\$'}[1+\mathbf{s}]-\mathbf{V} \\ -\mathbf{\$'f}[1+\mathbf{s}]-\mathbf{\$'i}[1+\mathbf{s}]\}[1-\mathbf{t}]-\mathbf{D})-1$$

Steve Asikin ISBN 13: **978-1564912361**, ISBN 10: **1564912363**

Rule-56196:

If both (**l**), (**N**), (**S'**), (**s**), (**V**), (**f**), (**i**), (**t**), (**D**), (**K**), (**S**), (**j**) and (**p**) are known, then its Utilized or Starting Capital must be:

$$U = ([N+K+Sj/360+Vp/360]-[1+l]\{S'[1+s]-V$$
$$-S'f[1+s]-S'i[1+s]\}[1-t]-D)/[1+l]$$

Rule-56197:

If both (**l**), (**U**), (**N**), (**s**), (**V**), (**f**), (**i**), (**t**), (**D**), (**K**), (**S**), (**j**) and (**p**) are known, then its Sales Past must be:

$$S' = (\{[N+K+Sj/360+Vp/360]/[1+l]-U+D\}$$
$$/[1-t]+V)/\{[1+s][1-f-i]\}$$
$$\text{or}$$
$$S' = S[1+s]$$

Rule-56198:

If both (**l**), (**U**), (**S'**), (**N**), (**V**), (**f**), (**i**), (**t**), (**D**), (**K**), (**S**), (**j**) and (**p**) are known, then its Sales Growth planned is:

$$s = (\{[N+K+Sj/360+Vp/360]/[1+l]-U+D\}$$
$$/[1-t]+V)/\{S'[1-f-i]\}-1$$
$$\text{or}$$
$$s = S/S'-1$$

Steve Asikin ISBN 13: **978-1564912361**, ISBN 10: **1564912363**

Rule-56199:

If both (**/**), (**U**), (**S'**), (**s**), (**N**), (**f**), (**i**), (**t**), (**D**), (**K**), (**S**), (**j**) and (**p**) are known, then its Variable Cost planned is:

$$V= [[1+\text{/}(U+\{S'[1+s]\ S'f[1+s]\text{-}S'i[1+s]\}[1\text{-}t]\text{-}D)$$
$$-[N+K+Sj/360]]/\{p/360+[1+\text{/}[1\text{-}t]\}$$

Rule-56200:

If both (**/**), (**U**), (**S'**), (**s**), (**V**), (**N**), (**i**), (**t**), (**D**), (**K**), (**S**), (**j**) and (**p**) are known, then its Fixed Portion planned is:

$$f= [[1+\text{/}(U+\{S'[1+s]\text{-}V\text{-}S'i[1+s]\}[1\text{-}t]\text{-}D)$$
$$-[N+K+Sj/360+Vp/360]]$$
$$/\{S'[1+s][1+\text{/}[1\text{-}t]\}$$

Rule-56201:

If both (**/**), (**U**), (**S'**), (**s**), (**V**), (**f**), (**N**), (**t**), (**D**), (**K**), (**S**), (**j**) and (**p**) are known, then its Interest Portion planned is:

$$i= [[1+\text{/}(U+\{S'[1+s]\text{-}V\text{-}S'f[1+s]\}[1\text{-}t]\text{-}D)$$
$$-[N+K+Sj/360+Vp/360]]$$
$$/\{S'[1+s][1+\text{/}[1\text{-}t]\}$$

Rule-56202:

If both (**/**), (**U**), (**S'**), (**s**), (**V**), (**f**), (**i**), (**N**), (**D**), (**K**), (**S**), (**j**) and (**p**) are known, then its Tsx Rate planned is:

$$t= 1\text{-}\{[N+K+Sj/360+Vp/360]/[1+\text{/}\text{-}U+D\}$$
$$/\{S'[1+s]\text{-}V\text{-}S'f[1+s]\text{-}S'i[1+s]\}$$

Steve Asikin ISBN 13: **978-1564912361**, ISBN 10: **1564912363**

<u>Rule-56203</u>:

If both (**/**), (**U**), (**S'**), (**s**), (**V**), (**f**), (**i**), (**t**), (**N**), (**K**), (**S**), (**j**) and (**p**) are known, then its Dividend planned is:

$$D= U+\{S'[1+s]-V-S'f[1+s]-S'i[1+s]\}[1-t]$$
$$-[N+K+Sj/360+Vp/360]/[1+/]$$

<u>Rule-56204</u>:

If both (**/**), (**U**), (**S'**), (**s**), (**V**), (**f**), (**i**), (**t**), (**D**), (**N**), (**S**), (**j**) and (**p**) are known, then its Kind of Cash planned is:

$$K= [1+/](U+\{S'[1+s]-V-S'f[1+s]-S'i[1+s]\}$$
$$[1-t]-D)-[N+Sj/360+Vp/360]$$

<u>Rule-56205</u>:

If both (**/**), (**U**), (**S'**), (**s**), (**V**), (**f**), (**i**), (**t**), (**D**), (**K**), (**N**), (**j**) and (**p**) are known, then its Sales or Revenue planned is:

$$S= 360[[1+/](U+\{S'[1+s]-V-S'f[1+s]-S'i[1+s]\}$$
$$[1-t]-D)-[N+K+Vp/360]]/j$$
or
$$S= S'[1+s]$$

<u>Rule-56206</u>:

If both (**/**), (**U**), (**S'**), (**s**), (**V**), (**f**), (**i**), (**t**), (**D**), (**K**), (**S**), (**N**) and (**p**) are known, then its Job or Trade Receivable Days planned is:

$$j= 360[[1+/](U+\{S'[1+s]-V-S'f[1+s]-S'i[1+s]\}$$
$$[1-t]-D)-[N+K+Vp/360]]/S$$

27

Steve Asikin ISBN 13: **978-1564912361**, ISBN 10: **1564912363**

Rule-56207:

If both (**/**), (**U**), (**$'**), (**s**), (**V**), (**f**), (**i**), (**t**), (**D**), (**K**), (**$**), (**j**) and (**N**) are known, then its Procured Inventory Days planned is:

$$\mathbf{p}= 360[[1+\mathbf{/}(\mathbf{U}+\{\mathbf{\$'}[1+\mathbf{s}]-\mathbf{V}-\mathbf{\$'f}[1+\mathbf{s}]-\mathbf{\$'i}[1+\mathbf{s}]\}$$
$$[1-\mathbf{t}]-\mathbf{D})-[\mathbf{N}+\mathbf{K}+\mathbf{\$j}/360]]/\mathbf{V}$$

Rule-56208:

If both (**/**), (**U**), (**$'**), (**s**), (**V**), (**f**), (**i**), (**t**), (**D**), (**K**), (**$**), (**j**), (**v**) and (**p**) are known, then its Non Current Asset planned is:

$$\mathbf{N}= [1+\mathbf{/}(\mathbf{U}+\{\mathbf{\$'}[1+\mathbf{s}]-\mathbf{V}-\mathbf{\$'f}[1+\mathbf{s}]-\mathbf{\$'i}[1+\mathbf{s}]\}$$
$$[1-\mathbf{t}]-\mathbf{D})-[\mathbf{K}+\mathbf{\$j}/360+\mathbf{\$vp}/360]$$

Rule-56209:

If both (**N**), (**U**), (**$'**), (**s**), (**V**), (**f**), (**i**), (**t**), (**D**), (**K**), (**$**), (**j**), (**v**) and (**p**) are known, then its Leverage or Gearing Ratio planned is:

$$\mathbf{/}= [\mathbf{N}+\mathbf{K}+\mathbf{\$j}/360+\mathbf{\$vp}/360]/(\mathbf{U}+\{\mathbf{\$'}[1+\mathbf{s}]-\mathbf{V}$$
$$-\mathbf{\$'f}[1+\mathbf{s}]-\mathbf{\$'i}[1+\mathbf{s}]\}[1-\mathbf{t}]-\mathbf{D})-1$$

Rule-56210:

If both (**/**), (**N**), (**$'**), (**s**), (**V**), (**f**), (**i**), (**t**), (**D**), (**K**), (**$**), (**j**), (**v**) and (**p**) are known, then its Utilized or Starting Capital must be:

$$\mathbf{U}= (\{\mathbf{N}+\mathbf{K}+\mathbf{\$j}/360+\mathbf{\$vp}/360\}-[1+\mathbf{/}\{\mathbf{\$'}[1+\mathbf{s}]-\mathbf{V}$$
$$-\mathbf{\$'f}[1+\mathbf{s}]-\mathbf{\$'i}[1+\mathbf{s}]\}[1-\mathbf{t}]-\mathbf{D})/[1+\mathbf{/}]$$

Steve Asikin ISBN 13: **978-1564912361**, ISBN 10: **1564912363**

Rule-56211:

If both (**/**), (**U**), (**N**), (**s**), (**V**), (**f**), (**i**), (**t**), (**D**), (**K**), (**S**), (**j**), (**v**) and (**p**) are known, then its Sales Past must be:

$$S'= (\{[N+K+Sj/360+Svp/360]/[1+/-U+D\}$$
$$/[1-t]+V)/\{[1+s][1-f-i]\}$$
or
$$S'= S/[1+s]$$


Rule-56212:

If both (**/**), (**U**), (**S'**), (**N**), (**V**), (**f**), (**i**), (**t**), (**D**), (**K**), (**S**), (**j**), (**v**) and (**p**) are known, then its Sales Growth planned is:

$$s= (\{[N+K+Sj/360+Svp/360]/[1+/-U+D\}$$
$$/[1-t]+V)/\{S'[1-f-i]\}-1$$
or
$$s= S/S'-1$$


Rule-56213:

If both (**/**), (**U**), (**S'**), (**s**), (**N**), (**f**), (**i**), (**t**), (**D**), (**K**), (**S**), (**j**), (**v**) and (**p**) are known, then its Variable Cost planned is:

$$V= [[1+/(U+\{S'[1+s]-S'f[1+s]-S'i[1+s]\}[1-t]-D)$$
$$-[N+K+Sj/360+Svp/360]]/\{[1+/][1-t]\}$$
or
$$V= Sv$$

Steve Asikin ISBN 13: **978-1564912361**, ISBN 10: **1564912363**

Rule-56214:

If both (**/**), (**U**), (**S'**), (**s**), (**V**), (**N**), (**i**), (**t**), (**D**), (**K**), (**S**), (**j**), (**v**) and (**p**) are known, then its Fixed Portion planned is:

$$f= [[1+/](U+\{S'[1+s]-V-S'i[1+s]\}[1-t]-D) \\ -[N+K+Sj/360+Svp/360]] \\ /\{S'[1+s][1+/][1-t]\}$$

Rule-56215:

If both (**/**), (**U**), (**S'**), (**s**), (**V**), (**f**), (**N**), (**t**), (**D**), (**K**), (**S**), (**j**), (**v**) and (**p**) are known, then its Interest Portion planned is:

$$i= [[1+/](U+\{S'[1+s]-V-S'f[1+s]\}[1-t]-D) \\ -[N+K+Sj/360+Svp/360]] \\ /\{S'[1+s][1+/][1-t]\}$$

Rule-56216:

If both (**/**), (**U**), (**S'**), (**s**), (**V**), (**f**), (**i**), (**N**), (**D**), (**K**), (**S**), (**j**), (**v**) and (**p**) are known, then its Tax Rate planned is:

$$t= 1-\{[N+K+Sj/360+Svp/360]/[1+/]-U+D\} \\ /\{S'[1+s]-V-S'f[1+s]-S'i[1+s]\}$$

Rule-56217:

If both (**/**), (**U**), (**S'**), (**s**), (**V**), (**f**), (**i**), (**t**), (**N**), (**K**), (**S**), (**j**), (**v**) and (**p**) are known, then its Dividend planned is:

$$D= U+\{S'[1+s]-V-S'f[1+s]-S'i[1+s]\}[1-t] \\ -[N+K+Sj/360+Svp/360]/[1+/]$$

Steve Asikin ISBN 13: **978-1564912361**, ISBN 10: **1564912363**

Rule-56218:

If both (**/**), (**U**), (**S'**), (**s**), (**V**), (**f**), (**i**), (**t**), (**D**), (**N**), (**S**), (**j**), (**v**) and (**p**) are known, then its Kind of Cash planned is:

$$K= [1+/](U+\{S'[1+s]-V-S'f[1+s]-S'i[1+s]\}$$
$$[1-t]-D)-[N+Sj/360+Svp/360]$$


Rule-56219:

If both (**/**), (**U**), (**S'**), (**s**), (**V**), (**f**), (**i**), (**t**), (**D**), (**K**), (**N**), (**j**), (**v**) and (**p**) are known, then its Sales or Revenue planned is:

$$S= [[1+/](U+\{S'[1+s]-V-S'f[1+s]-S'i[1+s]\}$$
$$[1-t]-D)-[N+K]]/[j/360+vp/360]$$

or

$$S= V/v$$

or

$$S= S'[1+s]$$


Rule-56220:

If both (**/**), (**U**), (**S'**), (**s**), (**V**), (**f**), (**i**), (**t**), (**D**), (**K**), (**S**), (**N**), (**v**) and (**p**) are known, then its Job or Trade Receivable Days planned is:

$$j= 360[[1+/](U+\{S'[1+s]-V-S'f[1+s]-S'i[1+s]\}$$
$$[1-t]-D)-\{N+K+Svp/360\}]/S$$

Steve Asikin ISBN 13: **978-1564912361**, ISBN 10: **1564912363**

Rule-56221:

If both (**/**), (**U**), (**S'**), (**s**), (**V**), (**f**), (**i**), (**t**), (**D**), (**K**), (**S**), (**j**), (**N**) and (**p**) are known, then its Variable Portion planned is:

$$\textbf{v}= 360[[1+\textbf{/}(\textbf{U}+\{\textbf{S'}[1+\textbf{s}]-\textbf{V}-\textbf{S'f}[1+\textbf{s}]-\textbf{S'i}[1+\textbf{s}]\}$$
$$[1-\textbf{t}]-\textbf{D})-[\textbf{N}+\textbf{K}+\textbf{Sj}/360]]/[\textbf{Sp}]$$

or

$$\textbf{v}= \textbf{V}/\textbf{S}$$

Rule-56222:

If both (**/**), (**U**), (**S'**), (**s**), (**V**), (**f**), (**i**), (**t**), (**D**), (**K**), (**S**), (**j**), (**v**) and (**N**) are known, then its Non Current Asset planned is:

$$\textbf{p}= 360[[1+\textbf{/}(\textbf{U}+\{\textbf{S'}[1+\textbf{s}]-\textbf{V}-\textbf{Sf}-\textbf{S'f}[1+\textbf{s}]$$
$$-\textbf{S'i}[1+\textbf{s}]\}[1-\textbf{t}]-\textbf{D})-[\textbf{N}+\textbf{K}+\textbf{Sj}/360]]/[\textbf{Sv}]$$

Rule-56223:

If both (**/**), (**U**), (**S'**), (**s**), (**V**), (**f**), (**i**), (**t**), (**D**), (**K**), (**S**), (**j**), (**v**) and (**p**) are known, then its Non Current Asset planned is:

$$\textbf{N}= [1+\textbf{/}(\textbf{U}+\{\textbf{S'}[1+\textbf{s}]-\textbf{V}-\textbf{S'f}[1+\textbf{s}]-\textbf{S'i}[1+\textbf{s}]\}$$
$$[1-\textbf{t}]-\textbf{D})-\{\textbf{K}+\textbf{Sj}/360+\textbf{S'vp}[1+\textbf{s}]/360\}$$

Rule-56224:

If both (**N**), (**U**), (**S'**), (**s**), (**V**), (**f**), (**i**), (**t**), (**D**), (**K**), (**S**), (**j**), (**v**) and (**p**) are known, then its Leverage or Gearing Ratio planned is:

$$\textbf{/}= \{\textbf{N}+\textbf{K}+\textbf{Sj}/360+\textbf{S'vp}[1+\textbf{s}]/360\}/(\textbf{U}+\{\textbf{S'}[1+\textbf{s}]-\textbf{V}$$
$$-\textbf{S'f}[1+\textbf{s}]-\textbf{S'i}[1+\textbf{s}]\}[1-\textbf{t}]-\textbf{D})-1$$

`

Steve Asikin ISBN 13: **978-1564912361**, ISBN 10: **1564912363**

Rule-56225:

If both (**/**), (**N**), (**S'**), (**s**), (**V**), (**f**), (**i**), (**t**), (**D**), (**K**), (**S**), (**j**), (**v**) and (**p**) are known, then its Utilized or Starting Capital must be:

$$U = (\{N+K+Sj/360+S'vp[1+s]/360\}-[1+/]\{S'[1+s]$$
$$-V-S'f[1+s]-S'i[1+s]\}[1-t]-D)/\{[1+/]$$

Rule-56226:

If both (**/**), (**U**), (**N**), (**s**), (**V**), (**f**), (**i**), (**t**), (**D**), (**K**), (**S**), (**j**), (**v**) and (**p**) are known, then its Sales Past must be:

$$S' = [(\{N+K+Sj/360+S'vp[1+s]/360\}/[1+/]$$
$$-U+D)/[1-t]+V]/\{[1+s][1-f-i]\}$$

or

$$S' = S/[1+s]$$

Rule-56227:

If both (**/**), (**U**), (**S'**), (**N**), (**V**), (**f**), (**i**), (**t**), (**D**), (**K**), (**S**), (**j**), (**v**) and (**p**) are known, then its Sales Growth planned is:

$$s = [(\{N+K+Sj/360+S'vp[1+s]/360\}/[1+/]-U+D)$$
$$/[1-t]+V]/\{S'[1-f-i]\}-1$$

or

$$s = S/S'-1$$

Steve Asikin ISBN 13: **978-1564912361**, ISBN 10: **1564912363**

Rule-56228:

If both (**/**), (**U**), (**$'**), (**s**), (**N**), (**f**), (**i**), (**t**), (**D**), (**K**), (**$**), (**j**), (**v**) and (**p**) are known, then its Variable Cost planned is:

$$V= [[1+/(U+\{S'[1+s]-S'f[1+s]-S'i[1+s]\}[1-t]-D) -\{N+K+Sj/360+Svp/360\}]/\{[1+/][1-t]\}$$

or

$$V= Sv$$

Rule-56229:

If both (**/**), (**U**), (**$'**), (**s**), (**V**), (**N**), (**I**), (**t**), (**D**), (**K**), (**$**), (**j**), (**v**) and (**p**) are known, then its Fixed Portion planned is:

$$f= [[1+/(U+\{S'[1+s]-V-S'i[1+s]\}[1-t]-D) -\{N+K+Sj/360+S'vp[1+s]/360\}] /\{S'[1+s][1+/][1-t]\}$$

Rule-56230:

If both (**/**), (**U**), (**$'**), (**s**), (**V**), (**f**), (**N**), (**t**), (**D**), (**K**), (**$**), (**j**), (**v**) and (**p**) are known, then its Interest Portion planned is:

$$i= [[1+/(U+\{S'[1+s]-V-S'f[1+s]\}[1-t]-D) -\{N+K+Sj/360+S'vp[1+s]/360\}] /\{S'[1+s][1+/][1-t]\}$$

Steve Asikin ISBN 13: **978-1564912361**, ISBN 10: **1564912363**

Rule-56231:

If both (**/**), (**U**), (**S'**), (**s**), (**V**), (**f**), (**i**), (**N**), (**D**), (**K**), (**S**), (**j**), (**v**) and (**p**) are known, then its Tax Rate planned is:

$$t = 1 - (\{N+K+Sj/360+S'vp[1+s]/360\}/[1+/]-U+D)$$
$$/\{S'[1+s]-V-S'f[1+s]-S'i[1+s]\}$$

Rule-56232:

If both (**/**), (**U**), (**S'**), (**s**), (**V**), (**f**), (**i**), (**t**), (**N**), (**K**), (**S**), (**j**), (**v**) and (**p**) are known, then its Dividend planned is:

$$D = (U+\{S'[1+s]-V-S'f[1+s]-S'i[1+s]\}[1-t]$$
$$-\{N+K+Sj/360+S'vp[1+s]/360\}/[1+/]$$

Rule-56233:

If both (**/**), (**U**), (**S'**), (**s**), (**V**), (**f**), (**i**), (**t**), (**D**), (**N**), (**S**), (**j**), (**v**) and (**p**) are known, then its Kind of Cash planned is:

$$K = [1+/](U+\{S'[1+s]-V-S'f[1+s]-S'i[1+s]\}[1-t]-D)$$
$$-\{N+Sj/360+S'vp[1+s]/360\}$$

Steve Asikin ISBN 13: **978-1564912361**, ISBN 10: **1564912363**

Rule-56234:

If both (**/**), (**U**), (**$'**), (**s**), (**V**), (**f**), (**i**), (**t**), (**D**), (**K**), (**$**), (**j**), (**v**) and (**p**) are known, then its Non Current Asset planned is:

$= 360**[**[1+**/**](**U**+{**$'**[1+s]-**V**-**$'f**[1+s]-**$'i**[1+s]}
[1-**t**]-**D**)-{**N**+**K**+**$'vp**[1+s]/360}**]/j**

or

$= **V**/**v**

or

$= **$'**[1+s]

Rule-56235:

If both (**/**), (**U**), (**$'**), (**s**), (**V**), (**f**), (**i**), (**t**), (**D**), (**K**), (**$**), (**N**), (**v**) and (**p**) are known, then its Job or Trade Receivable Days planned is:

**j**= 360**[**[1+**/**](**U**+{**$'**[1+s]-**V**-**$'f**[1+s]-**$'i**[1+s]}
[1-**t**]-**D**)-{**N**+**K**+**$'vp**[1+s]/360}**]/$**

Rule-56236:

If both (**/**), (**U**), (**$'**), (**s**), (**V**), (**f**), (**i**), (**t**), (**D**), (**K**), (**$**), (**j**), (**N**) and (**p**) are known, then its Variable Portion planned is:

**v**= 360**[**[1+**/**](**U**+{**$'**[1+s]-**V**-**$'f**[1+s]-**$'i**[1+s]}
[1-**t**]-**D**)-[**N**+**K**+**$j**/360]]/{**$'p**[1+s]}

or

**v**= **V**/**$**

Steve Asikin ISBN 13: **978-1564912361**, ISBN 10: **1564912363**

Rule-56237:

If both $(I)$, $(U)$, $(S')$, $(s)$, $(V)$, $(f)$, $(i)$, $(t)$, $(D)$, $(K)$, $(S)$, $(j)$, $(v)$ and $(N)$ are known, then its Procured Inventory Days planned is:

$$p = 360[[1+I(U+\{S'[1+s]-V-S'f[1+s]-S'i[1+s]\}$$
$$[1-t]-D)-[N+K+Sj/360]]/\{S'v[1+s]\}$$

Rule-56238:

If both $(I)$, $(U)$, $(S')$, $(s)$, $(V)$, $(f)$, $(i)$, $(t)$, $(D)$, $(K)$, $(j)$ and $(P)$ are known, then its Non Current Asset planned is:

$$N = [1+I(U+\{S'[1+s]-V-S'f[1+s]-S'i[1+s]\}$$
$$[1-t]-D)-\{K+S'j[1+s]/360+P\}$$

Rule-56239:

If both $(N)$, $(U)$, $(S')$, $(s)$, $(V)$, $(f)$, $(i)$, $(t)$, $(D)$, $(K)$, $(j)$ and $(P)$ are known, then its Leverage or Gearing Ratio planned is:

$$I = \{N+K+S'j[1+s]/360+P\}/(U+\{S'[1+s]-V$$
$$-S'f[1+s]-S'i[1+s]\}[1-t]-D)-1$$

Rule-56240:

If both $(I)$, $(N)$, $(S')$, $(s)$, $(V)$, $(f)$, $(i)$, $(t)$, $(D)$, $(K)$, $(j)$ and $(P)$ are known, then its Utilized or Starting Capital must be:

$$U = (\{N+K+S'j[1+s]/360+P\}-[1+I]\{S'[1+s]-V$$
$$-S'f[1+s]-S'i[1+s]\}[1-t]-D)/[1+I]$$

Steve Asikin ISBN 13: **978-1564912361**, ISBN 10: **1564912363**

Rule-56241:

If both (**/**), (**U**), (**N**), (**s**), (**V**), (**f**), (**i**), (**t**), (**D**), (**K**), (**j**) and (**P**) are known, then its Sales Past must be:

$$S' = [(\{N+K+S'j[1+s]/360+P\}/[1+/]-U+D)/[1-t] + V]/\{[1+s][1-f-i]\}$$

Rule-56242:

If both (**/**), (**U**), (**S'**), (**N**), (**V**), (**f**), (**i**), (**t**), (**D**), (**K**), (**j**) and (**P**) are known, then its Sales Growth planned is:

$$s = [(\{N+K+S'j[1+s]/360+P\}/[1+/]-U+D)/[1-t] + V]/\{S'[1+s][1-f-i]\} - 1$$

Rule-56243:

If both (**/**), (**U**), (**S'**), (**s**), (**N**), (**f**), (**i**), (**t**), (**D**), (**K**), (**j**) and (**P**) are known, then its Variable Cost planned is:

$$V = [[1+/](U+\{S'-S'f[1+s]-S'i[1+s]-T\}[1-d]) - \{N+K+S'j[1+s]/360+P\}]/\{[1+/][1-d]\}$$

Rule-56244:

If both (**/**), (**U**), (**S'**), (**s**), (**V**), (**N**), (**i**), (**t**), (**D**), (**K**), (**j**) and (**P**) are known, then its Fixed Portion planned is:

$$f = [[1+/](U+\{S'[1+s]-V-S'i[1+s]\}[1-t]-D) - \{N+K+S'j[1+s]/360+P\}] / \{S'[1+s][1+/][1-t]\}$$

`

Steve Asikin ISBN 13: **978-1564912361**, ISBN 10: **1564912363**

Rule-56245:

If both (**/**), (**U**), (**S'**), (**s**), (**V**), (**f**), (**N**), (**t**), (**D**), (**K**), (**j**) and (**P**) are known, then its Interest Portion planned is:

$$i= [[1+/](U+\{S'[1+s]-V-S'f[1+s]\}[1-t]-D)-\{N+K$$
$$+S'j[1+s]/360+P\}]/\{S'[1+s][1+/][1-t]\}$$

Rule-56246:

If both (**/**), (**U**), (**S'**), (**s**), (**V**), (**f**), (**i**), (**N**), (**D**), (**K**), (**j**) and (**P**) are known, then its Tax Rate planned is:

$$t= 1-(\{N+K+S'j[1+s]/360+P\}/[1+/]-U+D)$$
$$/\{S'[1+s]-V-S'f[1+s]-S'i[1+s]\}$$

Rule-56247:

If both (**/**), (**U**), (**S'**), (**s**), (**V**), (**f**), (**i**), (**t**), (**N**), (**K**), (**j**) and (**P**) are known, then its Dividend planned is:

$$D= U+\{S'[1+s]-V-S'f[1+s]-S'i[1+s]\}[1-t]$$
$$-\{N+K+S'j[1+s]/360+P\}/[1+/]$$

Rule-56248:

If both (**/**), (**U**), (**S'**), (**s**), (**V**), (**f**), (**i**), (**t**), (**D**), (**N**), (**j**) and (**P**) are known, then its Kind of Cash planned is:

$$K= [1+/](U+\{S'[1+s]-V-S'f[1+s]-S'i[1+s]\}[1-t]-D)$$
$$-\{N+S'j[1+s]/360+P\}$$

Steve Asikin ISBN 13: **978-1564912361**, ISBN 10: **1564912363**

Rule-56249:

If both (**I**), (**U**), (**S'**), (**s**), (**V**), (**f**), (**i**), (**t**), (**D**), (**K**), (**N**) and (**P**) are known, then its Job or Trade Receivable Days planned is:

$$j= 360[[1+I(U+\{S'[1+s]-V-S'f[1+s]-S'i[1+s]\} \\ [1-t]-D)-\{N+K+P\}]/\{S'[1+s]\}$$

Rule-56250:

If both (**I**), (**U**), (**S'**), (**s**), (**V**), (**f**), (**i**), (**t**), (**D**), (**K**), (**j**) and (**N**) are known, then its Procured Inventory planned is:

$$P= [1+I(U+\{S'[1+s]-V-S'f[1+s]-S'i[1+s]\} \\ [1-t]-D)-\{N+K+S'j[1+s]/360\}$$

Rule-56251:

If both (**I**), (**U**), (**S'**), (**s**), (**V**), (**f**), (**i**), (**t**), (**D**), (**K**), (**j**) and (**p**) are known, then its Non Current Asset planned is:

$$N= [1+I(U+\{S'[1+s]-V-S'f[1+s]-S'i[1+s]\} \\ [1-t]-D)-\{K+S'j[1+s]/360+Vp/360\}$$

Rule-56252:

If both (**N**), (**U**), (**S'**), (**s**), (**V**), (**f**), (**i**), (**t**), (**D**), (**K**), (**j**) and (**p**) are known, then its Leverage or Gearing Ratio planned is:

$$I= \{N+K+S'j[1+s]/360+Vp/360\}/(U+\{S'[1+s]-V \\ -S'f[1+s]-S'i[1+s]\}[1-t]-D)-1$$

`

Steve Asikin ISBN 13: **978-1564912361**, ISBN 10: **1564912363**

Rule-56253:

If both (**/**), (**U**), (**S'**), (**s**), (**V**), (**f**), (**i**), (**t**), (**D**), (**K**), (**j**) and (**p**) are known, then its Non Current Asset planned is:

$$U = (\{N+K+S'j[1+s]/360+Vp/360\}-[1+/]\{S'[1+s]$$
$$-V-S'f[1+s]-S'i[1+s]\}[1-t]-D)/[1+/]$$

Rule-56254:

If both (**/**), (**U**), (**N**), (**s**), (**V**), (**f**), (**i**), (**t**), (**D**), (**K**), (**j**) and (**p**) are known, then its Sales Past must be:

$$S' = [(\{N+K+S'j[1+s]/360+Vp/360\}/[1+/]-U+D)$$
$$/[1-t]+V]/\{S'[1-f-i]\}$$
or
$$S' = S/[1+s]$$

Rule-56255:

If both (**/**), (**U**), (**S'**), (**N**), (**V**), (**f**), (**i**), (**t**), (**D**), (**K**), (**j**) and (**p**) are known, then its Sales Growth planned is:

$$s = [(\{N+K+S'j[1+s]/360+Vp/360\}/[1+/]-U+D)$$
$$/[1-t]+V]/\{S'[1-f-i]\}$$
or
$$s = S/S'-1$$

Rule-56256:

If both (**/**), (**U**), (**S'**), (**s**), (**N**), (**f**), (**i**), (**t**), (**D**), (**K**), (**j**) and (**p**) are known, then its Variable Cost planned is:

$$V = [[1+/](U+\{S'[1+s]-S'f[1+s]-S'i[1+s]\}[1-t]-D)$$
$$-\{N+K+S'j[1+s]/360\}]$$
$$/\{p/360+[1+/][1-t]\}$$

41

Steve Asikin ISBN 13: **978-1564912361**, ISBN 10: **1564912363**

Rule-56257:

If both (**/**), (**U**), (**S'**), (**s**), (**V**), (**N**), (**i**), (**t**), (**D**), (**K**), (**j**) and (**p**) are known, then its Fixed Portion planned is:

$$f = [[1+/(U+\{S'[1+s]-V-S'i[1+s]\}[1-t]-D) -\{N+K+S'j[1+s]/360+Vp/360\}] /\{S'[1+s][1+/][1-t]\}$$

Rule-56258:

If both (**/**), (**U**), (**S'**), (**s**), (**V**), (**f**), (**N**), (**t**), (**D**), (**K**), (**j**) and (**p**) are known, then its Interest Portion planned is:

$$i = [[1+/(U+\{S'[1+s]-V-S'f[1+s]\}[1-t]-D) -\{N+K+S'j[1+s]/360+Vp/360\}] /\{S'[1+s][1+/][1-t]\}$$

Rule-56259:

If both (**/**), (**U**), (**S'**), (**s**), (**V**), (**f**), (**i**), (**N**), (**D**), (**K**), (**j**) and (**p**) are known, then its Tax Rate planned is:

$$t = 1-(\{N+K+S'j[1+s]/360+Vp/360\}/[1+/ -U+D)/\{S'[1+s]-V-S'f[1+s]-S'i[1+s]\}$$

Rule-56260:

If both (**/**), (**U**), (**S'**), (**s**), (**V**), (**f**), (**i**), (**t**), (**N**), (**K**), (**j**) and (**p**) are known, then its Dividend planned is:

$$D = U+\{S'[1+s]-V-S'f[1+s]-S'i[1+s]\}[1-t] -\{N+K+S'j[1+s]/360+Vp/360\}/[1+/]$$

`

42

Steve Asikin ISBN 13: **978-1564912361**, ISBN 10: **1564912363**

Rule-56261:

If both (**/**), (**U**), (**S'**), (**s**), (**V**), (**f**), (**i**), (**t**), (**D**), (**N**), (**j**)
and (**p**) are known, then its Kind of Cash planned is:
$$\mathbf{K}= [1+\mathbf{/}(\mathbf{U}+\{\mathbf{S'}[1+\mathbf{s}]-\mathbf{V}-\mathbf{S'f}[1+\mathbf{s}]-\mathbf{S'i}[1+\mathbf{s}]\}$$
$$[1-\mathbf{t}]-\mathbf{D})-\{\mathbf{N}+\mathbf{S'j}[1+\mathbf{s}]/360+\mathbf{Vp}/360\}$$

Rule-56262:

If both (**/**), (**U**), (**S'**), (**s**), (**V**), (**f**), (**i**), (**t**), (**D**), (**K**), (**N**)
and (**p**) are known, then its Job or Trade Receivable
Days planned is:
$$\mathbf{j}= 360[[1+\mathbf{/}(\mathbf{U}+\{\mathbf{S'}[1+\mathbf{s}]-\mathbf{V}-\mathbf{S'f}[1+\mathbf{s}]-\mathbf{S'i}[1+\mathbf{s}]\}$$
$$[1-\mathbf{t}]-\mathbf{D})-\{\mathbf{N}+\mathbf{K}+\mathbf{Vp}/360\}]/\{\mathbf{S'}[1+\mathbf{s}]\}$$

Rule-56263:

If both (**/**), (**U**), (**S'**), (**s**), (**V**), (**f**), (**i**), (**t**), (**D**), (**K**), (**j**)
and (**N**) are known, then its Procured Inventory Days
planned is:
$$\mathbf{p}= 360[[1+\mathbf{/}(\mathbf{U}+\{\mathbf{S'}[1+\mathbf{s}]-\mathbf{V}-\mathbf{S'f}[1+\mathbf{s}]-\mathbf{S'i}[1+\mathbf{s}]\}$$
$$[1-\mathbf{t}]-\mathbf{D})-\{\mathbf{N}+\mathbf{K}+\mathbf{S'j}[1+\mathbf{s}]/360\}]/\mathbf{V}$$

Rule-56264:

If both (**/**), (**U**), (**S'**), (**s**), (**V**), (**f**), (**i**), (**t**), (**D**), (**K**), (**j**),
(**S**), (**v**) and (**p**) are known, then its Non Current
Asset planned is:
$$\mathbf{N}= [1+\mathbf{/}(\mathbf{U}+\{\mathbf{S'}[1+\mathbf{s}]-\mathbf{V}-\mathbf{S'f}[1+\mathbf{s}]-\mathbf{S'i}[1+\mathbf{s}]\}$$
$$[1-\mathbf{t}]-\mathbf{D})-\{\mathbf{K}+\mathbf{S'j}[1+\mathbf{s}]/360+\mathbf{Svp}/360\}$$

Steve Asikin ISBN 13: **978-1564912361**, ISBN 10: **1564912363**

Rule-56265:

If both (**N**), (**U**), (**S'**), (**s**), (**V**), (**f**), (**i**), (**t**), (**D**), (**K**), (**j**), (**S**), (**v**) and (**p**) are known, then its Leverage or Gearing Ratio planned is:

$$I = \{N+K+S'j[1+s]/360+Svp/360\}/(U+\{S'[1+s]-V$$
$$-S'f[1+s]-S'i[1+s]\}[1-t]-D)-1$$

Rule-56266:

If both (**I**), (**N**), (**S'**), (**s**), (**V**), (**f**), (**i**), (**t**), (**D**), (**K**), (**j**), (**S**), (**v**) and (**p**) are known, then its Utilized or Starting Capital must br:

$$U = (\{N+K+S'j[1+s]/360+Svp/360\}$$
$$-[1+I]\{S'[1+s]-V-S'f[1+s]$$
$$-S'i[1+s]\}[1-t]-D)/\{[1+I]\}$$

Rule-56267:

If both (**I**), (**U**), (**N**), (**s**), (**V**), (**f**), (**i**), (**t**), (**D**), (**K**), (**j**), (**S**), (**v**) and (**p**) are known, then its Sales Past must be:

$$S' = [(\{N+K+S'j[1+s]/360+Svp/360\}/[1+I]$$
$$-U+D)/[1-t]]/\{[1+s][1-f-i]\}$$
$$\text{or}$$
$$S' = S/[1+s]$$

`

Steve Asikin ISBN 13: **978-1564912361**, ISBN 10: **1564912363**

Rule-56268:

If both (**/**), (**U**), (**S'**), (**N**), (**V**), (**f**), (**i**), (**t**), (**D**), (**K**), (**j**), (**S**), (**v**) and (**p**) are known, then its Sales Growth planned is:

$$s= [(\{N+K+S'j[1+s]/360+Svp/360\}/[1+/]$$
$$-U+D)/[1-t]]/\{S'[1-f-i]\}-1$$

or

$$s= S/S'-1$$

Rule-56269:

If both (**/**), (**U**), (**S'**), (**s**), (**N**), (**f**), (**i**), (**t**), (**D**), (**K**), (**j**), (**S**), (**v**) and (**p**) are known, then its Variable Cost planned is:

$$V= [[1+/](U+\{S'[1+s]-S'f[1+s]-S'i[1+s]\}[1-t]-D)$$
$$-\{N+K+S'j[1+s]/360+Svp/360\}]$$
$$/\{[1+/][1-t]\}$$

or

$$V= Sv$$

Rule-56270:

If both (**/**), (**U**), (**S'**), (**s**), (**V**), (**N**), (**i**), (**t**), (**D**), (**K**), (**j**), (**S**), (**v**) and (**p**) are known, then its Fixed Portion planned is:

$$f= ([1+/](U+\{S'[1+s]-V-S'i[1+s]\}[1-t]-D$$
$$-\{N+K+S'j[1+s]/360+Svp/360\})$$
$$/\{S'[1+s][1+/][1-t]\}$$

Steve Asikin ISBN 13: **978-1564912361**, ISBN 10: **1564912363**

Rule-56271:

If both (**/**), (**U**), (**S'**), (**s**), (**V**), (**f**), (**N**), (**t**), (**D**), (**K**), (**j**), (**S**), (**v**) and (**p**) are known, then its Interest Portion planned is:

$$i= ([1+/](U+\{S'[1+s]-V-S'f[1+s]\}[1-t]-D$$
$$-\{N+K+S'j[1+s]/360+Svp/360\})$$
$$/\{S'[1+s][1+/][1-t]\}$$

Rule-56272:

If both (**/**), (**U**), (**S'**), (**s**), (**V**), (**f**), (**i**), (**N**), (**D**), (**K**), (**j**), (**S**), (**v**) and (**p**) are known, then its Tax Rate planned is:

$$t= 1-(\{N+K+S'j[1+s]/360+Svp/360\}/[1+/]-U+D)$$
$$/\{S'[1+s]-V-S'f[1+s]-S'i[1+s]\}$$

Rule-56273:

If both (**/**), (**U**), (**S'**), (**s**), (**V**), (**f**), (**i**), (**t**), (**N**), (**K**), (**j**), (**S**), (**v**) and (**p**) are known, then its Dividend planned is:

$$D= U+\{S'[1+s]-V-S'f[1+s]-S'i[1+s]\}[1-t]$$
$$-\{N+K+S'j[1+s]/360+Svp/360\}/[1+/]$$

Rule-56274:

If both (**/**), (**U**), (**S'**), (**s**), (**V**), (**f**), (**i**), (**t**), (**D**), (**N**), (**j**), (**S**), (**v**) and (**p**) are known, then its Kind of Cash planned is:

$$K= [1+/](U+\{S'[1+s]-V-S'f[1+s]-S'i[1+s]\}$$
$$[1-t]-D)-\{N+S'j[1+s]/360+Svp/360\}$$

46

Rule-56275:

    If both (**/**), (**U**), (**S'**), (**s**), (**V**), (**f**), (**i**), (**t**), (**D**), (**K**), (**N**), (**S**), (**v**) and (**p**) are known, then its Job or Trade Receivable Days planned is:

$$j = 360[[1+/(U+\{S'[1+s]-V-S'f[1+s]-S'i[1+s]\}$$
$$[1-t]-D)-\{N+K+Svp/360\}]/\{S'[1+s]\}$$

Rule-56276:

    If both (**/**), (**U**), (**S'**), (**s**), (**V**), (**f**), (**i**), (**t**), (**D**), (**K**), (**j**), (**N**), (**v**) and (**p**) are known, then its Sales or Revenue planned is:

$$S = [[1+/(U+\{S'[1+s]-V-S'f[1+s]-S'i[1+s]\}$$
$$[1-t]-D)-\{N+K+S'j[1+s]/360\}]/[vp/360]$$

        or

$$S = S'[1+s]$$

        or

$$S = V/v$$

Rule-56277:

    If both (**/**), (**U**), (**S'**), (**s**), (**V**), (**f**), (**i**), (**t**), (**D**), (**K**), (**j**), (**S**), (**N**) and (**p**) are known, then its Variable Portion planned is:

$$v = 360[[1+/(U+\{S'[1+s]-V-S'f[1+s]-S'i[1+s]\}$$
$$[1-t]-D)-\{N+K+S'j[1+s]/360\}]/[Sp]$$

        or

$$v = V/S$$

47

Steve Asikin ISBN 13: **978-1564912361**, ISBN 10: **1564912363**

Rule-56278:

If both (**/**), (**U**), (**S'**), (**s**), (**V**), (**f**), (**i**), (**t**), (**D**), (**K**), (**j**), (**S**), (**v**) and (**N**) are known, then its Procured Inventory Days planned is:

$$\mathbf{p} = 360[[1+\mathbf{/}(\mathbf{U}+\{\mathbf{S'}[1+\mathbf{s}]-\mathbf{V}-\mathbf{S'f}[1+\mathbf{s}]-\mathbf{S'i}[1+\mathbf{s}]\}$$
$$[1-\mathbf{t}]-\mathbf{D})-\{\mathbf{N}+\mathbf{K}+\mathbf{S'j}[1+\mathbf{s}]/360\}]/[\mathbf{Sv}]$$

Rule-56279:

If both (**/**), (**U**), (**S'**), (**s**), (**V**), (**f**), (**i**), (**t**), (**D**), (**K**), (**j**), (**v**) and (**p**) are known, then its Non Current Asset planned is:

$$\mathbf{N} = [1+\mathbf{/}(\mathbf{U}+\{\mathbf{S'}[1+\mathbf{s}]-\mathbf{V}-\mathbf{S'f}[1+\mathbf{s}]-\mathbf{S'i}[1+\mathbf{s}]\}$$
$$[1-\mathbf{t}]-\mathbf{D})-\mathbf{K}+\mathbf{S'j}[1+\mathbf{s}]/360+\mathbf{S'vp}[1+\mathbf{s}]/360\}$$

Rule-56280:

If both (**N**), (**U**), (**S'**), (**s**), (**V**), (**f**), (**i**), (**t**), (**D**), (**K**), (**j**), (**v**) and (**p**) are known, then its Leverage or Gearing Ratio planned is:

$$\mathbf{/} = \{\mathbf{N}+\mathbf{K}+\mathbf{S'j}[1+\mathbf{s}]/360+\mathbf{S'vp}[1+\mathbf{s}]/360\}$$
$$/(\mathbf{U}+\{\mathbf{S'}[1+\mathbf{s}]-\mathbf{V}-\mathbf{S'f}[1+\mathbf{s}]$$
$$-\mathbf{S'i}[1+\mathbf{s}]\}[1-\mathbf{t}]-\mathbf{D})-1$$

Rule-56281:

If both (**/**), (**N**), (**S'**), (**s**), (**V**), (**f**), (**i**), (**t**), (**D**), (**K**), (**j**), (**v**) and (**p**) are known, then its Utilized or Starting Capital must be:

$$\mathbf{U} = (\{\mathbf{N}+\mathbf{K}+\mathbf{S'j}[1+\mathbf{s}]/360+\mathbf{S'vp}[1+\mathbf{s}]/360\}$$
$$-[1+\mathbf{/}]\{\mathbf{S'}[1+\mathbf{s}]-\mathbf{V}-\mathbf{S'f}[1+\mathbf{s}]$$
$$-\mathbf{S'i}[1+\mathbf{s}]\}[1-\mathbf{t}]-\mathbf{D})/[1+\mathbf{/}]$$

48

Steve Asikin ISBN 13: **978-1564912361**, ISBN 10: **1564912363**

Rule-56282:

If both (**∫**), (**U**), (**N**), (**s**), (**V**), (**f**), (**i**), (**t**), (**D**), (**K**), (**j**), (**v**) and (**p**) are known, then its Sales Past must be:

$$S' = [(\{N+K+S'j[1+s]/360+S'vp[1+s]/360\}/[1+∫]$$
$$-U+D)/[1-t]-V]/\{[1+s][1-f-i]\}$$

or

$$S' = S/[1+s]$$

Rule-56283:

If both (**∫**), (**U**), (**S'**), (**N**), (**V**), (**f**), (**i**), (**t**), (**D**), (**K**), (**j**), (**v**) and (**p**) are known, then its Sales Growth planned is:

$$s = [(\{N+K+S'j[1+s]/360+S'vp[1+s]/360\}/[1+∫]$$
$$-U+D)/[1-t]-V]/\{S'[1-f-i]\}-1$$

or

$$s = S/S'-1$$

Rule-56284:

If both (**∫**), (**U**), (**S'**), (**s**), (**N**), (**f**), (**i**), (**t**), (**D**), (**K**), (**j**), (**v**) and (**p**) are known, then its Variable Cost planned is:

$$V = [[1+∫](U+\{S'[1+s]-S'f[1+s]-S'i[1+s]\}[1-t]-D)$$
$$-\{N+K+S'j[1+s]/360+S'vp[1+s]/360\}]$$
$$/\{[1+∫][1-t]\}$$

or

$$V = Sv$$

Steve Asikin ISBN 13: **978-1564912361**, ISBN 10: **1564912363**

Rule-56285:

If both ($f$), (**U**), (**S'**), (**s**), (**V**), (**N**), (**i**), (**t**), (**D**), (**K**), (**j**), (**v**) and (**p**) are known, then its Fixed Portion planned is:

$$f= [[1+f](U+\{S'[1+s]-S'i[1+s]\}[1-t]-D)$$
$$-\{N+K+S'j[1+s]/360+S'vp[1+s]/360\}]$$
$$/\{S'[1+s][1+f][1-t]\}$$

Rule-56286:

If both ($f$), (**U**), (**S'**), (**s**), (**V**), (**f**), (**N**), (**t**), (**D**), (**K**), (**j**), (**v**) and (**p**) are known, then its Interest Portion planned is:

$$i= [[1+f](U+\{S'[1+s]-S'f[1+s]\}[1-t]-D)$$
$$-\{N+K+S'j[1+s]/360+S'vp[1+s]/360\}]$$
$$/\{S'[1+s][1+f][1-t]\}$$

Rule-56287:

If both ($f$), (**U**), (**S'**), (**s**), (**V**), (**f**), (**i**), (**N**), (**D**), (**K**), (**j**), (**v**) and (**p**) are known, then its Tax Rate planned is:

$$t= 1-\{N+K+S'j[1+s]/360+S'vp[1+s]/360\}/[1+f$$
$$-U+D)/\{S'[1+s]-V-S'f[1+s]-S'i[1+s]\}$$

Rule-56288:

If both ($f$), (**U**), (**S'**), (**s**), (**V**), (**f**), (**i**), (**t**), (**N**), (**K**), (**j**), (**v**) and (**p**) are known, then its Dividend planned is:

$$D= U+\{S'[1+s]-V-S'f[1+s]-S'i[1+s]\}[1-t]$$
$$- \{N+K+S'j[1+s]/360$$
$$+S'vp[1+s]/360\}/[1+f$$

Steve Asikin ISBN 13: **978-1564912361**, ISBN 10: **1564912363**

## Rule-56289:

If both (**/**), (**U**), (**$'**), (**s**), (**V**), (**f**), (**i**), (**t**), (**D**), (**N**), (**j**), (**v**) and (**p**) are known, then its Kind of Cash planned is:

$$K= [1+/(U+\{\$'[1+s]-V-\$'f[1+s]-\$'i[1+s]\}$$
$$[1-t]-D)-N+\$'j[1+s]/360+\$'vp[1+s]/360\}$$

## Rule-56290:

If both (**/**), (**U**), (**$'**), (**s**), (**V**), (**f**), (**i**), (**t**), (**D**), (**K**), (**N**), (**v**) and (**p**) are known, then its Job or Trade Receivable Days planned is:

$$j= 360[[1+/(U+\{\$'[1+s]-V-\$'f[1+s]-\$'i[1+s]\}$$
$$[1-t]-D)-\{N+K+\$'vp[1+s]/360\}]$$
$$/\{\$'[1+s]\}$$

## Rule-56291:

If both (**/**), (**U**), (**$'**), (**s**), (**V**), (**f**), (**i**), (**t**), (**D**), (**K**), (**j**), (**N**) and (**p**) are known, then its Variable Portion planned is:

$$v= 360[[1+/(U+\{\$'[1+s]-V-\$'f[1+s]-\$'i[1+s]\}$$
$$[1-t]-D)-\{N+K+\$'j[1+s]/360\}]$$
$$/\{\$'p[1+s]\}$$

or

$$v= V/\$$$

Steve Asikin ISBN 13: **978-1564912361**, ISBN 10: **1564912363**

Rule-56292:

If both $(\textbf{/})$, $(\textbf{U})$, $(\textbf{S'})$, $(\textbf{s})$, $(\textbf{V})$, $(\textbf{f})$, $(\textbf{i})$, $(\textbf{t})$, $(\textbf{D})$, $(\textbf{K})$, $(\textbf{j})$, $(\textbf{v})$ and $(\textbf{N})$ are known, then its Procured Inventory Days planned is:

$$p = 360[[1+/(\textbf{U}+\{\textbf{S'}[1+s]-\textbf{V}-\textbf{S'f}[1+s]-\textbf{S'i}[1+s]\}$$
$$[1-\textbf{t}]-\textbf{D})-\{\textbf{N}+\textbf{K}+\textbf{S'j}[1+s]/360\}]$$
$$/\{\textbf{S'v}[1+s]\}]$$

Rule-56293:

If both $(\textbf{/})$, $(\textbf{U})$, $(\textbf{S'})$, $(\textbf{s})$, $(\textbf{V})$, $(\textbf{f})$, $(\textbf{i})$, $(\textbf{t})$, $(\textbf{d})$, $(\textbf{K})$, $(\textbf{J})$ and $(\textbf{P})$ are known, then its Non Current Asset planned is:

$$\textbf{N} = [1+/(\textbf{U}+\{\textbf{S'}[1+s]-\textbf{V}-\textbf{S'f}[1+s]$$
$$-\textbf{S'i}[1+s]\}[1-t][1-d])-\textbf{K}-\textbf{J}-\textbf{P}$$

Rule-56294:

If both $(\textbf{N})$, $(\textbf{U})$, $(\textbf{S'})$, $(\textbf{s})$, $(\textbf{V})$, $(\textbf{f})$, $(\textbf{i})$, $(\textbf{t})$, $(\textbf{d})$, $(\textbf{K})$, $(\textbf{J})$ and $(\textbf{P})$ are known, then its Leverage or Gearing Ratio planned is:

$$\textbf{/} = \{\textbf{N}+\textbf{K}+\textbf{J}+\textbf{P}\}/(\textbf{U}+\{\textbf{S'}[1+s]-\textbf{V}-\textbf{S'f}[1+s]$$
$$-\textbf{S'i}[1+s]\}[1-\textbf{t}][1-\textbf{d}])-1$$

Rule-56295:

If both $(\textbf{/})$, $(\textbf{N})$, $(\textbf{S'})$, $(\textbf{s})$, $(\textbf{V})$, $(\textbf{f})$, $(\textbf{i})$, $(\textbf{t})$, $(\textbf{d})$, $(\textbf{K})$, $(\textbf{J})$ and $(\textbf{P})$ are known, then its Utilized or Starting Capital must be:

$$\textbf{U} = (\{\textbf{N}+\textbf{K}+\textbf{J}+\textbf{P}\}-[1+/\{\textbf{S'}[1+s]-\textbf{V}-\textbf{S'f}[1+s]$$
$$-\textbf{S'i}[1+s]\}[1-\textbf{t}][1-\textbf{d}])/[1+/]$$

`

Steve Asikin ISBN 13: **978-1564912361**, ISBN 10: **1564912363**

Rule-56296:

If both $(\textit{\textbf{I}})$, $(\textbf{U})$, $(\textbf{N})$, $(\textbf{s})$, $(\textbf{V})$, $(\textbf{f})$, $(\textbf{i})$, $(\textbf{t})$, $(\textbf{d})$, $(\textbf{K})$, $(\textbf{J})$ and $(\textbf{P})$ are known, then its Sales Past must be:

$$\textbf{S'} = (\{[\textbf{N}+\textbf{K}+\textbf{J}+\textbf{P}]/[1+\textit{\textbf{I}}]\}/\{[1-\textbf{t}][1-\textbf{d}]\}-\textbf{U}+\textbf{V})$$
$$-\{[1+\textbf{s}][1-\textbf{f}-\textbf{i}]\}$$


Rule-56297:

If both $(\textit{\textbf{I}})$, $(\textbf{U})$, $(\textbf{S'})$, $(\textbf{N})$, $(\textbf{V})$, $(\textbf{f})$, $(\textbf{i})$, $(\textbf{t})$, $(\textbf{d})$, $(\textbf{K})$, $(\textbf{J})$ and $(\textbf{P})$ are known, then its Sales Growth planned is:

$$\textbf{s} = (\{[\textbf{N}+\textbf{K}+\textbf{J}+\textbf{P}]/[1+\textit{\textbf{I}}]\}/\{[1-\textbf{t}][1-\textbf{d}]\}-\textbf{U}+\textbf{V})$$
$$-\{\textbf{S'}[1-\textbf{f}-\textbf{i}]\}-1$$


Rule-56298:

If both $(\textit{\textbf{I}})$, $(\textbf{U})$, $(\textbf{S'})$, $(\textbf{s})$, $(\textbf{N})$, $(\textbf{f})$, $(\textbf{i})$, $(\textbf{t})$, $(\textbf{d})$, $(\textbf{K})$, $(\textbf{J})$ and $(\textbf{P})$ are known, then its Variable Cost planned is:

$$\textbf{V} = [[1+\textit{\textbf{I}}](\textbf{U}+\{\textbf{S'}[1+\textbf{s}]-\textbf{S'f}[1+\textbf{s}]-\textbf{S'i}[1+\textbf{s}]\}$$
$$[1-\textbf{t}][1-\textbf{d}])-\{\textbf{N}+\textbf{K}+\textbf{J}+\textbf{P}\}]$$
$$/\{[1+\textit{\textbf{I}}][1-\textbf{t}][1-\textbf{d}]\}$$


Rule-56299:

If both $(\textit{\textbf{I}})$, $(\textbf{U})$, $(\textbf{S'})$, $(\textbf{s})$, $(\textbf{V})$, $(\textbf{N})$, $(\textbf{i})$, $(\textbf{t})$, $(\textbf{d})$, $(\textbf{K})$, $(\textbf{J})$ and $(\textbf{P})$ are known, then its Fixed Portion planned is:

$$\textbf{f} = [[1+\textit{\textbf{I}}](\textbf{U}+\{\textbf{S'}[1+\textbf{s}]-\textbf{V}-\textbf{S'i}[1+\textbf{s}]\}[1-\textbf{t}][1-\textbf{d}])$$
$$-\{\textbf{N}+\textbf{K}+\textbf{J}+\textbf{P}\}]/\{\textbf{S'}[1+\textbf{s}][1+\textit{\textbf{I}}][1-\textbf{t}][1-\textbf{d}]\}$$

Steve Asikin ISBN 13: **978-1564912361**, ISBN 10: **1564912363**

Rule-56300:

If both (**I**), (**U**), (**S'**), (**s**), (**V**), (**f**), (**N**), (**t**), (**d**), (**K**), (**J**) and (**P**) are known, then its Interest Portion planned is:

$$i= [[1+I(U+\{S'[1+s]-V-S'f[1+s]\}[1-t][1-d])$$
$$-\{N+K+J+P\}]/\{S'[1+s][1+I][1-t][1-d]\}$$

Rule-56301:

If both (**I**), (**U**), (**S'**), (**s**), (**V**), (**f**), (**i**), (**N**), (**d**), (**K**), (**J**) and (**P**) are known, then its Tax Rate planned is:

$$t= 1-\{[N+K+J+P]/[1+I]\}/(\{U+\{S'[1+s]-V$$
$$-S'f[1+s]-S'i[1+s]\}[1-d]\})$$

Rule-56302:

If both (**I**), (**U**), (**S'**), (**s**), (**V**), (**f**), (**i**), (**t**), (**N**), (**K**), (**J**) and (**P**) are known, then its Dividend Payout planned is:

$$d= 1-\{[N+K+J+P]/[1+I]\}/(\{U+\{S'[1+s]-V$$
$$-S'f[1+s]-S'i[1+s]\}[1-t]\})$$

Rule-56303:

If both (**I**), (**U**), (**S'**), (**s**), (**V**), (**f**), (**i**), (**t**), (**d**), (**N**), (**J**) and (**P**) are known, then its Kind of Cash planned is:

$$K= [1+I(U+\{S'[1+s]-V-S'f[1+s]-S'i[1+s]\}$$
$$[1-t][1-d])-\{N+J+P\}$$

Steve Asikin ISBN 13: **978-1564912361**, ISBN 10: **1564912363**

Rule-56304:

If both $(\textbf{\textit{I}})$, $(\textbf{U})$, $(\textbf{S'})$, $(\textbf{s})$, $(\textbf{V})$, $(\textbf{f})$, $(\textbf{i})$, $(\textbf{t})$, $(\textbf{d})$, $(\textbf{K})$, $(\textbf{N})$ and $(\textbf{P})$ are known, then its Job or Trade Account Receivable planned is:

$$J = [1 + I(U + \{S'[1+s] - V - S'f[1+s] - S'i[1+s]\}$$
$$[1-t][1-d]) - \{N+K+P\}$$

Rule-56305:

If both $(\textbf{\textit{I}})$, $(\textbf{U})$, $(\textbf{S'})$, $(\textbf{s})$, $(\textbf{V})$, $(\textbf{f})$, $(\textbf{i})$, $(\textbf{t})$, $(\textbf{d})$, $(\textbf{K})$, $(\textbf{J})$ and $(\textbf{N})$ are known, then its Procured Inventory planned is:

$$P = [1 + I(U + \{S'[1+s] - V - S'f[1+s] - S'i[1+s]\}[1-t][1-d]) - \{N+K+J\}$$

Rule-56306:

If both $(\textbf{\textit{I}})$, $(\textbf{U})$, $(\textbf{S'})$, $(\textbf{s})$, $(\textbf{V})$, $(\textbf{f})$, $(\textbf{i})$, $(\textbf{t})$, $(\textbf{d})$, $(\textbf{K})$, $(\textbf{J})$ and $(\textbf{p})$ are known, then its Non Current Asset planned is:

$$N = [1 + I(U + \{S'[1+s] - V - S'f[1+s] - S'i[1+s]\}$$
$$[1-t][1-d]) - K - J - Vp/360$$

Rule-56307:

If both $(\textbf{N})$, $(\textbf{U})$, $(\textbf{S'})$, $(\textbf{s})$, $(\textbf{V})$, $(\textbf{f})$, $(\textbf{i})$, $(\textbf{t})$, $(\textbf{d})$, $(\textbf{K})$, $(\textbf{J})$ and $(\textbf{p})$ are known, then its Leverage or Gearing Ratio planned is:

$$I = [N+K+J+Vp/360]/(U + \{S'[1+s] - V - S'f[1+s]$$
$$-S'i[1+s]\}[1-t][1-d]) - 1$$

Steve Asikin ISBN 13: **978-1564912361**, ISBN 10: **1564912363**

Rule-56308:

If both $(\textbf{\textit{I}})$, $(\textbf{N})$, $(\textbf{S'})$, $(\textbf{s})$, $(\textbf{V})$, $(\textbf{f})$, $(\textbf{i})$, $(\textbf{t})$, $(\textbf{d})$, $(\textbf{K})$, $(\textbf{J})$ and $(\textbf{p})$ are known, then its Utilized or Starting Capital must be:

$$U= ([N+K+J+Vp/360]-[1+I]\{S'[1+s]-V-S'f[1+s]$$
$$-S'i[1+s]\}[1-t][1-d])/[1+I]$$

Rule-56309:

If both $(\textbf{\textit{I}})$, $(\textbf{U})$, $(\textbf{N})$, $(\textbf{s})$, $(\textbf{V})$, $(\textbf{f})$, $(\textbf{i})$, $(\textbf{t})$, $(\textbf{d})$, $(\textbf{K})$, $(\textbf{J})$ and $(\textbf{p})$ are known, then its Sales Past must be:

$$S'= (\{[N+K+J+Vp/360]/[1+I]-U/\{[1-t][1-d]\}+V)$$
$$/\{[1+s][1-f-i]\}$$

Rule-56310:

If both $(\textbf{\textit{I}})$, $(\textbf{U})$, $(\textbf{S'})$, $(\textbf{N})$, $(\textbf{V})$, $(\textbf{f})$, $(\textbf{i})$, $(\textbf{t})$, $(\textbf{d})$, $(\textbf{K})$, $(\textbf{J})$ and $(\textbf{p})$ are known, then its Sales Growth planned is:

$$s= (\{[N+K+J+Vp/360]/[1+I]-U/\{[1-t][1-d]\}+V)$$
$$/\{S'[1-f-i]\}-1$$

Rule-56311:

If both $(\textbf{\textit{I}})$, $(\textbf{U})$, $(\textbf{S'})$, $(\textbf{s})$, $(\textbf{N})$, $(\textbf{f})$, $(\textbf{i})$, $(\textbf{t})$, $(\textbf{d})$, $(\textbf{K})$, $(\textbf{J})$ and $(\textbf{p})$ are known, then its Variable Cost planned is:

$$V= [[1+I](U+\{S'[1+s]-S'f[1+s]-S'i[1+s]\}$$
$$[1-t][1-d])-[N+K+J]]$$
$$/\{p/360+[1+I][1-t][1-d]\}$$

Steve Asikin ISBN 13: **978-1564912361**, ISBN 10: **1564912363**

Rule-56312:

If both (**/**), (**U**), (**S'**), (**s**), (**V**), (**N**), (**i**), (**t**), (**d**), (**K**), (**J**) and (**p**) are known, then its Fixed Portion planned is:

$$f= [[1+\text{/}(U+\{S'[1+s]-V-S'i[1+s]\}[1-t][1-d])$$
$$-[N+K+J+Vp/360]]$$
$$/\{S'[1+s][1+\text{/}][1-t][1-d]\}$$

Rule-56313:

If both (**/**), (**U**), (**S'**), (**s**), (**V**), (**f**), (**N**), (**t**), (**d**), (**K**), (**J**) and (**p**) are known, then its Interest Portion planned is:

$$i= [[1+\text{/}(U+\{S'[1+s]-V-S'f[1+s]\}[1-t][1-d])$$
$$-[N+K+J+Vp/360]]$$
$$/\{S'[1+s][1+\text{/}][1-t][1-d]\}$$

Rule-56314:

If both (**/**), (**U**), (**S'**), (**s**), (**V**), (**f**), (**i**), (**N**), (**d**), (**K**), (**J**) and (**p**) are known, then its Tax Rate planned is:

$$t= 1-\{[N+K+J+Vp/360]/[1+\text{/}-U]\}$$
$$/(\{S'[1+s]-V-S'f[1+s]-S'i[1+s]\}[1-d])$$

Rule-56315:

If both (**/**), (**U**), (**S'**), (**s**), (**V**), (**f**), (**i**), (**t**), (**N**), (**K**), (**J**) and (**p**) are known, then its Dividend Payout planned is:

$$d= 1-\{[N+K+J+Vp/360]/[1+\text{/}-U]\}$$
$$/(\{S'[1+s]-V-S'f[1+s]-S'i[1+s]\}[1-t])$$

Steve Asikin ISBN 13: **978-1564912361**, ISBN 10: **1564912363**

Rule-56316:

If both (**/**), (**U**), (**S'**), (**s**), (**V**), (**f**), (**i**), (**t**), (**d**), (**N**), (**J**) and (**p**) are known, then its Kind of Cash planned is:

$$K = [1 + / (U + \{S'[1+s] - V - S'f[1+s] - S'i[1+s]\}]$$
$$[1-t][1-d]) - [N+J+Vp/360]$$

Rule-56317:

If both (**/**), (**U**), (**S'**), (**s**), (**V**), (**f**), (**i**), (**t**), (**d**), (**K**), (**N**) and (**p**) are known, then its Job or Trade Account Receivable planned is:

$$J = [1 + / (U + \{S'[1+s] \ V \ S'f[1+s] - S'i[1+s]\}]$$
$$[1-t][1-d]) - \{N+K+Vp/360\}$$

Rule-56318:

If both (**/**), (**U**), (**S'**), (**s**), (**V**), (**f**), (**i**), (**t**), (**d**), (**K**), (**J**) and (**N**) are known, then its Procured Inventory Days planned is:

$$p = 360[[1 + / (U + \{S'[1+s] - V - S'f[1+s] - S'i[1+s]\}[1-t][1-d]) - [N+K+J]]/V$$

Rule-56319:

If both (**/**), (**U**), (**S'**), (**s**), (**V**), (**f**), (**i**), (**t**), (**d**), (**K**), (**J**), (**S**), (**v**) and (**p**) are known, then its Non Current Asset planned is:

$$N = [1 + / (U + \{S'[1+s] - V - S'f[1+s] - S'i[1+s]\}]$$
$$[1-t][1-d]) - K - J - Svp/360$$

`

58

Rule-56320:

If both **(N)**, **(U)**, **(S')**, **(s)**, **(V)**, **(f)**, **(i)**, **(t)**, **(d)**, **(K)**, **(J)**, **(S)**, **(v)** and **(p)** are known, then its Leverage or Gearing Ratio planned is:

$$\vDash [\mathbf{N+K+J+Sup}/360]/(\mathbf{U}+\{\mathbf{S'}[1+\mathbf{s}]-\mathbf{V}-\mathbf{S'f}[1+\mathbf{s}]$$
$$-\mathbf{S'i}[1+\mathbf{s}]\}[1-\mathbf{t}][1-\mathbf{d}])-1$$

Rule-56321:

If both **(/)**, **(N)**, **(S')**, **(s)**, **(V)**, **(f)**, **(i)**, **(t)**, **(d)**, **(K)**, **(J)**, **(S)**, **(v)** and **(p)** are known, then its Utilized or Starting Capital must be:

$$\mathbf{U}= ([\mathbf{N+K+Sj}/360+\mathbf{Sup}/360]-[1+\mathbf{/}]\{\mathbf{S'}[1+\mathbf{s}]-\mathbf{V}$$
$$-\mathbf{S'f}[1+\mathbf{s}]-\mathbf{S'i}[1+\mathbf{s}]\}[1-\mathbf{t}][1-\mathbf{d}])/[1+\mathbf{/}]$$

Rule-56322:

If both **(/)**, **(U)**, **(N)**, **(s)**, **(V)**, **(f)**, **(i)**, **(t)**, **(d)**, **(K)**, **(J)**, **(S)**, **(v)** and **(p)** are known, then its Sales Past must be:

$$\mathbf{S'}= (\{[\mathbf{N+K+J+Sup}/360]/[1+\mathbf{/}]-\mathbf{U}\}$$
$$/\{[1-\mathbf{t}][1-\mathbf{d}]\}-\mathbf{V})/\{[1+\mathbf{s}][1-\mathbf{f}-\mathbf{i}]\}$$
$$\text{or}$$
$$\mathbf{S'}= \mathbf{S}/[1+\mathbf{s}]$$

Steve Asikin ISBN 13: **978-1564912361**, ISBN 10: **1564912363**

Rule-56323:

If both (**I**), (**U**), (**S'**), (**N**), (**V**), (**f**), (**i**), (**t**), (**d**), (**K**), (**J**), (**S**), (**v**) and (**p**) are known, then its Sales Growth planned is:

$$s = (\{[\mathbf{N+K+J+Svp}/360]/[1+\mathbf{I}-\mathbf{U}\} / \{[1-\mathbf{t}][1-\mathbf{d}]\} - \mathbf{V})/\{\mathbf{S'}[1-\mathbf{f}-\mathbf{i}]\} - 1$$

or

$$s = \mathbf{S/S'} - 1$$

Rule-56324:

If both (**I**), (**U**), (**S'**), (**s**), (**N**), (**f**), (**i**), (**t**), (**d**), (**K**), (**J**), (**S**), (**v**) and (**p**) are known, then its Variable Cost planned is:

$$\mathbf{V} - [[1+\mathbf{I}](\mathbf{U}+\{\mathbf{S'}[1+s]-\mathbf{S'f}[1+s]-\mathbf{S'l}[1+s]\}$$
$$[1-\mathbf{t}][1-\mathbf{d}]) - [\mathbf{N+K+J+Svp}/360]]$$
$$/\{[1+\mathbf{I}][1-\mathbf{t}][1-\mathbf{d}]\}$$

or

$$\mathbf{V} = \mathbf{Sv}$$

Rule-56325:

If both (**I**), (**U**), (**S'**), (**s**), (**V**), (**N**), (**i**), (**t**), (**d**), (**K**), (**J**), (**S**), (**v**) and (**p**) are known, then its Fixed Portion planned is:

$$\mathbf{f} = [[1+\mathbf{I}](\mathbf{U}+\{\mathbf{S'}[1+s]-\mathbf{V}-\mathbf{S'i}[1+s]\}[1-\mathbf{t}][1-\mathbf{d}])$$
$$- [\mathbf{N+K+J+Svp}/360]]$$
$$/\{\mathbf{S'}[1+s][1+\mathbf{I}][1-\mathbf{t}][1-\mathbf{d}]\}$$

`

Steve Asikin ISBN 13: **978-1564912361**, ISBN 10: **1564912363**

Rule-56326:

If both (**/**), (**U**), (**S'**), (**s**), (**V**), (**f**), (**N**), (**t**), (**d**), (**K**), (**J**), (**S**), (**v**) and (**p**) are known, then its Interest Portion planned is:

$$i= [[1+/(U+\{S'[1+s]-V-S'f[1+s]\}[1-t][1-d])$$
$$-[N+K+J+Svp/360]]$$
$$/\{S'[1+s][1+/[1-t][1-d]\}$$

Rule-56327:

If both (**/**), (**U**), (**S'**), (**s**), (**V**), (**f**), (**i**), (**N**), (**d**), (**K**), (**J**), (**S**), (**v**) and (**p**) are known, then its Tax Rate planned is:

$$t= 1-\{[N+K+J+Svp/360]/[1+/-U\}$$
$$/(\{S'[1+s]-V-S'f[1+s]-S'i[1+s]\}[1-d])$$

Rule-56328:

If both (**/**), (**U**), (**S'**), (**s**), (**V**), (**f**), (**i**), (**t**), (**N**), (**K**), (**J**), (**S**), (**v**) and (**p**) are known, then its Dividend Payout planned is:

$$d= 1-\{[N+K+J+Svp/360]/[1+/-U\}$$
$$/(\{S'[1+s]-V-S'f[1+s]-S'i[1+s]\}[1-t])$$

Rule-56329:

If both (**/**), (**U**), (**S'**), (**s**), (**V**), (**f**), (**i**), (**t**), (**d**), (**N**), (**J**), (**S**), (**v**) and (**p**) are known, then its Kind of Cash planned is:

$$K= [1+/(U+\{S'[1+s]-V-S'f[1+s]-S'i[1+s]\}$$
$$[1-t][1-d])-[N+J+Svp/360]$$

Steve Asikin ISBN 13: **978-1564912361**, ISBN 10: **1564912363**

Rule-56330:

If both (**/**), (**U**), (**S'**), (**s**), (**V**), (**f**), (**i**), (**t**), (**d**), (**K**), (**N**), (**S**), (**v**) and (**p**) are known, then its Job or Trade Account Receivable planned is:

$$J= [1+/(U+\{S'[1+s]-V-S'f[1+s]-S'i[1+s]\}$$
$$[1-t][1-d])-\{N+K+Svp/360\}$$

Rule-56331:

If both (**/**), (**U**), (**S'**), (**s**), (**V**), (**f**), (**i**), (**t**), (**d**), (**K**), (**J**), (**N**), (**v**) and (**p**) are known, then its Sales or Revenue planned is:

$$S= 360[[1+/(U+\{S'[1+s]-V-S'f[1+s]-S'i[1+s]\}$$
$$[1-t][1-d])-[N+K+J]]/[vp]$$
or
$$S= S'[1+s]$$
or
$$S= V/v$$

Rule-56332:

If both (**/**), (**U**), (**S'**), (**s**), (**V**), (**f**), (**i**), (**t**), (**d**), (**K**), (**J**), (**S**), (**N**) and (**p**) are known, then its Variable Portion planned is:

$$v= 360[[1+/(U+\{S'[1+s]-V-S'f[1+s]-S'i[1+s]\}$$
$$[1-t][1-d])-[N+K+J]]/[Sp]$$
or
$$v= V/S$$

Steve Asikin ISBN 13: **978-1564912361**, ISBN 10: **1564912363**

Rule-56333:

If both (**/**), (**U**), (**S'**), (**s**), (**V**), (**f**), (**i**), (**t**), (**d**), (**K**), (**J**), (**S**), (**v**) and (**N**) are known, then its Procured Inventory Days planned is:

$$\mathbf{p}= 360[[1+\mathbf{/}(\mathbf{U}+\{\mathbf{S'}[1+\mathbf{s}]-\mathbf{V}-\mathbf{S'f}[1+\mathbf{s}]-\mathbf{S'i}[1+\mathbf{s}]\}$$
$$[1-\mathbf{t}][1-\mathbf{d}])-[\mathbf{N}+\mathbf{K}+\mathbf{J}]]/[\mathbf{Sv}]$$

Rule-56334:

If both (**/**), (**U**), (**S'**), (**s**), (**V**), (**f**), (**i**), (**t**), (**d**), (**K**), (**J**), (**v**) and (**p**) are known, then its Non Current Asset planned is:

$$\mathbf{N}= [1+\mathbf{/}(\mathbf{U}+\{\mathbf{S'}[1+\mathbf{s}]-\mathbf{V}-\mathbf{S'f}[1+\mathbf{s}]-\mathbf{S'i}[1+\mathbf{s}]\}$$
$$[1-\mathbf{t}][1-\mathbf{d}])-\mathbf{K}-\mathbf{J}-\mathbf{S'vp}[1+\mathbf{s}]/360$$

Rule-56335:

If both (**N**), (**U**), (**S'**), (**s**), (**V**), (**f**), (**i**), (**t**), (**d**), (**K**), (**J**), (**v**) and (**p**) are known, then its Leverage or Gearing Ratio planned is:

$$\mathbf{/}= \{\mathbf{N}+\mathbf{K}+\mathbf{J}+\mathbf{S'vp}[1+\mathbf{s}]/360\}/(\mathbf{U}+\{\mathbf{S'}[1+\mathbf{s}]-\mathbf{V}$$
$$-\mathbf{S'f}[1+\mathbf{s}]-\mathbf{S'i}[1+\mathbf{s}]\}[1-\mathbf{t}][1-\mathbf{d}])-1$$

Rule-56336:

If both (**/**), (**N**), (**S'**), (**s**), (**V**), (**f**), (**i**), (**t**), (**d**), (**K**), (**J**), (**v**) and (**p**) are known, then its Utilized or Starting Capital must be:

$$\mathbf{U}= (\{\mathbf{N}+\mathbf{K}+\mathbf{J}+\mathbf{S'vp}[1+\mathbf{s}]/360\}-[1+\mathbf{/}\{\mathbf{S'}[1+\mathbf{s}]-\mathbf{V}$$
$$-\mathbf{S'f}[1+\mathbf{s}]-\mathbf{S'i}[1+\mathbf{s}]\}[1-\mathbf{t}][1-\mathbf{d}])/[1+\mathbf{/}]$$

Steve Asikin ISBN 13: **978-1564912361**, ISBN 10: **1564912363**

Rule-56337:

If both ($I$), ($U$), ($N$), ($s$), ($V$), ($f$), ($i$), ($t$), ($d$), ($K$), ($J$), ($v$) and ($p$) are known, then its Sales Past must be:

$$S' = [(\{N+K+J+S'vp[1+s]/360\}/[1+I-U) \\ /\{[1-t][1-d]\}-V]/\{[1+s][1-f-i]\})$$

Rule-56338:

If both ($I$), ($U$), ($S'$), ($N$), ($V$), ($f$), ($i$), ($t$), ($d$), ($K$), ($J$), ($v$) and ($p$) are known, then its Sales Growth planned is:

$$s = [(\{N+K+J+S'vp[1+s]/360\}/[1+I-U) \\ /\{[1-t][1-d]\}-V]/\{S'[1-f-i]\})-1$$

Rule-56339:

If both ($I$), ($U$), ($S'$), ($s$), ($N$), ($f$), ($i$), ($t$), ($d$), ($K$), ($J$), ($v$) and ($p$) are known, then its Variable Cost planned is:

$$V = [[1+I(U+\{S'[1+s]-S'f[1+s]-S'i[1+s]\} \\ [1-t][1-d])-\{N+K+J+S'vp[1+s]/360\}] \\ /\{[1+I][1-t][1-d]\}$$

Rule-56340:

If both ($I$), ($U$), ($S'$), ($s$), ($V$), ($N$), ($i$), ($t$), ($d$), ($K$), ($J$), ($v$) and ($p$) are known, then its Fixed Portion planned is:

$$f = [[1+I(U+\{S'[1+s]-V-S'i[1+s]\}[1-t][1-d]) \\ -\{N+K+J+S'vp[1+s]/360\}] \\ /\{S'[1+s][1+I][1-t][1-d]\}$$

`

Steve Asikin ISBN 13: **978-1564912361**, ISBN 10: **1564912363**

Rule-56341:

If both (**/**), (**U**), (**S'**), (**s**), (**V**), (**f**), (**N**), (**t**), (**d**), (**K**), (**J**), (**v**) and (**p**) are known, then its Interest Portion planned is:

$$i= [[1+/(U+\{S'[1+s]-V-S'f[1+s]\}[1-t][1-d])$$
$$-\{N+K+J+S'vp[1+s]/360\}]$$
$$/\{S'[1+s][1+/[1-t][1-d]\}$$

Rule-56342:

If both (**/**), (**U**), (**S'**), (**s**), (**V**), (**f**), (**i**), (**N**), (**d**), (**K**), (**J**), (**v**) and (**p**) are known, then its Tax Rate planned is:

$$t= 1-(\{N+K+J+S'vp[1+s]/360\}/[1+/]-U)$$
$$/(\{S'[1+s]-V-S'f[1+s]-S'i[1+s]\}[1-d])$$

Rule-56343:

If both (**/**), (**U**), (**S'**), (**s**), (**V**), (**f**), (**i**), (**t**), (**N**), (**K**), (**J**), (**v**) and (**p**) are known, then its Dividend Payout planned is:

$$d= 1-(\{N+K+J+S'vp[1+s]/360\}/[1+/]-U)$$
$$/(\{S'[1+s]-V-S'f[1+s]-S'i[1+s]\}[1-t])$$

Rule-56344:

If both (**/**), (**U**), (**S'**), (**s**), (**V**), (**f**), (**i**), (**t**), (**d**), (**N**), (**J**), (**v**) and (**p**) are known, then its Kind of Cash planned is:

$$K= [1+/(U+\{S'[1+s]-V-S'f[1+s]-S'i[1+s]\}$$
$$[1-t][1-d])-\{N+J+S'vp[1+s]/360\}$$

Steve Asikin ISBN 13: **978-1564912361**, ISBN 10: **1564912363**

Rule-56345:

If both $(\textbf{\textit{I}})$, $(\textbf{U})$, $(\textbf{S'})$, $(\textbf{s})$, $(\textbf{V})$, $(\textbf{f})$, $(\textbf{i})$, $(\textbf{t})$, $(\textbf{d})$, $(\textbf{K})$, $(\textbf{N})$, $(\textbf{v})$ and $(\textbf{p})$ are known, then its Job or Trade Account Receivable planned is:

$$J= [[1+I](U+\{S'[1+s]-V-S'f[1+s]-S'i[1+s]\}$$
$$[1-t][1-d])-\{N+K+S'vp[1+s]/360\}$$

Rule-56346:

If both $(\textbf{\textit{I}})$, $(\textbf{U})$, $(\textbf{S'})$, $(\textbf{s})$, $(\textbf{V})$, $(\textbf{f})$, $(\textbf{i})$, $(\textbf{t})$, $(\textbf{d})$, $(\textbf{K})$, $(\textbf{J})$, $(\textbf{N})$ and $(\textbf{p})$ are known, then its Variable Portion planned is:

$$v= 360[[1+I](U+\{S'[1+s]-V-S'f[1+s]-S'i[1+s]\}$$
$$[1-t][1-d])-[N+K+J]]/\{S'p[1+s]\}$$

Rule-56347:

If both $(\textbf{\textit{I}})$, $(\textbf{U})$, $(\textbf{S'})$, $(\textbf{s})$, $(\textbf{V})$, $(\textbf{f})$, $(\textbf{i})$, $(\textbf{t})$, $(\textbf{d})$, $(\textbf{K})$, $(\textbf{J})$, $(\textbf{v})$ and $(\textbf{N})$ are known, then its Procured Inventory Days planned is:

$$p= 360[[1+I](U+\{S'[1+s]-V-S'f[1+s]-S'i[1+s]\}$$
$$[1-t][1-d])-[N+K+J]]/\{S'v[1+s]\}$$

Rule-56348:

If both $(\textbf{\textit{I}})$, $(\textbf{U})$, $(\textbf{S'})$, $(\textbf{s})$, $(\textbf{V})$, $(\textbf{f})$, $(\textbf{i})$, $(\textbf{t})$, $(\textbf{d})$, $(\textbf{K})$, $(\textbf{S})$, $(\textbf{j})$ and $(\textbf{P})$ are known, then its Non Current Asset planned is:

$$N= [1+I](U+\{S'[1+s]-V-S'f[1+s]-S'i[1+s]\}$$
$$[1-t][1-d])-K-Sj/360-P$$

Steve Asikin ISBN 13: **978-1564912361**, ISBN 10: **1564912363**

Rule-56349:

If both **(N)**, **(U)**, **(S')**, **(s)**, **(V)**, **(f)**, **(i)**, **(t)**, **(d)**, **(K)**, **(S)**, **(j)** and **(P)** are known, then its Leverage or Gearing Ratio planned is:

$$\textit{I} = \{N+K+Sj/360+P\}/(U+\{S'[1+s]-V-S'f[1+s] -S'i[1+s]\}[1-t][1-d])-1$$

Rule-56350:

If both **(I)**, **(N)**, **(S')**, **(s)**, **(V)**, **(f)**, **(i)**, **(t)**, **(d)**, **(K)**, **(S)**, **(j)** and **(P)** are known, then its Utilized or Starting Capital must be:

$$U = (\{N+K+Sj/360+P\}-[1+\textit{I}]\{S'[1+s]-V-S'f[1+s] -S'i[1+s]\}[1-t][1-d])/[1+\textit{I}]$$

Rule-56351:

If both **(I)**, **(U)**, **(N)**, **(s)**, **(V)**, **(f)**, **(i)**, **(t)**, **(d)**, **(K)**, **(S)**, **(j)** and **(P)** are known, then its Sales Past must be:

$$S' = (\{[N+K+Sj/360+P]/[1+\textit{I}]-U\} /\{[1-t][1-d]\}-V)/\{[1+s][1-f-i]\}$$

Rule-56352:

If both **(I)**, **(U)**, **(S')**, **(N)**, **(V)**, **(f)**, **(i)**, **(t)**, **(d)**, **(K)**, **(S)**, **(j)** and **(P)** are known, then its Sales Growth planned is:

$$s = (\{[N+K+Sj/360+P]/[1+\textit{I}]-U\} /\{[1-t][1-d]\}-V)/\{S'[1-f-i]\}-1$$

Steve Asikin ISBN 13: **978-1564912361**, ISBN 10: **1564912363**

Rule-56353:

If both (**I**), (**U**), (**S'**), (**s**), (**N**), (**f**), (**i**), (**t**), (**d**), (**K**), (**S**), (**j**) and (**P**) are known, then its Variable Cost planned is:

$$V= [[1+I](U+\{S'[1+s]-S'f[1+s]-S'i[1+s]\}$$
$$[1-t][1-d])-\{N+K+Sj/360+P\}]$$
$$/\{[1+I][1-t][1-d]\}$$

Rule-56354:

If both (**I**), (**U**), (**S'**), (**s**), (**V**), (**N**), (**i**), (**t**), (**d**), (**K**), (**S**), (**j**) and (**P**) are known, then its Fixed Portion planned is:

$$f= [[1+I](U+\{S'[1+s]-V-S'i[1+s]\}[1-t][1-d])$$
$$-\{N+K+Sj/360+P\}]$$
$$/\{S'[1+s][1+I][1-t][1-d]\}$$

Rule-56355:

If both (**I**), (**U**), (**S'**), (**s**), (**V**), (**f**), (**N**), (**t**), (**d**), (**K**), (**S**), (**j**) and (**P**) are known, then its Interest Portion planned is:

$$i= [[1+I](U+\{S'[1+s]-V-S'f[1+s]\}[1-t][1-d])$$
$$-\{N+K+Sj/360+P\}]$$
$$/\{S'[1+s][1+I][1-t][1-d]\}$$

Rule-56356:

If both (**I**), (**U**), (**S'**), (**s**), (**V**), (**f**), (**i**), (**N**), (**d**), (**K**), (**S**), (**j**) and (**P**) are known, then its Tax Rate planned is:

$$t= 1-\{[N+K+Sj/360+P]/[1+I]-U\}$$
$$/(\{S'[1+s]-V-S'f[1+s]-S'i[1+s]\}[1-d])$$

`

68

Rule-56357:

If both **(/)**, **(U)**, **(S')**, **(s)**, **(V)**, **(f)**, **(i)**, **(t)**, **(N)**, **(K)**, **(S)**, **(j)** and **(P)** are known, then its Dividend Payout planned is:

$$\mathbf{d}= 1-\{[\mathbf{N}+\mathbf{K}+\mathbf{S}j/360+\mathbf{P}]/[1+\mathbf{/}-\mathbf{U}\}$$
$$/(\{\mathbf{S'}[1+s]-\mathbf{V}-\mathbf{S'f}[1+s]-\mathbf{S'i}[1+s]\}[1-\mathbf{t}])$$

Rule-56358:

If both **(/)**, **(U)**, **(S')**, **(s)**, **(V)**, **(f)**, **(i)**, **(t)**, **(d)**, **(N)**, **(S)**, **(j)** and **(P)** are known, then its Kind of Cash planned is:

$$\mathbf{K}= [1+\mathbf{/}](\mathbf{U}+\{\mathbf{S'}[1+s]-\mathbf{V}-\mathbf{S'f}[1+s]-\mathbf{S'i}[1+s]\}$$
$$[1-\mathbf{t}][1-\mathbf{d}])-[\mathbf{N}+\mathbf{S}j/360+\mathbf{P}]$$

Rule-56359:

If both **(/)**, **(U)**, **(S')**, **(s)**, **(V)**, **(f)**, **(i)**, **(t)**, **(d)**, **(K)**, **(N)**, **(j)** and **(P)** are known, then its Sales or Revenue planned is:

$$\mathbf{S}= 360[[1+\mathbf{/}](\mathbf{U}+\{\mathbf{S'}[1+s]-\mathbf{V}-\mathbf{S'i}[1+s]\}$$
$$[1-\mathbf{t}][1-\mathbf{d}])-\{\mathbf{N}+\mathbf{K}+\mathbf{P}\}]/\mathbf{j}$$

or

$$\mathbf{S}= \mathbf{S'}[1+s]$$

Steve Asikin ISBN 13: **978-1564912361**, ISBN 10: **1564912363**

## Rule-56360:

If both $(I)$, $(U)$, $(S')$, $(s)$, $(V)$, $(f)$, $(i)$, $(t)$, $(d)$, $(K)$, $(S)$, $(N)$ and $(P)$ are known, then its Job or Trade Receivable Days planned is:

$$j = 360[[1+I(U+\{S'[1+s]-V-S'f[1+s]-S'i[1+s]\}[1-t][1-d])-\{N+K+P\}]/S$$

## Rule-56361:

If both $(I)$, $(U)$, $(S')$, $(s)$, $(V)$, $(f)$, $(i)$, $(t)$, $(d)$, $(K)$, $(S)$, $(j)$ and $(N)$ are known, then its Procured Inventory Days planned is:

$$P = [1+I(U+\{S'[1+s]-V-S'f[1+s]-S'i[1+s]\}[1-t][1-d])-\{N+K+Sj/360\}$$

## Rule-56362:

If both $(I)$, $(U)$, $(S')$, $(s)$, $(V)$, $(f)$, $(i)$, $(t)$, $(d)$, $(K)$, $(S)$, $(j)$ and $(p)$ are known, then its Non Current Asset planned is:

$$N = [1+I(U+\{S'[1+s]-V-S'f[1+s]-S'i[1+s]\}[1-t][1-d])-K-Sj/360-Vp/360$$

## Rule-56363:

If both $(N)$, $(U)$, $(S')$, $(s)$, $(V)$, $(f)$, $(i)$, $(t)$, $(d)$, $(K)$, $(S)$, $(j)$ and $(p)$ are known, then its Leverage or Gearing Ratio planned is:

$$I = [N+K+Sj/360+Vp/360]/(U+\{S'[1+s]-V-S'f[1+s]-S'i[1+s]\}[1-t][1-d])-1$$

Steve Asikin ISBN 13: **978-1564912361**, ISBN 10: **1564912363**

Rule-56364:

If both (**/**), (**N**), (**S'**), (**s**), (**V**), (**f**), (**i**), (**t**), (**d**), (**K**), (**S**), (**j**) and (**p**) are known, then its Utilized or Starting Capital must be:

$$U= ([N+K+Sj/360+Vp/360]-[1+/]\{S'[1+s]-V$$
$$-S'f[1+s]-S'i[1+s]\}[1-t][1-d])/[1+/]$$

Rule-56365:

If both (**/**), (**U**), (**N**), (**s**), (**V**), (**f**), (**i**), (**t**), (**d**), (**K**), (**S**), (**j**) and (**p**) are known, then its Sales Past must be:

$$S'= (\{[N+K+Sj/360+Vp/360]/[1+/]-U\}$$
$$/\{[1-t][1-d]\}-V)/\{[1+s][1-f-i]\}$$

or

$$S'= S[1+s]$$

Rule-56366:

If both (**/**), (**U**), (**S'**), (**N**), (**V**), (**f**), (**i**), (**t**), (**d**), (**K**), (**S**), (**j**) and (**p**) are known, then its Sales Growth planned is:

$$s= (\{[N+K+Sj/360+Vp/360]/[1+/]-U\}$$
$$/\{[1-t][1-d]\}-V)/\{S'[1-f-i]\}-1$$

or

$$s= S/S'-1$$

Steve Asikin ISBN 13: **978-1564912361**, ISBN 10: **1564912363**

<u>Rule-56367</u>:

If both (**/**), (**U**), (**S'**), (**s**), (**N**), (**f**), (**i**), (**t**), (**d**), (**K**), (**S**), (**j**) and (**p**) are known, then its Variable Cost planned is:

$$V= [[1+/](U+\{S'[1+s]\ S'f[1+s]-S'i[1+s]\}$$
$$[1-t][1-d])-[N+K+Sj/360]]$$
$$/\{p/360+[1+/][1-t][1-d]\}$$

<u>Rule-56368</u>:

If both (**/**), (**U**), (**S'**), (**s**), (**V**), (**N**), (**i**), (**t**), (**d**), (**K**), (**S**), (**j**) and (**p**) are known, then its Fixed Portion planned is:

$$f= [[1+/](U+\{S'[1+s]-V-S'i[1+s]\}[1-t][1-d])$$
$$-[N+K+Sj/360+Vp/360]]$$
$$/\{S'[1+s][1+/][1-t][1-d]\}$$

<u>Rule-56369</u>:

If both (**/**), (**U**), (**S'**), (**s**), (**V**), (**f**), (**N**), (**t**), (**d**), (**K**), (**S**), (**j**) and (**p**) are known, then its Interest Portion planned is:

$$i= [[1+/](U+\{S'[1+s]-V-S'f[1+s]\}[1-t][1-d])$$
$$-[N+K+Sj/360+Vp/360]]$$
$$/\{S'[1+s][1+/][1-t][1-d]\}$$

`

Steve Asikin ISBN 13: **978-1564912361**, ISBN 10: **1564912363**

Rule-56370:

If both (**∫**), (**U**), (**S'**), (**s**), (**V**), (**f**), (**i**), (**N**), (**d**), (**K**), (**S**), (**j**) and (**p**) are known, then its Tax Rate planned is:

$$t = 1 - \{[N+K+Sj/360+Vp/360]/[1+\text{∫}-U\}$$
$$/(\{S'[1+s]-V-S'f[1+s]-S'i[1+s]\}[1-d])$$

Rule-56371:

If both (**∫**), (**U**), (**S'**), (**s**), (**V**), (**f**), (**i**), (**t**), (**N**), (**K**), (**S**), (**j**) and (**p**) are known, then its Dividend Payout planned is:

$$d = 1 - \{[N+K+Sj/360+Vp/360]/[1+\text{∫}-U\}$$
$$/(\{S'[1+s]-V-S'f[1+s]-S'i[1+s]\}[1-t])$$

Rule-56372:

If both (**∫**), (**U**), (**S'**), (**s**), (**V**), (**f**), (**i**), (**t**), (**d**), (**N**), (**S**), (**j**) and (**p**) are known, then its Kind of Cash planned is:

$$K = [1+\text{∫}(U+\{S'[1+s]-V-S'f[1+s]-S'i[1+s]\}$$
$$[1-t][1-d])-[N+Sj/360+Vp/360]$$

Rule-56373:

If both (**∫**), (**U**), (**S'**), (**s**), (**V**), (**f**), (**i**), (**t**), (**d**), (**K**), (**N**), (**j**) and (**p**) are known, then its Sales or Revenue planned is:

$$S = 360[[1+\text{∫}(U+\{S'[1+s]-V-S'f[1+s]-S'i[1+s]\}$$
$$[1-t][1-d])-[N+K+Vp/360]]/j$$

or

$$S = S'[1+s]$$

Steve Asikin ISBN 13: **978-1564912361**, ISBN 10: **1564912363**

## Rule-56374:

If both ($\int$), (**U**), (**S'**), (**s**), (**V**), (**f**), (**i**), (**t**), (**d**), (**K**), (**S**), (**N**) and (**p**) are known, then its Job or Trade Receivable Days planned is:

$$j = 360[[1+\int(U+\{S'[1+s]-V-S'f[1+s]-S'i[1+s]\} [1-t][1-d])-[N+K+Vp/360]]/S$$

## Rule-56375:

If both ($\int$), (**U**), (**S'**), (**s**), (**V**), (**f**), (**i**), (**t**), (**d**), (**K**), (**S**), (**j**) and (**N**) are known, then its Procured Inventory Days planned is:

$$p = 360[[1+\int(U+\{S'[1+s]-V-S'f[1+s]-S'i[1+s]\} [1-t][1-d])-[N+K+Sj/360]]/V$$

## Rule-56376:

If both ($\int$), (**U**), (**S'**), (**s**), (**V**), (**f**), (**i**), (**t**), (**d**), (**K**), (**S**), (**j**), (**v**) and (**p**) are known, then its Non Current Asset planned is:

$$N = [1+\int(U+\{S'[1+s]-V-S'f[1+s]-S'i[1+s]\} [1-t][1-d])-K-Sj/360-Svp/360$$

## Rule-56377:

If both (**N**), (**U**), (**S'**), (**s**), (**V**), (**f**), (**i**), (**t**), (**d**), (**K**), (**S**), (**j**), (**v**) and (**p**) are known, then its Leverage or Gearing Ratio planned is:

$$\int = [N+K+Sj/360+Svp/360]/(U+\{S'[1+s]-V -S'f[1+s]-S'i[1+s]\}[1-t][1-d])-1$$

Steve Asikin ISBN 13: **978-1564912361**, ISBN 10: **1564912363**

Rule-56378:

If both (**/**), (**N**), (**S'**), (**s**), (**V**), (**f**), (**i**), (**t**), (**d**), (**K**), (**S**), (**j**), (**v**) and (**p**) are known, then its Utilized or Starting Capital must be:

$$U= [N+K+Sj/360+Svp/360]/[1+/-\{S'[1+s]-V$$
$$-S'f[1+s]-S'i[1+s]\}[1-t][1-d]$$

Rule-56379:

If both (**/**), (**U**), (**N**), (**s**), (**V**), (**f**), (**i**), (**t**), (**d**), (**K**), (**S**), (**j**), (**v**) and (**p**) are known, then its Sales Past must be:

$$S'= ([N+K+Sj/360+Svp/360]/[1+/-U\}$$
$$/\{[1-t][1-d]\}-V)/\{[1+s][1-f-i]\}$$
$$\text{or}$$
$$S'= S/[1+s]$$

Rule-56380:

If both (**/**), (**U**), (**S'**), (**N**), (**V**), (**f**), (**i**), (**t**), (**d**), (**K**), (**S**), (**j**), (**v**) and (**p**) are known, then its Sales Growth planned is:

$$s= ([N+K+Sj/360+Svp/360]/[1+/-U\}$$
$$/\{[1-t][1-d]\}-V)/\{S'[1-f-i]\}-1$$
$$\text{or}$$
$$s= S/S'-1$$

Steve Asikin ISBN 13: **978-1564912361**, ISBN 10: **1564912363**

Rule-56381:

If both (**/**), (**U**), (**S'**), (**s**), (**N**), (**f**), (**i**), (**t**), (**d**), (**K**), (**S**), (**j**), (**v**) and (**p**) are known, then its Variable Cost planned is:

$$V= [[1+/](U+\{S'[1+s]-S'f[1+s]-S'i[1+s]\}$$
$$[1-t][1-d])-[N+K+Sj/360+Svp/360]]$$
$$/\{[1+/][1-t][1-d]\}$$

or

$$V= Sv$$

Rule-56382:

If both (**/**), (**U**), (**S'**), (**s**), (**V**), (**N**), (**i**), (**t**), (**d**), (**K**), (**S**), (**j**), (**v**) and (**p**) are known, then its Fixed Portion planned is:

$$f= [[1+/](U+\{S'[1+s]-V-S'i[1+s]\}[1-t][1-d])$$
$$-[N+K+Sj/360+Svp/360]]$$
$$/\{S'[1+s][1+/][1-t][1-d]\}$$

Rule-56383:

If both (**/**), (**U**), (**S'**), (**s**), (**V**), (**f**), (**N**), (**t**), (**d**), (**K**), (**S**), (**j**), (**v**) and (**p**) are known, then its Interest Portion planned is:

$$i= [[1+/](U+\{S'[1+s]-V-S'f[1+s]\}[1-t][1-d])$$
$$-[N+K+Sj/360+Svp/360]]$$
$$/\{S'[1+s][1+/][1-t][1-d]\}$$

Steve Asikin ISBN 13: **978-1564912361**, ISBN 10: **1564912363**

Rule-56384:

If both **(I)**, **(U)**, **(S')**, **(s)**, **(V)**, **(f)**, **(i)**, **(N)**, **(d)**, **(K)**, **(S)**, **(j)**, **(v)** and **(p)** are known, then its Tax Rate planned is:

$$t= 1-[N+K+Sj/360+Svp/360]/[1+I-U\}$$
$$/(\{S'[1+s]-V-S'f[1+s]-S'i[1+s]\}[1-t])$$

Rule-56385:

If both **(I)**, **(U)**, **(S')**, **(s)**, **(V)**, **(f)**, **(i)**, **(t)**, **(N)**, **(K)**, **(S)**, **(j)**, **(v)** and **(p)** are known, then its Dividend Payout planned is:

$$d= 1-[N+K+Sj/360+Svp/360]/[1+I-U\}$$
$$/(\{S'[1+s]-V-S'f[1+s]-S'i[1+s]\}[1-t])$$

Rule-56386:

If both **(I)**, **(U)**, **(S')**, **(s)**, **(V)**, **(f)**, **(i)**, **(t)**, **(d)**, **(N)**, **(S)**, **(j)**, **(v)** and **(p)** are known, then its Kind of Cash planned is:

$$K= [1+I](U+\{S'[1+s]-V-S'f[1+s]-S'i[1+s]\}$$
$$[1-t][1-d])-[N+Sj/360+Svp/360]$$

Steve Asikin ISBN 13: **978-1564912361**, ISBN 10: **1564912363**

Rule-56387:

If both (**/**), (**U**), (**$'**), (**s**), (**V**), (**f**), (**i**), (**t**), (**d**), (**K**), (**N**), (**j**), (**v**) and (**p**) are known, then its Sales or Revenue planned is:

$$\$= [[1+\text{/}(\textbf{U}+\{\$'[1+s]-\textbf{V}-\$'\textbf{f}[1+s]-\$'\textbf{i}[1+s]\}$$
$$[1-\textbf{t}][1-\textbf{d}])-[\textbf{N}+\textbf{K}]]/[\textbf{j}/360+\textbf{vp}/360]$$

or

$$\$= \textbf{V}/\textbf{v}$$

or

$$\$= \$'[1+s]$$

Rule-56388:

If both (**/**), (**U**), (**$'**), (**s**), (**V**), (**f**), (**i**), (**t**), (**d**), (**K**), (**$**), (**N**), (**v**) and (**p**) are known, then its Job or Trade Receivable Days planned is:

$$\textbf{j}= 360[[1+\text{/}(\textbf{U}+\{\$'[1+s]-\textbf{V}-\$'\textbf{f}[1+s]-\$'\textbf{i}[1+s]\}$$
$$[1-\textbf{t}][1-\textbf{d}])-\{\textbf{N}+\textbf{K}+\$\textbf{vp}/360\}]/\$$$

Rule-56389:

If both (**/**), (**U**), (**$'**), (**s**), (**V**), (**f**), (**i**), (**t**), (**d**), (**K**), (**$**), (**j**), (**N**) and (**p**) are known, then its Variable Portion planned is:

$$\textbf{v}= 360[[1+\text{/}(\textbf{U}+\{\$'[1+s]-\textbf{V}-\$'\textbf{f}[1+s]$$
$$-\$'\textbf{i}[1+s]\}[1-\textbf{t}][1-\textbf{d}])-[\textbf{N}+\textbf{K}+\$\textbf{j}/360]]/[\$\textbf{p}]$$

or

$$\textbf{v}= \textbf{V}/\$$$

Steve Asikin ISBN 13: **978-1564912361**, ISBN 10: **1564912363**

Rule-56390:

If both (**/**), (**U**), (**S'**), (**s**), (**V**), (**f**), (**i**), (**t**), (**d**), (**K**), (**S**), (**j**), (**v**) and (**N**) are known, then its Procured Inventory Days planned is:

$$\mathbf{p}= 360[[1+\mathbf{/}(\mathbf{U}+\{\mathbf{S'}[1+\mathbf{s}]\text{-}\mathbf{V}\text{-}\mathbf{Sf}\text{-}\mathbf{S'f}[1+\mathbf{s}]$$
$$\text{-}\mathbf{S'i}[1+\mathbf{s}]\}[1\text{-}\mathbf{t}][1\text{-}\mathbf{d}])\text{-}[\mathbf{N}+\mathbf{K}+\mathbf{Sj}/360]]/[\mathbf{Sv}]$$

Rule-56391:

If both (**/**), (**U**), (**S'**), (**s**), (**V**), (**f**), (**i**), (**t**), (**d**), (**K**), (**S**), (**j**), (**v**) and (**p**) are known, then its Non Current Asset planned is:

$$\mathbf{N}= [1+\mathbf{/}(\mathbf{U}+\{\mathbf{S'}[1+\mathbf{s}]\text{-}\mathbf{V}\text{-}\mathbf{S'f}[1+\mathbf{s}]\text{-}\mathbf{S'i}[1+\mathbf{s}]\}$$
$$[1\text{-}\mathbf{t}][1\text{-}\mathbf{d}])\text{-}\mathbf{K}\text{-}\mathbf{Sj}/360\text{-}\mathbf{S'vp}[1+\mathbf{s}]/360$$

Rule-56392:

If both (**N**), (**U**), (**S'**), (**s**), (**V**), (**f**), (**i**), (**t**), (**d**), (**K**), (**S**), (**j**), (**v**) and (**p**) are known, then its Leverage or Gearing Ratio planned is:

$$\mathbf{/}= \{\mathbf{N}+\mathbf{K}+\mathbf{Sj}/360+\mathbf{S'vp}[1+\mathbf{s}]/360\}/(\mathbf{U}+\{\mathbf{S'}[1+\mathbf{s}]\text{-}\mathbf{V}$$
$$\text{-}\mathbf{S'f}[1+\mathbf{s}]\text{-}\mathbf{S'i}[1+\mathbf{s}]\}[1\text{-}\mathbf{t}][1\text{-}\mathbf{d}])\text{-}1$$

Rule-56393:

If both (**/**), (**N**), (**S'**), (**s**), (**V**), (**f**), (**i**), (**t**), (**d**), (**K**), (**S**), (**j**), (**v**) and (**p**) are known, then its Utilized or Starting Capital must be:

$$\mathbf{U}= (\{\mathbf{N}+\mathbf{K}+\mathbf{Sj}/360+\mathbf{S'vp}[1+\mathbf{s}]/360\}$$
$$\text{-}[1+\mathbf{/}\{\mathbf{S'}[1+\mathbf{s}]\text{-}\mathbf{V}\text{-}\mathbf{S'f}[1+\mathbf{s}]\text{-}\mathbf{S'i}[1+\mathbf{s}]\}$$
$$[1\text{-}\mathbf{t}][1\text{-}\mathbf{d}])/\{[1+\mathbf{/}$$

Steve Asikin ISBN 13: **978-1564912361**, ISBN 10: **1564912363**

Rule-56394:

If both (**/**), (**U**), (**N**), (**s**), (**V**), (**f**), (**i**), (**t**), (**d**), (**K**), (**$**), (**j**), (**v**) and (**p**) are known, then its Sales Past must be:

$$S' = [(\{N+K+Sj/360+S'vp[1+s]/360\}/[1+/-U)$$
$$/\{[1-t][1-d]\}+V]/\{[1+s][1-f-i]\}$$

or

$$S' = S/[1+s]$$

Rule-56395:

If both (**/**), (**U**), (**$'**), (**N**), (**V**), (**f**), (**i**), (**t**), (**d**), (**K**), (**$**), (**j**), (**v**) and (**p**) are known, then its Sales Growth planned is:

$$s = [(\{N+K+Sj/360+S'vp[1+s]/360\}/[1+/-U)$$
$$/\{[1-t][1-d]\}+V]/\{S'[1-f-i]\}-1$$

or

$$s = S/S'-1$$

Rule-56396:

If both (**/**), (**U**), (**$'**), (**s**), (**N**), (**f**), (**i**), (**t**), (**d**), (**K**), (**$**), (**j**), (**v**) and (**p**) are known, then its Variable Cost planned is:

$$V = [[1+/(U+\{S'[1+s]-S'f[1+s]-S'i[1+s]\}$$
$$[1-t][1-d])-\{N+K+Sj/360+Svp/360\}]$$
$$/\{[1+/][1-t][1-d]\}$$

or

$$V = Sv$$

80

Rule-56397:

If both (**/**), (**U**), (**S'**), (**s**), (**V**), (**N**), (**i**), (**t**), (**d**), (**K**), (**S**), (**j**), (**v**) and (**p**) are known, then its Fixed Portion planned is:

$$f= [[1+/(U+\{S'[1+s]-V-S'i[1+s]\}[1-t][1-d])$$
$$-\{N+K+Sj/360+S'vp[1+s]/360\}]$$
$$/\{S'[1+s][1+/][1-t][1-d]\}$$

Rule-56398:

If both (**/**), (**U**), (**S'**), (**s**), (**V**), (**f**), (**N**), (**t**), (**d**), (**K**), (**S**), (**j**), (**v**) and (**p**) are known, then its Interest Portion planned is:

$$i= [[1+/(U+\{S'[1+s]-V-S'f[1+s]\}[1-t][1-d])$$
$$-\{N+K+Sj/360+S'vp[1+s]/360\}]$$
$$/\{S'[1+s][1+/][1-t][1-d]\}$$

Rule-56399:

If both (**/**), (**U**), (**S'**), (**s**), (**V**), (**f**), (**i**), (**N**), (**d**), (**K**), (**S**), (**j**), (**v**) and (**p**) are known, then its Tax Rate planned is:

$$t= 1-(\{N+K+Sj/360+S'vp[1+s]/360\}/[1+/]-U)$$
$$/(\{S'[1+s]-V-S'f[1+s]-S'i[1+s]\}[1-d])$$

Rule-56400:

If both (**/**), (**U**), (**S'**), (**s**), (**V**), (**f**), (**i**), (**t**), (**N**), (**K**), (**S**), (**j**), (**v**) and (**p**) are known, then its Dividend Payout planned is:

$$d= 1-(\{N+K+Sj/360+S'vp[1+s]/360\}/[1+/]-U)$$
$$/(\{S'[1+s]-V-S'f[1+s]-S'i[1+s]\}[1-t])$$

Steve Asikin ISBN 13: **978-1564912361**, ISBN 10: **1564912363**

Rule-56401:

If both ($I$), ($U$), ($S'$), ($s$), ($V$), ($f$), ($i$), ($t$), ($d$), ($N$), ($S$), ($j$), ($v$) and ($p$) are known, then its Kind of Cash planned is:

$$K= [1+I(U+\{S'[1+s]-V-S'f[1+s]-S'i[1+s]\}$$
$$[1-t][1-d])-\{N+Sj/360+S'vp[1+s]/360\}$$

Rule-56402:

If both ($I$), ($U$), ($S'$), ($s$), ($V$), ($f$), ($i$), ($t$), ($d$), ($K$), ($N$), ($j$), ($v$) and ($p$) are known, then its Sales or Revenue planned is:

$$S= 360[[1+I(U+\{S'[1+s]-V-S'f[1+s]-S'i[1+s]\}$$
$$[1-t][1-d])-\{N+K+S'vp[1+s]/360\}]/j$$

or

$$S= V/v$$

or

$$S= S'[1+s]$$

Rule-56403:

If both ($I$), ($U$), ($S'$), ($s$), ($V$), ($f$), ($i$), ($t$), ($d$), ($K$), ($S$), ($N$), ($v$) and ($p$) are known, then its Job or Trade Receivable Days planned is:

$$j= 360[[1+I(U+\{S'[1+s]-V-S'f[1+s]-S'i[1+s]\}$$
$$[1-t][1-d])-\{N+K+S'vp[1+s]/360\}]/S$$

Steve Asikin ISBN 13: **978-1564912361**, ISBN 10: **1564912363**

Rule-56404:

If both (**/**), (**U**), (**S'**), (**s**), (**V**), (**f**), (**i**), (**t**), (**d**), (**K**), (**S**), (**j**), (**v**) and (**p**) are known, then its Non Current Asset planned is:

$$v= 360[[1+/(U+\{S'[1+s]-V-S'f[1+s]-S'i[1+s]\}$$
$$[1-t][1-d])-[N+K+Sj/360]]/\{S'p[1+s]\}$$
or
$$v= V/S$$

Rule-56405:

If both (**/**), (**U**), (**S'**), (**s**), (**V**), (**f**), (**i**), (**t**), (**d**), (**K**), (**S**), (**j**), (**v**) and (**N**) are known, then its Procured Inventory Days planned is:

$$p= 360[[1+/(U+\{S'[1+s]-V-S'f[1+s]-S'i[1+s]\}$$
$$[1-t][1-d])-[N+K+Sj/360]]/\{S'v[1+s]\}$$

Rule-56406:

If both (**/**), (**U**), (**S'**), (**s**), (**V**), (**f**), (**i**), (**t**), (**d**), (**K**), (**j**) and (**P**) are known, then its Non Current Asset planned is:

$$N= [1+/(U+\{S'[1+s]-V-S'f[1+s]-S'i[1+s]\}$$
$$[1-t][1-d])-K-S'j[1+s]/360-P$$

Rule-56407:

If both (**N**), (**U**), (**S'**), (**s**), (**V**), (**f**), (**i**), (**t**), (**d**), (**K**), (**j**) and (**P**) are known, then its Leverage or Gearing Ratio planned is:

$$/= \{N+K+S'j[1+s]/360+P\}/(U+\{S'[1+s]-V$$
$$-S'f[1+s]-S'i[1+s]\}[1-t][1-d])-1$$

Steve Asikin ISBN 13: **978-1564912361**, ISBN 10: **1564912363**

Rule-56408:

If both (**I**), (**N**), (**S'**), (**s**), (**V**), (**f**), (**i**), (**t**), (**d**), (**K**), (**j**) and (**P**) are known, then its Utilized or Starting Capital must be:

$$U = (\{N+K+S'j[1+s]/360+P\}-[1+I]\{S'[1+s]-V$$
$$-S'f[1+s]-S'i[1+s]\}[1-t][1-d])/[1+I]$$

Rule-56409:

If both (**I**), (**U**), (**N**), (**s**), (**V**), (**f**), (**i**), (**t**), (**d**), (**K**), (**j**) and (**P**) are known, then its Sales Past must be:

$$S' = [(\{N+K+S'j[1+s]/360+P\}/[1+I]-U)$$
$$/\{[1-t][1-d]\}+V]/\{[1+s][1-f-i]\}$$

Rule-56410:

If both (**I**), (**U**), (**S'**), (**N**), (**V**), (**f**), (**i**), (**t**), (**d**), (**K**), (**j**) and (**P**) are known, then its Sales Growth planned is:

$$s = [(\{N+K+S'j[1+s]/360+P\}/[1+I]-U)$$
$$/\{[1-t][1-d]\}+V]/\{S'[1-f-i]\}-1$$

Rule-56411:

If both (**I**), (**U**), (**S'**), (**s**), (**N**), (**f**), (**i**), (**t**), (**d**), (**K**), (**j**) and (**P**) are known, then its Variable Cost planned is:

$$V = [[1+I](U+\{S'-S'f[1+s]-S'i[1+s]\}[1-t][1-d])$$
$$-\{N+K+S'j[1+s]/360+P\}]$$
$$/\{[1+I][1-t][1-d]\}$$

`

84

Steve Asikin ISBN 13: **978-1564912361**, ISBN 10: **1564912363**

Rule-56412:

If both (**/**), (**U**), (**S'**), (**s**), (**V**), (**N**), (**i**), (**t**), (**d**), (**K**), (**j**) and (**P**) are known, then its Fixed Portion planned is:

$$f= [[1+/(U+\{S'[1+s]-V-S'i[1+s]\}[1-t][1-d])$$
$$-\{N+K+S'j[1+s]/360+P\}]$$
$$/\{S'[1+s][1+/[1-t][1-d]\}$$

Rule-56413:

If both (**/**), (**U**), (**S'**), (**s**), (**V**), (**f**), (**N**), (**t**), (**d**), (**K**), (**j**) and (**P**) are known, then its Interest Portion planned is:

$$i= [[1+/(U+\{S'[1+s]-V-S'f[1+s]\}[1-t][1-d])$$
$$-\{N+K+S'j[1+s]/360+P\}]$$
$$/\{S'[1+s][1+/[1-t][1-d]\}$$

Rule-56414:

If both (**/**), (**U**), (**S'**), (**s**), (**V**), (**f**), (**i**), (**N**), (**d**), (**K**), (**j**) and (**P**) are known, then its Tax Rate planned is:

$$t= 1-(\{N+K+S'j[1+s]/360+P\}/[1+/-U)$$
$$/(\{S'[1+s]-V-S'f[1+s]-S'i[1+s]\}[1-d])$$

Rule-56415:

If both (**/**), (**U**), (**S'**), (**s**), (**V**), (**f**), (**i**), (**t**), (**N**), (**K**), (**j**) and (**P**) are known, then its Dividend Payout planned is:

$$d= 1-(\{N+K+S'j[1+s]/360+P\}/[1+/-U)$$
$$/(\{S'[1+s]-V-S'f[1+s]-S'i[1+s]\}[1-t])$$

Steve Asikin ISBN 13: **978-1564912361**, ISBN 10: **1564912363**

Rule-56416:

If both (**/**), (**U**), (**S'**), (**s**), (**V**), (**f**), (**i**), (**t**), (**d**), (**N**), (**j**) and (**P**) are known, then its Kind of Cash planned is:

$$K= [1+/](U+\{S'[1+s]-V-S'f[1+s]-S'i[1+s]\}$$
$$[1-t][1-d])-\{N+S'j[1+s]/360+P\}$$

Rule-56417:

If both (**/**), (**U**), (**S'**), (**s**), (**V**), (**f**), (**i**), (**t**), (**d**), (**K**), (**N**) and (**P**) are known, then its Job or Trade Receivable Days planned is:

$$j= 360[[1+/](U+\{S'[1+s]-V-S'f[1+s]-S'i[1+s]\}$$
$$[1-t][1-d])-\{N+K+P\}]/\{S'[1+s]\}$$

Rule-56418:

If both (**/**), (**U**), (**S'**), (**s**), (**V**), (**f**), (**i**), (**t**), (**d**), (**K**), (**j**) and (**N**) are known, then its Procured Inventory planned is:

$$P= [1+/](U+\{S'[1+s]-V-S'f[1+s]-S'i[1+s]\}$$
$$[1-t][1-d])-\{N+K+S'j[1+s]/360\}$$

Rule-56419:

If both (**/**), (**U**), (**S'**), (**s**), (**V**), (**f**), (**i**), (**t**), (**d**), (**K**), (**j**) and (**p**) are known, then its Non Current Asset planned is:

$$N= [1+/](U+\{S'[1+s]-V-S'f[1+s]-S'i[1+s]\}$$
$$[1-t][1-d])-K-S'j[1+s]/360-Vp/360$$

Steve Asikin ISBN 13: **978-1564912361**, ISBN 10: **1564912363**

Rule-56420:

If both **(N)**, **(U)**, **(S')**, **(s)**, **(V)**, **(f)**, **(i)**, **(t)**, **(d)**, **(K)**, **(j)** and **(p)** are known, then its Leverage or Gearing Ratio planned is:

$$\textit{I} = \{N+K+S\text{'}j[1+s]/360+Vp/360\}/(U+\{S\text{'}[1+s]-V$$
$$-S\text{'}f[1+s]-S\text{'}i[1+s]\}[1-t][1-d])-1$$

Rule-56421:

If both **(I)**, **(N)**, **(S')**, **(s)**, **(V)**, **(f)**, **(i)**, **(t)**, **(d)**, **(K)**, **(j)** and **(p)** are known, then its Utilized or Starting Capital must be:

$$U = (\{N+K+S\text{'}j[1+s]/360+Vp/360\}-[1+\textit{I}]\{S\text{'}[1+s]$$
$$-V-S\text{'}f[1+s]-S\text{'}i[1+s]\}[1-t][1-d])/[1+\textit{I}]$$

Rule-56422:

If both **(I)**, **(U)**, **(N)**, **(s)**, **(V)**, **(f)**, **(i)**, **(t)**, **(d)**, **(K)**, **(j)** and **(p)** are known, then its Sales Past must be:

$$S\text{'} = [(\{N+K+S\text{'}j[1+s]/360+Vp/360\}/[1+\textit{I}]-U)$$
$$/\{[1-t][1-d]\}+V]/\{[1+s][1-f-i]\}$$

or

$$S\text{'} = S/[1+s]$$

Rule-56423:

If both **(I)**, **(U)**, **(S')**, **(N)**, **(V)**, **(f)**, **(i)**, **(t)**, **(d)**, **(K)**, **(j)** and **(p)** are known, then its Sales Growth planned is:

$$s = [(\{N+K+S\text{'}j[1+s]/360+Vp/360\}/[1+\textit{I}]-U)$$
$$/\{[1-t][1-d]\}+V]/\{S\text{'}[1-f-i]\}-1$$

or

$$s = S/S\text{'}-1$$

Steve Asikin ISBN 13: **978-1564912361**, ISBN 10: **1564912363**

Rule-56424:

If both (**/**), (**U**), (**$'**), (**s**), (**N**), (**f**), (**i**), (**t**), (**d**), (**K**), (**j**)
and (**p**) are known, then its Variable Cost planned is:

$$V= [[1+/(U+\{\$'[1+s] -\$'f[1+s]-\$'i[1+s]\}$$
$$[1-t][1-d])-\{N+K+\$'j[1+s]/360\}]$$
$$/\{p/360+[1+/][1-t][1-d]\}$$

Rule-56425:

If both (**/**), (**U**), (**$'**), (**s**), (**V**), (**N**), (**i**), (**t**), (**d**), (**K**), (**j**)
and (**p**) are known, then its Fixed Portion planned is:

$$f= [[1+/(U+\{\$'[1+s]-V-\$'i[1+s]\}[1-t][1-d])$$
$$-\{N+K+\$'j[1+s]/360+Vp/360\}]$$
$$/\{\$'[1+s][1+/][1-t][1-d]\}$$

Rule-56426:

If both (**/**), (**U**), (**$'**), (**s**), (**V**), (**f**), (**N**), (**t**), (**d**), (**K**), (**j**)
and (**p**) are known, then its Interest Portion planned
is:

$$i= [[1+/(U+\{\$'[1+s]-V-\$'f[1+s]\}[1-t][1-d])$$
$$-\{N+K+\$'j[1+s]/360+Vp/360\}]$$
$$/\{\$'[1+s][1+/][1-t][1-d]\}$$

Rule-56427:

If both (**/**), (**U**), (**$'**), (**s**), (**V**), (**f**), (**i**), (**N**), (**d**), (**K**), (**j**)
and (**p**) are known, then its Tax Rate planned is:

$$t= 1-(\{N+K+\$'j[1+s]/360+Vp/360\}/[1+/]-U)$$
$$/(\{\$'[1+s]-V-\$'f[1+s]-\$'i[1+s]\}[1-d])$$

Steve Asikin ISBN 13: **978-1564912361**, ISBN 10: **1564912363**

Rule-56428:

If both $(\textit{l})$, $(\mathbf{U})$, $(\mathbf{S'})$, $(\mathbf{s})$, $(\mathbf{V})$, $(\mathbf{f})$, $(\mathbf{i})$, $(\mathbf{t})$, $(\mathbf{N})$, $(\mathbf{K})$, $(\mathbf{j})$ and $(\mathbf{p})$ are known, then its Dividend Payout planned is:

$$d= 1-(\{N+K+S'j[1+s]/360+Vp/360\}/[1+\textit{l}-U)$$
$$/(\{S'[1+s]-V-S'f[1+s]-S'i[1+s]\}[1-t])$$

Rule-56429:

If both $(\textit{l})$, $(\mathbf{U})$, $(\mathbf{S'})$, $(\mathbf{s})$, $(\mathbf{V})$, $(\mathbf{f})$, $(\mathbf{i})$, $(\mathbf{t})$, $(\mathbf{d})$, $(\mathbf{N})$, $(\mathbf{j})$ and $(\mathbf{p})$ are known, then its Kind of Cash planned is:

$$K= [1+\textit{l}](U+\{S'[1+s]-V-S'f[1+s]-S'i[1+s]\}$$
$$[1-t][1-d])-\{N+S'j[1+s]/360+Vp/360\}$$

Rule-56430:

If both $(\textit{l})$, $(\mathbf{U})$, $(\mathbf{S'})$, $(\mathbf{s})$, $(\mathbf{V})$, $(\mathbf{f})$, $(\mathbf{i})$, $(\mathbf{t})$, $(\mathbf{d})$, $(\mathbf{K})$, $(\mathbf{N})$ and $(\mathbf{p})$ are known, then its Job or Trade Receivable Days planned is:

$$j= 360[[1+\textit{l}](U+\{S'[1+s]-V-S'f[1+s]-S'i[1+s]\}$$
$$[1-t][1-d])-\{N+K+Vp/360\}]/\{S'[1+s]\}$$

Rule-56431:

If both $(\textit{l})$, $(\mathbf{U})$, $(\mathbf{S'})$, $(\mathbf{s})$, $(\mathbf{V})$, $(\mathbf{f})$, $(\mathbf{i})$, $(\mathbf{t})$, $(\mathbf{d})$, $(\mathbf{K})$, $(\mathbf{j})$ and $(\mathbf{N})$ are known, then its Procured Inventory Days planned is:

$$p= 360[[1+\textit{l}](U+\{S'[1+s]-V-S'f[1+s]-S'i[1+s]\}$$
$$[1-t][1-d])-\{N+K+S'j[1+s]/360\}]/V$$

Steve Asikin ISBN 13: **978-1564912361**, ISBN 10: **1564912363**

<u>Rule-56432</u>:

If both (**/**), (**U**), (**$'**), (**s**), (**V**), (**f**), (**i**), (**t**), (**d**), (**K**), (**j**), (**$**), (**v**) and (**p**) are known, then its Non Current Asset planned is:

$$\mathbf{N} = [1 + \mathbf{/}(\mathbf{U} + \{\mathbf{\$'}[1+\mathbf{s}] - \mathbf{V} - \mathbf{\$'f}[1+\mathbf{s}] - \mathbf{\$'i}[1+\mathbf{s}]\}$$
$$[1 - \mathbf{t}][1 - \mathbf{d}]) - \mathbf{K} - \mathbf{\$'j}[1+\mathbf{s}]/360 - \mathbf{\$vp}/360$$

<u>Rule-56433</u>:

If both (**N**), (**U**), (**$'**), (**s**), (**V**), (**f**), (**i**), (**t**), (**d**), (**K**), (**j**), (**$**), (**v**) and (**p**) are known, then its Leverage or Gearing Ratio planned is:

$$\mathbf{/} = \{\mathbf{N} + \mathbf{K} + \mathbf{\$'j}[1+\mathbf{s}]/360 + \mathbf{\$vp}/360\}/(\mathbf{U} + \{\mathbf{\$'}[1+\mathbf{s}] - \mathbf{V}$$
$$- \mathbf{\$'f}[1+\mathbf{s}] - \mathbf{\$'i}[1+\mathbf{s}]\}[1 - \mathbf{t}][1 - \mathbf{d}]) - 1$$

<u>Rule-56434</u>:

If both (**/**), (**N**), (**$'**), (**s**), (**V**), (**f**), (**i**), (**t**), (**d**), (**K**), (**j**), (**$**), (**v**) and (**p**) are known, then its Utilized or Starting Capital must be:

$$\mathbf{U} = (\{\mathbf{N} + \mathbf{K} + \mathbf{\$'j}[1+\mathbf{s}]/360 + \mathbf{\$vp}/360\}$$
$$- [1 + \mathbf{/}]\{\mathbf{\$'}[1+\mathbf{s}] - \mathbf{V} - \mathbf{\$'f}[1+\mathbf{s}] - \mathbf{\$'i}[1+\mathbf{s}]\}$$
$$[1 - \mathbf{t}][1 - \mathbf{d}])/\{[1 + \mathbf{/}]\}$$

`

Steve Asikin ISBN 13: **978-1564912361**, ISBN 10: **1564912363**

Rule-56435:

If both $(\textbf{\textit{I}})$, $(\textbf{U})$, $(\textbf{N})$, $(\textbf{s})$, $(\textbf{V})$, $(\textbf{f})$, $(\textbf{i})$, $(\textbf{t})$, $(\textbf{d})$, $(\textbf{K})$, $(\textbf{j})$, $(\textbf{\$})$, $(\textbf{v})$ and $(\textbf{p})$ are known, then its Sales Past must be:

$$\textbf{\$'}= [(\{\textbf{N}+\textbf{K}+\textbf{\$'j}[1+\textbf{s}]/360+\textbf{\$vp}/360\}/[1+\textbf{\textit{I}}-\textbf{U})$$
$$/\{[1-\textbf{t}][1-\textbf{d}]\}-\textbf{V}]/\{[1+\textbf{s}][1-\textbf{f}-\textbf{i}]\}$$

or

$$\textbf{\$'}= \textbf{\$}/[1+\textbf{s}]$$

Rule-56436:

If both $(\textbf{\textit{I}})$, $(\textbf{U})$, $(\textbf{\$'})$, $(\textbf{N})$, $(\textbf{V})$, $(\textbf{f})$, $(\textbf{i})$, $(\textbf{t})$, $(\textbf{d})$, $(\textbf{K})$, $(\textbf{j})$, $(\textbf{\$})$, $(\textbf{v})$ and $(\textbf{p})$ are known, then its Sales Growth planned is:

$$\textbf{s}= [(\{\textbf{N}+\textbf{K}+\textbf{\$'j}[1+\textbf{s}]/360+\textbf{\$vp}/360\}/[1+\textbf{\textit{I}}-\textbf{U})$$
$$/\{[1-\textbf{t}][1-\textbf{d}]\}-\textbf{V}]/\{\textbf{\$'}[1-\textbf{f}-\textbf{i}]\}-1$$

or

$$\textbf{s}= \textbf{\$}/\textbf{\$'}-1$$

Rule-56437:

If both $(\textbf{\textit{I}})$, $(\textbf{U})$, $(\textbf{\$'})$, $(\textbf{s})$, $(\textbf{N})$, $(\textbf{f})$, $(\textbf{i})$, $(\textbf{t})$, $(\textbf{d})$, $(\textbf{K})$, $(\textbf{j})$, $(\textbf{\$})$, $(\textbf{v})$ and $(\textbf{p})$ are known, then its Variable Cost planned is:

$$\textbf{V}= [[1+\textbf{\textit{I}}](\textbf{U}+\{\textbf{\$'}[1+\textbf{s}]-\textbf{\$'f}[1+\textbf{s}]-\textbf{\$'i}[1+\textbf{s}]\}[1-\textbf{t}]$$
$$[1-\textbf{d}])-\{\textbf{N}+\textbf{K}+\textbf{\$'j}[1+\textbf{s}]/360+\textbf{\$vp}/360\}]$$
$$/\{[1+\textbf{\textit{I}}][1-\textbf{t}][1-\textbf{d}]\}$$

or

$$\textbf{V}= \textbf{\$v}$$

Steve Asikin ISBN 13: **978-1564912361**, ISBN 10: **1564912363**

Rule-56438:

If both (**/**), (**U**), (**$'**), (**s**), (**V**), (**N**), (**i**), (**t**), (**d**), (**K**), (**j**), (**$**), (**v**) and (**p**) are known, then its Fixed Portion planned is:

$$f= ([1+/](U+\{\$'[1+s]-V-\$'i[1+s]\}[1-t][1-d]$$
$$-\{N+K+\$'j[1+s]/360+\$vp/360\})$$
$$/\{\$'[1+s][1+/][1-t][1-d]\}$$

Rule-56439:

If both (**/**), (**U**), (**$'**), (**s**), (**V**), (**f**), (**N**), (**t**), (**d**), (**K**), (**j**), (**$**), (**v**) and (**p**) are known, then its Interest Portion planned is:

$$i= ([1+/](U+\{\$'[1+s]-V-\$'f[1+s]\}[1-t][1-d]$$
$$-\{N+K+\$'j[1+s]/360+\$vp/360\})$$
$$/\{\$'[1+s][1+/][1-t][1-d]\}$$

Rule-56440:

If both (**/**), (**U**), (**$'**), (**s**), (**V**), (**f**), (**i**), (**N**), (**d**), (**K**), (**j**), (**$**), (**v**) and (**p**) are known, then its Tax Rate planned is:

$$t= 1-(\{N+K+\$'j[1+s]/360+\$vp/360\}/[1+/]-U)$$
$$/(\{\$'[1+s]-V-\$'f[1+s]-\$'i[1+s]\}[1-d])$$

Rule-56441:

If both (**/**), (**U**), (**$'**), (**s**), (**V**), (**f**), (**i**), (**t**), (**N**), (**K**), (**j**), (**$**), (**v**) and (**p**) are known, then its Dividend Payout planned is:

$$d= 1-(\{N+K+\$'j[1+s]/360+\$vp/360\}/[1+/]-U)$$
$$/(\{\$'[1+s]-V-\$'f[1+s]-\$'i[1+s]\}[1-t])$$

Steve Asikin ISBN 13: **978-1564912361**, ISBN 10: **1564912363**

Rule-56442:

If both (**/**), (**U**), (**S'**), (**s**), (**V**), (**f**), (**i**), (**t**), (**d**), (**N**), (**j**), (**S**), (**v**) and (**p**) are known, then its Kind of Cash planned is:

$$K= [1+/](U+\{S'[1+s]-V-S'f[1+s]-S'i[1+s]\}$$
$$[1-t][1-d])-\{N+S'j[1+s]/360+Svp/360\}$$

Rule-56443:

If both (**/**), (**U**), (**S'**), (**s**), (**V**), (**f**), (**i**), (**t**), (**d**), (**K**), (**N**), (**S**), (**v**) and (**p**) are known, then its Job or Trade Receivable Days planned is:

$$j= 360[[1+/](U+\{S'[1+s]-V-S'f[1+s]-S'i[1+s]\}$$
$$[1-t][1-d])-\{N+K+Svp/360\}]/\{S'[1+s]\}$$

Rule-56444:

If both (**/**), (**U**), (**S'**), (**s**), (**V**), (**f**), (**i**), (**t**), (**d**), (**K**), (**j**), (**N**), (**v**) and (**p**) are known, then its Sales or Revenue planned is:

$$S= [[1+/](U+\{S'[1+s]-V-S'f[1+s]-S'i[1+s]\}[1-t]$$
$$[1-d])-\{N+K+S'j[1+s]/360\}]/[vp/360]$$

or

$$S= S'[1+s]$$

or

$$S= V/v$$

Steve Asikin ISBN 13: **978-1564912361**, ISBN 10: **1564912363**

Rule-56445:

If both (**/**), (**U**), (**S'**), (**s**), (**V**), (**f**), (**i**), (**t**), (**d**), (**K**), (**j**), (**S**), (**N**) and (**p**) are known, then its Variable Portion planned is:

$$v= 360[[1+/(U+\{S'[1+s]-V-S'f[1+s]-S'i[1+s]\}$$
$$[1-t][1-d])-\{N+K+S'j[1+s]/360\}]/[Sp]$$

or

$$v= V/S$$

Rule-56446:

If both (**/**), (**U**), (**S'**), (**s**), (**V**), (**f**), (**i**), (**t**), (**d**), (**K**), (**j**), (**S**), (**v**) and (**N**) are known, then its Procured Inventory Days planned is:

$$p= 360[[1+/(U+\{S'[1+s]-V-S'f[1+s]-S'i[1+s]\}$$
$$[1-t][1-d])-\{N+K+S'j[1+s]/360\}]/[Sv]$$

Rule-56447:

If both (**/**), (**U**), (**S'**), (**s**), (**V**), (**f**), (**i**), (**t**), (**d**), (**K**), (**j**), (**v**) and (**p**) are known, then its Non Current Asset planned is:

$$N= [1+/(U+\{S'[1+s]-V-S'f[1+s]-S'i[1+s]\}$$
$$[1-t][1-d])-K-S'j[1+s]/360-S'vp[1+s]/360$$

Steve Asikin ISBN 13: **978-1564912361**, ISBN 10: **1564912363**

Rule-56448:

If both **(N)**, **(U)**, **(S')**, **(s)**, **(V)**, **(f)**, **(i)**, **(t)**, **(d)**, **(K)**, **(j)**, **(v)** and **(p)** are known, then its Leverage or Gearing Ratio planned is:

$$I = \{N+K+S'j[1+s]/360+S'vp[1+s]/360\}$$
$$/(U+\{S'[1+s]-V-S'f[1+s]-S'i[1+s]\}$$
$$[1-t][1-d])-1$$

Rule-56449:

If both **(I)**, **(N)**, **(S')**, **(s)**, **(V)**, **(f)**, **(i)**, **(t)**, **(d)**, **(K)**, **(j)**, **(v)** and **(p)** are known, then its Utilized or Starting Capital must be:

$$U = (\{N+K+S'j[1+s]/360+S'vp[1+s]/360\}$$
$$-[1+I]\{S'[1+s]-V-S'f[1+s]-S'i[1+s]\}$$
$$[1-t][1-d])/[1+I]$$

Rule-56450:

If both **(I)**, **(U)**, **(N)**, **(s)**, **(V)**, **(f)**, **(i)**, **(t)**, **(d)**, **(K)**, **(j)**, **(v)** and **(p)** are known, then its Sales Past must be:

$$S' = [(\{N+K+S'j[1+s]/360+S'vp[1+s]/360\}$$
$$/[1-I-U)/\{[1-t][1-d]\}-V]/\{[1+s][1-f-i]\}$$
$$\text{or}$$
$$S' = S/[1+s]$$

Steve Asikin ISBN 13: **978-1564912361**, ISBN 10: **1564912363**

Rule-56451:

If both (**/**), (**U**), (**S'**), (**N**), (**V**), (**f**), (**i**), (**t**), (**d**), (**K**), (**j**), (**v**) and (**p**) are known, then its Sales Growth planned is:

$$s= [(\{N+K+S'j[1+s]/360+S'vp[1+s]/360\}$$
$$/[1-/-U)/\{[1-t][1-d]\}-V]/\{S'[1-f-i]\}-1$$

or

$$s= S/S'-1$$

Rule-56452:

If both (**/**), (**U**), (**S'**), (**s**), (**N**), (**f**), (**i**), (**t**), (**d**), (**K**), (**j**), (**v**) and (**p**) are known, then its Variable Cost planned is:

$$V= [[1+/](U+\{S'[1+s]-S'f[1+s]-S'i[1+s]\}[1-t]$$
$$[1-d])-\{N+K+S'j[1+s]/360$$
$$+S'vp[1+s]/360]$$
$$/\{[1+/][1-t][1-d]\}$$

or

$$V= Sv$$

Rule-56453:

If both (**/**), (**U**), (**S'**), (**s**), (**V**), (**N**), (**i**), (**t**), (**d**), (**K**), (**j**), (**v**) and (**p**) are known, then its Fixed Portion planned is:

$$f= [[1+/](U+\{S'[1+s]-S'i[1+s]\}[1-t][1-d])-\{N+K$$
$$+S'j[1+s]/360+S'vp[1+s]/360\}]$$
$$/\{S'[1+s][1+/][1-t][1-d]\}$$

Steve Asikin ISBN 13: **978-1564912361**, ISBN 10: **1564912363**

Rule-56454:

If both ($I$), (**U**), (**S'**), (**s**), (**V**), (**f**), (**N**), (**t**), (**d**), (**K**), (**j**), (**v**) and (**p**) are known, then its Interest Portion planned is:

$$i= [[1+I(U+\{S'[1+s]-S'f[1+s]\}[1-t][1-d])-\{N+K$$
$$+S'j[1+s]/360+S'vp[1+s]/360\}]$$
$$/\{S'[1+s][1+I[1-t][1-d]\}$$

Rule-56455:

If both ($I$), (**U**), (**S'**), (**s**), (**V**), (**f**), (**i**), (**N**), (**d**), (**K**), (**j**), (**v**) and (**p**) are known, then its Tax Rate planned is:

$$t= 1-(\{N+K+S'j[1+s]/360+S'vp[1+s]/360\}$$
$$/[1+I-U)/(\{S'[1+s]-V-S'f[1+s]$$
$$-S'i[1+s]\}[1-d])$$

Rule-56456:

If both ($I$), (**U**), (**S'**), (**s**), (**V**), (**f**), (**i**), (**t**), (**N**), (**K**), (**j**), (**v**) and (**p**) are known, then its Dividend Payout planned is:

$$d= 1-(\{N+K+S'j[1+s]/360+S'vp[1+s]/360\}$$
$$/[1+I-U)/(\{S'[1+s]-V-S'f[1+s]$$
$$-S'i[1+s]\}[1-t])$$

Rule-56457:

If both ($I$), (**U**), (**S'**), (**s**), (**V**), (**f**), (**i**), (**t**), (**d**), (**N**), (**j**), (**v**) and (**p**) are known, then its Kind of Cash planned is:

$$K= [1+I(U+\{S'[1+s]-V-S'f[1+s]-S'i[1+s]\}[1-t]$$
$$[1-d])-\{N+S'j[1+s]/360+S'vp[1+s]/360\}$$

Steve Asikin ISBN 13: **978-1564912361**, ISBN 10: **1564912363**

Rule-56458:

If both (**/**), (**U**), (**S'**), (**s**), (**V**), (**f**), (**i**), (**t**), (**d**), (**K**), (**N**), (**v**) and (**p**) are known, then its Job or Trade Receivable Days planned is:

$$j = 360\big[[1+\mathbf{/}(\mathbf{U}+\{\mathbf{S'}[1+\mathbf{s}]-\mathbf{V}-\mathbf{S'f}[1+\mathbf{s}]-\mathbf{S'i}[1+\mathbf{s}]\}$$
$$[1-\mathbf{t}][1-\mathbf{d}])-\{\mathbf{N}+\mathbf{K}+\mathbf{S'vp}[1+\mathbf{s}]/360\}\big]$$
$$/\{\mathbf{S'}[1+\mathbf{s}]\}$$

Rule-56459:

If both (**/**), (**U**), (**S'**), (**s**), (**V**), (**f**), (**i**), (**t**), (**d**), (**K**), (**j**), (**N**) and (**p**) are known, then its Variable Portion planned is:

$$v = 360\big[[1+\mathbf{/}(\mathbf{U}+\{\mathbf{S'}[1+\mathbf{s}]-\mathbf{V}-\mathbf{S'f}[1+\mathbf{s}]-\mathbf{S'i}[1+\mathbf{s}]\}$$
$$[1-\mathbf{t}][1-\mathbf{d}])-\{\mathbf{N}+\mathbf{K}+\mathbf{S'j}[1+\mathbf{s}]/360\}\big]$$
$$/\{\mathbf{S'p}[1+\mathbf{s}]\}$$

or

$$v = \mathbf{V}/\mathbf{S}$$

Rule-56460:

If both (**/**), (**U**), (**S'**), (**s**), (**V**), (**f**), (**i**), (**t**), (**d**), (**K**), (**j**), (**v**) and (**N**) are known, then its Procured Inventory Days planned is:

$$p = 360\big[[1+\mathbf{/}(\mathbf{U}+\{\mathbf{S'}[1+\mathbf{s}]-\mathbf{V}-\mathbf{S'f}[1+\mathbf{s}]-\mathbf{S'i}[1+\mathbf{s}]\}$$
$$[1-\mathbf{t}][1-\mathbf{d}])-\{\mathbf{N}+\mathbf{K}+\mathbf{S'j}[1+\mathbf{s}]/360\}\big]$$
$$/\{\mathbf{S'v}[1+\mathbf{s}]\}$$

`

Steve Asikin ISBN 13: **978-1564912361**, ISBN 10: **1564912363**

Rule-56461:

If both (**/**), (**U**), (**S'**), (**s**), (**S**), (**v**), (**F**), (**I**), (**T**), (**D**), (**K**), (**J**) and (**P**) are known, then its Non Current Asset planned is:

$$N= [1+/]\{U+S'[1+s]-Sv-F-I-T-D\}-K-J-P$$

Rule-56462:

If both (**N**), (**U**), (**S'**), (**s**), (**S**), (**v**), (**F**), (**I**), (**T**), (**D**), (**K**), (**J**) and (**P**) are known, then its Leverage or Gearing Ratio planned is:

$$/= \{N+K+J+P\}/(U+\{S'[1+s]-Sv-F-I\}-T-D)-1$$

Rule-56463:

If both (**/**), (**N**), (**S'**), (**s**), (**S**), (**v**), (**F**), (**I**), (**T**), (**D**), (**K**), (**J**) and (**P**) are known, then its Utilized or Starting Capital must be:

$$U= (\{N+K+J+P\}-[1+/]\{S'[1+s]-Sv-F-I\}-T-D)$$
$$/[1+/]$$

Rule-56464:

If both (**/**), (**U**), (**N**), (**s**), (**S**), (**v**), (**F**), (**I**), (**T**), (**D**), (**K**), (**J**) and (**P**) are known, then its Sales Past must be:

$$S'= \{[N+K+J+P]/[1+/]-U+T+D+F+I+Sv\}/[1+s]$$
$$\text{or}$$
$$S'= S/[1+s]$$

Steve Asikin ISBN 13: **978-1564912361**, ISBN 10: **1564912363**

Rule-56465:

If both ($I$), (**U**), (**S'**), (**N**), (**S**), (**v**), (**F**), (**I**), (**T**), (**D**), (**K**), (**J**) and (**P**) are known, then its Sales Growth planned is:

$$s= \{[\mathbf{N+K+J+P}]/[1+I\text{-}\mathbf{U+T+D+F+I+Sv}\}/\mathbf{S'}\text{-}1$$

or

$$s= \mathbf{S}/\mathbf{S'}\text{-}1$$

Rule-56466:

If both ($I$), (**U**), (**S'**), (**s**), (**N**), (**v**), (**F**), (**I**), (**T**), (**D**), (**K**), (**J**) and (**P**) are known, then its Sales or Revenue planned is:

$$\mathbf{S}= [[1+I](\mathbf{U}+\{\mathbf{S'}[1+s]\text{-}\mathbf{F}\text{-}\mathbf{I}\}\text{-}\mathbf{T}\text{-}\mathbf{D})\text{-}\{\mathbf{N+K+J+P}\}]/\mathbf{v}$$

or

$$\mathbf{S}= \mathbf{S'}[1+s]$$

Rule-56467:

If both ($I$), (**U**), (**S'**), (**s**), (**S**), (**N**), (**F**), (**I**), (**T**), (**D**), (**K**), (**J**) and (**P**) are known, then its Variable Portion planned is:

$$\mathbf{v}= [[1+I](\mathbf{U}+\{\mathbf{S'}[1+s]\text{-}\mathbf{F}\text{-}\mathbf{I}\}\text{-}\mathbf{T}\text{-}\mathbf{D})\text{-}\{\mathbf{N+K+J+P}\}]/\mathbf{S}$$

Rule-56468:

If both ($I$), (**U**), (**S'**), (**s**), (**S**), (**v**), (**N**), (**I**), (**T**), (**D**), (**K**), (**J**) and (**P**) are known, then its Fixed Cost planned is:

$$\mathbf{F}= \mathbf{U}+\{\mathbf{S'}[1+s]\text{-}\mathbf{Sv}\text{-}\mathbf{I}\}\text{-}\mathbf{T}\text{-}\mathbf{D}\text{-}[\mathbf{N+K+J+P}]/[1+I]$$

`

Steve Asikin ISBN 13: **978-1564912361**, ISBN 10: **1564912363**

Rule-56469:

If both (**/**), (**U**), (**S'**), (**s**), (**S**), (**v**), (**F**), (**N**), (**T**), (**D**), (**K**), (**J**) and (**P**) are known, then its Interest Portion planned is:

**I**= **U**+{**S'**[1+**s**]-**Sv**-**F**}-**T**-**D**-[**N**+**K**+**J**+**P**]/[1+**/**]

Rule-56470:

If both (**/**), (**U**), (**S'**), (**s**), (**S**), (**v**), (**F**), (**I**), (**N**), (**D**), (**K**), (**J**) and (**P**) are known, then its Tax planned is:

**T**= **U**+{**S'**[1+**s**]-**Sv**-**F**-**I**}-**D**-[**N**+**K**+**J**+**P**]/[1+**/**]

Rule-56471:

If both (**/**), (**U**), (**S'**), (**s**), (**S**), (**v**), (**F**), (**I**), (**T**), (**N**), (**K**), (**J**) and (**P**) are known, then its Dividend planned is:

**D**= **U**+{**S'**[1+**s**]-**Sv**-**F**-**I**}-**T**-[**N**+**K**+**J**+**P**]/[1+**/**]

Rule-56472:

If both (**/**), (**U**), (**S'**), (**s**), (**S**), (**v**), (**F**), (**I**), (**T**), (**D**), (**N**), (**J**) and (**P**) are known, then its Kind of Cash planned is:

**K**= [1+**/**](**U**+{**S'**[1+**s**]-**Sv**-**F**-**I**}-**T**-**D**)-[**N**+**J**+**P**]

Rule-56473:

If both (**/**), (**U**), (**S'**), (**s**), (**S**), (**v**), (**F**), (**I**), (**T**), (**D**), (**K**), (**N**) and (**P**) are known, then its Job or Trade Account Receivable planned is:

**J**= [1+**/**](**U**+{**S'**[1+**s**]-**Sv**-**F**-**I**}-**T**-**D**)-[**N**+**K**+**P**]

Steve Asikin ISBN 13: **978-1564912361**, ISBN 10: **1564912363**

Rule-56474:

If both (**/**), (**U**), (**S'**), (**s**), (**S**), (**v**), (**F**), (**I**), (**T**), (**D**), (**K**), (**J**) and (**N**) are known, then its Procured Inventory planned is:

$$\mathbf{P}= [1+\mathbf{/}](\mathbf{U}+\{\mathbf{S'}[1+\mathbf{s}]-\mathbf{Sv}-\mathbf{F}-\mathbf{I}\}-\mathbf{T}-\mathbf{D})-[\mathbf{N}+\mathbf{K}+\mathbf{J}]$$

Rule-56475:

If both (**/**), (**U**), (**S'**), (**s**), (**S**), (**v**), (**F**), (**I**), (**T**), (**D**), (**K**), (**J**), (**V**) and (**p**) are known, then its Non Current Asset planned is:

$$\mathbf{N}= [1+\mathbf{/}]\{\mathbf{U}+\mathbf{S'}[1+\mathbf{s}]-\mathbf{Sv}-\mathbf{F}-\mathbf{I}-\mathbf{T}-\mathbf{D}\}-\mathbf{K} \ \mathbf{J} \ \mathbf{Vp}/360$$

Rule-56476:

If both (**N**), (**U**), (**S'**), (**s**), (**S**), (**v**), (**F**), (**I**), (**T**), (**D**), (**K**), (**J**), (**V**) and (**p**) are known, then its Leverage or Gearing Ratio planned is:

$$\mathbf{/}= [\mathbf{N}+\mathbf{K}+\mathbf{J}+\mathbf{Vp}/360]/(\mathbf{U}+\{\mathbf{S'}[1+\mathbf{s}]-\mathbf{Sv}-\mathbf{F}-\mathbf{I}\}-\mathbf{T}-\mathbf{D})-1$$

Rule-56477:

If both (**/**), (**N**), (**S'**), (**s**), (**S**), (**v**), (**F**), (**I**), (**T**), (**D**), (**K**), (**J**), (**V**) and (**p**) are known, then its Utilized or Starting Capital must be:

$$\mathbf{U}= ([\mathbf{N}+\mathbf{K}+\mathbf{J}+\mathbf{Vp}/360]-[1+\mathbf{/}]\{\mathbf{S'}[1+\mathbf{s}]-\mathbf{Sv}-\mathbf{F}-\mathbf{I}\}-\mathbf{T}-\mathbf{D})/[1+\mathbf{/}]$$

Steve Asikin ISBN 13: **978-1564912361**, ISBN 10: **1564912363**

Rule-56478:

If both (**/**), (**U**), (**N**), (**s**), (**$**), (**v**), (**F**), (**I**), (**T**), (**D**), (**K**), (**J**), (**V**) and (**p**) are known, then its Sales Past must be:

$$\$' = \{[N+K+J+Vp/360]/[1+/]-U$$
$$+T+D+I+F+\$v\}/[1+s]$$

or

$$\$' = \$/[1+s]$$

Rule-56479:

If both (**/**), (**U**), (**$'**), (**N**), (**$**), (**v**), (**F**), (**I**), (**T**), (**D**), (**K**), (**J**), (**V**) and (**p**) are known, then its Sales Growth planned is:

$$s = \{[N+K+J+Vp/360]/[1+/]-U$$
$$+T+D+I+F+\$v\}/\$'-1$$

or

$$s = \$/\$'-1$$

Rule-56480:

If both (**/**), (**U**), (**$'**), (**s**), (**N**), (**v**), (**F**), (**I**), (**T**), (**D**), (**K**), (**J**), (**V**) and (**p**) are known, then its Sales or Revenue planned is:

$$\$ = [[1+/](U+\{\$'[1+s]-F-I\}-T-D)$$
$$-\{N+K+J+Vp/360\}]/v$$

or

$$\$ = \$'[1+s]$$

or

$$\$ = V/v$$

Steve Asikin ISBN 13: **978-1564912361**, ISBN 10: **1564912363**

Rule-56481:

If both $(\textit{I})$, $(\textbf{U})$, $(\textbf{S'})$, $(\textbf{s})$, $(\textbf{S})$, $(\textbf{N})$, $(\textbf{F})$, $(\textbf{I})$, $(\textbf{T})$, $(\textbf{D})$, $(\textbf{K})$, $(\textbf{J})$, $(\textbf{V})$ and $(\textbf{p})$ are known, then its Variable Portion planned is:

$$v= [[1+\textit{I}(U+\{S'[1+s]\text{-}F\text{-}I\}\text{-}T\text{-}D)$$
$$-\{N+K+J+Vp/360\}]/S$$

or

$$v= V/S$$

Rule-56482:

If both $(\textit{I})$, $(\textbf{U})$, $(\textbf{S'})$, $(\textbf{s})$, $(\textbf{S})$, $(\textbf{v})$, $(\textbf{N})$, $(\textbf{I})$, $(\textbf{T})$, $(\textbf{D})$, $(\textbf{K})$, $(\textbf{J})$, $(\textbf{V})$ and $(\textbf{p})$ are known, then its Fixed Cost planned is:

$$F= U+S'[1+s]\text{-}Sv\text{-}I\text{-}T\text{-}D\text{-}[N+K+J+Vp/360]/[1+\textit{I}]$$

Rule-56483:

If both $(\textit{I})$, $(\textbf{U})$, $(\textbf{S'})$, $(\textbf{s})$, $(\textbf{S})$, $(\textbf{v})$, $(\textbf{F})$, $(\textbf{N})$, $(\textbf{T})$, $(\textbf{D})$, $(\textbf{K})$, $(\textbf{J})$, $(\textbf{V})$ and $(\textbf{p})$ are known, then its Interest Expense planned is:

$$I= U+S'[1+s]\text{-}Sv\text{-}F\text{-}T\text{-}D\text{-}[N+K+J+Vp/360]/[1+\textit{I}]$$

Rule-56484:

If both $(\textit{I})$, $(\textbf{U})$, $(\textbf{S'})$, $(\textbf{s})$, $(\textbf{S})$, $(\textbf{v})$, $(\textbf{F})$, $(\textbf{I})$, $(\textbf{N})$, $(\textbf{D})$, $(\textbf{K})$, $(\textbf{J})$, $(\textbf{V})$ and $(\textbf{p})$ are known, then its Tax planned is:

$$T= U+S'[1+s]\text{-}Sv\text{-}F\text{-}I\text{-}D\text{-}[N+K+J+Vp/360]/[1+\textit{I}]$$

Steve Asikin ISBN 13: **978-1564912361**, ISBN 10: **1564912363**

Rule-56485:

If both $(I)$, $(U)$, $(S')$, $(s)$, $(S)$, $(v)$, $(F)$, $(I)$, $(T)$, $(N)$, $(K)$, $(J)$, $(V)$ and $(p)$ are known, then its Dividend planned is:

$$D= U+S'[1+s]-Sv-F-I-T-[N+K+J+Vp/360]/[1+I]$$

Rule-56486:

If both $(I)$, $(U)$, $(S')$, $(s)$, $(S)$, $(v)$, $(F)$, $(I)$, $(T)$, $(D)$, $(N)$, $(J)$, $(V)$ and $(p)$ are known, then its Kind of Cash planned is:

$$K= [1+I](U+\{S'[1+s]-Sv-F-I\}-T-D)$$
$$-[N+J+Vp/360]$$

Rule-56487:

If both $(I)$, $(U)$, $(S')$, $(s)$, $(S)$, $(v)$, $(F)$, $(I)$, $(T)$, $(D)$, $(K)$, $(N)$ and $(p)$ are known, then its Job or Trade Account Receivable planned is:

$$J= [1+I](U+\{S'[1+s]-Sv-F-I\}-T-D)$$
$$-\{N+K+Vp/360\}$$

Rule-56488:

If both $(I)$, $(U)$, $(S')$, $(s)$, $(S)$, $(v)$, $(F)$, $(I)$, $(T)$, $(D)$, $(K)$, $(J)$, $(N)$ and $(p)$ are known, then its Variable Cost planned is:

$$V= 360[[1+I](U+\{S'[1+s]-Sv-F-I\}-T-D)$$
$$-[N+K+J]]/p$$

or

$$V= Sv$$

Steve Asikin ISBN 13: **978-1564912361**, ISBN 10: **1564912363**

Rule-56489:

    If both $(I)$, $(U)$, $(S')$, $(s)$, $(S)$, $(v)$, $(F)$, $(I)$, $(T)$, $(D)$, $(K)$, $(J)$, $(V)$ and $(N)$ are known, then its Procured Inventory Days planned is:

$$p = 360[[1+I(U+\{S'[1+s]-Sv-F-I\}-T-D) \\ -[N+K+J]]/V$$

Rule-56490:

    If both $(I)$, $(U)$, $(S')$, $(s)$, $(S)$, $(v)$, $(F)$, $(I)$, $(T)$, $(D)$, $(K)$, $(J)$ and $(p)$ are known, then its Non Current Asset planned is:

$$N = [1+I\{U+S'[1+s]-Sv-F-I-T-D\}-K-J-Svp/360$$

Rule-56491:

    If both $(N)$, $(U)$, $(S')$, $(s)$, $(S)$, $(v)$, $(F)$, $(I)$, $(T)$, $(D)$, $(K)$, $(J)$ and $(p)$ are known, then its Leverage or Gearing Ratio planned is:

$$I = [N+K+J+Svp/360] \\ /(U+\{S'[1+s]-Sv-F-I\}-T-D)-1$$

Rule-56492:

    If both $(I)$, $(N)$, $(S')$, $(s)$, $(S)$, $(v)$, $(F)$, $(I)$, $(T)$, $(D)$, $(K)$, $(J)$ and $(p)$ are known, then its Utilized or Starting Capital must be:

$$U = ([N+K+J+Svp/360] \\ -[1+I\{S'[1+s]-Sv-F-I\}-T-D)/[1+I]$$

Steve Asikin ISBN 13: **978-1564912361**, ISBN 10: **1564912363**

Rule-56493:

If both **(/)**, **(U)**, **(N)**, **(s)**, **(S)**, **(v)**, **(F)**, **(I)**, **(T)**, **(D)**, **(K)**, **(J)** and **(p)** are known, then its Sales Past must be:

$$S' = \{[N+K+J+Svp/360]/[1+/]-U$$
$$+T+D+I+F+Sv\}/[1+s]$$

or

$$S' = S/[1+s]$$

Rule-56494:

If both **(/)**, **(U)**, **(S')**, **(N)**, **(S)**, **(v)**, **(F)**, **(I)**, **(T)**, **(D)**, **(K)**, **(J)** and **(p)** are known, then its Sales Growth planned is:

$$s = \{[N+K+J+Svp/360]/[1+/]-U$$
$$+T+D+I+F+Sv\}/S'-1$$

or

$$s = S/S'-1$$

Rule-56495:

If both **(/)**, **(U)**, **(S')**, **(s)**, **(N)**, **(v)**, **(F)**, **(I)**, **(T)**, **(D)**, **(K)**, **(J)** and **(p)** are known, then its Sales or Revenue planned is:

$$S = [[1+/](U+\{S'[1+s]-F-I\}-T-D)$$
$$-[N+K+J]]/(v\{p/360+[1+/]\})$$

or

$$S = S'[1+s]$$

Steve Asikin ISBN 13: **978-1564912361**, ISBN 10: **1564912363**

Rule-56496:

If both (**/**), (**U**), (**S'**), (**s**), (**S**), (**N**), (**F**), (**I**), (**T**), (**D**), (**K**), (**J**) and (**p**) are known, then its Variable Portion planned is:

$$v= [[1+/(U+\{S'[1+s]-F-I\}-T-D)-[N+K+J]] /(S\{p/360-[1+/]\})$$

Rule-56497:

If both (**/**), (**U**), (**S'**), (**s**), (**S**), (**v**), (**N**), (**I**), (**T**), (**D**), (**K**), (**J**) and (**p**) are known, then its Fixed Cost planned is:

$$F= U+S'[1+s]-Sv-I-T-D-[N+K+J+Svp/360]/[1+/]$$

Rule-56498:

If both (**/**), (**U**), (**S'**), (**s**), (**S**), (**v**), (**F**), (**N**), (**T**), (**D**), (**K**), (**J**) and (**p**) are known, then its Interest Expense planned is:

$$I= U+S'[1+s]-Sv-F-T-D-[N+K+J+Svp/360]/[1+/]$$

Rule-56499:

If both (**/**), (**U**), (**S'**), (**s**), (**S**), (**v**), (**F**), (**I**), (**N**), (**D**), (**K**), (**J**) and (**p**) are known, then its Tax planned is:

$$T= U+S'[1+s]-Sv-F-I-D-[N+K+J+Svp/360]/[1+/]$$

Rule-56500:

If both (**/**), (**U**), (**S'**), (**s**), (**S**), (**v**), (**F**), (**I**), (**T**), (**N**), (**K**), (**J**) and (**p**) are known, then its Dividend planned is:

$$D= U+S'[1+s]-Sv-F-I-T-[N+K+J+Svp/360]/[1+/]$$

Steve Asikin ISBN 13: **978-1564912361**, ISBN 10: **1564912363**

Rule-56501:

If both (**/**), (**U**), (**S'**), (**s**), (**S**), (**v**), (**F**), (**I**), (**T**), (**D**), (**N**), (**J**) and (**p**) are known, then its Kind of Cash planned is:

$$K= [1+/](U+\{S'[1+s]-Sv-F-I\}-T-D)$$
$$-[N+J+Svp/360]$$

Rule-56502:

If both (**/**), (**U**), (**S'**), (**s**), (**S**), (**v**), (**F**), (**I**), (**T**), (**D**), (**K**), (**N**) and (**p**) are known, then its Job or Trade Account Receivable planned is:

$$J= [1+/](U+\{S'[1+s]-Sv-F-I\}-T-D)$$
$$-\{N+K+Svp/360\}$$

Rule-56503:

If both (**/**), (**U**), (**S'**), (**s**), (**S**), (**v**), (**F**), (**I**), (**T**), (**D**), (**K**), (**J**) and (**N**) are known, then its Procured Inventory Days planned is:

$$p= 360[[1+/](U+\{S'[1+s]-Sv-F-I\}-T-D)$$
$$-[N+K+J]]/[Sv]$$

Rule-56504:

If both (**/**), (**U**), (**S'**), (**s**), (**S**), (**v**), (**F**), (**I**), (**T**), (**D**), (**K**), (**J**) and (**p**) are known, then its Non Current Asset planned is:

$$N= [1+/]\{U+S'[1+s]-Sv-F-I-T-D\}$$
$$-K-J-S'vp[1+s]/360$$

Steve Asikin ISBN 13: **978-1564912361**, ISBN 10: **1564912363**

Rule-56505:

If both (**N**), (**U**), (**S'**), (**s**), (**S**), (**v**), (**F**), (**I**), (**T**), (**D**), (**K**), (**J**) and (**p**) are known, then its Leverage or Gearing Ratio planned is:

$$I = \{N+K+J+S'vp[1+s]/360\}$$
$$/(U+\{S'[1+s]-Sv-F-I\}-T-D)-1$$

Rule-56506:

If both (**I**), (**N**), (**S'**), (**s**), (**S**), (**v**), (**F**), (**I**), (**T**), (**D**), (**K**), (**J**) and (**p**) are known, then its Utilized or Starting Capital must be:

$$U = (\{N+K+J+S'vp[1+s]/360\}$$
$$-[1+I]\{S'[1+s]-Sv-F-I\}-T-D)/[1+I]$$

Rule-56507:

If both (**I**), (**U**), (**N**), (**s**), (**S**), (**v**), (**F**), (**I**), (**T**), (**D**), (**K**), (**J**) and (**p**) are known, then its Sales Past must be:

$$S' = (\{N+K+J+S'vp[1+s]/360\}/[1+I]-U$$
$$+T+D+F+Sv)/[1+s]$$

or

$$S' = S/[1+s]$$

`

Steve Asikin ISBN 13: **978-1564912361**, ISBN 10: **1564912363**

Rule-56508:

If both $(\textbf{\textit{I}})$, $(\textbf{U})$, $(\textbf{S'})$, $(\textbf{N})$, $(\textbf{S})$, $(\textbf{v})$, $(\textbf{F})$, $(\textbf{I})$, $(\textbf{T})$, $(\textbf{D})$, $(\textbf{K})$, $(\textbf{J})$ and $(\textbf{p})$ are known, then its Sales Growth planned is:

$$s = (\{\textbf{N}+\textbf{K}+\textbf{J}+\textbf{S'vp}[1+s]/360\}/[1+\textbf{\textit{I}}]-\textbf{U}$$
$$+\textbf{T}+\textbf{D}+\textbf{F}+\textbf{Sv})/\textbf{S'}-1$$

or

$$s = \textbf{S}/\textbf{S'}-1$$

Rule-56509:

If both $(\textbf{\textit{I}})$, $(\textbf{U})$, $(\textbf{S'})$, $(\textbf{s})$, $(\textbf{N})$, $(\textbf{v})$, $(\textbf{F})$, $(\textbf{I})$, $(\textbf{T})$, $(\textbf{D})$, $(\textbf{K})$, $(\textbf{J})$ and $(\textbf{p})$ are known, then its Sales or Revenue planned is:

$$S = (\textbf{U}+\{\textbf{S'}[1+s]-\textbf{F}-\textbf{I}\}-\textbf{T}-\textbf{D}$$
$$-\{\textbf{N}+\textbf{K}+\textbf{J}+\textbf{S'vp}[1+s]/360\}/[1+\textbf{\textit{I}}])/\textbf{v}$$

or

$$S = \textbf{S'}[1+s]$$

Rule-56510:

If both $(\textbf{\textit{I}})$, $(\textbf{U})$, $(\textbf{S'})$, $(\textbf{s})$, $(\textbf{S})$, $(\textbf{N})$, $(\textbf{F})$, $(\textbf{I})$, $(\textbf{T})$, $(\textbf{D})$, $(\textbf{K})$, $(\textbf{J})$ and $(\textbf{p})$ are known, then its Variable Portion planned is:

$$v = (\textbf{U}+\{\textbf{S'}[1+s]-\textbf{F}-\textbf{I}\}-\textbf{T}-\textbf{D}$$
$$-\{\textbf{N}+\textbf{K}+\textbf{J}+\textbf{S'vp}[1+s]/360\}/[1+\textbf{\textit{I}}])/\textbf{S}$$

Steve Asikin ISBN 13: **978-1564912361**, ISBN 10: **1564912363**

Rule-56511:

If both (**I**), (**U**), (**S'**), (**s**), (**S**), (**v**), (**N**), (**I**), (**T**), (**D**), (**K**), (**J**) and (**p**) are known, then its Fixed Cost planned is:

$$F = U + \{S'\lfloor 1+s \rfloor - Sv - I\} - T - D$$
$$- \{N+K+J+S'vp[1+s]/360\}/[1+I]$$

Rule-56512:

If both (**I**), (**U**), (**S'**), (**s**), (**S**), (**v**), (**F**), (**N**), (**T**), (**D**), (**K**), (**J**) and (**p**) are known, then its Interest Expense planned is:

$$I = U + \{S'[1+s] - Sv - F\} - T - D$$
$$- \{N+K+J+S'vp[1+s]/360\}/[1+I]$$

Rule-56513:

If both (**I**), (**U**), (**S'**), (**s**), (**S**), (**v**), (**F**), (**I**), (**N**), (**D**), (**K**), (**J**) and (**p**) are known, then its Tax planned is:

$$T = U + \{S'[1+s] - Sv - F - I\} - D$$
$$- \{N+K+J+S'vp[1+s]/360\}/[1+I]$$

Rule-56514:

If both (**I**), (**U**), (**S'**), (**s**), (**S**), (**v**), (**F**), (**I**), (**T**), (**N**), (**K**), (**J**) and (**p**) are known, then its Dividend planned is:

$$D = U + \{S'[1+s] - Sv - F - I\} - T$$
$$- \{N+K+J+S'vp[1+s]/360\}/[1+I]$$

Steve Asikin ISBN 13: **978-1564912361**, ISBN 10: **1564912363**

Rule-56515:

If both (**/**), (**U**), (**S'**), (**s**), (**S**), (**v**), (**F**), (**I**), (**T**), (**D**), (**N**), (**J**) and (**p**) are known, then its Kind of Cash planned is:

$$\mathbf{K} = [1 + \mathbf{/}(\mathbf{U} + \{\mathbf{S'}[1+\mathbf{s}] - \mathbf{Sv} - \mathbf{F} - \mathbf{I}\} - \mathbf{T} - \mathbf{D})$$
$$- \{\mathbf{N} + \mathbf{J} + \mathbf{S'vp}[1+\mathbf{s}]/360\}$$

Rule-56516:

If both (**/**), (**U**), (**S'**), (**s**), (**S**), (**v**), (**F**), (**I**), (**T**), (**D**), (**K**), (**N**) and (**p**) are known, then its Job or Trade Account Receivable planned is:

$$\mathbf{J} = [[1 + \mathbf{/}(\mathbf{U} + \{\mathbf{S'}[1+\mathbf{s}] - \mathbf{Sv} - \mathbf{F} - \mathbf{I}\} - \mathbf{T} - \mathbf{D})$$
$$- \{\mathbf{N} + \mathbf{K} + \mathbf{S'vp}[1+\mathbf{s}]/360\}$$

Rule-56517:

If both (**/**), (**U**), (**S'**), (**s**), (**S**), (**v**), (**F**), (**I**), (**T**), (**D**), (**K**), (**J**) and (**N**) are known, then its Procured Inventory Days planned is:

$$\mathbf{p} = 360[[1 + \mathbf{/}(\mathbf{U} + \{\mathbf{S'}[1+\mathbf{s}] - \mathbf{Sv} - \mathbf{F} - \mathbf{I}\} - \mathbf{T} - \mathbf{D})$$
$$- [\mathbf{N} + \mathbf{K} + \mathbf{J}]] / \{\mathbf{S'v}[1+\mathbf{s}]\}$$

Rule-56518:

If both (**/**), (**U**), (**S'**), (**s**), (**S**), (**v**), (**F**), (**I**), (**T**), (**D**), (**K**), (**j**) and (**P**) are known, then its Non Current Asset planned is:

$$\mathbf{N} = [1 + \mathbf{/}\{\mathbf{U} + \mathbf{S'}[1+\mathbf{s}] - \mathbf{Sv} - \mathbf{F} - \mathbf{I} - \mathbf{T} - \mathbf{D}\} - \mathbf{K} - \mathbf{Sj}/360 - \mathbf{P}$$

Steve Asikin ISBN 13: **978-1564912361**, ISBN 10: **1564912363**

<u>Rule-56519</u>:

If both (**N**), (**U**), (**S'**), (**s**), (**S**), (**v**), (**F**), (**I**), (**T**), (**D**), (**K**), (**j**) and (**P**) are known, then its Leverage or Gearing Ratio planned is:

$$l = \{\mathbf{N+K+Sj}/360+\mathbf{P}\}/(\mathbf{U}+\{\mathbf{S'}[1+\mathbf{s}]-\mathbf{Sv-F-I}\}-\mathbf{T-D})-1$$

<u>Rule-56520</u>:

If both (**l**), (**N**), (**S'**), (**s**), (**S**), (**v**), (**F**), (**I**), (**T**), (**D**), (**K**), (**j**) and (**P**) are known, then its Utilized or Starting Capital must be:

$$\mathbf{U} = (\{\mathbf{N+K+Sj}/360+\mathbf{P}\}$$
$$-[1+l]\{\mathbf{S'}[1+\mathbf{s}]-\mathbf{Sv-F-I}\}-\mathbf{T-D})/[1+l]$$

<u>Rule-56521</u>:

If both (**l**), (**U**), (**N**), (**s**), (**S**), (**v**), (**F**), (**I**), (**T**), (**D**), (**K**), (**j**) and (**P**) are known, then its Sales Past must be:

$$\mathbf{S'} = \{[\mathbf{N+K+Sj}/360+\mathbf{P}]/[1+l]-\mathbf{U}$$
$$+\mathbf{Sv+F+I+T+D}\}/[1+\mathbf{s}]$$
$$\text{or}$$
$$\mathbf{S'} = \mathbf{S}/[1+\mathbf{s}]$$

<u>Rule-56522</u>:

If both (**l**), (**U**), (**S'**), (**N**), (**S**), (**v**), (**F**), (**I**), (**T**), (**D**), (**K**), (**j**) and (**P**) are known, then its Sales Growth planned is:

$$\mathbf{s} = \{[\mathbf{N+K+Sj}/360+\mathbf{P}]/[1+l]-\mathbf{U}$$
$$+\mathbf{Sv+F+I+T+D}\}/\mathbf{S'}-1$$
$$\text{or}$$
$$\mathbf{s} = \mathbf{S}/\mathbf{S'}-1$$

Steve Asikin ISBN 13: **978-1564912361**, ISBN 10: **1564912363**

Rule-56523:

If both (**∫**), (**U**), (**$'**), (**s**), (**N**), (**v**), (**F**), (**I**), (**T**), (**D**), (**K**), (**j**) and (**P**) are known, then its Sales or Revenue planned is:

$= [[1+∫(U+{$'[1+s]-F-I}-T-D)-[N+K+P]]
            /{j/360+v[1+∫]}

            or

$= $'[1+s]

Rule-56524:

If both (**∫**), (**U**), (**$'**), (**s**), (**$**), (**N**), (**F**), (**I**), (**T**), (**D**), (**K**), (**j**) and (**P**) are known, then its Variable Portion planned is:

v= (U+{$'[1+s]-I-F}-T-D
        -[N+K+$j/360+P]/[1+∫])/$

Rule-56525:

If both (**∫**), (**U**), (**$'**), (**s**), (**$**), (**v**), (**N**), (**I**), (**T**), (**D**), (**K**), (**j**) and (**P**) are known, then its Fixed Cost planned is:
F= U+{$'[1+s]-$v-I}-T-D-[N+K+$j/360+P]/[1+∫]

Rule-56526:

If both (**∫**), (**U**), (**$'**), (**s**), (**$**), (**v**), (**F**), (**N**), (**T**), (**D**), (**K**), (**j**) and (**P**) are known, then its Interest Expense planned is:

I= U+{$'[1+s]-$v-F}-T-D-[N+K+$j/360+P]/[1+∫]

Steve Asikin ISBN 13: **978-1564912361**, ISBN 10: **1564912363**

Rule-56527:

    If both $(I)$, $(U)$, $(S')$, $(s)$, $(S)$, $(v)$, $(F)$, $(I)$, $(N)$, $(D)$, $(K)$, $(j)$ and $(P)$ are known, then its Tax planned is:

$$T = U + \{S'[1+s] - Sv - F - I\} - D - [N + K + Sj/360 + P]/[1+I]$$

Rule-56528:

    If both $(I)$, $(U)$, $(S')$, $(s)$, $(S)$, $(v)$, $(F)$, $(I)$, $(T)$, $(N)$, $(K)$, $(j)$ and $(P)$ are known, then its Dividend planned is:

$$D = U + \{S'[1+s] - Sv - F - I\} - T - [N + K + Sj/360 + P]/[1+I]$$

Rule-56529:

    If both $(I)$, $(U)$, $(S')$, $(s)$, $(S)$, $(v)$, $(F)$, $(I)$, $(T)$, $(D)$, $(N)$, $(j)$ and $(P)$ are known, then its Kind of Cash planned is:

$$K = [1+I](U + \{S'[1+s] - Sv - F - I\} - T - D) - [N + Sj/360 + P]$$

Rule-56530:

    If both $(I)$, $(U)$, $(S')$, $(s)$, $(S)$, $(v)$, $(F)$, $(I)$, $(T)$, $(D)$, $(K)$, $(N)$ and $(P)$ are known, then its Job or Trade Receivable Days planned is:

$$j = 360[[1+I](U + \{S'[1+s] - Sv - F - I\} - T - D) - \{N + K + P\}]/S$$

Steve Asikin ISBN 13: **978-1564912361**, ISBN 10: **1564912363**

Rule-56531:

If both (**/**), (**U**), (**S'**), (**s**), (**S**), (**v**), (**F**), (**I**), (**T**), (**D**), (**K**), (**j**) and (**N**) are known, then its Procured Inventory planned is:

$$P= [1+/](U+\{S'[1+s]-Sv-F-I\}-T-D)$$
$$-\{N+K+Sj/360\}$$

Rule-56532:

If both (**/**), (**U**), (**S'**), (**s**), (**S**), (**v**), (**F**), (**I**), (**T**), (**D**), (**K**), (**j**), (**V**) and (**p**) are known, then its Non Current Asset planned is:

$$N= [1+/]\{U+S'[1+s]-Sv-F-I-T-D\}$$
$$-K-Sj/360-Vp/360$$

Rule-56533:

If both (**N**), (**U**), (**S'**), (**s**), (**S**), (**v**), (**F**), (**I**), (**T**), (**D**), (**K**), (**j**), (**V**) and (**p**) are known, then its Leverage or Gearing Ratio planned is:

$$/= [N+K+Sj/360+Vp/360]$$
$$/(U+\{S'[1+s]-Sv-F-I\}-T-D)-1$$

Rule-56534:

If both (**/**), (**N**), (**S'**), (**s**), (**S**), (**v**), (**F**), (**I**), (**T**), (**D**), (**K**), (**j**), (**V**) and (**p**) are known, then its Utilized or Starting Capital must be:

$$U= ([N+K+Sj/360+Vp/360]$$
$$-[1+/]\{S'[1+s]-Sv-F-I\}-T-D)/[1+/]$$

Steve Asikin ISBN 13: **978-1564912361**, ISBN 10: **1564912363**

Rule-56535:

If both (**/**), (**U**), (**N**), (**s**), (**S**), (**v**), (**F**), (**I**), (**T**), (**D**), (**K**), (**j**), (**V**) and (**p**) are known, then its Sales Past must be:

$$S'= [N+K+Sj/360+Vp/360]/[1+/-U$$
$$+D+T+I+F+Sv)/[1+s]$$

or

$$S'= S[1+s]$$

Rule-56536:

If both (**/**), (**U**), (**S'**), (**N**), (**S**), (**v**), (**F**), (**I**), (**T**), (**D**), (**K**), (**j**), (**V**) and (**p**) are known, then its Sales Growth planned is:

$$s= [N+K+Sj/360+Vp/360]/[1+/-U$$
$$+D+T+I+F+Sv)/S'-1$$

or

$$s= S/S'-1$$

Rule-56537:

If both (**/**), (**U**), (**S'**), (**s**), (**N**), (**v**), (**F**), (**I**), (**T**), (**D**), (**K**), (**j**), (**V**) and (**p**) are known, then its Sales or Revenue planned is:

$$S= [[1+/(U+\{S'[1+s]-F-I\}-T-D)$$
$$-[N+K+Vp/360]]/\{j/360+v[1+/]\}$$

or

$$S= S'[1+s]$$

`

Steve Asikin ISBN 13: **978-1564912361**, ISBN 10: **1564912363**

Rule-56538:

If both $(I)$, $(U)$, $(S')$, $(s)$, $(S)$, $(N)$, $(F)$, $(I)$, $(T)$, $(D)$, $(K)$, $(j)$, $(V)$ and $(p)$ are known, then its Variable Portion planned is:

$$v = (U + \{S'[1+s] - F - I\} - T - D$$
$$- [N + K + Sj/360 + Vp/360]/[1+I])/S$$

or

$$v = V/S$$


Rule-56539:

If both $(I)$, $(U)$, $(S')$, $(s)$, $(S)$, $(v)$, $(N)$, $(I)$, $(T)$, $(D)$, $(K)$, $(j)$, $(V)$ and $(p)$ are known, then its Fixed Cost planned is:

$$F = U + \{S'[1+s] - Sv - I\} - T - D$$
$$- [N + K + Sj/360 + Vp/360]/[1+I]$$


Rule-56540:

If both $(I)$, $(U)$, $(S')$, $(s)$, $(S)$, $(v)$, $(F)$, $(N)$, $(T)$, $(D)$, $(K)$, $(j)$, $(V)$ and $(p)$ are known, then its Interest Expense planned is:

$$I = U + \{S'[1+s] - Sv - F\} - T - D$$
$$- [N + K + Sj/360 + Vp/360]/[1+I]$$


Rule-56541:

If both $(I)$, $(U)$, $(S')$, $(s)$, $(S)$, $(v)$, $(F)$, $(I)$, $(N)$, $(D)$, $(K)$, $(j)$, $(V)$ and $(p)$ are known, then its Tax planned is:

$$T = U + \{S'[1+s] - Sv - F - I\} - D$$
$$- [N + K + Sj/360 + Vp/360]/[1+I]$$

Steve Asikin ISBN 13: **978-1564912361**, ISBN 10: **1564912363**

Rule-56542:

If both (**I**), (**U**), (**S'**), (**s**), (**S**), (**v**), (**F**), (**I**), (**T**), (**N**), (**K**), (**j**), (**V**) and (**p**) are known, then its Dividend planned is:

$$D= U+\{S'[1+s]-Sv-F-I\}-T$$
$$-[N+K+Sj/360+Vp/360]/[1+I]$$

Rule-56543:

If both (**I**), (**U**), (**S'**), (**s**), (**S**), (**v**), (**F**), (**I**), (**T**), (**D**), (**N**), (**j**), (**V**) and (**p**) are known, then its Kind of Cash planned is:

$$K= [1+I](U+\{S'[1+s]-Sv-F-I\}-T-D)$$
$$-[N+Sj/360+Vp/360]$$

Rule-56544:

If both (**I**), (**U**), (**S'**), (**s**), (**S**), (**v**), (**F**), (**I**), (**T**), (**D**), (**K**), (**N**), (**V**) and (**p**) are known, then its Job or Trade Receivable Days planned is:

$$j= 360[[1+I](U+\{S'[1+s]-Sv-F-I\}-T-D)$$
$$-[N+K+Vp/360]]/S$$

Steve Asikin ISBN 13: **978-1564912361**, ISBN 10: **1564912363**

Rule-56545:

If both (**/**), (**U**), (**S'**), (**s**), (**S**), (**v**), (**F**), (**I**), (**T**), (**D**), (**K**), (**j**), (**N**) and (**p**) are known, then its Variable Cost planned is:

$$V= 360[[1+/(U+\{S'[1+s]-Sv-F-I\}-T-D)$$
$$-[N+K+Sj/360]]/p$$

or

$$V= Sv$$

Rule-56546:

If both (**/**), (**U**), (**S'**), (**s**), (**S**), (**v**), (**F**), (**I**), (**T**), (**D**), (**K**), (**j**), (**V**) and (**N**) are known, then its Procured Inventory Days planned is:

$$p= 360[[1+/(U+\{S'[1+s]-Sv-F-I\}-T-D)$$
$$-[N+K+Sj/360]]/V$$

Rule-56547:

If both (**/**), (**U**), (**S'**), (**s**), (**S**), (**v**), (**F**), (**I**), (**T**), (**D**), (**K**), (**j**) and (**p**) are known, then its Non Current Asset planned is:

$$N= [1+/\{U+S'[1+s]-Sv-F-I-T-D\}$$
$$-K-Sj/360-Svp/360$$

Rule-56548:

If both (**N**), (**U**), (**S'**), (**s**), (**S**), (**v**), (**F**), (**I**), (**T**), (**D**), (**K**), (**j**) and (**p**) are known, then its Leverage or Gearing Ratio planned is:

$$/= [N+K+Sj/360+Svp/360]$$
$$/(U+\{S'[1+s]-Sv-F-I\}-T-D)-1$$

Steve Asikin ISBN 13: **978-1564912361**, ISBN 10: **1564912363**

Rule-56549:

If both (**∫**), (**N**), (**$'**), (**s**), (**$**), (**v**), (**F**), (**I**), (**T**), (**D**), (**K**), (**j**) and (**p**) are known, then its Utilized or Strarting Capital must be:

$$\mathbf{U} = ([\mathbf{N+K+\$j}/360+\mathbf{\$vp}/360]$$
$$-[1+\mathbf{\int}]\{\mathbf{\$'}[1+\mathbf{s}]-\mathbf{\$v-F-I-T-D})/[1+\mathbf{\int}]$$

Rule-56550:

If both (**∫**), (**U**), (**N**), (**s**), (**$**), (**v**), (**F**), (**I**), (**T**), (**D**), (**K**), (**j**) and (**p**) are known, then its Sales Past must be:

$$\mathbf{\$'} = \{[\mathbf{N+K+\$j}/360+\mathbf{\$vp}/360]/[1+\mathbf{\int}]-\mathbf{U}$$
$$+\mathbf{\$v+F+I+T+D}\}/[1+\mathbf{s}]$$

or

$$\mathbf{\$'} = \mathbf{\$}/[1+\mathbf{s}]$$

Rule-56551:

If both (**∫**), (**U**), (**$'**), (**N**), (**$**), (**v**), (**F**), (**I**), (**T**), (**D**), (**K**), (**j**) and (**p**) are known, then its Sales Growth planned is:

$$\mathbf{s} = \{[\mathbf{N+K+\$j}/360+\mathbf{\$vp}/360]/[1+\mathbf{\int}]-\mathbf{U}$$
$$+\mathbf{\$v+F+I+T+D}\}/\mathbf{\$'}-1$$

or

$$\mathbf{s} = \mathbf{\$}/\mathbf{\$'}-1$$

Steve Asikin ISBN 13: **978-1564912361**, ISBN 10: **1564912363**

Rule-56552:

If both (**/**), (**U**), (**S'**), (**s**), (**N**), (**v**), (**F**), (**I**), (**T**), (**D**), (**K**), (**j**) and (**p**) are known, then its Sales or Revenue planned is:

$$S= [[1+/](U+\{S'[1+s]-F-I\}-T-D)-[N+K]]$$
$$/\{j/360+vp/360+v[1+/]\}$$

or

$$S= S'[1+s]$$

Rule-56553:

If both (**/**), (**U**), (**S'**), (**s**), (**S**), (**N**), (**F**), (**I**), (**T**), (**D**), (**K**), (**j**) and (**p**) are known, then its Variable Portion planned is:

$$v= [[1+/](U+\{S'[1+s]-F-I\}-T-D)-[N+K+Sj/360]]$$
$$/(S\{p/360+[1+/]\})$$

Rule-56554:

If both (**/**), (**U**), (**S'**), (**s**), (**S**), (**v**), (**N**), (**I**), (**T**), (**D**), (**K**), (**j**) and (**p**) are known, then its Fixed Cost planned is:

$$F= U+\{S'[1+s]-Sv-I\}-T-D$$
$$-[N+K+Sj/360+Svp/360]/[1+/]$$

Rule-56555:

If both (**/**), (**U**), (**S'**), (**s**), (**S**), (**v**), (**F**), (**N**), (**T**), (**D**), (**K**), (**j**) and (**p**) are known, then its Interest Expense planned is:

$$I= U+\{S'[1+s]-Sv-F\}-T-D$$
$$-[N+K+Sj/360+Svp/360]/[1+/]$$

Steve Asikin ISBN 13: **978-1564912361**, ISBN 10: **1564912363**

Rule-56556:

If both $(I)$, $(U)$, $(S')$, $(s)$, $(\$)$, $(v)$, $(F)$, $(I)$, $(N)$, $(D)$, $(K)$, $(j)$ and $(p)$ are known, then its Tax planned is:

$$T = U + \{S'[1+s]-Sv-F-I\}-D$$
$$-[N+K+Sj/360+Svp/360]/[1+I]$$

Rule-56557:

If both $(I)$, $(U)$, $(S')$, $(s)$, $(\$)$, $(v)$, $(F)$, $(I)$, $(T)$, $(N)$, $(K)$, $(j)$ and $(p)$ are known, then its Dividend planned is:

$$D = U + \{S'[1+s]-Sv-F-I\}-T$$
$$-[N+K+Sj/360+Svp/360]/[1+I]$$

Rule-56558:

If both $(I)$, $(U)$, $(S')$, $(s)$, $(\$)$, $(v)$, $(F)$, $(I)$, $(T)$, $(D)$, $(N)$, $(j)$ and $(p)$ are known, then its Kind of Cash planned is:

$$K = [1+I](U+\{S'[1+s]-Sv-F-I\}-T-D)$$
$$-[N+Sj/360+Svp/360]$$

Rule-56559:

If both $(I)$, $(U)$, $(S')$, $(s)$, $(\$)$, $(v)$, $(F)$, $(I)$, $(T)$, $(D)$, $(K)$, $(N)$ and $(p)$ are known, then its Job or Trade Receivable Days planned is:

$$j = 360[[1+I](U+\{S'[1+s]-Sv-F-I\}-T-D)$$
$$-\{N+K+Svp/360\}]/\$$$

Steve Asikin ISBN 13: **978-1564912361**, ISBN 10: **1564912363**

Rule-56560:

If both $(I)$, $(U)$, $(S')$, $(s)$, $(S)$, $(v)$, $(F)$, $(I)$, $(T)$, $(D)$, $(K)$, $(j)$ and $(N)$ are known, then its Procured Inventory Days planned is:

$$p = 360[[1+I](U+\{S'[1+s]-Sv-F-I\}-T-D)$$
$$-[N+K+Sj/360]]/[Sv]$$

Rule-56561:

If both $(I)$, $(U)$, $(S')$, $(s)$, $(S)$, $(v)$, $(F)$, $(I)$, $(T)$, $(D)$, $(K)$, $(j)$ and $(p)$ are known, then its Non Current Asset planned is:

$$N = [1+I]\{U+S'[1+s]-Sv-F-I-T-D\}-K-Sj/360$$
$$-S'vp[1+s]/360$$

Rule-56562:

If both $(N)$, $(U)$, $(S')$, $(s)$, $(S)$, $(v)$, $(F)$, $(I)$, $(T)$, $(D)$, $(K)$, $(j)$ and $(p)$ are known, then its Leverage or Gearing Ratio planned is:

$$I = \{N+K+Sj/360+S'vp[1+s]/360\}$$
$$/(U+\{S'[1+s]-Sv-F-I\}-T-D)-1$$

Rule-56563:

If both $(I)$, $(N)$, $(S')$, $(s)$, $(S)$, $(v)$, $(F)$, $(I)$, $(T)$, $(D)$, $(K)$, $(j)$ and $(p)$ are known, then its Utilized or Starting Capital must be:

$$U = (\{N+K+Sj/360+S'vp[1+s]/360\}$$
$$-[1+I]\{S'[1+s]-Sv-F-I\}-T-D)/\{[1+I]$$

Steve Asikin ISBN 13: **978-1564912361**, ISBN 10: **1564912363**

<u>Rule-56564</u>:

If both (**/**), (**U**), (**S'**), (**s**), (**S**), (**v**), (**F**), (**I**), (**T**), (**D**), (**K**), (**j**) and (**p**) are known, then its Non Current Asset planned is:

$$\textbf{S'}= [[1+\textbf{/}](\textbf{U}-\{\textbf{Sv}+\textbf{F}+\textbf{I}\}-\textbf{T}-\textbf{D})-[\textbf{N}+\textbf{K}+\textbf{Sj}/360]]$$
$$/([1+\textbf{s}]\{\textbf{vp}/360-[1+\textbf{/}]\})$$

or

$$\textbf{S'}= \textbf{S}/[1+\textbf{s}]$$

<u>Rule-56565</u>:

If both (**/**), (**U**), (**S'**), (**N**), (**S**), (**v**), (**F**), (**I**), (**T**), (**D**), (**K**), (**j**) and (**p**) are known, then its Sales Growth planned is:

$$\textbf{s}= [[1+\textbf{/}](\textbf{U}-\{\textbf{Sv}+\textbf{F}+\textbf{I}\}-\textbf{T}-\textbf{D})-[\textbf{N}+\textbf{K}+\textbf{Sj}/360]]$$
$$/(\textbf{S'}\{\textbf{vp}/360-[1+\textbf{/}]\})-1$$

or

$$\textbf{s}= \textbf{S}/\textbf{S'}-1$$

<u>Rule-56566</u>:

If both (**/**), (**U**), (**S'**), (**s**), (**N**), (**v**), (**F**), (**I**), (**T**), (**D**), (**K**), (**j**) and (**p**) are known, then its Sales or Revenue planned is:

$$\textbf{S}= [\textbf{U}+\{\textbf{S'}[1+\textbf{s}]-\textbf{I}-\textbf{F}\}-\textbf{T}-\textbf{D}$$
$$-\{\textbf{N}+\textbf{K}+\textbf{Sj}/360+\textbf{S'vp}[1+\textbf{s}]/360\}/[1+\textbf{/}]/\textbf{v}$$

or

$$\textbf{S}= \textbf{S'}[1+\textbf{s}]$$

`

126

Steve Asikin ISBN 13: **978-1564912361**, ISBN 10: **1564912363**

Rule-56567:

If both ($\pmb{\jmath}$), (**U**), (**S'**), (**s**), (**S**), (**N**), (**F**), (**I**), (**T**), (**D**), (**K**), (**j**) and (**p**) are known, then its Variable Portion planned is:

$$\pmb{v}= [[1+\pmb{\jmath}(\pmb{U}+\{\pmb{S'}[1+\pmb{s}]\text{-}\pmb{F}\text{-}\pmb{I}\}\text{-}\pmb{T}\text{-}\pmb{D})\text{-}[\pmb{N}+\pmb{K}+\pmb{Sj}/360]]$$
$$/\{\pmb{S'p}[1+\pmb{s}]/360+\pmb{S}[1+\pmb{\jmath}]\}$$

Rule-56568:

If both ($\pmb{\jmath}$), (**U**), (**S'**), (**s**), (**S**), (**v**), (**N**), (**I**), (**T**), (**D**), (**K**), (**j**) and (**p**) are known, then its Fixed Cost planned is:

$$\pmb{F}= \pmb{U}+\{\pmb{S'}[1+\pmb{s}]\text{-}\pmb{Sv}\text{-}\pmb{I}\}\text{-}\pmb{T}\text{-}\pmb{D}$$
$$\text{-}\{\pmb{N}+\pmb{K}+\pmb{Sj}/360+\pmb{S'vp}[1+\pmb{s}]/360\}/[1+\pmb{\jmath}]$$

Rule-56569:

If both ($\pmb{\jmath}$), (**U**), (**S'**), (**s**), (**S**), (**v**), (**F**), (**N**), (**T**), (**D**), (**K**), (**j**) and (**p**) are known, then its Interest Expense planned is:

$$\pmb{I}= \pmb{U}+\{\pmb{S'}[1+\pmb{s}]\text{-}\pmb{Sv}\text{-}\pmb{F}\}\text{-}\pmb{T}\text{-}\pmb{D}$$
$$\text{-}\{\pmb{N}+\pmb{K}+\pmb{Sj}/360+\pmb{S'vp}[1+\pmb{s}]/360\}/[1+\pmb{\jmath}]$$

Rule-56570:

If both ($\pmb{\jmath}$), (**U**), (**S'**), (**s**), (**S**), (**v**), (**F**), (**I**), (**N**), (**D**), (**K**), (**j**) and (**p**) are known, then its Tax planned is:

$$\pmb{T}= \pmb{U}+\{\pmb{S'}[1+\pmb{s}]\text{-}\pmb{Sv}\text{-}\pmb{F}\text{-}\pmb{I}\}\text{-}\pmb{D}$$
$$\text{-}\{\pmb{N}+\pmb{K}+\pmb{Sj}/360+\pmb{S'vp}[1+\pmb{s}]/360\}/[1+\pmb{\jmath}]$$

Steve Asikin ISBN 13: **978-1564912361**, ISBN 10: **1564912363**

Rule-56571:

If both (**/**), (**U**), (**S'**), (**s**), (**S**), (**v**), (**F**), (**I**), (**T**), (**N**), (**K**), (**j**) and (**p**) are known, then its Dividend planned is:

$$D= U+\{S'[1+s]-Sv-F-I\}-T$$
$$-\{N+K+Sj/360+S'vp[1+s]/360\}/[1+\text{/}]$$

Rule-56572:

If both (**/**), (**U**), (**S'**), (**s**), (**S**), (**v**), (**F**), (**I**), (**T**), (**D**), (**N**), (**j**) and (**p**) are known, then its Kind of Cash planned is:

$$K= [1+\text{/}](U+\{S'[1+s]-Sv-F-I\}-T-D)$$
$$-\{N+Sj/360+S'vp[1+s]/360\}$$

Rule-56573:

If both (**/**), (**U**), (**S'**), (**s**), (**S**), (**v**), (**F**), (**I**), (**T**), (**D**), (**K**), (**N**) and (**p**) are known, then its Job or Trade Receivable Days planned is:

$$j= 360[[1+\text{/}](U+\{S'[1+s]-Sv-F-I\}-T-D)$$
$$-\{N+K+S'vp[1+s]/360\}]/S$$

Rule-56575:

If both (**/**), (**U**), (**S'**), (**s**), (**S**), (**v**), (**F**), (**I**), (**T**), (**D**), (**K**), (**j**) and (**N**) are known, then its Procured Inventory Days planned is:

$$p= 360[[1+\text{/}](U+\{S'[1+s]-Sv-F-I\}-T-D)$$
$$-[N+K+Sj/360]]/\{S'v[1+s]\}$$

Steve Asikin ISBN 13: **978-1564912361**, ISBN 10: **1564912363**

Rule-56576:

If both (**/**), (**U**), (**S'**), (**s**), (**S**), (**v**), (**F**), (**I**), (**T**), (**D**), (**K**), (**j**) and (**P**) are known, then its Non Current Asset planned is:

$$N = [1+/]\{U+S'[1+s]-Sv-F-I-T-D\}$$
$$-K-S'j[1+s]/360-P$$

Rule-56577:

If both (**N**), (**U**), (**S'**), (**s**), (**S**), (**v**), (**F**), (**I**), (**T**), (**D**), (**K**), (**j**) and (**P**) are known, then its Leverage or Gearing Ratio planned is:

$$/= \{N+K+S'j[1+s]/360+P\}$$
$$/(U+\{S'[1+s]-Sv-F-I\}-T-D)-1$$

Rule-56578:

If both (**/**), (**N**), (**S'**), (**s**), (**S**), (**v**), (**F**), (**I**), (**T**), (**D**), (**K**), (**j**) and (**P**) are known, then its Utilized or Starting Capital must be:

$$U = (\{N+K+S'j[1+s]/360+P\}$$
$$-[1+/]\{S'[1+s]-Sv-F-I\}-T-D)/[1+/]$$

Rule-56579:

If both (**/**), (**U**), (**N**), (**s**), (**S**), (**v**), (**F**), (**I**), (**T**), (**D**), (**K**), (**j**) and (**P**) are known, then its Sales Past must be:

$$S' = [[1+/](U-\{Sv+F+I\}-T-D)-[N+K+P]]$$
$$/([1+s]\{j/360-[1+/]\})$$
or
$$S' = S/[1+s]$$

Steve Asikin ISBN 13: **978-1564912361**, ISBN 10: **1564912363**

Rule-56580:

If both (**/**), (**U**), (**$'**), (**N**), (**$**), (**v**), (**F**), (**I**), (**T**), (**D**), (**K**), (**j**) and (**P**) are known, then its Sales Growth planned is:

$$s= [[1+/](U-\{$v+F+I\}-T-D)-[N+K+P]]$$
$$/($'\{j/360-[1+/]\})-1$$

or

$$s= $/$'-1$$

Rule-56581:

If both (**/**), (**U**), (**$'**), (**s**), (**N**), (**v**), (**F**), (**I**), (**T**), (**D**), (**K**), (**j**) and (**P**) are known, then its Sales or Revenue planned is:

$$$= [U+\{$'[1+s]-F-I\}-T-D$$
$$-\{N+K+$'j[1+s]/360+P\}/[1+/]]/v$$

or

$$$= $'[1+s]$$

Rule-56582:

If both (**/**), (**U**), (**$'**), (**s**), (**$**), (**N**), (**F**), (**I**), (**T**), (**D**), (**K**), (**j**) and (**P**) are known, then its Variable Portion planned is:

$$v= [U+\{$'[1+s]-F-I\}-T-D$$
$$-\{N+K+$'j[1+s]/360+P\}/[1+/]]/$$$

Steve Asikin ISBN 13: **978-1564912361**, ISBN 10: **1564912363**

Rule-56583:

If both $(I)$, $(U)$, $(S')$, $(s)$, $(S)$, $(v)$, $(N)$, $(I)$, $(T)$, $(D)$, $(K)$, $(j)$ and $(P)$ are known, then its Fixed Cost planned is:

$$F= U+\{S'[1+s]-Sv-I\}-T-D$$
$$-\{N+K+S'j[1+s]/360+P\}/[1+I]$$

Rule-56584:

If both $(I)$, $(U)$, $(S')$, $(s)$, $(S)$, $(v)$, $(F)$, $(N)$, $(T)$, $(D)$, $(K)$, $(j)$ and $(P)$ are known, then its Interest Expense planned is:

$$I= U+\{S'[1+s]-Sv-F\}-T-D$$
$$-\{N+K+S'j[1+s]/360+P\}/[1+I]$$

Rule-56585:

If both $(I)$, $(U)$, $(S')$, $(s)$, $(S)$, $(v)$, $(F)$, $(I)$, $(N)$, $(D)$, $(K)$, $(j)$ and $(P)$ are known, then its Tax planned is:

$$T= U+\{S'[1+s]-Sv-F-I\}-D$$
$$-\{N+K+S'j[1+s]/360+P\}/[1+I]$$

Rule-56586:

If both $(I)$, $(U)$, $(S')$, $(s)$, $(S)$, $(v)$, $(F)$, $(I)$, $(T)$, $(N)$, $(K)$, $(j)$ and $(P)$ are known, then its Dividend planned is:

$$D= U+\{S'[1+s]-Sv-F-I\}-T$$
$$-\{N+K+S'j[1+s]/360+P\}/[1+I]$$

Steve Asikin ISBN 13: **978-1564912361**, ISBN 10: **1564912363**

Rule-56587:
If both (**/**), (**U**), (**S'**), (**s**), (**S**), (**v**), (**F**), (**I**), (**T**), (**D**), (**N**), (**j**) and (**P**) are known, then its Kind of Cash planned is:

$$K = [1 + \text{/}(U + \{S'[1+s]-Sv-F-I\}-T-D) - \{N+S'j[1+s]/360+P\}$$

Rule-56588:
If both (**/**), (**U**), (**S'**), (**s**), (**S**), (**v**), (**F**), (**I**), (**T**), (**D**), (**K**), (**N**) and (**P**) are known, then its Sales or Revenue planned is:

$$j = 360[[1 + \text{/}(U + \{S'[1+s]-Sv-F-I\}-T-D) - \{N+K+P\}]/\{S'[1+s]\}$$

Rule-56589:
If both (**/**), (**U**), (**S'**), (**s**), (**S**), (**v**), (**F**), (**I**), (**T**), (**D**), (**K**), (**j**) and (**N**) are known, then its Procured Inventory planned is:

$$P = [1 + \text{/}(U + \{S'[1+s]-Sv-F-I\}-T-D) - \{N+K+S'j[1+s]/360\}$$

Rule-56590:
If both (**/**), (**U**), (**S'**), (**s**), (**S**), (**v**), (**F**), (**I**), (**T**), (**D**), (**K**), (**j**), (**V**) and (**p**) are known, then its Non Current Asset planned is:

$$N = [1 + \text{/}\{U+S'[1+s]-Sv-F-I-T-D\} - K-S'j[1+s]/360-Vp/360$$

Steve Asikin ISBN 13: **978-1564912361**, ISBN 10: **1564912363**

Rule-56591:

If both **(N)**, **(U)**, **(S')**, **(s)**, **(S)**, **(v)**, **(F)**, **(I)**, **(T)**, **(D)**, **(K)**, **(j)**, **(V)** and **(p)** are known, then its Leverage or Gearing Ratio planned is:

$$\not{l}= \{N+K+S'j[1+s]/360+Vp/360\}$$
$$/(U+\{S'[1+s]-Sv-F-I\}-T-D)-1$$


Rule-56592:

If both **(l)**, **(N)**, **(S')**, **(s)**, **(S)**, **(v)**, **(F)**, **(I)**, **(T)**, **(D)**, **(K)**, **(j)**, **(V)** and **(p)** are known, then its Utilized or Starting Capital must be:

$$U= (\{N+K+S'j[1+s]/360+Vp/360\}$$
$$-[1+l]\{S'[1+s]-Sv-F-I\}-T-D)/[1+l]$$


Rule-56593:

If both **(l)**, **(U)**, **(N)**, **(s)**, **(S)**, **(v)**, **(F)**, **(I)**, **(T)**, **(D)**, **(K)**, **(j)**, **(V)** and **(p)** are known, then its Sales Past must be:

$$S'= [[1+l](U-\{Sv+F+I\}-T-D)-\{N+K+Vp/360\}]$$
$$/([1+s]\{j/360-[1+l]\})$$
$$\text{or}$$
$$S'= S/[1+s]$$

Steve Asikin ISBN 13: **978-1564912361**, ISBN 10: **1564912363**

Rule-56594:

If both ($\pmb{/}$), (**U**), (**S'**), (**N**), (**S**), (**v**), (**F**), (**I**), (**T**), (**D**), (**K**), (**j**), (**V**) and (**p**) are known, then its Sales Growth planned is:

$$s = [[1+\pmb{/}(\mathbf{U}-\{\mathbf{Sv}+\mathbf{F}+\mathbf{I}\}-\mathbf{T}-\mathbf{D})-\{\mathbf{N}+\mathbf{K}+\mathbf{Vp}/360\}]$$
$$/(\mathbf{S'}\{\mathbf{j}/360-[1+\pmb{/}]\})-1$$

or

$$s = \mathbf{S}/\mathbf{S'}-1$$

Rule-56595:

If both ($\pmb{/}$), (**U**), (**S'**), (**s**), (**N**), (**v**), (**F**), (**I**), (**T**), (**D**), (**K**), (**j**), (**V**) and (**p**) are known, then its Sales or Revenue planned is:

$$\mathbf{S} = [\mathbf{U}+\{\mathbf{S'}[1+\mathbf{s}]-\mathbf{I}-\mathbf{F}\}-\mathbf{T}-\mathbf{D}$$
$$-\{\mathbf{N}+\mathbf{K}+\mathbf{S'j}[1+\mathbf{s}]/360+\mathbf{Vp}/360\}/[1+\pmb{/}]/\mathbf{v}$$

or

$$\mathbf{S} = \mathbf{S'}[1+\mathbf{s}]$$

or

$$\mathbf{S} = \mathbf{V}/\mathbf{v}$$

Rule-56596:

If both ($\pmb{/}$), (**U**), (**S'**), (**s**), (**S**), (**N**), (**F**), (**I**), (**T**), (**D**), (**K**), (**j**), (**V**) and (**p**) are known, then its Variable Portion planned is:

$$\mathbf{v} = [\mathbf{U}+\{\mathbf{S'}[1+\mathbf{s}]-\mathbf{I}-\mathbf{F}\}-\mathbf{T}-\mathbf{D}-\{\mathbf{N}+\mathbf{K}+\mathbf{S'j}[1+\mathbf{s}]/360$$
$$+\mathbf{Vp}/360\}/[1+\pmb{/}]/\mathbf{S}$$

or

$$\mathbf{v} = \mathbf{V}/\mathbf{S}$$

`

Steve Asikin ISBN 13: **978-1564912361**, ISBN 10: **1564912363**

Rule-56597:

If both $(I)$, $(U)$, $(S')$, $(s)$, $(S)$, $(v)$, $(N)$, $(I)$, $(T)$, $(D)$, $(K)$, $(j)$, $(V)$ and $(p)$ are known, then its Fixed Cost planned is:

$$F= U+\{S'[1+s]-Sv-I\}-T-D$$
$$-\{N+K+S'j[1+s]/360+Vp/360\}/[1+I]$$

Rule-56598:

If both $(I)$, $(U)$, $(S')$, $(s)$, $(S)$, $(v)$, $(F)$, $(N)$, $(T)$, $(D)$, $(K)$, $(j)$, $(V)$ and $(p)$ are known, then its Interest Expense planned is:

$$I= U+\{S'[1+s]-Sv-F\}-T-D$$
$$-\{N+K+S'j[1+s]/360+Vp/360\}/[1+I]$$

Rule-56599:

If both $(I)$, $(U)$, $(S')$, $(s)$, $(S)$, $(v)$, $(F)$, $(I)$, $(N)$, $(D)$, $(K)$, $(j)$, $(V)$ and $(p)$ are known, then its Tax planned is:

$$T= U+\{S'[1+s]-Sv-F-I\}-D$$
$$-\{N+K+S'j[1+s]/360+Vp/360\}/[1+I]$$

Rule-56600:

If both $(I)$, $(U)$, $(S')$, $(s)$, $(S)$, $(v)$, $(F)$, $(I)$, $(T)$, $(N)$, $(K)$, $(j)$, $(V)$ and $(p)$ are known, then its Dividend planned is:

$$D= U+\{S'[1+s]-Sv-F-I\}-T$$
$$-\{N+K+S'j[1+s]/360+Vp/360\}/[1+I]$$

Steve Asikin ISBN 13: **978-1564912361**, ISBN 10: **1564912363**

<u>Rule-56601</u>:

If both (**/**), (**U**), (**S'**), (**s**), (**$**), (**v**), (**F**), (**I**), (**T**), (**D**), (**N**), (**j**), (**V**) and (**p**) are known, then its Kind of Cash planned is:

$$K= [1+/(U+\{S'[1+s]-Sv-F-I\}-T-D)$$
$$-\{N+S'j[1+s]/360+Vp/360\}$$

<u>Rule-56602</u>:

If both (**/**), (**U**), (**S'**), (**s**), (**$**), (**v**), (**F**), (**I**), (**T**), (**D**), (**K**), (**N**), (**V**) and (**p**) are known, then its Job or Trade Receivable Days planned is:

$$j= 360[[1+/(U+\{S'[1+s]-Sv-F-I\}-T-D)$$
$$-\{N+K+Vp/360\}]/\{S'[1+s]\}$$

<u>Rule-56603</u>:

If both (**/**), (**U**), (**S'**), (**s**), (**$**), (**v**), (**F**), (**I**), (**T**), (**D**), (**K**), (**j**), (**N**) and (**p**) are known, then its Variable Cost planned is:

$$V= 360[[1+/(U+\{S'[1+s]-Sv-F-I\}-T-D)$$
$$-\{N+K+S'j[1+s]/360\}]/p$$
$$or$$
$$V= Sv$$

<u>Rule-56604</u>:

If both (**/**), (**U**), (**S'**), (**s**), (**$**), (**v**), (**F**), (**I**), (**T**), (**D**), (**K**), (**j**), (**V**) and (**N**) are known, then its Procured Inventory Days planned is:

$$p= 360[[1+/(U+\{S'[1+s]-Sv-F-I\}-T-D)$$
$$-\{N+K+S'j[1+s]/360\}]/V$$

136

Steve Asikin ISBN 13: **978-1564912361**, ISBN 10: **1564912363**

Rule-56605:

> If both ($f$), (**U**), (**S'**), (**s**), (**S**), (**v**), (**F**), (**I**), (**T**), (**D**), (**K**),
> (**j**) and (**p**) are known, then its Non Current Asset
> planned is:
> $$N= [1+f]\{U+S'[1+s]-Sv-F-I-T-D\}$$
> $$-K-S'j[1+s]/360-Svp/360$$

Rule-56606:

> If both (**N**), (**U**), (**S'**), (**s**), (**S**), (**v**), (**F**), (**I**), (**T**), (**D**),
> (**K**), (**j**) and (**p**) are known, then its Leverage or
> Gearing Ratio planned is:
> $$f= \{N+K+S'j[1+s]/360+Svp/360\}$$
> $$/(U+\{S'[1+s]-Sv-F-I\}-T-D)-1$$

Rule-56607:

> If both ($f$), (**N**), (**S'**), (**s**), (**S**), (**v**), (**F**), (**I**), (**T**), (**D**), (**K**),
> (**j**) and (**p**) are known, then its Utilized or Starting
> Capital must be:
> $$U= (\{N+K+S'j[1+s]/360+Svp/360\}$$
> $$-[1+f]\{S'[1+s]-Sv-F-I\}-T-D)/\{[1+f]\}$$

Rule-56608:

> If both ($f$), (**U**), (**N**), (**s**), (**S**), (**v**), (**F**), (**I**), (**T**), (**D**), (**K**),
> (**j**) and (**p**) are known, then its Sales Past must be:
> $$S'= [[1+f](U-\{Sv+F+I\}-T-D)-[N+K+Svp/360]]$$
> $$/([1+s]\{j/360-[1+f]\})$$
> or
> $$S'= S/[1+s]$$

Steve Asikin ISBN 13: **978-1564912361**, ISBN 10: **1564912363**

Rule-56609:

If both (**/**), (**U**), (**S'**), (**N**), (**S**), (**v**), (**F**), (**I**), (**T**), (**D**), (**K**), (**j**) and (**p**) are known, then its Sales Growth planned is:

$$s= [[1+\textbf{/}](\textbf{U}-\{\textbf{Sv}+\textbf{F}+\textbf{I}\}-\textbf{T}-\textbf{D})-[\textbf{N}+\textbf{K}+\textbf{Svp}/360]]$$
$$/(\textbf{S'}\{\textbf{j}/360-[1+\textbf{/}]\})-1$$

or

$$s= \textbf{S}/\textbf{S'}-1$$

Rule-56610:

If both (**/**), (**U**), (**S'**), (**s**), (**N**), (**v**), (**F**), (**I**), (**T**), (**D**), (**K**), (**j**) and (**p**) are known, then its Sales or Revenue planned is:

$$\textbf{S}= [[1+\textbf{/}](\textbf{U}+\{\textbf{S'}[1+\textbf{s}]-\textbf{F}-\textbf{I}\}-\textbf{T}-\textbf{D})$$
$$-\{\textbf{N}+\textbf{K}+\textbf{S'j}[1+\textbf{s}]/360\}]/(\textbf{v}\{\textbf{p}/360+[1+\textbf{/}]\})$$

or

$$\textbf{S}= \textbf{S'}[1+\textbf{s}]$$

Rule-56611:

If both (**/**), (**U**), (**S'**), (**s**), (**S**), (**N**), (**F**), (**I**), (**T**), (**D**), (**K**), (**j**) and (**p**) are known, then its Variable Portion planned is:

$$\textbf{v}= [[1+\textbf{/}](\textbf{U}+\{\textbf{S'}[1+\textbf{s}]-\textbf{F}-\textbf{I}\}-\textbf{T}-\textbf{D})$$
$$-\{\textbf{N}+\textbf{K}+\textbf{S'j}[1+\textbf{s}]/360\}]/(\textbf{S}\{\textbf{p}/360+[1+\textbf{/}]\})$$

Steve Asikin ISBN 13: **978-1564912361**, ISBN 10: **1564912363**

Rule-56612:

If both $(I)$, $(U)$, $(S')$, $(s)$, $(S)$, $(v)$, $(N)$, $(I)$, $(T)$, $(D)$, $(K)$, $(j)$ and $(p)$ are known, then its Fixed Cost planned is:

$$F = U + \{S'[1+s] - Sv - I\} - T - D$$
$$- \{N + K + S'j[1+s]/360 + Svp/360\}/[1+I]$$

Rule-56613:

If both $(I)$, $(U)$, $(S')$, $(s)$, $(S)$, $(v)$, $(F)$, $(N)$, $(T)$, $(D)$, $(K)$, $(j)$ and $(p)$ are known, then its Interest Expense planned is:

$$I = U + \{S'[1+s] - Sv - F\} - T - D$$
$$- \{N + K + S'j[1+s]/360 + Svp/360\}/[1+I]$$

Rule-56614:

If both $(I)$, $(U)$, $(S')$, $(s)$, $(S)$, $(v)$, $(F)$, $(I)$, $(N)$, $(D)$, $(K)$, $(j)$ and $(p)$ are known, then its Tax planned is:

$$T = U + \{S'[1+s] - Sv - F - I\} - D$$
$$- \{N + K + S'j[1+s]/360 + Svp/360\}/[1+I]$$

Rule-56615:

If both $(I)$, $(U)$, $(S')$, $(s)$, $(S)$, $(v)$, $(F)$, $(I)$, $(T)$, $(N)$, $(K)$, $(j)$ and $(p)$ are known, then its Dividend planned is:

$$D = U + \{S'[1+s] - Sv - F - I\} - T$$
$$- \{N + K + S'j[1+s]/360 + Svp/360\}/[1+I]$$

Steve Asikin ISBN 13: **978-1564912361**, ISBN 10: **1564912363**

Rule-56616:

If both (**/**), (**U**), (**$'**), (**s**), (**$**), (**v**), (**F**), (**I**), (**T**), (**D**), (**N**), (**j**) and (**p**) are known, then its Kind of Cash planned is:

$$\mathbf{K} = [1 + \mathbf{/}(\mathbf{U} + \{\mathbf{\$'}[1+\mathbf{s}] - \mathbf{\$v} - \mathbf{F} - \mathbf{I}\} - \mathbf{T} - \mathbf{D})$$
$$- \{\mathbf{N} + \mathbf{\$'j}[1+\mathbf{s}]/360 + \mathbf{\$vp}/360\}$$

Rule-56617:

If both (**/**), (**U**), (**$'**), (**s**), (**$**), (**v**), (**F**), (**I**), (**T**), (**D**), (**K**), (**N**) and (**p**) are known, then its Job or Trade Receivable Days planned is:

$$\mathbf{j} = 360[[1 + \mathbf{/}(\mathbf{U} + \{\mathbf{\$'}[1+\mathbf{s}] - \mathbf{\$v} - \mathbf{F} - \mathbf{I}\} - \mathbf{T} - \mathbf{D})$$
$$- \{\mathbf{N} + \mathbf{K} + \mathbf{\$vp}/360\}]/\{\mathbf{\$'}[1+\mathbf{s}]\}$$

Rule-56618:

If both (**/**), (**U**), (**$'**), (**s**), (**$**), (**v**), (**F**), (**I**), (**T**), (**D**), (**K**), (**j**) and (**N**) are known, then its Procured Inventory Days planned is:

$$\mathbf{p} = 360[[1 + \mathbf{/}(\mathbf{U} + \{\mathbf{\$'}[1+\mathbf{s}] - \mathbf{\$v} - \mathbf{F} - \mathbf{I}\} - \mathbf{T} - \mathbf{D})$$
$$- \{\mathbf{N} + \mathbf{K} + \mathbf{\$'j}[1+\mathbf{s}]/360\}]/[\mathbf{\$v}]$$

Rule-56619:

If both (**/**), (**U**), (**$'**), (**s**), (**$**), (**v**), (**F**), (**I**), (**T**), (**D**), (**K**), (**j**) and (**p**) are known, then its Non Current Asset planned is:

$$\mathbf{N} = [1 + \mathbf{/}\{\mathbf{U} + \mathbf{\$'}[1+\mathbf{s}] - \mathbf{\$v} - \mathbf{F} - \mathbf{I} - \mathbf{T} - \mathbf{D}\}$$
$$- \mathbf{K} - \mathbf{\$'j}[1+\mathbf{s}]/360 - \mathbf{\$'vp}[1+\mathbf{s}]/360$$

Steve Asikin ISBN 13: **978-1564912361**, ISBN 10: **1564912363**

Rule-56620:

If both (**N**), (**U**), (**S'**), (**s**), (**S**), (**v**), (**F**), (**I**), (**T**), (**D**), (**K**), (**j**) and (**p**) are known, then its Leverage or Gearing Ratio planned is:

$$I = \{N+K+S'j[1+s]/360+S'vp[1+s]/360\}$$
$$/(U+\{S'[1+s]-Sv-F-I\}-T-D)-1$$

Rule-56621:

If both (**I**), (**N**), (**S'**), (**s**), (**S**), (**v**), (**F**), (**I**), (**T**), (**D**), (**K**), (**j**) and (**p**) are known, then its Utilized or Starting Capital must be:

$$U = (\{N+K+S'j[1+s]/360+S'vp[1+s]/360\}$$
$$-[1+I]\{S'[1+s]-Sv-F-I\}-T-D)/[1+I]$$

Rule-56622:

If both (**I**), (**U**), (**N**), (**s**), (**S**), (**v**), (**F**), (**I**), (**T**), (**D**), (**K**), (**j**) and (**p**) are known, then its Sales Past must be:

$$S' = [[1+I](U-\{Sv+F+I\}-T-D)-[N+K]]$$
$$/([1+s]\{j/360+vp/360-[1+s]\})$$
$$\text{or}$$
$$S' = S/[1+s]$$

Steve Asikin ISBN 13: **978-1564912361**, ISBN 10: **1564912363**

<u>Rule-56623</u>:

If both (**/**), (**U**), (**$'**), (**N**), (**$**), (**v**), (**F**), (**I**), (**T**), (**D**), (**K**), (**j**) and (**p**) are known, then its Sales Growth planned is:

$$s= [[1+/(U-\{\$v+F+I\}-T-D)-[N+K]]$$
$$/(\$'\{j/360+vp/360-[1+s]\})-1$$

or

$$s= \$/\$'-1$$

<u>Rule-56624</u>:

If both (**/**), (**U**), (**$'**), (**s**), (**N**), (**v**), (**F**), (**I**), (**T**), (**D**), (**K**), (**j**) and (**p**) are known, then its Sales or Revenue planned is:

$$\$= [U+\{\$'[1+s]-F-I\}-T-D-\{N+K+\$'][1+s]/360$$
$$+\$'vp[1+s]/360\}/[1+/]/v$$

or

$$\$= \$'[1+s]$$

<u>Rule-56625</u>:

If both (**/**), (**U**), (**$'**), (**s**), (**$**), (**N**), (**F**), (**I**), (**T**), (**D**), (**K**), (**j**) and (**p**) are known, then its Variable Portion planned is:

$$v= [[1+/(U+\{\$'[1+s]-F-I\}-T-D)$$
$$-\{N+K+\$'j[1+s]/360\}]$$
$$/\{\$'p[1+s]/360+\$[1+/]\}$$

Steve Asikin ISBN 13: **978-1564912361**, ISBN 10: **1564912363**

Rule-56626:

If both (*I*), (**U**), (**S'**), (**s**), (**S**), (**v**), (**N**), (**I**), (**T**), (**D**), (**K**), (**j**) and (**p**) are known, then its Fixed Cost planned is:

$$F= U+\{S'[1+s]-Sv-I\}-T-D-\{N+K+S'j[1+s]/360 \\ +S'vp[1+s]/360\}/[1+I]$$

Rule-56627:

If both (*I*), (**U**), (**S'**), (**s**), (**S**), (**v**), (**F**), (**N**), (**T**), (**D**), (**K**), (**j**) and (**p**) are known, then its Interest Expense planned is:

$$I= U+\{S'[1+s]-Sv-F\}-T-D-\{N+K+S'j[1+s]/360 \\ +S'vp[1+s]/360\}/[1+I]$$

Rule-56628:

If both (*I*), (**U**), (**S'**), (**s**), (**S**), (**v**), (**F**), (**I**), (**N**), (**D**), (**K**), (**j**) and (**p**) are known, then its Tax planned is:

$$T= U+\{S'[1+s]-Sv-F-I\}-D-\{N+K+S'j[1+s]/360 \\ +S'vp[1+s]/360\}/[1+I]$$

Rule-56629:

If both (*I*), (**U**), (**S'**), (**s**), (**S**), (**v**), (**F**), (**I**), (**T**), (**N**), (**K**), (**j**) and (**p**) are known, then its Dividend planned is:

$$D= U+\{S'[1+s]-Sv-F-I\}-T-\{N+K+S'j[1+s]/360 \\ +S'vp[1+s]/360\}/[1+I]$$

Steve Asikin ISBN 13: **978-1564912361**, ISBN 10: **1564912363**

Rule-56630:

If both (**I**), (**U**), (**S'**), (**s**), (**S**), (**v**), (**F**), (**I**), (**T**), (**D**), (**N**), (**j**) and (**p**) are known, then its Kind of Cash planned is:

$$K= [1+I](U+\{S'[1+s]-Sv-F-I\}-T-D)$$
$$-\{N+S'j[1+s]/360+S'vp[1+s]/360\}$$

Rule-56631:

If both (**I**), (**U**), (**S'**), (**s**), (**S**), (**v**), (**F**), (**I**), (**T**), (**D**), (**K**), (**N**) and (**p**) are known, then its Job or Trade Receivable Days planned is:

$$j= 360[[1+I](U+\{S'[1+s]-Sv-F-I\}-T-D)$$
$$-\{N+K+S'vp[1+s]/360\}]/\{S'[1+s]\}$$

Rule-56632:

If both (**I**), (**U**), (**S'**), (**s**), (**S**), (**v**), (**F**), (**I**), (**T**), (**D**), (**K**), (**j**) and (**N**) are known, then its Procured Inventory Days planned is:

$$p= 360[[1+I](U+\{S'[1+s]-Sv-F-I\}-T-D)-$$
$$\{N+K+S'j[1+s]/360\}]/\{S'v[1+s]\}$$

Rule-56633:

If both (**I**), (**U**), (**S'**), (**s**), (**S**), (**v**), (**F**), (**I**), (**T**), (**d**), (**K**), (**J**) and (**P**) are known, then its Non Current Asset planned is:

$$N= [1+I](U+\{S'[1+s]-Sv-F-I-T\}[1-d])-K-J-P$$

Steve Asikin ISBN 13: **978-1564912361**, ISBN 10: **1564912363**

Rule-56634:

If both $(\mathbf{N})$, $(\mathbf{U})$, $(\mathbf{S'})$, $(\mathbf{s})$, $(\mathbf{S})$, $(\mathbf{v})$, $(\mathbf{F})$, $(\mathbf{I})$, $(\mathbf{T})$, $(\mathbf{d})$, $(\mathbf{K})$, $(\mathbf{J})$ and $(\mathbf{P})$ are known, then its Leverage or Gearing Ratio planned is:

$$\mathit{I} = \{\mathbf{N+K+J+P}\}/(\mathbf{U}+\{\mathbf{S'}[1+\mathbf{s}]-\mathbf{Sv-F-I-T}\}[1-\mathbf{d}])-1$$

Rule-56635:

If both $(\mathit{I})$, $(\mathbf{N})$, $(\mathbf{S'})$, $(\mathbf{s})$, $(\mathbf{S})$, $(\mathbf{v})$, $(\mathbf{F})$, $(\mathbf{I})$, $(\mathbf{T})$, $(\mathbf{d})$, $(\mathbf{K})$, $(\mathbf{J})$ and $(\mathbf{P})$ are known, then its Utilized or Starting Capital must be:

$$\mathbf{U} = (\{\mathbf{N+K+J+P}\}-[1+\mathit{I}]\{\mathbf{S'}[1+\mathbf{s}]-\mathbf{Sv-F-I-T}\}[1-\mathbf{d}])/[1+\mathit{I}]$$

Rule-56636:

If both $(\mathit{I})$, $(\mathbf{U})$, $(\mathbf{N})$, $(\mathbf{s})$, $(\mathbf{S})$, $(\mathbf{v})$, $(\mathbf{F})$, $(\mathbf{I})$, $(\mathbf{T})$, $(\mathbf{d})$, $(\mathbf{K})$, $(\mathbf{J})$ and $(\mathbf{P})$ are known, then its Sales Past must be:

$$\mathbf{S'} = (\{[\mathbf{N+K+J+P}]/[1+\mathit{I}]-\mathbf{U}\}/[1-\mathbf{d}]+\mathbf{Sv+F+I+T})/[1+\mathbf{s}]$$

or

$$\mathbf{S'} = \mathbf{S}/[1+\mathbf{s}]$$

Rule-56637:

If both $(\mathit{I})$, $(\mathbf{U})$, $(\mathbf{S'})$, $(\mathbf{N})$, $(\mathbf{S})$, $(\mathbf{v})$, $(\mathbf{F})$, $(\mathbf{I})$, $(\mathbf{T})$, $(\mathbf{d})$, $(\mathbf{K})$, $(\mathbf{J})$ and $(\mathbf{P})$ are known, then its Sales Growth planned is:

$$\mathbf{s} = (\{[\mathbf{N+K+J+P}]/[1+\mathit{I}]-\mathbf{U}\}/[1-\mathbf{d}]+\mathbf{Sv+F+I+T})/\mathbf{S'}-1$$

or

$$\mathbf{s} = \mathbf{S}/\mathbf{S'}-1$$

145

Steve Asikin ISBN 13: **978-1564912361**, ISBN 10: **1564912363**

Rule-56638:

If both (**/**), (**U**), (**S'**), (**s**), (**N**), (**v**), (**F**), (**I**), (**T**), (**d**), (**K**), (**J**) and (**P**) are known, then its Sales or Revenue planned is:

$$S= [S'[1+s]-F-I-T-\{[N+K+J+P]/[1+/]-U\}/[1-d]]/v$$

or

$$S= S'[1+s]$$

Rule-56639:

If both (**/**), (**U**), (**S'**), (**s**), (**S**), (**N**), (**F**), (**I**), (**T**), (**d**), (**K**), (**J**) and (**P**) are known, then its Variable Portion planned is:

$$v= [S'[1+s]-F-I-T-\{[N+K+J+P]/[1+/]-U\}/[1-d]]/S$$

Rule-56640:

If both (**/**), (**U**), (**S'**), (**s**), (**S**), (**v**), (**N**), (**I**), (**T**), (**d**), (**K**), (**J**) and (**P**) are known, then its Fixed Portion planned is:

$$F= S'[1+s]-Sv-I-T-\{[N+K+J+P]/[1+/]-U\}/[1-d]$$

Rule-56641:

If both (**/**), (**U**), (**S'**), (**s**), (**S**), (**v**), (**F**), (**N**), (**T**), (**d**), (**K**), (**J**) and (**P**) are known, then its Interest Expense planned is:

$$I= S'[1+s]-Sv-F-T-\{[N+K+J+P]/[1+/]-U\}/[1-d]$$

Steve Asikin ISBN 13: **978-1564912361**, ISBN 10: **1564912363**

Rule-56642:

If both (**/**), (**U**), (**S'**), (**s**), (**S**), (**v**), (**F**), (**I**), (**N**), (**d**), (**K**), (**J**) and (**P**) are known, then its Tax planned is:

$$T= S'[1+s]-Sv-F-I-\{[N+K+J+P]/[1+/-U\}/[1-d]$$

Rule-56643:

If both (**/**), (**U**), (**S'**), (**s**), (**S**), (**v**), (**F**), (**I**), (**T**), (**N**), (**K**), (**J**) and (**P**) are known, then its Dividend Payout planned is:

$$d= 1-\{[N+K+J+P]/[1+/-U\}/\{S'[1+s]-Sv-F-I-T\}$$

Rule-56644:

If both (**/**), (**U**), (**S'**), (**s**), (**S**), (**v**), (**F**), (**I**), (**T**), (**d**), (**N**), (**J**) and (**P**) are known, then its Kind of Cash planned is:

$$K= [1+/(U+\{S'[1+s]-Sv-F-I-T\}[1-d])-[N+J+P]$$

Rule-56645:

If both (**/**), (**U**), (**S'**), (**s**), (**S**), (**v**), (**F**), (**I**), (**T**), (**d**), (**K**), (**N**) and (**P**) are known, then its Job or Trade Account Receivable planned is:

$$J= [1+/(U+\{S'[1+s]-Sv-F-I-T\}[1-d])-[N+K+P]$$

Rule-56646:

If both (**/**), (**U**), (**S'**), (**s**), (**S**), (**v**), (**F**), (**I**), (**T**), (**d**), (**K**), (**J**) and (**N**) are known, then its Procured Inventory planned is:

$$P= [1+/(U+\{S'[1+s]-Sv-F-I-T\}[1-d])-[N+K+J]$$

Steve Asikin ISBN 13: **978-1564912361**, ISBN 10: **1564912363**

<u>Rule-56647</u>:

If both (**/**), (**U**), (**S'**), (**s**), (**S**), (**v**), (**F**), (**I**), (**T**), (**d**), (**K**), (**J**), (**V**) and (**p**) are known, then its Non Current Asset planned is:

$$N= [1+/(U+\{S'[1+s]-Sv-F-I-T\}[1-d])$$
$$-K-J-Vp/360$$

<u>Rule-56648</u>:

If both (**N**), (**U**), (**S'**), (**s**), (**S**), (**v**), (**F**), (**I**), (**T**), (**d**), (**K**), (**J**), (**V**) and (**p**) are known, then its Leverage or Gearing Ratio planned is:

$$/= [N+K+J+Vp/360]$$
$$/(U+\{S'[1+s]-Sv-F-I-T\}[1-d])-1$$

<u>Rule-56649</u>:

If both (**/**), (**N**), (**S'**), (**s**), (**S**), (**v**), (**F**), (**I**), (**T**), (**d**), (**K**), (**J**), (**V**) and (**p**) are known, then its Utilized or Starting Capital must be:

$$U= ([N+K+J+Vp/360]-[1+/\{S'[1+s]-Sv-F-I-T\}$$
$$[1-d])/[1+/]$$

<u>Rule-56650</u>:

If both (**/**), (**U**), (**N**), (**s**), (**S**), (**v**), (**F**), (**I**), (**T**), (**d**), (**K**), (**J**), (**V**) and (**p**) are known, then its Sales Past must be:

$$S'= (\{[N+K+J+Vp/360]/[1+/-U\}/[1-d]$$
$$+Sv+F+I+T)/[1+s]$$

or

$$S'= S/[1+s]$$

148

Steve Asikin ISBN 13: **978-1564912361**, ISBN 10: **1564912363**

Rule-56651:

If both $(\textit{\textbf{I}})$, $(\textbf{U})$, $(\textbf{S'})$, $(\textbf{N})$, $(\textbf{S})$, $(\textbf{v})$, $(\textbf{F})$, $(\textbf{I})$, $(\textbf{T})$, $(\textbf{d})$, $(\textbf{K})$, $(\textbf{J})$, $(\textbf{V})$ and $(\textbf{p})$ are known, then its Sales Growth planned is:

$$s = (\{[\textbf{N}+\textbf{K}+\textbf{J}+\textbf{Vp}/360]/[1+\textit{\textbf{I}}-\textbf{U}\}/[1-\textbf{d}]$$
$$+\textbf{Sv}+\textbf{F}+\textbf{I}+\textbf{T})/\textbf{S'}-1$$

or

$$s = \textbf{S}/\textbf{S'}-1$$

Rule-56652:

If both $(\textit{\textbf{I}})$, $(\textbf{U})$, $(\textbf{S'})$, $(\textbf{s})$, $(\textbf{N})$, $(\textbf{v})$, $(\textbf{F})$, $(\textbf{I})$, $(\textbf{T})$, $(\textbf{d})$, $(\textbf{K})$, $(\textbf{J})$, $(\textbf{V})$ and $(\textbf{p})$ are known, then its Sales or Revenue planned is:

$$\textbf{S} = (\textbf{S'}[1+s]-\textbf{F}-\textbf{I}-\textbf{T}-\{[\textbf{N}+\textbf{K}+\textbf{J}+\textbf{Vp}/360]/[1+\textit{\textbf{I}}-\textbf{U}\}$$
$$/[1-\textbf{d}])/\textbf{v}$$

or

$$\textbf{S} = \textbf{S'}[1+s]$$

or

$$\textbf{S} = \textbf{V}/\textbf{v}$$

Rule-56653:

If both $(\textit{\textbf{I}})$, $(\textbf{U})$, $(\textbf{S'})$, $(\textbf{s})$, $(\textbf{S})$, $(\textbf{N})$, $(\textbf{F})$, $(\textbf{I})$, $(\textbf{T})$, $(\textbf{d})$, $(\textbf{K})$, $(\textbf{J})$, $(\textbf{V})$ and $(\textbf{p})$ are known, then its Variable Portion planned is:

$$v = (\textbf{S'}[1+s]-\textbf{F}-\textbf{I}-\textbf{T}-\{[\textbf{N}+\textbf{K}+\textbf{J}+\textbf{Vp}/360]/[1+\textit{\textbf{I}}-\textbf{U}\}$$
$$/[1-\textbf{d}])/\textbf{S}$$

or

$$v = \textbf{V}/\textbf{S}$$

Steve Asikin ISBN 13: **978-1564912361**, ISBN 10: **1564912363**

Rule-56654:

If both (**/**), (**U**), (**S'**), (**s**), (**S**), (**v**), (**N**), (**I**), (**T**), (**d**), (**K**), (**J**), (**V**) and (**p**) are known, then its Fixed Cost planned is:

$$\mathbf{F} = \mathbf{S'}[1+\mathbf{s}]-\mathbf{Sv}-\mathbf{I}-\mathbf{T}-\{[\mathbf{N}+\mathbf{K}+\mathbf{J}+\mathbf{Vp}/360]/[1+\mathbf{/}]-\mathbf{U}\}/[1-\mathbf{d}]$$

Rule-56655:

If both (**/**), (**U**), (**S'**), (**s**), (**S**), (**v**), (**F**), (**N**), (**T**), (**d**), (**K**), (**J**), (**V**) and (**p**) are known, then its Interest Expense planned is:

$$\mathbf{I} = \mathbf{S'}[1+\mathbf{s}]-\mathbf{Sv}-\mathbf{F}-\mathbf{T}-\{[\mathbf{N}+\mathbf{K}+\mathbf{J}+\mathbf{Vp}/360]/[1+\mathbf{/}]-\mathbf{U}\}/[1-\mathbf{d}]$$

Rule-56656:

If both (**/**), (**U**), (**S'**), (**s**), (**S**), (**v**), (**F**), (**I**), (**N**), (**d**), (**K**), (**J**), (**V**) and (**p**) are known, then its Tax planned is:

$$\mathbf{T} = \mathbf{S'}[1+\mathbf{s}]-\mathbf{Sv}-\mathbf{F}-\mathbf{I}-\{[\mathbf{N}+\mathbf{K}+\mathbf{J}+\mathbf{Vp}/360]/[1+\mathbf{/}]-\mathbf{U}\}/[1-\mathbf{d}]$$

Rule-56657:

If both (**/**), (**U**), (**S'**), (**s**), (**S**), (**v**), (**F**), (**I**), (**T**), (**N**), (**K**), (**J**), (**V**) and (**p**) are known, then its Dividend Payout planned is:

$$\mathbf{d} = 1-\{[\mathbf{N}+\mathbf{K}+\mathbf{J}+\mathbf{Vp}/360]/[1+\mathbf{/}]-\mathbf{U}\}/\{\mathbf{S'}[1+\mathbf{s}]-\mathbf{Sv}-\mathbf{F}-\mathbf{I}-\mathbf{T}\}$$

Steve Asikin ISBN 13: **978-1564912361**, ISBN 10: **1564912363**

Rule-56658:

If both $(\textit{I})$, $(\mathbf{U})$, $(\mathbf{S'})$, $(\mathbf{s})$, $(\mathbf{S})$, $(\mathbf{v})$, $(\mathbf{F})$, $(\mathbf{I})$, $(\mathbf{T})$, $(\mathbf{d})$, $(\mathbf{N})$, $(\mathbf{J})$, $(\mathbf{V})$ and $(\mathbf{p})$ are known, then its Kind of Cash planned is:

$$\mathbf{K} = [1+\textit{I}(\mathbf{U}+\{\mathbf{S'}[1+\mathbf{s}]\text{-}\mathbf{Sv}\text{-}\mathbf{F}\text{-}\mathbf{I}\text{-}\mathbf{T}\}[1\text{-}\mathbf{d}])$$
$$-[\mathbf{N}+\mathbf{J}+\mathbf{Vp}/360]$$

Rule-56659:

If both $(\textit{I})$, $(\mathbf{U})$, $(\mathbf{S'})$, $(\mathbf{s})$, $(\mathbf{S})$, $(\mathbf{v})$, $(\mathbf{F})$, $(\mathbf{I})$, $(\mathbf{T})$, $(\mathbf{d})$, $(\mathbf{K})$, $(\mathbf{N})$, $(\mathbf{V})$ and $(\mathbf{p})$ are known, then its Job or Trade Account Receivable planned is:

$$\mathbf{J} = [1+\textit{I}(\mathbf{U}+\{\mathbf{S'}[1+\mathbf{s}]\text{-}\mathbf{Sv}\text{-}\mathbf{F}\text{-}\mathbf{I}\text{-}\mathbf{T}\}[1\text{-}\mathbf{d}])$$
$$-\{\mathbf{N}+\mathbf{K}+\mathbf{Vp}/360\}$$

Rule-56660:

If both $(\textit{I})$, $(\mathbf{U})$, $(\mathbf{S'})$, $(\mathbf{s})$, $(\mathbf{S})$, $(\mathbf{v})$, $(\mathbf{F})$, $(\mathbf{I})$, $(\mathbf{T})$, $(\mathbf{d})$, $(\mathbf{K})$, $(\mathbf{J})$, $(\mathbf{N})$ and $(\mathbf{p})$ are known, then its Variable Cost planned is:

$$\mathbf{V} = 360[[1+\textit{I}(\mathbf{U}+\{\mathbf{S'}[1+\mathbf{s}]\text{-}\mathbf{Sv}\text{-}\mathbf{F}\text{-}\mathbf{I}\text{-}\mathbf{T}\}[1\text{-}\mathbf{d}])$$
$$-[\mathbf{N}+\mathbf{K}+\mathbf{J}]]/\mathbf{p}$$
$$\text{or}$$
$$\mathbf{V} = \mathbf{Sv}$$

Rule-56661:

If both $(\textit{I})$, $(\mathbf{U})$, $(\mathbf{S'})$, $(\mathbf{s})$, $(\mathbf{S})$, $(\mathbf{v})$, $(\mathbf{F})$, $(\mathbf{I})$, $(\mathbf{T})$, $(\mathbf{d})$, $(\mathbf{K})$, $(\mathbf{J})$, $(\mathbf{V})$ and $(\mathbf{N})$ are known, then its Procured Inventory Days planned is:

$$\mathbf{p} = 360[[1+\textit{I}(\mathbf{U}+\{\mathbf{S'}[1+\mathbf{s}]\text{-}\mathbf{Sv}\text{-}\mathbf{F}\text{-}\mathbf{I}\text{-}\mathbf{T}\}[1\text{-}\mathbf{d}])$$
$$-[\mathbf{N}+\mathbf{K}+\mathbf{J}]]/\mathbf{V}$$

Steve Asikin ISBN 13: **978-1564912361**, ISBN 10: **1564912363**

Rule-56662:

If both ($I$), (**U**), (**S'**), (**s**), (**S**), (**v**), (**F**), (**I**), (**T**), (**d**), (**K**), (**J**) and (**p**) are known, then its Non Current Asset planned is:

$$N= [1+I](U+\{S'[1+s]-Sv-F-I-T\}[1-d]) -K-J-Sup/360$$

Rule-56663:

If both (**N**), (**U**), (**S'**), (**s**), (**S**), (**v**), (**F**), (**I**), (**T**), (**d**), (**K**), (**J**) and (**p**) are known, then its Leverage or Gearing Ratio planned is:

$$I= [N+K+J+Sup/360] /(U+\{S'[1+s]-Sv-F-I-T\}[1-d])-1$$

Rule-56664:

If both ($I$), (**N**), (**S'**), (**s**), (**S**), (**v**), (**F**), (**I**), (**T**), (**d**), (**K**), (**J**) and (**p**) are known, then its Utilized or Starting Capital must be:

$$U= ([N+K+J+Sup/360] -[1+I]\{S'[1+s]-Sv-F-I-T\}[1-d])/[1+I]$$

Rule-56665:

If both ($I$), (**U**), (**N**), (**s**), (**S**), (**v**), (**F**), (**I**), (**T**), (**d**), (**K**), (**J**) and (**p**) are known, then its Sales Past must be:

$$S'= (\{[N+K+J+Sup/360]/[1+I]-U\}/[1-d] +Sv+F+I+T\})/[1+s]$$

or

$$S'= S/[1+s]$$

Steve Asikin ISBN 13: **978-1564912361**, ISBN 10: **1564912363**

Rule-56666:

If both (**/**), (**U**), (**S'**), (**N**), (**S**), (**v**), (**F**), (**I**), (**T**), (**d**), (**K**), (**J**) and (**p**) are known, then its Sales Growth planned is:

$$s= (\{[N+K+J+Svp/360]/[1+/-U\}/[1-d]$$
$$+Sv+F+I+T\})/S'-1$$

or

$$s= S/S'-1$$

Rule-56667:

If both (**/**), (**U**), (**S'**), (**s**), (**S**), (**v**), (**F**), (**I**), (**T**), (**d**), (**K**), (**J**) and (**p**) are known, then its Non Current Asset planned is:

$$S= [[1+/(U+\{S'[1+s]-F-I-T\}[1-d])-[N+K+J]]$$
$$/(v\{p/360+[1+/]\})$$

or

$$S= S'[1+s]$$

Rule-56668:

If both (**/**), (**U**), (**S'**), (**s**), (**S**), (**N**), (**F**), (**I**), (**T**), (**d**), (**K**), (**J**) and (**p**) are known, then its Variable Portion planned is:

$$v= [[1+/(U+\{S'[1+s]-F-I-T\}[1-d])-[N+K+J]]$$
$$/(S\{p/360+[1+/]\})$$

Steve Asikin ISBN 13: **978-1564912361**, ISBN 10: **1564912363**

Rule-56669:

If both (**I**), (**U**), (**S'**), (**s**), (**S**), (**v**), (**F**), (**I**), (**T**), (**d**), (**K**), (**J**) and (**p**) are known, then its Non Current Asset planned is:

$$\mathbf{F}= \mathbf{S'}[1+\mathbf{s}]-\mathbf{Sv}-\mathbf{I}-\mathbf{T}-\{[\mathbf{N}+\mathbf{K}+\mathbf{J}+\mathbf{Svp}/360]/[1+\mathbf{I}-\mathbf{U}\}/[1-\mathbf{d}]$$

Rule-56670:

If both (**I**), (**U**), (**S'**), (**s**), (**S**), (**v**), (**F**), (**N**), (**T**), (**d**), (**K**), (**J**) and (**p**) are known, then its Interest Expense planned is:

$$\mathbf{I}= \mathbf{S'}[1+\mathbf{s}]-\mathbf{Sv}-\mathbf{F}-\mathbf{T}-\{[\mathbf{N}+\mathbf{K}+\mathbf{J}+\mathbf{Svp}/360]/[1+\mathbf{I}-\mathbf{U}\}/[1-\mathbf{d}]$$

Rule-56671:

If both (**I**), (**U**), (**S'**), (**s**), (**S**), (**v**), (**F**), (**I**), (**N**), (**d**), (**K**), (**J**) and (**p**) are known, then its Tax planned is:

$$\mathbf{T}= \mathbf{S'}[1+\mathbf{s}]-\mathbf{Sv}-\mathbf{F}-\mathbf{I}-\{[\mathbf{N}+\mathbf{K}+\mathbf{J}+\mathbf{Svp}/360]/[1+\mathbf{I}-\mathbf{U}\}/[1-\mathbf{d}]$$

Rule-56672:

If both (**I**), (**U**), (**S'**), (**s**), (**S**), (**v**), (**F**), (**I**), (**T**), (**N**), (**K**), (**J**) and (**p**) are known, then its Dividend Payout planned is:

$$\mathbf{d}= 1-\{[\mathbf{N}+\mathbf{K}+\mathbf{J}+\mathbf{Svp}/360]/[1+\mathbf{I}-\mathbf{U}\}/\{\mathbf{S'}[1+\mathbf{s}]-\mathbf{Sv}-\mathbf{F}-\mathbf{I}-\mathbf{T}\}$$

`

Steve Asikin ISBN 13: **978-1564912361**, ISBN 10: **1564912363**

Rule-56673:

If both $(\mathit{I})$, $(\mathbf{U})$, $(\mathbf{S'})$, $(\mathbf{s})$, $(\mathbf{S})$, $(\mathbf{v})$, $(\mathbf{F})$, $(\mathbf{I})$, $(\mathbf{T})$, $(\mathbf{d})$, $(\mathbf{N})$, $(\mathbf{J})$ and $(\mathbf{p})$ are known, then its Kind of Cash planned is:

$$\mathbf{K} = [1 + \mathit{I}(\mathbf{U} + \{\mathbf{S'}[1+\mathbf{s}] - \mathbf{Sv} - \mathbf{F} - \mathbf{I} - \mathbf{T}\}[1-\mathbf{d}]) - [\mathbf{N} + \mathbf{J} + \mathbf{Svp}/360]$$

Rule-56674:

If both $(\mathit{I})$, $(\mathbf{U})$, $(\mathbf{S'})$, $(\mathbf{s})$, $(\mathbf{S})$, $(\mathbf{v})$, $(\mathbf{F})$, $(\mathbf{I})$, $(\mathbf{T})$, $(\mathbf{d})$, $(\mathbf{K})$, $(\mathbf{N})$ and $(\mathbf{p})$ are known, then its Job or Trade Account Receivable planned is:

$$\mathbf{J} = [1 + \mathit{I}(\mathbf{U} + \{\mathbf{S'}[1+\mathbf{s}] - \mathbf{Sv} - \mathbf{F} - \mathbf{I} - \mathbf{T}\}[1-\mathbf{d}]) - \{\mathbf{N} + \mathbf{K} + \mathbf{Svp}/360\}$$

Rule-56675:

If both $(\mathit{I})$, $(\mathbf{U})$, $(\mathbf{S'})$, $(\mathbf{s})$, $(\mathbf{S})$, $(\mathbf{v})$, $(\mathbf{F})$, $(\mathbf{I})$, $(\mathbf{T})$, $(\mathbf{d})$, $(\mathbf{K})$, $(\mathbf{J})$ and $(\mathbf{N})$ are known, then its Procured Inventory Days planned is:

$$\mathbf{p} = 360[[1 + \mathit{I}(\mathbf{U} + \{\mathbf{S'}[1+\mathbf{s}] - \mathbf{Sv} - \mathbf{F} - \mathbf{I} - \mathbf{T}\}[1-\mathbf{d}]) - [\mathbf{N} + \mathbf{K} + \mathbf{J}]]/[\mathbf{Sv}]$$

Rule-56676:

If both $(\mathit{I})$, $(\mathbf{U})$, $(\mathbf{S'})$, $(\mathbf{s})$, $(\mathbf{S})$, $(\mathbf{v})$, $(\mathbf{F})$, $(\mathbf{I})$, $(\mathbf{T})$, $(\mathbf{d})$, $(\mathbf{K})$, $(\mathbf{J})$ and $(\mathbf{p})$ are known, then its Non Current Asset planned is:

$$\mathbf{N} = [1 + \mathit{I}(\mathbf{U} + \{\mathbf{S'}[1+\mathbf{s}] - \mathbf{Sv} - \mathbf{F} - \mathbf{I} - \mathbf{T}\}[1-\mathbf{d}]) - \mathbf{K} - \mathbf{J} - \mathbf{S'vp}[1+\mathbf{s}]/360$$

Steve Asikin ISBN 13: **978-1564912361**, ISBN 10: **1564912363**

Rule-56677:

If both $(\mathbf{N})$, $(\mathbf{U})$, $(\mathbf{S'})$, $(\mathbf{s})$, $(\mathbf{S})$, $(\mathbf{v})$, $(\mathbf{F})$, $(\mathbf{I})$, $(\mathbf{T})$, $(\mathbf{d})$, $(\mathbf{K})$, $(\mathbf{J})$ and $(\mathbf{p})$ are known, then its Leverage or Gearing Ratio planned is:

$$I = \{\mathbf{N+K+J+S'vp}[1+\mathbf{s}]/360\} / (\mathbf{U}+\{\mathbf{S'}[1+\mathbf{s}]-\mathbf{Sv-F-I-T}\}[1-\mathbf{d}])-1$$

Rule-56678:

If both $(I)$, $(\mathbf{N})$, $(\mathbf{S'})$, $(\mathbf{s})$, $(\mathbf{S})$, $(\mathbf{v})$, $(\mathbf{F})$, $(\mathbf{I})$, $(\mathbf{T})$, $(\mathbf{d})$, $(\mathbf{K})$, $(\mathbf{J})$ and $(\mathbf{p})$ are known, then its Utilized or Starting Capital must be:

$$\mathbf{U} = (\{\mathbf{N+K+J+S'vp}[1+\mathbf{s}]/360\} -[1+I]\{\mathbf{S'}[1+\mathbf{s}]-\mathbf{Sv-F-I-T}\}[1-\mathbf{d}])/[1+I]$$

Rule-56679:

If both $(I)$, $(\mathbf{U})$, $(\mathbf{N})$, $(\mathbf{s})$, $(\mathbf{S})$, $(\mathbf{v})$, $(\mathbf{F})$, $(\mathbf{I})$, $(\mathbf{T})$, $(\mathbf{d})$, $(\mathbf{K})$, $(\mathbf{J})$ and $(\mathbf{p})$ are known, then its Sales Past must be:

$$\mathbf{S'} = [[1+I](\mathbf{U}-\{\mathbf{Sv+F+I+T}\}[1-\mathbf{d}])-[\mathbf{N+K+J}]] /([1+\mathbf{s}]\{\mathbf{vp}/360-[1+I]\})$$

or

$$\mathbf{S'} = \mathbf{S}/[1+\mathbf{s}]$$

Steve Asikin ISBN 13: **978-1564912361**, ISBN 10: **1564912363**

Rule-56680:

If both (**/**), (**U**), (**S'**), (**N**), (**S**), (**v**), (**F**), (**I**), (**T**), (**d**), (**K**), (**J**) and (**p**) are known, then its Sales Growth planned is:

$$s= [[1+\text{/}(U-\{Sv+F+I+T\}[1-d])-[N+K+J]]$$
$$/(S'\{vp/360-[1+\text{/}]\})-1$$

or

$$s= S/S'-1$$


Rule-56681:

If both (**/**), (**U**), (**S'**), (**s**), (**N**), (**v**), (**F**), (**I**), (**T**), (**d**), (**K**), (**J**) and (**p**) are known, then its Sales or Revenue planned is:

$$S= (S'[1+s]-F-I-T-([N+K+J+S'vp[1+s]/360\}$$
$$/[1+\text{/}-U)/[1-d])/v$$

or

$$S= S'[1+s]$$


Rule-56682:

If both (**/**), (**U**), (**S'**), (**s**), (**S**), (**N**), (**F**), (**I**), (**T**), (**d**), (**K**), (**J**) and (**p**) are known, then its Variable Portion planned is:

$$v= (S'[1+s]-F-I-T-([N+K+J+S'vp[1+s]/360\}$$
$$/[1+\text{/}-U)/[1-d])/S$$

Steve Asikin ISBN 13: **978-1564912361**, ISBN 10: **1564912363**

Rule-56683:

If both ($I$), ($U$), ($S'$), ($s$), ($S$), ($v$), ($N$), ($I$), ($T$), ($d$), ($K$), ($J$) and ($p$) are known, then its Fixed Cost planned is:

$$F = S'[1+s]-Sv-I-T-(\{N+K+J+S'vp[1+s]/360\}/[1+I-U)/[1-d]$$

Rule-56684:

If both ($I$), ($U$), ($S'$), ($s$), ($S$), ($v$), ($F$), ($N$), ($T$), ($d$), ($K$), ($J$) and ($p$) are known, then its Interest Expense planned is:

$$I = S'[1+s]-Sv-F-T-(\{N+K+J+S'vp[1+s]/360\}/[1+I-U)/[1-d]$$

Rule-56685:

If both ($I$), ($U$), ($S'$), ($s$), ($S$), ($v$), ($F$), ($I$), ($N$), ($d$), ($K$), ($J$) and ($p$) are known, then its Tax planned is:

$$T = S'[1+s]-Sv-F-I-(\{N+K+J+S'vp[1+s]/360\}/[1+I-U)/[1-d]$$

Rule-56686:

If both ($I$), ($U$), ($S'$), ($s$), ($S$), ($v$), ($F$), ($I$), ($T$), ($N$), ($K$), ($J$) and ($p$) are known, then its Dividend Payout planned is:

$$d = 1-(\{N+K+J+S'vp[1+s]/360\}/[1+I-U)/\{S'[1+s]-Sv-F-I-T\}$$

`

Steve Asikin ISBN 13: **978-1564912361**, ISBN 10: **1564912363**

Rule-56687:

If both (**/**), (**U**), (**S'**), (**s**), (**S**), (**v**), (**F**), (**I**), (**T**), (**d**), (**N**), (**J**) and (**p**) are known, then its Kind of Cash planned is:

$$K= [1+/(U+\{S'[1+s]-Sv-F-I-T\}[1-d])$$
$$-\{N+J+S'vp[1+s]/360\}$$

Rule-56688:

If both (**/**), (**U**), (**S'**), (**s**), (**S**), (**v**), (**F**), (**I**), (**T**), (**d**), (**K**), (**N**) and (**p**) are known, then its Job or Trade Account Receivable planned is:

$$J= [[1+/(U+\{S'[1+s]-Sv-F-I-T\}[1-d])$$
$$-\{N+K+S'vp[1+s]/360\}$$

Rule-56689:

If both (**/**), (**U**), (**S'**), (**s**), (**S**), (**v**), (**F**), (**I**), (**T**), (**d**), (**K**), (**J**) and (**N**) are known, then its Procured Inventory planned is:

$$p= 360[[1+/(U+\{S'[1+s]-Sv-F-I-T\}[1-d])$$
$$-[N+K+J]]/\{S'v[1+s]\}$$

Rule-56690:

If both (**/**), (**U**), (**S'**), (**s**), (**S**), (**v**), (**F**), (**I**), (**T**), (**d**), (**K**), (**j**) and (**P**) are known, then its Non Current Asset planned is:

$$N= [1+/(U+\{S'[1+s]-Sv-F-I-T\}[1-d])-K-Sj/360-P$$

Steve Asikin ISBN 13: **978-1564912361**, ISBN 10: **1564912363**

<u>Rule-56691</u>:

If both (**N**), (**U**), (**S'**), (**s**), (**S**), (**v**), (**F**), (**I**), (**T**), (**d**), (**K**), (**j**) and (**P**) are known, then its Leverage oir Gearing Ratio planned is:

$$I= \{N+K+Sj/360+P\}/(U+\{S'[1+s]-Sv-F-I-T\}[1-d])-1$$

<u>Rule-56692</u>:

If both (**I**), (**N**), (**S'**), (**s**), (**S**), (**v**), (**F**), (**I**), (**T**), (**d**), (**K**), (**j**) and (**P**) are known, then its Utilized or Starting Capital must be:

$$U= (\{N+K+Sj/360+P\}-[1+I]\{S'[1+s]-Sv-F-I-T\}[1-d])/[1+I]$$

<u>Rule-56693</u>:

If both (**I**), (**U**), (**N**), (**s**), (**S**), (**v**), (**F**), (**I**), (**T**), (**d**), (**K**), (**j**) and (**P**) are known, then its Sales Past must be:

$$S'= (\{[N+K+Sj/360+P]/[1+I]-U\}/[1-d]+Sv+F+I+T)/[1+s]$$

or

$$S'= S/[1+s]$$

Steve Asikin ISBN 13: **978-1564912361**, ISBN 10: **1564912363**

Rule-56694:

If both $(I)$, $(U)$, $(S')$, $(N)$, $(S)$, $(v)$, $(F)$, $(I)$, $(T)$, $(d)$, $(K)$, $(j)$ and $(P)$ are known, then its Sales Growth planned is:

$$s = (\{[N+K+Sj/360+P]/[1+I-U\}/[1-d]$$
$$+Sv+F+I+T)/S'-1$$

or

$$s = S/S'-1$$

Rule-56695:

If both $(I)$, $(U)$, $(S')$, $(s)$, $(N)$, $(v)$, $(F)$, $(I)$, $(T)$, $(d)$, $(K)$, $(j)$ and $(P)$ are known, then its Sales or Revenue planned is:

$$S = (S'[1+s]-F-I-T-\{[N+K+Sj/360+P]/[1+I-U\}$$
$$/[1-d])/v$$

or

$$S = S'[1+s]$$

Rule-56696:

If both $(I)$, $(U)$, $(S')$, $(s)$, $(S)$, $(N)$, $(F)$, $(I)$, $(T)$, $(d)$, $(K)$, $(j)$ and $(P)$ are known, then its Variable Portion planned is:

$$v = (S'[1+s]-F-I-T-\{[N+K+Sj/360+P]/[1+I-U\}$$
$$/[1-d])/S$$

Steve Asikin ISBN 13: **978-1564912361**, ISBN 10: **1564912363**

Rule-56697:

If both (**I**), (**U**), (**S'**), (**s**), (**S**), (**v**), (**N**), (**I**), (**T**), (**d**), (**K**),
(**j**) and (**P**) are known, then its Fixed Cost planned is:

$$F = S'[1+s]-Sv-I-T-\{[N+K+Sj/360+P]/[1+I]-U\}$$
$$/[1-d]$$

Rule-56698:

If both (**I**), (**U**), (**S'**), (**s**), (**S**), (**v**), (**F**), (**N**), (**T**), (**d**), (**K**),
(**j**) and (**P**) are known, then its Interest Portion
planned is:

$$I = S'[1+s]-Sv-F-T-\{[N+K+Sj/360+P]/[1+I]-U\}$$
$$/[1-d]$$

Rule-56699:

If both (**I**), (**U**), (**S'**), (**s**), (**S**), (**v**), (**F**), (**I**), (**N**), (**d**), (**K**),
(**j**) and (**P**) are known, then its Tax planned is:

$$T = S'[1+s]-Sv-F-I-\{[N+K+Sj/360+P]/[1+I]-U\}$$
$$/[1-d]$$

Rule-56700:

If both (**I**), (**U**), (**S'**), (**s**), (**S**), (**v**), (**F**), (**I**), (**T**), (**N**), (**K**),
(**j**) and (**P**) are known, then its Dividend Payout
planned is:

$$d = 1-\{[N+K+Sj/360+P]/[1+I]-U\}$$
$$/\{S'[1+s]-Sv-F-I-T\}$$

`

Steve Asikin ISBN 13: **978-1564912361**, ISBN 10: **1564912363**

Rule-56701:

If both $(I)$, $(U)$, $(S')$, $(s)$, $(S)$, $(v)$, $(F)$, $(I)$, $(T)$, $(d)$, $(N)$, $(j)$ and $(P)$ are known, then its Kind of Cash planned is:

$$K= [1+I](U+\{S'[1+s]-Sv-F-I-T\}[1-d])$$
$$-[N+Sj/360+P]$$

Rule-56702:

If both $(I)$, $(U)$, $(S')$, $(s)$, $(S)$, $(v)$, $(F)$, $(I)$, $(T)$, $(d)$, $(K)$, $(N)$ and $(P)$ are known, then its Job or Trade Receivable Days planned is:

$$j= 360[[1+I](U+\{S'[1+s]-Sv-F-I-T\}[1-d])$$
$$-\{N+K+P\}]/S$$

Rule-56703:

If both $(I)$, $(U)$, $(S')$, $(s)$, $(S)$, $(v)$, $(F)$, $(I)$, $(T)$, $(d)$, $(K)$, $(j)$ and $(N)$ are known, then its Procured Inventory planned is:

$$P= [1+I](U+\{S'[1+s]-Sv-F-I-T\}[1-d])$$
$$-\{N+K+Sj/360\}$$

Rule-56704:

If both $(I)$, $(U)$, $(S')$, $(s)$, $(S)$, $(v)$, $(F)$, $(I)$, $(T)$, $(d)$, $(K)$, $(j)$, $(V)$ and $(p)$ are known, then its Non Current Asset planned is:

$$N= [1+I](U+\{S'[1+s]-Sv-F-I-T\}[1-d])-K-Sj/360$$
$$-Vp/360$$

Steve Asikin ISBN 13: **978-1564912361**, ISBN 10: **1564912363**

Rule-56705:

If both (**N**), (**U**), (**S'**), (**s**), (**$**), (**v**), (**F**), (**I**), (**T**), (**d**), (**K**), (**j**), (**V**) and (**p**) are known, then its Leverage or Gearing Ratio planned is:

$$\textit{l} = [\mathbf{N+K+Sj}/360+\mathbf{Vp}/360]$$
$$/(\mathbf{U}+\{\mathbf{S'}[1+\mathbf{s}]-\mathbf{Sv-F-I-T}\}[1-\mathbf{d}])-1$$

Rule-56706:

If both (**l**), (**N**), (**S'**), (**s**), (**$**), (**v**), (**F**), (**I**), (**T**), (**d**), (**K**), (**j**), (**V**) and (**p**) are known, then its Utilized or Starting Capital must be:

$$\mathbf{U} = ([\mathbf{N+K+Sj}/360+\mathbf{Vp}/360]$$
$$-[1+\textit{l}]\{\mathbf{S'}[1+\mathbf{s}]-\mathbf{Sv-F-I-T}\}[1-\mathbf{d}])/[1+\textit{l}]$$

Rule-56707:

If both (**l**), (**U**), (**S'**), (**s**), (**$**), (**v**), (**F**), (**I**), (**T**), (**d**), (**K**), (**j**), (**V**) and (**p**) are known, then its Sales Past must be:

$$\mathbf{S'} = (\{[\mathbf{N+K+Sj}/360+\mathbf{Vp}/360]/[1+\textit{l}]-\mathbf{U}\}[1-\mathbf{d}]$$
$$+\mathbf{Sv+F+I+T})/[1+\mathbf{s}]$$
$$\text{or}$$
$$\mathbf{S'} = \mathbf{S}[1+\mathbf{s}]$$

`

Steve Asikin ISBN 13: **978-1564912361**, ISBN 10: **1564912363**

Rule-56708:

If both (**/**), (**U**), (**S'**), (**N**), (**S**), (**v**), (**F**), (**I**), (**T**), (**d**), (**K**), (**j**), (**V**) and (**p**) are known, then its Sales Growth planned is:

$$s = (\{[\mathbf{N}+\mathbf{K}+\mathbf{Sj}/360+\mathbf{Vp}/360]/[1+\mathbf{/}]-\mathbf{U}\}/[1-\mathbf{d}]$$
$$+\mathbf{Sv}+\mathbf{F}+\mathbf{I}+\mathbf{T})/\mathbf{S'}$$

or

$$s = \mathbf{S}/\mathbf{S'} - 1$$

Rule-56709:

If both (**/**), (**U**), (**S'**), (**s**), (**N**), (**v**), (**F**), (**I**), (**T**), (**d**), (**K**), (**j**), (**V**) and (**p**) are known, then its Sales or Revenue planned is:

$$\mathbf{S} = [[1+\mathbf{/}](\mathbf{U}+\{\mathbf{S'}[1+\mathbf{s}]-\mathbf{F}-\mathbf{I}-\mathbf{T}\}[1-\mathbf{d}])$$
$$-[\mathbf{N}+\mathbf{K}+\mathbf{Vp}/360]]/\{\mathbf{j}/360+\mathbf{v}[1+\mathbf{/}]\}$$

or

$$\mathbf{S} = \mathbf{S'}[1+\mathbf{s}]$$

Rule-56710:

If both (**/**), (**U**), (**S'**), (**s**), (**S**), (**N**), (**F**), (**I**), (**T**), (**d**), (**K**), (**j**), (**V**) and (**p**) are known, then its Variable Portion planned is:

$$\mathbf{v} = [\mathbf{S'}[1+\mathbf{s}]-\mathbf{F}-\mathbf{I}-\mathbf{T}-\{[\mathbf{N}+\mathbf{K}+\mathbf{Sj}/360+\mathbf{Vp}/360]$$
$$/[1+\mathbf{/}]-\mathbf{U}\}/[1-\mathbf{d}]]/\mathbf{S}$$

or

$$\mathbf{v} = \mathbf{V}/\mathbf{S}$$

Steve Asikin ISBN 13: **978-1564912361**, ISBN 10: **1564912363**

Rule-56711:

If both (**/**), (**U**), (**S'**), (**s**), (**S**), (**v**), (**N**), (**I**), (**T**), (**d**), (**K**), (**j**), (**V**) and (**p**) are known, then its Fixed Cost planned is:

$$F= S'[1+s]-Sv-I-T-\{[N+K+Sj/360+Vp/360]\ /[1+/-U\}/[1-d]$$

Rule-56712:

If both (**/**), (**U**), (**S'**), (**s**), (**S**), (**v**), (**F**), (**N**), (**T**), (**d**), (**K**), (**j**), (**V**) and (**p**) are known, then its Interest Expense planned is:

$$I= S'[1+s]-Sv-F-T-\{[N+K+Sj/360+Vp/360]\ /[1+/-U\}/[1-d]$$

Rule-56713:

If both (**/**), (**U**), (**S'**), (**s**), (**S**), (**v**), (**F**), (**I**), (**N**), (**d**), (**K**), (**j**), (**V**) and (**p**) are known, then its Tax planned is:

$$T= S'[1+s]-Sv-F-I-\{[N+K+Sj/360+Vp/360]\ /[1+/-U\}/[1-d]$$

Rule-56714:

If both (**/**), (**U**), (**S'**), (**s**), (**S**), (**v**), (**F**), (**I**), (**T**), (**N**), (**K**), (**j**), (**V**) and (**p**) are known, then its Dividend Payout planned is:

$$d= 1-\{[N+K+Sj/360+Vp/360]/[1+/-U\}\ /\{S'[1+s]-Sv-F-I-T\}$$

Steve Asikin ISBN 13: **978-1564912361**, ISBN 10: **1564912363**

Rule-56715:

If both (**/**), (**U**), (**S'**), (**s**), (**S**), (**v**), (**F**), (**I**), (**T**), (**d**), (**N**), (**j**), (**V**) and (**p**) are known, then its Kind of Cash planned is:

$$K= [1+/(U+\{S'[1+s]-Sv-F-I-T\}[1-d])$$
$$-[N+Sj/360+Vp/360]$$

Rule-56716:

If both (**/**), (**U**), (**S'**), (**s**), (**S**), (**v**), (**F**), (**I**), (**T**), (**d**), (**K**), (**N**), (**V**) and (**p**) are known, then its Job or Trade Receivable Days planned is:

$$j= 360[[1+/(U+\{S'[1+s]-Sv-F-I-T\}[1-d])$$
$$-[N+K+Vp/360]]/S$$

Rule-56717:

If both (**/**), (**U**), (**S'**), (**s**), (**S**), (**v**), (**F**), (**I**), (**T**), (**d**), (**K**), (**j**), (**N**) and (**p**) are known, then its Variable Cost planned is:

$$V= 360[[1+/(U+\{S'[1+s]-Sv-F-I-T\}[1-d])$$
$$-[N+K+Sj/360]]/p$$
$$\text{or}$$
$$V= Sv$$

Rule-56718:

If both (**/**), (**U**), (**S'**), (**s**), (**S**), (**v**), (**F**), (**I**), (**T**), (**d**), (**K**), (**j**), (**V**) and (**N**) are known, then its Procured Inventory Days planned is:

$$p= 360[[1+/(U+\{S'[1+s]-Sv-F-I-T\}[1-d])$$
$$-[N+K+Sj/360]]/V$$

Steve Asikin ISBN 13: **978-1564912361**, ISBN 10: **1564912363**

Rule-56719:

If both $(I)$, $(U)$, $(S')$, $(s)$, $(S)$, $(v)$, $(F)$, $(I)$, $(T)$, $(d)$, $(K)$, $(j)$ and $(p)$ are known, then its Non Current Asset planned is:

$$N = [1 + I(U + \{S'[1+s] - Sv - F - I - T\}[1-d]) - K - Sj/360 - Svp/360$$

Rule-56720:

If both $(N)$, $(U)$, $(S')$, $(s)$, $(S)$, $(v)$, $(F)$, $(I)$, $(T)$, $(d)$, $(K)$, $(j)$ and $(p)$ are known, then its Leverage or Gearing Ratio planned is:

$$I = [N + K + Sj/360 + Svp/360] / (U + \{S'[1+s] - Sv - F - I - T\}[1-d]) - 1$$

Rule-56721:

If both $(I)$, $(N)$, $(S')$, $(s)$, $(S)$, $(v)$, $(F)$, $(I)$, $(T)$, $(d)$, $(K)$, $(j)$ and $(p)$ are known, then its Utilized or Starting Capital must be:

$$U = ([N + K + Sj/360 + Svp/360] - [1 + I]\{S'[1+s] - Sv - F - I - T\}[1-d])/[1+I]$$

Rule-56722:

If both $(I)$, $(U)$, $(N)$, $(s)$, $(S)$, $(v)$, $(F)$, $(I)$, $(T)$, $(d)$, $(K)$, $(j)$ and $(p)$ are known, then its Sales Past must be:

$$S' = (\{[N + K + Sj/360 + Svp/360]/[1+I] - U\}/[1-d] + Sv + F + I + T)/S'[1+s]$$

or

$$S' = S/[1+s]$$

Steve Asikin ISBN 13: **978-1564912361**, ISBN 10: **1564912363**

Rule-56723:

If both $(\textit{I})$, $(\textbf{U})$, $(\textbf{S'})$, $(\textbf{N})$, $(\textbf{S})$, $(\textbf{v})$, $(\textbf{F})$, $(\textbf{I})$, $(\textbf{T})$, $(\textbf{d})$, $(\textbf{K})$, $(\textbf{j})$ and $(\textbf{p})$ are known, then its Sales Growth planned is:

$$s= (\{[\textbf{N}+\textbf{K}+\textbf{Sj}/360+\textbf{Svp}/360]/[1+\textit{I}]-\textbf{U}\}/[1-\textbf{d}]$$
$$+\textbf{Sv}+\textbf{F}+\textbf{I}+\textbf{T})/\textbf{S'}-1$$

or

$$s= \textbf{S}/\textbf{S'}-1$$

Rule-56724:

If both $(\textit{I})$, $(\textbf{U})$, $(\textbf{S'})$, $(\textbf{s})$, $(\textbf{S})$, $(\textbf{v})$, $(\textbf{F})$, $(\textbf{I})$, $(\textbf{T})$, $(\textbf{d})$, $(\textbf{K})$, $(\textbf{j})$ and $(\textbf{p})$ are known, then its Non Current Asset planned is:

$$S= \{[1+\textit{I}](\textbf{U}+\{\textbf{S'}[1+\textbf{s}]-\textbf{F}-\textbf{I}-\textbf{T}\}[1-\textbf{d}])-[\textbf{N}+\textbf{K}]\}$$
$$/[(\textbf{j}/360+\textbf{v}\{\textbf{p}/360+[1+\textit{I}]\})]$$

or

$$S= \textbf{S'}[1+\textbf{s}]$$

Rule-56725:

If both $(\textit{I})$, $(\textbf{U})$, $(\textbf{S'})$, $(\textbf{s})$, $(\textbf{S})$, $(\textbf{N})$, $(\textbf{F})$, $(\textbf{I})$, $(\textbf{T})$, $(\textbf{d})$, $(\textbf{K})$, $(\textbf{j})$ and $(\textbf{p})$ are known, then its Variable Portion planned is:

$$v= [\{[1+\textit{I}](\textbf{U}+\{\textbf{S'}[1+\textbf{s}]-\textbf{F}-\textbf{I}-\textbf{T}\}[1-\textbf{d}])-[\textbf{N}+\textbf{K}]\}/\textbf{S}$$
$$-\textbf{j}/360]/\{\textbf{p}/360+[1+\textit{I}]\}$$

Steve Asikin ISBN 13: **978-1564912361**, ISBN 10: **1564912363**

Rule-56726:

If both (**/**), (**U**), (**S'**), (**s**), (**S**), (**v**), (**N**), (**I**), (**T**), (**d**), (**K**), (**j**) and (**p**) are known, then its Fixed Cost planned is:

$$F= S'[1+s]-Sv-I-T-\{[N+K+Sj]/360+Svp/360]$$
$$/[1+/-U\}/[1-d]$$

Rule-56727:

If both (**/**), (**U**), (**S'**), (**s**), (**S**), (**v**), (**F**), (**N**), (**T**), (**d**), (**K**), (**j**) and (**p**) are known, then its Interest Expense planned is:

$$I= S'[1+s]-Sv-F-T-\{[N+K+Sj]/360+Svp/360]$$
$$/[1+/-U\}/[1-d]$$

Rule-56728:

If both (**/**), (**U**), (**S'**), (**s**), (**S**), (**v**), (**F**), (**I**), (**N**), (**d**), (**K**), (**j**) and (**p**) are known, then its Tax planned is:

$$T= S'[1+s]-Sv-F-I-\{[N+K+Sj]/360+Svp/360]$$
$$/[1+/-U\}/[1-d]$$

Rule-56729:

If both (**/**), (**U**), (**S'**), (**s**), (**S**), (**v**), (**F**), (**I**), (**T**), (**N**), (**K**), (**j**) and (**p**) are known, then its Dividend Payout planned is:

$$d= 1-\{[N+K+Sj]/360+Svp/360]/[1+/-U\}/\{S'[1+s]$$
$$-Sv-F-I-T\}$$

`

Steve Asikin ISBN 13: **978-1564912361**, ISBN 10: **1564912363**

Rule-56730:

If both (**/**), (**U**), (**S'**), (**s**), (**S**), (**v**), (**F**), (**I**), (**T**), (**d**), (**N**), (**j**) and (**p**) are known, then its Kind of Cash planned is:

$$K= [1+\text{/}(U+\{S'[1+s]-Sv-F-I-T\}[1-d])$$
$$-[N+Sj/360+Svp/360]$$

Rule-56731:

If both (**/**), (**U**), (**S'**), (**s**), (**S**), (**v**), (**F**), (**I**), (**T**), (**d**), (**K**), (**N**) and (**p**) are known, then its Job or Trade Receivable Days planned is:

$$j= 360[[1+\text{/}(U+\{S'[1+s]-Sv-F-I-T\}[1-d])$$
$$-\{N+K+Svp/360\}]/S$$

Rule-56732:

If both (**/**), (**U**), (**S'**), (**s**), (**S**), (**v**), (**F**), (**I**), (**T**), (**d**), (**K**), (**j**) and (**N**) are known, then its Procured Inventory Days planned is:

$$p= 360[[1+\text{/}(U+\{S'[1+s]-Sv-F-I-T\}[1-d])$$
$$-[N+K+Sj/360]]/[Sv]$$

Rule-56733:

If both (**/**), (**U**), (**S'**), (**s**), (**S**), (**v**), (**F**), (**I**), (**T**), (**d**), (**K**), (**j**) and (**p**) are known, then its Non Current Asset planned is:

$$N= [1+\text{/}(U+\{S'[1+s]-Sv-F-I-T\}[1-d])$$
$$-K-Sj/360-S'vp[1+s]/360$$

Steve Asikin ISBN 13: **978-1564912361**, ISBN 10: **1564912363**

Rule-56734:

If both (**N**), (**U**), (**S'**), (**s**), (**S**), (**v**), (**F**), (**I**), (**T**), (**d**), (**K**), (**j**) and (**p**) are known, then its Leverage or Gearing Ratio planned is:

$$I= \{\textbf{N}+\textbf{K}+\textbf{Sj}/360+\textbf{S'vp}[1+\textbf{s}]/360\}$$
$$/(\textbf{U}+\{\textbf{S'}[1+\textbf{s}]-\textbf{Sv}-\textbf{F}-\textbf{I}-\textbf{T}\}[1-\textbf{d}])-1$$

Rule-56735:

If both (**I**), (**N**), (**S'**), (**s**), (**S**), (**v**), (**F**), (**I**), (**T**), (**d**), (**K**), (**j**) and (**p**) are known, then its Utilized or Starting Capital must be:

$$\textbf{U}= (\{\textbf{N}+\textbf{K}+\textbf{Sj}/360+\textbf{S'vp}[1+\textbf{s}]/360\}$$
$$-[1+\textbf{I}]\{\textbf{S'}[1+\textbf{s}]-\textbf{Sv}-\textbf{F}-\textbf{I}-\textbf{T}\}[1-\textbf{d}])/\{[1+\textbf{I}]$$

Rule-56736:

If both (**I**), (**U**), (**N**), (**s**), (**S**), (**v**), (**F**), (**I**), (**T**), (**d**), (**K**), (**j**) and (**p**) are known, then its Sales Past must be:

$$\textbf{S'}= [[1+\textbf{I}](\textbf{U}-\{\textbf{Sv}+\textbf{F}+\textbf{I}+\textbf{T}\}[1-\textbf{d}])-[\textbf{N}+\textbf{K}+\textbf{Sj}/360]]$$
$$/([1+\textbf{s}]\{\textbf{vp}/360-[1+\textbf{I}]\})$$

or

$$\textbf{S'}= \textbf{S}/[1+\textbf{s}]$$

Steve Asikin ISBN 13: **978-1564912361**, ISBN 10: **1564912363**

Rule-56737:

If both $(\textit{\textbf{l}})$, $(\mathbf{U})$, $(\mathbf{S'})$, $(\mathbf{N})$, $(\mathbf{S})$, $(\mathbf{v})$, $(\mathbf{F})$, $(\mathbf{I})$, $(\mathbf{T})$, $(\mathbf{d})$, $(\mathbf{K})$, $(\mathbf{j})$ and $(\mathbf{p})$ are known, then its Sales Growth planned is:

$$s= [[1+\textit{\textbf{l}}(U-\{Sv+F+I+T\}[1-d])-[N+K+Sj/360]]$$
$$/(S'\{vp/360-[1+\textit{\textbf{l}}]\})-1$$

or

$$s= S/S'-1$$


Rule-56738:

If both $(\textit{\textbf{l}})$, $(\mathbf{U})$, $(\mathbf{S'})$, $(\mathbf{s})$, $(\mathbf{N})$, $(\mathbf{v})$, $(\mathbf{F})$, $(\mathbf{I})$, $(\mathbf{T})$, $(\mathbf{d})$, $(\mathbf{K})$, $(\mathbf{j})$ and $(\mathbf{p})$ are known, then its Sales or Revenue planned is:

$$S= [[1+\textit{\textbf{l}}(U+\{S'[1+s]-F-I-T\}[1-d])$$
$$-\{N+K+S'vp[1+s]/360\}]/\{j/360+v[1+\textit{\textbf{l}}]\}$$

or

$$S= S'[1+s]$$


Rule-56739:

If both $(\textit{\textbf{l}})$, $(\mathbf{U})$, $(\mathbf{S'})$, $(\mathbf{s})$, $(\mathbf{S})$, $(\mathbf{N})$, $(\mathbf{F})$, $(\mathbf{I})$, $(\mathbf{T})$, $(\mathbf{d})$, $(\mathbf{K})$, $(\mathbf{j})$ and $(\mathbf{p})$ are known, then its Variable Portion planned is:

$$v= [[1+\textit{\textbf{l}}(U+\{S'[1+s]-F-I-T\}[1-d])$$
$$-[N+K+Sj/360]]/\{S'p[1+s]/360+S[1+\textit{\textbf{l}}]\}$$

Steve Asikin ISBN 13: **978-1564912361**, ISBN 10: **1564912363**

Rule-56740:

> If both (**I**), (**U**), (**S'**), (**s**), (**S**), (**v**), (**N**), (**I**), (**T**), (**d**), (**K**),
> (**j**) and (**p**) are known, then its Fixed Cost planned is:
>
> $$F= S'[1+s]-Sv-I-T-(\{N+K+Sj/360+S'vp[1+s]/360\}$$
> $$/[1+I-U)/[1-d]$$

Rule-56741:

> If both (**I**), (**U**), (**S'**), (**s**), (**S**), (**v**), (**F**), (**N**), (**T**), (**d**), (**K**),
> (**j**) and (**p**) are known, then its Interest Expense
> planned is:
>
> $$I= S'[1+s]-Sv-F-T-(\{N+K+Sj/360+S'vp[1+s]/360\}$$
> $$/[1+I-U)/[1-d]$$

Rule-56742:

> If both (**I**), (**U**), (**S'**), (**s**), (**S**), (**v**), (**F**), (**I**), (**N**), (**d**), (**K**),
> (**j**) and (**p**) are known, then its Tax planned is:
>
> $$T= S'[1+s]-Sv-F-I-(\{N+K+Sj/360+S'vp[1+s]/360\}$$
> $$/[1+I-U)/[1-d]$$

Rule-56743:

> If both (**I**), (**U**), (**S'**), (**s**), (**S**), (**v**), (**F**), (**I**), (**T**), (**N**), (**K**),
> (**j**) and (**p**) are known, then its Dividend Payout
> planned is:
>
> $$d= 1-(\{N+K+Sj/360+S'vp[1+s]/360\}/[1+I-U)$$
> $$/\{S'[1+s]-Sv-F-I-T\}$$

Steve Asikin ISBN 13: **978-1564912361**, ISBN 10: **1564912363**

Rule-56744:

If both ($\mathit{l}$), (**U**), (**S'**), (**s**), (**S**), (**v**), (**F**), (**I**), (**T**), (**d**), (**N**), (**j**) and (**p**) are known, then its Kind of Cash planned is:

$$K= [1+\mathit{l}(U+\{S'[1+s]-Sv-F-I-T\}[1-d])$$
$$-\{N+Sj/360+S'vp[1+s]/360\}$$

Rule-56745:

If both ($\mathit{l}$), (**U**), (**S'**), (**s**), (**S**), (**v**), (**F**), (**I**), (**T**), (**d**), (**K**), (**N**) and (**p**) are known, then its Job or Trade Receivable Days planned is:

$$j= 360[[1+\mathit{l}(U+\{S'[1+s]-Sv-F-I-T\}[1-d])$$
$$-\{N+K+S'vp[1+s]/360\}]/S$$

Rule-56746:

If both ($\mathit{l}$), (**U**), (**S'**), (**s**), (**S**), (**v**), (**F**), (**I**), (**T**), (**d**), (**K**), (**j**) and (**N**) are known, then its Procured Inventory Days planned is:

$$p= 360[[1+\mathit{l}(U+\{S'[1+s]-Sv-F-I-T\}[1-d])$$
$$-[N+K+Sj/360]]/\{S'v[1+s]\}$$

Rule-56747:

If both ($\mathit{l}$), (**U**), (**S'**), (**s**), (**S**), (**v**), (**F**), (**I**), (**T**), (**d**), (**K**), (**j**) and (**P**) are known, then its Non Current Asset planned is:

$$N= [1+\mathit{l}(U+\{S'[1+s]-Sv-F-I-T\}[1-d])$$
$$-K-S'j[1+s]/360-P$$

Steve Asikin ISBN 13: **978-1564912361**, ISBN 10: **1564912363**

Rule-56748:

If both (**N**), (**U**), (**S'**), (**s**), (**S**), (**v**), (**F**), (**I**), (**T**), (**d**), (**K**), (**j**) and (**P**) are known, then its Leverage or Gearing Ratio planned is:

$$\textbf{\textit{l}}= \{\textbf{N}+\textbf{K}+\textbf{S'}\textbf{j}[1+\textbf{s}]/360+\textbf{P}\}$$
$$/(\textbf{U}+\{\textbf{S'}[1+\textbf{s}]-\textbf{Sv}-\textbf{F}-\textbf{I}-\textbf{T}\}[1-\textbf{d}])-1$$

Rule-56749:

If both (**l**), (**N**), (**S'**), (**s**), (**S**), (**v**), (**F**), (**I**), (**T**), (**d**), (**K**), (**j**) and (**P**) are known, then its Utilized or Starting Capital must be:

$$\textbf{U}= (\{\textbf{N}+\textbf{K}+\textbf{S'}\textbf{j}[1+\textbf{s}]/360+\textbf{P}\}$$
$$-[1+\textbf{\textit{l}}]\{\textbf{S'}[1+\textbf{s}]-\textbf{Sv}-\textbf{F}-\textbf{I}-\textbf{T}\}[1-\textbf{d}])/[1+\textbf{\textit{l}}]$$

Rule-56750:

If both (**l**), (**U**), (**N**), (**s**), (**S**), (**v**), (**F**), (**I**), (**T**), (**d**), (**K**), (**j**) and (**P**) are known, then its Sales Past must be:

$$\textbf{S'}= [[1+\textbf{\textit{l}}](\textbf{U}-\{\textbf{Sv}+\textbf{F}+\textbf{I}+\textbf{T}\}[1-\textbf{d}])-[\textbf{N}+\textbf{K}+\textbf{P}]]$$
$$/(\textbf{S'}[1+\textbf{s}]\{\textbf{j}/360-[1+\textbf{\textit{l}}]\})$$

or

$$\textbf{S'}= \textbf{S}/[1+\textbf{s}]$$

Steve Asikin ISBN 13: **978-1564912361**, ISBN 10: **1564912363**

Rule-56751:

If both $(\textbf{\textit{l}})$, $(\textbf{U})$, $(\textbf{S'})$, $(\textbf{N})$, $(\textbf{S})$, $(\textbf{v})$, $(\textbf{F})$, $(\textbf{I})$, $(\textbf{T})$, $(\textbf{d})$, $(\textbf{K})$, $(\textbf{j})$ and $(\textbf{P})$ are known, then its Sales Growth planned is:

$$s= [[1+\textit{l}(\textbf{U}-\{\textbf{Sv}+\textbf{F}+\textbf{I}+\textbf{T}\}[1-\textbf{d}])-[\textbf{N}+\textbf{K}+\textbf{P}]]$$
$$/(\textbf{S'}\{\textbf{j}/360-[1+\textit{l}]\})-1$$

or

$$s= \textbf{S}/\textbf{S'}-1$$


Rule-56752:

If both $(\textbf{\textit{l}})$, $(\textbf{U})$, $(\textbf{S'})$, $(\textbf{s})$, $(\textbf{N})$, $(\textbf{v})$, $(\textbf{F})$, $(\textbf{I})$, $(\textbf{T})$, $(\textbf{d})$, $(\textbf{K})$, $(\textbf{j})$ and $(\textbf{P})$ are known, then its Sales or Revenue planned is:

$$S= [\textbf{S'}[1+\textbf{s}]-\textbf{F}-\textbf{I}-\textbf{T}-(\{\textbf{N}+\textbf{K}+\textbf{S'}\textbf{j}[1+\textbf{s}]/360+\textbf{P}\}$$
$$/[1+\textit{l}-\textbf{U})/[1-\textbf{d}]]/\textbf{v}$$

or

$$S= \textbf{S'}[1+\textbf{s}]$$


Rule-56753:

If both $(\textbf{\textit{l}})$, $(\textbf{U})$, $(\textbf{S'})$, $(\textbf{s})$, $(\textbf{S})$, $(\textbf{N})$, $(\textbf{F})$, $(\textbf{I})$, $(\textbf{T})$, $(\textbf{d})$, $(\textbf{K})$, $(\textbf{j})$ and $(\textbf{P})$ are known, then its Variable Portion planned is:

$$v= [\textbf{S'}[1+\textbf{s}]-\textbf{F}-\textbf{I}-\textbf{T}-(\{\textbf{N}+\textbf{K}+\textbf{S'}\textbf{j}[1+\textbf{s}]/360+\textbf{P}\}$$
$$/[1+\textit{l}-\textbf{U})/[1-\textbf{d}]]/\textbf{S}$$

Steve Asikin ISBN 13: **978-1564912361**, ISBN 10: **1564912363**

Rule-56754:

If both (**/**), (**U**), (**S'**), (**s**), (**S**), (**v**), (**N**), (**I**), (**T**), (**d**), (**K**), (**j**) and (**P**) are known, then its Fixed Cost planned is:

$$F= S'[1+s]-Sv-I-T-(\{N+K+S'j[1+s]/360+P\}$$
$$/[1+/-U)/[1-d]$$

Rule-56755:

If both (**/**), (**U**), (**S'**), (**s**), (**S**), (**v**), (**F**), (**N**), (**T**), (**d**), (**K**), (**j**) and (**P**) are known, then its Interest Expense planned is:

$$I= S'[1+s]-Sv-F-T-(\{N+K+S'j[1+s]/360+P\}$$
$$/[1+/-U)/[1-d]$$

Rule-56756:

If both (**/**), (**U**), (**S'**), (**s**), (**S**), (**v**), (**F**), (**I**), (**N**), (**d**), (**K**), (**j**) and (**P**) are known, then its Tax planned is:

$$T= S'[1+s]-Sv-F-I-(\{N+K+S'j[1+s]/360+P\}$$
$$/[1+/-U)/[1-d]$$

Rule-56757:

If both (**/**), (**U**), (**S'**), (**s**), (**S**), (**v**), (**F**), (**I**), (**T**), (**N**), (**K**), (**j**) and (**P**) are known, then its Dividend Payout planned is:

$$d= 1-(\{N+K+S'j[1+s]/360+P\}/[1+/-U)$$
$$/\{S'[1+s]-Sv-F-I-T\}$$

Steve Asikin ISBN 13: **978-1564912361**, ISBN 10: **1564912363**

Rule-56758:

If both $(\pmb{f})$, $(\pmb{U})$, $(\pmb{S'})$, $(\pmb{s})$, $(\pmb{\$})$, $(\pmb{v})$, $(\pmb{F})$, $(\pmb{I})$, $(\pmb{T})$, $(\pmb{d})$, $(\pmb{N})$, $(\pmb{j})$ and $(\pmb{P})$ are known, then its Kind of Cash planned is:

$$\pmb{K} = [1 + \pmb{f}(\pmb{U} + \{\pmb{S'}[1+\pmb{s}] - \pmb{Sv} - \pmb{F} - \pmb{I} - \pmb{T}\}[1-\pmb{d}])$$
$$- \{\pmb{N} + \pmb{S'j}[1+\pmb{s}]/360 + \pmb{P}\}$$

Rule-56759:

If both $(\pmb{f})$, $(\pmb{U})$, $(\pmb{S'})$, $(\pmb{s})$, $(\pmb{\$})$, $(\pmb{v})$, $(\pmb{F})$, $(\pmb{I})$, $(\pmb{T})$, $(\pmb{d})$, $(\pmb{K})$, $(\pmb{N})$ and $(\pmb{P})$ are known, then its Job or Trade Receivable Days planned is:

$$\pmb{j} = 360[[1 + \pmb{f}(\pmb{U} + \{\pmb{S'}[1+\pmb{s}] - \pmb{Sv} - \pmb{F} - \pmb{I} - \pmb{T}\}[1-\pmb{d}])$$
$$- \{\pmb{N} + \pmb{K} + \pmb{P}\}]/\{\pmb{S'}[1+\pmb{s}]\}$$

Rule-56760:

If both $(\pmb{f})$, $(\pmb{U})$, $(\pmb{S'})$, $(\pmb{s})$, $(\pmb{\$})$, $(\pmb{v})$, $(\pmb{F})$, $(\pmb{I})$, $(\pmb{T})$, $(\pmb{d})$, $(\pmb{K})$, $(\pmb{j})$ and $(\pmb{N})$ are known, then its Procured Inventory planned is:

$$\pmb{P} = [1 + \pmb{f}(\pmb{U} + \{\pmb{S'}[1+\pmb{s}] - \pmb{Sv} - \pmb{F} - \pmb{I} - \pmb{T}\}[1-\pmb{d}])$$
$$- \{\pmb{N} + \pmb{K} + \pmb{S'j}[1+\pmb{s}]/360\}$$

Rule-56761:

If both $(\pmb{f})$, $(\pmb{U})$, $(\pmb{S'})$, $(\pmb{s})$, $(\pmb{\$})$, $(\pmb{v})$, $(\pmb{F})$, $(\pmb{I})$, $(\pmb{T})$, $(\pmb{d})$, $(\pmb{K})$, $(\pmb{j})$, $(\pmb{V})$ and $(\pmb{p})$ are known, then its Non Current Asset planned is:

$$\pmb{N} = [1 + \pmb{f}(\pmb{U} + \{\pmb{S'}[1+\pmb{s}] - \pmb{Sv} - \pmb{F} - \pmb{I} - \pmb{T}\}[1-\pmb{d}])$$
$$- \pmb{K} - \pmb{S'j}[1+\pmb{s}]/360 - \pmb{Vp}/360$$

Steve Asikin ISBN 13: **978-1564912361**, ISBN 10: **1564912363**

Rule-56762:

If both (**N**), (**U**), (**S'**), (**s**), (**S**), (**v**), (**F**), (**I**), (**T**), (**d**), (**K**), (**j**), (**V**) and (**p**) are known, then its Leverage or Gearing Ratio planned is:

$$I = \{N+K+S'j[1+s]/360+Vp/360\}$$
$$/(U+\{S'[1+s]-Sv-F-I-T\}[1-d])-1$$

Rule-56763:

If both (**I**), (**N**), (**S'**), (**s**), (**S**), (**v**), (**F**), (**I**), (**T**), (**d**), (**K**), (**j**), (**V**) and (**p**) are known, then its Utilized or Starting Capital must be:

$$U= (\{N+K+S'j[1+s]/360+Vp/360\}$$
$$-[1+I]\{S'[1+s]-Sv-F-I-T\}[1-d])/[1+I]$$

Rule-56764:

If both (**I**), (**U**), (**N**), (**s**), (**S**), (**v**), (**F**), (**I**), (**T**), (**d**), (**K**), (**j**), (**V**) and (**p**) are known, then its Sales Past must be:

$$S'= [[1+I](U-\{Sv+F+I+T\}[1-d])-[N+K+Vp/360]]$$
$$/(S'[1+s]\{j/360-[1+I]\})$$

or

$$S'= S/[1+s]$$

Steve Asikin ISBN 13: **978-1564912361**, ISBN 10: **1564912363**

Rule-56765:

> If both $(I)$, $(U)$, $(S')$, $(N)$, $(S)$, $(v)$, $(F)$, $(I)$, $(T)$, $(d)$, $(K)$, $(j)$, $(V)$ and $(p)$ are known, then its Sales Growth planned is:
>
> $s= [[1+I(U-\{Sv+F+I+T\}[1-d])-[N+K+Vp/360]]$
> $/(S'\{j/360-[1+I]\})-1$
>
> or
>
> $s= S/S'-1$

Rule-56766:

> If both $(I)$, $(U)$, $(S')$, $(s)$, $(N)$, $(v)$, $(F)$, $(I)$, $(T)$, $(d)$, $(K)$, $(j)$, $(V)$ and $(p)$ are known, then its Sales or Revenue planned is:
>
> $S= [S'[1+s]-F-I-T-(\{N+K+S'j[1+s]/360+Vp/360\}$
> $/[1+I-U)/[1-d]]/v$
>
> or
>
> $S= S'[1+s]$
>
> or
>
> $S= V/v$

Rule-56767:

> If both $(I)$, $(U)$, $(S')$, $(s)$, $(S)$, $(N)$, $(F)$, $(I)$, $(T)$, $(d)$, $(K)$, $(j)$, $(V)$ and $(p)$ are known, then its Variable Portion planned is:
>
> $v= [S'[1+s]-F-I-T-(\{N+K+S'j[1+s]/360+Vp/360\}$
> $/[1+I-U)/[1-d]]/S$
>
> or
>
> $v= V/S$

Steve Asikin ISBN 13: **978-1564912361**, ISBN 10: **1564912363**

Rule-56768:

If both (**/**), (**U**), (**S'**), (**s**), (**S**), (**v**), (**N**), (**I**), (**T**), (**d**), (**K**), (**j**), (**V**) and (**p**) are known, then its Fixed Cost planned is:

$$F= S'[1+s]-Sv-I-T-(\{N+K+S'j[1+s]/360+Vp/360\}/[1+/-U)/[1-d]$$

Rule-56769:

If both (**/**), (**U**), (**S'**), (**s**), (**S**), (**v**), (**F**), (**N**), (**T**), (**d**), (**K**), (**j**), (**V**) and (**p**) are known, then its Interest Expense planned is:

$$I= S'[1+s]-Sv-F-T-(\{N+K+S'j[1+s]/360+Vp/360\}/[1+/-U)/[1-d]$$

Rule-56770:

If both (**/**), (**U**), (**S'**), (**s**), (**S**), (**v**), (**F**), (**I**), (**N**), (**d**), (**K**), (**j**), (**V**) and (**p**) are known, then its Tax planned is:

$$T= S'[1+s]-Sv-F-I-(\{N+K+S'j[1+s]/360+Vp/360\}/[1+/-U)/[1-d]$$

Rule-56771:

If both (**/**), (**U**), (**S'**), (**s**), (**S**), (**v**), (**F**), (**I**), (**T**), (**N**), (**K**), (**j**), (**V**) and (**p**) are known, then its Dividend Payout planned is:

$$d= 1-(\{N+K+S'j[1+s]/360+Vp/360\}/[1+/-U)/\{S'[1+s]-Sv-F-I-T\}$$

`

Steve Asikin ISBN 13: **978-1564912361**, ISBN 10: **1564912363**

Rule-56772:

If both (**/**), (**U**), (**S'**), (**s**), (**S**), (**v**), (**F**), (**I**), (**T**), (**d**), (**N**), (**j**), (**V**) and (**p**) are known, then its Kind of Cash planned is:

$$K = [1+/(U+\{S'[1+s]-Sv-F-I-T\}[1-d]) \\ -\{N+S'j[1+s]/360+Vp/360\}$$

Rule-56773:

If both (**/**), (**U**), (**S'**), (**s**), (**S**), (**v**), (**F**), (**I**), (**T**), (**d**), (**K**), (**N**), (**V**) and (**p**) are known, then its Job or Trade Receivable Days planned is:

$$j = 360[[1+/(U+\{S'[1+s]-Sv-F-I-T\}[1-d]) \\ -\{N+K+Vp/360\}]/\{S'[1+s]\}$$

Rule-56774:

If both (**/**), (**U**), (**S'**), (**s**), (**S**), (**v**), (**F**), (**I**), (**T**), (**d**), (**K**), (**j**), (**N**) and (**p**) are known, then its Variable Cost planned is:

$$V = 360[[1+/(U+\{S'[1+s]-Sv-F-I-T\}[1-d]) \\ -\{N+K+S'j[1+s]/360\}]/p$$

or

$$V = Sv$$

Rule-56775:

If both (**/**), (**U**), (**S'**), (**s**), (**S**), (**v**), (**F**), (**I**), (**T**), (**d**), (**K**), (**j**), (**V**) and (**N**) are known, then its Procured Inventory Days planned is:

$$p = 360[[1+/(U+\{S'[1+s]-Sv-F-I-T\}[1-d]) \\ -\{N+K+S'j[1+s]/360\}]/V$$

Steve Asikin ISBN 13: **978-1564912361**, ISBN 10: **1564912363**

<u>Rule-56776</u>:

If both ($I$), (**U**), (**S'**), (**s**), (**S**), (**v**), (**F**), (**I**), (**T**), (**d**), (**K**), (**j**) and (**p**) are known, then its Non Current Asset planned is:

$$N= [1+I(U+\{S'[1+s]-Sv-F-I-T\}[1-d])-K$$
$$-S'j[1+s]/360-Svp/360$$

<u>Rule-56777</u>:

If both (**N**), (**U**), (**S'**), (**s**), (**S**), (**v**), (**F**), (**I**), (**T**), (**d**), (**K**), (**j**) and (**p**) are known, then its Leverage or Gearing Ratio planned is:

$$I= \{N+K+S'j[1+s]/360+Svp/360\}$$
$$/(U+\{S'[1+s]-Sv-F-I-T\}[1-d])-1$$

<u>Rule-56778</u>:

If both ($I$), (**N**), (**S'**), (**s**), (**S**), (**v**), (**F**), (**I**), (**T**), (**d**), (**K**), (**j**) and (**p**) are known, then its Utilized or Starting Capital must be:

$$U= (\{N+K+S'j[1+s]/360+Svp/360\}$$
$$-[1+I]\{S'[1+s]-Sv-F-I-T\}[1-d])/\{[1+I]\}$$

<u>Rule-56779</u>:

If both ($I$), (**U**), (**N**), (**s**), (**S**), (**v**), (**F**), (**I**), (**T**), (**d**), (**K**), (**j**) and (**p**) are known, then its Sales Past must be:

$$S'= [[1+I(U-\{Sv+F+I+T\}[1-d])-[N+K$$
$$+Svp/360]]/(S'[1+s]\{j/360-[1+I]\})$$

or

$$S'= S/[1+s]$$

Steve Asikin ISBN 13: **978-1564912361**, ISBN 10: **1564912363**

Rule-56780:

If both (**/**), (**U**), (**S'**), (**N**), (**S**), (**v**), (**F**), (**I**), (**T**), (**d**), (**K**), (**j**) and (**p**) are known, then its Sales Growth planned is:

$$s= [[1+/](U-\{Sv+F+I+T\}[1-d])-[N+K+Svp/360]]$$
$$/(S'\{j/360-[1+/]\})-1$$

or

$$s= S/S'-1$$

Rule-56781:

If both (**/**), (**U**), (**S'**), (**s**), (**N**), (**v**), (**F**), (**I**), (**T**), (**d**), (**K**), (**j**) and (**p**) are known, then its Sales or Revenue planned is:

$$S= [[1+/](U+\{S'[1+s]-F-I-T\}[1-d])$$
$$-\{N+K+S'j[1+s]/360\}]/(v\{p/360+[1+/]\})$$

or

$$S= S'[1+s]$$

Rule-56782:

If both (**/**), (**U**), (**S'**), (**s**), (**S**), (**N**), (**F**), (**I**), (**T**), (**d**), (**K**), (**j**) and (**p**) are known, then its Variable Portion planned is:

$$v= [[1+/](U+\{S'[1+s]-F-I-T\}[1-d])$$
$$-\{N+K+S'j[1+s]/360\}]/(S\{p/360+[1+/]\})$$

Steve Asikin ISBN 13: **978-1564912361**, ISBN 10: **1564912363**

<u>Rule-56783</u>:

If both (**l**), (**U**), (**S'**), (**s**), (**S**), (**v**), (**N**), (**I**), (**T**), (**d**), (**K**), (**j**) and (**p**) are known, then its Fixed Cost planned is:

$$F = S'[1+s] - Sv - I - T - (\{N+K+S'j[1+s]/360 + Svp/360\} /[1+l-U)/[1-d]$$

<u>Rule-56784</u>:

If both (**l**), (**U**), (**S'**), (**s**), (**S**), (**v**), (**F**), (**N**), (**T**), (**d**), (**K**), (**j**) and (**p**) are known, then its Interest Expense planned is:

$$I = S'[1+s] - Sv - F - T - (\{N+K+S'j[1+s]/360 + Svp/360\} /[1+l-U)/[1-d]$$

<u>Rule-56785</u>:

If both (**l**), (**U**), (**S'**), (**s**), (**S**), (**v**), (**F**), (**I**), (**N**), (**d**), (**K**), (**j**) and (**p**) are known, then its Tax planned is:

$$T = S'[1+s] - Sv - F - I - (\{N+K+S'j[1+s]/360 + Svp/360\} /[1+l-U)/[1-d]$$

<u>Rule-56786</u>:

If both (**l**), (**U**), (**S'**), (**s**), (**S**), (**v**), (**F**), (**I**), (**T**), (**N**), (**K**), (**j**) and (**p**) are known, then its Dividend Payout planned is:

$$d = 1 - (\{N+K+S'j[1+s]/360 + Svp/360\}/[1+l-U) /\{S'[1+s] - Sv - F - I - T\}$$

Steve Asikin ISBN 13: **978-1564912361**, ISBN 10: **1564912363**

Rule-56787:

If both ($\mathit{f}$), (**U**), (**$'**), (**s**), (**$**), (**v**), (**F**), (**I**), (**T**), (**d**), (**N**), (**j**) and (**p**) are known, then its Kind of Cash planned is:

$$K= [1+\mathit{f}(U+\{\$'[1+s]-\$v-F-I-T\}[1-d])$$
$$-\{N+\$j[1+s]/360+\$vp/360\}$$

Rule-56788:

If both ($\mathit{f}$), (**U**), (**$'**), (**s**), (**$**), (**v**), (**F**), (**I**), (**T**), (**d**), (**K**), (**N**) and (**p**) are known, then its Job or Trade Receivable Days planned is:

$$j= 360[[1+\mathit{f}(U+\{\$'[1+s]-\$v-F-I-T\}[1-d])$$
$$-\{N+K+\$vp/360\}]/\{\$'[1+s]\}$$

Rule-56789:

If both ($\mathit{f}$), (**U**), (**$'**), (**s**), (**$**), (**v**), (**F**), (**I**), (**T**), (**d**), (**K**), (**j**) and (**N**) are known, then its Procured Inventory Days planned is:

$$p= 360[[1+\mathit{f}(U+\{\$'[1+s]-\$v-F-I-T\}[1-d])$$
$$-\{N+K+\$j[1+s]/360\}]/[\$v]$$

Rule-56790:

If both ($\mathit{f}$), (**U**), (**$'**), (**s**), (**$**), (**v**), (**F**), (**I**), (**T**), (**d**), (**K**), (**j**) and (**p**) are known, then its Non Current Asset planned is:

$$N= [1+\mathit{f}(U+\{\$'[1+s]-\$v-F-I-T\}[1-d])$$
$$-K-\$j[1+s]/360-\$'vp[1+s]/360$$

Steve Asikin ISBN 13: **978-1564912361**, ISBN 10: **1564912363**

Rule-56791:

If both (**N**), (**U**), (**S'**), (**s**), (**S**), (**v**), (**F**), (**I**), (**T**), (**d**), (**K**), (**j**) and (**p**) are known, then its Leverage or Gearing Ratio planned is:

$$I= \{N+K+S'j[1+s]/360+S'vp[1+s]/360\}$$
$$/(U+\{S'[1+s]-Sv-F-I-T\}[1-d])-1$$

Rule-56792:

If both (**I**), (**N**), (**S'**), (**s**), (**S**), (**v**), (**F**), (**I**), (**T**), (**d**), (**K**), (**j**) and (**p**) are known, then its Utilized or Starting Capital must be:

$$U= (\{N+K+S'j[1+s]/360+S'vp[1+s]/360\}$$
$$-[1+I]\{S'[1+s]-Sv-F-I-T\}[1-d])/[1+I]$$

Rule-56793:

If both (**I**), (**U**), (**N**), (**s**), (**S**), (**v**), (**F**), (**I**), (**T**), (**d**), (**K**), (**j**) and (**p**) are known, then its Sales Past must be:

$$S'= [[1+I](U-\{Sv+F+I+T\}[1-d])-[N+K]]$$
$$/([1+s]\{j/360+vp/360-[1+I]\})$$

or

$$S'= S/[1+s]$$

Steve Asikin ISBN 13: **978-1564912361**, ISBN 10: **1564912363**

Rule-56794:

If both (**/**), (**U**), (**S'**), (**N**), (**S**), (**v**), (**F**), (**I**), (**T**), (**d**), (**K**), (**j**) and (**p**) are known, then its Sales Growth planned is:

$$s= [[1+/(U-\{Sv+F+I+T\}[1-d])-[N+K]]$$
$$/(S'\{j/360+vp/360-[1+/]\})-1$$
$$\text{or}$$
$$s= S/S'-1$$


Rule-56795:

If both (**/**), (**U**), (**S'**), (**s**), (**S**), (**v**), (**F**), (**I**), (**T**), (**d**), (**K**), (**j**) and (**p**) are known, then its Non Current Asset planned is:

$$S= [\{S'[1+s]-F-I-T-(\{N+K+S'j[1+s]/360$$
$$+S'vp[1+s]/360\}/[1+/-U)/[1-d]]/v$$
$$\text{or}$$
$$S= S'[1+s]$$


Rule-56796:

If both (**/**), (**U**), (**S'**), (**s**), (**S**), (**N**), (**F**), (**I**), (**T**), (**d**), (**K**), (**j**) and (**p**) are known, then its Variable Portion planned is:

$$v= [\{S'[1+s]-F-I-T-(\{N+K+S'j[1+s]/360$$
$$+S'vp[1+s]/360\}/[1+/-U)/[1-d]]/S$$

Steve Asikin ISBN 13: **978-1564912361**, ISBN 10: **1564912363**

## Rule-56797:

If both (**/**), (**U**), (**$'**), (**s**), (**$**), (**v**), (**N**), (**I**), (**T**), (**d**), (**K**), (**j**) and (**p**) are known, then its Fixed Cost planned is:

$$F= \{\$'[1+s]-\$v-I-T-(\{N+K+\$'j[1+s]/360$$
$$+\$'vp[1+s]/360\}/[1+/-U)/[1-d]$$

## Rule-56798:

If both (**N**), (**U**), (**$'**), (**s**), (**$**), (**v**), (**F**), (**I**), (**T**), (**d**), (**K**), (**j**) and (**p**) are known, then its Interest Expense planned is:

$$I= \{\$'[1+s]-\$v-F-T-(\{N+K+\$'j[1+s]/360$$
$$+\$'vp[1+s]/360\}/[1+/-U)/[1-d]$$

## Rule-56799:

If both (**/**), (**U**), (**$'**), (**s**), (**$**), (**v**), (**F**), (**I**), (**N**), (**d**), (**K**), (**j**) and (**p**) are known, then its Tax planned is:

$$T= \{\$'[1+s]-\$v-F-I-(\{N+K+\$'j[1+s]/360$$
$$+\$'vp[1+s]/360\}/[1+/-U)/[1-d]$$

## Rule-56800:

If both (**/**), (**U**), (**$'**), (**s**), (**$**), (**v**), (**F**), (**I**), (**T**), (**N**), (**K**), (**j**) and (**p**) are known, then its Dividend Payout planned is:

$$d= 1-(\{N+K+\$'j[1+s]/360+\$'vp[1+s]/360\}$$
$$/[1+/-U)/\{\$'[1+s]-\$v-F-I-T\}$$

Steve Asikin ISBN 13: **978-1564912361**, ISBN 10: **1564912363**

Rule-56801:

If both (**/**), (**U**), (**S'**), (**s**), (**S**), (**v**), (**F**), (**I**), (**T**), (**d**), (**N**), (**j**) and (**p**) are known, then its Kind of Cash planned is:

$$K = [1 + /(U + \{S'[1+s] - Sv - F - I - T\}[1-d])$$
$$- \{N + S'j[1+s]/360 + S'vp[1+s]/360\}$$

Rule-56802:

If both (**/**), (**U**), (**S'**), (**s**), (**S**), (**v**), (**F**), (**I**), (**T**), (**d**), (**K**), (**N**) and (**p**) are known, then its Job or Trade Receivable Days planned is:

$$j = 360[[1 + /(U + \{S'[1+s] - Sv - F - I - T\}[1-d])$$
$$- \{N + K + S'vp[1+s]/360\}]/\{S'[1+s]\}]$$

Rule-56803:

If both (**/**), (**U**), (**S'**), (**s**), (**S**), (**v**), (**F**), (**I**), (**T**), (**d**), (**K**), (**j**) and (**N**) are known, then its Procured Inventory Days planned is:

$$p = 360[[1 + /(U + \{S'[1+s] - Sv - F - I - T\}[1-d])$$
$$- \{N + K + S'j[1+s]/360\}]/\{S'v[1+s]\}]$$

Rule-56804:

If both (**/**), (**U**), (**S'**), (**s**), (**S**), (**v**), (**F**), (**I**), (**t**), (**D**), (**K**), (**J**) and (**P**) are known, then its Non Current Asset planned is:

$$N = [1 + /(U + \{S'[1+s] - Sv - F - I\}[1-t] - D) - K - J - P$$

Steve Asikin ISBN 13: **978-1564912361**, ISBN 10: **1564912363**

<u>Rule-56805</u>:

If both (**N**), (**U**), (**S'**), (**s**), (**S**), (**v**), (**F**), (**I**), (**t**), (**D**), (**K**), (**J**) and (**P**) are known, then its Leverage or Gearing Ratio planned is:

$$I = \{N+K+J+P\}/(U+\{S'[1+s]-Sv-F-I\}[1-t]-D)-1$$

<u>Rule-56806</u>:

If both (**I**), (**N**), (**S'**), (**s**), (**S**), (**v**), (**F**), (**I**), (**t**), (**D**), (**K**), (**J**) and (**P**) are known, then its Utilized or Starting Capital must be:

$$U = (\{N+K+J+P\}-[1+I]\{S'[1+s]-Sv-I\}[1-t]-D)$$
$$/[1+I]$$

<u>Rule-56807</u>:

If both (**I**), (**U**), (**N**), (**s**), (**S**), (**v**), (**F**), (**I**), (**t**), (**D**), (**K**), (**J**) and (**P**) are known, then its Sales Past must be:

$$S' = (\{[N+K+J+P]/[1+I]-U+D\}/[1-t]+F+I+Sv)$$
$$/[1+s]$$
$$\text{or}$$
$$S' = S/[1+s]$$

<u>Rule-56808</u>:

If both (**I**), (**U**), (**S'**), (**N**), (**S**), (**v**), (**F**), (**I**), (**t**), (**D**), (**K**), (**J**) and (**P**) are known, then its Sales Growth planned is:

$$s = (\{[N+K+J+P]/[1+I]-U+D\}/[1-t]+F+I+Sv)/S'-1$$
$$\text{or}$$
$$s = S/S'-1$$

`

Steve Asikin ISBN 13: **978-1564912361**, ISBN 10: **1564912363**

Rule-56809:

If both (**/**), (**U**), (**S'**), (**s**), (**N**), (**v**), (**F**), (**I**), (**t**), (**D**), (**K**), (**J**) and (**P**) are known, then its Sales or Revenue planned is:

$$S= [S'[1+s]-F-I-\{[N+K+J+P]/[1+/]-U+D\}$$
$$/[1-t]]/v$$

or

$$S= S'[1+s]$$

Rule-56810:

If both (**/**), (**U**), (**S'**), (**s**), (**S**), (**N**), (**F**), (**I**), (**t**), (**D**), (**K**), (**J**) and (**P**) are known, then its Variable Portion planned is:

$$v= [S'[1+s]-F-I-\{[N+K+J+P]/[1+/]-U+D\}$$
$$/[1-t]]/S$$

Rule-56811:

If both (**/**), (**U**), (**S'**), (**s**), (**S**), (**v**), (**N**), (**I**), (**t**), (**D**), (**K**), (**J**) and (**P**) are known, then its Fixed Cost planned is:

$$F= S'[1+s]-Sv-I-\{[N+K+J+P]/[1+/]-U+D\}/[1-t]$$

Rule-56812:

If both (**/**), (**U**), (**S'**), (**s**), (**S**), (**v**), (**F**), (**N**), (**t**), (**D**), (**K**), (**J**) and (**P**) are known, then its Interest Expense planned is:

$$I= S'[1+s]-Sv-F-\{[N+K+J+P]/[1+/]-U+D\}/[1-t]$$

Steve Aoikin ISBN 13: 078 1564012361, ISBN 10: 1564012363

Rule-56813:

If both (**/**), (**U**), (**S'**), (**s**), (**S**), (**v**), (**F**), (**I**), (**N**), (**D**), (**K**), (**J**) and (**P**) are known, then its Tax Rate planned is:

$$t= 1-\{[N+K+J+P]/[1+/]-U+D\}/\{S'[1+s]-Sv-F-I\}$$

Rule-56814:

If both (**/**), (**U**), (**S'**), (**s**), (**S**), (**v**), (**F**), (**I**), (**t**), (**N**), (**K**), (**J**) and (**P**) are known, then its Dividend planned is:

$$D= U+\{S'[1+s]-Sv-F-I\}[1-t]-[N+K+J+P]/[1+/]$$

Rule-56815:

If both (**/**), (**U**), (**S'**), (**s**), (**S**), (**v**), (**F**), (**I**), (**t**), (**D**), (**N**), (**J**) and (**P**) are known, then its Kind of Cash planned is:

$$K= [1+/](U+\{S'[1+s]-Sv-F-I\}[1-t]-D)-[N+J+P]$$

Rule-56816:

If both (**/**), (**U**), (**S'**), (**s**), (**S**), (**v**), (**F**), (**I**), (**t**), (**D**), (**K**), (**N**) and (**P**) are known, then its Job or Trade Account Receivable planned is:

$$J= [1+/](U+\{S'[1+s]-Sv-F-I\}[1-t]-D)-[N+K+P]$$

Rule-56817:

If both (**/**), (**U**), (**S'**), (**s**), (**S**), (**v**), (**F**), (**I**), (**t**), (**D**), (**K**), (**J**) and (**N**) are known, then its Procured Inventory planned is:

$$P= [1+/](U+\{S'[1+s]-Sv-F-I\}[1-t]-D)-[N+K+J]$$

Steve Asikin ISBN 13: **978-1564912361**, ISBN 10: **1564912363**

Rule-56818:

If both $(I)$, $(U)$, $(S')$, $(s)$, $(S)$, $(v)$, $(F)$, $(I)$, $(t)$, $(D)$, $(K)$, $(J)$, $(V)$ and $(p)$ are known, then its Non Current Asset planned is:

$$N= [1+I](U+\{S'[1+s]-Sv-F-I\}[1-t]-D)$$
$$-K-J-Vp/360$$

Rule-56819:

If both $(N)$, $(U)$, $(S')$, $(s)$, $(S)$, $(v)$, $(F)$, $(I)$, $(t)$, $(D)$, $(K)$, $(J)$, $(V)$ and $(p)$ are known, then its Leverage or Gearing Ratio planned is:

$$I= [N+K+J+Vp/360]/(U+\{S'[1+s]-Sv-F-I\}[1-t]-D)-1$$

Rule-56820:

If both $(I)$, $(N)$, $(S')$, $(s)$, $(S)$, $(v)$, $(F)$, $(I)$, $(t)$, $(D)$, $(K)$, $(J)$, $(V)$ and $(p)$ are known, then its Utilized or Starting Capital must be:

$$U= ([N+K+J+Vp/360]-[1+I]\{S'[1+s]-Sv-F-I\}[1-t]-D)/[1+I]$$

Rule-56821:

If both $(I)$, $(U)$, $(N)$, $(s)$, $(S)$, $(v)$, $(F)$, $(I)$, $(t)$, $(D)$, $(K)$, $(J)$, $(V)$ and $(p)$ are known, then its Sales Past must be:

$$S'= (\{[N+K+J+Vp/360]/[1+I]-U+D\}/[1-t]+Sv+F+I)/[1+s]$$
or
$$S'= S/[1+s]$$

Steve Aoikin ISBN 13: 978 1564912361, ISBN 10: 1564912363

Rule-56822:

If both (**∕**), (**U**), (**$'**), (**N**), (**$**), (**ʋ**), (**F**), (**I**), (**t**), (**D**), (**K**), (**J**), (**V**) and (**p**) are known, then its Sales Growth planned is:

$$s= (\{[\mathbf{N}+\mathbf{K}+\mathbf{J}+\mathbf{Vp}/360]/[1+\mathbf{\textit{∕}}-\mathbf{U}+\mathbf{D}\}/[1-\mathbf{t}]$$
$$+\mathbf{\$ʋ}+\mathbf{F}+\mathbf{I})/\mathbf{\$'}-1$$

or

$$s= \mathbf{\$}/\mathbf{\$'}-1$$

Rule-56823:

If both (**∕**), (**U**), (**$'**), (**s**), (**N**), (**ʋ**), (**F**), (**I**), (**t**), (**D**), (**K**), (**J**), (**V**) and (**p**) are known, then its Sales or Revenue planned is:

$$\$= [\mathbf{\$'}[1+s]-\mathbf{\$ʋ}-\mathbf{I}-\{[\mathbf{N}+\mathbf{K}+\mathbf{J}+\mathbf{Vp}/360]/[1+\mathbf{\textit{∕}}-\mathbf{U}+\mathbf{D}\}$$
$$/[1-\mathbf{t}]]/\mathbf{ʋ}$$

or

$$\$= \mathbf{\$'}[1+s]$$

or

$$\$= \mathbf{V}/\mathbf{ʋ}$$

Rule-56824:

If both (**∕**), (**U**), (**$'**), (**s**), (**$**), (**N**), (**F**), (**I**), (**t**), (**D**), (**K**), (**J**), (**V**) and (**p**) are known, then its Variable Portion planned is:

$$ʋ= [\mathbf{\$'}[1+s]-\mathbf{\$ʋ}-\mathbf{I}-\{[\mathbf{N}+\mathbf{K}+\mathbf{J}+\mathbf{Vp}/360]/[1+\mathbf{\textit{∕}}$$
$$-\mathbf{U}+\mathbf{D}\}/[1-\mathbf{t}]]/\mathbf{\$}$$

or

$$ʋ= \mathbf{V}/\mathbf{\$}$$

Steve Asikin ISBN 13: **978-1564912361**, ISBN 10: **1564912363**

Rule-56825:

If both (**/**), (**U**), (**S'**), (**s**), (**S**), (**v**), (**N**), (**I**), (**t**), (**D**), (**K**), (**J**), (**V**) and (**p**) are known, then its Fixed Cost planned is:

$$F= S'[1+s]-Sv-I-\{[N+K+J+Vp/360]/[1+/]$$
$$-U+D\}/[1-t]$$

Rule-56826:

If both (**/**), (**U**), (**S'**), (**s**), (**S**), (**v**), (**F**), (**N**), (**t**), (**D**), (**K**), (**J**), (**V**) and (**p**) are known, then its Interest Portion planned is:

$$I= S'[1+s]-Sv-F-\{[N+K+J+Vp/360]/[1+/]$$
$$-U+D\}/[1-t]$$

Rule-56827:

If both (**/**), (**U**), (**S'**), (**s**), (**S**), (**v**), (**F**), (**I**), (**N**), (**D**), (**K**), (**J**), (**V**) and (**p**) are known, then its Tax Rate planned is:

$$t= 1-\{[N+K+J+Vp/360]/[1+/]-U+D\}$$
$$/\{S'[1+s]-Sv-F-I\}$$

Rule-56828:

If both (**/**), (**U**), (**S'**), (**s**), (**S**), (**v**), (**F**), (**I**), (**t**), (**N**), (**K**), (**J**), (**V**) and (**p**) are known, then its Dividend planned is:

$$D= U+\{S'[1+s]-Sv-F-I\}[1-t]$$
$$-[N+K+J+Vp/360]/[1+/]$$

Steve Asikin ISBN 13: 978-1564912361, ISBN 10: 1564912363

Rule-56829:

If both (**/**), (**U**), (**S'**), (**s**), (**S**), (**v**), (**F**), (**I**), (**t**), (**D**), (**N**), (**J**), (**V**) and (**p**) are known, then its Kind of Cash planned is:

$$\mathbf{K} = [1 + \mathbf{/}(\mathbf{U} + \{\mathbf{S'}[1+\mathbf{s}] - \mathbf{Sv} - \mathbf{F} - \mathbf{I}\}[1-\mathbf{t}] - \mathbf{D})$$
$$- [\mathbf{N} + \mathbf{J} + \mathbf{Vp}/360]$$

Rule-56830:

If both (**/**), (**U**), (**S'**), (**s**), (**S**), (**v**), (**F**), (**I**), (**t**), (**D**), (**K**), (**N**), (**V**) and (**p**) are known, then its Job or Trade Account Receivable planned is:

$$\mathbf{J} = [1 + \mathbf{/}(\mathbf{U} + \{\mathbf{S'}[1+\mathbf{s}] - \mathbf{Sv} - \mathbf{F} - \mathbf{I}\}[1-\mathbf{t}] - \mathbf{D})$$
$$- \{\mathbf{N} + \mathbf{K} + \mathbf{Vp}/360\}$$

Rule-56831:

If both (**/**), (**U**), (**S'**), (**s**), (**S**), (**v**), (**F**), (**I**), (**t**), (**D**), (**K**), (**J**), (**N**) and (**p**) are known, then its Variable Cost planned is:

$$\mathbf{V} = 360[[1 + \mathbf{/}(\mathbf{U} + \{\mathbf{S'}[1+\mathbf{s}] - \mathbf{Sv} - \mathbf{F} - \mathbf{I}\}[1-\mathbf{t}] - \mathbf{D})$$
$$- [\mathbf{N} + \mathbf{K} + \mathbf{J}]]/\mathbf{p}$$

or

$$\mathbf{V} = \mathbf{Sv}$$

Rule-56832:

If both (**/**), (**U**), (**S'**), (**s**), (**S**), (**v**), (**F**), (**I**), (**t**), (**D**), (**K**), (**J**), (**V**) and (**N**) are known, then its Procured Inventory Days planned is:

$$\mathbf{p} = 360[[1 + \mathbf{/}(\mathbf{U} + \{\mathbf{S'}[1+\mathbf{s}] - \mathbf{Sv} - \mathbf{F} - \mathbf{I}\}[1-\mathbf{t}] - \mathbf{D})$$
$$- [\mathbf{N} + \mathbf{K} + \mathbf{J}]]/\mathbf{V}$$

Steve Asikin ISBN 13: **978-1564912361**, ISBN 10: **1564912363**

Rule-56833:

If both $(I)$, $(U)$, $(S')$, $(s)$, $(S)$, $(v)$, $(F)$, $(I)$, $(t)$, $(D)$, $(K)$, $(J)$ and $(p)$ are known, then its Non Current Asset planned is:

$$N= [1+I](U+\{S'[1+s]-Sv-F-I\}[1-t]-D)$$
$$-K-J-Svp/360$$

Rule-56834:

If both $(N)$, $(U)$, $(S')$, $(s)$, $(S)$, $(v)$, $(F)$, $(I)$, $(t)$, $(D)$, $(K)$, $(J)$ and $(p)$ are known, then its Leverage or Gearing Ratio planned is:

$$I= [N+K+J+Svp/360]/(U+\{S'[1+s]-Sv-F-I\}$$
$$[1-t]-D)-1$$

Rule-56835:

If both $(I)$, $(N)$, $(S')$, $(s)$, $(S)$, $(v)$, $(F)$, $(I)$, $(t)$, $(D)$, $(K)$, $(J)$ and $(p)$ are known, then its Utilized or Starting capital must be:

$$U= ([N+K+J+Svp/360]-[1+I]\{S'[1+s]-Sv-F-I\}$$
$$[1-t]-D)/[1+I]$$

Rule-56836:

If both $(I)$, $(U)$, $(N)$, $(s)$, $(S)$, $(v)$, $(F)$, $(I)$, $(t)$, $(D)$, $(K)$, $(J)$ and $(p)$ are known, then its Sales Past planned is:

$$S'= (\{[N+K+J+Svp/360]/[1+I]-U+D\}/[1-t]$$
$$+Sv+F+I)/[1+s]$$
$$\text{or}$$
$$S'= S/[1+s]$$

Steve Asikin ISBN 13: **978-1564912361**, ISBN 10: **1564912363**

Rule-56837:

If both (**/**), (**U**), (**S'**), (**N**), (**S**), (**v**), (**F**), (**I**), (**t**), (**D**), (**K**), (**J**) and (**p**) are known, then its Sales Growthnplanned is:

$$s = (\{[\mathbf{N+K+J+Svp}/360]/[1+\mathbf{/-U+D}\}/[1-\mathbf{t}]$$
$$+\mathbf{Sv+F+I})/\mathbf{S'}-1$$

or

$$s = \mathbf{S}/\mathbf{S'}-1$$

Rule-56838:

If both (**/**), (**U**), (**S'**), (**s**), (**N**), (**v**), (**F**), (**I**), (**t**), (**D**), (**K**), (**J**) and (**p**) are known, then its Sales or Revenue planned is:

$$S = [[1+\mathbf{/}(\mathbf{U}+\{\mathbf{S'}[1+\mathbf{s}]-\mathbf{F-I}\}[1-\mathbf{t}]-\mathbf{D})$$
$$-[\mathbf{N+K+J}]]/(\mathbf{v}\{\mathbf{p}/360+[1+\mathbf{/}]\})$$

or

$$S = \mathbf{S'}[1+\mathbf{s}]$$

Rule-56839:

If both (**/**), (**U**), (**S'**), (**s**), (**S**), (**N**), (**F**), (**I**), (**t**), (**D**), (**K**), (**J**) and (**p**) are known, then its Variable Portion planned is:

$$v = [[1+\mathbf{/}(\mathbf{U}+\{\mathbf{S'}[1+\mathbf{s}]-\mathbf{F-I}\}[1-\mathbf{t}]-\mathbf{D})$$
$$-[\mathbf{N+K+J}]]/(\mathbf{S}\{\mathbf{p}/360+[1+\mathbf{/}]\})$$

Steve Asikin ISBN 13: **978-1564912361**, ISBN 10: **1564912363**

Rule-56840:

If both $(\mathit{l})$, $(\mathbf{U})$, $(\mathbf{S'})$, $(\mathbf{s})$, $(\mathbf{\$})$, $(\mathbf{v})$, $(\mathbf{N})$, $(\mathbf{I})$, $(\mathbf{t})$, $(\mathbf{D})$, $(\mathbf{K})$, $(\mathbf{J})$ and $(\mathbf{p})$ are known, then its Fixed Cost planned is:

$$F= S'[1+s]-Sv-I-\{[N+K+J+Svp/360]/[1+\mathit{l}]\\-U+D\}/[1-t]$$

Rule-56841:

If both $(\mathit{l})$, $(\mathbf{U})$, $(\mathbf{S'})$, $(\mathbf{s})$, $(\mathbf{\$})$, $(\mathbf{v})$, $(\mathbf{F})$, $(\mathbf{N})$, $(\mathbf{t})$, $(\mathbf{D})$, $(\mathbf{K})$, $(\mathbf{J})$ and $(\mathbf{p})$ are known, then its Interest Portion planned is:

$$I= S'[1+s]-Sv-F-\{[N+K+J+Svp/360]/[1+\mathit{l}]\\-U+D\}/[1-t]$$

Rule-56842:

If both $(\mathit{l})$, $(\mathbf{U})$, $(\mathbf{S'})$, $(\mathbf{s})$, $(\mathbf{\$})$, $(\mathbf{v})$, $(\mathbf{F})$, $(\mathbf{I})$, $(\mathbf{N})$, $(\mathbf{D})$, $(\mathbf{K})$, $(\mathbf{J})$ and $(\mathbf{p})$ are known, then its Tax Rate planned is:

$$t= 1-\{[N+K+J+Svp/360]/[1+\mathit{l}]-U+D\}\\/\{S'[1+s]-Sv-F-I\}$$

Rule-56843:

If both $(\mathit{l})$, $(\mathbf{U})$, $(\mathbf{S'})$, $(\mathbf{s})$, $(\mathbf{\$})$, $(\mathbf{v})$, $(\mathbf{F})$, $(\mathbf{I})$, $(\mathbf{t})$, $(\mathbf{N})$, $(\mathbf{K})$, $(\mathbf{J})$ and $(\mathbf{p})$ are known, then its Dividend planned is:

$$D= U+\{S'[1+s]-Sv-F-I\}[1-t]\\-[N+K+J+Svp/360]/[1+\mathit{l}]$$

Steve Asikin ISBN 13: **978-1564912361**, ISBN 10: **1564912363**

Rule-56844:

If both $(I)$, $(U)$, $(S')$, $(s)$, $(S)$, $(v)$, $(F)$, $(I)$, $(t)$, $(D)$, $(N)$, $(J)$ and $(p)$ are known, then its Kind of Cash planned is:

$$K = [1+I(U+\{S'[1+s]-Sv-F-I\}[1-t]-D) -[N+J+Svp/360]$$

Rule-56845:

If both $(I)$, $(U)$, $(S')$, $(s)$, $(S)$, $(v)$, $(F)$, $(I)$, $(t)$, $(D)$, $(K)$, $(N)$ and $(p)$ are known, then its Job or Trade Account Receivable planned is:

$$J = [1+I(U+\{S'[1+s]-Sv-F-I\}[1-t]-D) -\{N+K+Svp/360\}$$

Rule-56846:

If both $(I)$, $(U)$, $(S')$, $(s)$, $(S)$, $(v)$, $(F)$, $(I)$, $(t)$, $(D)$, $(K)$, $(J)$ and $(N)$ are known, then its Procured Inventory Days planned is:

$$p = 360[[1+I(U+\{S'[1+s]-Sv-F-I\}[1-t]-D) -[N+K+J]]/[Sv]$$

Rule-56847:

If both $(I)$, $(U)$, $(S')$, $(s)$, $(S)$, $(v)$, $(F)$, $(I)$, $(t)$, $(D)$, $(K)$, $(J)$ and $(p)$ are known, then its Non Current Asset planned is:

$$N = [1+I(U+\{S'[1+s]-Sv-F-I\}[1-t]-D) -K-J-S'vp[1+s]/360$$

Steve Asikin ISBN 13: **978-1564912361**, ISBN 10: **1564912363**

Rule-56848:

If both **(N)**, **(U)**, **(S')**, **(s)**, **(S)**, **(v)**, **(F)**, **(I)**, **(t)**, **(D)**, **(K)**, **(J)** and **(p)** are known, then its Leverage or Gearing Ratio planned is:

$$\textit{I} = \{\mathbf{N+K+J+S'vp}[1+\mathbf{s}]/360\}$$
$$/(\mathbf{U}+\{\mathbf{S'}[1+\mathbf{s}]-\mathbf{Sv-F-I}\}[1-\mathbf{t}]-\mathbf{D})-1$$

Rule-56849:

If both **(I)**, **(N)**, **(S')**, **(s)**, **(S)**, **(v)**, **(F)**, **(I)**, **(t)**, **(D)**, **(K)**, **(J)** and **(p)** are known, then its Utilized or Starting Capital must be:

$$\mathbf{U} = (\{\mathbf{N+K+J+S'vp}[1+\mathbf{s}]/360\}$$
$$-[1+\textit{I}]\{\mathbf{S'}[1+\mathbf{s}]-\mathbf{Sv-F-I}\}[1-\mathbf{t}]-\mathbf{D})/[1+\textit{I}]$$

Rule-56850:

If both **(I)**, **(U)**, **(N)**, **(s)**, **(S)**, **(v)**, **(F)**, **(I)**, **(t)**, **(D)**, **(K)**, **(J)** and **(p)** are known, then its Sales Past must be:

$$\mathbf{S'} = [(\{\mathbf{N+K+J+S'vp}[1+\mathbf{s}]/360\}/[1+\textit{I}]-\mathbf{U+D})$$
$$/[1-\mathbf{t}]+\mathbf{Sv+F+I}\}]/[1+\mathbf{s}]$$
$$\text{or}$$
$$\mathbf{S'} = \mathbf{S}/[1+\mathbf{s}]$$

Steve Asikin ISBN 13: **978-1564912361**, ISBN 10: **1564912363**

Rule-56851:

If both (**/**), (**U**), (**S'**), (**N**), (**S**), (**v**), (**F**), (**I**), (**t**), (**D**), (**K**), (**J**) and (**p**) are known, then its Sales Growth planned is:

$$s= [(\{N+K+J+S'vp[1+s]/360\}/[1+/]-U+D)$$
$$/[1-t]+Sv+F+I\}]/S'-1$$

or

$$s= S/S'-1$$

Rule-56852:

If both (**/**), (**U**), (**S'**), (**s**), (**N**), (**v**), (**F**), (**I**), (**t**), (**D**), (**K**), (**J**) and (**p**) are known, then its Sales or Revenue planned is:

$$S= [S'[1+s]-F-I-(\{N+K+J+S'vp[1+s]/360\}/[1+/]$$
$$-U+D)/[1-t]]/v$$

or

$$S= S'[1+s]$$

Rule-56853:

If both (**/**), (**U**), (**S'**), (**s**), (**S**), (**N**), (**F**), (**I**), (**t**), (**D**), (**K**), (**J**) and (**p**) are known, then its Variable Portion planned is:

$$v= [S'[1+s]-F-I-(\{N+K+J+S'vp[1+s]/360\}/[1+/]$$
$$-U+D)/[1-t]]/S$$

Steve Asikin ISBN 13: **978-1564912361**, ISBN 10: **1564912363**

Rule-56854:

If both (**/**), (**U**), (**S'**), (**s**), (**S**), (**v**), (**N**), (**I**), (**t**), (**D**), (**K**),
(**J**) and (**p**) are known, then its Fixed Cost planned is:

$$F = S'[1+s] - Sv - I - (\{N+K+J+S'vp[1+s]/360\}/[1+\ell] - U + D)/[1-t]$$

Rule-56855:

If both (**/**), (**U**), (**S'**), (**s**), (**S**), (**v**), (**F**), (**N**), (**t**), (**D**), (**K**),
(**J**) and (**p**) are known, then its Interest Expense
planned is:

$$I = S'[1+s] - Sv - F - (\{N+K+J+S'vp[1+s]/360\}/[1+\ell] - U + D)/[1-t]$$

Rule-56856:

If both (**/**), (**U**), (**S'**), (**s**), (**S**), (**v**), (**F**), (**I**), (**N**), (**D**), (**K**),
(**J**) and (**p**) are known, then its Tax Rate planned is:

$$t = 1 - (\{N+K+J+S'vp[1+s]/360\}/[1+\ell] - U + D) / \{S'[1+s] - Sv - F - I\}$$

Rule-56857:

If both (**/**), (**U**), (**S'**), (**s**), (**S**), (**v**), (**F**), (**I**), (**t**), (**N**), (**K**),
(**J**) and (**p**) are known, then its Dividend planned is:

$$D = (U + \{S'[1+s] - Sv - F - I\}[1-t] - \{N+K+J+S'vp[1+s]/360\}/[1+\ell]$$

Steve Asikin ISBN 13: **978-1564912361**, ISBN 10: **1564912363**

Rule-56858:

If both ($I$), (**U**), (**$'**), (**s**), (**$**), (**v**), (**F**), (**I**), (**t**), (**D**), (**N**), (**J**) and (**p**) are known, then its Kind of Cash planned is:

$$K= [1+I(U+\{\$'[1+s]-\$v-F-I\}[1-t]-D)$$
$$-\{N+J+\$'vp[1+s]/360\}$$

Rule-56859:

If both ($I$), (**U**), (**$'**), (**s**), (**$**), (**v**), (**F**), (**I**), (**t**), (**D**), (**K**), (**N**) and (**p**) are known, then its Job or Trade Account Receivable planned is:

$$J= [[1+I(U+\{\$'[1+s]-\$v-F-I\}[1-t]-D)$$
$$-\{N+K+\$'vp[1+s]/360\}$$

Rule-56860:

If both ($I$), (**U**), (**$'**), (**s**), (**$**), (**v**), (**F**), (**I**), (**t**), (**D**), (**K**), (**J**) and (**N**) are known, then its Procured Inventory Days planned is:

$$p= 360[[1+I(U+\{\$'[1+s]-\$v-F-I\}[1-t]-D)$$
$$-[N+K+J]]/\{\$'v[1+s]\}$$

Rule-56861:

If both ($I$), (**U**), (**$'**), (**s**), (**$**), (**v**), (**F**), (**I**), (**t**), (**D**), (**K**), (**j**) and (**P**) are known, then its Non Current Asset planned is:

$$N= [1+I(U+\{\$'[1+s]-\$v-F-I\}[1-t]-D)-K-\$j/360-P$$

Steve Asikin ISBN 13: **978-1564912361**, ISBN 10: **1564912363**

Rule-56862:

If both (**N**), (**U**), (**S'**), (**s**), (**S**), (**v**), (**F**), (**I**), (**t**), (**D**), (**K**), (**j**) and (**P**) are known, then its Leverage or Gearing Ratio planned is:

$$I = \{N+K+Sj/360+P\}/(U+\{S'[1+s]-Sv-F-I\}$$
$$[1-t]-D)-1$$

Rule-56863:

If both (**I**), (**N**), (**S'**), (**s**), (**S**), (**v**), (**F**), (**I**), (**t**), (**D**), (**K**), (**j**) and (**P**) are known, then its Utilized or Starting Capital must be:

$$U = (\{N+K+Sj/360+P\}-[1+I]\{S'[1+s]-Sv-F-I\}$$
$$[1-t]-D)/[1+I]$$

Rule-56864:

If both (**I**), (**U**), (**N**), (**s**), (**S**), (**v**), (**F**), (**I**), (**t**), (**D**), (**K**), (**j**) and (**P**) are known, then its Sales Past must be:

$$S' = (\{[N+K+Sj/360+P]/[1+I]-U+D\}/[1-t]$$
$$+Sv+F+I)/[1+s]$$

or

$$S' = S/[1+s]$$

Steve Asikin ISBN 13: **978-1564912361**, ISBN 10: **1564912363**

<u>Rule-56865</u>:

If both (**/**), (**U**), (**$'**), (**N**), (**$**), (**v**), (**F**), (**I**), (**t**), (**D**), (**K**), (**j**) and (**P**) are known, then its Sales Growth planned is:

$$s = (\{[N+K+\$j/360+P]/[1+/-U+D\}/[1-t]$$
$$+\$v+F+I)/\ \$'-1$$

or

$$s = \$/\$'-1$$

<u>Rule-56866</u>:

If both (**/**), (**U**), (**$'**), (**s**), (**N**), (**v**), (**F**), (**I**), (**t**), (**D**), (**K**), (**j**) and (**P**) are known, then its Sales or Revenue planned is:

$$\$ = [[1+/](U+\{\$'[1+s]-F-I\}[1-t]-D)$$
$$-[N+K+P]]/\{j/360+v[1+/]\}$$

or

$$\$ = \$'[1+s]$$

<u>Rule-56867</u>:

If both (**/**), (**U**), (**$'**), (**s**), (**$**), (**N**), (**F**), (**I**), (**t**), (**D**), (**K**), (**j**) and (**P**) are known, then its Variable Portion planned is:

$$v = (\$'[1+s]-F-I-\{[N+K+\$j/360+P]/[1+/]$$
$$-U+D\}/[1-t])/\$$$

Steve Asikin ISBN 13: **978-1564912361**, ISBN 10: **1564912363**

Rule-56868:

If both $(I)$, $(U)$, $(S')$, $(s)$, $(S)$, $(v)$, $(N)$, $(I)$, $(t)$, $(D)$, $(K)$, $(j)$ and $(P)$ are known, then its Fixed Cost planned is:

$$F= S'[1+s]-Sv-I-\{[N+K+Sj/360+P]/[1+I] -U+D\}/[1-t]$$

Rule-56869:

If both $(I)$, $(U)$, $(S')$, $(s)$, $(S)$, $(v)$, $(F)$, $(N)$, $(t)$, $(D)$, $(K)$, $(j)$ and $(P)$ are known, then its Interest Expense planned is:

$$I= S'[1+s]-Sv-F-\{[N+K+Sj/360+P]/[1+I] -U+D\}/[1-t]$$

Rule-56870:

If both $(I)$, $(U)$, $(S')$, $(s)$, $(S)$, $(v)$, $(F)$, $(I)$, $(N)$, $(D)$, $(K)$, $(j)$ and $(P)$ are known, then its Tax Rate planned is:

$$t= 1-\{[N+K+Sj/360+P]/[1+I]-U+D\} /\{S'[1+s]-Sv-F-I\}$$

Rule-56871:

If both $(I)$, $(U)$, $(S')$, $(s)$, $(S)$, $(v)$, $(F)$, $(I)$, $(t)$, $(N)$, $(K)$, $(j)$ and $(P)$ are known, then its Dividend planned is:

$$D= U+\{S'[1+s]-Sv-F-I\}[1-t] - [N+K+Sj/360+P]/[1+I]$$

Steve Asikin ISBN 13: **978-1564912361**, ISBN 10: **1564912363**

Rule-56872:

If both (**I**), (**U**), (**S'**), (**s**), (**S**), (**v**), (**F**), (**I**), (**t**), (**D**), (**N**), (**j**) and (**P**) are known, then its Kind of Cash planned is:

$$K= [1+I](U+\{S'[1+s]-Sv-F-I\}[1-t]-D)$$
$$-[N+Sj/360+P]$$

Rule-56873:

If both (**I**), (**U**), (**S'**), (**s**), (**S**), (**v**), (**F**), (**I**), (**t**), (**D**), (**K**), (**N**) and (**P**) are known, then its Job or Trade Receivable Days planned is:

$$j= 360[[1+I](U+\{S'[1+s]-Sv-F-I\}[1-t]-D)$$
$$-\{N+K+P\}]/S$$

Rule-56874:

If both (**I**), (**U**), (**S'**), (**s**), (**S**), (**v**), (**F**), (**I**), (**t**), (**D**), (**K**), (**j**) and (**N**) are known, then its Procured Inventory planned is:

$$P= [1+I](U+\{S'[1+s]-Sv-F-I\}[1-t]-D)$$
$$-\{N+K+Sj/360\}$$

Rule-56875:

If both (**I**), (**U**), (**S'**), (**s**), (**S**), (**v**), (**F**), (**I**), (**t**), (**D**), (**K**), (**j**), (**V**) and (**p**) are known, then its Non Current Asset planned is:

$$N= [1+I](U+\{S'[1+s]-Sv-F-I\}[1-t]-D)-K-Sj/360$$
$$-Vp/360$$

Steve Asikin ISBN 13: **978-1564912361**, ISBN 10: **1564912363**

Rule-56876:

If both **(N)**, **(U)**, **(S')**, **(s)**, **(S)**, **(v)**, **(F)**, **(I)**, **(t)**, **(D)**, **(K)**, **(j)**, **(V)** and **(p)** are known, then its Leverage or Gearing Ratio planned is:

$$\textit{I} = [\mathbf{N+K+Sj}/360+\mathbf{Vp}/360]$$
$$/(\mathbf{U}+\{\mathbf{S'}[1+\mathbf{s}]-\mathbf{Sv-F-I}\}[1-\mathbf{t}]-\mathbf{D})-1$$


Rule-56877:

If both **(***I***)**, **(N)**, **(S')**, **(s)**, **(S)**, **(v)**, **(F)**, **(I)**, **(t)**, **(D)**, **(K)**, **(j)**, **(V)** and **(p)** are known, then its Utilized or Starting Capital must be:

$$\mathbf{U} = ([\mathbf{N+K+Sj}/360+\mathbf{Vp}/360]$$
$$-[1+\textit{I}]\{\mathbf{S'}[1+\mathbf{s}]-\mathbf{Sv-F-I}\}[1-\mathbf{t}]-\mathbf{D})/[1+\textit{I}]$$


Rule-56878:

If both **(***I***)**, **(U)**, **(N)**, **(s)**, **(S)**, **(v)**, **(F)**, **(I)**, **(t)**, **(D)**, **(K)**, **(j)**, **(V)** and **(p)** are known, then its Sales Past must be:

$$\mathbf{S'} = (\{[\mathbf{N+K+Sj}/360+\mathbf{Vp}/360]/[1+\textit{I}]-\mathbf{U+D}\}$$
$$/[1-\mathbf{t}]+\mathbf{Sv+F+I})/[1+\mathbf{s}]$$
$$\text{or}$$
$$\mathbf{S'} = \mathbf{S}[1+\mathbf{s}]$$

Steve Asikin ISBN 13: **978-1564912361**, ISBN 10: **1564912363**

Rule-56879:

If both (**/**), (**U**), (**S'**), (**N**), (**S**), (**v**), (**F**), (**I**), (**t**), (**D**), (**K**), (**j**), (**V**) and (**p**) are known, then its Sales Growth planned is:

$$s= (\{[\mathbf{N+K+Sj}/360+\mathbf{Vp}/360]/[1+\mathbf{/}\mathbf{-U+D}\}$$
$$/[1\mathbf{-t}]+\mathbf{Sv+F+I})/\mathbf{S'}-1$$

or

$$s= \mathbf{S/S'}-1$$


Rule-56880:

If both (**/**), (**U**), (**S'**), (**s**), (**N**), (**v**), (**F**), (**I**), (**t**), (**D**), (**K**), (**j**), (**V**) and (**p**) are known, then its Sales or Revenue planned is:

$$S= [[1+\mathbf{/}(\mathbf{U}+\{\mathbf{S'}[1+s]\mathbf{-F-I}\}[1\mathbf{-t}]\mathbf{-D})$$
$$-[\mathbf{N+K+Vp}/360]]/\{\mathbf{j}/360+\mathbf{v}[1+\mathbf{/}]\}$$

or

$$S= \mathbf{S'}[1+s]$$


Rule-56881:

If both (**/**), (**U**), (**S'**), (**s**), (**S**), (**N**), (**F**), (**I**), (**t**), (**D**), (**K**), (**j**), (**V**) and (**p**) are known, then its Variable Portion planned is:

$$v= (\mathbf{S'}[1+s]\mathbf{-F-I}-\{[\mathbf{N+K+Sj}/360+\mathbf{Vp}/360]/[1+\mathbf{/}$$
$$\mathbf{-U+D}\}/[1\mathbf{-t}])/\mathbf{S}$$

or

$$v= \mathbf{V/S}$$

Steve Asikin ISBN 13: **978-1564912361**, ISBN 10: **1564912363**

Rule-56882:

If both (**/**), (**U**), (**S'**), (**s**), (**S**), (**v**), (**N**), (**I**), (**t**), (**D**), (**K**), (**j**), (**V**) and (**p**) are known, then its Fixed Cost planned is:

$$F= S'[1+s]-Sv-I-\{[N+K+Sj/360+Vp/360]/[1+/]$$
$$-U+D\}/[1-t]$$

Rule-56883:

If both (**/**), (**U**), (**S'**), (**s**), (**S**), (**v**), (**F**), (**N**), (**t**), (**D**), (**K**), (**j**), (**V**) and (**p**) are known, then its Interest Expense planned is:

$$I= S'[1+s]-Sv-F-\{[N+K+Sj/360+Vp/360]/[1+/]$$
$$-U+D\}/[1-t]$$

Rule-56884:

If both (**/**), (**U**), (**S'**), (**s**), (**S**), (**v**), (**F**), (**I**), (**N**), (**D**), (**K**), (**j**), (**V**) and (**p**) are known, then its Tax Rate planned is:

$$t= 1-\{[N+K+Sj/360+Vp/360]/[1+/]-U+D\}$$
$$/\{S'[1+s]-Sv-F-I\}$$

Rule-56885:

If both (**/**), (**U**), (**S'**), (**s**), (**S**), (**v**), (**F**), (**I**), (**t**), (**N**), (**K**), (**j**), (**V**) and (**p**) are known, then its Dividend planned is:

$$D= U+\{S'[1+s]-Sv-F-I\}[1-t]$$
$$- [N+K+Sj/360+Vp/360]/[1+/]$$

Steve Asikin ISBN 13: **978-1564912361**, ISBN 10: **1564912363**

Rule-56886:

If both (**/**), (**U**), (**S'**), (**s**), (**S**), (**v**), (**F**), (**I**), (**t**), (**D**), (**N**), (**j**), (**V**) and (**p**) are known, then its Kind of Cash planned is:

$$K = [1+/(U+\{S'[1+s]-Sv-F-I\}[1-t]-D)$$
$$-[N+Sj/360+Vp/360]$$

Rule-56887:

If both (**/**), (**U**), (**S'**), (**s**), (**S**), (**v**), (**F**), (**I**), (**t**), (**D**), (**K**), (**N**), (**V**) and (**p**) are known, then its Job or Trade Receivable Days planned is:

$$j = 360[[1+/(U+\{S'[1+s]-Sv-F-I\}[1-t]-D)$$
$$-[N+K+Vp/360]]/S$$

Rule-56888:

If both (**/**), (**U**), (**S'**), (**s**), (**S**), (**v**), (**F**), (**I**), (**t**), (**D**), (**K**), (**j**), (**N**) and (**p**) are known, then its Variable Cost planned is:

$$V = 360[[1+/(U+\{S'[1+s]-Sv-F-I\}[1-t]-D)$$
$$-[N+K+Sj/360]]/p$$
$$\text{or}$$
$$V = Sv$$

Rule-56889:

If both (**/**), (**U**), (**S'**), (**s**), (**S**), (**v**), (**F**), (**I**), (**t**), (**D**), (**K**), (**j**), (**V**) and (**N**) are known, then its Procured Inventory Days planned is:

$$p = 360[[1+/(U+\{S'[1+s]-Sv-F-I\}[1-t]-D)$$
$$-[N+K+Sj/360]]/V$$

Steve Asikin ISBN 13: **978-1564912361**, ISBN 10: **1564912363**

Rule-56890:

If both $(\textit{I})$, $(\textbf{U})$, $(\textbf{S'})$, $(\textbf{s})$, $(\textbf{S})$, $(\textbf{v})$, $(\textbf{F})$, $(\textbf{I})$, $(\textbf{t})$, $(\textbf{D})$, $(\textbf{K})$, $(\textbf{j})$ and $(\textbf{p})$ are known, then its Non Current Asset planned is:

$$N= [1+\textit{I}](U+\{S'[1+s]-Sv-F-I\}[1-t]-D)$$
$$-K-Sj/360-Svp/360$$

Rule-56891:

If both $(\textbf{N})$, $(\textbf{U})$, $(\textbf{S'})$, $(\textbf{s})$, $(\textbf{S})$, $(\textbf{v})$, $(\textbf{F})$, $(\textbf{I})$, $(\textbf{t})$, $(\textbf{D})$, $(\textbf{K})$, $(\textbf{j})$ and $(\textbf{p})$ are known, then its Leverage or Gearing Ratio planned is:

$$\textit{I}= [N+K+Sj/360+Svp/360]$$
$$/(U+\{S'[1+s]-Sv-F-I\}[1-t]-D)-1$$

Rule-56892:

If both $(\textit{I})$, $(\textbf{N})$, $(\textbf{S'})$, $(\textbf{s})$, $(\textbf{S})$, $(\textbf{v})$, $(\textbf{F})$, $(\textbf{I})$, $(\textbf{t})$, $(\textbf{D})$, $(\textbf{K})$, $(\textbf{j})$ and $(\textbf{p})$ are known, then its Utilized or Starting Capital must be:

$$U= ([N+K+Sj/360+Svp/360]$$
$$-[1+\textit{I}]\{S'[1+s]-Sv-F-I\}[1-t]-D)/[1+\textit{I}]$$

Rule-56893:

If both $(\textit{I})$, $(\textbf{U})$, $(\textbf{N})$, $(\textbf{s})$, $(\textbf{S})$, $(\textbf{v})$, $(\textbf{F})$, $(\textbf{I})$, $(\textbf{t})$, $(\textbf{D})$, $(\textbf{K})$, $(\textbf{j})$ and $(\textbf{p})$ are known, then its Sales Past must be:

$$S'=[N+K+Sj/360+Svp/360]/[1+\textit{I}]-U+D\}/[1-t]$$
$$+Sv+F+I)/[1+s]$$
$$\text{or}$$
$$S'= S/[1+s]$$

Steve Asikin ISBN 13: **978-1564912361**, ISBN 10: **1564912363**

Rule-56894:

If both (**/**), (**U**), (**S'**), (**N**), (**S**), (**v**), (**F**), (**I**), (**t**), (**D**), (**K**), (**j**) and (**p**) are known, then its Sales Growth planned is:

$$s = [N + K + Sj/360 + Svp/360] / [1 + /-U + D] / [1-t]$$
$$+ Sv + F + I) / S' - 1$$

or

$$s = S/S' - 1$$

Rule-56895:

If both (**/**), (**U**), (**S'**), (**s**), (**N**), (**v**), (**F**), (**I**), (**t**), (**D**), (**K**), (**j**) and (**p**) are known, then its Sales or Revenue planned is:

$$S = [[1 + /](U + \{S'[1+s] - F - I\}[1-t] - D) [N+K]]$$
$$/ (j/360 + v\{p/360 + [1 + /]\})$$

or

$$S = S'[1+s]$$

Rule-56896:

If both (**/**), (**U**), (**S'**), (**s**), (**S**), (**N**), (**F**), (**I**), (**t**), (**D**), (**K**), (**j**) and (**p**) are known, then its Variable Portion planned is:

$$v = \{[[1 + /](U + \{S'[1+s] - F - I\}[1-t] - D) - [N+K]]/S$$
$$- j/360\}/\{p/360 + [1 + /]\})$$

Steve Asikin ISBN 13: **978-1564912361**, ISBN 10: **1564912363**

Rule-56897:

If both $(\textit{I})$, $(\textbf{U})$, $(\textbf{S'})$, $(\textbf{s})$, $(\textbf{S})$, $(\textbf{v})$, $(\textbf{N})$, $(\textbf{I})$, $(\textbf{t})$, $(\textbf{D})$, $(\textbf{K})$, $(\textbf{j})$ and $(\textbf{p})$ are known, then its Fixed Cost planned is:

$$F = S'[1+s] - Sv - I - [N+K+Sj/360+Svp/360]/[1+I] -U+D\}/[1-t]$$

Rule-56898:

If both $(\textit{I})$, $(\textbf{U})$, $(\textbf{S'})$, $(\textbf{s})$, $(\textbf{S})$, $(\textbf{v})$, $(\textbf{F})$, $(\textbf{N})$, $(\textbf{t})$, $(\textbf{D})$, $(\textbf{K})$, $(\textbf{j})$ and $(\textbf{p})$ are known, then its Interest Expense planned is:

$$I = S'[1+s] - Sv - F - [N+K+Sj/360+Svp/360]/[1+I] -U+D\}/[1-t]$$

Rule-56899:

If both $(\textit{I})$, $(\textbf{U})$, $(\textbf{S'})$, $(\textbf{s})$, $(\textbf{S})$, $(\textbf{v})$, $(\textbf{F})$, $(\textbf{I})$, $(\textbf{N})$, $(\textbf{D})$, $(\textbf{K})$, $(\textbf{j})$ and $(\textbf{p})$ are known, then its Tax Rate planned is:

$$t = 1 - [N+K+Sj/360+Svp/360]/[1+I-U+D\} /\{S'[1+s]-Sv-F-I\}$$

Rule-56900:

If both $(\textit{I})$, $(\textbf{U})$, $(\textbf{S'})$, $(\textbf{s})$, $(\textbf{S})$, $(\textbf{v})$, $(\textbf{F})$, $(\textbf{I})$, $(\textbf{t})$, $(\textbf{N})$, $(\textbf{K})$, $(\textbf{j})$ and $(\textbf{p})$ are known, then its Dividend planned is:

$$D = U + \{S'[1+s]-Sv-F-I\}[1-t] -[N+K+Sj/360+Svp/360]/[1+I]$$

Steve Asikin ISBN 13: **978-1564912361**, ISBN 10: **1564912363**

Rule-56901:

If both (**/**), (**U**), (**S'**), (**s**), (**S**), (**v**), (**F**), (**I**), (**t**), (**D**), (**N**), (**j**) and (**p**) are known, then its Kind of Cash planned is:

$$K= [1+/](U+\{S'[1+s]-Sv-F-I\}[1-t]-D)$$
$$-[N+Sj/360+Svp/360]$$

Rule-56902:

If both (**/**), (**U**), (**S'**), (**s**), (**S**), (**v**), (**F**), (**I**), (**t**), (**D**), (**K**), (**N**) and (**p**) are known, then its Job or Trade Receivable Days planned is:

$$j= 360[[1+/](U+\{S'[1+s]-Sv-F-I\}[1-t]-D)$$
$$-\{N+K+Svp/360\}]/S$$

Rule-56903:

If both (**/**), (**U**), (**S'**), (**s**), (**S**), (**v**), (**F**), (**I**), (**t**), (**D**), (**K**), (**j**) and (**N**) are known, then its Procured Inventory Days planned is:

$$p= 360[[1+/](U+\{S'[1+s]-Sv-F-I\}[1-t]-D)$$
$$-[N+K+Sj/360]]/[Sv]$$

Rule-56904:

If both (**/**), (**U**), (**S'**), (**s**), (**S**), (**v**), (**F**), (**I**), (**t**), (**D**), (**K**), (**j**) and (**p**) are known, then its Non Current Asset planned is:

$$N= [1+/](U+\{S'[1+s]-Sv-F-I\}[1-t]-D)$$
$$-K-Sj/360-S'vp[1+s]/360$$

Steve Asikin ISBN 13: **978-1564912361**, ISBN 10: **1564912363**

Rule-56905:

If both (**N**), (**U**), (**S'**), (**s**), (**S**), (**v**), (**F**), (**I**), (**t**), (**D**), (**K**), (**j**) and (**p**) are known, then its Leverage or Gearing Ratio planned is:

$$\mathit{l} = \{N+K+Sj/360+S'vp[1+s]/360\}$$
$$/(U+\{S'[1+s]-Sv-F-I\}[1-t]-D)-1$$

Rule-56906:

If both (**l**), (**N**), (**S'**), (**s**), (**S**), (**v**), (**F**), (**I**), (**t**), (**D**), (**K**), (**j**) and (**p**) are known, then its Utilized or Starting Capital must be:

$$U = (\{N+K+Sj/360+S'vp[1+s]/360\}$$
$$-[1+\mathit{l}]\{S'[1+s]-Sv-F-I\}[1-t]-D)/\{[1+\mathit{l}]$$

Rule-56907:

If both (**l**), (**U**), (**N**), (**s**), (**S**), (**v**), (**F**), (**I**), (**t**), (**D**), (**K**), (**j**) and (**p**) are known, then its Sales Past must be:

$$S' = [[1+\mathit{l}](U-\{Sv+F+I\}[1-t]-D)-[N+K+Sj/360]]$$
$$/([1+s]\{vp/360-[1+\mathit{l}]\})$$
$$\text{or}$$
$$S' = S/[1+s]$$

Steve Asikin ISBN 13: **978-1564912361**, ISBN 10: **1564912363**

Rule-56908:

If both (**/**), (**U**), (**$'**), (**N**), (**$**), (**v**), (**F**), (**I**), (**t**), (**D**), (**K**), (**j**) and (**p**) are known, then its Sales Growth planned is:

$$s= [[1+/](U-\{\$v+F+I\}[1-t]-D)-[N+K+\$j/360]]$$
$$/(\$'\{vp/360-[1+/]\})-1$$

or

$$s= \$/\$'-1$$

Rule-56909:

If both (**/**), (**U**), (**$'**), (**s**), (**N**), (**v**), (**F**), (**I**), (**t**), (**D**), (**K**), (**j**) and (**p**) are known, then its Sales or Revenue planned is:

$$\$= [[1+/](U+\{\$'[1+s]-F-I\}[1-t]-D)-\{N+K$$
$$+\$'vp[1+s]/360\}]/\{j/360+v[1+/]\}$$

or

$$\$= \$'[1+s]$$

Rule-56910:

If both (**/**), (**U**), (**$'**), (**s**), (**$**), (**N**), (**F**), (**I**), (**t**), (**D**), (**K**), (**j**) and (**p**) are known, then its Variable Portion planned is:

$$v= [[1+/](U+\{\$'[1+s]-F-I\}[1-t]-D)$$
$$-[N+K+\$j/360]]/\{\$'p[1+s]/360\}+\$[1+/]\}$$

Steve Asikin ISBN 13: **978-1564912361**, ISBN 10: **1564912363**

Rule-56911:

If both $(\textit{l})$, $(\textbf{U})$, $(\textbf{S'})$, $(\textbf{s})$, $(\textbf{S})$, $(\textbf{v})$, $(\textbf{N})$, $(\textbf{I})$, $(\textbf{t})$, $(\textbf{D})$, $(\textbf{K})$, $(\textbf{j})$ and $(\textbf{p})$ are known, then its Fixed Cost planned is:

$$F = \{S'[1+s]-Sv-I-(\{N+K+Sj/360+S'vp[1+s]/360\}$$
$$/[1+\textit{l}]-U+D)/[1-t]$$


Rule-56912:

If both $(\textit{l})$, $(\textbf{U})$, $(\textbf{S'})$, $(\textbf{s})$, $(\textbf{S})$, $(\textbf{v})$, $(\textbf{F})$, $(\textbf{N})$, $(\textbf{t})$, $(\textbf{D})$, $(\textbf{K})$, $(\textbf{j})$ and $(\textbf{p})$ are known, then its Interest Expense planned is:

$$I = \{S'[1+s]-Sv-F-(\{N+K+Sj/360+S'vp[1+s]/360\}$$
$$/[1+\textit{l}]-U+D)/[1-t]$$


Rule-56913:

If both $(\textit{l})$, $(\textbf{U})$, $(\textbf{S'})$, $(\textbf{s})$, $(\textbf{S})$, $(\textbf{v})$, $(\textbf{F})$, $(\textbf{I})$, $(\textbf{N})$, $(\textbf{D})$, $(\textbf{K})$, $(\textbf{j})$ and $(\textbf{p})$ are known, then its Tax Rate planned is:

$$t = 1-(\{N+K+Sj/360+S'vp[1+s]/360\}/[1+\textit{l}]-U+D)$$
$$/\{S'[1+s]-Sv-F-I\}$$


Rule-56914:

If both $(\textit{l})$, $(\textbf{U})$, $(\textbf{S'})$, $(\textbf{s})$, $(\textbf{S})$, $(\textbf{v})$, $(\textbf{F})$, $(\textbf{I})$, $(\textbf{t})$, $(\textbf{N})$, $(\textbf{K})$, $(\textbf{j})$ and $(\textbf{p})$ are known, then its Dividend planned is:

$$D = U+\{S'[1+s]-Sv-F-I\}[1-t]$$
$$-\{N+K+Sj/360+S'vp[1+s]/360\}/[1+\textit{l}]$$

Steve Asikin ISBN 13: **978-1564912361**, ISBN 10: **1564912363**

Rule-56915:

If both (**I**), (**U**), (**S'**), (**s**), (**S**), (**v**), (**F**), (**I**), (**t**), (**D**), (**N**), (**j**) and (**p**) are known, then its Kind of Cash planned is:

$$K= [1+I](U+\{S'[1+s]-Sv-F-I\}[1-t]-D)$$
$$-\{N+Sj/360+S'vp[1+s]/360\}$$

Rule-56916:

If both (**I**), (**U**), (**S'**), (**s**), (**S**), (**v**), (**F**), (**I**), (**t**), (**D**), (**K**), (**N**) and (**p**) are known, then its Job or Trade Receivable Days planned is:

$$j= 360[[1+I](U+\{S'[1+s]-Sv-F-I\}[1-t]-D)$$
$$-\{N+K+S'vp[1+s]/360\}]/S$$

Rule-56917:

If both (**I**), (**U**), (**S'**), (**s**), (**S**), (**v**), (**F**), (**I**), (**t**), (**D**), (**K**), (**j**) and (**N**) are known, then its Procured Inventory Days planned is:

$$p= 360[[1+I](U+\{S'[1+s]-Sv-F-I\}[1-t]-D)$$
$$-[N+K+Sj/360]]/\{S'v[1+s]\}$$

Rule-56918:

If both (**I**), (**U**), (**S'**), (**s**), (**S**), (**v**), (**F**), (**I**), (**t**), (**D**), (**K**), (**j**) and (**P**) are known, then its Non Current Asset planned is:

$$N= [1+I](U+\{S'[1+s]-Sv-F-I\}[1-t]-D)$$
$$-K-S'j[1+s]/360-P$$

Steve Asikin ISBN 13: **978-1564912361**, ISBN 10: **1564912363**

Rule-56919:

If both **(N)**, **(U)**, **(S')**, **(s)**, **(S)**, **(v)**, **(F)**, **(I)**, **(t)**, **(D)**, **(K)**, **(j)** and **(P)** are known, then its Leverage or Gearing Ratio planned is:

$$\text{I} = \{\textbf{N}+\textbf{K}+\textbf{S'j}[1+\textbf{s}]/360+\textbf{P}\}/(\textbf{U}+\{\textbf{S'}[1+\textbf{s}]-\textbf{Sv}-\textbf{F}-\textbf{I}\}[1-\textbf{t}]-\textbf{D})-1$$

Rule-56920:

If both **(I)**, **(N)**, **(S')**, **(s)**, **(S)**, **(v)**, **(F)**, **(I)**, **(t)**, **(D)**, **(K)**, **(j)** and **(P)** are known, then its Utilized or Starting Capital must be:

$$\textbf{U} = (\{\textbf{N}+\textbf{K}+\textbf{S'j}[1+\textbf{s}]/360+\textbf{P}\}$$
$$-[1+\textbf{I}]\{\textbf{S'}[1+\textbf{s}]-\textbf{Sv}-\textbf{F}-\textbf{I}\}[1-\textbf{t}]-\textbf{D})/[1+\textbf{I}]$$

Rule-56921:

If both **(I)**, **(U)**, **(N)**, **(s)**, **(S)**, **(v)**, **(F)**, **(I)**, **(t)**, **(D)**, **(K)**, **(j)** and **(P)** are known, then its Sales Past must be:

$$\textbf{S'} = [[1+\textbf{I}](\textbf{U}-\{\textbf{Sv}+\textbf{F}+\textbf{I}\}[1-\textbf{t}]-\textbf{D})-[\textbf{N}+\textbf{K}+\textbf{P}]]$$
$$/([1+\textbf{s}]\{\textbf{j}/360-[1+\textbf{I}]\})$$

or

$$\textbf{S'} = \textbf{S}/[1+\textbf{s}]$$

Steve Asikin ISBN 13: **978-1564912361**, ISBN 10: **1564912363**

Rule-56922:

If both (**/**), (**U**), (**S'**), (**N**), (**S**), (**v**), (**F**), (**I**), (**t**), (**D**), (**K**), (**j**) and (**P**) are known, then its Sales Growth planned is:

$$s = [[1+/](U-\{Sv+F+I\}[1-t]-D)-[N+K+P]]$$
$$/(S'\{j/360-[1+/]\})-1$$

or

$$s = S/S'-1$$

Rule-56923:

If both (**/**), (**U**), (**S'**), (**s**), (**N**), (**v**), (**F**), (**I**), (**t**), (**D**), (**K**), (**j**) and (**P**) are known, then its Sales or Revenue planned is:

$$S = [S'[1+s]-F-I-(\{N+K+S'j[1+s]/360+P\}/[1+/]$$
$$-U+D)/[1-t]]/v$$

or

$$S = S'[1+s]$$

Rule-56924:

If both (**/**), (**U**), (**S'**), (**s**), (**S**), (**N**), (**F**), (**I**), (**t**), (**D**), (**K**), (**j**) and (**P**) are known, then its Variable Portion planned is:

$$v = [S'[1+s]-F-I-(\{N+K+S'j[1+s]/360+P\}/[1+/]$$
$$-U+D)/[1-t]]/S$$

Steve Asikin ISBN 13: **978-1564912361**, ISBN 10: **1564912363**

Rule-56925:

If both (**/**), (**U**), (**S'**), (**s**), (**S**), (**v**), (**N**), (**I**), (**t**), (**D**), (**K**), (**j**) and (**P**) are known, then its Fixed Cost planned is:

$$F= S'[1+s]-Sv-I-(\{N+K+S'j[1+s]/360+P\}/[1+/] -U+D)/[1-t]$$

Rule-56926:

If both (**/**), (**U**), (**S'**), (**s**), (**S**), (**v**), (**F**), (**N**), (**t**), (**D**), (**K**), (**j**) and (**P**) are known, then its Interest Expense planned is:

$$I= S'[1+s]-Sv-F-(\{N+K+S'j[1+s]/360+P\}/[1+/] -U+D)/[1-t]$$

Rule-56927:

If both (**/**), (**U**), (**S'**), (**s**), (**S**), (**v**), (**F**), (**I**), (**N**), (**D**), (**K**), (**j**) and (**P**) are known, then its Tax Rate planned is:

$$t= 1-(\{N+K+S'j[1+s]/360+P\}/[1+/]-U+D) /\{S'[1+s]-Sv-F-I\}$$

Rule-56928:

If both (**/**), (**U**), (**S'**), (**s**), (**S**), (**v**), (**F**), (**I**), (**t**), (**N**), (**K**), (**j**) and (**P**) are known, then its Dividend planned is:

$$D= U+\{S'[1+s]-Sv-F-I\}[1-t] - \{N+K+S'j[1+s]/360+P\}/[1+/]$$

Steve Asikin ISBN 13: **978-1564912361**, ISBN 10: **1564912363**

## Rule-56929:

If both (**/**), (**U**), (**S'**), (**s**), (**S**), (**v**), (**F**), (**I**), (**t**), (**D**), (**N**), (**j**) and (**P**) are known, then its Kind of Cash planned is:

$$K = [1+/(U+\{S'[1+s]-Sv-F-I\}[1-t]-D) -\{N+S'j[1+s]/360+P\}$$

## Rule-56930:

If both (**/**), (**U**), (**S'**), (**s**), (**S**), (**v**), (**F**), (**I**), (**t**), (**D**), (**K**), (**N**) and (**P**) are known, then its Job or Trade Receivable Days planned is:

$$j = 360[[1+/(U+\{S'[1+s]-Sv-F-I\}[1-t]-D) -\{N+K+P\}]/\{S'[1+s]\}$$

## Rule-56931:

If both (**/**), (**U**), (**S'**), (**s**), (**S**), (**v**), (**F**), (**I**), (**t**), (**D**), (**K**), (**j**) and (**N**) are known, then its Procured Inventory planned is:

$$P = [1+/(U+\{S'[1+s]-Sv-F-I\}[1-t]-D)- \{N+K+S'j[1+s]/360\}$$

## Rule-56932:

If both (**/**), (**U**), (**S'**), (**s**), (**S**), (**v**), (**F**), (**I**), (**t**), (**D**), (**K**), (**j**), (**V**) and (**p**) are known, then its Non Current Asset planned is:

$$N = [1+/(U+\{S'[1+s]-Sv/360-F-I\}[1-t]-D) -K-S'j[1+s]/360-Vp/360$$

Steve Asikin ISBN 13: **978-1564912361**, ISBN 10: **1564912363**

Rule-56933:

If both $(\mathbf{N})$, $(\mathbf{U})$, $(\mathbf{S'})$, $(\mathbf{s})$, $(\mathbf{S})$, $(\mathbf{v})$, $(\mathbf{F})$, $(\mathbf{I})$, $(\mathbf{t})$, $(\mathbf{D})$, $(\mathbf{K})$, $(\mathbf{j})$, $(\mathbf{V})$ and $(\mathbf{p})$ are known, then its Leverage or Gearing Ratio planned is:

$$\mathbf{I} = \{\mathbf{N}+\mathbf{K}+\mathbf{S'j}[1+\mathbf{s}]/360+\mathbf{Vp}/360\}$$
$$/(\mathbf{U}+\{\mathbf{S'}[1+\mathbf{s}]-\mathbf{Sv}-\mathbf{F}-\mathbf{I}\}[1-\mathbf{t}]-\mathbf{D})-1$$

Rule-56934:

If both $(\mathbf{I})$, $(\mathbf{N})$, $(\mathbf{S'})$, $(\mathbf{s})$, $(\mathbf{S})$, $(\mathbf{v})$, $(\mathbf{F})$, $(\mathbf{I})$, $(\mathbf{t})$, $(\mathbf{D})$, $(\mathbf{K})$, $(\mathbf{j})$, $(\mathbf{V})$ and $(\mathbf{p})$ are known, then its Utilized or Starting Capital must be:

$$\mathbf{U} = (\{\mathbf{N}+\mathbf{K}+\mathbf{S'j}[1+\mathbf{s}]/360+\mathbf{Vp}/360\}$$
$$-[1+\mathbf{I}]\{\mathbf{S'}[1+\mathbf{s}]-\mathbf{Sv}-\mathbf{F}-\mathbf{I}\}[1-\mathbf{t}]-\mathbf{D})/[1+\mathbf{I}]$$

Rule-56935:

If both $(\mathbf{I})$, $(\mathbf{U})$, $(\mathbf{N})$, $(\mathbf{s})$, $(\mathbf{S})$, $(\mathbf{v})$, $(\mathbf{F})$, $(\mathbf{I})$, $(\mathbf{t})$, $(\mathbf{D})$, $(\mathbf{K})$, $(\mathbf{j})$, $(\mathbf{V})$ and $(\mathbf{p})$ are known, then its Sales Past must be:

$$\mathbf{S'} = [[1+\mathbf{I}](\mathbf{U}-\{\mathbf{Sv}+\mathbf{F}+\mathbf{I}\}[1-\mathbf{t}]-\mathbf{D})$$
$$-[\mathbf{N}+\mathbf{K}+\mathbf{Vp}/360]]/([1+\mathbf{s}]\{\mathbf{j}/360-[1+\mathbf{I}]\})$$
$$\text{or}$$
$$\mathbf{S'} = \mathbf{S}/[1+\mathbf{s}]$$

Steve Asikin ISBN 13: **978-1564912361**, ISBN 10: **1564912363**

Rule-56936:

If both (**/**), (**U**), (**$'**), (**N**), (**$**), (**ʋ**), (**F**), (**I**), (**t**), (**D**), (**K**), (**j**), (**V**) and (**p**) are known, then its Sales Growth planned is:

$$s= [[1+/(U-\{\$\nu+F+I\}[1-t]-D)-[N+K+Vp/360]]$$
$$/(\$'\{j/360-[1+/]\})-1$$

or

$$s= \$/\$'-1$$

Rule-56937:

If both (**/**), (**U**), (**$'**), (**$**), (**N**), (**ʋ**), (**F**), (**I**), (**t**), (**D**), (**K**), (**j**), (**V**) and (**p**) are known, then its Sales or Revenue planned is:

$$\$= [\{\$'[1+s]-F-I-(\{N+K+\$'j[1+s]/360$$
$$+Vp/360\}/[1+/-U+D)/[1-t]]/\nu$$

or

$$\$= \$'[1+s]$$

or

$$\$= V/\nu$$

Rule-56938:

If both (**/**), (**U**), (**$'**), (**$**), (**$**), (**N**), (**F**), (**I**), (**t**), (**D**), (**K**), (**j**), (**V**) and (**p**) are known, then its Variable Portion planned is:

$$\nu= [\{\$'[1+s]-F-I-(\{N+K+\$'j[1+s]/360+Vp/360\}$$
$$/[1+/-U+D)/[1-t]]/\$$$

or

$$\nu= V/\$$$

Steve Asikin ISBN 13: **978-1564912361**, ISBN 10: **1564912363**

Rule-56939:

If both (**/**), (**U**), (**S'**), (**s**), (**S**), (**v**), (**N**), (**I**), (**t**), (**D**), (**K**), (**j**), (**V**) and (**p**) are known, then its Fixed Cost planned is:

$$F= \{S'[1+s]-Sv-I-(\{N+K+S'j[1+s]/360+Vp/360\} /[1+/]-U+D)/[1-t]$$

Rule-56940:

If both (**/**), (**U**), (**S'**), (**s**), (**S**), (**v**), (**F**), (**N**), (**t**), (**D**), (**K**), (**j**), (**V**) and (**p**) are known, then its Interest Expense planned is:

$$I= \{S'[1+s]-Sv-F-(\{N+K+S'j[1+s]/360+Vp/360\} /[1+/]-U+D)/[1-t]$$

Rule-56941:

If both (**/**), (**U**), (**S'**), (**s**), (**S**), (**v**), (**F**), (**I**), (**N**), (**D**), (**K**), (**j**), (**V**) and (**p**) are known, then its Tax Rate planned is:

$$t= 1-(\{N+K+S'j[1+s]/360+Vp/360\}/[1+/] -U+D)/\{S'[1+s]-Sv-F-I\}$$

Rule-56942:

If both (**/**), (**U**), (**S'**), (**s**), (**S**), (**v**), (**F**), (**I**), (**t**), (**N**), (**K**), (**j**), (**V**) and (**p**) are known, then its Dividend planned is:

$$D= U+\{S'[1+s]-Sv-F-I\}[1-t] -\{N+K+S'j[1+s]/360+Vp/360\}/[1+/]$$

Steve Asikin ISBN 13: **978-1564912361**, ISBN 10: **1564912363**

Rule-56943:

If both (**/**), (**U**), (**S'**), (**s**), (**S**), (**v**), (**F**), (**I**), (**t**), (**D**), (**N**), (**j**), (**V**) and (**p**) are known, then its Kind of Cash planned is:

$$K= [1+/(U+\{S'[1+s]-Sv-F-I\}[1-t]-D)$$
$$-\{N+S'j[1+s]/360+Vp/360\}$$

Rule-56944:

If both (**/**), (**U**), (**S'**), (**s**), (**S**), (**v**), (**F**), (**I**), (**t**), (**D**), (**K**), (**N**), (**V**) and (**p**) are known, then its Job or Trade Receivable Days planned is:

$$j= 360[[1+/(U+\{S'[1+s]-Sv-F-I\}[1-t]-D)$$
$$-\{N+K+Vp/360\}]/\{S'[1+s]\}$$

Rule-56945:

If both (**/**), (**U**), (**S'**), (**s**), (**S**), (**v**), (**F**), (**I**), (**t**), (**D**), (**K**), (**j**), (**N**) and (**p**) are known, then its Variable Cost planned is:

$$V= 360[[1+/(U+\{S'[1+s]-Sv-F-I\}[1-t]-D)$$
$$-\{N+K+S'j[1+s]/360\}]/p$$

or

$$V= Sv$$

Steve Asikin ISBN 13: **978-1564912361**, ISBN 10: **1564912363**

Rule-56946:

If both $(I)$, $(U)$, $(S')$, $(s)$, $(S)$, $(v)$, $(F)$, $(I)$, $(t)$, $(D)$, $(K)$, $(j)$, $(V)$ and $(N)$ are known, then its Procured Inventory Days planned is:

$$p = 360[[1+I(U+\{S'[1+s]-Sv-F-I\}[1-t]-D) -\{N+K+S'j[1+s]/360\}]/V$$

Rule-56947:

If both $(I)$, $(U)$, $(S')$, $(s)$, $(S)$, $(v)$, $(F)$, $(I)$, $(t)$, $(D)$, $(K)$, $(j)$ and $(p)$ are known, then its Non Current Asset planned is:

$$N = [1+I(U+\{S'[1+s]-Sv-F-I\}[1-t]-D) -K-S'j[1+s]/360-Svp/360$$

Rule-56948:

If both $(N)$, $(U)$, $(S')$, $(s)$, $(S)$, $(v)$, $(F)$, $(I)$, $(t)$, $(D)$, $(K)$, $(j)$ and $(p)$ are known, then its Leverage or Gearing Ratio planned is:

$$I = \{N+K+S'j[1+s]/360+Svp/360\} /(U+\{S'[1+s]-Sv-F-I\}[1-t]-D)-1$$

Rule-56949:

If both $(I)$, $(N)$, $(S')$, $(s)$, $(S)$, $(v)$, $(F)$, $(I)$, $(t)$, $(D)$, $(K)$, $(j)$ and $(p)$ are known, then its Utilized or Starting Capital must be:

$$U = (\{N+K+S'j[1+s]/360+Svp/360\}-[1+I] \{S'[1+s]-Sv-F-I\}[1-t]-D)/\{[1+I]\}$$

Steve Asikin ISBN 13: **978-1564912361**, ISBN 10: **1564912363**

Rule-56950:

If both (**/**), (**U**), (**N**), (**s**), (**S**), (**v**), (**F**), (**I**), (**t**), (**D**), (**K**),
(**j**) and (**p**) are known, then its Sales Past must be:

$$S'= [[1+/](U-\{Sv+F+I\}[1-t]-D)-[N+K+Svp/360]]$$
$$/([1+s]\{j/360-[1+/]\})$$

or

$$S'= S/[1+s]$$


Rule-56951:

If both (**/**), (**U**), (**S'**), (**N**), (**S**), (**v**), (**F**), (**I**), (**t**), (**D**), (**K**),
(**j**) and (**p**) are known, then its Sales Growth planned
is:

$$s= [[1+/](U-\{Sv+F+I\}[1-t]-D)-[N+K+Svp/360]]$$
$$/(S'\{j/360-[1+/]\})-1$$

or

$$s= S/S'-1$$


Rule-56952:

If both (**/**), (**U**), (**S'**), (**s**), (**N**), (**v**), (**F**), (**I**), (**t**), (**D**), (**K**),
(**j**) and (**p**) are known, then its Sales or Revenue
planned is:

$$S= [[1+/](U+\{S'[1+s]-F-I\}[1-t]-D)$$
$$-\{N+K+S'j[1+s]/360\}]/(v\{p/360+[1+/]\})$$

or

$$S= S'[1+s]$$

Steve Asikin ISBN 13: **978-1564912361**, ISBN 10: **1564912363**

Rule-56953:

> If both (**/**), (**U**), (**S'**), (**s**), (**S**), (**N**), (**F**), (**I**), (**t**), (**D**), (**K**), (**j**) and (**p**) are known, then its Variable Portion planned is:
>
> $$v= [[1+/](U+\{S'[1+s]-F-I\}[1-t]-D)$$
> $$-\{N+K+S'j[1+s]/360\}]/(S\{p/360+[1+/]\})$$

Rule-56954:

> If both (**/**), (**U**), (**S'**), (**s**), (**S**), (**v**), (**N**), (**I**), (**t**), (**D**), (**K**), (**j**) and (**p**) are known, then its Fixed Cost planned is:
>
> $$F= \{S'[1+s]-Sv-I-(\{N+K+S'j[1+s]/360+Svp/360\}$$
> $$/[1+/]-U+D)/[1-t]$$

Rule-56955:

> If both (**/**), (**U**), (**S'**), (**s**), (**S**), (**v**), (**F**), (**N**), (**t**), (**D**), (**K**), (**j**) and (**p**) are known, then its Interest Expense planned is:
>
> $$I= \{S'[1+s]-Sv-F-(\{N+K+S'j[1+s]/360+Svp/360\}$$
> $$/[1+/]-U+D)/[1-t]$$

Rule-56956:

> If both (**/**), (**U**), (**S'**), (**s**), (**S**), (**v**), (**F**), (**I**), (**N**), (**D**), (**K**), (**j**) and (**p**) are known, then its Tax Rate planned is:
>
> $$t= 1-(\{N+K+S'j[1+s]/360+Svp/360\}/[1+/]-U+D)$$
> $$/\{S'[1+s]-Sv-F-I\}$$

Steve Asikin ISBN 13: **978-1564912361**, ISBN 10: **1564912363**

Rule-56957:

If both (**/**), (**U**), (**S'**), (**s**), (**S**), (**v**), (**F**), (**I**), (**t**), (**N**), (**K**), (**j**) and (**p**) are known, then its Dividend planned is:

$$D= U+\{S'[1+s]-Sv-F-I\}[1-t]$$
$$-\{N+K+S'j[1+s]/360+Svp/360\}/[1+I]$$

Rule-56958:

If both (**/**), (**U**), (**S'**), (**s**), (**S**), (**v**), (**F**), (**I**), (**t**), (**D**), (**N**), (**j**) and (**p**) are known, then its Kind of Cash planned is:

$$K- [1+I(U+\{S'[1+s]-Sv-F-I\}[1-t]-D)$$
$$-\{N+S'j[1+s]/360+Svp/360\}$$

Rule-56959:

If both (**/**), (**U**), (**S'**), (**s**), (**S**), (**v**), (**F**), (**I**), (**t**), (**D**), (**K**), (**N**) and (**p**) are known, then its Job or Trade Receivable Days planned is:

$$j= 360[[1+I(U+\{S'[1+s]-Sv-F-I\}[1-t]-D)$$
$$-\{N+K+Svp/360\}]/\{S'[1+s]\}$$

Rule-56960:

If both (**/**), (**U**), (**S'**), (**s**), (**S**), (**v**), (**F**), (**I**), (**t**), (**D**), (**K**), (**j**) and (**N**) are known, then its Procured Inventory Days planned is:

$$p= 360[[1+I(U+\{S'[1+s]-Sv-F-I\}[1-t]-D)$$
$$-\{N+K+S'j[1+s]/360\}]/[Sv]$$

Steve Asikin ISBN 13: **978-1564912361**, ISBN 10: **1564912363**

Rule-56961:

If both $(I)$, $(U)$, $(S')$, $(s)$, $(S)$, $(v)$, $(F)$, $(I)$, $(t)$, $(D)$, $(K)$, $(j)$ and $(p)$ are known, then its Non Current Asset planned is:

$$N = [1 + I(U + \{S'[1+s] - Sv - F - I\}[1-t] - D) - K - S'j[1+s]/360 - S'vp[1+s]/360$$

Rule-56962:

If both $(N)$, $(U)$, $(S')$, $(s)$, $(S)$, $(v)$, $(F)$, $(I)$, $(t)$, $(D)$, $(K)$, $(j)$ and $(p)$ are known, then its Leverage or Gearing Ratio planned is:

$$I = \{N + K + S'j[1+s]/360 + S'vp[1+s]/360\} / (U + \{S'[1+s] - Sv - F - I\}[1-t] - D) - 1$$

Rule-56963:

If both $(I)$, $(N)$, $(S')$, $(s)$, $(S)$, $(v)$, $(F)$, $(I)$, $(t)$, $(D)$, $(K)$, $(j)$ and $(p)$ are known, then its Utilized or Starting Capital must be:

$$U = (\{N + K + S'j[1+s]/360 + S'vp[1+s]/360\} - [1 + I]\{S'[1+s] - Sv - F - I\}[1-t] - D)/[1+I]$$

Rule-56964:

If both $(I)$, $(U)$, $(N)$, $(s)$, $(S)$, $(v)$, $(F)$, $(I)$, $(t)$, $(D)$, $(K)$, $(j)$ and $(p)$ are known, then its Sales Past must be:

$$S' = [[1 + I(U - \{Sv + F + I\}[1-t] - D) - [N + K]] / ([1+s]\{j/360 + vp/360 - [1+I]\})$$

or

$$S' = S/[1+s]$$

Steve Asikin ISBN 13: **978-1564912361**, ISBN 10: **1564912363**

Rule-56965:

If both (**/**), (**U**), (**S'**), (**N**), (**S**), (**v**), (**F**), (**I**), (**t**), (**D**), (**K**), (**j**) and (**p**) are known, then its Sales Growth planned is:

$$s= [[1+\textbf{/}](\textbf{U}-\{\textbf{Sv}+\textbf{F}+\textbf{I}\}[1-\textbf{t}]-\textbf{D})-[\textbf{N}+\textbf{K}]] /(\textbf{S'}\{\textbf{j}/360+\textbf{vp}/360-[1+\textbf{/}]\})-1$$

or

$$s= \textbf{S}/\textbf{S'}-1$$

Rule-56966:

If both (**/**), (**U**), (**S'**), (**s**), (**N**), (**v**), (**F**), (**I**), (**t**), (**D**), (**K**), (**j**) and (**p**) are known, then its Sales or Revenue planned is:

$$S- [\textbf{S'}[1+\textbf{s}] \textbf{F}-\textbf{I}-(\{\textbf{N}+\textbf{S'j}[1+\textbf{s}]/360 +\textbf{S'vp}[1+\textbf{s}]/360\}/[1+\textbf{/}]-\textbf{U}+\textbf{D})/[1-\textbf{t}]]/\textbf{v}$$

or

$$S= \textbf{S'}[1+\textbf{s}]$$

Rule-56967:

If both (**/**), (**U**), (**S'**), (**s**), (**S**), (**N**), (**F**), (**I**), (**t**), (**D**), (**K**), (**j**) and (**p**) are known, then its Variable Portion planned is:

$$v= [[1+\textbf{/}](\textbf{U}+\{\textbf{S'}[1+\textbf{s}]-\textbf{F}-\textbf{I}\}[1-\textbf{t}]-\textbf{D}) -\{\textbf{N}+\textbf{K}+\textbf{S'j}[1+\textbf{s}]/360\}] /\{\textbf{S'p}[1+\textbf{s}]/360+\textbf{S}[1+\textbf{/}]\}$$

Steve Asikin ISBN 13: **978-1564912361**, ISBN 10: **1564912363**

Rule-56968:

If both (**/**), (**U**), (**S'**), (**s**), (**S**), (**v**), (**N**), (**I**), (**t**), (**D**), (**K**), (**j**) and (**p**) are known, then its Fixed Cost planned is:

$$F= S'[1+s]-Sv-I-(\{N+S'j[1+s]/360$$
$$+S'vp[1+s]/360\}/[1+\text{/}-U+D)/[1-t]$$

Rule-56969:

If both (**/**), (**U**), (**S'**), (**s**), (**S**), (**v**), (**F**), (**N**), (**t**), (**D**), (**K**), (**j**) and (**p**) are known, then its Interest Expense planned is:

$$I= S'[1+s]-Sv-F-(\{N+S'j[1+s]/360$$
$$+S'vp[1+s]/360\}/[1+\text{/}-U+D)/[1-t]$$

Rule-56970:

If both (**/**), (**U**), (**S'**), (**s**), (**S**), (**v**), (**F**), (**I**), (**N**), (**D**), (**K**), (**j**) and (**p**) are known, then its Tax Rate planned is:

$$t= 1-(\{N+S'j[1+s]/360+S'vp[1+s]/360\}/[1+\text{/}$$
$$-U+D)/\{S'[1+s]-Sv-F-I\}$$

Rule-56971:

If both (**/**), (**U**), (**S'**), (**s**), (**S**), (**v**), (**F**), (**I**), (**t**), (**N**), (**K**), (**j**) and (**p**) are known, then its Dividend planned is:

$$D= U+\{S'[1+s]-Sv-F-I\}[1-t]-\{N+S'j[1+s]/360$$
$$+S'vp[1+s]/360\}/[1+\text{/}$$

Steve Asikin ISBN 13: **978-1564912361**, ISBN 10: **1564912363**

Rule-56972:

If both (**∕**), (**U**), (**$'**), (**s**), (**$**), (**ʋ**), (**F**), (**I**), (**t**), (**D**), (**N**), (**j**) and (**p**) are known, then its Kind of Cash planned is:

$$K= [1+\textit{∕}(U+\{\$'[1+s]-\$ʋ-F-I\}[1-t]-D)$$
$$-\{N+\$'j[1+s]/360+\$'ʋp[1+s]/360\}$$

Rule-56973:

If both (**∕**), (**U**), (**$'**), (**s**), (**$**), (**ʋ**), (**F**), (**I**), (**t**), (**D**), (**K**), (**N**) and (**p**) are known, then its Job or Trade Receivable Days planned is:

$$j= 360[[1+\textit{∕}(U+\{\$'[1+s]-\$ʋ-F-I\}[1-t]-D)$$
$$-\{N+K+\$'ʋp[1+s]/360\}]/\{\$'[1+s]\}$$

Rule-56974:

If both (**∕**), (**U**), (**$'**), (**s**), (**$**), (**ʋ**), (**F**), (**I**), (**t**), (**D**), (**K**), (**j**) and (**N**) are known, then its Procured Inventory Days planned is:

$$p= 360[[1+\textit{∕}(U+\{\$'[1+s]-\$ʋ-F-I\}[1-t]-D)$$
$$-\{N+K+\$'j[1+s]/360\}]/\{\$'ʋ[1+s]\}$$

Rule-56975:

If both (**∕**), (**U**), (**$'**), (**s**), (**$**), (**ʋ**), (**F**), (**I**), (**t**), (**d**), (**K**), (**J**) and (**P**) are known, then its Non Current Asset planned is:

$$N= [1+\textit{∕}(U+\{\$'[1+s]-\$ʋ-F-I\}[1-t][1-d])-K-J-P$$

Steve Asikin ISBN 13: **978-1564912361**, ISBN 10: **1564912363**

Rule-56976:

If both **(N)**, **(U)**, **(S')**, **(s)**, **($)**, **(v)**, **(F)**, **(I)**, **(t)**, **(d)**, **(K)**, **(J)** and **(P)** are known, then its Leverage or Gearing Ratio planned is:

$$I = \{N+K+J+P\}/(U+\{S'[1+s]-Sv-F-I\}[1-t][1-d])-1$$

Rule-56977:

If both **(I)**, **(N)**, **(S')**, **(s)**, **($)**, **(v)**, **(F)**, **(I)**, **(t)**, **(d)**, **(K)**, **(J)** and **(P)** are known, then its Utilized or Starting Capital must be:

$$U = (\{N+K+J+P\}-[1+I]\{S'[1+s]-Sv-I\}[1-t][1-d])$$
$$/[1+I]$$

Rule-56978:

If both **(I)**, **(U)**, **(N)**, **(s)**, **($)**, **(v)**, **(F)**, **(I)**, **(t)**, **(d)**, **(K)**, **(J)** and **(P)** are known, then its Sales Past must be:

$$S' = (\{[N+K+J+P]/[1+I]-U\}/\{[1-t][1-d]\}$$
$$+Sv+F+I)/[1+s]$$
$$\text{or}$$
$$S' = S/[1+s]$$

Rule-56979:

If both **(I)**, **(U)**, **(S')**, **(s)**, **($)**, **(v)**, **(F)**, **(I)**, **(t)**, **(d)**, **(K)**, **(J)** and **(P)** are known, then its Non Current Asset planned is:

$$s = (\{[N+K+J+P]/[1+I]-U\}/\{[1-t][1-d]\}$$
$$+Sv+F+I)/S'-1$$
$$\text{or}$$
$$s = S/S'-1$$

Steve Asikin ISBN 13: **978-1564912361**, ISBN 10: **1564912363**

Rule-56980:

If both (**/**), (**U**), (**S'**), (**s**), (**N**), (**v**), (**F**), (**I**), (**t**), (**d**), (**K**), (**J**) and (**P**) are known, then its Sales or Revenue planned is:

$$S = (S'[1+s]-F-I-\{[N+K+J+P]/[1+/]-U\}$$
$$/\{[1-t][1-d]\})/v$$

or

$$S = S'[1+s]$$

Rule-56981:

If both (**/**), (**U**), (**S'**), (**s**), (**S**), (**N**), (**F**), (**I**), (**t**), (**d**), (**K**), (**J**) and (**P**) are known, then its Variable Portion planned is:

$$v = (S'[1+s]-F-I-\{[N+K+J+P]/[1+/]-U\}$$
$$/\{[1-t][1-d]\})/S$$

Rule-56982:

If both (**/**), (**U**), (**S'**), (**s**), (**S**), (**v**), (**N**), (**I**), (**t**), (**d**), (**K**), (**J**) and (**P**) are known, then its Fixed Cost planned is:

$$F = S'[1+s]-Sv-I-\{[N+K+J+P]/[1+/]-U\}$$
$$/\{[1-t][1-d]\}$$

Rule-56983:

If both (**/**), (**U**), (**S'**), (**s**), (**S**), (**v**), (**F**), (**N**), (**t**), (**d**), (**K**), (**J**) and (**P**) are known, then its Interest Espense planned is:

$$I = S'[1+s]-Sv-F-\{[N+K+J+P]/[1+/]-U\}$$
$$/\{[1-t][1-d]\}$$

Steve Asikin ISBN 13: **978-1564912361**, ISBN 10: **1564912363**

Rule-56984:

If both $(\textbf{I}), (\textbf{U}), (\textbf{S'}), (\textbf{s}), (\textbf{S}), (\textbf{v}), (\textbf{F}), (\textbf{I}), (\textbf{N}), (\textbf{d}), (\textbf{K}),$ $(\textbf{J})$ and $(\textbf{P})$ are known, then its Tax Rate planned is:

$$\textbf{t}= 1-\{[\textbf{N}+\textbf{K}+\textbf{J}+\textbf{P}]/[1+\textbf{I}]\}/(\textbf{U}+\{\textbf{S'}[1+\textbf{s}]-\textbf{Sv}-\textbf{F}-\textbf{I}\}[1-\textbf{d}])$$

Rule-56985:

If both $(\textbf{I}), (\textbf{U}), (\textbf{S'}), (\textbf{s}), (\textbf{S}), (\textbf{v}), (\textbf{F}), (\textbf{I}), (\textbf{t}), (\textbf{N}), (\textbf{K}),$ $(\textbf{J})$ and $(\textbf{P})$ are known, then its Dividend Payout planned is:

$$\textbf{d}= 1-\{[\textbf{N}+\textbf{K}+\textbf{J}+\textbf{P}]/[1+\textbf{I}]\}/(\textbf{U}+\{\textbf{S'}[1+\textbf{s}]-\textbf{Sv}-\textbf{F}-\textbf{I}\}[1-\textbf{t}])$$

Rule-56986:

If both $(\textbf{I}), (\textbf{U}), (\textbf{S'}), (\textbf{s}), (\textbf{S}), (\textbf{v}), (\textbf{F}), (\textbf{I}), (\textbf{t}), (\textbf{d}), (\textbf{N}),$ $(\textbf{J})$ and $(\textbf{P})$ are known, then its Kind of Cash planned is:

$$\textbf{K}= [1+\textbf{I}](\textbf{U}+\{\textbf{S'}[1+\textbf{s}]-\textbf{Sv}-\textbf{F}-\textbf{I}\}[1-\textbf{t}][1-\textbf{d}])-[\textbf{N}+\textbf{J}+\textbf{P}]$$

Rule-56987:

If both $(\textbf{I}), (\textbf{U}), (\textbf{S'}), (\textbf{s}), (\textbf{S}), (\textbf{v}), (\textbf{F}), (\textbf{I}), (\textbf{t}), (\textbf{d}), (\textbf{K}),$ $(\textbf{N})$ and $(\textbf{P})$ are known, then its Job or Trade Account Receivable planned is:

$$\textbf{J}= [1+\textbf{I}](\textbf{U}+\{\textbf{S'}[1+\textbf{s}]-\textbf{Sv}-\textbf{F}-\textbf{I}\}[1-\textbf{t}][1-\textbf{d}])-[\textbf{N}+\textbf{K}+\textbf{P}]$$

Steve Asikin ISBN 13: **978-1564912361**, ISBN 10: **1564912363**

Rule-56988:
 If both (**/**), (**U**), (**$'**), (**s**), (**$**), (**v**), (**F**), (**I**), (**t**), (**d**), (**K**), (**J**) and (**N**) are known, then its Procured Inventory planned is:
 **P**= [1+**/**](**U**+{**$'**[1+**s**]-**$v**-**F**-**I**}[1-**t**][1-**d**])-[**N**+**K**+**J**]

Rule-56989:
 If both (**/**), (**U**), (**$'**), (**s**), (**$**), (**v**), (**F**), (**I**), (**t**), (**d**), (**K**), (**J**), (**V**) and (**p**) are known, then its Non Current Asset planned is:
 **N**= [1+**/**](**U**+{**$'**[1+**s**]-**$v**-**F**-**I**}[1-**t**][1-**d**])
  -**K**-**J**-**Vp**/360

Rule-56990:
 If both (**N**), (**U**), (**$'**), (**s**), (**$**), (**v**), (**F**), (**I**), (**t**), (**d**), (**K**), (**J**), (**V**) and (**p**) are known, then its Leverage or Gearing Ratio planned is:
 **/**= [**N**+**K**+**J**+**Vp**/360]/(**U**+{**$'**[1+**s**]-**$v**-**F**-**I**}
  [1-**t**][1-**d**])-1

Rule-56991:
 If both (**/**), (**N**), (**$'**), (**s**), (**$**), (**v**), (**F**), (**I**), (**t**), (**d**), (**K**), (**J**), (**V**) and (**p**) are known, then its Utilized or Starting Capital must be:
 **U**= ([**N**+**K**+**J**+**Vp**/360]-[1+**/**]{**$'**[1+**s**]-**$v**-**F**-**I**}
  [1-**t**][1-**d**])/[1+**/**]

Steve Asikin ISBN 13: **978-1564912361**, ISBN 10: **1564912363**

Rule-56992:

If both (**/**), (**U**), (**N**), (**s**), (**S**), (**v**), (**F**), (**I**), (**t**), (**d**), (**K**), (**J**), (**V**) and (**p**) are known, then its Sales Past must be:

$$S'= ([N+K+J+Vp/360]/[1+/\text{-}U]/\{[1\text{-}t][1\text{-}d]\}$$
$$+Sv+F+I)/[1+s]$$

or

$$S'= S/[1+s]$$

Rule-56993:

If both (**/**), (**U**), (**S'**), (**N**), (**S**), (**v**), (**F**), (**I**), (**t**), (**d**), (**K**), (**J**), (**V**) and (**p**) are known, then its Sales Growth planned is:

$$s= ([N+K+J+Vp/360]/[1+/\text{-}U]/\{[1\text{-}t][1\text{-}d]\}$$
$$+Sv+F+I)/S'\text{-}1$$

or

$$s= S/S'\text{-}1$$

Rule-56994:

If both (**/**), (**U**), (**S'**), (**s**), (**S**), (**v**), (**F**), (**I**), (**t**), (**d**), (**K**), (**J**), (**V**) and (**p**) are known, then its Non Current Asset planned is:

$$S= (S'[1+s]\text{-}F\text{-}I\text{-}[N+K+J+Vp/360]/[1+/\text{-}U]$$
$$/\{[1\text{-}t][1\text{-}d]\})/v$$

or

$$S= S'[1+s]$$

or

$$S= V/v$$

Steve Asikin ISBN 13: **978-1564912361**, ISBN 10: **1564912363**

Rule-56995:

If both (**/**), (**U**), (**S'**), (**s**), (**S**), (**v**), (**F**), (**I**), (**t**), (**d**), (**K**), (**J**), (**V**) and (**p**) are known, then its Non Current Asset planned is:

$$v= (S'[1+s]-F-I-[N+K+J+Vp/360]/[1+/-U]$$
$$/\{[1-t][1-d]\})/S$$

or

$$v= V/S$$

Rule-56996:

If both (**/**), (**U**), (**S'**), (**s**), (**S**), (**v**), (**N**), (**I**), (**t**), (**d**), (**K**), (**J**), (**V**) and (**p**) are known, then its Fixed Cost planned is:

$$F= S'[1+s]-Sv-I-[N+K+J+Vp/360]/[1+/-U]$$
$$/\{[1-t][1-d]\}$$

Rule-56997:

If both (**/**), (**U**), (**S'**), (**s**), (**S**), (**v**), (**F**), (**N**), (**t**), (**d**), (**K**), (**J**), (**V**) and (**p**) are known, then its Interest Expense planned is:

$$I= S'[1+s]-Sv-F-[N+K+J+Vp/360]/[1+/-U]$$
$$/\{[1-t][1-d]\}$$

Rule-56998:

If both (**/**), (**U**), (**S'**), (**s**), (**S**), (**v**), (**F**), (**I**), (**N**), (**d**), (**K**), (**J**), (**V**) and (**p**) are known, then its Tax Rate planned is:

$$t= 1-[N+K+J+Vp/360]/[1+/-U]$$
$$/(\{S'[1+s]-Sv-F-I\}[1-d])$$

`

Steve Asikin ISBN 13: **978-1564912361**, ISBN 10: **1564912363**

Rule-56999:

If both (**/**), (**U**), (**S'**), (**s**), (**S**), (**v**), (**F**), (**I**), (**t**), (**N**), (**K**), (**J**), (**V**) and (**p**) are known, then its Dividend Payout planned is:

$$d = 1-[\mathbf{N+K+J+Vp}/360]/[1+\text{/}\text{-}\mathbf{U}\}$$
$$/(\{\mathbf{S'}[1+\mathbf{s}]\text{-}\mathbf{Sv\text{-}F\text{-}I}\}[1\text{-}\mathbf{t}])$$

Rule-57000:

If both (**/**), (**U**), (**S'**), (**s**), (**S**), (**v**), (**F**), (**I**), (**t**), (**d**), (**N**), (**J**), (**V**) and (**p**) are known, then its Kind of Cash planned is:

$$\mathbf{K} = [1+\text{/}(\mathbf{U}+\{\mathbf{S'}[1+\mathbf{s}]\text{-}\mathbf{Sv\text{-}F\text{-}I}\}[1\text{-}\mathbf{t}][1\text{-}\mathbf{d}])$$
$$-[\mathbf{N+J+Vp}/360]$$

Rule-57001:

If both (**/**), (**U**), (**S'**), (**s**), (**S**), (**v**), (**F**), (**I**), (**t**), (**d**), (**K**), (**N**), (**V**) and (**p**) are known, then its Job or Trade Account Receivable planned is:

$$\mathbf{J} = [1+\text{/}(\mathbf{U}+\{\mathbf{S'}[1+\mathbf{s}]\text{-}\mathbf{Sv\text{-}F\text{-}I}\}[1\text{-}\mathbf{t}][1\text{-}\mathbf{d}])$$
$$-\{\mathbf{N+K+Vp}/360\}$$

Rule-57002:

If both (**/**), (**U**), (**S'**), (**s**), (**S**), (**v**), (**F**), (**I**), (**t**), (**d**), (**K**), (**J**), (**N**) and (**p**) are known, then its Variable Cost planned is:

$$\mathbf{V} = 360[[1+\text{/}(\mathbf{U}+\{\mathbf{S'}[1+\mathbf{s}]\text{-}\mathbf{Sv\text{-}F\text{-}I}\}[1\text{-}\mathbf{t}][1\text{-}\mathbf{d}])$$
$$-[\mathbf{N+K+J}]]/\mathbf{p}$$

or

$$\mathbf{V} = \mathbf{Sv}$$

Steve Asikin ISBN 13: **978-1564912361**, ISBN 10: **1564912363**

Rule-57003:

If both (**/**), (**U**), (**S'**), (**s**), (**S**), (**v**), (**F**), (**I**), (**t**), (**d**), (**K**), (**J**), (**V**) and (**N**) are known, then its Procured Inventory Days planned is:

$$\mathbf{p}= 360[[1+\textbf{/}(\mathbf{U}+\{\mathbf{S'}[1+\mathbf{s}]-\mathbf{Sv}-\mathbf{F}-\mathbf{I}\}[1-\mathbf{t}][1-\mathbf{d}])$$
$$-[\mathbf{N}+\mathbf{K}+\mathbf{J}]]/\mathbf{V}$$

Rule-57004:

If both (**/**), (**U**), (**S'**), (**s**), (**S**), (**v**), (**F**), (**I**), (**t**), (**d**), (**K**), (**J**) and (**p**) are known, then its Non Current Asset planned is:

$$\mathbf{N}= [1+\textbf{/}(\mathbf{U}+\{\mathbf{S'}[1+\mathbf{s}]-\mathbf{Sv}-\mathbf{F}-\mathbf{I}\}[1-\mathbf{t}][1-\mathbf{d}])$$
$$-\mathbf{K}-\mathbf{J}-\mathbf{Svp}/360$$

Rule-57005:

If both (**N**), (**U**), (**S'**), (**s**), (**S**), (**v**), (**F**), (**I**), (**t**), (**d**), (**K**), (**J**) and (**p**) are known, then its Leverage or Gearing Ratio planned is:

$$\textbf{/}= [\mathbf{N}+\mathbf{K}+\mathbf{J}+\mathbf{Svp}/360]/(\mathbf{U}+\{\mathbf{S'}[1+\mathbf{s}]-\mathbf{Sv}-\mathbf{F}-\mathbf{I}\}$$
$$[1-\mathbf{t}][1-\mathbf{d}])-1$$

Rule-57006:

If both (**/**), (**N**), (**S'**), (**s**), (**S**), (**v**), (**F**), (**I**), (**t**), (**d**), (**K**), (**J**) and (**p**) are known, then its Utilized or Starting Capital must be:

$$\mathbf{U}= ([\mathbf{N}+\mathbf{K}+\mathbf{J}+\mathbf{Svp}/360]-[1+\textbf{/}\{\mathbf{S'}[1+\mathbf{s}]-\mathbf{Sv}-\mathbf{F}-\mathbf{I}\}$$
$$[1-\mathbf{t}][1-\mathbf{d}])/[1+\textbf{/}$$

Steve Asikin ISBN 13: **978-1564912361**, ISBN 10: **1564912363**

Rule-57007:

> If both $(\textit{I})$, $(\textbf{U})$, $(\textbf{N})$, $(\textbf{s})$, $(\textbf{S})$, $(\textbf{v})$, $(\textbf{F})$, $(\textbf{I})$, $(\textbf{t})$, $(\textbf{d})$, $(\textbf{K})$, $(\textbf{J})$ and $(\textbf{p})$ are known, then its Sales Past must be:
>
> $$S' = (\{[N+K+J+Svp/360]/[1+\textit{I}-U\}$$
> $$/\{[1-t][1-d]\}+Sv+F+I)/[1+s]$$
>
> or
>
> $$S' = S/[1+s]$$

Rule-57008:

> If both $(\textit{I})$, $(\textbf{U})$, $(\textbf{S'})$, $(\textbf{N})$, $(\textbf{S})$, $(\textbf{v})$, $(\textbf{F})$, $(\textbf{I})$, $(\textbf{t})$, $(\textbf{d})$, $(\textbf{K})$, $(\textbf{J})$ and $(\textbf{p})$ are known, then its Sales Growth planned is:
>
> $$s = (\{[N+K+J+Svp/360]/[1+\textit{I}-U\}$$
> $$/\{[1-t][1-d]\}+Sv+F+I)/S'-1$$
>
> or
>
> $$s = S/S'-1$$

Rule-57009:

> If both $(\textit{I})$, $(\textbf{U})$, $(\textbf{S'})$, $(\textbf{s})$, $(\textbf{S})$, $(\textbf{v})$, $(\textbf{F})$, $(\textbf{I})$, $(\textbf{t})$, $(\textbf{d})$, $(\textbf{K})$, $(\textbf{J})$ and $(\textbf{p})$ are known, then its Sales or Revenue planned is:
>
> $$S = [[1+\textit{I}(U+\{S'[1+s]-F-I\}[1-t][1-d])-[N+K+J]]$$
> $$/(v\{p/360+[1+\textit{I}]\})$$
>
> or
>
> $$S = S'[1+s]$$

Steve Asikin ISBN 13: **978-1564912361**, ISBN 10: **1564912363**

Rule-57010:

If both ($\Lambda$), (**U**), (**S'**), (**s**), (**S**), (**N**), (**F**), (**I**), (**t**), (**d**), (**K**), (**J**) and (**p**) are known, then its Variable Portioin planned is:

$$v= [[1+\Lambda(U+\{S'[1+s]-F-I\}[1-t][1-d])-[N+K+J]] /(S\{p/360+[1+\Lambda]\})$$

Rule-57011:

If both ($\Lambda$), (**U**), (**S'**), (**s**), (**S**), (**v**), (**N**), (**I**), (**t**), (**d**), (**K**), (**J**) and (**p**) are known, then its Fixed Cost planned is:

$$F= S'[1+s]-Sv-I-\{[N+K+J+Svp/360]/[1+\Lambda]-U\} /\{[1-t][1-d]\}$$

Rule-57012:

If both ($\Lambda$), (**U**), (**S'**), (**s**), (**S**), (**v**), (**F**), (**N**), (**t**), (**d**), (**K**), (**J**) and (**p**) are known, then its Interest Expense planned is:

$$I= S'[1+s]-Sv-F-\{[N+K+J+Svp/360]/[1+\Lambda]-U\} /\{[1-t][1-d]\}$$

Rule-57013:

If both ($\Lambda$), (**U**), (**S'**), (**s**), (**S**), (**v**), (**F**), (**I**), (**N**), (**d**), (**K**), (**J**) and (**p**) are known, then its Tax Rate planned is:

$$t= 1-\{[N+K+J+Svp/360]/[1+\Lambda]-U\} /(\{S'[1+s]-Sv-F-I\}[1-d])$$

Steve Asikin ISBN 13: **978-1564912361**, ISBN 10: **1564912363**

Rule-57014:

If both (**/**), (**U**), (**S'**), (**s**), (**S**), (**v**), (**F**), (**I**), (**t**), (**N**), (**K**), (**J**) and (**p**) are known, then its Dividend Payout planned is:

$$d= 1-\{[N+K+J+Svp/360]/[1+/-U\}$$
$$/(\{S'[1+s]-Sv-F-I\}[1-t])$$

Rule-57015:

If both (**/**), (**U**), (**S'**), (**s**), (**S**), (**v**), (**F**), (**I**), (**t**), (**d**), (**N**), (**J**) and (**p**) are known, then its Kind of Cash planned is:

$$K= [1+/](U+\{S'[1+s]-Sv-F-I\}[1-t][1-d])$$
$$-[N+J+Svp/360]$$

Rule-57016:

If both (**/**), (**U**), (**S'**), (**s**), (**S**), (**v**), (**F**), (**I**), (**t**), (**d**), (**K**), (**N**) and (**p**) are known, then its Job or Trade Account Receivable planned is:

$$J= [1+/](U+\{S'[1+s]-Sv-F-I\}[1-t][1-d])$$
$$-\{N+K+Svp/360\}$$

Rule-57017:

If both (**/**), (**U**), (**S'**), (**s**), (**S**), (**v**), (**F**), (**I**), (**t**), (**d**), (**K**), (**J**) and (**N**) are known, then its Procured Inventory Days planned is:

$$p= 360[[1+/](U+\{S'[1+s]-Sv-F-I\}[1-t][1-d])$$
$$-[N+K+J]]/[Sv]$$

Steve Asikin ISBN 13: **978-1564912361**, ISBN 10: **1564912363**

<u>Rule-57018</u>:

If both (**/**), (**U**), (**S'**), (**s**), (**S**), (**v**), (**F**), (**I**), (**t**), (**d**), (**K**), (**J**) and (**p**) are known, then its Non Current Asset planned is:

$$N = [1 + /(U + \{S'[1+s] - Sv - F - I\}[1-t][1-d]) - K - J - S'vp[1+s]/360$$

<u>Rule-57019</u>:

If both (**N**), (**U**), (**S'**), (**s**), (**S**), (**v**), (**F**), (**I**), (**t**), (**d**), (**K**), (**J**) and (**p**) are known, then its Leverage or Gearing Ratio planned is:

$$/ = \{N + K + J + S'vp[1+s]/360\} / (U + \{S'[1+s] - Sv - F - I\}[1-t][1-d]) - 1$$

<u>Rule-57020</u>:

If both (**/**), (**N**), (**S'**), (**s**), (**S**), (**v**), (**F**), (**I**), (**t**), (**d**), (**K**), (**J**) and (**p**) are known, then its Utilized or Starting Capital must be:

$$U = (\{N + K + J + S'vp[1+s]/360\} - [1 + /]\{S'[1+s] - Sv - F - I\}[1-t][1-d]) / [1 + /$$

<u>Rule-57021</u>:

If both (**/**), (**U**), (**S'**), (**s**), (**S**), (**v**), (**F**), (**I**), (**t**), (**d**), (**K**), (**J**) and (**p**) are known, then its Non Current Asset planned is:

$$S' = [[1 + /(U - \{Sv + F + I\}[1-t][1-d]) - [N + K + J]] / ([1+s]\{vp/360 + [1 + /]\})$$

or

$$S' = S/[1+s]$$

Steve Asikin ISBN 13: **978-1564912361**, ISBN 10: **1564912363**

Rule-57022:

If both (**/**), (**U**), (**$'**), (**N**), (**$**), (**v**), (**F**), (**I**), (**t**), (**d**), (**K**), (**J**) and (**p**) are known, then its Sales Growth planned is:

$$s= [[1+\text{/}](U-\{\$v+F+I\}[1-t][1-d])-[N+K+J]]$$
$$/(\$'\{vp/360+[1+\text{/}]\})-1$$

or

$$s= \$/\$'-1$$

Rule-57023:

If both (**/**), (**U**), (**$'**), (**s**), (**$**), (**v**), (**F**), (**I**), (**t**), (**d**), (**K**), (**J**) and (**p**) are known, then its Non Current Asset planned is:

$$\$= [\$'[1+s]-F-I-(\{N+K+J+\$'vp[1+s]/360\}$$
$$/[1+\text{/}-U)/\{ [1-t][1-d]\}]/v$$

or

$$\$= \$'[1+s]$$

Rule-57024:

If both (**/**), (**U**), (**$'**), (**s**), (**$**), (**N**), (**F**), (**I**), (**t**), (**d**), (**K**), (**J**) and (**p**) are known, then its Variable Portion planned is:

$$v= [\$'[1+s]-F-I-(\{N+K+J+\$'vp[1+s]/360\}$$
$$/[1+\text{/}-U)/\{ [1-t][1-d]\}]/\$$$

Steve Asikin ISBN 13: **978-1564912361**, ISBN 10: **1564912363**

Rule-57025:

If both (**/**), (**U**), (**S'**), (**s**), (**S**), (**v**), (**N**), (**I**), (**t**), (**d**), (**K**),
(**J**) and (**p**) are known, then its Fixed Cost planned is:

$$F= S'[1+s]-Sv-I-(\{N+K+J+S'vp[1+s]/360\}$$
$$/[1+/-U)/\{ [1-t][1-d]\}$$


Rule-57026:

If both (**/**), (**U**), (**S'**), (**s**), (**S**), (**v**), (**F**), (**N**), (**t**), (**d**), (**K**),
(**J**) and (**p**) are known, then its Interest Expense
planned is:

$$I= S'[1+s]-Sv-F-(\{N+K+J+S'vp[1+s]/360\}$$
$$/[1+/-U)/\{ [1-t][1-d]\}$$


Rule-57027:

If both (**/**), (**U**), (**S'**), (**s**), (**S**), (**v**), (**F**), (**I**), (**N**), (**d**), (**K**),
(**J**) and (**p**) are known, then its Tax Rate planned is:

$$t= 1-(\{N+K+J+S'vp[1+s]/360\}/[1+/-U)$$
$$/(\{S'[1+s]-Sv-F-I\}[1-d])$$


Rule-57028:

If both (**/**), (**U**), (**S'**), (**s**), (**S**), (**v**), (**F**), (**I**), (**t**), (**N**), (**K**),
(**J**) and (**p**) are known, then its Dividend Payout
planned is:

$$d= 1-(\{N+K+J+S'vp[1+s]/360\}/[1+/-U)$$
$$/(\{S'[1+s]-Sv-F-I\}[1-t])$$

Steve Asikin ISBN 13: **978-1564912361**, ISBN 10: **1564912363**

Rule-57029:

If both (**/**), (**U**), (**S'**), (**s**), (**S**), (**v**), (**F**), (**I**), (**t**), (**d**), (**N**), (**J**) and (**p**) are known, then its Kind of Cash planned is:

$$K= [1+/(U+\{S'[1+s]-Sv-F-I\}[1-t][1-d])$$
$$-\{N+J+S'vp[1+s]/360\}$$

Rule-57030:

If both (**/**), (**U**), (**S'**), (**s**), (**S**), (**v**), (**F**), (**I**), (**t**), (**d**), (**K**), (**N**) and (**p**) are known, then its Job or Trade Account Receivable planned is:

$$J= [[1+/(U+\{S'[1+s]-Sv-F-I\}[1-t][1-d])$$
$$-\{N+K+S'vp[1+s]/360\}$$

Rule-57031:

If both (**/**), (**U**), (**S'**), (**s**), (**S**), (**v**), (**F**), (**I**), (**t**), (**d**), (**K**), (**J**) and (**N**) are known, then its Procured Inventory Days planned is:

$$p= 360[[1+/(U+\{S'[1+s]-Sv-F-I\}[1-t][1-d])$$
$$-[N+K+J]]/\{S'v[1+s]\}$$

Rule-57032:

If both (**/**), (**U**), (**S'**), (**s**), (**S**), (**v**), (**F**), (**I**), (**t**), (**d**), (**K**), (**j**) and (**P**) are known, then its Non Current Asset planned is:

$$N= [1+/(U+\{S'[1+s]-Sv-F-I\}[1-t][1-d])$$
$$-K-Sj/360-P$$

Steve Asikin ISBN 13: **978-1564912361**, ISBN 10: **1564912363**

Rule-57033:

If both (**N**), (**U**), (**S'**), (**s**), (**S**), (**v**), (**F**), (**I**), (**t**), (**d**), (**K**), (**j**) and (**P**) are known, then its Leverage or Gearing Ratio planned is:

$$I = \{N+K+Sj/360+P\}/(U+\{S'[1+s]-Sv-F-I\}[1-t][1-d])-1$$

Rule-57034:

If both (**I**), (**N**), (**S'**), (**s**), (**S**), (**v**), (**F**), (**I**), (**t**), (**d**), (**K**), (**j**) and (**P**) are known, then its Utilized or Starting Capital must be:

$$U = (\{N+K+Sj/360+P\}-[1+I]\{S'[1+s]-Sv-F-I\}[1-t][1-d])/[1+I]$$

Rule-57035:

If both (**I**), (**U**), (**S'**), (**s**), (**S**), (**v**), (**F**), (**I**), (**t**), (**d**), (**K**), (**j**) and (**P**) are known, then its Sales Past must be:

$$S' = (\{[N+K+Sj/360+P]/[1+I]-U\}/\{[1-t][1-d]\}+Sv+F+I)/[1+s]$$

or

$$S' = S/[1+s]$$

Steve Asikin ISBN 13: **978-1564912361**, ISBN 10: **1564912363**

Rule-57036:

If both (**/**), (**U**), (**S'**), (**N**), (**S**), (**v**), (**F**), (**I**), (**t**), (**d**), (**K**), (**j**) and (**P**) are known, then its Sales Growth planned is:

$$s= (\{[N+K+Sj/360+P]/[1+/]-U\}/\{[1-t][1-d]\}$$
$$+Sv+F+I)/ S'-1$$

or

$$s= S/S'-1$$

Rule-57037:

If both (**/**), (**U**), (**S'**), (**s**), (**N**), (**v**), (**F**), (**I**), (**t**), (**d**), (**K**), (**j**) and (**P**) are known, then its Sales or Revenue planned is:

$$S= [[1+/](U+\{S'[1+s]-F-I\}[1-t][1-d])-[N+K+P]]$$
$$/\{j/360+v[1+/]\}$$

or

$$S= S'[1+s]$$

Rule-57038:

If both (**/**), (**U**), (**S'**), (**s**), (**S**), (**N**), (**F**), (**I**), (**t**), (**d**), (**K**), (**j**) and (**P**) are known, then its Variable Portion planned is:

$$v= [S'[1+s]-F-I-\{[N+K+Sj/360+P]/[1+/]-U\}$$
$$/[1-t][1-d]]/S$$

Steve Asikin ISBN 13: **978-1564912361**, ISBN 10: **1564912363**

<u>Rule-57039</u>:

If both ($I$), (**U**), (**S'**), (**s**), (**S**), (**v**), (**N**), (**I**), (**t**), (**d**), (**K**), (**j**) and (**P**) are known, then its Fixed Cost planned is:

$$F= S'[1+s]-Sv-I-\{[N+K+Sj/360+P]/[1+I-U\}$$
$$/[1-t][1-d]$$

<u>Rule-57040</u>:

If both ($I$), (**U**), (**S'**), (**s**), (**S**), (**v**), (**F**), (**N**), (**t**), (**d**), (**K**), (**j**) and (**P**) are known, then its Interest Expense planned is:

$$I= S'[1+s]-Sv-F-\{[N+K+Sj/360+P]/[1+I-U\}$$
$$/[1-t][1-d]$$

<u>Rule-57041</u>:

If both ($I$), (**U**), (**S'**), (**s**), (**S**), (**v**), (**F**), (**I**), (**N**), (**d**), (**K**), (**j**) and (**P**) are known, then its Tax Rate planned is:

$$t= 1-\{[N+K+Sj/360+P]/[1+I-U\}$$
$$/(\{S'[1+s]-Sv-F-I\}[1-d])$$

<u>Rule-57042</u>:

If both ($I$), (**U**), (**S'**), (**s**), (**S**), (**v**), (**F**), (**I**), (**t**), (**N**), (**K**), (**j**) and (**P**) are known, then its Dividend Payout planned is:

$$d= 1-\{[N+K+Sj/360+P]/[1+I-U\}$$
$$/(\{S'[1+s]-Sv-F-I\}[1-t])$$

Steve Asikin ISBN 13: **978-1564912361**, ISBN 10: **1564912363**

Rule-57043:

If both ($\textit{l}$), (**U**), (**$'**), (**s**), (**$**), (**v**), (**F**), (**I**), (**t**), (**d**), (**N**), (**j**) and (**P**) are known, then its Kind of Cash planned is:

$$K= [1+\textit{l}(U+\{\$'[1+s]-\$v-F-I\}[1-t][1-d])$$
$$-[N+\$j/360+P]$$

Rule-57044:

If both ($\textit{l}$), (**U**), (**$'**), (**s**), (**$**), (**v**), (**F**), (**I**), (**t**), (**d**), (**K**), (**N**) and (**P**) are known, then its Job or Trade Receivable Days planned is:

$$j= 360[[1+\textit{l}(U+\{\$'[1+s]-\$v-F-I\}[1-t][1-d])$$
$$-\{N+K+P\}]/\$$$

Rule-57045:

If both ($\textit{l}$), (**U**), (**$'**), (**s**), (**$**), (**v**), (**F**), (**I**), (**t**), (**d**), (**K**), (**j**) and (**N**) are known, then its Procured Inventory planned is:

$$P= [1+\textit{l}(U+\{\$'[1+s]-\$v-F-I\}[1-t][1-d])$$
$$-\{N+K+\$j/360\}$$

Rule-57046:

If both ($\textit{l}$), (**U**), (**$'**), (**s**), (**$**), (**v**), (**F**), (**I**), (**t**), (**d**), (**K**), (**j**), (**V**) and (**p**) are known, then its Non Current Asset planned is:

$$N= [1+\textit{l}(U+\{\$'[1+s]-\$v-F-I\}[1-t][1-d])$$
$$-K-\$j/360-Vp/360$$

Steve Asikin ISBN 13: **978-1564912361**, ISBN 10: **1564912363**

Rule-57047:

If both (**N**), (**U**), (**S'**), (**s**), (**S**), (**v**), (**F**), (**I**), (**t**), (**d**), (**K**), (**j**), (**V**) and (**p**) are known, then its Leverage or Gearing Ratio planned is:

$$I = [\mathbf{N}+\mathbf{K}+\mathbf{Sj}/360+\mathbf{Vp}/360]/(\mathbf{U}+\{\mathbf{S'}[1+\mathbf{s}]-\mathbf{Sv}-\mathbf{F}-\mathbf{I}\}$$
$$[1-\mathbf{t}][1-\mathbf{d}])-1$$

Rule-57048:

If both (**I**), (**N**), (**S'**), (**s**), (**S**), (**v**), (**F**), (**I**), (**t**), (**d**), (**K**), (**j**), (**V**) and (**p**) are known, then its Utilized or Starting Capital must be:

$$\mathbf{U} = ([\mathbf{N}+\mathbf{K}+\mathbf{Sj}/360+\mathbf{Vp}/360]-[1+I]\{\mathbf{S'}[1+\mathbf{s}]$$
$$-\mathbf{Sv}-\mathbf{F}-\mathbf{I}\}[1-\mathbf{t}][1-\mathbf{d}])/[1+I]$$

Rule-57049:

If both (**I**), (**U**), (**N**), (**s**), (**S**), (**v**), (**F**), (**I**), (**t**), (**d**), (**K**), (**j**), (**V**) and (**p**) are known, then its Sales Past must be:

$$\mathbf{S'} = (\{[\mathbf{N}+\mathbf{K}+\mathbf{Sj}/360+\mathbf{Vp}/360]/[1+I]-\mathbf{U}\}$$
$$/\{[1-\mathbf{t}][1-\mathbf{d}]\}+\mathbf{Sv}+\mathbf{F}+\mathbf{I})/[1+\mathbf{s}]$$

or

$$\mathbf{S'} = \mathbf{S}[1+\mathbf{s}]$$

Steve Asikin ISBN 13: **978-1564912361**, ISBN 10: **1564912363**

Rule-57050:

If both $(\textbf{\textit{l}})$, $(\textbf{U})$, $(\textbf{S'})$, $(\textbf{N})$, $(\textbf{S})$, $(\textbf{v})$, $(\textbf{F})$, $(\textbf{I})$, $(\textbf{t})$, $(\textbf{d})$, $(\textbf{K})$, $(\textbf{j})$, $(\textbf{V})$ and $(\textbf{p})$ are known, then its Sales Growth planned is:

$$s= (\{[\textbf{N}+\textbf{K}+\textbf{S}\textbf{j}/360+\textbf{V}\textbf{p}/360]/[1+\textbf{\textit{l}}\text{-}\textbf{U}\}$$
$$/\{[1\text{-}\textbf{t}][1\text{-}\textbf{d}]\}+\textbf{S}\textbf{v}+\textbf{F}+\textbf{I})/\textbf{S'}\text{-}1$$

or

$$s= \textbf{S}/\textbf{S'}\text{-}1$$

Rule-57051:

If both $(\textbf{\textit{l}})$, $(\textbf{U})$, $(\textbf{S'})$, $(\textbf{s})$, $(\textbf{N})$, $(\textbf{v})$, $(\textbf{F})$, $(\textbf{I})$, $(\textbf{t})$, $(\textbf{d})$, $(\textbf{K})$, $(\textbf{j})$, $(\textbf{V})$ and $(\textbf{p})$ are known, then its Sales or Revenue planned is:

$$S= [[1+\textbf{\textit{l}}(\textbf{U}+\{\textbf{S'}[1+\textbf{s}]\text{-}\textbf{F}\text{-}\textbf{I}\}[1\text{-}\textbf{t}][1\text{-}\textbf{d}])$$
$$\text{-}[\textbf{N}+\textbf{K}+\textbf{V}\textbf{p}/360]]/\{\textbf{j}/360+\textbf{v}[1+\textbf{\textit{l}}]\}$$

or

$$S= \textbf{S'}[1+\textbf{s}]$$

Rule-57052:

If both $(\textbf{\textit{l}})$, $(\textbf{U})$, $(\textbf{S'})$, $(\textbf{s})$, $(\textbf{S})$, $(\textbf{N})$, $(\textbf{F})$, $(\textbf{I})$, $(\textbf{t})$, $(\textbf{d})$, $(\textbf{K})$, $(\textbf{j})$, $(\textbf{V})$ and $(\textbf{p})$ are known, then its Variable Portion planned is:

$$v= [\textbf{S'}[1+\textbf{s}]\text{-}\textbf{F}\text{-}\textbf{I}\text{-}\{[\textbf{N}+\textbf{K}+\textbf{S}\textbf{j}/360+\textbf{V}\textbf{p}/360]/[1+\textbf{\textit{l}}\text{-}\textbf{U}\}$$
$$/\{[1\text{-}\textbf{t}][1\text{-}\textbf{d}]\}]/\textbf{S}$$

or

$$v= \textbf{V}/\textbf{S}$$

259

Steve Asikin ISBN 13: **978-1564912361**, ISBN 10: **1564912363**

Rule-57053:

If both (**/**), (**U**), (**S'**), (**s**), (**S**), (**v**), (**N**), (**I**), (**t**), (**d**), (**K**), (**j**), (**V**) and (**p**) are known, then its Fixed Cost planned is:

$$\mathbf{F} = \mathbf{S'}[1+\mathbf{s}]-\mathbf{Sv}-\mathbf{I}-\{[\mathbf{N}+\mathbf{K}+\mathbf{Sj}/360+\mathbf{Vp}/360]$$
$$/[1+\mathbf{/}-\mathbf{U}]/\{[1-\mathbf{t}][1-\mathbf{d}]\}$$

Rule-57054:

If both (**/**), (**U**), (**S'**), (**s**), (**S**), (**v**), (**F**), (**N**), (**t**), (**d**), (**K**), (**j**), (**V**) and (**p**) are known, then its Interest Expense planned is:

$$\mathbf{I} = \mathbf{S'}[1+\mathbf{s}]-\mathbf{Sv}-\mathbf{F}-\{[\mathbf{N}+\mathbf{K}+\mathbf{Sj}/360+\mathbf{Vp}/360]$$
$$/[1+\mathbf{/}-\mathbf{U}]/\{[1-\mathbf{t}][1-\mathbf{d}]\}$$

Rule-57055:

If both (**/**), (**U**), (**S'**), (**s**), (**S**), (**v**), (**F**), (**I**), (**N**), (**d**), (**K**), (**j**), (**V**) and (**p**) are known, then its Tax Rate planned is:

$$\mathbf{t} = 1-\{[\mathbf{N}+\mathbf{K}+\mathbf{Sj}/360+\mathbf{Vp}/360]/[1+\mathbf{/}-\mathbf{U}]$$
$$/(\{\mathbf{S'}[1+\mathbf{s}]-\mathbf{Sv}-\mathbf{F}-\mathbf{I}\}[1-\mathbf{d}])$$

Rule-57056:

If both (**/**), (**U**), (**S'**), (**s**), (**S**), (**v**), (**F**), (**I**), (**t**), (**N**), (**K**), (**j**), (**V**) and (**p**) are known, then its Dividend Payout planned is:

$$\mathbf{d} = 1-\{[\mathbf{N}+\mathbf{K}+\mathbf{Sj}/360+\mathbf{Vp}/360]/[1+\mathbf{/}-\mathbf{U}]$$
$$/(\{\mathbf{S'}[1+\mathbf{s}]-\mathbf{Sv}-\mathbf{F}-\mathbf{I}\}[1-\mathbf{t}])$$

Steve Asikin ISBN 13: **978-1564912361**, ISBN 10: **1564912363**

Rule-57057:

If both (**/**), (**U**), (**S'**), (**s**), (**S**), (**v**), (**F**), (**I**), (**t**), (**d**), (**N**), (**j**), (**V**) and (**p**) are known, then its Kind of Cash planned is:

$$K= [1+\text{/}(U+\{S'[1+s]-Sv-F-I\}[1-t][1-d])$$
$$-[N+Sj/360+Vp/360]$$

Rule-57058:

If both (**/**), (**U**), (**S'**), (**s**), (**S**), (**v**), (**F**), (**I**), (**t**), (**d**), (**K**), (**N**), (**V**) and (**p**) are known, then its Job or Trade Receivable Days planned is:

$$j= 360[[1+\text{/}(U+\{S'[1+s]-Sv-F-I\}[1-t][1-d])$$
$$-[N+K+Vp/360]]/S$$

Rule-57059:

If both (**/**), (**U**), (**S'**), (**s**), (**S**), (**v**), (**F**), (**I**), (**t**), (**d**), (**K**), (**j**), (**N**) and (**p**) are known, then its Variable Cost planned is:

$$V= 360[[1+\text{/}(U+\{S'[1+s]-Sv-F-I\}[1-t][1-d])$$
$$-[N+K+Sj/360]]/p$$
$$\text{or}$$
$$V= Sv$$

Rule-57060:

If both (**/**), (**U**), (**S'**), (**s**), (**S**), (**v**), (**F**), (**I**), (**t**), (**d**), (**K**), (**j**), (**V**) and (**N**) are known, then its Procured Inventory Days planned is:

$$p= 360[[1+\text{/}(U+\{S'[1+s]-Sv-F-I\}[1-t][1-d])$$
$$-[N+K+Sj/360]]/V$$

Steve Asikin ISBN 13: **978-1564912361**, ISBN 10: **1564912363**

Rule-57061:

If both ($I$), (**U**), (**S'**), (**s**), (**S**), (**v**), (**F**), (**I**), (**t**), (**d**), (**K**), (**j**) and (**p**) are known, then its Non Current Asset planned is:

$$N = [1+I(U+\{S'[1+s]-Sv-F-I\}[1-t][1-d]) \\ -K-Sj/360-Svp/360$$

Rule-57062:

If both (**N**), (**U**), (**S'**), (**s**), (**S**), (**v**), (**F**), (**I**), (**t**), (**d**), (**K**), (**j**) and (**p**) are known, then its Leverage or Gearing Ratio planned is:

$$I = [N+K+Sj/360+Svp/360] \\ /(U+\{S'[1+s]-Sv-F-I\}[1-t][1-d])-1$$

Rule-57063:

If both ($I$), (**N**), (**S'**), (**s**), (**S**), (**v**), (**F**), (**I**), (**t**), (**d**), (**K**), (**j**) and (**p**) are known, then its Utilized or Starting Capital must be:

$$U = ([N+K+Sj/360+Svp/360]-[1+I\{S'[1+s] \\ -Sv-F-I\}[1-t][1-d])/[1+I$$

Rule-57064:

If both ($I$), (**U**), (**N**), (**s**), (**S**), (**v**), (**F**), (**I**), (**t**), (**d**), (**K**), (**j**) and (**p**) are known, then its Sales Past must be:

$$S' = (\{[N+K+Sj/360+Svp/360]/[1+I-U\} \\ /\{[1-t][1-d]\}+Sv+F+I)/[1+s]$$
$$\text{or}$$
$$S' = S/[1+s]$$

Steve Asikin ISBN 13: **978-1564912361**, ISBN 10: **1564912363**

Rule-57065:

If both (**/**), (**U**), (**S'**), (**s**), (**S**), (**v**), (**F**), (**I**), (**t**), (**d**), (**K**), (**j**) and (**p**) are known, then its Non Current Asset planned is:

$$s= (\{[N+K+Sj/360+Svp/360]/[1+/]-U\}$$
$$/\{[1-t][1-d]\}+Sv+F+I)/S'-1$$

or

$$s= S/S'-1$$


Rule-57066:

If both (**/**), (**U**), (**S'**), (**s**), (**S**), (**v**), (**F**), (**I**), (**t**), (**d**), (**K**), (**j**) and (**p**) are known, then its Sales or Revenue planned is:

$$S= [[1+/](U+\{S'[1+s]-F-I\}[1-t][1-d])-[N+K]]$$
$$/(j/360+v\{p/360]+[1+/]\})$$

or

$$S= S'[1+s]$$


Rule-57067:

If both (**/**), (**U**), (**S'**), (**s**), (**S**), (**N**), (**F**), (**I**), (**t**), (**d**), (**K**), (**j**) and (**p**) are known, then its Variable Portion planned is:

$$v= [S'[1+s]-F-I-\{[N+K+Sj/360+Svp/360]$$
$$/[1+/]-U\}/\{[1-t][1-d]\}]/S$$

Steve Asikin ISBN 13: **978-1564912361**, ISBN 10: **1564912363**

Rule-57068:

If both (**/**), (**U**), (**S'**), (**s**), (**S**), (**v**), (**N**), (**I**), (**t**), (**d**), (**K**), (**j**) and (**p**) are known, then its Fixed Cost planned is:

$$F = S'[1+s] - Sv - I - \{[N+K+Sj/360 + Svp/360]$$
$$/[1+/-U\} / \{[1-t][1-d]\}$$

Rule-57069:

If both (**/**), (**U**), (**S'**), (**s**), (**S**), (**v**), (**F**), (**N**), (**t**), (**d**), (**K**), (**j**) and (**p**) are known, then its Interest Expense planned is:

$$I = S'[1+s] \; Sv \; F \; \{[N+K+Sj/360 + Svp/360]$$
$$/[1+/-U\} / \{[1-t][1-d]\}$$

Rule-57070:

If both (**/**), (**U**), (**S'**), (**s**), (**S**), (**v**), (**F**), (**I**), (**N**), (**d**), (**K**), (**j**) and (**p**) are known, then its Tax Rate planned is:

$$t = 1 - \{[N+K+Sj/360 + Svp/360]/[1+/-U\}$$
$$/(\{S'[1+s] - Sv - F - I\}[1-d])$$

Rule-57071:

If both (**/**), (**U**), (**S'**), (**s**), (**S**), (**v**), (**F**), (**I**), (**t**), (**N**), (**K**), (**j**) and (**p**) are known, then its Dividend Payout planned is:

$$d = 1 - \{[N+K+Sj/360 + Svp/360]/[1+/-U\}$$
$$/(\{S'[1+s] - Sv - F - I\}[1-t])$$

Steve Asikin ISBN 13: **978-1564912361**, ISBN 10: **1564912363**

Rule-57072:

If both (**∫**), (**U**), (**$'**), (**s**), (**$**), (**v**), (**F**), (**I**), (**t**), (**d**), (**N**), (**j**) and (**p**) are known, then its Kind of Cash planned is:

$$K= [1+ʃ(U+\{\$'[1+s]-\$v-F-I\}[1-t][1-d])$$
$$-[N+\$j/360+\$vp/360]$$

Rule-57073:

If both (**∫**), (**U**), (**$'**), (**s**), (**$**), (**v**), (**F**), (**I**), (**t**), (**d**), (**K**), (**N**) and (**p**) are known, then its Job or Trade Receivable Days planned is:

$$j= 360[[1+ʃ(U+\{\$'[1+s]-\$v-F-I\}[1-t][1-d])$$
$$-\{N+K+\$vp/360\}]/\$$$

Rule-57074:

If both (**∫**), (**U**), (**$'**), (**s**), (**$**), (**v**), (**F**), (**I**), (**t**), (**d**), (**K**), (**j**) and (**N**) are known, then its Procured Inventory Days planned is:

$$p= 360[[1+ʃ(U+\{\$'[1+s]-\$v-F-I\}[1-t][1-d])$$
$$-[N+K+\$j/360]]/[\$v]$$

Rule-57075:

If both (**∫**), (**U**), (**$'**), (**s**), (**$**), (**v**), (**F**), (**I**), (**t**), (**d**), (**K**), (**j**) and (**p**) are known, then its Non Current Asset planned is:

$$N= [1+ʃ(U+\{\$'[1+s]-\$v-F-I\}[1-t][1-d])$$
$$-K-\$j/360-\$'vp[1+s]/360$$

Steve Asikin ISBN 13: **978-1564912361**, ISBN 10: **1564912363**

Rule-57076:

If both (**N**), (**U**), (**S'**), (**s**), (**S**), (**v**), (**F**), (**I**), (**t**), (**d**), (**K**), (**j**) and (**p**) are known, then its Leverage or Gearing Ratio planned is:

$$\textit{l} = \{\mathbf{N}+\mathbf{K}+\mathbf{S}\mathbf{j}/360+\mathbf{S'}\mathbf{v}\mathbf{p}[1+\mathbf{s}]/360\}$$
$$/(\mathbf{U}+\{\mathbf{S'}[1+\mathbf{s}]-\mathbf{S}\mathbf{v}-\mathbf{F}-\mathbf{I}\}[1-\mathbf{t}][1-\mathbf{d}])-1$$

Rule-57077:

If both (**l**), (**N**), (**S'**), (**s**), (**S**), (**v**), (**F**), (**I**), (**t**), (**d**), (**K**), (**j**) and (**p**) are known, then its Utilized or Starting Capital must be:

$$\mathbf{U} = (\{\mathbf{N}+\mathbf{K}+\mathbf{S}\mathbf{j}/360+\mathbf{S'}\mathbf{v}\mathbf{p}[1+\mathbf{s}]/360\}$$
$$-[1+\textit{l}]\{\mathbf{S'}[1+\mathbf{s}]-\mathbf{S}\mathbf{v}-\mathbf{F}-\mathbf{I}\}[1-\mathbf{t}][1-\mathbf{d}])/\{[1+\textit{l}]$$

Rule-57078:

If both (**l**), (**U**), (**N**), (**s**), (**S**), (**v**), (**F**), (**I**), (**t**), (**d**), (**K**), (**j**) and (**p**) are known, then its Sales Past must be:

$$\mathbf{S'} = [[1+\textit{l}](\mathbf{U}-\{\mathbf{S}\mathbf{v}+\mathbf{F}+\mathbf{I}\}[1-\mathbf{t}][1-\mathbf{d}])-[\mathbf{N}+\mathbf{K}+\mathbf{S}\mathbf{j}/360]]$$
$$/([1+\mathbf{s}]\{\mathbf{v}\mathbf{p}/360-[1+\textit{l}]\})$$

or

$$\mathbf{S'} = \mathbf{S}/[1+\mathbf{s}]$$

Steve Asikin ISBN 13: **978-1564912361**, ISBN 10: **1564912363**

Rule-57079:

If both **(/)**, **(U)**, **(S')**, **(N)**, **(S)**, **(v)**, **(F)**, **(I)**, **(t)**, **(d)**, **(K)**, **(j)** and **(p)** are known, then its Sales Growth planned is:

$$s= [[1+/](U-\{Sv+F+I\}[1-t][1-d])-[N+K+Sj/360]]$$
$$/(S'\{vp/360-[1+/]\})-1$$

or

$$s= S/S'-1$$

Rule-57080:

If both **(/)**, **(U)**, **(S')**, **(s)**, **(N)**, **(v)**, **(F)**, **(I)**, **(t)**, **(d)**, **(K)**, **(j)** and **(p)** are known, then its Sales or Revenue planned is:

$$S= [[1+/](U+\{S'[1+s]-F-I\}[1-t][1-d])$$
$$-\{N+K+S'vp[1+s]/360\}]/\{j/360+v[1+/]\}$$

or

$$S= S'[1+s]$$

Rule-57081:

If both **(/)**, **(U)**, **(S')**, **(s)**, **(S)**, **(N)**, **(F)**, **(I)**, **(t)**, **(d)**, **(K)**, **(j)** and **(p)** are known, then its Variable Portion planned is:

$$v= [S'[1+s]-Sv-I-(\{N+K+Sj/360+S'vp[1+s]/360\}$$
$$/[1+/]-U)/\{[1-t][1-d]\}]/S$$

Steve Asikin ISBN 13: **978-1564912361**, ISBN 10: **1564912363**

<u>Rule-57082</u>:

If both ($I$), (**U**), (**S'**), (**s**), (**S**), (**v**), (**N**), (**I**), (**t**), (**d**), (**K**), (**j**) and (**p**) are known, then its Fixed Cost planned is:

$$F= S'[1+s]-Sv-I-(\{N+K+Sj/360+S'vp[1+s]/360\}/[1+I-U)/\{[1-t][1-d]\}$$

<u>Rule-57083</u>:

If both ($I$), (**U**), (**S'**), (**s**), (**S**), (**v**), (**F**), (**N**), (**t**), (**d**), (**K**), (**j**) and (**p**) are known, then its Interest Expense planned is:

$$I= S'[1+s]-Sv-F-(\{N+K+Sj/360+S'vp[1+s]/360\}/[1+I-U)/\{[1-t][1-d]\}$$

<u>Rule-57084</u>:

If both ($I$), (**U**), (**S'**), (**s**), (**S**), (**v**), (**F**), (**I**), (**N**), (**d**), (**K**), (**j**) and (**p**) are known, then its Tax Rate planned is:

$$t= 1-(\{N+K+Sj/360+S'vp[1+s]/360\}/[1+I-U)/(\{S'[1+s]-Sv-F-I\}[1-d])$$

<u>Rule-57085</u>:

If both ($I$), (**U**), (**S'**), (**s**), (**S**), (**v**), (**F**), (**I**), (**t**), (**N**), (**K**), (**j**) and (**p**) are known, then its Dividend Payout planned is:

$$d= 1-(\{N+K+Sj/360+S'vp[1+s]/360\}/[1+I-U)/(\{S'[1+s]-Sv-F-I\}[1-t])$$

`

268

Steve Asikin ISBN 13: **978-1564912361**, ISBN 10: **1564912363**

Rule-57086:

If both (**/**), (**U**), (**S'**), (**s**), (**S**), (**v**), (**F**), (**I**), (**t**), (**d**), (**N**), (**j**) and (**p**) are known, then its Kind of Cash planned is:

$$K= [1+/(U+\{S'[1+s]-Sv-F-I\}[1-t][1-d])$$
$$-\{N+Sj/360+S'vp[1+s]/360\}$$

Rule-57087:

If both (**/**), (**U**), (**S'**), (**s**), (**S**), (**v**), (**F**), (**I**), (**t**), (**d**), (**K**), (**N**) and (**p**) are known, then its Job or Trade Receivable Days planned is:

$$j= 360[[1+/(U+\{S'[1+s]-Sv-F-I\}[1-t][1-d])$$
$$-\{N+K+S'vp[1+s]/360\}]/S$$

Rule-57088:

If both (**/**), (**U**), (**S'**), (**s**), (**S**), (**v**), (**F**), (**I**), (**t**), (**d**), (**K**), (**j**) and (**N**) are known, then its Procured Inventory Days planned is:

$$p= 360[[1+/(U+\{S'[1+s]-Sv-F-I\}[1-t][1-d])$$
$$-[N+K+Sj/360]]/\{S'v[1+s]\}$$

Rule-57089:

If both (**/**), (**U**), (**S'**), (**s**), (**S**), (**v**), (**F**), (**I**), (**t**), (**d**), (**K**), (**j**) and (**p**) are known, then its Non Current Asset planned is:

$$N= [1+/(U+\{S'[1+s]-Sv-F-I\}[1-t][1-d])$$
$$-K-S'j[1+s]/360-P$$

Steve Asikin ISBN 13: **978-1564912361**, ISBN 10: **1564912363**

Rule-57090:

If both (**N**), (**U**), (**S'**), (**s**), (**S**), (**v**), (**F**), (**I**), (**t**), (**d**), (**K**), (**j**) and (**p**) are known, then its Leverage or Gearing Ratio planned is:

$$I = \{N+K+S'j[1+s]/360+P\} \\ /(U+\{S'[1+s]-Sv-F-I\}[1-t][1-d])-1$$

Rule-57091:

If both (**I**), (**N**), (**S'**), (**s**), (**S**), (**v**), (**F**), (**I**), (**t**), (**d**), (**K**), (**j**) and (**p**) are known, then its Utilized or Starting Capital must be:

$$U = (\{N+K+S'j[1+s]/360+P\}-[1+I]\{S'[1+s] \\ -Sv-F-I\}[1-t][1-d])/[1+I]$$

Rule-57092:

If both (**I**), (**U**), (**N**), (**s**), (**S**), (**v**), (**F**), (**I**), (**t**), (**d**), (**K**), (**j**) and (**p**) are known, then its Sales Past must be:

$$S' = [[1+I](U-\{Sv+F+I\}[1-t][1-d])-[N+K+P]] \\ /([1+s]\{j/360-[1+I]\})$$

or

$$S' = S/[1+s]$$

Steve Asikin ISBN 13: **978-1564912361**, ISBN 10: **1564912363**

Rule-57093:

If both (**/**), (**U**), (**S'**), (**N**), (**S**), (**v**), (**F**), (**I**), (**t**), (**d**), (**K**), (**j**) and (**p**) are known, then its Sales Growth planned is:

$$s= [[1+\textbf{/}](\textbf{U}-\{\textbf{Sv}+\textbf{F}+\textbf{I}\}[1-\textbf{t}][1-\textbf{d}])-[\textbf{N}+\textbf{K}+\textbf{P}]]$$
$$/(\textbf{S'}\{\textbf{j}/360-[1+\textbf{/}]\})-1$$

or

$$s= \textbf{S}/\textbf{S'}-1$$


Rule-57094:

If both (**/**), (**U**), (**S'**), (**s**), (**N**), (**v**), (**F**), (**I**), (**t**), (**d**), (**K**), (**j**) and (**p**) are known, then its Sales or Revenue planned is:

$$\textbf{S}= [\textbf{S'}[1+\textbf{s}]-\textbf{F}-\textbf{I}-(\{\textbf{N}+\textbf{K}+\textbf{S'j}[1+\textbf{s}]/360+\textbf{P}\}$$
$$/[1+\textbf{/}-\textbf{U}]/\{[1-\textbf{t}][1-\textbf{d}]\}]/\textbf{v}$$

or

$$\textbf{S}= \textbf{S'}[1+\textbf{s}]$$


Rule-57095:

If both (**/**), (**U**), (**S'**), (**s**), (**S**), (**N**), (**F**), (**I**), (**t**), (**d**), (**K**), (**j**) and (**p**) are known, then its Variable Portion planned is:

$$\textbf{v}= [\textbf{S'}[1+\textbf{s}]-\textbf{F}-\textbf{I}-(\{\textbf{N}+\textbf{K}+\textbf{S'j}[1+\textbf{s}]/360+\textbf{P}\}$$
$$/[1+\textbf{/}-\textbf{U}]/\{[1-\textbf{t}][1-\textbf{d}]\}]/\textbf{S}$$

Steve Asikin ISBN 13: **978-1564912361**, ISBN 10: **1564912363**

Rule-57096:

If both (**/**), (**U**), (**S'**), (**s**), (**S**), (**v**), (**N**), (**I**), (**t**), (**d**), (**K**), (**j**) and (**p**) are known, then its Fixed Cost planned is:

$$F= S'[1+s]-Sv-I-(\{N+K+S'j[1+s]/360+P\}$$
$$/[1+/-U)/\{[1-t][1-d]\}$$

Rule-57097:

If both (**/**), (**U**), (**S'**), (**s**), (**S**), (**v**), (**F**), (**N**), (**t**), (**d**), (**K**), (**j**) and (**p**) are known, then its Interest Portion planned is:

$$I= S'[1+s]-Sv-F-(\{N+K+S'j[1+s]/360+P\}$$
$$/[1+/-U)/\{[1-t][1-d]\}$$

Rule-57098:

If both (**/**), (**U**), (**S'**), (**s**), (**S**), (**v**), (**F**), (**I**), (**N**), (**d**), (**K**), (**j**) and (**p**) are known, then its Tax Rate planned is:

$$t= 1-(\{N+K+S'j[1+s]/360+P\}/[1+/-U)/(\{S'[1+s]$$
$$-Sv-F-I\}[1-d])$$

Rule-57099:

If both (**/**), (**U**), (**S'**), (**s**), (**S**), (**v**), (**F**), (**I**), (**t**), (**N**), (**K**), (**j**) and (**p**) are known, then its Dividend Payout planned is:

$$d= 1-(\{N+K+S'j[1+s]/360+P\}/[1+/-U)$$
$$/(\{S'[1+s]-Sv-F-I\}[1-t])$$

Steve Asikin ISBN 13: **978-1564912361**, ISBN 10: **1564912363**

Rule-57100:

If both (**/**), (**U**), (**S'**), (**s**), (**S**), (**v**), (**F**), (**I**), (**t**), (**d**), (**N**), (**j**) and (**p**) are known, then its Kind of Cash planned is:

$$K= [1+\text{/}](U+\{S'[1+s]-Sv-F-I\}[1-t][1-d])$$
$$-\{N+S'j[1+s]/360+P\}$$


Rule-57101:

If both (**/**), (**U**), (**S'**), (**s**), (**S**), (**v**), (**F**), (**I**), (**t**), (**d**), (**K**), (**N**) and (**p**) are known, then its Job or Trade Receivable Days planned is:

$$j= 360[[1+\text{/}](U+\{S'[1+s]-Sv-F-I\}[1-t][1-d])$$
$$-\{N+K+P\}]/\{S'[1+s]\}$$


Rule-57102:

If both (**/**), (**U**), (**S'**), (**s**), (**S**), (**v**), (**F**), (**I**), (**t**), (**d**), (**K**), (**j**) and (**N**) are known, then its Procured Inventory planned is:

$$P= [1+\text{/}](U+\{S'[1+s]-Sv-F-I\}[1-t][1-d])$$
$$-\{N+K+S'j[1+s]/360\}$$


Rule-57103:

If both (**/**), (**U**), (**S'**), (**s**), (**S**), (**v**), (**F**), (**I**), (**t**), (**d**), (**K**), (**j**), (**V**) and (**p**) are known, then its Non Current Asset planned is:

$$N= [1+\text{/}](U+\{S'[1+s]-Sv-F-I\}[1-t][1-d])$$
$$-K-S'j[1+s]/360-Vp/360$$

Steve Asikin ISBN 13: **978-1564912361**, ISBN 10: **1564912363**

Rule-57104:

If both (**N**), (**U**), (**S'**), (**s**), (**S**), (**v**), (**F**), (**I**), (**t**), (**d**), (**K**), (**j**), (**V**) and (**p**) are known, then its Leverage or Gearing Ratio planned is:

$$I= \{N+K+S'j[1+s]/360+Vp/360\}$$
$$/(U+\{S'[1+s]-Sv-F-I\}[1-t][1-d])-1$$

Rule-57105:

If both (**I**), (**N**), (**S'**), (**s**), (**S**), (**v**), (**F**), (**I**), (**t**), (**d**), (**K**), (**j**), (**V**) and (**p**) are known, then its Utilized or Starting Capital must be:

$$U= (\{N+K+S'j[1+s]/360+Vp/360\}-[1+I]\{S'[1+s]$$
$$-Sv-F-I\}[1-t][1-d])/[1+I]$$

Rule-57106:

If both (**I**), (**U**), (**N**), (**s**), (**S**), (**v**), (**F**), (**I**), (**t**), (**d**), (**K**), (**j**), (**V**) and (**p**) are known, then its Sales Past must be:

$$S'= [[1+I](U-\{Sv+F+I\}[1-t][1-d])$$
$$-\{N+K+Vp/360\}]/([1+s]\{j/360-[1+I]\})$$

or

$$S'= S/[1+s]$$

Steve Asikin ISBN 13: **978-1564912361**, ISBN 10: **1564912363**

Rule-57107:

If both (**/**), (**U**), (**S'**), (**N**), (**S**), (**v**), (**F**), (**I**), (**t**), (**d**), (**K**), (**j**), (**V**) and (**p**) are known, then its Sales Growth planned is:

$s= [[1+/](U-\{Sv+F+I\}[1-t][1-d])$
$\qquad -\{N+K+Vp/360\}]/(S'\{j/360-[1+/]\})-1$

or

$s= S/S'-1$

Rule-57108:

If both (**/**), (**U**), (**S'**), (**s**), (**N**), (**v**), (**F**), (**I**), (**t**), (**d**), (**K**), (**j**), (**V**) and (**p**) are known, then its Sales or Revenue planned is:

$S= [S'[1+s]-F-I-(\{N+K+S'j[1+s]/360+Vp/360\}$
$\qquad /[1+/-U)/\{[1-t][1-d]\}]/v$

or

$S= S'[1+s]$

or

$S= V/v$

Rule-57109:

If both (**/**), (**U**), (**S'**), (**s**), (**S**), (**N**), (**F**), (**I**), (**t**), (**d**), (**K**), (**j**), (**V**) and (**p**) are known, then its Variable Portion planned is:

$v= [S'[1+s]-F-I-(\{N+K+S'j[1+s]/360+Vp/360\}$
$\qquad /[1+/-U)/\{[1-t][1-d]\}]/S$

or

$v= V/S$

Steve Asikin ISBN 13: **978-1564912361**, ISBN 10: **1564912363**

Rule-57110:

If both (**/**), (**U**), (**S'**), (**s**), (**S**), (**v**), (**N**), (**I**), (**t**), (**d**), (**K**), (**j**), (**V**) and (**p**) are known, then its Fixed Cost planned is:

$$F= S'[1+s]-Sv-I-(\{N+K+S'j[1+s]/360+Vp/360\} \\ /[1+\text{/}-U)/\{[1-t][1-d]\}$$

Rule-57111:

If both (**/**), (**U**), (**S'**), (**s**), (**S**), (**v**), (**F**), (**N**), (**t**), (**d**), (**K**), (**j**), (**V**) and (**p**) are known, then its Interest Portion planned is:

$$I= S'[1+s]-Sv-F-(\{N+K+S'j[1+s]/360+Vp/360\} \\ /[1+\text{/}-U)/\{[1-t][1-d]\}$$

Rule-57112:

If both (**/**), (**U**), (**S'**), (**s**), (**S**), (**v**), (**F**), (**I**), (**N**), (**d**), (**K**), (**j**), (**V**) and (**p**) are known, then its Tax Rate planned is:

$$t= 1-(\{N+K+S'j[1+s]/360+Vp/360\}/[1+\text{/}-U) \\ /(\{S'[1+s]-Sv-F-I\}[1-d])$$

Rule-57113:

If both (**/**), (**U**), (**S'**), (**s**), (**S**), (**v**), (**F**), (**I**), (**t**), (**N**), (**K**), (**j**), (**V**) and (**p**) are known, then its Dividend Payout planned is:

$$d= 1-(\{N+K+S'j[1+s]/360+Vp/360\}/[1+\text{/}-U) \\ /(\{S'[1+s]-Sv-F-I\}[1-t])$$

Steve Asikin ISBN 13: **978-1564912361**, ISBN 10: **1564912363**

Rule-57114:

If both $(I)$, $(U)$, $(S')$, $(s)$, $(S)$, $(v)$, $(F)$, $(I)$, $(t)$, $(d)$, $(N)$, $(j)$, $(V)$ and $(p)$ are known, then its Kind of Cash planned is:

$$K = [1 + I(U + \{S'[1+s] - Sv - F - I\}[1-t][1-d])$$
$$- \{N + S'j[1+s]/360 + Vp/360\}$$

Rule-57115:

If both $(I)$, $(U)$, $(S')$, $(s)$, $(S)$, $(v)$, $(F)$, $(I)$, $(t)$, $(d)$, $(K)$, $(N)$, $(V)$ and $(p)$ are known, then its Job or Trade Receivable Days planned is:

$$j = 360[[1 + I(U + \{S'[1+s] - Sv - F - I\}[1-t][1-d])$$
$$- \{N + K + Vp/360\}]/\{S'[1+s]\}$$

Rule-57116:

If both $(I)$, $(U)$, $(S')$, $(s)$, $(S)$, $(v)$, $(F)$, $(I)$, $(t)$, $(d)$, $(K)$, $(j)$, $(N)$ and $(p)$ are known, then its Variable Cost planned is:

$$V = 360[[1 + I(U + \{S'[1+s] - Sv - F - I\}[1-t][1-d])$$
$$- \{N + K + S'j[1+s]/360\}]/p$$

or

$$V = Sv$$

Rule-57117:

If both $(I)$, $(U)$, $(S')$, $(s)$, $(S)$, $(v)$, $(F)$, $(I)$, $(t)$, $(d)$, $(K)$, $(j)$, $(V)$ and $(N)$ are known, then its Procured Inventory Days planned is:

$$p = 360[[1 + I(U + \{S'[1+s] - Sv - F - I\}[1-t][1-d])$$
$$- \{N + K + S'j[1+s]/360\}]/V$$

Steve Asikin ISBN 13: **978-1564912361**, ISBN 10: **1564912363**

<u>Rule-57118</u>:

If both ($I$), (**U**), (**S'**), (**s**), (**S**), (**v**), (**F**), (**I**), (**t**), (**d**), (**K**), (**j**) and (**p**) are known, then its Non Current Asset planned is:

$$N = [1+I](U+\{S'[1+s]-Sv-F-I\}[1-t][1-d])$$
$$-K-S'j[1+s]/360-Svp/360$$

<u>Rule-57119</u>:

If both (**N**), (**U**), (**S'**), (**s**), (**S**), (**v**), (**F**), (**I**), (**t**), (**d**), (**K**), (**j**) and (**p**) are known, then its Leverage or Gearing Ratio planned is:

$$I = \{N+K+S'j[1+s]/360+Svp/360\}$$
$$/(U+\{S'[1+s]-Sv-F-I\}[1-t][1-d])-1$$

<u>Rule-57120</u>:

If both ($I$), (**N**), (**S'**), (**s**), (**S**), (**v**), (**F**), (**I**), (**t**), (**d**), (**K**), (**j**) and (**p**) are known, then its Utilized or Starting Capital must be:

$$U = (\{N+K+S'j[1+s]/360+Svp/360\}$$
$$-[1+I]\{S'[1+s]-Sv-F-I\}[1-t][1-d])/\{[1+I]\}$$

<u>Rule-57121</u>:

If both ($I$), (**U**), (**S'**), (**s**), (**S**), (**v**), (**F**), (**I**), (**t**), (**d**), (**K**), (**j**) and (**p**) are known, then its Utilized or starting Capital must be:

$$S' = [[1+I](U-\{Sv+F+I\}[1-t][1-d])$$
$$-[N+K+Svp/360]]/([1+s]\{j/360-[1+I]\})$$
$$\text{or}$$
$$S' = S/[1+s]$$

278

Steve Asikin ISBN 13: **978-1564912361**, ISBN 10: **1564912363**

Rule-57122:

If both (**/**), (**U**), (**S'**), (**N**), (**S**), (**v**), (**F**), (**I**), (**t**), (**d**), (**K**), (**j**) and (**p**) are known, then its Sales Growth planned is:

$$s= [[1+/(U-\{Sv+F+I\}[1-t][1-d])$$
$$-[N+K+Svp/360]]/(S'\{j/360-[1+/]\})-1$$

or

$$s= S/S'-1$$


Rule-57123:

If both (**/**), (**U**), (**S'**), (**s**), (**N**), (**v**), (**F**), (**I**), (**t**), (**d**), (**K**), (**j**) and (**p**) are known, then its Sales or Revenue planned is:

$$S= [[1+/(U+\{S'[1+s]-F-I\}[1-t][1-d])$$
$$-\{N+K+S'j[1+s]/360\}]/(v\{p/360+[1+/]\})$$

or

$$S= S'[1+s]$$


Rule-57124:

If both (**/**), (**U**), (**S'**), (**s**), (**S**), (**N**), (**F**), (**I**), (**t**), (**d**), (**K**), (**j**) and (**p**) are known, then its Variable Portion planned is:

$$v= [[1+/(U+\{S'[1+s]-F-I\}[1-t][1-d])$$
$$-\{N+K+S'j[1+s]/360\}]/(S\{p/360+[1+/]\})$$

Steve Asikin ISBN 13: **978-1564912361**, ISBN 10: **1564912363**

<u>Rule-57125</u>:

If both (**/**), (**U**), (**S'**), (**s**), (**S**), (**v**), (**N**), (**I**), (**t**), (**d**), (**K**), (**j**) and (**p**) are known, then its Fixed Cost planned is:

$$F= S'[1+s]-Sv-I-(\{N+K+S'j[1+s]/360+Svp/360\}$$
$$/[1+/]-U)/\{[1-t][1-d]\}$$

<u>Rule-57126</u>:

If both (**/**), (**U**), (**S'**), (**s**), (**S**), (**v**), (**F**), (**N**), (**t**), (**d**), (**K**), (**j**) and (**p**) are known, then its Interest Expense planned is:

$$I= S'[1+s]-Sv-F-(\{N+K+S'j[1+s]/360+Svp/360\}$$
$$/[1+/]-U)/\{[1-t][1-d]\}$$

<u>Rule-57127</u>:

If both (**/**), (**U**), (**S'**), (**s**), (**S**), (**v**), (**F**), (**I**), (**N**), (**d**), (**K**), (**j**) and (**p**) are known, then its Tax Rate planned is:

$$t= 1-(\{N+K+S'j[1+s]/360+Svp/360\}/[1+/]-U)$$
$$/(\{S'[1+s]-Sv-F-I\}[1-d])$$

<u>Rule-57128</u>:

If both (**/**), (**U**), (**S'**), (**s**), (**S**), (**v**), (**F**), (**I**), (**t**), (**N**), (**K**), (**j**) and (**p**) are known, then its Dividend Payout planned is:

$$d= 1-(\{N+K+S'j[1+s]/360+Svp/360\}/[1+/]-U)$$
$$/(\{S'[1+s]-Sv-F-I\}[1-t])$$

Steve Asikin ISBN 13: **978-1564912361**, ISBN 10: **1564912363**

Rule-57129:

If both (**/**), (**U**), (**S'**), (**s**), (**$**), (**v**), (**F**), (**I**), (**t**), (**d**), (**N**), (**j**) and (**p**) are known, then its Kind of Cash planned is:

$$K= [1+/(U+\{S'[1+s]-Sv-F-I\}[1-t][1-d])$$
$$-\{N+S'j[1+s]/360+Svp/360\}$$

Rule-57130:

If both (**/**), (**U**), (**S'**), (**s**), (**$**), (**v**), (**F**), (**I**), (**t**), (**d**), (**K**), (**N**) and (**p**) are known, then its Job or Trade Receivable Days planned is:

$$j= 360[[1+/(U+\{S'[1+s]-Sv-F-I\}[1-t][1-d])$$
$$-\{N+K+Svp/360\}]/\{S'[1+s]\}$$

Rule-57131:

If both (**/**), (**U**), (**S'**), (**s**), (**$**), (**v**), (**F**), (**I**), (**t**), (**d**), (**K**), (**j**) and (**N**) are known, then its Procured Inventory Days planned is:

$$p= 360[[1+/(U+\{S'[1+s]-Sv-F-I\}[1-t][1-d])$$
$$-\{N+K+S'j[1+s]/360\}]/[Sv]$$

Rule-57132:

If both (**/**), (**U**), (**S'**), (**s**), (**$**), (**v**), (**F**), (**I**), (**t**), (**d**), (**K**), (**j**) and (**p**) are known, then its Non Current Asset planned is:

$$N= [1+/(U+\{S'[1+s]-Sv-F-I\}[1-t][1-d])$$
$$-K-S'j[1+s]/360-S'vp[1+s]/360$$

Steve Asikin ISBN 13: **978-1564912361**, ISBN 10: **1564912363**

Rule-57133:

If both $(\mathbf{N})$, $(\mathbf{U})$, $(\mathbf{S'})$, $(\mathbf{s})$, $(\mathbf{S})$, $(\mathbf{v})$, $(\mathbf{F})$, $(\mathbf{I})$, $(\mathbf{t})$, $(\mathbf{d})$, $(\mathbf{K})$, $(\mathbf{j})$ and $(\mathbf{p})$ are known, then its Leverage or Gearing Ratio planned is:

$$I = \{\mathbf{N}+\mathbf{K}+\mathbf{S'j}[1+\mathbf{s}]/360+\mathbf{S'vp}[1+\mathbf{s}]/360\}$$
$$/(\mathbf{U}+\{\mathbf{S'}[1+\mathbf{s}]-\mathbf{Sv}-\mathbf{F}-\mathbf{I}\}[1-\mathbf{t}][1-\mathbf{d}])-1$$

Rule-57134:

If both $(I)$, $(\mathbf{N})$, $(\mathbf{S'})$, $(\mathbf{s})$, $(\mathbf{S})$, $(\mathbf{v})$, $(\mathbf{F})$, $(\mathbf{I})$, $(\mathbf{t})$, $(\mathbf{d})$, $(\mathbf{K})$, $(\mathbf{j})$ and $(\mathbf{p})$ are known, then its Utilized or Starting Capital must be:

$$\mathbf{U}= (\{\mathbf{N}+\mathbf{K}+\mathbf{S'j}[1+\mathbf{s}]/360+\mathbf{S'vp}[1+\mathbf{s}]/360\}$$
$$-[1+I]\{\mathbf{S'}[1+\mathbf{s}]-\mathbf{Sv}-\mathbf{F}-\mathbf{I}\}[1-\mathbf{t}][1-\mathbf{d}])/[1+I]$$

Rule-57135:

If both $(I)$, $(\mathbf{U})$, $(\mathbf{N})$, $(\mathbf{s})$, $(\mathbf{S})$, $(\mathbf{v})$, $(\mathbf{F})$, $(\mathbf{I})$, $(\mathbf{t})$, $(\mathbf{d})$, $(\mathbf{K})$, $(\mathbf{j})$ and $(\mathbf{p})$ are known, then its Sales Past must be:

$$\mathbf{S'}= [[1+I](\mathbf{U}-\{\mathbf{Sv}+\mathbf{F}+\mathbf{I}\}[1-\mathbf{t}][1-\mathbf{d}])-[\mathbf{N}+\mathbf{K}]]$$
$$/([1+\mathbf{s}]\{\mathbf{j}/360+\mathbf{vp}/360-[1+I]\})$$

or

$$\mathbf{S'}= \mathbf{S}/[1+\mathbf{s}]$$

Steve Asikin ISBN 13: **978-1564912361**, ISBN 10: **1564912363**

Rule-57136:

If both (**∫**), (**U**), (**S'**), (**N**), (**S**), (**v**), (**F**), (**I**), (**t**), (**d**), (**K**), (**j**) and (**p**) are known, then its Sales Growth planned is:

$$s= [[1+\text{∫}](U-\{Sv+F+I\}[1-t][1-d])-[N+K]]$$
$$/(S'\{j/360+vp/360-[1+\text{∫}]\})-1$$

or

$$s= S/S'-1$$

Rule-57137:

If both (**∫**), (**U**), (**S'**), (**s**), (**N**), (**v**), (**F**), (**I**), (**t**), (**d**), (**K**), (**j**) and (**p**) are known, then its Sales or Revenue planned is:

$$S= [S'[1+s]-F-I-(\{N+K+S'j[1+s]/360$$
$$+S'vp[1+s]/360\}/[1+\text{∫}-U)$$
$$/\{[1-t][1-d]\}]/v$$

or

$$S= S'[1+s]$$

Rule-57138:

If both (**∫**), (**U**), (**S'**), (**s**), (**S**), (**N**), (**F**), (**I**), (**t**), (**d**), (**K**), (**j**) and (**p**) are known, then its Variable Portion planned is:

$$v= [S'[1+s]-F-I-(\{N+K+S'j[1+s]/360$$
$$+S'vp[1+s]/360\}/[1+\text{∫}-U)$$
$$/\{[1-t][1-d]\}]/S$$

Steve Asikin ISBN 13: **978-1564912361**, ISBN 10: **1564912363**

Rule-57139:

If both (**/**), (**U**), (**S'**), (**s**), (**S**), (**v**), (**N**), (**I**), (**t**), (**d**), (**K**), (**j**) and (**p**) are known, then its Fixed Cost planned is:

$$F= S'[1+s]-Sv-I-(\{N+K+S'j[1+s]/360 +S'vp[1+s]/360\}/[1+/-U)/\{[1-t][1-d]\}$$

Rule-57140:

If both (**/**), (**U**), (**S'**), (**s**), (**S**), (**v**), (**F**), (**N**), (**t**), (**d**), (**K**), (**j**) and (**p**) are known, then its Interest Expense planned is:

$$I= S'[1+s]-Sv-F-(\{N+K+S'j[1+s]/360 +S'vp[1+s]/360\}/[1+/-U)/\{[1-t][1-d]\}$$

Rule-57141:

If both (**/**), (**U**), (**S'**), (**s**), (**S**), (**v**), (**F**), (**I**), (**t**), (**N**), (**K**), (**j**) and (**p**) are known, then its Dividend Payout planned is:

$$d= 1-(\{N+K+S'j[1+s]/360+S'vp[1+s]/360\} /[1+/-U)/(\{S'[1+s]-Sv-F-I\}[1-t])$$

Rule-57142:

If both (**/**), (**U**), (**S'**), (**s**), (**S**), (**v**), (**F**), (**I**), (**N**), (**d**), (**K**), (**j**) and (**p**) are known, then its Tax Rate planned is:

$$t= 1-(\{N+K+S'j[1+s]/360+S'vp[1+s]/360\} /[1+/-U)/(\{S'[1+s]-Sv-F-I\}[1-d])$$

Steve Asikin ISBN 13: **978-1564912361**, ISBN 10: **1564912363**

Rule-57143:
If both (**/**), (**U**), (**S'**), (**s**), (**S**), (**v**), (**F**), (**I**), (**t**), (**d**), (**N**), (**j**) and (**p**) are known, then its Kind of Cash planned is:

$$\mathbf{K}= [1+\mathbf{/}(\mathbf{U}+\{\mathbf{S'}[1+\mathbf{s}]-\mathbf{Sv}-\mathbf{F}-\mathbf{I}\}[1-\mathbf{t}][1-\mathbf{d}])$$
$$-\{\mathbf{N}+\mathbf{S'j}[1+\mathbf{s}]/360+\mathbf{S'vp}[1+\mathbf{s}]/360\}$$

Rule-57144:
If both (**/**), (**U**), (**S'**), (**s**), (**S**), (**v**), (**F**), (**I**), (**t**), (**d**), (**K**), (**N**) and (**p**) are known, then its Job or Trade Receivable Days planned is:

$$\mathbf{j}= 360[[1+\mathbf{/}(\mathbf{U}+\{\mathbf{S'}[1+\mathbf{s}]-\mathbf{Sv}-\mathbf{F}-\mathbf{I}\}[1-\mathbf{t}][1-\mathbf{d}])$$
$$-\{\mathbf{N}+\mathbf{K}+\mathbf{S'vp}[1+\mathbf{s}]/360\}]/\{\mathbf{S'}[1+\mathbf{s}]\}$$

Rule-57145:
If both (**/**), (**U**), (**S'**), (**s**), (**S**), (**v**), (**F**), (**I**), (**t**), (**d**), (**K**), (**j**) and (**N**) are known, then its Procured Inventory Days planned is:

$$\mathbf{p}= 360[[1+\mathbf{/}(\mathbf{U}+\{\mathbf{S'}[1+\mathbf{s}]-\mathbf{Sv}-\mathbf{F}-\mathbf{I}\}[1-\mathbf{t}][1-\mathbf{d}])$$
$$-\{\mathbf{N}+\mathbf{K}+\mathbf{S'j}[1+\mathbf{s}]/360\}]/\{\mathbf{S'v}[1+\mathbf{s}]\}$$

Rule-57146:
If both (**/**), (**U**), (**S'**), (**s**), (**S**), (**v**), (**F**), (**i**), (**T**), (**D**), (**K**), (**J**) and (**P**) are known, then its Non Current Asset planned is:

$$\mathbf{N}= [1+\mathbf{/}\{\mathbf{U}+\mathbf{S'}[1+\mathbf{s}]-\mathbf{Sv}-\mathbf{F}-\mathbf{Si}-\mathbf{T}-\mathbf{D}\}-\mathbf{K}-\mathbf{J}-\mathbf{P}$$

Steve Asikin ISBN 13: **978-1564912361**, ISBN 10: **1564912363**

Rule-57147:

If both $(\mathbf{N})$, $(\mathbf{U})$, $(\mathbf{S'})$, $(\mathbf{s})$, $(\mathbf{S})$, $(\mathbf{v})$, $(\mathbf{F})$, $(\mathbf{i})$, $(\mathbf{T})$, $(\mathbf{D})$, $(\mathbf{K})$, $(\mathbf{J})$ and $(\mathbf{P})$ are known, then its Leverage or Gearing Ratio planned is:

$$l = \{\mathbf{N+K+J+P}\}/(\mathbf{U}+\{\mathbf{S'}[1+\mathbf{s}]-\mathbf{Sv-F-Si}\}-\mathbf{T-D})-1$$

Rule-57148:

If both $(\mathbf{l})$, $(\mathbf{N})$, $(\mathbf{S'})$, $(\mathbf{s})$, $(\mathbf{S})$, $(\mathbf{v})$, $(\mathbf{F})$, $(\mathbf{i})$, $(\mathbf{T})$, $(\mathbf{D})$, $(\mathbf{K})$, $(\mathbf{J})$ and $(\mathbf{P})$ are known, then its Utilized or Starting Capital must be:

$$\mathbf{U} = (\{\mathbf{N+K+J+P}\}-[1+l]\{\mathbf{S'}[1+\mathbf{s}]-\mathbf{Sv-Si}\}-\mathbf{T-D})/[1+l]$$

Rule-57149:

If both $(\mathbf{l})$, $(\mathbf{U})$, $(\mathbf{N})$, $(\mathbf{s})$, $(\mathbf{S})$, $(\mathbf{v})$, $(\mathbf{F})$, $(\mathbf{i})$, $(\mathbf{T})$, $(\mathbf{D})$, $(\mathbf{K})$, $(\mathbf{J})$ and $(\mathbf{P})$ are known, then its Sales Past must be:

$$\mathbf{S'} = \{[\mathbf{N+K+J+P}]/[1+l]-\mathbf{U+Sv+F+Si+T+D}\}/[1+\mathbf{s}]$$

or

$$\mathbf{S'} = \mathbf{S}/[1+\mathbf{s}]$$

Rule-57150:

If both $(\mathbf{l})$, $(\mathbf{U})$, $(\mathbf{S'})$, $(\mathbf{N})$, $(\mathbf{S})$, $(\mathbf{v})$, $(\mathbf{F})$, $(\mathbf{i})$, $(\mathbf{T})$, $(\mathbf{D})$, $(\mathbf{K})$, $(\mathbf{J})$ and $(\mathbf{P})$ are known, then its Sales Growth planned is:

$$\mathbf{s} = \{[\mathbf{N+K+J+P}]/[1+l]-\mathbf{U+Sv+F+Si+T+D}\}/\mathbf{S'}-1$$

or

$$\mathbf{s} = \mathbf{S}/\mathbf{S'}-1$$

Steve Asikin ISBN 13: **978-1564912361**, ISBN 10: **1564912363**

Rule-57151:

If both $(\textit{I})$, $(\mathbf{U})$, $(\mathbf{S'})$, $(\mathbf{s})$, $(\mathbf{S})$, $(\mathbf{v})$, $(\mathbf{F})$, $(\mathbf{i})$, $(\mathbf{T})$, $(\mathbf{D})$, $(\mathbf{K})$, $(\mathbf{J})$ and $(\mathbf{P})$ are known, then its Non Current Asset planned is:

$$S= (U+\{S'[1+s]-F-T\}-D-[N+K+J+P]/[1+\textit{I}])/[v+i]$$

or

$$S= S'[1+s]$$

Rule-57152:

If both $(\textit{I})$, $(\mathbf{U})$, $(\mathbf{S'})$, $(\mathbf{s})$, $(\mathbf{S})$, $(\mathbf{N})$, $(\mathbf{F})$, $(\mathbf{i})$, $(\mathbf{T})$, $(\mathbf{D})$, $(\mathbf{K})$, $(\mathbf{J})$ and $(\mathbf{P})$ are known, then its Variable Portion planned is:

$$v= (U+\{S'[1+s]-Si-F-T\}-D-[N+K+J+P]/[1+\textit{I}])/S$$

Rule-57153:

If both $(\textit{I})$, $(\mathbf{U})$, $(\mathbf{S'})$, $(\mathbf{s})$, $(\mathbf{S})$, $(\mathbf{v})$, $(\mathbf{N})$, $(\mathbf{i})$, $(\mathbf{T})$, $(\mathbf{D})$, $(\mathbf{K})$, $(\mathbf{J})$ and $(\mathbf{P})$ are known, then its Fixed Cost planned is:

$$F= U+\{S'[1+s]-Sv-Si-T\}-D-[N+K+J+P]/[1+\textit{I}]$$

Rule-57154:

If both $(\textit{I})$, $(\mathbf{U})$, $(\mathbf{S'})$, $(\mathbf{s})$, $(\mathbf{S})$, $(\mathbf{v})$, $(\mathbf{F})$, $(\mathbf{N})$, $(\mathbf{T})$, $(\mathbf{D})$, $(\mathbf{K})$, $(\mathbf{J})$ and $(\mathbf{P})$ are known, then its Interest Portion planned is:

$$i= (U+\{S'[1+s]-Sv-F-T\}-D-[N+K+J+P]/[1+\textit{I}])/S$$

Steve Asikin ISBN 13: **978-1564912361**, ISBN 10: **1564912363**

Rule-57155:

If both $(I)$, $(U)$, $(S')$, $(s)$, $(S)$, $(v)$, $(F)$, $(i)$, $(N)$, $(D)$, $(K)$, $(J)$ and $(P)$ are known, then its Tax planned is:

$$T= U+\{S'[1+s]-Sv-F-Si\}-D-[N+K+J+P]/[1+I]$$

Rule-57156:

If both $(I)$, $(U)$, $(S')$, $(s)$, $(S)$, $(v)$, $(F)$, $(i)$, $(T)$, $(N)$, $(K)$, $(J)$ and $(P)$ are known, then its Dividend planned is:

$$D= U+\{S'[1+s]-Sv-F-Si\}-T-[N+K+J+P]/[1+I]$$

Rule-57157:

If both $(I)$, $(U)$, $(S')$, $(s)$, $(S)$, $(v)$, $(F)$, $(i)$, $(T)$, $(D)$, $(N)$, $(J)$ and $(P)$ are known, then its Kind of Cash planned is:

$$K= [1+I](U+\{S'[1+s]-Sv-F-Si\}-T-D)-[N+J+P]$$

Rule-57158:

If both $(I)$, $(U)$, $(S')$, $(s)$, $(S)$, $(v)$, $(F)$, $(i)$, $(T)$, $(D)$, $(K)$, $(N)$ and $(P)$ are known, then its Job or Trade Account Receivable planned is:

$$J= [1+I](U+\{S'[1+s]-Sv-F-Si\}-T-D)-[N+K+P]$$

Rule-57159:

If both $(I)$, $(U)$, $(S')$, $(s)$, $(S)$, $(v)$, $(F)$, $(i)$, $(T)$, $(D)$, $(K)$, $(J)$ and $(N)$ are known, then its Procured Inventory planned is:

$$P= [1+I](U+\{S'[1+s]-Sv-F-Si\}-T-D)-[N+K+J]$$

Steve Asikin ISBN 13: **978-1564912361**, ISBN 10: **1564912363**

Rule-57160:

If both $(\textit{I})$, $(\textbf{U})$, $(\textbf{S'})$, $(\textbf{s})$, $(\textbf{S})$, $(\textbf{v})$, $(\textbf{F})$, $(\textbf{i})$, $(\textbf{T})$, $(\textbf{D})$, $(\textbf{K})$, $(\textbf{J})$, $(\textbf{V})$ and $(\textbf{p})$ are known, then its Non Current Asset planned is:

$$\textbf{N} = [1 + \textit{I}\{\textbf{U} + \textbf{S'}[1+\textbf{s}] - \textbf{Sv} - \textbf{F} - \textbf{Si} - \textbf{T} - \textbf{D}\} - \textbf{K} - \textbf{J} - \textbf{Vp}/360$$

Rule-57161:

If both $(\textbf{N})$, $(\textbf{U})$, $(\textbf{S'})$, $(\textbf{s})$, $(\textbf{S})$, $(\textbf{v})$, $(\textbf{F})$, $(\textbf{i})$, $(\textbf{T})$, $(\textbf{D})$, $(\textbf{K})$, $(\textbf{J})$, $(\textbf{V})$ and $(\textbf{p})$ are known, then its Leverage or Gearing Ratio planned is:

$$\textit{I} = [\textbf{N} + \textbf{K} + \textbf{J} + \textbf{Vp}/360]$$
$$/(\textbf{U} + \{\textbf{S'}[1+\textbf{s}] - \textbf{Sv} - \textbf{F} - \textbf{Si}\} - \textbf{T} - \textbf{D}) - 1$$

Rule-57162:

If both $(\textit{I})$, $(\textbf{N})$, $(\textbf{S'})$, $(\textbf{s})$, $(\textbf{S})$, $(\textbf{v})$, $(\textbf{F})$, $(\textbf{i})$, $(\textbf{T})$, $(\textbf{D})$, $(\textbf{K})$, $(\textbf{J})$, $(\textbf{V})$ and $(\textbf{p})$ are known, then its Utilized or Starting Capital must be:

$$\textbf{U} = ([\textbf{N} + \textbf{K} + \textbf{J} + \textbf{Vp}/360] - [1 + \textit{I}\{\textbf{S'}[1+\textbf{s}] - \textbf{Sv} - \textbf{F} - \textbf{Si}\}$$
$$- \textbf{T} - \textbf{D})/[1 + \textit{I}]$$

Rule-57163:

If both $(\textit{I})$, $(\textbf{U})$, $(\textbf{N})$, $(\textbf{s})$, $(\textbf{S})$, $(\textbf{v})$, $(\textbf{F})$, $(\textbf{i})$, $(\textbf{T})$, $(\textbf{D})$, $(\textbf{K})$, $(\textbf{J})$, $(\textbf{V})$ and $(\textbf{p})$ are known, then its Sales Past must be:

$$\textbf{S'} = (\{[\textbf{N} + \textbf{K} + \textbf{J} + \textbf{Vp}/360]/[1 + \textit{I}] - \textbf{U}\}$$
$$+ \textbf{Sv} + \textbf{F} + \textbf{Si} + \textbf{T} + \textbf{D})/[1 + \textbf{s}]$$
$$\text{or}$$
$$\textbf{S'} = \textbf{S}/[1 + \textbf{s}]$$

Steve Asikin ISBN 13: **978-1564912361**, ISBN 10: **1564912363**

Rule-57164:

If both (**/**), (**U**), (**S'**), (**N**), (**S**), (**v**), (**F**), (**i**), (**T**), (**D**), (**K**), (**J**), (**V**) and (**p**) are known, then its Sales Growth planned is:

$$s = (\{[N+K+J+Vp/360]/[1+/]-U\}$$
$$+Sv+F+Si+T+D)/S'-1$$

or

$$s = S/S'-1$$

Rule-57165:

If both (**/**), (**U**), (**S'**), (**s**), (**N**), (**v**), (**F**), (**i**), (**T**), (**D**), (**K**), (**J**), (**V**) and (**p**) are known, then its Sales or Revenue planned is:

$$S = (U+\{S'[1+s]-F\}-T-D$$
$$-[N+K+J+Vp/360]/[1+/])/[v+i]$$

or

$$S = S'[1+s]$$

or

$$S = V/v$$

Rule-57166:

If both (**/**), (**U**), (**S'**), (**s**), (**S**), (**N**), (**F**), (**i**), (**T**), (**D**), (**K**), (**J**), (**V**) and (**p**) are known, then its Variable Portion planned is:

$$v = (U+\{S'[1+s]-F-Si\}-T-D$$
$$-[N+K+J+Vp/360]/[1+/])/S$$

or

$$v = V/S$$

Steve Asikin ISBN 13: **978-1564912361**, ISBN 10: **1564912363**

Rule-57167:

If both (**/**), (**U**), (**S'**), (**s**), (**S**), (**v**), (**N**), (**i**), (**T**), (**D**), (**K**), (**J**), (**V**) and (**p**) are known, then its Fixed Cost planned is:

$$F = (U + \{S'[1+s] - Sv - Si\} - T - D$$
$$- [N + K + J + Vp/360]/[1 + I]$$

Rule-57168:

If both (**/**), (**U**), (**S'**), (**s**), (**S**), (**v**), (**F**), (**N**), (**T**), (**D**), (**K**), (**J**), (**V**) and (**p**) are known, then its Interest Portion planned is:

$$i = (U + \{S'[1+s] - Sv - F\} - T - D$$
$$- [N + K + J + Vp/360]/[1 + I])/S$$

Rule-57169:

If both (**/**), (**U**), (**S'**), (**s**), (**S**), (**v**), (**F**), (**i**), (**N**), (**D**), (**K**), (**J**), (**V**) and (**p**) are known, then its Tax planned is:

$$T = U + \{S'[1+s] - Sv - F - Si\} - D$$
$$- [N + K + J + Vp/360]/[1 + I]$$

Rule-57170:

If both (**/**), (**U**), (**S'**), (**s**), (**S**), (**v**), (**F**), (**i**), (**T**), (**N**), (**K**), (**J**), (**V**) and (**p**) are known, then its Dividend planned is:

$$D = U + \{S'[1+s] - Sv - F - Si\} - T$$
$$- [N + K + J + Vp/360]/[1 + I]$$

Steve Asikin ISBN 13: **978-1564912361**, ISBN 10: **1564912363**

Rule-57171:

If both (**/**), (**U**), (**S'**), (**s**), (**S**), (**v**), (**F**), (**i**), (**T**), (**D**), (**N**), (**J**), (**V**) and (**p**) are known, then its Kind of Cash planned is:

$$K = [1 + /(U + \{S'[1+s] - Sv - F - Si\} - T - D) - [N + J + Vp/360]$$

Rule-57172:

If both (**/**), (**U**), (**S'**), (**s**), (**S**), (**v**), (**F**), (**i**), (**T**), (**D**), (**K**), (**N**), (**V**) and (**p**) are known, then its Job or Trade Account Receivable planned is:

$$J = [1 + /(U + \{S'[1+s] - Sv - F - Si\} - T - D) - \{N + K + Vp/360\}$$

Rule-57173:

If both (**/**), (**U**), (**S'**), (**s**), (**S**), (**v**), (**F**), (**i**), (**T**), (**D**), (**K**), (**J**), (**N**) and (**p**) are known, then its Variable Cost planned is:

$$V = 360[[1 + /(U + \{S'[1+s] - Sv - F - Si\} - T - D) - [N + K + J]]/p$$

or

$$V = Sv$$

Rule-57174:

If both (**/**), (**U**), (**S'**), (**s**), (**S**), (**v**), (**F**), (**i**), (**T**), (**D**), (**K**), (**J**), (**V**) and (**N**) are known, then its Procured Inventory Days planned is:

$$p = 360[[1 + /(U + \{S'[1+s] - Sv - F - Si\} - T - D) - [N + K + J]]/V$$

Steve Asikin ISBN 13: **978-1564912361**, ISBN 10: **1564912363**

Rule-57175:

   If both $(\boldsymbol{I})$, $(\mathbf{U})$, $(\mathbf{S'})$, $(\mathbf{s})$, $(\mathbf{S})$, $(\mathbf{v})$, $(\mathbf{F})$, $(\mathbf{i})$, $(\mathbf{T})$, $(\mathbf{D})$, $(\mathbf{K})$, $(\mathbf{J})$ and $(\mathbf{p})$ are known, then its Non Current Asset planned is:

$$\mathbf{N} = [1+\boldsymbol{I}]\{\mathbf{U}+\mathbf{S'}[1+\mathbf{s}]-\mathbf{Sv}-\mathbf{F}-\mathbf{Si}-\mathbf{T}-\mathbf{D}\}-\mathbf{K}-\mathbf{J}-\mathbf{Svp}/360$$

Rule-57176:

   If both $(\mathbf{N})$, $(\mathbf{U})$, $(\mathbf{S'})$, $(\mathbf{s})$, $(\mathbf{S})$, $(\mathbf{v})$, $(\mathbf{F})$, $(\mathbf{i})$, $(\mathbf{T})$, $(\mathbf{D})$, $(\mathbf{K})$, $(\mathbf{J})$ and $(\mathbf{p})$ are known, then its Leverage or Gearing Ratio planned is:

$$\boldsymbol{I} = [\mathbf{N}+\mathbf{K}+\mathbf{J}+\mathbf{Svp}/360]$$
$$/(\mathbf{U}+\{\mathbf{S'}[1+\mathbf{s}]-\mathbf{Sv}-\mathbf{F}-\mathbf{Si}\}-\mathbf{T}-\mathbf{D})-1$$

Rule-57177:

   If both $(\boldsymbol{I})$, $(\mathbf{N})$, $(\mathbf{S'})$, $(\mathbf{s})$, $(\mathbf{S})$, $(\mathbf{v})$, $(\mathbf{F})$, $(\mathbf{i})$, $(\mathbf{T})$, $(\mathbf{D})$, $(\mathbf{K})$, $(\mathbf{J})$ and $(\mathbf{p})$ are known, then its Utilized or Starting Capital must be:

$$\mathbf{U} = ([\mathbf{N}+\mathbf{K}+\mathbf{J}+\mathbf{Svp}/360]$$
$$-[1+\boldsymbol{I}]\{\mathbf{S'}[1+\mathbf{s}]-\mathbf{Sv}-\mathbf{F}-\mathbf{Si}-\mathbf{T}-\mathbf{D})/[1+\boldsymbol{I}]$$

Rule-57178:

   If both $(\boldsymbol{I})$, $(\mathbf{U})$, $(\mathbf{N})$, $(\mathbf{s})$, $(\mathbf{S})$, $(\mathbf{v})$, $(\mathbf{F})$, $(\mathbf{i})$, $(\mathbf{T})$, $(\mathbf{D})$, $(\mathbf{K})$, $(\mathbf{J})$ and $(\mathbf{p})$ are known, then its Sales Past must be:

$$\mathbf{S'} = ([\mathbf{N}+\mathbf{K}+\mathbf{J}+\mathbf{Svp}/360]/[1+\boldsymbol{I}]$$
$$-\mathbf{U}+\mathbf{Sv}+\mathbf{F}+\mathbf{Si}+\mathbf{T}+\mathbf{D})/[1+\mathbf{s}]$$

        or

$$\mathbf{S'} = \mathbf{S}/[1+\mathbf{s}]$$

Steve Asikin ISBN 13: **978-1564912361**, ISBN 10: **1564912363**

Rule-57179:

If both (**/**), (**U**), (**$'**), (**N**), (**$**), (**v**), (**F**), (**i**), (**T**), (**D**), (**K**), (**J**) and (**p**) are known, then its Sales Growth planned is:

$$s = ([\mathbf{N}+\mathbf{K}+\mathbf{J}+\mathbf{Svp}/360]/[1+\mathbf{/}]$$
$$-\mathbf{U}+\mathbf{Sv}+\mathbf{F}+\mathbf{Si}+\mathbf{T}+\mathbf{D})/\mathbf{\$'}-1$$

or

$$s = \mathbf{\$}/\mathbf{\$'}-1$$

Rule-57180:

If both (**/**), (**U**), (**$'**), (**s**), (**N**), (**v**), (**F**), (**i**), (**T**), (**D**), (**K**), (**J**) and (**p**) are known, then its Sales or Revenue planned is:

$$\mathbf{\$} = [[1+\mathbf{/}](\mathbf{U}+\{\mathbf{\$'}[1+\mathbf{s}]-\mathbf{F}\}-\mathbf{T}-\mathbf{D})$$
$$-[\mathbf{N}+\mathbf{K}+\mathbf{J}]]/(\{\mathbf{p}/360+[1+\mathbf{/}][\mathbf{v}+\mathbf{i}]\})$$

or

$$\mathbf{\$} = \mathbf{\$'}[1+\mathbf{s}]$$

Rule-57181:

If both (**/**), (**U**), (**$'**), (**s**), (**$**), (**N**), (**F**), (**i**), (**T**), (**D**), (**K**), (**J**) and (**p**) are known, then its Variable Portion planned is:

$$\mathbf{v} = [[1+\mathbf{/}](\mathbf{U}+\{\mathbf{\$'}[1+\mathbf{s}]-\mathbf{F}-\mathbf{Si}\}-\mathbf{T}-\mathbf{D})-[\mathbf{N}+\mathbf{K}+\mathbf{J}]]$$
$$/(\mathbf{\$}\{\mathbf{p}/360+[1+\mathbf{/}]\})$$

Steve Asikin ISBN 13: **978-1564912361**, ISBN 10: **1564912363**

Rule-57182:

If both (**/**), (**U**), (**S'**), (**s**), (**S**), (**v**), (**N**), (**i**), (**T**), (**D**), (**K**), (**J**) and (**p**) are known, then its Fixed Cost planned is:
$$F= (U+\{S'[1+s]-Sv-Si\}-T-D$$
$$-[N+K+J+Svp]/360]/[1+/]$$

Rule-57183:

If both (**/**), (**U**), (**S'**), (**s**), (**S**), (**v**), (**F**), (**N**), (**T**), (**D**), (**K**), (**J**) and (**p**) are known, then its Interest Portion planned is:
$$i= (U+\{S'[1+s]-Sv-F\}-T-D$$
$$-[N+K+J+Svp]/360]/[1+/])/S$$

Rule-57184:

If both (**/**), (**U**), (**S'**), (**s**), (**S**), (**v**), (**F**), (**i**), (**N**), (**D**), (**K**), (**J**) and (**p**) are known, then its Tax planned is:
$$T= U+\{S'[1+s]-Sv-F-Si\}-D$$
$$-[N+K+J+Svp]/360]/[1+/]$$

Rule-57185:

If both (**/**), (**U**), (**S'**), (**s**), (**S**), (**v**), (**F**), (**i**), (**T**), (**N**), (**K**), (**J**) and (**p**) are known, then its Dividend planned is:
$$D= U+\{S'[1+s]-Sv-F-Si\}-T$$
$$-[N+K+J+Svp]/360]/[1+/]$$

Steve Asikin ISBN 13: **978-1564912361**, ISBN 10: **1564912363**

Rule-57186:

If both (**/**), (**U**), (**S'**), (**s**), (**S**), (**v**), (**F**), (**i**), (**T**), (**D**), (**N**), (**J**) and (**p**) are known, then its Kind of Cash planned is:

$$K= [1+/(U+\{S'[1+s]-Sv-F-Si\}-T-D)$$
$$-[N+J+Svp/360]$$

Rule-57187:

If both (**/**), (**U**), (**S'**), (**s**), (**S**), (**v**), (**F**), (**i**), (**T**), (**D**), (**K**), (**N**) and (**p**) are known, then its Job or Trade Account Receivable planned is:

$$J= [1+/(U+\{S'[1+s]-Sv-F-Si\}-T-D)$$
$$-\{N+K+Svp/360\}$$

Rule-57188:

If both (**/**), (**U**), (**S'**), (**s**), (**S**), (**v**), (**F**), (**i**), (**T**), (**D**), (**K**), (**J**) and (**N**) are known, then its Procured Inventory Days planned is:

$$p= 360[[1+/(U+\{S'[1+s]-Sv-F-Si\}-T-D)$$
$$-[N+K+J]]/[Sv]$$

Rule-57189:

If both (**/**), (**U**), (**S'**), (**s**), (**S**), (**v**), (**F**), (**i**), (**T**), (**D**), (**K**), (**J**) and (**p**) are known, then its Non Current Asset planned is:

$$N= [1+/\{U+S'[1+s]-Sv-F-Si-T-D\}$$
$$-K-J-S'vp[1+s]/360$$

Steve Asikin ISBN 13: **978-1564912361**, ISBN 10: **1564912363**

Rule-57190:

If both (**N**), (**U**), (**S'**), (**s**), (**S**), (**v**), (**F**), (**i**), (**T**), (**D**), (**K**), (**J**) and (**p**) are known, then its Leverage or Gearing Ratio planned is:

$$I = \{N+K+J+S'vp[1+s]/360\}$$
$$/(U+\{S'[1+s]-Sv-F-Si\}-T-D)-1$$


Rule-57191:

If both (**I**), (**N**), (**S'**), (**s**), (**S**), (**v**), (**F**), (**i**), (**T**), (**D**), (**K**), (**J**) and (**p**) are known, then its Utilized or Starting Capital must be:

$$U = (\{N+K+J+S'vp[1+s]/360\}$$
$$-[1+I]\{S'[1+s]-Sv-F-Si\}-T-D)/[1+I]$$


Rule-57192:

If both (**I**), (**U**), (**S'**), (**s**), (**S**), (**v**), (**F**), (**i**), (**T**), (**D**), (**K**), (**J**) and (**p**) are known, then its Non Current Asset planned is:

$$S' = [[1+I](U-\{Sv+F+Si\}-T-D)-[N+K+J]]$$
$$/([1+s]\{vp/360-[1+I]\})$$

or

$$S' = S/[1+s]$$

Steve Asikin ISBN 13: **978-1564912361**, ISBN 10: **1564912363**

Rule-57193:

If both (**/**), (**U**), (**S'**), (**N**), (**S**), (**v**), (**F**), (**i**), (**T**), (**D**), (**K**), (**J**) and (**p**) are known, then its Sales Growth planned is:

$$s= [[1+/](U-\{Sv+F+Si\}-T-D)-[N+K+J]]$$
$$/(S'\{vp/360-[1+/]\})-1$$

or

$$s= S/S'-1$$

Rule-57194:

If both (**/**), (**U**), (**S'**), (**s**), (**N**), (**v**), (**F**), (**i**), (**T**), (**D**), (**K**), (**J**) and (**p**) are known, then its Sales or Revenue planned is:

$$S=( U+\{S'[1+s]-F\}-T-D-\{N+K+J$$
$$+S'vp[1+s]/360\}/[1+/])/[v+i]$$

or

$$S= S'[1+s]$$

Rule-57195:

If both (**/**), (**U**), (**S'**), (**s**), (**S**), (**N**), (**F**), (**i**), (**T**), (**D**), (**K**), (**J**) and (**p**) are known, then its Variable Portion planned is:

$$v= [[1+/](U+\{S'[1+s]-F-Si\}-T-D)-[N+J+K]]$$
$$/\{S'p[1+s]/360+S[1+/]\}$$

Steve Asikin ISBN 13: **978-1564912361**, ISBN 10: **1564912363**

Rule-57196:

If both (**$I$**), (**U**), (**S'**), (**s**), (**$**), (**v**), (**N**), (**i**), (**T**), (**D**), (**K**), (**J**) and (**p**) are known, then its Fixed Cost planned is:

$$F= U+\{S'[1+s]-Sv-Si\}-T-D$$
$$-\{N+K+J+S'vp[1+s]/360\}/[1+I]$$


Rule-57197:

If both (**$I$**), (**U**), (**S'**), (**s**), (**$**), (**v**), (**F**), (**N**), (**T**), (**D**), (**K**), (**J**) and (**p**) are known, then its Interest Portion planned is:

$$i=( U+\{S'[1+s]-Sv-F\}-T-D$$
$$-\{N+K+J+S'vp[1+s]/360\}/[1+I])/S$$


Rule-57198:

If both (**$I$**), (**U**), (**S'**), (**s**), (**$**), (**v**), (**F**), (**i**), (**N**), (**D**), (**K**), (**J**) and (**p**) are known, then its Tax planned is:

$$T= U+\{S'[1+s]-Sv-F-Si\}-D$$
$$-\{N+K+J+S'vp[1+s]/360\}/[1+I]$$


Rule-57199:

If both (**$I$**), (**U**), (**S'**), (**s**), (**$**), (**v**), (**F**), (**i**), (**T**), (**N**), (**K**), (**J**) and (**p**) are known, then its Dividend planned is:

$$D= U+\{S'[1+s]-Sv-F-Si\}-T$$
$$-\{N+K+J+S'vp[1+s]/360\}/[1+I]$$

Steve Asikin ISBN 13: **978-1564912361**, ISBN 10: **1564912363**

Rule-57200:

If both (**/**), (**U**), (**S'**), (**s**), (**S**), (**v**), (**F**), (**i**), (**T**), (**D**), (**N**), (**J**) and (**p**) are known, then its Kind of Cash planned is:

$$K= [1+/](U+\{S'[1+s]-Sv-F-Si\}-T-D)$$
$$-\{N+J+S'vp[1+s]/360\}$$

Rule-57201:

If both (**/**), (**U**), (**S'**), (**s**), (**S**), (**v**), (**F**), (**i**), (**T**), (**D**), (**K**), (**N**) and (**p**) are known, then its Job or Trade Account Receivable planned is:

$$J= [[1+/](U+\{S'[1+s]-Sv-F-Si\}-T-D)$$
$$-\{N+K+S'vp[1+s]/360\}$$

Rule-57202:

If both (**/**), (**U**), (**S'**), (**s**), (**S**), (**v**), (**F**), (**i**), (**T**), (**D**), (**K**), (**J**) and (**N**) are known, then its Procured Inventory Days planned is:

$$p= 360[[1+/](U+\{S'[1+s]-Sv-F-Si\}-T-D)$$
$$-[N+K+J]]/\{S'v[1+s]\}$$

Rule-57203:

If both (**/**), (**U**), (**S'**), (**s**), (**S**), (**v**), (**F**), (**i**), (**T**), (**D**), (**K**), (**j**) and (**P**) are known, then its Non Current Asset planned is:

$$N= [1+/]\{U+S'[1+s]-Sv-F-Si-T-D\}-K-Sj/360-P$$

Steve Asikin ISBN 13: **978-1564912361**, ISBN 10: **1564912363**

Rule-57204:

If both $(\mathbf{N})$, $(\mathbf{U})$, $(\mathbf{S'})$, $(\mathbf{s})$, $(\mathbf{S})$, $(\mathbf{v})$, $(\mathbf{F})$, $(\mathbf{i})$, $(\mathbf{T})$, $(\mathbf{D})$, $(\mathbf{K})$, $(\mathbf{j})$ and $(\mathbf{P})$ are known, then its Leverage or Gearing Ratio planned is:

$$l = \{\mathbf{N} + \mathbf{K} + \mathbf{Sj}/360 + \mathbf{P}\}$$
$$/(\mathbf{U} + \{\mathbf{S'}[1+\mathbf{s}] - \mathbf{Sv} - \mathbf{F} - \mathbf{Si}\} - \mathbf{T} - \mathbf{D}) - 1$$


Rule-57205:

If both $(l)$, $(\mathbf{N})$, $(\mathbf{S'})$, $(\mathbf{s})$, $(\mathbf{S})$, $(\mathbf{v})$, $(\mathbf{F})$, $(\mathbf{i})$, $(\mathbf{T})$, $(\mathbf{D})$, $(\mathbf{K})$, $(\mathbf{j})$ and $(\mathbf{P})$ are known, then its Utilized or Starting Capital must be:

$$\mathbf{U} = (\{\mathbf{N} + \mathbf{K} + \mathbf{Sj}/360 + \mathbf{P}\} - [1 + l]\{\mathbf{S'}[1+\mathbf{s}] - \mathbf{Sv} - \mathbf{F} - \mathbf{Si}\}$$
$$- \mathbf{T} - \mathbf{D})/[1 + l]$$


Rule-57206:

If both $(l)$, $(\mathbf{U})$, $(\mathbf{S'})$, $(\mathbf{s})$, $(\mathbf{S})$, $(\mathbf{v})$, $(\mathbf{F})$, $(\mathbf{i})$, $(\mathbf{T})$, $(\mathbf{D})$, $(\mathbf{K})$, $(\mathbf{j})$ and $(\mathbf{P})$ are known, then its Sales Past must be:

$$\mathbf{S'} = [\mathbf{N} + \mathbf{K} + \mathbf{Sj}/360 + \mathbf{P}]/[1 + l] - \mathbf{U}$$
$$+ \mathbf{Sv} + \mathbf{F} + \mathbf{Si} + \mathbf{T} + \mathbf{D})/[1+\mathbf{s}]$$

or

$$\mathbf{S'} = \mathbf{S}/[1+\mathbf{s}]$$


Rule-57207:

If both $(l)$, $(\mathbf{U})$, $(\mathbf{S'})$, $(\mathbf{N})$, $(\mathbf{S})$, $(\mathbf{v})$, $(\mathbf{F})$, $(\mathbf{i})$, $(\mathbf{T})$, $(\mathbf{D})$, $(\mathbf{K})$, $(\mathbf{j})$ and $(\mathbf{P})$ are known, then its Sales Growth planned is:

$$\mathbf{s} = [\mathbf{N} + \mathbf{K} + \mathbf{Sj}/360 + \mathbf{P}]/[1 + l] - \mathbf{U} + \mathbf{Sv} + \mathbf{F} + \mathbf{Si} + \mathbf{T} + \mathbf{D})/\mathbf{S'} - 1$$

or

$$\mathbf{s} = \mathbf{S}/\mathbf{S'} - 1$$

301

Steve Asikin ISBN 13: **978-1564912361**, ISBN 10: **1564912363**

Rule-57208:

If both (**/**), (**U**), (**$'**), (**s**), (**$**), (**v**), (**F**), (**i**), (**T**), (**D**), (**K**), (**j**) and (**P**) are known, then its Sales or Revenue planned is:

$$\$= [[1+/](U+\{\$'[1+s]-F\}-T-D)-[N+K+P]]$$
$$/\{j/360+[1+/][v+i]\}$$

or

$$\$= \$'[1+s]$$

Rule-57209:

If both (**/**), (**U**), (**$'**), (**s**), (**$**), (**N**), (**F**), (**i**), (**T**), (**D**), (**K**), (**j**) and (**P**) are known, then its Variable Portion planned is:

$$v= (U+\{\$'[1+s]-\$i-F\}-T-D$$
$$-[N+K+\$j/360+P]/[1+/])/\$$$

Rule-57210:

If both (**/**), (**U**), (**$'**), (**s**), (**$**), (**v**), (**N**), (**i**), (**T**), (**D**), (**K**), (**j**) and (**P**) are known, then its Fixed Cost planned is:

$$F= (U+\{\$'[1+s]-\$v-\$i\}-T-D$$
$$-[N+K+\$j/360+P]/[1+/])/\$$$

Rule-57211:

If both (**/**), (**U**), (**$'**), (**s**), (**$**), (**v**), (**F**), (**N**), (**T**), (**D**), (**K**), (**j**) and (**P**) are known, then its Interest Portion planned is:

$$i= (U+\{\$'[1+s]-\$v-F\}-T-D$$
$$-[N+K+\$j/360+P]/[1+/])/\$$$

Steve Asikin ISBN 13: **978-1564912361**, ISBN 10: **1564912363**

Rule-57212:
If both (**/**), (**U**), (**S'**), (**s**), (**S**), (**v**), (**F**), (**i**), (**N**), (**D**), (**K**), (**j**) and (**P**) are known, then its Tax planned is:

$$T = U + \{S'[1+s] - Sv - F - Si\} - D - [N + K + Sj/360 + P] / [1 + /]$$

Rule-57213:
If both (**/**), (**U**), (**S'**), (**s**), (**S**), (**v**), (**F**), (**i**), (**T**), (**N**), (**K**), (**j**) and (**P**) are known, then its Dividend planned is:

$$D = U + \{S'[1+s] - Sv - F - Si\} - T - [N + K + Sj/360 + P] / [1 + /]$$

Rule-57214:
If both (**/**), (**U**), (**S'**), (**s**), (**S**), (**v**), (**F**), (**i**), (**T**), (**D**), (**N**), (**j**) and (**P**) are known, then its Kind of Cash planned is:

$$K = [1 + /](U + \{S'[1+s] - Sv - F - Si\} - T - D) - [N + Sj/360 + P]$$

Rule-57215:
If both (**/**), (**U**), (**S'**), (**s**), (**S**), (**v**), (**F**), (**i**), (**T**), (**D**), (**K**), (**N**) and (**P**) are known, then its Job or Trade Receivable Days planned is:

$$j = 360[[1 + /](U + \{S'[1+s] - Sv - F - Si\} - T - D) - \{N + K + P\}] / S$$

303

Steve Asikin ISBN 13: **978-1564912361**, ISBN 10: **1564912363**

Rule-57216:

If both (**/**), (**U**), (**S'**), (**s**), (**S**), (**v**), (**F**), (**i**), (**T**), (**D**), (**K**), (**j**) and (**P**) are known, then its Non Current Asset planned is:

$$\mathbf{P} = [1 + \textit{/}](\mathbf{U} + \{\mathbf{S'}[1+\mathbf{s}] - \mathbf{Sv} - \mathbf{F} - \mathbf{Si}\} - \mathbf{T} - \mathbf{D})$$
$$- \{\mathbf{N} + \mathbf{K} + \mathbf{Sj}/360\}$$

Rule-57217:

If both (**/**), (**U**), (**S'**), (**s**), (**S**), (**v**), (**F**), (**i**), (**T**), (**D**), (**K**), (**j**), (**V**) and (**p**) are known, then its Non Current Asset planned is:

$$\mathbf{N} = [1 + \textit{/}]\{\mathbf{U} + \mathbf{S'}[1+\mathbf{s}] - \mathbf{Sv} - \mathbf{F} - \mathbf{Si} - \mathbf{T} - \mathbf{D}\}$$
$$- \mathbf{K} - \mathbf{Sj}/360 - \mathbf{Vp}/360$$

Rule-57218:

If both (**N**), (**U**), (**S'**), (**s**), (**S**), (**v**), (**F**), (**i**), (**T**), (**D**), (**K**), (**j**), (**V**) and (**p**) are known, then its Leverage or Gearing Ratio planned is:

$$\textit{/} = [\mathbf{N} + \mathbf{K} + \mathbf{Sj}/360 + \mathbf{Vp}/360]$$
$$/(\mathbf{U} + \{\mathbf{S'}[1+\mathbf{s}] - \mathbf{Sv} - \mathbf{F} - \mathbf{Si}\} - \mathbf{T} - \mathbf{D}) - 1$$

Rule-57219:

If both (**/**), (**N**), (**S'**), (**s**), (**S**), (**v**), (**F**), (**i**), (**T**), (**D**), (**K**), (**j**), (**V**) and (**p**) are known, then its Utilized or Starting Capital must be:

$$\mathbf{U} = ([\mathbf{N} + \mathbf{K} + \mathbf{Sj}/360 + \mathbf{Vp}/360] - [1 + \textit{/}]\{\mathbf{S'}[1+\mathbf{s}]$$
$$- \mathbf{Sv} - \mathbf{F} - \mathbf{Si}\} - \mathbf{T} - \mathbf{D})/[1 + \textit{/}]$$

Steve Asikin ISBN 13: **978-1564912361**, ISBN 10: **1564912363**

Rule-57220:

If both (**/**), (**U**), (**S'**), (**s**), (**S**), (**v**), (**F**), (**i**), (**T**), (**D**), (**K**), (**j**), (**V**) and (**p**) are known, then its Sales Past must be:

$$S' = ([N+K+Sj/360+Vp/360]/[1+/]-U$$
$$+Sv+F+Si+T+D)/[1+s]$$

or

$$S' = S[1+s]$$

Rule-57221:

If both (**/**), (**U**), (**S'**), (**N**), (**S**), (**v**), (**F**), (**i**), (**T**), (**D**), (**K**), (**j**), (**V**) and (**p**) are known, then its Sales Growth planned is:

$$s = ([N+K+Sj/360+Vp/360]/[1+/]-U$$
$$+Sv+F+Si+T+D)/S'-1$$

or

$$s = S/S'-1$$

Rule-57222:

If both (**/**), (**U**), (**S'**), (**s**), (**N**), (**v**), (**F**), (**i**), (**T**), (**D**), (**K**), (**j**), (**V**) and (**p**) are known, then its Sales or Revenue planned is:

$$S = [[1+/](U+\{S'[1+s]-F\}-T-D)-[N+K+Vp/360]]$$
$$/\{j/360+[1+/][v+i]\}$$

or

$$S = S'[1+s]$$

Steve Asikin ISBN 13: **978-1564912361**, ISBN 10: **1564912363**

Rule-57223:

If both (**/**), (**U**), (**S'**), (**s**), (**S**), (**N**), (**F**), (**i**), (**T**), (**D**), (**K**), (**j**), (**V**) and (**p**) are known, then its Variable Portion planned is:

$$\mathbf{v} = (\mathbf{U} + \{\mathbf{S'}[1+\mathbf{s}] - \mathbf{F} - \mathbf{Si}\} - \mathbf{T} - \mathbf{D}$$
$$- [\mathbf{N} + \mathbf{K} + \mathbf{Sj}/360 + \mathbf{Vp}/360]/[1+\mathbf{/}])/\mathbf{S}$$

or

$$\mathbf{v} = \mathbf{V}/\mathbf{S}$$

Rule-57224:

If both (**/**), (**U**), (**S'**), (**s**), (**S**), (**v**), (**N**), (**i**), (**T**), (**D**), (**K**), (**j**), (**V**) and (**p**) are known, then its Fixed Cost planned is:

$$\mathbf{F} = (\mathbf{U} + \{\mathbf{S'}[1+\mathbf{s}] - \mathbf{Sv} - \mathbf{Si}\} - \mathbf{T} - \mathbf{D}$$
$$- [\mathbf{N} + \mathbf{K} + \mathbf{Sj}/360 + \mathbf{Vp}/360]/[1+\mathbf{/}]$$

Rule-57225:

If both (**/**), (**U**), (**S'**), (**s**), (**S**), (**v**), (**F**), (**N**), (**T**), (**D**), (**K**), (**j**), (**V**) and (**p**) are known, then its Interest Portion planned is:

$$\mathbf{i} = (\mathbf{U} + \{\mathbf{S'}[1+\mathbf{s}] - \mathbf{Sv} - \mathbf{F}\} - \mathbf{T} - \mathbf{D}$$
$$- [\mathbf{N} + \mathbf{K} + \mathbf{Sj}/360 + \mathbf{Vp}/360]/[1+\mathbf{/}])/\mathbf{S}$$

Rule-57226:

If both (**/**), (**U**), (**S'**), (**s**), (**S**), (**v**), (**F**), (**i**), (**N**), (**D**), (**K**), (**j**), (**V**) and (**p**) are known, then its Tax planned is:

$$\mathbf{T} = \mathbf{U} + \{\mathbf{S'}[1+\mathbf{s}] - \mathbf{Sv} - \mathbf{F} - \mathbf{Si}\} - \mathbf{D}$$
$$- [\mathbf{N} + \mathbf{K} + \mathbf{Sj}/360 + \mathbf{Vp}/360]/[1+\mathbf{/}]$$

`

Steve Asikin ISBN 13: **978-1564912361**, ISBN 10: **1564912363**

Rule-57227:

If both (**ƒ**), (**U**), (**S'**), (**s**), (**S**), (**v**), (**F**), (**i**), (**T**), (**N**), (**K**), (**j**), (**V**) and (**p**) are known, then its Dividend planned is:

$$\mathbf{D}= \mathbf{U}+\{\mathbf{S'}[1+\mathbf{s}]-\mathbf{Sv}-\mathbf{F}-\mathbf{Si}\}-\mathbf{T}$$
$$-[\mathbf{N}+\mathbf{K}+\mathbf{Sj}/360+\mathbf{Vp}/360]/[1+\mathbf{ƒ}]$$

Rule-57228:

If both (**ƒ**), (**U**), (**S'**), (**s**), (**S**), (**v**), (**F**), (**i**), (**T**), (**D**), (**N**), (**j**), (**V**) and (**p**) are known, then its Kind of Cash planned is:

$$\mathbf{K}= [1+\mathbf{ƒ}](\mathbf{U}+\{\mathbf{S'}[1+\mathbf{s}]-\mathbf{Sv}-\mathbf{F}-\mathbf{Si}\}-\mathbf{T}-\mathbf{D})$$
$$-[\mathbf{N}+\mathbf{K}+\mathbf{Sj}/360+\mathbf{Vp}/360]$$

Rule-57229:

If both (**ƒ**), (**U**), (**S'**), (**s**), (**S**), (**v**), (**F**), (**i**), (**T**), (**D**), (**K**), (**N**), (**V**) and (**p**) are known, then its Job or Trade Receivable Days planned is:

$$\mathbf{j}= 360[[1+\mathbf{ƒ}](\mathbf{U}+\{\mathbf{S'}[1+\mathbf{s}]-\mathbf{Sv}-\mathbf{F}-\mathbf{Si}\}-\mathbf{T}-\mathbf{D})$$
$$-[\mathbf{N}+\mathbf{K}+\mathbf{Vp}/360]]/\mathbf{S}$$

Rule-57230:

If both (**ƒ**), (**U**), (**S'**), (**s**), (**S**), (**v**), (**F**), (**i**), (**T**), (**D**), (**K**), (**j**), (**N**) and (**p**) are known, then its Variable Cost planned is:

$$\mathbf{V}= 360[[1+\mathbf{ƒ}](\mathbf{U}+\{\mathbf{S'}[1+\mathbf{s}]-\mathbf{Sv}-\mathbf{F}-\mathbf{Si}\}-\mathbf{T}-\mathbf{D})$$
$$-[\mathbf{N}+\mathbf{K}+\mathbf{Sj}/360]]/\mathbf{p}$$

or

$$\mathbf{V}= \mathbf{Sv}$$

Steve Asikin ISBN 13: **978-1564912361**, ISBN 10: **1564912363**

Rule-57231:

If both (**/**), (**U**), (**S'**), (**s**), (**S**), (**v**), (**F**), (**i**), (**T**), (**D**), (**K**), (**j**), (**V**) and (**N**) are known, then its Procured Inventory Days planned is:

$$\mathbf{p}= 360[[1+\mathbf{/}](\mathbf{U}+\{\mathbf{S'}[1+\mathbf{s}]-\mathbf{Sv}-\mathbf{F}-\mathbf{Si}\}-\mathbf{T}-\mathbf{D})$$
$$-[\mathbf{N}+\mathbf{K}+\mathbf{Sj}/360]]/\mathbf{V}$$

Rule-57232:

If both (**/**), (**U**), (**S'**), (**s**), (**S**), (**v**), (**F**), (**i**), (**T**), (**D**), (**K**), (**j**) and (**p**) are known, then its Non Current Asset planned is:

$$\mathbf{N}= [1+\mathbf{/}]\{\mathbf{U}+\mathbf{S'}[1+\mathbf{s}]-\mathbf{Sv}-\mathbf{F}-\mathbf{Si}-\mathbf{T}-\mathbf{D}\}$$
$$-\mathbf{K}-\mathbf{Sj}/360-\mathbf{Svp}/360$$

Rule-57233:

If both (**N**), (**U**), (**S'**), (**s**), (**S**), (**v**), (**F**), (**i**), (**T**), (**D**), (**K**), (**j**) and (**p**) are known, then its Leverage or Gearing Ratio planned is:

$$\mathbf{/}= [\mathbf{N}+\mathbf{K}+\mathbf{Sj}/360+\mathbf{Svp}/360]$$
$$/(\mathbf{U}+\{\mathbf{S'}[1+\mathbf{s}]-\mathbf{Sv}-\mathbf{F}-\mathbf{Si}\}-\mathbf{T}-\mathbf{D})-1$$

Rule-57234:

If both (**/**), (**N**), (**S'**), (**s**), (**S**), (**v**), (**F**), (**i**), (**T**), (**D**), (**K**), (**j**) and (**p**) are known, then its Utilized or Starting Capital must be:

$$\mathbf{U}= ([\mathbf{N}+\mathbf{K}+\mathbf{Sj}/360+\mathbf{Svp}/360]$$
$$-[1+\mathbf{/}]\{\mathbf{S'}[1+\mathbf{s}]-\mathbf{Sv}-\mathbf{F}-\mathbf{Si}\}-\mathbf{T}-\mathbf{D})/[1+\mathbf{/}]$$

`

Steve Asikin ISBN 13: **978-1564912361**, ISBN 10: **1564912363**

Rule-57235:

If both $(\textit{I})$, $(\mathbf{U})$, $(\mathbf{N})$, $(\mathbf{s})$, $(\mathbf{S})$, $(\mathbf{v})$, $(\mathbf{F})$, $(\mathbf{i})$, $(\mathbf{T})$, $(\mathbf{D})$, $(\mathbf{K})$, $(\mathbf{j})$ and $(\mathbf{p})$ are known, then its Sales Past must be:

$$S' = ([\mathbf{N+K+Sj}/360+\mathbf{Sup}/360]/[1+\textit{I}]-\mathbf{U}$$
$$+\mathbf{Sv+F+Si+T+D})/[1+\mathbf{s}]$$

or

$$S' = S/[1+\mathbf{s}]$$

Rule-57236:

If both $(\textit{I})$, $(\mathbf{U})$, $(\mathbf{S'})$, $(\mathbf{N})$, $(\mathbf{S})$, $(\mathbf{v})$, $(\mathbf{F})$, $(\mathbf{i})$, $(\mathbf{T})$, $(\mathbf{D})$, $(\mathbf{K})$, $(\mathbf{j})$ and $(\mathbf{p})$ are known, then its Sales Growth planned is:

$$s = ([\mathbf{N+K+Sj}/360+\mathbf{Sup}/360]/[1+\textit{I}]-\mathbf{U}$$
$$+\mathbf{Sv+F+Si+T+D})/\mathbf{S'}-1$$

or

$$s = S/S'-1$$

Rule-57237:

If both $(\textit{I})$, $(\mathbf{U})$, $(\mathbf{S'})$, $(\mathbf{s})$, $(\mathbf{N})$, $(\mathbf{v})$, $(\mathbf{F})$, $(\mathbf{i})$, $(\mathbf{T})$, $(\mathbf{D})$, $(\mathbf{K})$, $(\mathbf{j})$ and $(\mathbf{p})$ are known, then its Sales or Revenue planned is:

$$S = [[1+\textit{I}](\mathbf{U}+\{\mathbf{S'}[1+\mathbf{s}]-\mathbf{F}\}-\mathbf{T}-\mathbf{D})-[\mathbf{N+K}]]$$
$$/\{\mathbf{j}/360+\mathbf{up}/360+[1+\textit{I}][\mathbf{v+i}]\}$$

or

$$S = S'[1+\mathbf{s}]$$

Steve Asikin ISBN 13: **978-1564912361**, ISBN 10: **1564912363**

Rule-57238:

If both ($I$), (**U**), (**S'**), (**s**), (**S**), (**N**), (**F**), (**i**), (**T**), (**D**), (**K**), (**j**) and (**p**) are known, then its Variable Portion planned is:

$$v= [[1+I(U+\{S'[1+s]-Sv-F-Si\}-T-D)$$
$$-[N+K+Sj/360]]/(S\{p/360]+[1+I]\})$$

Rule-57239:

If both ($I$), (**U**), (**S'**), (**s**), (**S**), (**v**), (**N**), (**i**), (**T**), (**D**), (**K**), (**j**) and (**p**) are known, then its Fixed Cost planned is:

$$F= U+\{S'[1+s]-Sv-Si\}-T-D$$
$$-[N+K+Sj/360+Svp/360]/[1+I]$$

Rule-57240:

If both ($I$), (**U**), (**S'**), (**s**), (**S**), (**v**), (**F**), (**N**), (**T**), (**D**), (**K**), (**j**) and (**p**) are known, then its Interest Portion planned is:

$$i= (U+\{S'[1+s]-Si-F\}-T-D$$
$$-[N+K+Sj/360+Svp/360]/[1+I])/S$$

Rule-57241:

If both ($I$), (**U**), (**S'**), (**s**), (**S**), (**v**), (**F**), (**i**), (**N**), (**D**), (**K**), (**j**) and (**p**) are known, then its Tax planned is:

$$T= U+\{S'[1+s]-Sv-F-Si\}-D$$
$$-[N+K+Sj/360+Svp/360]/[1+I]$$

Steve Asikin ISBN 13: **978-1564912361**, ISBN 10: **1564912363**

Rule-57242:

If both (**/**), (**U**), (**S'**), (**s**), (**S**), (**v**), (**F**), (**i**), (**T**), (**N**), (**K**), (**j**) and (**p**) are known, then its Dividend planned is:

$$D= U+\{S'[1+s]-Sv-F-Si\}-T$$
$$-[N+K+Sj/360+Svp/360]/[1+/]$$


Rule-57243:

If both (**/**), (**U**), (**S'**), (**s**), (**S**), (**v**), (**F**), (**i**), (**T**), (**D**), (**N**), (**j**) and (**p**) are known, then its Kind of Cash planned is:

$$K= [1+/](U+\{S'[1+s]-Sv-F-Si\}-T-D)$$
$$-[N+Sj/360+Svp/360]$$


Rule-57244:

If both (**/**), (**U**), (**S'**), (**s**), (**S**), (**v**), (**F**), (**i**), (**T**), (**D**), (**K**), (**N**) and (**p**) are known, then its Job or Trade Receivable Days planned is:

$$j= 360[[1+/](U+\{S'[1+s]-Sv-F-Si\}-T-D)$$
$$-\{N+K+Svp/360\}]/S$$


Rule-57245:

If both (**/**), (**U**), (**S'**), (**s**), (**S**), (**v**), (**F**), (**i**), (**T**), (**D**), (**K**), (**j**) and (**N**) are known, then its Procured Inventory Days planned is:

$$p= 360[[1+/](U+\{S'[1+s]-Sv-F-Si\}-T-D)$$
$$-[N+K+Sj/360]]/[Sv]$$

Steve Asikin ISBN 13: **978-1564912361**, ISBN 10: **1564912363**

Rule-57246:

If both $(I)$, $(U)$, $(S')$, $(s)$, $(S)$, $(v)$, $(F)$, $(i)$, $(T)$, $(D)$, $(K)$, $(j)$ and $(p)$ are known, then its Non Current Asset planned is:

$$N = [1+I]\{U+S'[1+s]-Sv-F-Si-T-D\}$$
$$-K-Sj/360-S'vp[1+s]/360$$

Rule-57247:

If both $(N)$, $(U)$, $(S')$, $(s)$, $(S)$, $(v)$, $(F)$, $(i)$, $(T)$, $(D)$, $(K)$, $(j)$ and $(p)$ are known, then its Leverage or Gearing Ratio planned is:

$$I = \{N+K+Sj/360+S'vp[1+s]/360\}$$
$$/(U+\{S'[1+s]-Sv-F-Si\}-T-D)-1$$

Rule-57248:

If both $(I)$, $(N)$, $(S')$, $(s)$, $(S)$, $(v)$, $(F)$, $(i)$, $(T)$, $(D)$, $(K)$, $(j)$ and $(p)$ are known, then its Utilized or Starting Capital must be:

$$U = (\{N+K+Sj/360+S'vp[1+s]/360\}$$
$$-[1+I]\{S'[1+s]-Sv-F-Si\}-T-D)/\{[1+I]$$

Rule-57249:

If both $(I)$, $(U)$, $(S')$, $(s)$, $(S)$, $(v)$, $(F)$, $(i)$, $(T)$, $(D)$, $(K)$, $(j)$ and $(p)$ are known, then its Sales Past must be:

$$S' = ([1+I](U-\{Sv+F+Si\}-T-D)-[N+K+Sj/360])$$
$$/([1+s]\{vp/360-[1+I]\})$$
$$\text{or}$$
$$S' = S/[1+s]$$

Steve Asikin ISBN 13: **978-1564912361**, ISBN 10: **1564912363**

Rule-57250:

If both (*I*), (**U**), (**S'**), (**N**), (**S**), (**v**), (**F**), (**i**), (**T**), (**D**), (**K**), (**j**) and (**p**) are known, then its Sales Growth planned is:

$$s = ([1+I](U-\{Sv+F+Si\}-T-D)-[N+K+Sj/360])$$
$$/(S'\{vp/360-[1+I]\})-1$$

or

$$s = S/S'-1$$

Rule-57251:

If both (*I*), (**U**), (**S'**), (**s**), (**S**), (**v**), (**F**), (**i**), (**T**), (**D**), (**K**), (**j**) and (**p**) are known, then its Sales or Revenue planned is:

$$S = [[1+I](U+\{S'[1+s]-F\}-T-D)-\{N+K$$
$$+S'vp[1+s]/360\}]/\{j/360+[1+I][v+i]\}$$

or

$$S = S'[1+s]$$

Rule-57252:

If both (*I*), (**U**), (**S'**), (**s**), (**S**), (**N**), (**F**), (**i**), (**T**), (**D**), (**K**), (**j**) and (**p**) are known, then its Variable Portion planned is:

$$v = ([1+I](U+\{S'[1+s]-F-Si\}-T-D)/$$
$$-[N+K+Sj/360])/\{S'p[1+s]/360+S[1+I]\}$$

Steve Asikin ISBN 13: **978-1564912361**, ISBN 10: **1564912363**

Rule-57253:

    If both (**/**), (**U**), (**S'**), (**s**), (**S**), (**v**), (**N**), (**i**), (**T**), (**D**), (**K**), (**j**) and (**p**) are known, then its Fixed Cost planned is:

$$F = U + \{S'[1+s] - Sv - Si\} - T - D$$
$$- \{N + K + Sj/360 + S'vp[1+s]/360\}/[1+/]$$


Rule-57254:

    If both (**/**), (**U**), (**S'**), (**s**), (**S**), (**v**), (**F**), (**N**), (**T**), (**D**), (**K**), (**j**) and (**p**) are known, then its Interest Portion planned is:

$$i = (U + \{S'[1+s] - Sv - F\} - T - D$$
$$- \{N + K + Sj/360 + S'vp[1+s]/360\}/[1+/])/S$$


Rule-57255:

    If both (**/**), (**U**), (**S'**), (**s**), (**S**), (**v**), (**F**), (**i**), (**N**), (**D**), (**K**), (**j**) and (**p**) are known, then its Tax planned is:

$$T = U + \{S'[1+s] - Sv - F - Si\} - D$$
$$- \{N + K + Sj/360 + S'vp[1+s]/360\}/[1+/]$$


Rule-57256:

    If both (**/**), (**U**), (**S'**), (**s**), (**S**), (**v**), (**F**), (**i**), (**T**), (**N**), (**K**), (**j**) and (**p**) are known, then its Dividend planned is:

$$D = U + \{S'[1+s] - Sv - F - Si\} - T$$
$$- \{N + K + Sj/360 + S'vp[1+s]/360\}/[1+/]$$

Steve Asikin ISBN 13: **978-1564912361**, ISBN 10: **1564912363**

Rule-57257:

If both (**/**), (**U**), (**S'**), (**s**), (**S**), (**v**), (**F**), (**i**), (**T**), (**D**), (**N**), (**j**) and (**p**) are known, then its Kind of Cash planned is:

$$K= [1+/](U+\{S'[1+s]-Sv-F-Si\}-T-D)$$
$$-\{N+Sj/360+S'vp[1+s]/360\}$$

Rule-57258:

If both (**/**), (**U**), (**S'**), (**s**), (**S**), (**v**), (**F**), (**i**), (**T**), (**D**), (**K**), (**N**) and (**p**) are known, then its Job or Trade Receivable Days planned is:

$$j= 360[[1+/](U+\{S'[1+s]-Sv-F-Si\}-T-D)$$
$$-\{N+K+S'vp[1+s]/360\}]/S$$

Rule-57259:

If both (**/**), (**U**), (**S'**), (**s**), (**S**), (**v**), (**F**), (**i**), (**T**), (**D**), (**K**), (**j**) and (**N**) are known, then its Procured Inventory Days planned is:

$$p= 360[[1+/](U+\{S'[1+s]-Sv-F-Si\}-T-D)$$
$$-[N+K+Sj/360]]/\{S'v[1+s]\}$$

Rule-57260:

If both (**/**), (**U**), (**S'**), (**s**), (**S**), (**v**), (**F**), (**i**), (**T**), (**D**), (**K**), (**j**) and (**P**) are known, then its Non Current Asset planned is:

$$N= [1+/]\{U+S'[1+s]-Sv-F-Si-T-D\}$$
$$-K-S'j[1+s]/360-P$$

Steve Asikin ISBN 13: **978-1564912361**, ISBN 10: **1564912363**

Rule-57261:

If both $(\mathbf{N})$, $(\mathbf{U})$, $(\mathbf{S'})$, $(\mathbf{s})$, $(\mathbf{S})$, $(\mathbf{v})$, $(\mathbf{F})$, $(\mathbf{i})$, $(\mathbf{T})$, $(\mathbf{D})$, $(\mathbf{K})$, $(\mathbf{j})$ and $(\mathbf{P})$ are known, then its Leverage or Gearing Ratio planned is:

$$\mathbf{I} = \{\mathbf{N}+\mathbf{K}+\mathbf{S'j}[1+\mathbf{s}]/360+\mathbf{P}\}$$
$$/(\mathbf{U}+\{\mathbf{S'}[1+\mathbf{s}]-\mathbf{Sv}-\mathbf{F}-\mathbf{Si}\}-\mathbf{T}-\mathbf{D})-1$$

Rule-57262:

If both $(\mathbf{I})$, $(\mathbf{N})$, $(\mathbf{S'})$, $(\mathbf{s})$, $(\mathbf{S})$, $(\mathbf{v})$, $(\mathbf{F})$, $(\mathbf{i})$, $(\mathbf{T})$, $(\mathbf{D})$, $(\mathbf{K})$, $(\mathbf{j})$ and $(\mathbf{P})$ are known, then its Utilized or Starting Capital must be:

$$\mathbf{U} = (\{\mathbf{N}+\mathbf{K}+\mathbf{S'j}[1+\mathbf{s}]/360+\mathbf{P}\}-[1+\mathbf{I}]\{\mathbf{S'}[1+\mathbf{s}]$$
$$-\mathbf{Sv}-\mathbf{F}-\mathbf{Si}\}-\mathbf{T}-\mathbf{D})/[1+\mathbf{I}]$$

Rule-57263:

If both $(\mathbf{I})$, $(\mathbf{U})$, $(\mathbf{N})$, $(\mathbf{s})$, $(\mathbf{S})$, $(\mathbf{v})$, $(\mathbf{F})$, $(\mathbf{i})$, $(\mathbf{T})$, $(\mathbf{D})$, $(\mathbf{K})$, $(\mathbf{j})$ and $(\mathbf{P})$ are known, then its Sales Past must be:

$$\mathbf{S'} = [[1+\mathbf{I}](\mathbf{U}-\{\mathbf{Sv}+\mathbf{F}+\mathbf{Si}\}-\mathbf{T}-\mathbf{D})$$
$$-[\mathbf{N}+\mathbf{K}+\mathbf{P}]]/([1+\mathbf{s}]\{\mathbf{j}/360-[1+\mathbf{I}]\})$$

or

$$\mathbf{S'} = \mathbf{S}/[1+\mathbf{s}]$$

Steve Asikin ISBN 13: **978-1564912361**, ISBN 10: **1564912363**

Rule-57264:

If both (**/**), (**U**), (**S'**), (**s**), (**S**), (**v**), (**F**), (**i**), (**T**), (**D**), (**K**), (**j**) and (**P**) are known, then its Sales Growth planned is:

$$s= [[1+\textbf{/}(\textbf{U}-\{\textbf{Sv}+\textbf{F}+\textbf{Si}\}-\textbf{T}-\textbf{D})$$
$$-[\textbf{N}+\textbf{K}+\textbf{P}]]/(\textbf{S'}\{\textbf{j}/360-[1+\textbf{/}]\})-1$$

or

$$s= \textbf{S}/\textbf{S'}-1$$


Rule-57265:

If both (**/**), (**U**), (**S'**), (**s**), (**N**), (**v**), (**F**), (**i**), (**T**), (**D**), (**K**), (**j**) and (**P**) are known, then its Sales or Revenue planned is:

$$\textbf{S}= (\textbf{U}+\{\textbf{S'}[1+\textbf{s}]-\textbf{F}\}-\textbf{T}-\textbf{D}$$
$$-\{\textbf{N}+\textbf{K}+\textbf{S'j}[1+\textbf{s}]/360+\textbf{P}\}/[1+\textbf{/}])/[\textbf{v}+\textbf{i}]$$

or

$$\textbf{S}= \textbf{S'}[1+\textbf{s}]$$


Rule-57266:

If both (**/**), (**U**), (**S'**), (**s**), (**S**), (**N**), (**F**), (**i**), (**T**), (**D**), (**K**), (**j**) and (**P**) are known, then its Variable Portion planned is:

$$v= (\textbf{U}+\{\textbf{S'}[1+\textbf{s}]-\textbf{F}-\textbf{Si}\}-\textbf{T}-\textbf{D}$$
$$-\{\textbf{N}+\textbf{K}+\textbf{S'j}[1+\textbf{s}]/360+\textbf{P}\}/[1+\textbf{/}])/\textbf{S}$$

Steve Asikin ISBN 13: **978-1564912361**, ISBN 10: **1564912363**

Rule-57267:

If both (**I**), (**U**), (**S'**), (**s**), (**S**), (**v**), (**N**), (**i**), (**T**), (**D**), (**K**), (**j**) and (**P**) are known, then its Fixed Cost planned is:

$$F= U+\{S'[1+s]-Sv-Si\}-T-D$$
$$-\{N+K+S'j[1+s]/360+P\}/[1+I]$$

Rule-57268:

If both (**I**), (**U**), (**S'**), (**s**), (**S**), (**v**), (**F**), (**N**), (**T**), (**D**), (**K**), (**j**) and (**P**) are known, then its Interest Portion planned is:

$$i= (U+\{S'[1+s]-Sv-F\}-T-D$$
$$-\{N+K+S'j[1+s]/360+P\}/[1+I])/S$$

Rule-57269:

If both (**I**), (**U**), (**S'**), (**s**), (**S**), (**v**), (**F**), (**i**), (**N**), (**D**), (**K**), (**j**) and (**P**) are known, then its Tax planned is:

$$T= U+\{S'[1+s]-Sv-F-Si\}-D$$
$$-\{N+K+S'j[1+s]/360+P\}/[1+I]$$

Rule-57270:

If both (**I**), (**U**), (**S'**), (**s**), (**S**), (**v**), (**F**), (**i**), (**T**), (**N**), (**K**), (**j**) and (**P**) are known, then its Dividend planned is:

$$D= U+\{S'[1+s]-Sv-F-Si\}-T$$
$$-\{N+K+S'j[1+s]/360+P\}/[1+I]$$

Steve Asikin ISBN 13: **978-1564912361**, ISBN 10: **1564912363**

Rule-57271:

If both $(\mathit{I})$, $(\mathbf{U})$, $(\mathbf{S'})$, $(\mathbf{s})$, $(\mathbf{S})$, $(\mathbf{v})$, $(\mathbf{F})$, $(\mathbf{i})$, $(\mathbf{T})$, $(\mathbf{D})$, $(\mathbf{N})$, $(\mathbf{j})$ and $(\mathbf{P})$ are known, then its Kind of Cash planned is:

$$K = [1 + \mathit{I}(U + \{S'[1+s] - Sv - F - Si\} - T - D)$$
$$- \{N + S'j[1+s]/360 + P\}$$

Rule-57272:

If both $(\mathit{I})$, $(\mathbf{U})$, $(\mathbf{S'})$, $(\mathbf{s})$, $(\mathbf{S})$, $(\mathbf{v})$, $(\mathbf{F})$, $(\mathbf{i})$, $(\mathbf{T})$, $(\mathbf{D})$, $(\mathbf{K})$, $(\mathbf{N})$ and $(\mathbf{P})$ are known, then its Job or Trade Receivable Days planned is:

$$j = 360[[1 + \mathit{I}(U + \{S'[1+s] - Sv - F - Si\} - T - D)$$
$$- \{N + K + P\}] / \{S'[1+s]\}$$

Rule-57273:

If both $(\mathit{I})$, $(\mathbf{U})$, $(\mathbf{S'})$, $(\mathbf{s})$, $(\mathbf{S})$, $(\mathbf{v})$, $(\mathbf{F})$, $(\mathbf{i})$, $(\mathbf{T})$, $(\mathbf{D})$, $(\mathbf{K})$, $(\mathbf{j})$ and $(\mathbf{N})$ are known, then its Procured Inventory planned is:

$$P = [1 + \mathit{I}(U + \{S'[1+s] - Sv - F - Si\} - T - D)$$
$$- \{N + K + S'j[1+s]/360\}$$

Rule-57274:

If both $(\mathit{I})$, $(\mathbf{U})$, $(\mathbf{S'})$, $(\mathbf{s})$, $(\mathbf{S})$, $(\mathbf{v})$, $(\mathbf{F})$, $(\mathbf{i})$, $(\mathbf{T})$, $(\mathbf{D})$, $(\mathbf{K})$, $(\mathbf{j})$, $(\mathbf{V})$ and $(\mathbf{p})$ are known, then its Non Current Asset planned is:

$$N = [1 + \mathit{I}\{U + S'[1+s] - Sv - F - Si - T - D\}$$
$$- K - S'j[1+s]/360 - Vp/360$$

Steve Asikin ISBN 13: **978-1564912361**, ISBN 10: **1564912363**

Rule-57275:

If both (**N**), (**U**), (**S'**), (**s**), (**$**), (**v**), (**F**), (**i**), (**T**), (**D**), (**K**), (**j**), (**V**) and (**p**) are known, then its Leverage or Gearing Ratio planned is:

$$I = \{N+K+S'j[1+s]/360+Vp/360\}$$
$$/(U+\{S'[1+s]-Sv-F-Si\}-T-D)-1$$

Rule-57276:

If both (**I**), (**N**), (**S'**), (**s**), (**$**), (**v**), (**F**), (**i**), (**T**), (**D**), (**K**), (**j**), (**V**) and (**p**) are known, then its Utilized or Starting Capital must be:

$$U = (\{N+K+S'j[1+s]/360+Vp/360\}$$
$$-[1+I]\{S'[1+s]-Sv-F-Si\}-T-D)/[1+I]$$

Rule-57277:

If both (**I**), (**U**), (**N**), (**s**), (**$**), (**v**), (**F**), (**i**), (**T**), (**D**), (**K**), (**j**), (**V**) and (**p**) are known, then its Sales Past must be:

$$S' = [[1+I](U-\{Sv+F+Si\}-T-D)-[N+K+Vp/360]]$$
$$/([1+s]\{j/360-[1+I]\})$$
$$\text{or}$$
$$S' = S/[1+s]$$

Steve Asikin ISBN 13: **978-1564912361**, ISBN 10: **1564912363**

Rule-57278:

If both (**/**), (**U**), (**S'**), (**N**), (**S**), (**υ**), (**F**), (**i**), (**T**), (**D**), (**K**), (**j**), (**V**) and (**p**) are known, then its Sales Growth planned is:

$$s = [[1+/](U-\{Sυ+F+Si\}-T-D)-[N+K+Vp/360]]$$
$$/(S'\{j/360-[1+/]\})-1$$

or

$$s = S/S'-1$$

Rule-57279:

If both (**/**), (**U**), (**S'**), (**s**), (**N**), (**υ**), (**F**), (**i**), (**T**), (**D**), (**K**), (**j**), (**V**) and (**p**) are known, then its Sales or Revenue planned is:

$$S = (U+\{S'[1+s]-F\}-T-D-\{N+K+S'j[1+s]/360$$
$$+Vp/360\}/[1+/])/[υ+i]$$

or

$$S = S'[1+s]$$

or

$$S = V/υ$$

Rule-57280:

If both (**/**), (**U**), (**S'**), (**s**), (**S**), (**N**), (**F**), (**i**), (**T**), (**D**), (**K**), (**j**), (**V**) and (**p**) are known, then its Variable Portion planned is:

$$υ = (U+\{S'[1+s]-F-Si\}-T-D-\{N+K+S'j[1+s]/360$$
$$+Vp/360\}/[1+/])/S$$

or

$$υ = V/S$$

321

Steve Asikin ISBN 13: **978-1564912361**, ISBN 10: **1564912363**

Rule-57281:

If both (**I**), (**U**), (**S'**), (**s**), (**S**), (**v**), (**N**), (**i**), (**T**), (**D**), (**K**), (**j**), (**V**) and (**p**) are known, then its Fixed Cost planned is:

$$F= U+\{S'[1+s]-Sv-Si\}-T-D-\{N+K+S'j[1+s]/360 +Vp/360\}/[1+I]$$

Rule-57282:

If both (**I**), (**U**), (**S'**), (**s**), (**S**), (**v**), (**F**), (**N**), (**T**), (**D**), (**K**), (**j**), (**V**) and (**p**) are known, then its Interest Portion planned is:

$$i= (U+\{S'[1+s]-Sv-F\}-T-D-\{N+K+S'j[1+s]/360 +Vp/360\}/[1+I])/S$$

Rule-57283:

If both (**I**), (**U**), (**S'**), (**s**), (**S**), (**v**), (**F**), (**i**), (**N**), (**D**), (**K**), (**j**), (**V**) and (**p**) are known, then its Tax planned is:

$$T= U+\{S'[1+s]-Sv-F-Si\}-D-\{N+K+S'j[1+s]/360 +Vp/360\}/[1+I]$$

Rule-57284:

If both (**I**), (**U**), (**S'**), (**s**), (**S**), (**v**), (**F**), (**i**), (**T**), (**N**), (**K**), (**j**), (**V**) and (**p**) are known, then its Dividend planned is:

$$D= U+\{S'[1+s]-Sv-F-Si\}-T-\{N+K+S'j[1+s]/360 +Vp/360\}/[1+I]$$

Steve Asikin ISBN 13: **978-1564912361**, ISBN 10: **1564912363**

Rule-57285:

If both (**/**), (**U**), (**S'**), (**s**), (**S**), (**v**), (**F**), (**i**), (**T**), (**D**), (**N**), (**j**), (**V**) and (**p**) are known, then its Kind of Cash planned is:

$$K= [1+/](U+\{S'[1+s]-Sv-F-Si\}-T-D)$$
$$-\{N+S'j[1+s]/360+Vp/360\}$$

Rule-57286:

If both (**/**), (**U**), (**S'**), (**s**), (**S**), (**v**), (**F**), (**i**), (**T**), (**D**), (**K**), (**N**), (**V**) and (**p**) are known, then its Job or Trade Receivable Days planned is:

$$j= 360[[1+/](U+\{S'[1+s]-Sv-F-Si\}-T-D)$$
$$-\{N+K+Vp/360\}]/\{S'[1+s]\}$$

Rule-57287:

If both (**/**), (**U**), (**S'**), (**s**), (**S**), (**v**), (**F**), (**i**), (**T**), (**D**), (**K**), (**j**), (**N**) and (**p**) are known, then its Variable Cost planned is:

$$V= 360[[1+/](U+\{S'[1+s]-Sv-F-Si\}-T-D)$$
$$-\{N+K+S'j[1+s]/360\}]/p$$
$$or$$
$$V= Sv$$

Steve Asikin ISBN 13: **978-1564912361**, ISBN 10: **1564912363**

Rule-57288:

If both (**/**), (**U**), (**S'**), (**s**), (**S**), (**v**), (**F**), (**i**), (**T**), (**D**), (**K**), (**j**), (**V**) and (**N**) are known, then its Procured Inventory Days planned is:

$$\mathbf{p} = 360\mathbf{[}[1+\textbf{/}](\mathbf{U}+\{\mathbf{S'}[1+\mathbf{s}]-\mathbf{Sv}-\mathbf{F}-\mathbf{Si}\}-\mathbf{T}-\mathbf{D})$$
$$-\{\mathbf{N}+\mathbf{K}+\mathbf{S'j}[1+\mathbf{s}]/360\}\mathbf{]}/\mathbf{V}$$

Rule-57289:

If both (**/**), (**U**), (**S'**), (**s**), (**S**), (**v**), (**F**), (**i**), (**T**), (**D**), (**K**), (**j**) and (**p**) are known, then its Non Current Asset planned is:

$$\mathbf{N} = [1+\textbf{/}]\{\mathbf{U}+\mathbf{S'}[1+\mathbf{s}]-\mathbf{Sv}-\mathbf{F}-\mathbf{Si}-\mathbf{T}-\mathbf{D}\}$$
$$-\mathbf{K}-\mathbf{S'j}[1+\mathbf{s}]/360-\mathbf{Svp}/360$$

Rule-57290:

If both (**N**), (**U**), (**S'**), (**s**), (**S**), (**v**), (**F**), (**i**), (**T**), (**D**), (**K**), (**j**) and (**p**) are known, then its Leverage or Gearing Ratio planned is:

$$\textbf{/} = \{\mathbf{N}+\mathbf{K}+\mathbf{S'j}[1+\mathbf{s}]/360+\mathbf{Svp}/360\}$$
$$/(\mathbf{U}+\{\mathbf{S'}[1+\mathbf{s}]-\mathbf{Sv}-\mathbf{F}-\mathbf{Si}\}-\mathbf{T}-\mathbf{D})-1$$

Rule-57291:

If both (**/**), (**N**), (**S'**), (**s**), (**S**), (**v**), (**F**), (**i**), (**T**), (**D**), (**K**), (**j**) and (**p**) are known, then its Utilized or Starting Capital must be:

$$\mathbf{U} = (\{\mathbf{N}+\mathbf{K}+\mathbf{S'j}[1+\mathbf{s}]/360+\mathbf{Svp}/360\}$$
$$-[1+\textbf{/}]\{\mathbf{S'}[1+\mathbf{s}]-\mathbf{Sv}-\mathbf{F}-\mathbf{Si}\}-\mathbf{T}-\mathbf{D})/\{[1+\textbf{/}]\}$$

Steve Asikin ISBN 13: **978-1564912361**, ISBN 10: **1564912363**

Rule-57292:

If both (**/**), (**U**), (**N**), (**s**), (**$**), (**v**), (**F**), (**i**), (**T**), (**D**), (**K**), (**j**) and (**p**) are known, then its Sales Past must be:

$$S' = [[1+/(U-\{Sv+F+Si\}-T-D)-[N+K+Svp/360]]$$
$$/([1+s]\{j/360-[1+/]\})$$

or

$$S' = S/[1+s]$$

Rule-57293:

If both (**/**), (**U**), (**S'**), (**s**), (**$**), (**v**), (**F**), (**i**), (**T**), (**D**), (**K**), (**j**) and (**p**) are known, then its Non Current Asset planned is:

$$s = [[1+/(U-\{Sv+F+Si\}-T-D)-[N+K+Svp/360]]$$
$$/(S'\{j/360-[1+/]\})-1$$

or

$$s = S/S'-1$$

Rule-57294:

If both (**/**), (**U**), (**S'**), (**s**), (**N**), (**v**), (**F**), (**i**), (**T**), (**D**), (**K**), (**j**) and (**p**) are known, then its Sales or Revenue planned is:

$$S = [[1+/(U+\{S'[1+s]-F\}-T-D)-\{N+K$$
$$+S'j[1+s]/360\}]/\{vp/360+[1+/][v+i]\}$$

or

$$S = S'[1+s]$$

Steve Asikin ISBN 13: **978-1564912361**, ISBN 10: **1564912363**

Rule-57295:

If both (**/**), (**U**), (**S'**), (**s**), (**S**), (**N**), (**F**), (**i**), (**T**), (**D**), (**K**), (**j**) and (**p**) are known, then its Variable Portion planned is:

$$v= [[1+/](U+\{S'[1+s]-F-Si\}-T-D)$$
$$-\{N+K+S'j[1+s]/360\}]/\{Sp/360+S[1+/]\}$$

Rule-57296:

If both (**/**), (**U**), (**S'**), (**s**), (**S**), (**v**), (**N**), (**i**), (**T**), (**D**), (**K**), (**j**) and (**p**) are known, then its Fixed Cost planned is:

$$F= U+\{S'[1+s]-Sv-Si\}-T-D$$
$$-\{N+K+S'j[1+s]/360+Svp/360\}/[1+/]$$

Rule-57297:

If both (**/**), (**U**), (**S'**), (**s**), (**S**), (**v**), (**F**), (**N**), (**T**), (**D**), (**K**), (**j**) and (**p**) are known, then its Interest Portion planned is:

$$i= (U+\{S'[1+s]-Sv-F\}-T-D$$
$$-\{N+K+S'j[1+s]/360+Svp/360\}/[1+/])/S$$

Rule-57298:

If both (**/**), (**U**), (**S'**), (**s**), (**S**), (**v**), (**F**), (**i**), (**N**), (**D**), (**K**), (**j**) and (**p**) are known, then its Tax planned is:

$$T= U+\{S'[1+s]-Sv-F-Si\}-D$$
$$-\{N+K+S'j[1+s]/360+Svp/360\}/[1+/]$$

Steve Asikin ISBN 13: **978-1564912361**, ISBN 10: **1564912363**

Rule-57299:

If both $(I)$, $(U)$, $(S')$, $(s)$, $(S)$, $(v)$, $(F)$, $(i)$, $(T)$, $(N)$, $(K)$, $(j)$ and $(p)$ are known, then its Dividend planned is:

$$D= U+\{S'[1+s]-Sv-F-Si\}-T$$
$$-\{N+K+S'j[1+s]/360+Svp/360\}/[1+I]$$


Rule-57300:

If both $(I)$, $(U)$, $(S')$, $(s)$, $(S)$, $(v)$, $(F)$, $(i)$, $(T)$, $(D)$, $(N)$, $(j)$ and $(p)$ are known, then its Kind of Cash planned is:

$$K= [1+I](U+\{S'[1+s]-Sv-F-Si\}-T-D)$$
$$-\{N+S'j[1+s]/360+Svp/360\}$$


Rule-57301:

If both $(I)$, $(U)$, $(S')$, $(s)$, $(S)$, $(v)$, $(F)$, $(i)$, $(T)$, $(D)$, $(K)$, $(N)$ and $(p)$ are known, then its Job or Trade Receivable Days planned is:

$$j= 360[[1+I](U+\{S'[1+s]-Sv-F-Si\}-T-D)$$
$$-\{N+K+Svp/360\}]/\{S'[1+s]\}$$


Rule-57302:

If both $(I)$, $(U)$, $(S')$, $(s)$, $(S)$, $(v)$, $(F)$, $(i)$, $(T)$, $(D)$, $(K)$, $(j)$ and $(N)$ are known, then its Procured Inventory Days planned is:

$$p= 360[[1+I](U+\{S'[1+s]-Sv-F-Si\}-T-D)$$
$$-\{N+K+S'j[1+s]/360\}]/[Sv]$$

Steve Asikin ISBN 13: **978-1564912361**, ISBN 10: **1564912363**

Rule-57303:

If both (**I**), (**U**), (**S'**), (**s**), (**S**), (**v**), (**F**), (**i**), (**T**), (**D**), (**K**), (**j**) and (**p**) are known, then its Non Current Asset planned is:

$$N = [1+I]\{U+S'[1+s]-Sv-F-Si-T-D\}$$
$$-K-S'j[1+s]/360-S'vp[1+s]/360$$

Rule-57304:

If both (**N**), (**U**), (**S'**), (**s**), (**S**), (**v**), (**F**), (**i**), (**T**), (**D**), (**K**), (**j**) and (**p**) are known, then its Leverage or Gearing Ratio planned is:

$$I = \{N+K+S'j[1+s]/360+S'vp[1+s]/360\}$$
$$/(U+\{S'[1+s]-Sv-F-Si\}-T-D)-1$$

Rule-57305:

If both (**I**), (**N**), (**S'**), (**s**), (**S**), (**v**), (**F**), (**i**), (**T**), (**D**), (**K**), (**j**) and (**p**) are known, then its Utilized or Starting Capital must be:

$$U = (\{N+K+S'j[1+s]/360+S'vp[1+s]/360\}$$
$$-[1+I]\{S'[1+s]-Sv-F-Si\}-T-D)/[1+I]$$

Rule-57306:

If both (**I**), (**U**), (**N**), (**s**), (**S**), (**v**), (**F**), (**i**), (**T**), (**D**), (**K**), (**j**) and (**p**) are known, then its Sales Past must be:

$$S' = [[1+I](U-\{Sv+F+Si\}-T-D)-[N+K]]$$
$$/([1+s]\{j/360+vp/360-[1+I]\})$$
$$\text{or}$$
$$S' = S/[1+s]$$

Steve Asikin ISBN 13: **978-1564912361**, ISBN 10: **1564912363**

Rule-57307:

If both (*I*), (**U**), (**S'**), (**N**), (**S**), (**v**), (**F**), (**i**), (**T**), (**D**), (**K**), (**j**) and (**p**) are known, then its Sales Growth planned is:

$$s= [[1+I(U-\{Sv+F+Si\}-T-D)-[N+K]]$$
$$/(S'\{j/360+vp/360-[1+I]\})-1$$

or

$$s= S/S'-1$$


Rule-57308:

If both (*I*), (**U**), (**S'**), (**s**), (**N**), (**v**), (**F**), (**i**), (**T**), (**D**), (**K**), (**j**) and (**p**) are known, then its Sales or Revenue planned is:

$$S= (\{N+K+S'j[1+s]/360+S'vp[1+s]/360\}/[1+I$$
$$-U+\{S'[1+s]-F\}-T-D)/[v+i]$$

or

$$S= S'[1+s]$$


Rule-57309:

If both (*I*), (**U**), (**S'**), (**s**), (**S**), (**N**), (**F**), (**i**), (**T**), (**D**), (**K**), (**j**) and (**p**) are known, then its Variable Portion planned is:

$$v= [[1+I(U+\{S'[1+s]-F-Si\}-T-D)-\{N+K$$
$$+S'j[1+s]/360\}]/\{S'p[1+s]/360+S[1+I]\}$$

Steve Asikin ISBN 13: **978-1564912361**, ISBN 10: **1564912363**

Rule-57310:

If both (**/**), (**U**), (**S'**), (**s**), (**S**), (**v**), (**N**), (**i**), (**T**), (**D**), (**K**), (**j**) and (**p**) are known, then its Fixed Cost planned is:

$$F= U+\{S'[1+s]-Sv-Si\}-T-D-\{N+K$$
$$+S'j[1+s]/360+S'vp[1+s]/360\}/[1+/]$$

Rule-57311:

If both (**/**), (**U**), (**S'**), (**s**), (**S**), (**v**), (**F**), (**N**), (**T**), (**D**), (**K**), (**j**) and (**p**) are known, then its Interest Portion planned is:

$$i= (U+\{S'[1+s]-Sv-F\}-T-D-\{N+K+S'j[1+s]/360$$
$$+S'vp[1+s]/360\}/[1+/])/S$$

Rule-57312:

If both (**/**), (**U**), (**S'**), (**s**), (**S**), (**v**), (**F**), (**i**), (**N**), (**D**), (**K**), (**j**) and (**p**) are known, then its Tax planned is:

$$T= U+\{S'[1+s]-Sv-F-Si\}-D-\{N+K+S'j[1+s]/360$$
$$+S'vp[1+s]/360\}/[1+/]$$

Rule-57313:

If both (**/**), (**U**), (**S'**), (**s**), (**S**), (**v**), (**F**), (**i**), (**T**), (**N**), (**K**), (**j**) and (**p**) are known, then its Dividend planned is:

$$D= U+\{S'[1+s]-Sv-F-Si\}-T-\{N+K+S'j[1+s]/360$$
$$+S'vp[1+s]/360\}/[1+/]$$

Steve Asikin ISBN 13: **978-1564912361**, ISBN 10: **1564912363**

Rule-57314:

If both ($I$), (**U**), (**S'**), (**s**), (**S**), (**v**), (**F**), (**i**), (**T**), (**D**), (**N**), (**j**) and (**p**) are known, then its Kind of Cash planned is:

$$K= [1+I(U+\{S'[1+s]-Sv-F-Si\}-T-D)$$
$$-\{N+S'j[1+s]/360+S'vp[1+s]/360\}$$

Rule-57315:

If both ($I$), (**U**), (**S'**), (**s**), (**S**), (**v**), (**F**), (**i**), (**T**), (**D**), (**K**), (**N**) and (**p**) are known, then its Job or Trade Receivable Days planned is:

$$j= 360[[1+I(U+\{S'[1+s]-Sv-F-Si\}-T-D)$$
$$-\{N+K+S'vp[1+s]/360\}]/\{S'[1+s]\}$$

Rule-57316:

If both ($I$), (**U**), (**S'**), (**s**), (**S**), (**v**), (**F**), (**i**), (**T**), (**D**), (**K**), (**j**) and (**N**) are known, then its Procured Invenbtory Days planned is:

$$p= 360[[1+I(U+\{S'[1+s]-Sv-F-Si\}-T-D)$$
$$-\{N+K+S'j[1+s]/360\}]/\{S'v[1+s]\}$$

Rule-57317:

If both ($I$), (**U**), (**S'**), (**s**), (**S**), (**v**), (**F**), (**i**), (**T**), (**d**), (**K**), (**J**) and (**P**) are known, then its Non Current Asset planned is:

$$N= [1+I(U+\{S'[1+s]-Sv-F-Si-T\}[1-d])-K-J-P$$

Steve Asikin ISBN 13: **978-1564912361**, ISBN 10: **1564912363**

Rule-57318:

If both (**N**), (**U**), (**S'**), (**s**), (**S**), (**v**), (**F**), (**i**), (**T**), (**d**), (**K**), (**J**) and (**P**) are known, then its Leverage or Gearing Ratio planned is:

$$l = \{N+K+J+P\}/(U+\{S'[1+s]-Sv-F-Si-T\}[1-d])-1$$

Rule-57319:

If both (**l**), (**N**), (**S'**), (**s**), (**S**), (**v**), (**F**), (**i**), (**T**), (**d**), (**K**), (**J**) and (**P**) are known, then its Utilized or Starting Capital must be:

$$U = (\{N+K+J+P\}-[1+l]\{S'[1+s]-Sv-Si-T\}[1-d])$$
$$/[1+l]$$

Rule-57320:

If both (**l**), (**U**), (**N**), (**s**), (**S**), (**v**), (**F**), (**i**), (**T**), (**d**), (**K**), (**J**) and (**P**) are known, then its Sales Past must be:

$$S' = (\{[N+K+J+P]/[1+l]-U\}/[1-d]+Sv+F+Si+T)$$
$$/[1+s]$$
$$\text{or}$$
$$S' = S/[1+s]$$

Rule-57321:

If both (**l**), (**U**), (**S'**), (**N**), (**S**), (**v**), (**F**), (**i**), (**T**), (**d**), (**K**), (**J**) and (**P**) are known, then its Sales Growth planned is:

$$s = (\{[N+K+J+P]/[1+l]-U\}/[1-d]$$
$$+Sv+F+Si+T)/S'-1$$
$$\text{or}$$
$$s = S/S'-1$$

Steve Asikin ISBN 13: **978-1564912361**, ISBN 10: **1564912363**

Rule-57322:

If both (**/**), (**U**), (**S'**), (**s**), (**N**), (**v**), (**F**), (**i**), (**T**), (**d**), (**K**), (**J**) and (**P**) are known, then its Sales or Revenue planned is:

$$S= (S'[1+s]-F-T-\{[N+K+J+P]/[1+\text{/}-U\}/[1-d])$$
$$/[v+i]$$

or

$$S= S'[1+s]$$

Rule-57323:

If both (**/**), (**U**), (**S'**), (**s**), (**S**), (**N**), (**F**), (**i**), (**T**), (**d**), (**K**), (**J**) and (**P**) are known, then its Variable Portion planned is:

$$v= (S'[1+s]-F-Si-T-\{[N+K+J+P]/[1+\text{/}-U\}$$
$$/[1-d])/S$$

Rule-57324:

If both (**/**), (**U**), (**S'**), (**s**), (**S**), (**v**), (**N**), (**i**), (**T**), (**d**), (**K**), (**J**) and (**P**) are known, then its Fixed Cost planned is:

$$F= S'[1+s]-Sv-Si-T-\{[N+K+J+P]/[1+\text{/}-U\}/[1-d]$$

Rule-57325:

If both (**/**), (**U**), (**S'**), (**s**), (**S**), (**v**), (**F**), (**N**), (**T**), (**d**), (**K**), (**J**) and (**P**) are known, then its Interest Portion planned is:

$$i= (S'[1+s]-Sv-F-T-\{[N+K+J+P]/[1+\text{/}-U\}$$
$$/[1-d])/S$$

Steve Asikin ISBN 13: **978-1564912361**, ISBN 10: **1564912363**

Rule-57326:

If both (**/**), (**U**), (**S'**), (**s**), (**$**), (**v**), (**F**), (**i**), (**N**), (**d**), (**K**), (**J**) and (**P**) are known, then its Tax planned is:

$$T = S'[1+s] - Sv - F - Si - \{[N+K+J+P]/[1+/-U\}/[1-d]$$

Rule-57327:

If both (**/**), (**U**), (**S'**), (**s**), (**$**), (**v**), (**F**), (**i**), (**T**), (**N**), (**K**), (**J**) and (**P**) are known, then its Dividend Payout planned is:

$$d = 1 - \{[N+K+J+P]/[1+/-U\}/\{S'[1+s]-Sv-F-Si-T\}$$

Rule-57328:

If both (**/**), (**U**), (**S'**), (**s**), (**$**), (**v**), (**F**), (**i**), (**T**), (**D**), (**N**), (**J**) and (**P**) are known, then its Kind of Cash planned is:

$$K = [1+/(U+\{S'[1+s]-Sv-F-Si-T\}[1-d]) - [N+J+P]$$

Rule-57329:

If both (**/**), (**U**), (**S'**), (**s**), (**$**), (**v**), (**F**), (**i**), (**T**), (**D**), (**K**), (**N**) and (**P**) are known, then its Job or Trade Account Receivable planned is:

$$J = [1+/(U+\{S'[1+s]-Sv-F-Si-T\}[1-d]) - [N+K+P]$$

Rule-57330:

If both (**/**), (**U**), (**S'**), (**s**), (**$**), (**v**), (**F**), (**i**), (**T**), (**D**), (**K**), (**J**) and (**N**) are known, then its Procured Inventory planned is:

$$P = [1+/(U+\{S'[1+s]-Sv-F-Si-T\}[1-d]) - [N+K+J]$$

Steve Asikin ISBN 13: **978-1564912361**, ISBN 10: **1564912363**

Rule-57331:

If both (**/**), (**U**), (**S'**), (**s**), (**S**), (**v**), (**F**), (**i**), (**T**), (**d**), (**K**), (**J**), (**V**) and (**p**) are known, then its Non Current Asset planned is:

$$\mathbf{N}= [1+\mathbf{/}](\mathbf{U}+\{\mathbf{S'}[1+\mathbf{s}]-\mathbf{Sv}-\mathbf{F}-\mathbf{Si}-\mathbf{T}\}[1-\mathbf{d}])$$
$$-\mathbf{K}-\mathbf{J}-\mathbf{Vp}/360$$

Rule-57332:

If both (**N**), (**U**), (**S'**), (**s**), (**S**), (**v**), (**F**), (**i**), (**T**), (**d**), (**K**), (**J**), (**V**) and (**p**) are known, then its Leverage or Gearing Ratio planned is:

$$\mathbf{/}= [\mathbf{N}+\mathbf{K}+\mathbf{J}+\mathbf{Vp}/360]/(\mathbf{U}+\{\mathbf{S'}[1+\mathbf{s}]-\mathbf{Sv}-\mathbf{F}-\mathbf{Si}-\mathbf{T}\}$$
$$[1-\mathbf{d}])-1$$

Rule-57333:

If both (**/**), (**N**), (**S'**), (**s**), (**S**), (**v**), (**F**), (**i**), (**T**), (**d**), (**K**), (**J**), (**V**) and (**p**) are known, then its Utilized or Starting Capital must be:

$$\mathbf{U}= ([\mathbf{N}+\mathbf{K}+\mathbf{J}+\mathbf{Vp}/360]-[1+\mathbf{/}]\{\mathbf{S'}[1+\mathbf{s}]$$
$$-\mathbf{Sv}-\mathbf{F}-\mathbf{Si}-\mathbf{T}\}[1-\mathbf{d}])/[1+\mathbf{/}]$$

Rule-57334:

If both (**/**), (**U**), (**N**), (**s**), (**S**), (**v**), (**F**), (**i**), (**T**), (**d**), (**K**), (**J**), (**V**) and (**p**) are known, then its Sales Past must be:

$$\mathbf{S'}= (\{[\mathbf{N}+\mathbf{K}+\mathbf{J}+\mathbf{Vp}/360]/[1+\mathbf{/}]-\mathbf{U}\}/[1-\mathbf{d}]$$
$$+\mathbf{Sv}+\mathbf{F}+\mathbf{Si}+\mathbf{T})/[1+\mathbf{s}]$$

or

$$\mathbf{S'}= \mathbf{S}/[1+\mathbf{s}]$$

Steve Asikin ISBN 13: **978-1564912361**, ISBN 10: **1564912363**

Rule-57335:

If both (**/**), (**U**), (**S'**), (**N**), (**S**), (**v**), (**F**), (**i**), (**T**), (**d**), (**K**), (**J**), (**V**) and (**p**) are known, then its Sales Growth planned is:

$$s= (\{[N+K+J+Vp/360]/[1+/-U\}/[1-d]$$
$$+Sv+F+Si+T)/S'-1$$

or

$$s= S/S'-1$$

Rule-57336:

If both (**/**), (**U**), (**S'**), (**s**), (**N**), (**v**), (**F**), (**i**), (**T**), (**d**), (**K**), (**J**), (**V**) and (**p**) are known, then its Sales or Revenue planned is:

$$S- (S'[1+s]-F-T-\{[N+K+J+Vp/360]/[1+/-U\}$$
$$/[1-d])/[v+i]$$

or

$$S= S'[1+s]$$

or

$$S= V/v$$

Rule-57337:

If both (**/**), (**U**), (**S'**), (**s**), (**S**), (**N**), (**F**), (**i**), (**T**), (**d**), (**K**), (**J**), (**V**) and (**p**) are known, then its Variable Portion planned is:

$$v= (S'[1+s]-F-Si-T-\{[N+K+J+Vp/360]$$
$$/[1+/-U\}/[1-d])/S$$

or

$$v= V/S$$

Steve Asikin ISBN 13: **978-1564912361**, ISBN 10: **1564912363**

Rule-57338:

    If both (**ʃ**), (**U**), (**S'**), (**s**), (**S**), (**v**), (**N**), (**i**), (**T**), (**d**), (**K**), (**J**), (**V**) and (**p**) are known, then its Fixed Cost planned is:

$$F = S'[1+s] - Sv - Si - T - \{[N+K+J+Vp/360]/[1+ʃ-U\}/[1-d]$$

Rule-57339:

    If both (**ʃ**), (**U**), (**S'**), (**s**), (**S**), (**v**), (**F**), (**N**), (**T**), (**d**), (**K**), (**J**), (**V**) and (**p**) are known, then its Interest Portion planned is:

$$i = (S'[1+s] - Sv - F - T - \{[N+K+J+Vp/360]/[1+ʃ-U\}/[1-d])/S$$

Rule-57340:

    If both (**ʃ**), (**U**), (**S'**), (**s**), (**S**), (**v**), (**F**), (**i**), (**N**), (**d**), (**K**), (**J**), (**V**) and (**p**) are known, then its Tax planned is:

$$T = S'[1+s] - Sv - F - Si - \{[N+K+J+Vp/360]/[1+ʃ-U\}/[1-d]$$

Rule-57341:

    If both (**ʃ**), (**U**), (**S'**), (**s**), (**S**), (**v**), (**F**), (**i**), (**T**), (**N**), (**K**), (**J**), (**V**) and (**p**) are known, then its Dividend Portion planned is:

$$d = 1 - \{[N+K+J+Vp/360]/[1+ʃ-U\}/\{S'[1+s] - Sv - F - Si - T\}$$

Steve Asikin ISBN 13: **978-1564912361**, ISBN 10: **1564912363**

Rule-57342:

If both (**ƒ**), (**U**), (**S'**), (**s**), (**S**), (**v**), (**F**), (**i**), (**T**), (**d**), (**N**), (**J**), (**V**) and (**p**) are known, then its Kind of Cash planned is:

$$K = [1 + ƒ(U + \{S'[1+s] - Sv - F - Si - T\}[1-d]) - [N + J + Vp/360]$$

Rule-57343:

If both (**ƒ**), (**U**), (**S'**), (**s**), (**S**), (**v**), (**F**), (**i**), (**T**), (**d**), (**K**), (**N**), (**V**) and (**p**) are known, then its Job or Trade Account Receivable planned is:

$$J = [1 + ƒ(U + \{S'[1+s] - Sv - F - Si - T\}[1-d]) - \{N + K + Vp/360\}$$

Rule-57344:

If both (**ƒ**), (**U**), (**S'**), (**s**), (**S**), (**v**), (**F**), (**i**), (**T**), (**d**), (**K**), (**J**), (**N**) and (**p**) are known, then its Variable Cost planned is:

$$V = 360[[1 + ƒ(U + \{S'[1+s] - Sv - F - Si - T\}[1-d]) - [N + K + J]]/p$$

or

$$V = Sv$$

Rule-57345:

If both (**ƒ**), (**U**), (**S'**), (**s**), (**S**), (**v**), (**F**), (**i**), (**T**), (**d**), (**K**), (**J**), (**V**) and (**N**) are known, then its Procured Inventory Days planned is:

$$p = 360[[1 + ƒ(U + \{S'[1+s] - Sv - F - Si - T\}[1-d]) - [N + K + J]]/V$$

Steve Asikin ISBN 13: **978-1564912361**, ISBN 10: **1564912363**

Rule-57346:

If both (*I*), (**U**), (**S'**), (**s**), (**S**), (**v**), (**F**), (**i**), (**T**), (**d**), (**K**), (**J**) and (**p**) are known, then its Non Current Asset planned is:

$$N= [1+I](U+\{S'[1+s]-Sv-F-Si-T\}[1-d])$$
$$-K-J-Svp/360$$

Rule-57347:

If both (**N**), (**U**), (**S'**), (**s**), (**S**), (**v**), (**F**), (**i**), (**T**), (**d**), (**K**), (**J**) and (**p**) are known, then its Leverage or Gearing Ratio planned is:

$$I= [N+K+J+Svp/360]/(U+\{S'[1+s]-Sv-F-Si-T\}$$
$$[1-d])-1$$

Rule-57348:

If both (*I*), (**N**), (**S'**), (**s**), (**S**), (**v**), (**F**), (**i**), (**T**), (**d**), (**K**), (**J**) and (**p**) are known, then its Utilized or Starting Capital must be:

$$U= ([N+K+J+Svp/360]-[1+I]\{S'[1+s]$$
$$-Sv-F-Si-T\}[1-d])/[1+I]$$

Rule-57349:

If both (*I*), (**U**), (**N**), (**s**), (**S**), (**v**), (**F**), (**i**), (**T**), (**d**), (**K**), (**J**) and (**p**) are known, then its Sales Past must be:

$$S'= (\{[N+K+J+Svp/360]/[1+I]-U\}/[1-d]$$
$$+Sv+F+Si+T)/[1+s]$$

or

$$S'= S/[1+s]$$

Steve Asikin ISBN 13: **978-1564912361**, ISBN 10: **1564912363**

Rule-57350:

If both (**/**), (**U**), (**S'**), (**N**), (**S**), (**v**), (**F**), (**i**), (**T**), (**d**), (**K**), (**J**) and (**p**) are known, then its Sales Growth planned is:

$$s= (\{[N+K+J+Svp/360]/[1+/-U\}/[1-d]$$
$$+Sv+F+Si+T)/S'-1$$

or

$$s= S/S'-1$$

Rule-57351:

If both (**/**), (**U**), (**S'**), (**s**), (**N**), (**v**), (**F**), (**i**), (**T**), (**d**), (**K**), (**J**) and (**p**) are known, then its Sales or Revenue planned is:

$$S= [[1+/(U+\{S'[1+s]-F-Si-T\}[1-d])$$
$$-[N+K+J]]/(v\{p/360+[1+/[1-d]\})$$

or

$$S= S'[1+s]$$

Rule-57352:

If both (**/**), (**U**), (**S'**), (**s**), (**S**), (**N**), (**F**), (**i**), (**T**), (**d**), (**K**), (**J**) and (**p**) are known, then its Variable Portion planned is:

$$v= [[1+/(U+\{S'[1+s]-F-Si-T\}[1-d])-[N+K+J]]$$
$$/(S\{p/360+[1+/[1-d]\})$$

Steve Asikin ISBN 13: **978-1564912361**, ISBN 10: **1564912363**

Rule-57353:

> If both $(\textit{l})$, $(\textbf{U})$, $(\textbf{S'})$, $(\textbf{s})$, $(\textbf{S})$, $(\textbf{v})$, $(\textbf{N})$, $(\textbf{i})$, $(\textbf{T})$, $(\textbf{d})$, $(\textbf{K})$, $(\textbf{J})$ and $(\textbf{p})$ are known, then its Fixed Cost planned is:
>
> $$F = S'[1+s] - Sv - Si - T - \{[N+K+J+Svp/360]/[1+\textit{l}-U\} /[1-d]$$

Rule-57354:

> If both $(\textit{l})$, $(\textbf{U})$, $(\textbf{S'})$, $(\textbf{s})$, $(\textbf{S})$, $(\textbf{v})$, $(\textbf{F})$, $(\textbf{N})$, $(\textbf{T})$, $(\textbf{d})$, $(\textbf{K})$, $(\textbf{J})$ and $(\textbf{p})$ are known, then its Interest Portion planned is:
>
> $$i = (S'[1+s] - Sv - F - T - \{[N+K+J+Svp/360]/[1+\textit{l}-U\} /[1-d])/S$$

Rule-57355:

> If both $(\textit{l})$, $(\textbf{U})$, $(\textbf{S'})$, $(\textbf{s})$, $(\textbf{S})$, $(\textbf{v})$, $(\textbf{F})$, $(\textbf{i})$, $(\textbf{T})$, $(\textbf{d})$, $(\textbf{N})$, $(\textbf{J})$ and $(\textbf{p})$ are known, then its Tax planned is:
>
> $$T = S'[1+s] - Sv - F - Si - \{[N+K+J+Svp/360]/[1+\textit{l}-U\} /[1-d]$$

Rule-57356:

> If both $(\textit{l})$, $(\textbf{U})$, $(\textbf{S'})$, $(\textbf{s})$, $(\textbf{S})$, $(\textbf{v})$, $(\textbf{F})$, $(\textbf{i})$, $(\textbf{T})$, $(\textbf{N})$, $(\textbf{K})$, $(\textbf{J})$ and $(\textbf{p})$ are known, then its Dividend Payout planned is:
>
> $$d = 1 - \{[N+K+J+Svp/360]/[1+\textit{l}-U\} /\{S'[1+s] - Sv - F - Si - T\}$$

Steve Asikin ISBN 13: **978-1564912361**, ISBN 10: **1564912363**

Rule-57357:

If both ($I$), (**U**), (**S'**), (**s**), (**S**), (**v**), (**F**), (**i**), (**T**), (**d**), (**N**), (**J**) and (**p**) are known, then its Kind of Cash planned is:

$$K= [1+I(U+\{S'[1+s]-Sv-F-Si-T\}[1-d])$$
$$-[N+J+Svp/360]$$

Rule-57358:

If both ($I$), (**U**), (**S'**), (**s**), (**S**), (**v**), (**F**), (**i**), (**T**), (**d**), (**K**), (**N**) and (**p**) are known, then its Job or Trade Account Receivable planned is:

$$J= [1+I(U+\{S'[1+s]-Sv-F-Si-T\}[1-d])$$
$$-\{N+K+Svp/360\}$$

Rule-57359:

If both ($I$), (**U**), (**S'**), (**s**), (**S**), (**v**), (**F**), (**i**), (**T**), (**d**), (**K**), (**J**) and (**N**) are known, then its Procured Inventory Days planned is:

$$p= 360[[1+I(U+\{S'[1+s]-Sv-F-Si-T\}[1-d])$$
$$-[N+K+J]]/[Sv]$$

Rule-57360:

If both ($I$), (**U**), (**S'**), (**s**), (**S**), (**v**), (**F**), (**i**), (**T**), (**d**), (**K**), (**J**) and (**p**) are known, then its Non Current Asset planned is:

$$N= [1+I(U+\{S'[1+s]-Sv-F-Si-T\}[1-d])$$
$$-K-J-S'vp[1+s]/360$$

Steve Asikin ISBN 13: **978-1564912361**, ISBN 10: **1564912363**

Rule-57361:

If both **(N)**, **(U)**, **(S')**, **(s)**, **(S)**, **(v)**, **(F)**, **(i)**, **(T)**, **(d)**, **(K)**, **(J)** and **(p)** are known, then its Leverage or Gearing Ratio planned is:

$$I\!= \{N+K+J+S'vp[1+s]/360\}$$
$$/(U+\{S'[1+s]-Sv-F-Si-T\}[1-d])-1$$

Rule-57362:

If both **(I)**, **(N)**, **(S')**, **(s)**, **(S)**, **(v)**, **(F)**, **(i)**, **(T)**, **(d)**, **(K)**, **(J)** and **(p)** are known, then its Utilized or Starting Capital must be:

$$U= (\{N+K+J+S'vp[1+s]/360\}-[1+I]\{S'[1+s]$$
$$-Sv-F-Si-T\}[1-d])/[1+I]$$

Rule-57363:

If both **(I)**, **(U)**, **(N)**, **(s)**, **(S)**, **(v)**, **(F)**, **(i)**, **(T)**, **(d)**, **(K)**, **(J)** and **(p)** are known, then its Sales Past must be:

$$S'= [[1+I(U-\{Sv+F+Si+T\}[1-d])-[N+K+J]]$$
$$/([1+s]\{vp/360-[1+I][1-d]\})$$

or

$$S'= S/[1+s]$$

Steve Asikin ISBN 13: **978-1564912361**, ISBN 10: **1564912363**

Rule-57364:

If both (**∫**), (**U**), (**$'**), (**N**), (**$**), (**v**), (**F**), (**i**), (**T**), (**d**), (**K**), (**J**) and (**p**) are known, then its Sales Growth planned is:

$s$= [[1+**∫**](**U**-{**$v**+**F**+**$i**+**T**}[1-**d**])-[**N**+**K**+**J**]]
    /(**$'**{**vp**/360-[1+**∫**][1-**d**]})-1

or

$s$= **$**/**$'**-1


Rule-57365:

If both (**∫**), (**U**), (**$'**), (**s**), (**N**), (**v**), (**F**), (**i**), (**T**), (**d**), (**K**), (**J**) and (**p**) are known, then its Sales or Revenue planned is:

**$**= [**$'**[1+**s**]-**F**-**T**-({**N**+**K**+**J**}+**$'vp**[1+**s**]/360}
    /[1+**∫**-**U**)/[1-**d**]]/[**v**+**i**]

or

**$**= **$'**[1+**s**]


Rule-57366:

If both (**∫**), (**U**), (**$'**), (**s**), (**$**), (**N**), (**F**), (**i**), (**T**), (**d**), (**K**), (**J**) and (**p**) are known, then its Variable Portion planned is:

**v**= [[1+**∫**](**U**+{**$'**[1+**s**]-**F**-**$i**-**T**}[1-**d**])-[**N**+**K**+**J**]]
    /{**$'p**[1+**s**]/360+**$**[1+**∫**][1-**d**]}

Steve Asikin ISBN 13: **978-1564912361**, ISBN 10: **1564912363**

Rule-57367:

    If both $(\textit{I})$, $(\mathbf{U})$, $(\textbf{S'})$, $(\textbf{s})$, $(\textbf{S})$, $(\textbf{v})$, $(\mathbf{N})$, $(\textbf{i})$, $(\mathbf{T})$, $(\mathbf{d})$, $(\mathbf{K})$, $(\mathbf{J})$ and $(\mathbf{p})$ are known, then its Fixed Cost planned is:

$$F= S'[1+s]-Sv-Si-T-(\{N+K+J+S'vp[1+s]/360\}/[1+\textit{I}-U)/[1-d]$$

Rule-57368:

    If both $(\textit{I})$, $(\mathbf{U})$, $(\textbf{S'})$, $(\textbf{s})$, $(\textbf{S})$, $(\textbf{v})$, $(\mathbf{F})$, $(\mathbf{N})$, $(\mathbf{T})$, $(\mathbf{d})$, $(\mathbf{K})$, $(\mathbf{J})$ and $(\mathbf{p})$ are known, then its Interest Portion planned is:

$$i= [S'[1+s]-Sv-F-T-(\{N+K+J+S'vp[1+s]/360\}/[1+\textit{I}-U)/[1-d]]/S$$

Rule-57369:

    If both $(\textit{I})$, $(\mathbf{U})$, $(\textbf{S'})$, $(\textbf{s})$, $(\textbf{S})$, $(\textbf{v})$, $(\mathbf{F})$, $(\textbf{i})$, $(\mathbf{N})$, $(\mathbf{d})$, $(\mathbf{K})$, $(\mathbf{J})$ and $(\mathbf{p})$ are known, then its Tax planned is:

$$T= S'[1+s]-Sv-F-Si-(\{N+K+J+S'vp[1+s]/360\}/[1+\textit{I}-U)/[1-d]$$

Rule-57370:

    If both $(\textit{I})$, $(\mathbf{U})$, $(\textbf{S'})$, $(\textbf{s})$, $(\textbf{S})$, $(\textbf{v})$, $(\mathbf{F})$, $(\textbf{i})$, $(\mathbf{T})$, $(\mathbf{N})$, $(\mathbf{K})$, $(\mathbf{J})$ and $(\mathbf{p})$ are known, then its Dividend Payout planned is:

$$d= 1-(\{N+K+J+S'vp[1+s]/360\}/[1+\textit{I}-U)/\{S'[1+s]-Sv-F-Si-T\}$$

Steve Asikin ISBN 13: **978-1564912361**, ISBN 10: **1564912363**

## Rule-57371:

If both ($I$), (**U**), (**S'**), (**s**), (**S**), (**v**), (**F**), (**i**), (**T**), (**d**), (**N**), (**J**) and (**p**) are known, then its Kind of Cash planned is:

$$K= [1+I(U+\{S'[1+s]-Sv-F-Si-T\}[1-d]) \\ -\{N+J+S'vp[1+s]/360\}$$

## Rule-57372:

If both ($I$), (**U**), (**S'**), (**s**), (**S**), (**v**), (**F**), (**i**), (**T**), (**d**), (**K**), (**N**) and (**p**) are known, then its Job or Trade Account Receivable planned is:

$$J= [[1+I(U+\{S'[1+s]-Sv-F-Si-T\}[1-d]) \\ -\{N+K+S'vp[1+s]/360\}$$

## Rule-57373:

If both ($I$), (**U**), (**S'**), (**s**), (**S**), (**v**), (**F**), (**i**), (**T**), (**d**), (**K**), (**J**) and (**N**) are known, then its Procured Inventory Days planned is:

$$p= 360[[1+I(U+\{S'[1+s]-Sv-F-Si-T\}[1-d]) \\ -[N+K+J]]/\{S'v[1+s]\}$$

## Rule-57374:

If both ($I$), (**U**), (**S'**), (**s**), (**S**), (**v**), (**F**), (**i**), (**T**), (**d**), (**K**), (**j**) and (**P**) are known, then its Non Current Asset planned is:

$$N= [1+I(U+\{S'[1+s]-Sv-F-Si-T\}[1-d]) \\ -K-Sj/360-P$$

`

Steve Asikin ISBN 13: **978-1564912361**, ISBN 10: **1564912363**

Rule-57375:

If both **(N)**, **(U)**, **(S')**, **(s)**, **(S)**, **(v)**, **(F)**, **(i)**, **(T)**, **(d)**, **(K)**, **(j)** and **(P)** are known, then its Leverage or Gearing Ratio planned is:

$$I = \{N+K+Sj/360+P\}/(U+\{S'[1+s]-Sv-F-Si-T\}[1-d])-1$$

Rule-57376:

If both **(I)**, **(N)**, **(S')**, **(s)**, **(S)**, **(v)**, **(F)**, **(i)**, **(T)**, **(d)**, **(K)**, **(j)** and **(P)** are known, then its Utilized or Starting Capital must be:

$$U = (\{N+K+Sj/360+P\}-[1+I]\{S'[1+s]-Sv-F-Si-T\}[1-d])/[1+I]$$

Rule-57377:

If both **(I)**, **(U)**, **(N)**, **(s)**, **(S)**, **(v)**, **(F)**, **(i)**, **(T)**, **(d)**, **(K)**, **(j)** and **(P)** are known, then its Sales Past must be:

$$S' = (\{[N+K+Sj/360+P]/[1+I]-U\}/[1-d]+Sv+F+Si+T)/[1+s]$$

or

$$S' = S/[1+s]$$

Steve Asikin ISBN 13: **978-1564912361**, ISBN 10: **1564912363**

<u>Rule-57378</u>:

If both $(I)$, $(U)$, $(S')$, $(N)$, $(S)$, $(v)$, $(F)$, $(i)$, $(T)$, $(d)$, $(K)$, $(j)$ and $(P)$ are known, then its Sales Growth planned is:

$$s = (\{[N+K+Sj/360+P]/[1+I]-U\}/[1-d]$$
$$+Sv+F+Si+T)/S'-1$$

or

$$s = S/S'-1$$


<u>Rule-57379</u>:

If both $(I)$, $(U)$, $(S')$, $(s)$, $(S)$, $(v)$, $(F)$, $(i)$, $(T)$, $(d)$, $(K)$, $(j)$ and $(P)$ are known, then its Non Current Asset planned is:

$$S = [\lfloor 1+I \rfloor (U+\{S'[1+s]-F-T\}[1-d])-[N+K+P]]$$
$$/\{j/360+[1+I][1-d][v+i]\}$$

or

$$S = S'[1+s]$$


<u>Rule-57380</u>:

If both $(I)$, $(U)$, $(S')$, $(s)$, $(S)$, $(N)$, $(F)$, $(i)$, $(T)$, $(d)$, $(K)$, $(j)$ and $(P)$ are known, then its Variable Portion planned is:

$$v = (S'[1+s]-F-Si-T-\{[N+K+Sj/360+P]/[1+I]-U\}$$
$$/[1-d])/S$$

Steve Asikin ISBN 13: **978-1564912361**, ISBN 10: **1564912363**

Rule-57381:

If both $(\textit{l})$, $(\textbf{U})$, $(\textbf{S'})$, $(\textbf{s})$, $(\textbf{\$})$, $(\textbf{v})$, $(\textbf{N})$, $(\textbf{i})$, $(\textbf{T})$, $(\textbf{d})$, $(\textbf{K})$, $(\textbf{j})$ and $(\textbf{P})$ are known, then its Fixed Cost planned is:

$$F = S'[1+s] - Sv - Si - T - \{[N+K+Sj/360+P]/[1+\textit{l}-U\} /[1-d]$$

Rule-57382:

If both $(\textit{l})$, $(\textbf{U})$, $(\textbf{S'})$, $(\textbf{s})$, $(\textbf{\$})$, $(\textbf{v})$, $(\textbf{F})$, $(\textbf{N})$, $(\textbf{T})$, $(\textbf{d})$, $(\textbf{K})$, $(\textbf{j})$ and $(\textbf{P})$ are known, then its Interest Portion planned is:

$$i = (S'[1+s] - Sv - F - T - \{[N+K+Sj/360+P]/[1+\textit{l}-U\} /[1-d])/S$$

Rule-57383:

If both $(\textit{l})$, $(\textbf{U})$, $(\textbf{S'})$, $(\textbf{s})$, $(\textbf{\$})$, $(\textbf{v})$, $(\textbf{F})$, $(\textbf{i})$, $(\textbf{N})$, $(\textbf{d})$, $(\textbf{K})$, $(\textbf{j})$ and $(\textbf{P})$ are known, then its Tax planned is:

$$T = S'[1+s] - Sv - F - Si - \{[N+K+Sj/360+P]/[1+\textit{l}-U\} /[1-d]$$

Rule-57384:

If both $(\textit{l})$, $(\textbf{U})$, $(\textbf{S'})$, $(\textbf{s})$, $(\textbf{\$})$, $(\textbf{v})$, $(\textbf{F})$, $(\textbf{i})$, $(\textbf{T})$, $(\textbf{N})$, $(\textbf{K})$, $(\textbf{j})$ and $(\textbf{P})$ are known, then its Dividend Payout planned is:

$$d = 1 - \{[N+K+Sj/360+P]/[1+\textit{l}-U\} /\{S'[1+s] - Sv - F - Si - T\}$$

Steve Asikin ISBN 13: **978-1564912361**, ISBN 10: **1564912363**

Rule-57385:

If both (**/**), (**U**), (**$'**), (**s**), (**$**), (**v**), (**F**), (**i**), (**T**), (**d**), (**N**), (**j**) and (**P**) are known, then its Kind of Cash planned is:

$$K= [1+\textbf{/}(\textbf{U}+\{\textbf{\$'}[1+\textbf{s}]-\textbf{\$v}-\textbf{F}-\textbf{\$i}-\textbf{T}\}[1-\textbf{d}]) \\ -[\textbf{N}+\textbf{\$j}/360+\textbf{P}]$$

Rule-57386:

If both (**/**), (**U**), (**$'**), (**s**), (**$**), (**v**), (**F**), (**i**), (**T**), (**d**), (**K**), (**N**) and (**P**) are known, then its Job or Trade Receivable Days planned is:

$$j= 360[[1+\textbf{/}(\textbf{U}+\{\textbf{\$'}[1+\textbf{s}]-\textbf{\$v}-\textbf{F}-\textbf{\$i}-\textbf{T}\}[1-\textbf{d}]) \\ -\{\textbf{N}+\textbf{K}+\textbf{P}\}]/\textbf{\$}$$

Rule-57387:

If both (**/**), (**U**), (**$'**), (**s**), (**$**), (**v**), (**F**), (**i**), (**T**), (**d**), (**K**), (**j**) and (**N**) are known, then its Procured Inventory planned is:

$$P= [1+\textbf{/}(\textbf{U}+\{\textbf{\$'}[1+\textbf{s}]-\textbf{\$v}-\textbf{F}-\textbf{\$i}-\textbf{T}\}[1-\textbf{d}]) \\ -\{\textbf{N}+\textbf{K}+\textbf{\$j}/360\}$$

Rule-57388:

If both (**/**), (**U**), (**$'**), (**s**), (**$**), (**v**), (**F**), (**i**), (**T**), (**d**), (**K**), (**j**), (**V**) and (**p**) are known, then its Non Current Asset planned is:

$$N= [1+\textbf{/}(\textbf{U}+\{\textbf{\$'}[1+\textbf{s}]-\textbf{\$v}-\textbf{F}-\textbf{\$i}-\textbf{T}\}[1-\textbf{d}]) \\ -\textbf{K}-\textbf{\$j}/360-\textbf{Vp}/360$$

Steve Asikin ISBN 13: **978-1564912361**, ISBN 10: **1564912363**

Rule-57389:

If both $(\mathbf{N})$, $(\mathbf{U})$, $(\mathbf{S'})$, $(\mathbf{s})$, $(\mathbf{S})$, $(\mathbf{v})$, $(\mathbf{F})$, $(\mathbf{i})$, $(\mathbf{T})$, $(\mathbf{d})$, $(\mathbf{K})$, $(\mathbf{j})$, $(\mathbf{V})$ and $(\mathbf{p})$ are known, then its Leverage or Gearing Ratio planned is:

$$\mathbf{I} = [\mathbf{N+K+Sj}/360+\mathbf{Vp}/360]$$
$$/(\mathbf{U+\{S'}[1+\mathbf{s}]-\mathbf{Sv-F-Si-T}\}[1-\mathbf{d}])-1$$

Rule-57390:

If both $(\mathbf{I})$, $(\mathbf{N})$, $(\mathbf{S'})$, $(\mathbf{s})$, $(\mathbf{S})$, $(\mathbf{v})$, $(\mathbf{F})$, $(\mathbf{i})$, $(\mathbf{T})$, $(\mathbf{d})$, $(\mathbf{K})$, $(\mathbf{j})$, $(\mathbf{V})$ and $(\mathbf{p})$ are known, then its Utilized or Starting Capital must be:

$$\mathbf{U} = ([\mathbf{N+K+Sj}/360+\mathbf{Vp}/360]$$
$$-[1+\mathbf{I}]\{\mathbf{S'}[1+\mathbf{s}]-\mathbf{Sv-F-Si-T}\}[1-\mathbf{d}])/[1+\mathbf{I}]$$

Rule-57391:

If both $(\mathbf{I})$, $(\mathbf{U})$, $(\mathbf{N})$, $(\mathbf{s})$, $(\mathbf{S})$, $(\mathbf{v})$, $(\mathbf{F})$, $(\mathbf{i})$, $(\mathbf{T})$, $(\mathbf{d})$, $(\mathbf{K})$, $(\mathbf{j})$, $(\mathbf{V})$ and $(\mathbf{p})$ are known, then its Sales Past must be:

$$\mathbf{S'} = (\{[\mathbf{N+K+Sj}/360+\mathbf{Vp}/360]/[1+\mathbf{I}]-\mathbf{U}\}/[1-\mathbf{d}]$$
$$+\mathbf{Sv+F+Si+T})/[1+\mathbf{s}]$$

or

$$\mathbf{S'} = \mathbf{S}[1+\mathbf{s}]$$

Steve Asikin ISBN 13: **978-1564912361**, ISBN 10: **1564912363**

Rule-57392:

If both ($I$), ($U$), ($S'$), ($N$), ($S$), ($v$), ($F$), ($i$), ($T$), ($d$), ($K$), ($j$), ($V$) and ($p$) are known, then its Sales Growth planned is:

$$s = (\{[N+K+Sj/360+Vp/360]/[1+I-U\}/[1-d] + Sv+F+Si+T)/S'-1$$

or

$$s = S/S'-1$$


Rule-57393:

If both ($I$), ($U$), ($S'$), ($s$), ($N$), ($v$), ($F$), ($i$), ($T$), ($d$), ($K$), ($j$), ($V$) and ($p$) are known, then its Sales or Revenue planned is:

$$S = [[1+I](U+\{S'[1+s]-F-T\}[1-d]) -[N+K+Vp/360]] /\{j/360+[1+I][1-d][v+i]\}$$

or

$$S = S'[1+s]$$


Rule-57394:

If both ($I$), ($U$), ($S'$), ($s$), ($S$), ($N$), ($F$), ($i$), ($T$), ($d$), ($K$), ($j$), ($V$) and ($p$) are known, then its Variable Portion planned is:

$$v = (S'[1+s]-Si-F-T-\{[N+K+Sj/360+Vp/360] /[1+I-U\}/[1-d])/S$$

or

$$v = V/S$$

Steve Asikin ISBN 13: **978-1564912361**, ISBN 10: **1564912363**

Rule-57395:

If both (**/**), (**U**), (**S'**), (**s**), (**S**), (**v**), (**N**), (**i**), (**T**), (**d**), (**K**), (**j**), (**V**) and (**p**) are known, then its Fixed Cost planned is:

$$F= S'[1+s]-Sv-Si-T-\{[N+K+Sj/360+Vp/360] /[1+/-U\}/[1-d]$$

Rule-57396:

If both (**/**), (**U**), (**S'**), (**s**), (**S**), (**v**), (**F**), (**N**), (**T**), (**d**), (**K**), (**j**), (**V**) and (**p**) are known, then its Interesting Portion planned is:

$$i= (S'[1+s]-Sv-F-T-\{[N+K+Sj/360+Vp/360] /[1+/-U\}/[1-d])/S$$

Rule-57397:

If both (**/**), (**U**), (**S'**), (**s**), (**S**), (**v**), (**F**), (**i**), (**N**), (**d**), (**K**), (**j**), (**V**) and (**p**) are known, then its Tax planned is:

$$T= S'[1+s]-Sv-F-Si-\{[N+K+Sj/360+Vp/360] /[1+/-U\}/[1-d]$$

Rule-57398:

If both (**/**), (**U**), (**S'**), (**s**), (**S**), (**v**), (**F**), (**i**), (**T**), (**N**), (**K**), (**j**), (**V**) and (**p**) are known, then its Dividend Payout planned is:

$$d= 1-\{[N+K+Sj/360+Vp/360]/[1+/-U\} /\{S'[1+s]-Sv-F-Si-T\}$$

Steve Asikin ISBN 13: **978-1564912361**, ISBN 10: **1564912363**

Rule-57399:

If both $(I)$, $(U)$, $(S')$, $(s)$, $(S)$, $(v)$, $(F)$, $(i)$, $(T)$, $(d)$, $(N)$, $(j)$, $(V)$ and $(p)$ are known, then its Kind of Cash planned is:

$$K= [1+I(U+\{S'[1+s]-Sv-F-Si-T\}[1-d])$$
$$-[N+Sj/360+Vp/360]$$

Rule-57400:

If both $(I)$, $(U)$, $(S')$, $(s)$, $(S)$, $(v)$, $(F)$, $(i)$, $(T)$, $(d)$, $(K)$, $(N)$, $(V)$ and $(p)$ are known, then its Job or Trade Receivable planned is:

$$j= 360[[1+I(U+\{S'[1+s]-Sv-F-Si-T\}[1-d])$$
$$-[N+K+Vp/360]]/S$$

Rule-57401:

If both $(I)$, $(U)$, $(S')$, $(s)$, $(S)$, $(v)$, $(F)$, $(i)$, $(T)$, $(d)$, $(K)$, $(j)$, $(N)$ and $(p)$ are known, then its Variable Cost planned is:

$$V= 360[[1+I(U+\{S'[1+s]-Sv-F-Si-T\}[1-d])$$
$$-[N+K+Sj/360]]/p$$
$$\text{or}$$
$$V= Sv$$

Rule-57402:

If both $(I)$, $(U)$, $(S')$, $(s)$, $(S)$, $(v)$, $(F)$, $(i)$, $(T)$, $(d)$, $(K)$, $(j)$, $(V)$ and $(N)$ are known, then its Procured Inventory Days planned is:

$$p= 360[[1+I(U+\{S'[1+s]-Sv-F-Si-T\}[1-d])$$
$$-[N+K+Sj/360]]/V$$

Steve Asikin ISBN 13: **978-1564912361**, ISBN 10: **1564912363**

Rule-57403:

If both $(I)$, $(U)$, $(S')$, $(s)$, $(S)$, $(v)$, $(F)$, $(i)$, $(T)$, $(d)$, $(K)$, $(j)$ and $(p)$ are known, then its Non Current Asset planned is:

$$N= [1+I](U+\{S'[1+s]-Sv-F-Si-T\}[1-d])$$
$$-K-Sj/360-Svp/360$$

Rule-57404:

If both $(N)$, $(U)$, $(S')$, $(s)$, $(S)$, $(v)$, $(F)$, $(i)$, $(T)$, $(d)$, $(K)$, $(j)$ and $(p)$ are known, then its Leverage or Gearing Ratio planned is:

$$I= [N+K+Sj/360+Svp/360]$$
$$/(U+\{S'[1+s]-Sv-F-Si-T\}[1-d])-1$$

Rule-57405:

If both $(I)$, $(N)$, $(S')$, $(s)$, $(S)$, $(v)$, $(F)$, $(i)$, $(T)$, $(d)$, $(K)$, $(j)$ and $(p)$ are known, then its Utilized or Starting Capital must be:

$$U= ([N+K+Sj/360+Svp/360]$$
$$-[1+I]\{S'[1+s]-Sv-F-Si-T\}[1-d])/[1+I]$$

Rule-57406:

If both $(I)$, $(U)$, $(N)$, $(s)$, $(S)$, $(v)$, $(F)$, $(i)$, $(T)$, $(d)$, $(K)$, $(j)$ and $(p)$ are known, then its Sales Past must be:

$$S'= (\{[N+K+Sj/360+Svp/360]/[1+I]-U\}/[1-d]$$
$$+Sv+F+Si+T)/[1+s]$$
$$\text{or}$$
$$S'= S/[1+s]$$

Steve Asikin ISBN 13: **978-1564912361**, ISBN 10: **1564912363**

Rule-57407:

If both $(\textbf{\textit{I}})$, $(\textbf{U})$, $(\textbf{S'})$, $(\textbf{N})$, $(\textbf{S})$, $(\textbf{v})$, $(\textbf{F})$, $(\textbf{i})$, $(\textbf{T})$, $(\textbf{d})$, $(\textbf{K})$, $(\textbf{j})$ and $(\textbf{p})$ are known, then its Sales Growth planned is:

$$s= (\{[N+K+Sj/360+Svp/360]/[1+I-U\}/[1-d]$$
$$+Sv+F+Si+T)/S'-1$$

or

$$s= S/S'-1$$

Rule-57408:

If both $(\textbf{\textit{I}})$, $(\textbf{U})$, $(\textbf{S'})$, $(\textbf{s})$, $(\textbf{N})$, $(\textbf{v})$, $(\textbf{F})$, $(\textbf{i})$, $(\textbf{T})$, $(\textbf{d})$, $(\textbf{K})$, $(\textbf{j})$ and $(\textbf{p})$ are known, then its Sales or Revenue planned is:

$$S= [[1+I(U+\{S'[1+s]-Sv-F-Si-T\}[1\ d])-[N+K]]$$
$$/(j/360+v\{p/360+[1+I[1-d]\})$$

or

$$S= S'[1+s]$$

Rule-57409:

If both $(\textbf{\textit{I}})$, $(\textbf{U})$, $(\textbf{S'})$, $(\textbf{s})$, $(\textbf{S})$, $(\textbf{N})$, $(\textbf{F})$, $(\textbf{i})$, $(\textbf{T})$, $(\textbf{d})$, $(\textbf{K})$, $(\textbf{j})$ and $(\textbf{p})$ are known, then its Variable Portion planned is:

$$v= \{[[1+I(U+\{S'[1+s]-Sv-F-Si-T\}[1-d])-[N+K]]$$
$$/S-j/360\}/\{p/360+[1+I[1-d]\}$$

Steve Asikin ISBN 13: **978-1564912361**, ISBN 10: **1564912363**

Rule-57410:

If both $(\textbf{\textit{I}})$, $(\textbf{U})$, $(\textbf{S'})$, $(\textbf{s})$, $(\textbf{S})$, $(\textbf{v})$, $(\textbf{N})$, $(\textbf{i})$, $(\textbf{T})$, $(\textbf{d})$, $(\textbf{K})$, $(\textbf{j})$ and $(\textbf{p})$ are known, then its Fixed Cost planned is:

$$\textbf{F} = \textbf{S'}[1+\textbf{s}]-\textbf{Sv}-\textbf{Si}-\textbf{T}-\{[\textbf{N}+\textbf{K}+\textbf{Sj}/360+\textbf{Svp}/360]$$
$$/[1+\textbf{\textit{I}}-\textbf{U}\}/[1-\textbf{d}]$$

Rule-57411:

If both $(\textbf{\textit{I}})$, $(\textbf{U})$, $(\textbf{S'})$, $(\textbf{s})$, $(\textbf{S})$, $(\textbf{v})$, $(\textbf{F})$, $(\textbf{N})$, $(\textbf{T})$, $(\textbf{d})$, $(\textbf{K})$, $(\textbf{j})$ and $(\textbf{p})$ are known, then its Interest Portion planned is:

$$\textbf{i} = (\textbf{S'}[1+\textbf{s}]-\textbf{Sv}-\textbf{F}-\textbf{T}-\{[\textbf{N}+\textbf{K}+\textbf{Sj}/360+\textbf{Svp}/360]$$
$$/[1+\textbf{\textit{I}}-\textbf{U}\}/[1-\textbf{d}])/\textbf{S}$$

Rule-57412:

If both $(\textbf{\textit{I}})$, $(\textbf{U})$, $(\textbf{S'})$, $(\textbf{s})$, $(\textbf{S})$, $(\textbf{v})$, $(\textbf{F})$, $(\textbf{i})$, $(\textbf{N})$, $(\textbf{d})$, $(\textbf{K})$, $(\textbf{j})$ and $(\textbf{p})$ are known, then its Tax planned is:

$$\textbf{T} = \textbf{S'}[1+\textbf{s}]-\textbf{Sv}-\textbf{F}-\textbf{Si}-\{[\textbf{N}+\textbf{K}+\textbf{Sj}/360+\textbf{Svp}/360]$$
$$/[1+\textbf{\textit{I}}-\textbf{U}\}/[1-\textbf{d}]$$

Rule-57413:

If both $(\textbf{\textit{I}})$, $(\textbf{U})$, $(\textbf{S'})$, $(\textbf{s})$, $(\textbf{S})$, $(\textbf{v})$, $(\textbf{F})$, $(\textbf{i})$, $(\textbf{T})$, $(\textbf{N})$, $(\textbf{K})$, $(\textbf{j})$ and $(\textbf{p})$ are known, then its Dividend Payout planned is:

$$\textbf{d} = 1-\{[\textbf{N}+\textbf{K}+\textbf{Sj}/360+\textbf{Svp}/360]/[1+\textbf{\textit{I}}-\textbf{U}\}$$
$$/\{\textbf{S'}[1+\textbf{s}]-\textbf{Sv}-\textbf{F}-\textbf{Si}-\textbf{T}\}$$

Steve Asikin ISBN 13: **978-1564912361**, ISBN 10: **1564912363**

Rule-57414:

If both ($I$), ($U$), ($S'$), ($s$), ($S$), ($v$), ($F$), ($i$), ($T$), ($d$), ($N$), ($j$) and ($p$) are known, then its Kind of Cash planned is:

$$K= [1+I(U+\{S'[1+s]-Sv-F-Si-T\}[1-d])$$
$$-[N+Sj/360+Svp/360]$$

Rule-57415:

If both ($I$), ($U$), ($S'$), ($s$), ($S$), ($v$), ($F$), ($i$), ($T$), ($d$), ($K$), ($N$) and ($p$) are known, then its Job or Trade Receivable Days planned is:

$$j= 360[[1+I(U+\{S'[1+s]-Sv-F-Si-T\}[1-d])$$
$$-\{N+K+Svp/360\}]/S$$

Rule-57416:

If both ($I$), ($U$), ($S'$), ($s$), ($S$), ($v$), ($F$), ($i$), ($T$), ($d$), ($K$), ($j$) and ($N$) are known, then its Procured Inventory Days planned is:

$$p= 360[[1+I(U+\{S'[1+s]-Sv-F-Si-T\}[1-d])$$
$$-[N+K+Sj/360]]/[Sv]$$

Rule-57417:

If both ($I$), ($U$), ($S'$), ($s$), ($S$), ($v$), ($F$), ($i$), ($T$), ($d$), ($K$), ($j$) and ($p$) are known, then its Non Current Asset planned is:

$$N= [1+I(U+\{S'[1+s]-Sv-F-Si-T\}[1-d])$$
$$-K-Sj/360-S'vp[1+s]/360$$

`

Steve Asikin ISBN 13: **978-1564912361**, ISBN 10: **1564912363**

Rule-57418:

If both $(\mathbf{N})$, $(\mathbf{U})$, $(\mathbf{S'})$, $(\mathbf{s})$, $(\mathbf{S})$, $(\mathbf{v})$, $(\mathbf{F})$, $(\mathbf{i})$, $(\mathbf{T})$, $(\mathbf{d})$, $(\mathbf{K})$, $(\mathbf{j})$ and $(\mathbf{p})$ are known, then its Leverage or Gearing Ratio planned is:

$$\mathbf{\textit{l}} = \{\mathbf{N}+\mathbf{K}+\mathbf{Sj}/360+\mathbf{S'vp}[1+\mathbf{s}]/360\}$$
$$/(\mathbf{U}+\{\mathbf{S'}[1+\mathbf{s}]-\mathbf{Sv}-\mathbf{F}-\mathbf{Si}-\mathbf{T}\}[1-\mathbf{d}])-1$$

Rule-57419:

If both $(\mathbf{\textit{l}})$, $(\mathbf{N})$, $(\mathbf{S'})$, $(\mathbf{s})$, $(\mathbf{S})$, $(\mathbf{v})$, $(\mathbf{F})$, $(\mathbf{i})$, $(\mathbf{T})$, $(\mathbf{d})$, $(\mathbf{K})$, $(\mathbf{j})$ and $(\mathbf{p})$ are known, then its Utilized or Starting Capital must be:

$$\mathbf{U} = (\{\mathbf{N}+\mathbf{K}+\mathbf{Sj}/360+\mathbf{S'vp}[1+\mathbf{s}]/360\}$$
$$-[1+\mathbf{\textit{l}}]\{\mathbf{S'}[1+\mathbf{s}]-\mathbf{Sv}-\mathbf{F}-\mathbf{Si}-\mathbf{T}\}[1-\mathbf{d}])/\{[1+\mathbf{\textit{l}}]$$

Rule-57420:

If both $(\mathbf{\textit{l}})$, $(\mathbf{U})$, $(\mathbf{N})$, $(\mathbf{s})$, $(\mathbf{S})$, $(\mathbf{v})$, $(\mathbf{F})$, $(\mathbf{i})$, $(\mathbf{T})$, $(\mathbf{d})$, $(\mathbf{K})$, $(\mathbf{j})$ and $(\mathbf{p})$ are known, then its Sales Past must be:

$$\mathbf{S'} = [[1+\mathbf{\textit{l}}](\mathbf{U}-\{\mathbf{Sv}+\mathbf{F}+\mathbf{Si}+\mathbf{T}\}[1-\mathbf{d}])-[\mathbf{N}+\mathbf{K}+\mathbf{Sj}/360]]$$
$$/([1+\mathbf{s}]\{\mathbf{vp}/360-[1+\mathbf{\textit{l}}][1-\mathbf{d}]\})$$

or

$$\mathbf{S'} = \mathbf{S}/[1+\mathbf{s}]$$

Steve Asikin ISBN 13: **978-1564912361**, ISBN 10: **1564912363**

Rule-57421:

If both $(I)$, $(U)$, $(S')$, $(N)$, $(S)$, $(v)$, $(F)$, $(i)$, $(T)$, $(d)$, $(K)$, $(j)$ and $(p)$ are known, then its Sales Growth planned is:

$$s= [[1+I](U-\{Sv+F+Si+T\}[1-d])-[N+K+Sj/360]]$$
$$/(S'\{vp/360-[1+I][1-d]\})-1$$

or

$$s= S/S'-1$$


Rule-57422:

If both $(I)$, $(U)$, $(S')$, $(s)$, $(N)$, $(v)$, $(F)$, $(i)$, $(T)$, $(d)$, $(K)$, $(j)$ and $(p)$ are known, then its Sales or Revenue planned is:

$$S= [[1+I](U+\{S'[1+s]-F-T\}[1-d])$$
$$-\{N+K+S'vp[1+s]/360\}]$$
$$/\{j/360+[1+I][1-d][v+i]\}$$

or

$$S= S'[1+s]$$


Rule-57423:

If both $(I)$, $(U)$, $(S')$, $(s)$, $(S)$, $(N)$, $(F)$, $(i)$, $(T)$, $(d)$, $(K)$, $(j)$ and $(p)$ are known, then its Variable Portion planned is:

$$v= [[1+I](U+\{S'[1+s]-F-Si-T\}[1-d])$$
$$-[N+K+Sj/360]]$$
$$/\{S'p[1+s]/360+S[1+I][1-d]\}$$

Steve Asikin ISBN 13: **978-1564912361**, ISBN 10: **1564912363**

Rule-57424:

If both ($/$), (**U**), (**S'**), (**s**), (**S**), (**v**), (**N**), (**i**), (**T**), (**d**), (**K**), (**j**) and (**p**) are known, then its Fixed Cost planned is:

$$F= S'[1+s]-Sv-Si-T-(\{N+K+Sj/360$$
$$+S'vp[1+s]/360\}/[1+/]-U)/[1-d]$$

Rule-57425:

If both ($/$), (**U**), (**S'**), (**s**), (**S**), (**v**), (**F**), (**N**), (**T**), (**d**), (**K**), (**j**) and (**p**) are known, then its Interest Portion planned is:

$$i= [S'[1+s]-Sv-F-T-(\{N+K+Sj/360$$
$$+S'vp[1+s]/360\}/[1+/]-U)/[1-d]]/S$$

Rule-57426:

If both ($/$), (**U**), (**S'**), (**s**), (**S**), (**v**), (**F**), (**i**), (**N**), (**d**), (**K**), (**j**) and (**p**) are known, then its Tax planned is:

$$T= S'[1+s]-Sv-F-Si-(\{N+K+Sj/360$$
$$+S'vp[1+s]/360\}/[1+/]-U)/[1-d]$$

Rule-57427:

If both ($/$), (**U**), (**S'**), (**s**), (**S**), (**v**), (**F**), (**i**), (**T**), (**N**), (**K**), (**j**) and (**p**) are known, then its Dividend Payout planned is:

$$d= 1-(\{N+K+Sj/360+S'vp[1+s]/360\}/[1+/]-U)$$
$$/\{S'[1+s]-Sv-F-Si-T\}$$

Steve Asikin ISBN 13: **978-1564912361**, ISBN 10: **1564912363**

Rule-57428:

If both (**∫**), (**U**), (**$'**), (**s**), (**$**), (**v**), (**F**), (**i**), (**T**), (**d**), (**N**), (**j**) and (**p**) are known, then its Kind of Cash planned is:

$$K= [1+\text{∫}(U+\{\$'[1+s]-\$v-F-\$i-T\}[1-d])$$
$$-\{N+\$j/360+\$'vp[1+s]/360\}$$

Rule-57429:

If both (**∫**), (**U**), (**$'**), (**s**), (**$**), (**v**), (**F**), (**i**), (**T**), (**d**), (**K**), (**N**) and (**p**) are known, then its Job or Trade Receivable Days planned is:

$$j= 360[[1+\text{∫}(U+\{\$'[1+s]-\$v-F-\$i-T\}[1-d])$$
$$-\{N+K+\$'vp[1+s]/360\}]/\$$$

Rule-57430:

If both (**∫**), (**U**), (**$'**), (**s**), (**$**), (**v**), (**F**), (**i**), (**T**), (**d**), (**K**), (**j**) and (**N**) are known, then its Procured Inventory Days planned is:

$$p= 360[[1+\text{∫}(U+\{\$'[1+s]-\$v-F-\$i-T\}[1-d])$$
$$-[N+K+\$j/360]]/\{\$'v[1+s]\}$$

Rule-57431:

If both (**∫**), (**U**), (**$'**), (**s**), (**$**), (**v**), (**F**), (**i**), (**T**), (**d**), (**K**), (**j**) and (**P**) are known, then its Non Current Asset planned is:

$$N= [1+\text{∫}(U+\{\$'[1+s]-\$v-F-\$i-T\}[1-d])$$
$$-K-\$'j[1+s]/360-P$$

Steve Asikin ISBN 13: **978-1564912361**, ISBN 10: **1564912363**

Rule-57432:

If both (**N**), (**U**), (**S'**), (**s**), (**S**), (**v**), (**F**), (**i**), (**T**), (**d**), (**K**), (**j**) and (**P**) are known, then its Leverage or Gearing Ratio planned is:

$$\textit{I} = \{\textbf{N}+\textbf{K}+\textbf{S'j}[1+\textbf{s}]/360+\textbf{P}\}$$
$$/(\textbf{U}+\{\textbf{S'}[1+\textbf{s}]-\textbf{Sv}-\textbf{F}-\textbf{Si}-\textbf{T}\}[1-\textbf{d}])-1$$

Rule-57433:

If both (**I**), (**N**), (**S'**), (**s**), (**S**), (**v**), (**F**), (**i**), (**T**), (**d**), (**K**), (**j**) and (**P**) are known, then its Utilized or Starting Capital must be:

$$\textbf{U} = (\{\textbf{N}+\textbf{K}+\textbf{S'j}[1+\textbf{s}]/360+\textbf{P}\}$$
$$-[1+\textit{I}]\{\textbf{S'}[1+\textbf{s}]-\textbf{Sv}-\textbf{F}-\textbf{Si}-\textbf{T}\}[1-\textbf{d}])/[1+\textit{I}]$$

Rule-57434:

If both (**I**), (**U**), (**N**), (**s**), (**S**), (**v**), (**F**), (**i**), (**T**), (**d**), (**K**), (**j**) and (**P**) are known, then its Sales Past must be:

$$\textbf{S'} = [[1+\textit{I}](\textbf{U}-\{\textbf{Sv}+\textbf{F}+\textbf{Si}+\textbf{T}\}[1-\textbf{d}])-[\textbf{N}+\textbf{K}+\textbf{P}]]$$
$$/([1+\textbf{s}]\{\textbf{j}/360-[1+\textit{I}][1-\textbf{d}]\})$$

or

$$\textbf{S'} = \textbf{S}/[1+\textbf{s}]$$

Steve Asikin ISBN 13: **978-1564912361**, ISBN 10: **1564912363**

Rule-57435:

If both (**/**), (**U**), (**S'**), (**N**), (**S**), (**v**), (**F**), (**i**), (**T**), (**d**), (**K**), (**j**) and (**P**) are known, then its Sales Growth planned is:

$$s= [[1+/(U-\{Sv+F+Si+T\}[1-d])-[N+K+P]]$$
$$/(S'\{j/360-[1+/[1-d]\})-1$$

or

$$s= S/S'-1$$


Rule-57436:

If both (**/**), (**U**), (**S'**), (**s**), (**N**), (**v**), (**F**), (**i**), (**T**), (**d**), (**K**), (**j**) and (**P**) are known, then its Sales or Revenue planned is:

$$S= [S'[1+s]-F-T-(\{N+K+S'j[1+s]/360+P\}$$
$$/[1+/-U)/[1-d]]/[v+i]$$

or

$$S= S'[1+s]$$


Rule-57437:

If both (**/**), (**U**), (**S'**), (**s**), (**S**), (**N**), (**F**), (**i**), (**T**), (**d**), (**K**), (**j**) and (**P**) are known, then its Variable Portion planned is:

$$v= [S'[1+s]-F-Si-T-(\{N+K+S'j[1+s]/360+P\}$$
$$/[1+/-U)/[1-d]]/S$$

Steve Asikin ISBN 13: **978-1564912361**, ISBN 10: **1564912363**

Rule-57438:

If both $(I)$, $(U)$, $(S')$, $(s)$, $(S)$, $(v)$, $(N)$, $(i)$, $(T)$, $(d)$, $(K)$, $(j)$ and $(P)$ are known, then its Fixed Cost planned is:

$$F= S'[1+s]-Sv-Si-T-(\{N+K+S'j[1+s]/360+P\}$$
$$/[1+I-U)/[1-d]$$


Rule-57439:

If both $(I)$, $(U)$, $(S')$, $(s)$, $(S)$, $(v)$, $(F)$, $(N)$, $(T)$, $(d)$, $(K)$, $(j)$ and $(P)$ are known, then its Interest Portion planned is:

$$i= [S'[1+s]-Sv-F-T-(\{N+K+S'j[1+s]/360+P\}$$
$$/[1+I-U)/[1-d]]/S$$


Rule-57440:

If both $(I)$, $(U)$, $(S')$, $(s)$, $(S)$, $(v)$, $(F)$, $(i)$, $(N)$, $(d)$, $(K)$, $(j)$ and $(P)$ are known, then its Tax planned is:

$$T= S'[1+s]-Sv-F-Si-(\{N+K+S'j[1+s]/360+P\}$$
$$/[1+I-U)/[1-d]$$


Rule-57441:

If both $(I)$, $(U)$, $(S')$, $(s)$, $(S)$, $(v)$, $(F)$, $(i)$, $(T)$, $(N)$, $(K)$, $(j)$ and $(P)$ are known, then its Dividend Payout planned is:

$$d= 1-(\{N+K+S'j[1+s]/360+P\}/[1+I-U)$$
$$/\{S'[1+s]-Sv-F-Si-T\}$$

Steve Asikin ISBN 13: **978-1564912361**, ISBN 10: **1564912363**

Rule-57442:

If both (**/**), (**U**), (**S'**), (**s**), (**S**), (**v**), (**F**), (**i**), (**T**), (**d**), (**N**), (**j**) and (**P**) are known, then its Kind of Cash planned is:

$$K= [1+/(U+\{S'[1+s]-Sv-F-Si-T\}[1-d])$$
$$-\{N+S'j[1+s]/360+P\}$$

Rule-57443:

If both (**/**), (**U**), (**S'**), (**s**), (**S**), (**v**), (**F**), (**i**), (**T**), (**d**), (**K**), (**N**) and (**P**) are known, then its Job or Trade Receivable Days planned is:

$$j= 360[[1+/(U+\{S'[1+s]-Sv-F-Si-T\}[1-d])$$
$$-\{N+K+P\}]/\{S'[1+s]\}$$

Rule-57444:

If both (**/**), (**U**), (**S'**), (**s**), (**S**), (**v**), (**F**), (**i**), (**T**), (**d**), (**K**), (**j**) and (**N**) are known, then its Procured Inventory planned is:

$$P= [1+/(U+\{S'[1+s]-Sv-F-Si-T\}[1-d])$$
$$-\{N+K+S'j[1+s]/360\}$$

Rule-57445:

If both (**/**), (**U**), (**S'**), (**s**), (**S**), (**v**), (**F**), (**i**), (**T**), (**d**), (**K**), (**j**), (**V**) and (**p**) are known, then its Non Current Asset planned is:

$$N= [1+/(U+\{S'[1+s]-Sv-F-Si-T\}[1-d])$$
$$-K-S'j[1+s]/360-Vp/360$$

Steve Asikin ISBN 13: **978-1564912361**, ISBN 10: **1564912363**

<u>Rule-57446</u>:

If both $(\mathbf{N})$, $(\mathbf{U})$, $(\mathbf{S'})$, $(\mathbf{s})$, $(\mathbf{S})$, $(\mathbf{v})$, $(\mathbf{F})$, $(\mathbf{i})$, $(\mathbf{T})$, $(\mathbf{d})$, $(\mathbf{K})$, $(\mathbf{j})$, $(\mathbf{V})$ and $(\mathbf{p})$ are known, then its Leverage or Gearing Ratio planned is:

$$\mathbf{f} = \{\mathbf{N}+\mathbf{K}+\mathbf{S'j}[1+\mathbf{s}]/360+\mathbf{Vp}/360\}$$
$$/(\mathbf{U}+\{\mathbf{S'}[1+\mathbf{s}]-\mathbf{Sv}-\mathbf{F}-\mathbf{Si}-\mathbf{T}\}[1-\mathbf{d}])-1$$

<u>Rule-57447</u>:

If both $(\mathbf{f})$, $(\mathbf{N})$, $(\mathbf{S'})$, $(\mathbf{s})$, $(\mathbf{S})$, $(\mathbf{v})$, $(\mathbf{F})$, $(\mathbf{i})$, $(\mathbf{T})$, $(\mathbf{d})$, $(\mathbf{K})$, $(\mathbf{j})$, $(\mathbf{V})$ and $(\mathbf{p})$ are known, then its Utilized or Starting Capital must be:

$$\mathbf{U} = (\{\mathbf{N}+\mathbf{K}+\mathbf{S'j}[1+\mathbf{s}]/360+\mathbf{Vp}/360\}-[1+\mathbf{f}]\{\mathbf{S'}[1+\mathbf{s}]$$
$$-\mathbf{Sv}-\mathbf{F}-\mathbf{Si}-\mathbf{T}\}[1-\mathbf{d}])/[1+\mathbf{f}]$$

<u>Rule-57448</u>:

If both $(\mathbf{f})$, $(\mathbf{U})$, $(\mathbf{N})$, $(\mathbf{s})$, $(\mathbf{S})$, $(\mathbf{v})$, $(\mathbf{F})$, $(\mathbf{i})$, $(\mathbf{T})$, $(\mathbf{d})$, $(\mathbf{K})$, $(\mathbf{j})$, $(\mathbf{V})$ and $(\mathbf{p})$ are known, then its Sales Past must be:

$$\mathbf{S'} = [[1+\mathbf{f}(\mathbf{U}-\{\mathbf{Sv}+\mathbf{F}+\mathbf{Si}+\mathbf{T}\}[1-\mathbf{d}])-[\mathbf{N}+\mathbf{K}$$
$$+\mathbf{Vp}/360]]/([1+\mathbf{s}]\{\mathbf{j}/360-[1+\mathbf{f}][1-\mathbf{d}]\})$$

or

$$\mathbf{S'} = \mathbf{S}/[1+\mathbf{s}]$$

Steve Asikin ISBN 13: **978-1564912361**, ISBN 10: **1564912363**

Rule-57449:

If both (**∫**), (**U**), (**S'**), (**N**), (**S**), (**υ**), (**F**), (**i**), (**T**), (**d**), (**K**), (**j**), (**V**) and (**p**) are known, then its Sales Growth planned is:

$$s= [[1+\text{∫}](U-\{Sυ+F+Si+T\}[1-d])-[N+K+Vp/360]]$$
$$/(S'\{j/360-[1+\text{∫}][1-d]\})-1$$

or

$$s= S/S'-1$$

Rule-57450:

If both (**∫**), (**U**), (**S'**), (**s**), (**N**), (**υ**), (**F**), (**i**), (**T**), (**d**), (**K**), (**j**), (**V**) and (**p**) are known, then its Sales or Ratio planned is:

$$S= [S'[1+s]-F-Si-T-(\{N+K+S'j[1+s]/360$$
$$+Vp/360\}/[1+\text{∫}-U)/[1-d]]/υ$$

or

$$S= S'[1+s]$$

or

$$S= V/υ$$

Rule-57451:

If both (**∫**), (**U**), (**S'**), (**s**), (**S**), (**N**), (**F**), (**i**), (**T**), (**d**), (**K**), (**j**), (**V**) and (**p**) are known, then its Variable Portion planned is:

$$v= [S'[1+s]-F-Si-T-(\{N+K+S'j[1+s]/360+Vp/360\}$$
$$/[1+\text{∫}-U)/[1-d]]/S$$

or

$$v= V/S$$

Steve Asikin ISBN 13: **978-1564912361**, ISBN 10: **1564912363**

Rule-57452:

If both (**/**), (**U**), (**S'**), (**s**), (**S**), (**v**), (**N**), (**i**), (**T**), (**d**), (**K**), (**j**), (**V**) and (**p**) are known, then its Fixed Cost planned is:

$$F= S'[1+s]-Sv-Si-T-(\{N+K+S'j[1+s]/360+Vp/360\}$$
$$/[1+/-U)/[1-d]$$

Rule-57453:

If both (**/**), (**U**), (**S'**), (**s**), (**S**), (**v**), (**F**), (**i**), (**T**), (**d**), (**K**), (**j**), (**V**) and (**p**) are known, then its Non Current Asset planned is:

$$i= [S'[1+s]-Sv-F-T-(\{N+K+S'j[1+s]/360+Vp/360\}$$
$$/[1+/-U)/[1-d]]/S$$

Rule-57454:

If both (**/**), (**U**), (**S'**), (**s**), (**S**), (**v**), (**F**), (**i**), (**N**), (**d**), (**K**), (**j**), (**V**) and (**p**) are known, then its Tax planned is:

$$T= S'[1+s]-Sv-F-Si-(\{N+K+S'j[1+s]/360+Vp/360\}$$
$$/[1+/-U)/[1-d]$$

Rule-57455:

If both (**/**), (**U**), (**S'**), (**s**), (**S**), (**v**), (**F**), (**i**), (**T**), (**N**), (**K**), (**j**), (**V**) and (**p**) are known, then its Dividend Payout planned is:

$$d= 1-(\{N+K+S'j[1+s]/360+Vp/360\}/[1+/-U)$$
$$/\{S'[1+s]-Sv-F-Si-T\}$$

Steve Asikin ISBN 13: **978-1564912361**, ISBN 10: **1564912363**

Rule-57456:

If both (**ı**), (**U**), (**S'**), (**s**), (**S**), (**υ**), (**F**), (**i**), (**T**), (**d**), (**N**), (**j**), (**V**) and (**p**) are known, then its Kind of Cash planned is:

$$K = [1+ı(U+\{S'[1+s]-Sυ-F-Si-T\}[1-d])$$
$$-\{N+S'j[1+s]/360+Vp/360\}$$

Rule-57457:

If both (**ı**), (**U**), (**S'**), (**s**), (**S**), (**υ**), (**F**), (**i**), (**T**), (**d**), (**K**), (**N**), (**V**) and (**p**) are known, then its Job or Trade Receivable Days planned is:

$$j = 360[[1+ı(U+\{S'[1+s]-Sυ-F-Si-T\}[1-d])$$
$$-\{N+K+Vp/360\}]/\{S'[1+s]\}$$

Rule-57458:

If both (**ı**), (**U**), (**S'**), (**s**), (**S**), (**υ**), (**F**), (**i**), (**T**), (**d**), (**K**), (**j**), (**N**) and (**p**) are known, then its Variable Cost planned is:

$$V = 360[[1+ı(U+\{S'[1+s]-Sυ-F-Si-T\}[1-d])$$
$$-\{N+K+S'j[1+s]/360\}]/p$$

or

$$V = Sυ$$

Rule-57459:

If both (**ı**), (**U**), (**S'**), (**s**), (**S**), (**υ**), (**F**), (**i**), (**T**), (**d**), (**K**), (**j**), (**V**) and (**N**) are known, then its Procured Inventory Days planned is:

$$p = 360[[1+ı(U+\{S'[1+s]-Sυ-F-Si-T\}[1-d])$$
$$-\{N+K+S'j[1+s]/360\}]/V$$

Steve Asikin ISBN 13: **978-1564912361**, ISBN 10: **1564912363**

Rule-57460:

If both (*I*), (**U**), (**S'**), (**s**), (**S**), (**v**), (**F**), (**i**), (**T**), (**d**), (**K**), (**j**) and (**p**) are known, then its Non Current Asset planned is:

$$\mathbf{N} = [1 + I(\mathbf{U} + \{\mathbf{S'}[1+\mathbf{s}] - \mathbf{Sv} - \mathbf{F} - \mathbf{Si} - \mathbf{T}\}[1-\mathbf{d}])$$
$$- \mathbf{K} - \mathbf{S'j}[1+\mathbf{s}]/360 - \mathbf{Svp}/360$$

Rule-57461:

If both (**N**), (**U**), (**S'**), (**s**), (**S**), (**v**), (**F**), (**i**), (**T**), (**d**), (**K**), (**j**) and (**p**) are known, then its Leverage or Gearing Ratio planned is:

$$I = \{\mathbf{N} + \mathbf{K} + \mathbf{S'j}[1+\mathbf{s}]/360 + \mathbf{Svp}/360\}$$
$$/(\mathbf{U} + \{\mathbf{S'}[1+\mathbf{s}] - \mathbf{Sv} - \mathbf{F} - \mathbf{Si} - \mathbf{T}\}[1-\mathbf{d}]) - 1$$

Rule-57462:

If both (*I*), (**N**), (**S'**), (**s**), (**S**), (**v**), (**F**), (**i**), (**T**), (**d**), (**K**), (**j**) and (**p**) are known, then its Utilized or Starting Capital must be:

$$\mathbf{U} = (\{\mathbf{N} + \mathbf{K} + \mathbf{S'j}[1+\mathbf{s}]/360 + \mathbf{Svp}/360\}$$
$$- [1 + I\{\mathbf{S'}[1+\mathbf{s}] - \mathbf{Sv} - \mathbf{F} - \mathbf{Si} - \mathbf{T}\}[1-\mathbf{d}])/\{[1 + I]\}$$

Rule-57463:

If both (*I*), (**U**), (**N**), (**s**), (**S**), (**v**), (**F**), (**i**), (**T**), (**d**), (**K**), (**j**) and (**p**) are known, then its Sales Past must be:

$$\mathbf{S'} = [[1 + I(\mathbf{U} - \{\mathbf{Sv} + \mathbf{F} + \mathbf{Si} + \mathbf{T}\}[1-\mathbf{d}]) - \{\mathbf{N} + \mathbf{K}$$
$$+ \mathbf{Svp}/360\}]/([1+\mathbf{s}]\{\mathbf{j}/360 - [1 + I][1-\mathbf{d}]\})$$

or

$$\mathbf{S'} = \mathbf{S}/[1+\mathbf{s}]$$

Steve Asikin ISBN 13: **978-1564912361**, ISBN 10: **1564912363**

Rule-57464:

If both (**/**), (**U**), (**S'**), (**N**), (**S**), (**v**), (**F**), (**i**), (**T**), (**d**), (**K**), (**j**) and (**p**) are known, then its Sales Growth planned is:

$$s= [[1+/(U-\{Sv+F+Si+T\}[1-d])-\{N+K$$
$$+Svp/360\}]/(S'\{j/360-[1+/][1-d]\})-1$$

or

$$s= S/S'-1$$

Rule-57465:

If both (**/**), (**U**), (**S'**), (**s**), (**N**), (**v**), (**F**), (**i**), (**T**), (**d**), (**K**), (**j**) and (**p**) are known, then its Sales or Revenue planned is:

$$S= [[1+/(U+\{S'[1+s]-F-T\}[1\ d])$$
$$-\{N+K+S'j[1+s]/360\}]$$
$$/\{vp/360+[1+/][1-d][v+i]\}$$

or

$$S= S'[1+s]$$

Rule-57466:

If both (**/**), (**U**), (**S'**), (**s**), (**S**), (**N**), (**F**), (**i**), (**T**), (**d**), (**K**), (**j**) and (**p**) are known, then its Variable Portion planned is:

$$v= [[1+/(U+\{S'[1+s]-F-Si-T\}[1-d])$$
$$-\{N+K+S'j[1+s]/360\}]$$
$$/\{Sp/360+S[1+/][1-d]\}$$

Steve Asikin ISBN 13: **978-1564912361**, ISBN 10: **1564912363**

Rule-57467:

If both (**ʃ**), (**U**), (**S'**), (**s**), (**S**), (**v**), (**N**), (**i**), (**T**), (**d**), (**K**), (**j**) and (**p**) are known, then its Fixed Cost planned is:

$$F= S'[1+s]-Sv-Si-T-(\{N+K+S'j[1+s]/360$$
$$+Svp/360\}/[1+\text{ʃ}-U)/[1-d]$$


Rule-57468:

If both (**ʃ**), (**U**), (**S'**), (**s**), (**S**), (**v**), (**F**), (**N**), (**T**), (**d**), (**K**), (**j**) and (**p**) are known, then its Interest Portion planned is:

$$i= [S'[1+s]-Sv-F-T-(\{N+K+S'j[1+s]/360$$
$$+Svp/360\}/[1+\text{ʃ}-U)/[1-d]]/S$$


Rule-57469:

If both (**ʃ**), (**U**), (**S'**), (**s**), (**S**), (**v**), (**F**), (**i**), (**N**), (**d**), (**K**), (**j**) and (**p**) are known, then its Tax planned is:

$$T= S'[1+s]-Sv-F-Si-(\{N+K+S'j[1+s]/360$$
$$+Svp/360\}/[1+\text{ʃ}-U)/[1-d]$$


Rule-57470:

If both (**ʃ**), (**U**), (**S'**), (**s**), (**S**), (**v**), (**F**), (**i**), (**T**), (**N**), (**K**), (**j**) and (**p**) are known, then its Dividend Payout planned is:

$$d= 1-(\{N+K+S'j[1+s]/360+Svp/360\}/[1+\text{ʃ}-U)$$
$$/\{S'[1+s]-Sv-F-Si-T\}$$

Steve Asikin ISBN 13: **978-1564912361**, ISBN 10: **1564912363**

Rule-57471:

If both (**/**), (**U**), (**S'**), (**s**), (**$**), (**v**), (**F**), (**i**), (**T**), (**d**), (**N**), (**j**) and (**p**) are known, then its Kind of Cash planned is:

$$K = [1+/(U+\{S'[1+s]-Sv-F-Si-T\}[1-d])$$
$$-\{N+S'j[1+s]/360+Svp/360\}$$

Rule-57472:

If both (**/**), (**U**), (**S'**), (**s**), (**$**), (**v**), (**F**), (**i**), (**T**), (**d**), (**K**), (**N**) and (**p**) are known, then its Job or Trade Receivable Days planned is.

$$j = 360[[1+/(U+\{S'[1+s]-Sv-F-Si-T\}[1-d])$$
$$-\{N+K+Svp/360\}]/\{S'[1+s]\}$$

Rule-57473:

If both (**/**), (**U**), (**S'**), (**s**), (**$**), (**v**), (**F**), (**i**), (**T**), (**d**), (**K**), (**j**) and (**N**) are known, then its Procured Inventory Days planned is:

$$p = 360[[1+/(U+\{S'[1+s]-Sv-F-Si-T\}[1-d])$$
$$-\{N+K+S'j[1+s]/360\}]/[Sv]$$

Rule-57474:

If both (**/**), (**U**), (**S'**), (**s**), (**$**), (**v**), (**F**), (**i**), (**T**), (**d**), (**K**), (**j**) and (**p**) are known, then its Non Current Asset planned is:

$$N = [1+/(U+\{S'[1+s]-Sv-F-Si-T\}[1-d])$$
$$-K-S'j[1+s]/360-S'vp[1+s]/360$$

`

374

Steve Asikin ISBN 13: **978-1564912361**, ISBN 10: **1564912363**

Rule-57475:

If both **(N)**, **(U)**, **(S')**, **(s)**, **(S)**, **(v)**, **(F)**, **(i)**, **(T)**, **(d)**, **(K)**, **(j)** and **(p)** are known, then its Leverage or Gearing Ratio planned is:

$$l = \{N+K+S'j[1+s]/360+S'vp[1+s]/360\}$$
$$/(U+\{S'[1+s]-Sv-F-Si-T\}[1-d])-1$$

Rule-57476:

If both **(l)**, **(N)**, **(S')**, **(s)**, **(S)**, **(v)**, **(F)**, **(i)**, **(T)**, **(d)**, **(K)**, **(j)** and **(p)** are known, then its Utilized or Starting Capital must be:

$$U = (\{N+K+S'j[1+s]/360+S'vp[1+s]/360\}$$
$$-[1+l]\{S'[1+s]-Sv-F-Si-T\}[1-d])/[1+l]$$

Rule-57477:

If both **(l)**, **(U)**, **(N)**, **(s)**, **(S)**, **(v)**, **(F)**, **(i)**, **(T)**, **(d)**, **(K)**, **(j)** and **(p)** are known, then its Sales Growth must be:

$$S' = [[1+l](U-\{Sv+F+Si+T\}[1-d])-[N+K]]$$
$$/(S'[1+s]\{j/360+vp/360-[1+l][1-d]\})$$
$$\text{or}$$
$$S' = S/[1+s]$$

Steve Asikin ISBN 13: **978-1564912361**, ISBN 10: **1564912363**

Rule-57478:

If both (**/**), (**U**), (**S'**), (**N**), (**S**), (**v**), (**F**), (**i**), (**T**), (**d**), (**K**), (**j**) and (**p**) are known, then its Sales Growth planned is:

$$s= [[1+/(U-\{Sv+F+Si+T\}[1-d])-[N+K]]$$
$$/(S'\{j/360+vp/360-[1+/[1-d]\})-1$$

or

$$s= S/S'-1$$

Rule-57479:

If both (**/**), (**U**), (**S'**), (**s**), (**N**), (**v**), (**F**), (**i**), (**T**), (**d**), (**K**), (**j**) and (**p**) are known, then its Sales or Revenue planned is:

$$S- [S'[1+s]-F-Si-T-(\{N+K+S'j[1+s]/360$$
$$+S'vp[1+s]/360\}/[1+/]-U)/[1-d]]/v$$

or

$$S= S'[1+s]$$

Rule-57480:

If both (**/**), (**U**), (**S'**), (**s**), (**S**), (**N**), (**F**), (**i**), (**T**), (**d**), (**K**), (**j**) and (**p**) are known, then its Variable Portion planned is:

$$v= [[1+/(U+\{S'[1+s]-Sv-F-Si-T\}[1-d])$$
$$-\{N+S'j[1+s]/360+K\}]$$
$$/\{S'p[1+s]/360\}+S[1+/[1-d]\}$$

Steve Asikin ISBN 13: **978-1564912361**, ISBN 10: **1564912363**

Rule-57481:

If both (**/**), (**U**), (**S'**), (**s**), (**$**), (**v**), (**N**), (**i**), (**T**), (**d**), (**K**), (**j**) and (**p**) are known, then its Fixed Cost planned is:

$$F= S'[1+s]-Sv-Si-T-(\{N+K+S'j[1+s]/360$$
$$+S'vp[1+s]/360\}/[1+/-U)/[1-d]$$

Rule-57482:

If both (**/**), (**U**), (**S'**), (**s**), (**$**), (**v**), (**F**), (**N**), (**T**), (**d**), (**K**), (**j**) and (**p**) are known, then its Interest Portion planned is:

$$i= [S'[1+s]-Sv-F-T-(\{N+K+S'j[1+s]/360$$
$$+S'vp[1+s]/360\}/[1+/-U)/[1-d]]/S$$

Rule-57483:

If both (**/**), (**U**), (**S'**), (**s**), (**$**), (**v**), (**F**), (**i**), (**N**), (**d**), (**K**), (**j**) and (**p**) are known, then its Tax planned is:

$$T= S'[1+s]-Sv-F-Si-(\{N+K+S'j[1+s]/360$$
$$+S'vp[1+s]/360\}/[1+/-U)/[1-d]$$

Rule-57484:

If both (**/**), (**U**), (**S'**), (**s**), (**$**), (**v**), (**F**), (**i**), (**T**), (**N**), (**K**), (**j**) and (**p**) are known, then its Dividend Payout planned is:

$$d= 1-(\{N+K+S'j[1+s]/360+S'vp[1+s]/360\}$$
$$/[1+/-U)/\{S'[1+s]-Sv-F-Si-T\}$$

Steve Asikin ISBN 13: **978-1564912361**, ISBN 10: **1564912363**

## Rule-57485:

If both (**/**), (**U**), (**S'**), (**s**), (**S**), (**v**), (**F**), (**i**), (**T**), (**d**), (**N**), (**j**) and (**p**) are known, then its Kind of Cash planned is:

$$K= [1+\text{/}(U+\{S'[1+s]-Sv-F-Si-T\}[1-d])$$
$$-\{N+S'j[1+s]/360+S'vp[1+s]/360\}$$

## Rule-57486:

If both (**/**), (**U**), (**S'**), (**s**), (**S**), (**v**), (**F**), (**i**), (**T**), (**d**), (**K**), (**N**) and (**p**) are known, then its Job or Trade Receivable Days planned is:

$$j= 360[[1+\text{/}(U+\{S'[1+s]-Sv-F-Si-T\}[1-d])$$
$$-\{N+K+S'vp[1+s]/360\}]/\{S'[1+s]\}$$

## Rule-57487:

If both (**/**), (**U**), (**S'**), (**s**), (**S**), (**v**), (**F**), (**i**), (**T**), (**d**), (**K**), (**j**) and (**N**) are known, then its Procured Inventory Days planned is:

$$p= 360[[1+\text{/}(U+\{S'[1+s]-Sv-F-Si-T\}[1-d])$$
$$-\{N+K+S'j[1+s]/360\}]/\{S'v[1+s]\}$$

## Rule-57474:

If both (**/**), (**U**), (**S'**), (**s**), (**S**), (**v**), (**F**), (**i**), (**t**), (**D**), (**K**), (**J**) and (**P**) are known, then its Non Current Asset planned is:

$$N= [1+\text{/}(U+\{S'[1+s]-Sv-F-Si\}[1-t]-D)-K-J-P$$

Steve Asikin ISBN 13: **978-1564912361**, ISBN 10: **1564912363**

Rule-57475:

If both **(N)**, **(U)**, **(S')**, **(s)**, **(S)**, **(v)**, **(F)**, **(i)**, **(t)**, **(D)**, **(K)**, **(J)** and **(P)** are known, then its Leverage or Gearing Ratio planned is:

$$\text{\textipa{F}}= \{N+K+J+P\}/(U+\{S'[1+s]-Sv-F-Si\}[1-t]-D)-1$$

Rule-57476:

If both **(Ɫ)**, **(N)**, **(S')**, **(s)**, **(S)**, **(v)**, **(F)**, **(i)**, **(t)**, **(D)**, **(K)**, **(J)** and **(P)** are known, then its Utilized or Starting Capital must be:

$$U= (\{N+K+J+P\}-[1+Ɫ]\{S'[1+s]-Sv-Si\}[1-t]-D)$$
$$/[1+Ɫ]$$

Rule-57477:

If both **(Ɫ)**, **(U)**, **(N)**, **(s)**, **(S)**, **(v)**, **(F)**, **(i)**, **(t)**, **(D)**, **(K)**, **(J)** and **(P)** are known, then its Sales Past must be:

$$S'= \{[N+K+J+P]/[1+Ɫ]-U+D\}/[1-t]$$
$$+Sv+F+Si)/[1+s]$$

or

$$S'= S/[1+s]$$

Rule-57478:

If both **(Ɫ)**, **(U)**, **(S')**, **(N)**, **(S)**, **(v)**, **(F)**, **(i)**, **(t)**, **(D)**, **(K)**, **(J)** and **(P)** are known, then its Sales Growth planned is:

$$s= \{[N+K+J+P]/[1+Ɫ]-U+D\}/[1-t]$$
$$+Sv+F+Si)/S'-1$$

or

$$s= S/S'-1$$

Steve Asikin ISBN 13: **978-1564912361**, ISBN 10: **1564912363**

Rule-57479:

If both ($\mathit{I}$), (**U**), (**$'**), (**s**), (**$**), (**v**), (**F**), (**i**), (**t**), (**D**), (**K**), (**J**) and (**P**) are known, then its Non Current Asset planned is:

$$\$= (\{\$'[1+s]-F-\{[N+K+J+P]/[1+\mathit{I}-U+D\}/[1-t])$$
$$/[v+i]$$

or

$$\$= \$'[1+s]$$

Rule-57480:

If both ($\mathit{I}$), (**U**), (**$'**), (**s**), (**$**), (**N**), (**F**), (**i**), (**t**), (**D**), (**K**), (**J**) and (**P**) are known, then its Variable Portion planned is:

$$v= (\{\$'[1+s]-\$i-F-\{[N+K+J+P]/[1+\mathit{I}\ U+D\}$$
$$/[1-t])/\$$$

Rule-57481:

If both ($\mathit{I}$), (**U**), (**$'**), (**s**), (**$**), (**v**), (**N**), (**i**), (**t**), (**D**), (**K**), (**J**) and (**P**) are known, then its Fixed Cost planned is:

$$F= \{\$'[1+s]-\$v-\$i-\{[N+K+J+P]/[1+\mathit{I}-U+D\}/[1-t]$$

Rule-57482:

If both ($\mathit{I}$), (**U**), (**$'**), (**s**), (**$**), (**v**), (**F**), (**N**), (**t**), (**D**), (**K**), (**J**) and (**P**) are known, then its Interest Portion planned is:

$$i= (\{\$'[1+s]-\$v-F-\{[N+K+J+P]/[1+\mathit{I}-U+D\}$$
$$/[1-t])/\$$$

Steve Asikin ISBN 13: **978-1564912361**, ISBN 10: **1564912363**

Rule-57483:

If both (**/**), (**U**), (**S'**), (**s**), (**S**), (**v**), (**F**), (**i**), (**N**), (**D**), (**K**), (**J**) and (**P**) are known, then its Tax Rate planned is:

$$t= 1-\{[N+K+J+P]/[1+/-U+D]/\{S'[1+s]-Sv-F-Si\}$$

Rule-57484:

If both (**/**), (**U**), (**S'**), (**s**), (**S**), (**v**), (**F**), (**i**), (**t**), (**N**), (**K**), (**J**) and (**P**) are known, then its Dividend planned is:

$$D= U+\{S'[1+s]-Sv-F-Si\}[1-t]-[N+K+J+P]/[1+/]$$

Rule-57485:

If both (**/**), (**U**), (**S'**), (**s**), (**S**), (**v**), (**F**), (**i**), (**t**), (**D**), (**N**), (**J**) and (**P**) are known, then its Kind of Cash planned is:

$$K= [1+/](U+\{S'[1+s]-Sv-F-Si\}[1-t]-D)-[N+J+P]$$

Rule-57486:

If both (**/**), (**U**), (**S'**), (**s**), (**S**), (**v**), (**F**), (**i**), (**t**), (**D**), (**K**), (**N**) and (**P**) are known, then its Job or Trade Account Receivable planned is:

$$J= [1+/](U+\{S'[1+s]-Sv-F-Si\}[1-t]-D)-[N+K+P]$$

Rule-57487:

If both (**/**), (**U**), (**S'**), (**s**), (**S**), (**v**), (**F**), (**i**), (**t**), (**D**), (**K**), (**J**) and (**N**) are known, then its Procured Inventory planned is:

$$P= [1+/](U+\{S'[1+s]-Sv-F-Si\}[1-t]-D)-[N+K+J]$$

Steve Asikin ISBN 13: **978-1564912361**, ISBN 10: **1564912363**

Rule-57488:

If both (**/**), (**U**), (**S'**), (**s**), (**S**), (**v**), (**F**), (**i**), (**t**), (**D**), (**K**), (**J**), (**V**) and (**p**) are known, then its Non Current Asset planned is:

$$N= [1+/](U+\{S'[1+s]-Sv-F-Si\}[1-t]-D) -K-J-Vp/360$$

Rule-57489:

If both (**N**), (**U**), (**S'**), (**s**), (**S**), (**v**), (**F**), (**i**), (**t**), (**D**), (**K**), (**J**), (**V**) and (**p**) are known, then its Leverage or Gearing Ratio planned is:

$$/= [N+K+J+Vp/360] /(U+\{S'[1+s]-Sv-F-Si\}[1-t]-D)-1$$

Rule-57490:

If both (**/**), (**N**), (**S'**), (**s**), (**S**), (**v**), (**F**), (**i**), (**t**), (**D**), (**K**), (**J**), (**V**) and (**p**) are known, then its Utilized or Starting Capital must be:

$$U= ([N+K+J+Vp/360]-[1+/]\{S'[1+s]-Sv-F-Si\} [1-t]-D)/[1+/]$$

Rule-57491:

If both (**/**), (**U**), (**N**), (**s**), (**S**), (**v**), (**F**), (**i**), (**t**), (**D**), (**K**), (**J**), (**V**) and (**p**) are known, then its Sales Past must be:

$$S'= \{[N+K+J+Vp/360]/[1+/]-U+D\}/[1-t] +Sv+F+Si)/S'[1+s]$$

or

$$S'= S/[1+s]$$

Steve Asikin ISBN 13: **978-1564912361**, ISBN 10: **1564912363**

Rule-57492:

If both $(\textbf{/})$, $(\textbf{U})$, $(\textbf{S'})$, $(\textbf{N})$, $(\textbf{S})$, $(\textbf{v})$, $(\textbf{F})$, $(\textbf{i})$, $(\textbf{t})$, $(\textbf{D})$, $(\textbf{K})$, $(\textbf{J})$, $(\textbf{V})$ and $(\textbf{p})$ are known, then its Sales Growth planned is:

$$s= \{[\textbf{N}+\textbf{K}+\textbf{J}+\textbf{Vp}/360]/[1+\textbf{/}-\textbf{U}+\textbf{D}\}/[1-\textbf{t}]$$
$$+\textbf{Sv}+\textbf{F}+\textbf{Si})/\textbf{S'}-1$$

or

$$s= \textbf{S}/\textbf{S'}-1$$

Rule-57493:

If both $(\textbf{/})$, $(\textbf{U})$, $(\textbf{S'})$, $(\textbf{s})$, $(\textbf{N})$, $(\textbf{v})$, $(\textbf{F})$, $(\textbf{i})$, $(\textbf{t})$, $(\textbf{D})$, $(\textbf{K})$, $(\textbf{J})$, $(\textbf{V})$ and $(\textbf{p})$ are known, then its Sales or Revenue planned is:

$$\textbf{S}= (\textbf{S'}[1+\textbf{s}]-\textbf{F}-\{[\textbf{N}+\textbf{K}+\textbf{J}+\textbf{Vp}/360]/[1+\textbf{/}-\textbf{U}+\textbf{D}\}$$
$$/[1-\textbf{t}])/[\textbf{v}+\textbf{i}]$$

or

$$\textbf{S}= \textbf{S'}[1+\textbf{s}]$$

or

$$\textbf{S}= \textbf{V}/\textbf{v}$$

Rule-57494:

If both $(\textbf{/})$, $(\textbf{U})$, $(\textbf{S'})$, $(\textbf{s})$, $(\textbf{S})$, $(\textbf{N})$, $(\textbf{F})$, $(\textbf{i})$, $(\textbf{t})$, $(\textbf{D})$, $(\textbf{K})$, $(\textbf{J})$, $(\textbf{V})$ and $(\textbf{p})$ are known, then its Variable Portion planned is:

$$v= (\textbf{S'}[1+\textbf{s}]-\textbf{F}-\textbf{Si}-\{[\textbf{N}+\textbf{K}+\textbf{J}+\textbf{Vp}/360]/[1+\textbf{/}$$
$$-\textbf{U}+\textbf{D}\}/[1-\textbf{t}])/\textbf{S}$$

or

$$v= \textbf{V}/\textbf{S}$$

Steve Asikin ISBN 13: **978-1564912361**, ISBN 10: **1564912363**

<u>Rule-57495</u>:

If both (**I**), (**U**), (**S'**), (**s**), (**S**), (**v**), (**N**), (**i**), (**t**), (**D**), (**K**), (**J**), (**V**) and (**p**) are known, then its Fixed Cost planned is:

$$F = S'[1+s] - Sv - Si - \{[N+K+J+Vp/360]/[1+I] - U+D\}/[1-t]$$

<u>Rule-57496</u>:

If both (**I**), (**U**), (**S'**), (**s**), (**N**), (**v**), (**F**), (**i**), (**t**), (**D**), (**K**), (**J**), (**V**) and (**p**) are known, then its Sales or Revenue planned is:

$$i = (S'[1+s] - Sv - F - \{[N+K+J+Vp/360]/[1+I] - U+D\}/[1-t])/S$$

<u>Rule-57497</u>:

If both (**I**), (**U**), (**S'**), (**s**), (**S**), (**v**), (**F**), (**N**), (**t**), (**D**), (**K**), (**J**), (**V**) and (**p**) are known, then its Interest Portion planned is:

$$t = 1 - \{[N+K+J+Vp/360]/[1+I] - U+D\}/\{S'[1+s] - Sv - F - Si\}$$

<u>Rule-57498</u>:

If both (**I**), (**U**), (**S'**), (**s**), (**S**), (**v**), (**F**), (**i**), (**t**), (**N**), (**K**), (**J**), (**V**) and (**p**) are known, then its Dividend planned is:

$$D = U + \{S'[1+s] - Sv - F - Si\}[1-t] - [N+K+J+Vp/360]/[1+I]$$

`

Steve Asikin ISBN 13: **978-1564912361**, ISBN 10: **1564912363**

Rule-57499:

If both $(f)$, $(U)$, $(S')$, $(s)$, $(S)$, $(v)$, $(F)$, $(i)$, $(t)$, $(D)$, $(N)$, $(J)$, $(V)$ and $(p)$ are known, then its Kind of Cash planned is:

$$K= [1+f(U+\{S'[1+s]-Sv-F-Si\}[1-t]-D) -[N+J+Vp/360]$$

Rule-57500:

If both $(f)$, $(U)$, $(S')$, $(s)$, $(S)$, $(v)$, $(F)$, $(i)$, $(t)$, $(D)$, $(K)$, $(N)$, $(V)$ and $(p)$ are known, then its Job or Trade Account Receivable planned is:

$$J= [1+f(U+\{S'[1+s]-Sv-F-Si\}[1-t]-D) -\{N+K+Vp/360\}$$

Rule-57501:

If both $(f)$, $(U)$, $(S')$, $(s)$, $(S)$, $(v)$, $(F)$, $(i)$, $(t)$, $(D)$, $(K)$, $(J)$, $(V)$ and $(p)$ are known, then its Non Current Asset planned is:

$$V= 360[[1+f(U+\{S'[1+s]-Sv-F-Si\}[1-t]-D) -[N+K+J]]/p$$

or

$$V= Sv$$

Rule-57502:

If both $(f)$, $(U)$, $(S')$, $(s)$, $(S)$, $(v)$, $(F)$, $(i)$, $(t)$, $(D)$, $(K)$, $(J)$, $(V)$ and $(p)$ are known, then its Non Current Asset planned is:

$$p= 360[[1+f(U+\{S'[1+s]-Sv-F-Si\}[1-t]-D) -[N+K+J]]/V$$

Steve Asikin ISBN 13: **978-1564912361**, ISBN 10: **1564912363**

Rule-57503:

If both $(I)$, $(U)$, $(S')$, $(s)$, $(S)$, $(v)$, $(F)$, $(i)$, $(t)$, $(D)$, $(K)$, $(J)$ and $(p)$ are known, then its Non Current Asset planned is:

$$N = [1+I](U+\{S'[1+s]-Sv-F-Si\}[1-t]-D) -K-J-Svp/360$$

Rule-57504:

If both $(N)$, $(U)$, $(S')$, $(s)$, $(S)$, $(v)$, $(F)$, $(i)$, $(t)$, $(D)$, $(K)$, $(J)$ and $(p)$ are known, then its Leverage or Gearing Ratio planned is:

$$I = [N+K+J+Svp/360] /(U+\{S'[1+s]-Sv-F-Si\}[1-t]-D)-1$$

Rule-57505:

If both $(I)$, $(N)$, $(S')$, $(s)$, $(S)$, $(v)$, $(F)$, $(i)$, $(t)$, $(D)$, $(K)$, $(J)$ and $(p)$ are known, then its Utilized or Starting Capital must be:

$$U = ([N+K+J+Svp/360] -[1+I]\{S'[1+s]-Sv-F-Si\}[1-t]-D)/[1+I]$$

Rule-57506:

If both $(I)$, $(U)$, $(N)$, $(s)$, $(S)$, $(v)$, $(F)$, $(i)$, $(t)$, $(D)$, $(K)$, $(J)$ and $(p)$ are known, then its Sales Past must be:

$$S' = (F+Sv+Si+\{[N+K+J+Svp/360]/[1+I] -U+D\}/[1-t]\})/[1+s]$$

or

$$S' = S/[1+s]$$

Steve Asikin ISBN 13: **978-1564912361**, ISBN 10: **1564912363**

Rule-57507:

If both **(/)**, **(U)**, **(S')**, **(N)**, **(S)**, **(v)**, **(F)**, **(i)**, **(t)**, **(D)**, **(K)**, **(J)** and **(p)** are known, then its Sales Growth planned is:

$$s= (F+Sv+Si+\{[N+K+J+Svp/360]/[1+/]-U+D\}/[1-t]\})/S'-1$$

or

$$s= S/S'-1$$

Rule-57508:

If both **(/)**, **(U)**, **(S')**, **(s)**, **(N)**, **(v)**, **(F)**, **(i)**, **(t)**, **(D)**, **(K)**, **(J)** and **(p)** are known, then its Sales or Revenue planned is:

$$S= [[1+/](U+\{S'[1+s]-F\}[1-t]-D)-[N+K+J]]/\{vp/360+[1+/][1-t][v+i]\}$$

or

$$S= S'[1+s]$$

Rule-57509:

If both **(/)**, **(U)**, **(S')**, **(s)**, **(S)**, **(N)**, **(F)**, **(i)**, **(t)**, **(D)**, **(K)**, **(J)** and **(p)** are known, then its Variable Portion planned is:

$$v= [[1+/](U+\{S'[1+s]-F-Si\}[1-t]-D)-[N+K+J]]/(S\{p/360+[1+/][1-t]\})$$

Steve Asikin ISBN 13: **978-1564912361**, ISBN 10: **1564912363**

Rule-57510:

If both (**/**), (**U**), (**S'**), (**s**), (**S**), (**v**), (**N**), (**i**), (**t**), (**D**), (**K**), (**J**) and (**p**) are known, then its Fixed Cost planned is:

$$F= S'[1+s]-Sv-Si-\{[N+K+J+Svp/360]/[1+/]-U+D\}$$
$$/[1-t]$$

Rule-57511:

If both (**/**), (**U**), (**S'**), (**s**), (**S**), (**v**), (**F**), (**N**), (**t**), (**D**), (**K**), (**J**) and (**p**) are known, then its Interest Portion planned is:

$$i= (S'[1+s]-Sv-F-\{[N+K+J+Svp/360]/[1+/]$$
$$-U+D\}/[1-t])/S$$

Rule-57512:

If both (**/**), (**U**), (**S'**), (**s**), (**S**), (**v**), (**F**), (**i**), (**N**), (**D**), (**K**), (**J**) and (**p**) are known, then its Tax Rate planned is:

$$t= 1-\{[N+K+J+Svp/360]/[1+/]-U+D\}$$
$$/\{S'[1+s]-Sv-F-Si\}$$

Rule-57513:

If both (**/**), (**U**), (**S'**), (**s**), (**S**), (**v**), (**F**), (**i**), (**t**), (**N**), (**K**), (**J**) and (**p**) are known, then its Dividend planned is:

$$D= U+\{S'[1+s]-Sv-F-Si\}[1-t]-[N+K+J+Svp/360]$$
$$/[1+/]$$

Steve Asikin ISBN 13: **978-1564912361**, ISBN 10: **1564912363**

Rule-57514:

If both (**/**), (**U**), (**S'**), (**s**), (**S**), (**v**), (**F**), (**i**), (**t**), (**D**), (**N**), (**J**) and (**p**) are known, then its Kind of Cash planned is:

$$K= [1+/(U+\{S'[1+s]-Sv-F-Si\}[1-t]-D)$$
$$-[N+J+Svp/360]$$

Rule-57515:

If both (**/**), (**U**), (**S'**), (**s**), (**S**), (**v**), (**F**), (**i**), (**t**), (**D**), (**K**), (**N**) and (**p**) are known, then its Job or Trade Account Receivable planned is:

$$J= [1+/(U+\{S'[1+s]-Sv-F-Si\}[1-t]-D)$$
$$-\{N+K+Svp/360\}$$

Rule-57516:

If both (**/**), (**U**), (**S'**), (**s**), (**S**), (**v**), (**F**), (**i**), (**t**), (**D**), (**K**), (**J**) and (**N**) are known, then its Procured Inventory Days planned is:

$$p= 360[[1+/(U+\{S'[1+s]-Sv-F-Si\}[1-t]-D)$$
$$-[N+K+J]]/[Sv]$$

Rule-57517:

If both (**/**), (**U**), (**S'**), (**s**), (**S**), (**v**), (**F**), (**i**), (**t**), (**D**), (**K**), (**J**) and (**p**) are known, then its Non Current Asset planned is:

$$N= [1+/(U+\{S'[1+s]-Sv-F-Si\}[1-t]-D)$$
$$-K-J-S'vp[1+s]/360$$

Steve Asikin ISBN 13: **978-1564912361**, ISBN 10: **1564912363**

Rule-57518:

If both $(\mathbf{N})$, $(\mathbf{U})$, $(\mathbf{S'})$, $(\mathbf{s})$, $(\mathbf{S})$, $(\mathbf{v})$, $(\mathbf{F})$, $(\mathbf{i})$, $(\mathbf{t})$, $(\mathbf{D})$, $(\mathbf{K})$, $(\mathbf{J})$ and $(\mathbf{p})$ are known, then its Leverage or Gearing Ratio planned is:

$$I = \{\mathbf{N}+\mathbf{K}+\mathbf{J}+\mathbf{S'vp}[1+\mathbf{s}]/360\}$$
$$/(\mathbf{U}+\{\mathbf{S'}[1+\mathbf{s}]-\mathbf{Sv}-\mathbf{F}-\mathbf{Si}\}[1-\mathbf{t}]-\mathbf{D})-1$$

Rule-57519:

If both $(\mathbf{I})$, $(\mathbf{N})$, $(\mathbf{S'})$, $(\mathbf{s})$, $(\mathbf{S})$, $(\mathbf{v})$, $(\mathbf{F})$, $(\mathbf{i})$, $(\mathbf{t})$, $(\mathbf{D})$, $(\mathbf{K})$, $(\mathbf{J})$ and $(\mathbf{p})$ are known, then its Utilized or Starting Capital must be:

$$\mathbf{U} = (\{\mathbf{N}+\mathbf{K}+\mathbf{J}+\mathbf{S'vp}[1+\mathbf{s}]/360\}$$
$$-[1+\mathbf{I}]\{\mathbf{S'}[1+\mathbf{s}]-\mathbf{Sv}-\mathbf{F}-\mathbf{Si}\}[1-\mathbf{t}]-\mathbf{D})/[1+\mathbf{I}]$$

Rule-57520:

If both $(\mathbf{I})$, $(\mathbf{U})$, $(\mathbf{N})$, $(\mathbf{s})$, $(\mathbf{S})$, $(\mathbf{v})$, $(\mathbf{F})$, $(\mathbf{i})$, $(\mathbf{t})$, $(\mathbf{D})$, $(\mathbf{K})$, $(\mathbf{J})$ and $(\mathbf{p})$ are known, then its Sales Past must be:

$$\mathbf{S'} = [[1+\mathbf{I}](\mathbf{U}-\{\mathbf{Sv}+\mathbf{F}+\mathbf{Si}\}[1-\mathbf{t}]-\mathbf{D})-[\mathbf{N}+\mathbf{K}+\mathbf{J}]]$$
$$/([1+\mathbf{s}]\{\mathbf{vp}/360+[1+\mathbf{I}][1-\mathbf{t}]\})$$

or

$$\mathbf{S'} = \mathbf{S}/[1+\mathbf{s}]$$

Steve Asikin ISBN 13: **978-1564912361**, ISBN 10: **1564912363**

Rule-57521:

If both (**/**), (**U**), (**S'**), (**N**), (**S**), (**v**), (**F**), (**i**), (**t**), (**D**), (**K**), (**J**) and (**p**) are known, then its Sales Growth planned is:

$$s= [[1+/](U-\{Sv+F+Si\}[1-t]-D)-[N+K+J]]$$
$$/(S'\{vp/360+[1+/][1-t]\})-1$$

or

$$s= S/S'-1$$


Rule-57522:

If both (**/**), (**U**), (**S'**), (**s**), (**N**), (**v**), (**F**), (**i**), (**t**), (**D**), (**K**), (**J**) and (**p**) are known, then its Sales or Revenue planned is:

$$S= [S'[1+s]-F-(\{N+K+J+S'vp[1+s]/360\}/[1+/]$$
$$-U+D)/[1-t]]/[v+i]$$

or

$$S= S'[1+s]$$


Rule-57523:

If both (**/**), (**U**), (**S'**), (**s**), (**S**), (**N**), (**F**), (**i**), (**t**), (**D**), (**K**), (**J**) and (**p**) are known, then its Variable Portion planned is:

$$v= [[1+/](U+\{S'[1+s]-F-Si\}[1-t]-D)-[N+K+J]]$$
$$/\{S'p[1+s]/360+S[1+/][1-d]\}$$

Steve Asikin ISBN 13: **978-1564912361**, ISBN 10: **1564912363**

Rule-57524:

If both (**I**), (**U**), (**S'**), (**s**), (**S**), (**v**), (**N**), (**i**), (**t**), (**D**), (**K**), (**J**) and (**p**) are known, then its Fixed Cost planned is:

$$F= S'[1+s]-Sv-Si-(\{N+K+J+S'vp[1+s]/360\}/[1+I] -U+D)/[1-t]$$

Rule-57525:

If both (**I**), (**U**), (**S'**), (**s**), (**S**), (**v**), (**F**), (**N**), (**t**), (**D**), (**K**), (**J**) and (**p**) are known, then its Interest Portion planned is:

$$i= [S'[1+s]-Sv-F-(\{N+K+J+S'vp[1+s]/360\}/[1+I] -U+D)/[1-t]]/S$$

Rule-57526:

If both (**I**), (**U**), (**S'**), (**s**), (**S**), (**v**), (**F**), (**i**), (**N**), (**D**), (**K**), (**J**) and (**p**) are known, then its Tax Rate planned is:

$$t= 1-(\{N+K+J+S'vp[1+s]/360\}/[1+I] -U+D)/\{S'[1+s]-Sv-F-Si\}$$

Rule-57527:

If both (**I**), (**U**), (**S'**), (**s**), (**S**), (**v**), (**F**), (**i**), (**t**), (**N**), (**K**), (**J**) and (**p**) are known, then its Dividend planned is:

$$D= U+\{S'[1+s]-Sv-F-Si\}[1-t] -\{N+K+J+S'vp[1+s]/360\}/[1+I]$$

Steve Asikin ISBN 13: **978-1564912361**, ISBN 10: **1564912363**

Rule-57528:

If both (**/**), (**U**), (**S'**), (**s**), (**S**), (**v**), (**F**), (**i**), (**t**), (**D**), (**N**), (**J**) and (**p**) are known, then its Kind of Cash planned is:

$$K = [1 + /](U + \{S'[1+s] - Sv - F - Si\}[1-t] - D)$$
$$- \{N + J + S'vp[1+s]/360\}$$

Rule-57529:

If both (**/**), (**U**), (**S'**), (**s**), (**S**), (**v**), (**F**), (**i**), (**t**), (**D**), (**K**), (**N**) and (**p**) are known, then its Job or Trade Account Receivable planned is:

$$J = [[1 + /](U + \{S'[1+s] - Sv - F - Si\}[1-t] - D)$$
$$- \{N + K + S'vp[1+s]/360\}$$

Rule-57530:

If both (**/**), (**U**), (**S'**), (**s**), (**S**), (**v**), (**F**), (**i**), (**t**), (**D**), (**K**), (**J**) and (**N**) are known, then its Procured Inventory Days planned is:

$$p = 360[[1 + /](U + \{S'[1+s] - Sv - F - Si\}[1-t] - D)$$
$$- [N + K + J]]/\{S'v[1+s]\}$$

Rule-57531:

If both (**/**), (**U**), (**S'**), (**s**), (**S**), (**v**), (**F**), (**i**), (**t**), (**D**), (**K**), (**j**) and (**P**) are known, then its Non Current Asset planned is:

$$N = [1 + /](U + \{S'[1+s] - Sv - F - Si\}[1-t] - D)$$
$$- K - Sj/360 - P$$

Steve Asikin ISBN 13: **978-1564912361**, ISBN 10: **1564912363**

Rule-57532:

If both (**N**), (**U**), (**S'**), (**s**), (**S**), (**v**), (**F**), (**i**), (**t**), (**D**), (**K**), (**j**) and (**P**) are known, then its Leverage or Gearing Ratio planned is:

$$l = \{\mathbf{N}+\mathbf{K}+\mathbf{Sj}/360+\mathbf{P}\}$$
$$/(\mathbf{U}+\{\mathbf{S'}[1+\mathbf{s}]-\mathbf{Sv}-\mathbf{F}-\mathbf{Si}\}[1-\mathbf{t}]-\mathbf{D})-1$$


Rule-57533:

If both (**l**), (**N**), (**S'**), (**s**), (**S**), (**v**), (**F**), (**i**), (**t**), (**D**), (**K**), (**j**) and (**P**) are known, then its Utilized or Starting Capital must be:

$$\mathbf{U}= (\{\mathbf{N}+\mathbf{K}+\mathbf{Sj}/360+\mathbf{P}\}-[1+l]\{\mathbf{S'}[1+\mathbf{s}]-\mathbf{Sv}-\mathbf{F}-\mathbf{Si}\}$$
$$[1-\mathbf{t}]-\mathbf{D})/[1+l]$$


Rule-57534:

If both (**l**), (**U**), (**N**), (**s**), (**S**), (**v**), (**F**), (**i**), (**t**), (**D**), (**K**), (**j**) and (**P**) are known, then its Sales Past must be:

$$\mathbf{S'}= \{[\mathbf{N}+\mathbf{K}+\mathbf{Sj}/360+\mathbf{P}]/[1+l]-\mathbf{U}+\mathbf{D}\}/[1-\mathbf{t}]$$
$$+\mathbf{Sv}+\mathbf{F}+\mathbf{Si})/[1+\mathbf{s}]$$

or

$$\mathbf{S'}= \mathbf{S}/[1+\mathbf{s}]$$

Steve Asikin ISBN 13: **978-1564912361**, ISBN 10: **1564912363**

Rule-57535:

If both (**/**), (**U**), (**S'**), (**N**), (**S**), (**v**), (**F**), (**i**), (**t**), (**D**), (**K**), (**j**) and (**P**) are known, then its Sales Growth planned is:

$$s= \{[\mathbf{N}+\mathbf{K}+\mathbf{Sj}/360+\mathbf{P}]/[1+\mathbf{/}-\mathbf{U}+\mathbf{D}\}/[1-\mathbf{t}]$$
$$+\mathbf{Sv}+\mathbf{F}+\mathbf{Si})/\mathbf{S'}-1$$

or

$$s= \mathbf{S}/\mathbf{S'}-1$$

Rule-57536:

If both (**/**), (**U**), (**S'**), (**s**), (**N**), (**v**), (**F**), (**i**), (**t**), (**D**), (**K**), (**j**) and (**P**) are known, then its Sales or Revenue planned is:

$$\mathbf{S}= [[1+\mathbf{/}(\mathbf{U}+\{\mathbf{S'}[1+\mathbf{s}]-\mathbf{F}\}[1-\mathbf{t}]-\mathbf{D})-[\mathbf{N}+\mathbf{K}+\mathbf{P}]]$$
$$/\{\mathbf{j}/360+[1+\mathbf{/}][1-\mathbf{t}]\ [\mathbf{v}+\mathbf{i}]\}$$

or

$$\mathbf{S}= \mathbf{S'}[1+\mathbf{s}]$$

Rule-57537:

If both (**/**), (**U**), (**S'**), (**s**), (**S**), (**N**), (**F**), (**i**), (**t**), (**D**), (**K**), (**j**) and (**P**) are known, then its Variable Portion planned is:

$$\mathbf{v}= (\mathbf{S'}[1+\mathbf{s}]-\mathbf{F}-\mathbf{Si}-\{[\mathbf{N}+\mathbf{K}+\mathbf{Sj}/360+\mathbf{P}]/[1+\mathbf{/}-\mathbf{U}+\mathbf{D}\}$$
$$/[1-\mathbf{t}])/\mathbf{S}$$

Steve Asikin ISBN 13: **978-1564912361**, ISBN 10: **1564912363**

Rule-57538:

If both (**I**), (**U**), (**S'**), (**s**), (**S**), (**v**), (**N**), (**i**), (**t**), (**D**), (**K**), (**j**) and (**P**) are known, then its Fixed Cost planned is:

$$F= S'[1+s]-Sv-Si-\{[N+K+Sj/360+P]/[1+I]-U+D\}/[1-t]$$

Rule-57539:

If both (**I**), (**U**), (**S'**), (**s**), (**S**), (**v**), (**F**), (**N**), (**t**), (**D**), (**K**), (**j**) and (**P**) are known, then its Interest Portion planned is:

$$i= (S'[1+s]-Sv-F-\{[N+K+Sj/360+P]/[1+I]-U+D\}/[1-t])/S$$

Rule-57540:

If both (**I**), (**U**), (**S'**), (**s**), (**S**), (**v**), (**F**), (**i**), (**N**), (**D**), (**K**), (**j**) and (**P**) are known, then its Tax Rate planned is:

$$t= 1-\{[N+K+Sj/360+P]/[1+I]-U+D\}/\{S'[1+s]-Sv-F-Si\}$$

Rule-57541:

If both (**I**), (**U**), (**S'**), (**s**), (**S**), (**v**), (**F**), (**i**), (**t**), (**N**), (**K**), (**j**) and (**P**) are known, then its Dividend planned is:

$$D= U+\{S'[1+s]-Sv-F-Si\}[1-t] - [N+K+Sj/360+P]/[1+I]$$

Steve Asikin ISBN 13: **978-1564912361**, ISBN 10: **1564912363**

Rule-57542:

If both (**ʃ**), (**U**), (**S'**), (**s**), (**S**), (**v**), (**F**), (**i**), (**t**), (**D**), (**N**), (**j**) and (**P**) are known, then its Kind of Cash planned is:

$$K= [1+ʃ(U+\{S'[1+s]-Sv-F-Si\}[1-t]-D)$$
$$-[N+Sj/360+P]$$

Rule-57543:

If both (**ʃ**), (**U**), (**S'**), (**s**), (**S**), (**v**), (**F**), (**i**), (**t**), (**D**), (**K**), (**N**) and (**P**) are known, then its Job or Trade Receivable Days planned is:

$$j= 360[[1+ʃ(U+\{S'[1+s]-Sv-F-Si\}[1-t]-D)$$
$$-\{N+K+P\}]/S$$

Rule-57544:

If both (**ʃ**), (**U**), (**S'**), (**s**), (**S**), (**v**), (**F**), (**i**), (**t**), (**D**), (**K**), (**j**) and (**N**) are known, then its Procured Inventory planned is:

$$P= [1+ʃ(U+\{S'[1+s]-Sv-F-Si\}[1-t]-D)$$
$$-\{N+K+Sj/360\}$$

Rule-57545:

If both (**ʃ**), (**U**), (**S'**), (**s**), (**S**), (**v**), (**F**), (**i**), (**t**), (**D**), (**K**), (**j**), (**V**) and (**p**) are known, then its Non Current Asset planned is:

$$N= [1+ʃ(U+\{S'[1+s]-Sv-F-Si\}[1-t]-D)$$
$$-K-Sj/360-Vp/360$$

Steve Asikin ISBN 13: **978-1564912361**, ISBN 10: **1564912363**

Rule-57546:

If both (**N**), (**U**), (**S'**), (**s**), (**S**), (**v**), (**F**), (**i**), (**t**), (**D**), (**K**), (**j**), (**V**) and (**p**) are known, then its Leverage or Gearing Ratio planned is:

$$l = [N+K+Sj/360+Vp/360]$$
$$/(U+\{S'[1+s]-Sv-F-Si\}[1-t]-D)-1$$

Rule-57547:

If both (**l**), (**N**), (**S'**), (**s**), (**S**), (**v**), (**F**), (**i**), (**t**), (**D**), (**K**), (**j**), (**V**) and (**p**) are known, then its Utilized or Starting Capital must be:

$$U= ([N+K+Sj/360+Vp/360]$$
$$-[1+l]\{S'[1+s]-Sv-F-Si\}[1-t]-D)/[1+l]$$

Rule-57548:

If both (**l**), (**U**), (**N**), (**s**), (**S**), (**v**), (**F**), (**i**), (**t**), (**D**), (**K**), (**j**), (**V**) and (**p**) are known, then its Sales Past must be:

$$S'= (\{[N+K+Sj/360+Vp/360]/[1+l]-U+D\}/[1-t]$$
$$+Sv+F+Si)/[1+s]$$
or
$$S'= S[1+s]$$

Steve Asikin ISBN 13: **978-1564912361**, ISBN 10: **1564912363**

Rule-57549:

If both $(\mathit{l})$, $(\mathbf{U})$, $(\mathbf{S'})$, $(\mathbf{N})$, $(\mathbf{S})$, $(\mathbf{v})$, $(\mathbf{F})$, $(\mathbf{i})$, $(\mathbf{t})$, $(\mathbf{D})$, $(\mathbf{K})$, $(\mathbf{j})$, $(\mathbf{V})$ and $(\mathbf{p})$ are known, then its Sales Growth planned is:

$$s= (\{[\mathbf{N}+\mathbf{K}+\mathbf{Sj}/360+\mathbf{Vp}/360]/[1+\mathit{l}-\mathbf{U}+\mathbf{D}\}/[1-\mathbf{t}]$$
$$+\mathbf{Sv}+\mathbf{F}+\mathbf{Si})/\mathbf{S'}-1$$

or

$$s= \mathbf{S}/\mathbf{S'}-1$$


Rule-57550:

If both $(\mathit{l})$, $(\mathbf{U})$, $(\mathbf{S'})$, $(\mathbf{s})$, $(\mathbf{N})$, $(\mathbf{v})$, $(\mathbf{F})$, $(\mathbf{i})$, $(\mathbf{t})$, $(\mathbf{D})$, $(\mathbf{K})$, $(\mathbf{j})$, $(\mathbf{V})$ and $(\mathbf{p})$ are known, then its Sales or Revenue planned is:

$$S= [[1+\mathit{l}(\mathbf{U}+\{\mathbf{S'}[1+\mathbf{s}]-\mathbf{F}\}[1-\mathbf{t}]-\mathbf{D})$$
$$-[\mathbf{N}+\mathbf{K}+\mathbf{Vp}/360]]$$
$$/\{\mathbf{j}/360+[1+\mathit{l}][1-\mathbf{t}][\mathbf{v}+\mathbf{i}]\}$$

or

$$S= \mathbf{S'}[1+\mathbf{s}]$$


Rule-57551:

If both $(\mathit{l})$, $(\mathbf{U})$, $(\mathbf{S'})$, $(\mathbf{s})$, $(\mathbf{S})$, $(\mathbf{N})$, $(\mathbf{F})$, $(\mathbf{i})$, $(\mathbf{t})$, $(\mathbf{D})$, $(\mathbf{K})$, $(\mathbf{j})$, $(\mathbf{V})$ and $(\mathbf{p})$ are known, then its Variable Portion planned is:

$$v= [\mathbf{S'}[1+\mathbf{s}]-\mathbf{F}-\mathbf{Si}-\{[\mathbf{N}+\mathbf{K}+\mathbf{Sj}/360+\mathbf{Vp}/360]/[1+\mathit{l}$$
$$-\mathbf{U}+\mathbf{D}\}/[1-\mathbf{t}]]/\mathbf{S}$$

or

$$v= \mathbf{V}/\mathbf{S}$$

Steve Asikin ISBN 13: **978-1564912361**, ISBN 10: **1564912363**

Rule-57552:

If both (**/**), (**U**), (**S'**), (**s**), (**S**), (**v**), (**N**), (**i**), (**t**), (**D**), (**K**), (**j**), (**V**) and (**p**) are known, then its Fixed Cost planned is:

$$F= S'[1+s]-Sv-Si-\{[N+K+Sj/360+Vp/360]/[1+/\ -U+D\}/[1-t]$$

Rule-57553:

If both (**/**), (**U**), (**S'**), (**s**), (**S**), (**v**), (**F**), (**N**), (**t**), (**D**), (**K**), (**j**), (**V**) and (**p**) are known, then its Interest Portion planned is:

$$i= [S'[1+s]-Sv-F-\{[N+K+Sj/360+Vp/360]/[1+/\ -U+D\}/[1-t]]/S$$

Rule-57554:

If both (**/**), (**U**), (**S'**), (**s**), (**S**), (**v**), (**F**), (**i**), (**N**), (**D**), (**K**), (**j**), (**V**) and (**p**) are known, then its Tax Rate planned is:

$$t= 1-\{[N+K+Sj/360+Vp/360]/[1+/\ -U+D\}/\{S'[1+s]-Sv-F-Si\}$$

Rule-57555:

If both (**/**), (**U**), (**S'**), (**s**), (**S**), (**v**), (**F**), (**i**), (**t**), (**N**), (**K**), (**j**), (**V**) and (**p**) are known, then its Dividend planned is:

$$D= U+\{S'[1+s]-Sv-F-Si\}[1-t] - [N+K+Sj/360+Vp/360]/[1+/\ $$

Steve Asikin ISBN 13: **978-1564912361**, ISBN 10: **1564912363**

Rule-57556:

If both (**/**), (**U**), (**S'**), (**s**), (**S**), (**v**), (**F**), (**i**), (**t**), (**D**), (**N**), (**j**), (**V**) and (**p**) are known, then its Kind of Cash planned is:

$$K= [1+/](U+\{S'[1+s]-Sv-F-Si\}[1-t]-D)$$
$$-[N+Sj/360+Vp/360]$$

Rule-57557:

If both (**/**), (**U**), (**S'**), (**s**), (**S**), (**v**), (**F**), (**i**), (**t**), (**D**), (**K**), (**N**), (**V**) and (**p**) are known, then its Job or Trade Receivable Days planned is:

$$j= 360[[1+/](U+\{S'[1+s]-Sv-F-Si\}[1-t]-D)$$
$$-[N+K+Vp/360]]/S$$

Rule-57558:

If both (**/**), (**U**), (**S'**), (**s**), (**S**), (**v**), (**F**), (**i**), (**t**), (**D**), (**K**), (**j**), (**N**) and (**p**) are known, then its Variable Cost planned is:

$$V= 360[[1+/](U+\{S'[1+s]-Sv-F-Si\}[1-t]-D)$$
$$-[N+K+Sj/360]]/p$$
$$or$$
$$V= Sv$$

Rule-57559:

If both (**/**), (**U**), (**S'**), (**s**), (**S**), (**v**), (**F**), (**i**), (**t**), (**D**), (**K**), (**j**), (**V**) and (**N**) are known, then its Procured Inventory Days planned is:

$$p= 360[[1+/](U+\{S'[1+s]-Sv-F-Si\}[1-t]-D)$$
$$-[N+K+Sj/360]]/V$$

Steve Asikin ISBN 13: **978-1564912361**, ISBN 10: **1564912363**

Rule-57560:

If both (**/**), (**U**), (**S'**), (**s**), (**S**), (**v**), (**F**), (**i**), (**t**), (**D**), (**K**), (**j**) and (**p**) are known, then its Non Current Asset planned is:

$$\mathbf{N}= [1+\mathbf{/}](\mathbf{U}+\{\mathbf{S'}[1+\mathbf{s}]-\mathbf{Sv}-\mathbf{F}-\mathbf{Si}\}[1-\mathbf{t}]-\mathbf{D})$$
$$-\mathbf{K}-\mathbf{Sj}/360-\mathbf{Svp}/360$$

Rule-57561:

If both (**N**), (**U**), (**S'**), (**s**), (**S**), (**v**), (**F**), (**i**), (**t**), (**D**), (**K**), (**j**) and (**p**) are known, then its Leverage or Gearing Ratio planned is:

$$\mathbf{/}= [\mathbf{N}+\mathbf{K}+\mathbf{Sj}/360+\mathbf{Svp}/360]$$
$$/(\mathbf{U}+\{\mathbf{S'}[1+\mathbf{s}]-\mathbf{Sv}-\mathbf{F}-\mathbf{Si}\}[1-\mathbf{t}]-\mathbf{D})-1$$

Rule-57562:

If both (**/**), (**N**), (**S'**), (**s**), (**S**), (**v**), (**F**), (**i**), (**t**), (**D**), (**K**), (**j**) and (**p**) are known, then its Utilized or Starting Capital must be:

$$\mathbf{U}= ([\mathbf{N}+\mathbf{K}+\mathbf{Sj}/360+\mathbf{Svp}/360]$$
$$-[1+\mathbf{/}]\{\mathbf{S'}[1+\mathbf{s}]-\mathbf{Sv}-\mathbf{F}-\mathbf{Si}\}[1-\mathbf{t}]-\mathbf{D})/[1+\mathbf{/}]$$

Rule-57563:

If both (**/**), (**U**), (**N**), (**s**), (**S**), (**v**), (**F**), (**i**), (**t**), (**D**), (**K**), (**j**) and (**p**) are known, then its Sales Past must be:

$$\mathbf{S'}= (\mathbf{Sv}+\mathbf{Si}-\mathbf{F}+\{[\mathbf{N}+\mathbf{K}+\mathbf{Sj}/360+\mathbf{Svp}/360]/[1+\mathbf{/}]$$
$$-\mathbf{U}+\mathbf{D}\}/[1-\mathbf{t}])/[1+\mathbf{s}]$$
$$\text{or}$$
$$\mathbf{S'}= \mathbf{S}/[1+\mathbf{s}]$$

Steve Asikin ISBN 13: **978-1564912361**, ISBN 10: **1564912363**

Rule-57564:

If both $(I)$, $(U)$, $(S')$, $(N)$, $(S)$, $(v)$, $(F)$, $(i)$, $(t)$, $(D)$, $(K)$, $(j)$ and $(p)$ are known, then its Sales Growth planned is:

$$s= (Sv+Si-F+\{[N+K+Sj/360+Svp/360]/[1+I]$$
$$-U+D\}/[1-t])/S'-1$$

or

$$s= S/S'-1$$


Rule-57565:

If both $(I)$, $(U)$, $(S')$, $(s)$, $(N)$, $(v)$, $(F)$, $(i)$, $(t)$, $(D)$, $(K)$, $(j)$ and $(p)$ are known, then its Sales or Revenue planned is:

$$S= [[1+I](U+\{S'[1+s]-F\}[1-t]-D)-[N+K]]$$
$$/\{j/360+vp/360+[1+I][1-t][v+i]\}$$

or

$$S= S'[1+s]$$


Rule-57566:

If both $(I)$, $(U)$, $(S')$, $(s)$, $(S)$, $(N)$, $(F)$, $(i)$, $(t)$, $(D)$, $(K)$, $(j)$ and $(p)$ are known, then its Variable Portion planned is:

$$v= [[1+I](U+\{S'[1+s]-F-Si\}[1-t]-D)$$
$$-[N+K+Sj/360]]/(S\{p/360+[1+I][1-t]\})$$

Steve Asikin ISBN 13: **978-1564912361**, ISBN 10: **1564912363**

Rule-57567:

If both (**/**), (**U**), (**S'**), (**s**), (**S**), (**v**), (**N**), (**i**), (**t**), (**D**), (**K**), (**j**) and (**p**) are known, then its Fixed Cost planned is:

$$F= S'[1+s]-Sv-Si-\{[N+K+Sj/360+Svp/360]/[1+/]-U+D\}[1-t]/[1-t]$$

Rule-57568:

If both (**/**), (**U**), (**S'**), (**s**), (**S**), (**v**), (**F**), (**N**), (**t**), (**D**), (**K**), (**j**) and (**p**) are known, then its Interest Portion planned is:

$$i= [S'[1+s]-Sv-F-\{[N+K+Sj/360+Svp/360]/[1+/]-U+D\}[1-t]/[1-t]]/S$$

Rule-57569:

If both (**/**), (**U**), (**S'**), (**s**), (**S**), (**v**), (**F**), (**i**), (**N**), (**D**), (**K**), (**j**) and (**p**) are known, then its Tax Rate planned is:

$$t= 1-\{[N+K+Sj/360+Svp/360]/[1+/]-U+D\}/\{S'[1+s]-Sv-F-Si\}$$

Rule-57570:

If both (**/**), (**U**), (**S'**), (**s**), (**S**), (**v**), (**F**), (**i**), (**t**), (**N**), (**K**), (**j**) and (**p**) are known, then its Dividend planned is:

$$D= U+\{S'[1+s]-Sv-F-Si\}[1-t]-[N+K+Sj/360+Svp/360]/[1+/]$$

`

404

Steve Asikin ISBN 13: **978-1564912361**, ISBN 10: **1564912363**

Rule-57571:

If both (**/**), (**U**), (**$'**), (**s**), (**$**), (**v**), (**F**), (**i**), (**t**), (**D**), (**N**), (**j**) and (**p**) are known, then its Kind of Cash planned is:

$$K= [1+/](U+\{\$'[1+s]-\$v-F-\$i\}[1-t]-D)$$
$$-[N+\$j/360+\$vp/360]$$

Rule-57572:

If both (**/**), (**U**), (**$'**), (**s**), (**$**), (**v**), (**F**), (**i**), (**t**), (**D**), (**K**), (**N**) and (**p**) are known, then its Job or Trade Receivable Days planned is:

$$j= 360[[1+/](U+\{\$'[1+s]-\$v-F-\$i\}[1-t]-D)$$
$$-\{N+K+\$vp/360\}]/\$$$

Rule-57573:

If both (**/**), (**U**), (**$'**), (**s**), (**$**), (**v**), (**F**), (**i**), (**t**), (**D**), (**K**), (**j**) and (**N**) are known, then its Procured Inventory Days planned is:

$$p= 360[[1+/](U+\{\$'[1+s]-\$v-F-\$i\}[1-t]-D)$$
$$-[N+K+\$j/360]]/[\$v]$$

Rule-57574:

If both (**/**), (**U**), (**$'**), (**s**), (**$**), (**v**), (**F**), (**i**), (**t**), (**D**), (**K**), (**j**) and (**p**) are known, then its Non Current Asset planned is:

$$N= [1+/](U+\{\$'[1+s]-\$v-F-\$i\}[1-t]-D)$$
$$-K-\$j/360-\$'vp[1+s]/360$$

Steve Asikin ISBN 13: **978-1564912361**, ISBN 10: **1564912363**

<u>Rule-57575</u>:

If both (**N**), (**U**), (**S'**), (**s**), (**S**), (**v**), (**F**), (**i**), (**t**), (**D**), (**K**), (**j**) and (**p**) are known, then its Leverage or Gearing Ratio planned is:

$$I = \{N+K+Sj/360+S'vp[1+s]/360\}$$
$$/(U+\{S'[1+s]-Sv-F-Si\}[1-t]-D)-1$$

<u>Rule-57576</u>:

If both (**I**), (**N**), (**S'**), (**s**), (**S**), (**v**), (**F**), (**i**), (**t**), (**D**), (**K**), (**j**) and (**p**) are known, then its Utilized or Starting Capital must be:

$$U = (\{N+K+Sj/360+S'vp[1+s]/360\}$$
$$-[1+I]\{S'[1+s]-Sv-F-Si\}[1-t]-D)/\{[1+I]$$

<u>Rule-57577</u>:

If both (**I**), (**U**), (**S'**), (**s**), (**S**), (**v**), (**F**), (**i**), (**t**), (**D**), (**K**), (**j**) and (**p**) are known, then its Sales Past must be:

$$S' = [[1+I](U+\{S'[1+s]-Sv-F-Si\}[1-t]-D)$$
$$-[N+K+Sj/360]]$$
$$/([1+s]\{vp/360-[1+I][1-t]\})$$

or

$$S' = S/[1+s]$$

Steve Asikin ISBN 13: **978-1564912361**, ISBN 10: **1564912363**

Rule-57578:

If both (**/**), (**U**), (**S'**), (**N**), (**S**), (**v**), (**F**), (**i**), (**t**), (**D**), (**K**), (**j**) and (**p**) are known, then its Sales Growth planned is:

$$s= [[1+/(U+\{S'[1+s]-Sv-F-Si\}[1-t]-D)-[N+K$$
$$+Sj/360]]/(S'\{vp/360-[1+/[1-t]\})-1$$
$$\text{or}$$
$$s= S/S'-1$$

Rule-57579:

If both (**/**), (**U**), (**S'**), (**s**), (**N**), (**v**), (**F**), (**i**), (**t**), (**D**), (**K**), (**j**) and (**p**) are known, then its Sales or Revenue planned is:

$$S= [[1+/(U+\{S'[1+s]-F\}[1-t]-D)-\{N+K$$
$$+S'vp[1+s]/360\}]/\{j/360+v[1+/[1-t]\}$$
$$\text{or}$$
$$S= S'[1+s]$$

Rule-57580:

If both (**/**), (**U**), (**S'**), (**s**), (**S**), (**N**), (**F**), (**i**), (**t**), (**D**), (**K**), (**j**) and (**p**) are known, then its Variable Portion planned is:

$$v= [S'[1+s]-F-Si-(\{N+K+Sj/360+S'vp[1+s]/360\}$$
$$/[1+/-U+D)/[1-t]]/S$$

Steve Asikin ISBN 13: **978-1564912361**, ISBN 10: **1564912363**

Rule-57581:

If both ($I$), ($U$), ($S'$), ($s$), ($S$), ($v$), ($N$), ($i$), ($t$), ($D$), ($K$), ($j$) and ($p$) are known, then its Fixed Cost planned is:

$$F= S'[1+s]-Sv-Si-(\{N+K+Sj/360+S'vp[1+s]/360\}/[1+I]-U+D)/[1-t]$$

Rule-57582:

If both ($I$), ($U$), ($S'$), ($s$), ($S$), ($v$), ($F$), ($N$), ($t$), ($D$), ($K$), ($j$) and ($p$) are known, then its Interest Portion planned is:

$$i= [S'[1+s]-Sv-F-(\{N+K+Sj/360+S'vp[1+s]/360\}/[1+I]-U+D)/[1-t]]/S$$

Rule-57583:

If both ($I$), ($U$), ($S'$), ($s$), ($S$), ($v$), ($F$), ($i$), ($t$), ($D$), ($K$), ($j$) and ($p$) are known, then its Tax Rate planned is:

$$t= 1-(\{N+K+Sj/360+S'vp[1+s]/360/[1+I]-U+D)/\{S'[1+s]-Sv-F-Si\}$$

Rule-57584:

If both ($I$), ($U$), ($S'$), ($s$), ($S$), ($v$), ($F$), ($i$), ($t$), ($N$), ($K$), ($j$) and ($p$) are known, then its Dividend planned is:

$$D= U+\{S'[1+s]-Sv-F-Si\}[1-t]-\{N+K+Sj/360+S'vp[1+s]/360\}/[1+I]$$

`

Steve Asikin ISBN 13: **978-1564912361**, ISBN 10: **1564912363**

Rule-57585:

If both (**/**), (**U**), (**S'**), (**s**), (**S**), (**v**), (**F**), (**i**), (**t**), (**D**), (**N**), (**j**) and (**p**) are known, then its Kind of Cash planned is:

$$K= [1+/(U+\{S'[1+s]-Sv-F-Si\}[1-t]-D)$$
$$-\{N+Sj/360+S'vp[1+s]/360\}$$

Rule-57586:

If both (**/**), (**U**), (**S'**), (**s**), (**S**), (**v**), (**F**), (**i**), (**t**), (**D**), (**K**), (**N**) and (**p**) are known, then its Job or Trade Receivable Days planned is:

$$j= 360[[1+/(U+\{S'[1+s]-Sv-F-Si\}[1-t]-D)$$
$$-\{N+K+S'vp[1+s]/360\}]/S$$

Rule-57587:

If both (**/**), (**U**), (**S'**), (**s**), (**S**), (**v**), (**F**), (**i**), (**t**), (**D**), (**K**), (**j**) and (**N**) are known, then its Procured Inventory Days planned is:

$$p= 360[[1+/(U+\{S'[1+s]-Sv-F-Si\}[1-t]-D)$$
$$-[N+K+Sj/360]]/\{S'v[1+s]\}$$

Rule-57588:

If both (**/**), (**U**), (**S'**), (**s**), (**S**), (**v**), (**F**), (**i**), (**t**), (**D**), (**K**), (**j**) and (**P**) are known, then its Non Current Asset planned is:

$$N= [1+/(U+\{S'[1+s]-Sv-F-Si\}[1-t]-D)$$
$$-K-S'j[1+s]/360-P$$

Steve Asikin ISBN 13: **978-1564912361**, ISBN 10: **1564912363**

Rule-57589:

If both (**N**), (**U**), (**S'**), (**s**), (**S**), (**v**), (**F**), (**i**), (**t**), (**D**), (**K**), (**j**) and (**P**) are known, then its Leverage or Gearing Ratio planned is:

$$l = \{N+K+S'j[1+s]/360+P\}$$
$$/(U+\{S'[1+s]-Sv-F-Si\}[1-t]-D)-1$$

Rule-57590:

If both (**l**), (**N**), (**S'**), (**s**), (**S**), (**v**), (**F**), (**i**), (**t**), (**D**), (**K**), (**j**) and (**P**) are known, then its Utilized or Starting Capital must be:

$$U = (\{N+K+S'j[1+s]/360+P\}$$
$$-[1+l]\{S'[1+s]-Sv-F-Si\}[1-t]-D)/[1+l]$$

Rule-57591:

If both (**l**), (**U**), (**N**), (**s**), (**S**), (**v**), (**F**), (**i**), (**t**), (**D**), (**K**), (**j**) and (**P**) are known, then its Sales Past must be:

$$S' = [[1+l](U-\{Sv+F+Si\}[1-t]-D)-[N+K+P]]$$
$$/([1+s]\{j/360-[1+l][1-t]\})$$

$$\text{or}$$

$$S' = S/[1+s]$$

Steve Asikin ISBN 13: **978-1564912361**, ISBN 10: **1564912363**

Rule-57592:

If both (**/**), (**U**), (**S'**), (**s**), (**S**), (**v**), (**F**), (**i**), (**t**), (**D**), (**K**), (**j**) and (**P**) are known, then its Sales Growth planned is:

$$s= [[1+\text{/}(U-\{Sv+F+Si\}[1-t]-D)-[N+K+P]]$$
$$/(S'\{j/360-[1+\text{/}[1-t]\})-1$$

or

$$s= S/S'-1$$

Rule-57593:

If both (**/**), (**U**), (**S'**), (**s**), (**S**), (**v**), (**F**), (**i**), (**t**), (**D**), (**K**), (**j**) and (**P**) are known, then its Sales or Revenue planned is:

$$S= [S'[1+s]-F-(\{N+K+S'j[1+s]/360+P\}/[1+\text{/}$$
$$-U+D)/[1-t]]/[v+i]$$

or

$$S= S'[1+s]$$

Rule-57594:

If both (**/**), (**U**), (**S'**), (**s**), (**S**), (**N**), (**F**), (**i**), (**t**), (**D**), (**K**), (**j**) and (**P**) are known, then its Variable Portion planned is:

$$v= [S'[1+s]-F-Si-(\{N+K+S'j[1+s]/360+P\}/[1+\text{/}$$
$$-U+D)/[1-t]]/S$$

Steve Asikin ISBN 13: **978-1564912361**, ISBN 10: **1564912363**

Rule-57595:

If both (**I**), (**U**), (**S'**), (**s**), (**S**), (**v**), (**N**), (**i**), (**t**), (**D**), (**K**), (**j**) and (**P**) are known, then its Fixed Cost planned is:

$$F = S'[1+s] - Sv - Si - (\{N+K+S'j[1+s]/360+P\}/[1+I -U+D)/[1-t]$$

Rule-57596:

If both (**I**), (**U**), (**S'**), (**s**), (**S**), (**v**), (**F**), (**N**), (**t**), (**D**), (**K**), (**j**) and (**P**) are known, then its Interest Portion planned is:

$$i = [S'[1+s] - Sv - F - (\{N+K+S'j[1+s]/360+P\}/[1+I -U+D)/[1-t]]/S$$

Rule-57597:

If both (**I**), (**U**), (**S'**), (**s**), (**S**), (**v**), (**F**), (**i**), (**N**), (**D**), (**K**), (**j**) and (**P**) are known, then its Tax Rate planned is:

$$t = 1 - (\{N+K+S'j[1+s]/360+P\}/[1+I-U+D) /\{S'[1+s]-Sv-F-Si\}$$

Rule-57598:

If both (**I**), (**U**), (**S'**), (**s**), (**S**), (**v**), (**F**), (**i**), (**t**), (**N**), (**K**), (**j**) and (**P**) are known, then its Dividend planned is:

$$D = U + \{S'[1+s]-Sv-F-Si\}[1-t] -\{N+K+S'j[1+s]/360+P\}/[1+I]$$

Steve Asikin ISBN 13: **978-1564912361**, ISBN 10: **1564912363**

Rule-57599:

If both (**/**), (**U**), (**S'**), (**s**), (**S**), (**v**), (**F**), (**i**), (**t**), (**D**), (**N**), (**j**) and (**P**) are known, then its Kind of Cash planned is:

$$K= [1+/(U+\{S'[1+s]-Sv-F-Si\}[1-t]-D) \\ -\{N+S'j[1+s]/360+P\}$$

Rule-57600:

If both (**/**), (**U**), (**S'**), (**s**), (**S**), (**v**), (**F**), (**i**), (**t**), (**D**), (**K**), (**N**) and (**P**) are known, then its Job or Trade Receivable Days planned is:

$$j= 360[[1+/(U+\{S'[1+s]-Sv-F-Si\}[1-t]-D) \\ -\{N+K+P\}]/\{S'[1+s]\}$$

Rule-57601:

If both (**/**), (**U**), (**S'**), (**s**), (**S**), (**v**), (**F**), (**i**), (**t**), (**D**), (**K**), (**j**) and (**N**) are known, then its Procured Inventory Days planned is:

$$P= [1+/(U+\{S'[1+s]-Sv-F-Si\}[1-t]-D) \\ -\{N+K+S'j[1+s]/360\}$$

Rule-57602:

If both (**/**), (**U**), (**S'**), (**s**), (**S**), (**v**), (**F**), (**i**), (**t**), (**D**), (**K**), (**j**), (**V**) and (**p**) are known, then its Non Current Asset planned is:

$$N= [1+/(U+\{S'[1+s]-Sv-F-Si\}[1-t]-D) \\ -K-S'j[1+s]/360-Vp/360$$

Steve Asikin ISBN 13: **978-1564912361**, ISBN 10: **1564912363**

Rule-57603:

If both (**N**), (**U**), (**S'**), (**s**), (**S**), (**v**), (**F**), (**i**), (**t**), (**D**), (**K**), (**j**), (**V**) and (**p**) are known, then its Leverage or Gearing Ratio planned is:

$$l = \{N+K+S'j[1+s]/360+Vp/360\}$$
$$/(U+\{S'[1+s]-Sv-F-Si\}[1-t]-D)-1$$

Rule-57604:

If both (**l**), (**N**), (**S'**), (**s**), (**S**), (**v**), (**F**), (**i**), (**t**), (**D**), (**K**), (**j**), (**V**) and (**p**) are known, then its Utilized or Starting Capital must be:

$$U = (\{N+K+S'j[1+s]/360+Vp/360\}$$
$$-[1+l]\{S'[1+s]-Sv-F-Si\}[1-t]-D)/[1+l]$$

Rule-57605:

If both (**l**), (**U**), (**N**), (**s**), (**S**), (**v**), (**F**), (**i**), (**t**), (**D**), (**K**), (**j**), (**V**) and (**p**) are known, then its Sales Past must be:

$$S' = [[1+l](U-\{Sv+F+Si\}[1-t]-D])$$
$$-[N+K+Vp/360]]$$
$$/([1+s]\{j/360-[1+l][1-t]\})$$

or

$$S' = S/[1+s]$$

Steve Asikin ISBN 13: **978-1564912361**, ISBN 10: **1564912363**

Rule-57606:

If both (**/**), (**U**), (**S'**), (**N**), (**S**), (**v**), (**F**), (**i**), (**t**), (**D**), (**K**), (**j**), (**V**) and (**p**) are known, then its Sales Growth planned is:

$$s= [[1+/](U-\{Sv+F+Si\}[1-t]-D)-[N+K+Vp/360]]$$
$$/(S'\{j/360-[1+/][1-t]\})-1$$

or

$$s= S/S'-1$$


Rule-57607:

If both (**/**), (**U**), (**S'**), (**s**), (**N**), (**v**), (**F**), (**i**), (**t**), (**D**), (**K**), (**j**), (**V**) and (**p**) are known, then its Sales or Revenue planned is:

$$S= [S'[1+s]-F-(\{N+K+S'j[1+s]/360+Vp/360\}$$
$$/[1+/]-U+D)/[1-t]]/[v+i]$$

or

$$S= S'[1+s]$$

or

$$S= V/v$$


Rule-57608:

If both (**/**), (**U**), (**S'**), (**s**), (**S**), (**N**), (**F**), (**i**), (**t**), (**D**), (**K**), (**j**), (**V**) and (**p**) are known, then its Variable Portion planned is:

$$v= [S'[1+s]-F-Si-(\{N+K+S'j[1+s]/360+Vp/360\}$$
$$/[1+/]-U+D)/[1-t]]/S$$

or

$$v= V/S$$

Steve Asikin ISBN 13: **978-1564912361**, ISBN 10: **1564912363**

Rule-57609:

If both ($I$), (**U**), (**S'**), (**s**), (**S**), (**v**), (**N**), (**i**), (**t**), (**D**), (**K**), (**j**), (**V**) and (**p**) are known, then its Fixed Cost planned is:

$$F= S'[1+s]-Sv-Si-(\{N+K+S'j[1+s]/360+Vp/360\}/[1+I-U+D)/[1-t]$$

Rule-57610:

If both ($I$), (**U**), (**S'**), (**s**), (**S**), (**v**), (**F**), (**N**), (**t**), (**D**), (**K**), (**j**), (**V**) and (**p**) are known, then its Interest Portion planned is:

$$i= [S'[1+s]-Sv-F-(\{N+K+S'j[1+s]/360+Vp/360\}/[1+I-U+D)/[1-t]]/S$$

Rule-57611:

If both ($I$), (**U**), (**S'**), (**s**), (**S**), (**v**), (**F**), (**i**), (**N**), (**D**), (**K**), (**j**), (**V**) and (**p**) are known, then its Tax Rate planned is:

$$t= 1-(\{N+K+S'j[1+s]/360+Vp/360\}/[1+I-U+D)/\{S'[1+s]-Sv-F-Si\}$$

Rule-57612:

If both ($I$), (**U**), (**S'**), (**s**), (**S**), (**v**), (**F**), (**i**), (**t**), (**N**), (**K**), (**j**), (**V**) and (**p**) are known, then its Dividend planned is:

$$D= U+\{S'[1+s]-Sv-F-Si\}[1-t]-\{N+K+S'j[1+s]/360+Vp/360\}/[1+I]$$

Steve Asikin ISBN 13: **978-1564912361**, ISBN 10: **1564912363**

Rule-57613:

If both (**/**), (**U**), (**S'**), (**s**), (**S**), (**v**), (**F**), (**i**), (**t**), (**D**), (**N**), (**j**), (**V**) and (**p**) are known, then its Kind of Cash planned is:

$$K = [1 + /(U + \{S'[1+s]-Sv-F-Si\}[1-t]-D) -\{N+S'j[1+s]/360+Vp/360\}$$

Rule-57614:

If both (**/**), (**U**), (**S'**), (**s**), (**S**), (**v**), (**F**), (**i**), (**t**), (**D**), (**K**), (**N**), (**V**) and (**p**) are known, then its Job or Trade Receivable Days planned is:

$$j = 360[[1 + /(U + \{S'[1+s]-Sv-F-Si\}[1-t]-D) -\{N+K+Vp/360\}]/\{S'[1+s]\}$$

Rule-57615:

If both (**/**), (**U**), (**S'**), (**s**), (**S**), (**v**), (**F**), (**i**), (**t**), (**D**), (**K**), (**j**), (**N**) and (**p**) are known, then its Variable Cost planned is:

$$V = 360[[1 + /(U + \{S'[1+s]-Sv-F-Si\}[1-t]-D) -\{N+K+S'j[1+s]/360\}]/p$$

or

$$V = Sv$$

Rule-57616:

If both (**/**), (**U**), (**S'**), (**s**), (**S**), (**v**), (**F**), (**i**), (**t**), (**D**), (**K**), (**j**), (**V**) and (**N**) are known, then its Procured Inventory Days planned is:

$$p = 360[[1 + /(U + \{S'[1+s]-Sv-F-Si\}[1-t]-D) -\{N+K+S'j[1+s]/360\}]/V$$

417

Steve Asikin ISBN 13: **978-1564912361**, ISBN 10: **1564912363**

Rule-57617:

If both (**/**), (**U**), (**S'**), (**s**), (**S**), (**v**), (**F**), (**i**), (**t**), (**D**), (**K**), (**j**) and (**p**) are known, then its Non Current Asset planned is:

$$N = [1+/](U+\{S'[1+s]-Sv-F-Si\}[1-t]-D)$$
$$-K-S'j[1+s]/360-Svp/360$$

Rule-57618:

If both (**N**), (**U**), (**S'**), (**s**), (**S**), (**v**), (**F**), (**i**), (**t**), (**D**), (**K**), (**j**) and (**p**) are known, then its Leverage or Gearing Ratio planned is:

$$/ = \{N+K+S'j[1+s]/360+Svp/360\}$$
$$/(U+\{S'[1+s]-Sv-F-Si\}[1-t]-D)-1$$

Rule-57619:

If both (**/**), (**N**), (**S'**), (**s**), (**S**), (**v**), (**F**), (**i**), (**t**), (**D**), (**K**), (**j**) and (**p**) are known, then its Utilized or Starting Capital must be:

$$U = (\{N+K+S'j[1+s]/360+Svp/360\}$$
$$-[1+/]\{S'[1+s]-Sv-F-Si\}[1-t]-D)/\{[1+/]\}$$

Rule-57620:

If both (**/**), (**U**), (**N**), (**s**), (**S**), (**v**), (**F**), (**i**), (**t**), (**D**), (**K**), (**j**) and (**p**) are known, then its Sales Past must be:

$$S' = [[1+/](U-\{Sv+F+Si\{[1-t]-D)$$
$$-\{N+K+Svp/360\}]$$
$$/([1+s]\{j/360-[1+/][1-t]\})$$
or
$$S' = S/[1+s]$$

Steve Asikin ISBN 13: **978-1564912361**, ISBN 10: **1564912363**

Rule-57621:

If both (**/**), (**U**), (**S'**), (**N**), (**S**), (**v**), (**F**), (**i**), (**t**), (**D**), (**K**), (**j**) and (**p**) are known, then its Sales Growth planned is:

$$s= [[1+/(U-\{Sv+F+Si\}[1-t]-D)$$
$$-\{N+K+Svp/360\}]$$
$$/(S'\{j/360-[1+/][1-t]\})-1$$

or

$$s= S/S'-1$$

Rule-57622:

If both (**/**), (**U**), (**S'**), (**s**), (**S**), (**v**), (**F**), (**i**), (**t**), (**D**), (**K**), (**j**) and (**p**) are known, then its Non Current Asset planned is:

$$S= [[1+/(U+\{S'[1+s]-F\}[1-t]-D)$$
$$-\{N+K+S'j[1+s]/360\}]$$
$$/\{vp/360+[1+/][1-t][v+i]\}$$

or

$$S= S'[1+s]$$

Rule-57623:

If both (**/**), (**U**), (**S'**), (**s**), (**S**), (**N**), (**F**), (**i**), (**t**), (**D**), (**K**), (**j**) and (**p**) are known, then its Variable Portion planned is:

$$v= [[1+/(U+\{S'[1+s]-F-Si\}[1-t]-D)$$
$$-\{N+K+S'j[1+s]/360\}]$$
$$/\{Sp/360+S[1+/][1-t]\}$$

Steve Asikin ISBN 13: **978-1564912361**, ISBN 10: **1564912363**

Rule-57624:

If both (**/**), (**U**), (**S'**), (**s**), (**S**), (**v**), (**N**), (**i**), (**t**), (**D**), (**K**), (**j**) and (**p**) are known, then its Fixed Cost planned is:

$$F= S'[1+s]-Sv-Si-(\{N+K+S'j[1+s]/360+Svp/360\}$$
$$/[1+/]-U+D)/[1-t]$$

Rule-57625:

If both (**/**), (**U**), (**S'**), (**s**), (**S**), (**v**), (**F**), (**N**), (**t**), (**D**), (**K**), (**j**) and (**p**) are known, then its Interest Portion planned is:

$$i= [S'[1+s]\ Sv\ F\ (\{N+K+S'j[1+s]/360+Svp/360\}$$
$$/[1+/]-U+D)/[1-t]]/S$$

Rule-57626:

If both (**/**), (**U**), (**S'**), (**s**), (**S**), (**v**), (**F**), (**i**), (**N**), (**D**), (**K**), (**j**) and (**p**) are known, then its Tax Rate planned is:

$$t= 1-(\{N+K+S'j[1+s]/360+Svp/360\}/[1+/]$$
$$-U+D)/\{S'[1+s]-Sv-F-Si\}$$

Rule-57627:

If both (**/**), (**U**), (**S'**), (**s**), (**S**), (**v**), (**F**), (**i**), (**t**), (**N**), (**K**), (**j**) and (**p**) are known, then its Dividend planned is:

$$D= U+\{S'[1+s]-Sv-F-Si\}[1-t]-\{N+K+S'j[1+s]/360$$
$$+Svp/360\}/[1+/]$$

Steve Asikin ISBN 13: **978-1564912361**, ISBN 10: **1564912363**

Rule-57628:

If both (**/**), (**U**), (**S'**), (**s**), (**S**), (**v**), (**F**), (**i**), (**t**), (**D**), (**N**), (**j**) and (**p**) are known, then its Kimnd of Cash planned is:

$$K= [1+/(U+\{S'[1+s]-Sv-F-Si\}[1-t]-D) -\{N+S'j[1+s]/360+Svp/360\}$$

Rule-57629:

If both (**/**), (**U**), (**S'**), (**s**), (**S**), (**v**), (**F**), (**i**), (**t**), (**D**), (**K**), (**N**) and (**p**) are known, then its Job or Trade Receivable Days planned is:

$$j= 360[[1+/(U+\{S'[1+s]-Sv-F-Si\}[1-t]-D) -\{N+K+Svp/360\}]/\{S'[1+s]\}$$

Rule-57630:

If both (**/**), (**U**), (**S'**), (**s**), (**S**), (**v**), (**F**), (**i**), (**t**), (**D**), (**K**), (**j**) and (**N**) are known, then its Procured Inventory Days planned is:

$$p= 360[[1+/(U+\{S'[1+s]-Sv-F-Si\}[1-t]-D) -\{N+K+S'j[1+s]/360\}]/[Sv]$$

Rule-57631:

If both (**/**), (**U**), (**S'**), (**s**), (**S**), (**v**), (**F**), (**i**), (**t**), (**D**), (**K**), (**j**) and (**p**) are known, then its Non Current Asset planned is:

$$N= [1+/(U+\{S'[1+s]-Sv-F-Si\}[1-t]-D) -K-S'j[1+s]/360-S'vp[1+s]/360$$

Steve Asikin ISBN 13: **978-1564912361**, ISBN 10: **1564912363**

Rule-57632:

If both $(\mathbf{N})$, $(\mathbf{U})$, $(\mathbf{S'})$, $(\mathbf{s})$, $(\mathbf{S})$, $(\mathbf{v})$, $(\mathbf{F})$, $(\mathbf{i})$, $(\mathbf{t})$, $(\mathbf{D})$, $(\mathbf{K})$, $(\mathbf{j})$ and $(\mathbf{p})$ are known, then its Leverage or Gearing Ratio planned is:

$$I = \{\mathbf{N+K+S'j}[1+\mathbf{s}]/360+\mathbf{S'vp}[1+\mathbf{s}]/360\}$$
$$/(\mathbf{U}+\{\mathbf{S'}[1+\mathbf{s}]-\mathbf{Sv}-\mathbf{F}-\mathbf{Si}\}[1-\mathbf{t}]-\mathbf{D})-1$$

Rule-57633:

If both $(I)$, $(\mathbf{N})$, $(\mathbf{S'})$, $(\mathbf{s})$, $(\mathbf{S})$, $(\mathbf{v})$, $(\mathbf{F})$, $(\mathbf{i})$, $(\mathbf{t})$, $(\mathbf{D})$, $(\mathbf{K})$, $(\mathbf{j})$ and $(\mathbf{p})$ are known, then its Utilized or Starting Capital must be:

$$\mathbf{U} = (\{\mathbf{N+K+S'j}[1+\mathbf{s}]/360+\mathbf{S'vp}[1+\mathbf{s}]/360\}$$
$$-[1+I]\{\mathbf{S'}[1+\mathbf{s}]-\mathbf{Sv}-\mathbf{F}-\mathbf{Si}\}[1-\mathbf{t}]-\mathbf{D})/[1+I]$$

Rule-57634:

If both $(I)$, $(\mathbf{U})$, $(\mathbf{N})$, $(\mathbf{s})$, $(\mathbf{S})$, $(\mathbf{v})$, $(\mathbf{F})$, $(\mathbf{i})$, $(\mathbf{t})$, $(\mathbf{D})$, $(\mathbf{K})$, $(\mathbf{j})$ and $(\mathbf{p})$ are known, then its Sales Past must be:

$$\mathbf{S'} = [[1+I](\mathbf{U}-\{\mathbf{Sv+F+Si}\}[1-\mathbf{t}]-\mathbf{D})-[\mathbf{N+K}]]$$
$$/(\mathbf{S'}[1+\mathbf{s}]\{\mathbf{j}/360+\mathbf{vp}/360-[1+I][1-\mathbf{t}]\})$$
$$\text{or}$$
$$\mathbf{S'} = \mathbf{S}/[1+\mathbf{s}]$$

Steve Asikin ISBN 13: **978-1564912361**, ISBN 10: **1564912363**

Rule-57635:

If both (**/**), (**U**), (**S'**), (**N**), (**S**), (**v**), (**F**), (**i**), (**t**), (**D**), (**K**), (**j**) and (**p**) are known, then its Sales Growth planned is:

$$s= [[1+/(U-\{Sv+F+Si\}[1-t]-D)-[N+K]]$$
$$/(S'\{j/360+vp/360-[1+/[1-t]\})-1$$

or

$$s= S/S'-1$$


Rule-57636:

If both (**/**), (**U**), (**S'**), (**s**), (**N**), (**v**), (**F**), (**i**), (**t**), (**D**), (**K**), (**j**) and (**p**) are known, then its Sales or Revenue planned is:

$$S= [S'[1+s]-F-(\{N+K+S'j[1+s]/360$$
$$+S'vp[1+s]/360\}/[1+/-U+D)/[1-t]]/[v+i]$$

or

$$S= S'[1+s]$$


Rule-57637:

If both (**/**), (**U**), (**S'**), (**s**), (**S**), (**N**), (**F**), (**i**), (**t**), (**D**), (**K**), (**j**) and (**p**) are known, then its Variable Portion planned is:

$$v= [[1+/(U+\{S'[1+s]-Sv-F-Si\}[1-t]-D)$$
$$-\{N+S'j[1+s]/360+K\}]$$
$$/\{S'p[1+s]/360\}+S[1+/[1-t]\}$$

Steve Asikin ISBN 13: **978-1564912361**, ISBN 10: **1564912363**

Rule-57638:

If both $(I)$, $(U)$, $(S')$, $(s)$, $(S)$, $(v)$, $(N)$, $(i)$, $(t)$, $(D)$, $(K)$, $(j)$ and $(p)$ are known, then its Fixed Cost planned is:

$$F= S'[1+s]-Sv-Si-(\{N+K+S'j[1+s]/360 \\ +S'vp[1+s]/360\}/[1+I-U+D)/[1-t]$$

Rule-57639:

If both $(I)$, $(U)$, $(S')$, $(s)$, $(S)$, $(v)$, $(F)$, $(N)$, $(t)$, $(D)$, $(K)$, $(j)$ and $(p)$ are known, then its Interest Portion planned is:

$$i= [S'[1+s]-Sv-F-(\{N+K+S'j[1+s]/360 \\ +S'vp[1+s]/360\}/[1+I-U+D)/[1-t]]/S$$

Rule-57640:

If both $(I)$, $(U)$, $(S')$, $(s)$, $(S)$, $(v)$, $(F)$, $(i)$, $(N)$, $(D)$, $(K)$, $(j)$ and $(p)$ are known, then its Variable Portion planned is:

$$t= 1-(\{N+K+S'j[1+s]/360+S'vp[1+s]/360\}/[1+I \\ -U+D)/\{S'[1+s]-Sv-F-Si\}$$

Rule-57641:

If both $(I)$, $(U)$, $(S')$, $(s)$, $(S)$, $(v)$, $(F)$, $(i)$, $(t)$, $(N)$, $(K)$, $(j)$ and $(p)$ are known, then its Dividend planned is:

$$D= U+\{S'[1+s]-Sv-F-Si\}[1-t]-\{N+K+S'j[1+s]/360 \\ +S'vp[1+s]/360\}/[1+I]$$

Steve Asikin ISBN 13: **978-1564912361**, ISBN 10: **1564912363**

Rule-57642:

If both $(I)$, $(U)$, $(S')$, $(s)$, $(S)$, $(v)$, $(F)$, $(i)$, $(t)$, $(D)$, $(N)$, $(j)$ and $(p)$ are known, then its Kind of Cash planned is:

$$K = [1 + I(U + \{S'[1+s] - Sv - F - Si\}[1-t] - D) - \{N + S'j[1+s]/360 + S'vp[1+s]/360\}$$

Rule-57643:

If both $(I)$, $(U)$, $(S')$, $(s)$, $(S)$, $(v)$, $(F)$, $(i)$, $(t)$, $(D)$, $(K)$, $(N)$ and $(p)$ are known, then its Job or Trade Receivable Days planned is:

$$j = 360[[1 + I(U + \{S'[1+s] - Sv - F - Si\}[1-t] - D) - \{N + K + S'vp[1+s]/360\}]/\{S'[1+s]\}]$$

Rule-57644:

If both $(I)$, $(U)$, $(S')$, $(s)$, $(S)$, $(v)$, $(F)$, $(i)$, $(t)$, $(D)$, $(K)$, $(j)$ and $(N)$ are known, then its Procured Inventory Days planned is:

$$p = 360[[1 + I(U + \{S'[1+s] - Sv - F - Si\}[1-t] - D) - \{N + K + S'j[1+s]/360\}]/\{S'v[1+s]\}]$$

Rule-57645:

If both $(I)$, $(U)$, $(S')$, $(s)$, $(S)$, $(v)$, $(F)$, $(i)$, $(t)$, $(d)$, $(K)$, $(J)$ and $(P)$ are known, then its Non Current Asset planned is:

$$N = [1 + I(U + \{S'[1+s] - Sv - F - Si\}[1-t][1-d]) - K - J - P$$

Steve Asikin ISBN 13: **978-1564912361**, ISBN 10: **1564912363**

Rule-57646:

If both (**I**), (**U**), (**S'**), (**s**), (**S**), (**v**), (**F**), (**i**), (**t**), (**d**), (**K**), (**J**) and (**P**) are known, then its Non Current Asset planned is:

$$I = \{N+K+J+P\}$$
$$/(U+\{S'[1+s]-Sv-F-Si\}[1-t][1-d])-1$$

Rule-57647:

If both (**I**), (**N**), (**S'**), (**s**), (**S**), (**v**), (**F**), (**i**), (**t**), (**d**), (**K**), (**J**) and (**P**) are known, then its Utilized or Starting Capital must be:

$$U = (\{N+K+J+P\}-[1+I]\{S'[1+s]-Sv-F-Si\}[1-t][1-d])$$
$$/[1+I]$$

Rule-57648:

If both (**I**), (**U**), (**N**), (**s**), (**S**), (**v**), (**F**), (**i**), (**t**), (**d**), (**K**), (**J**) and (**P**) are known, then its Sales Past must be:

$$S' = \{[N+K+J+P]/[1+I]-U\}/\{[1-t][1-d]\}$$
$$+Sv+F+Si)/[1+s]$$

or

$$S' = S/[1+s]$$

Steve Asikin ISBN 13: **978-1564912361**, ISBN 10: **1564912363**

Rule-57649:

If both $(I)$, $(U)$, $(S')$, $(N)$, $(S)$, $(v)$, $(F)$, $(i)$, $(t)$, $(d)$, $(K)$, $(J)$ and $(P)$ are known, then its Sales Growth planned is:

$$s= \{[N+K+J+P]/[1+I-U]/\{[1-t][1-d]\}$$
$$+Sv+F+Si)/S'-1$$

or

$$s= S/S'-1$$

Rule-57650:

If both $(I)$, $(U)$, $(S')$, $(s)$, $(N)$, $(v)$, $(F)$, $(i)$, $(t)$, $(d)$, $(K)$, $(J)$ and $(P)$ are known, then its Sales or Revenue planned is:

$$S= (S'[1+s]-F-\{[N+K+J+P]/[1+I-U\}$$
$$/\{[1-t][1-d]\})/[v+i]$$

or

$$S= S'[1+s]$$

Rule-57651:

If both $(I)$, $(U)$, $(S')$, $(s)$, $(S)$, $(N)$, $(F)$, $(i)$, $(t)$, $(d)$, $(K)$, $(J)$ and $(P)$ are known, then its Variable Portion planned is:

$$v= [S'[1+s]-F-Si-\{[N+K+J+P]/[1+I-U\}$$
$$/\{[1-t][1-d]\}]/S$$

Steve Asikin ISBN 13: **978-1564912361**, ISBN 10: **1564912363**

Rule-57652:

If both ($I$), (**U**), (**S'**), (**s**), (**S**), (**v**), (**N**), (**i**), (**t**), (**d**), (**K**), (**J**) and (**P**) are known, then its Fixed Cost planned is:

$$F= S'[1+s]-Sv-Si-\{[N+K+J+P]/[1+I-U\}$$
$$/\{[1-t][1-d]\}$$

Rule-57653:

If both ($I$), (**U**), (**S'**), (**s**), (**S**), (**v**), (**F**), (**i**), (**t**), (**d**), (**K**), (**J**) and (**P**) are known, then its Non Current Asset planned is:

$$i= [S'[1+s]-Sv-F-\{[N+K+J+P]/[1+I-U\}$$
$$/\{[1-t][1-d]\}]/S$$

Rule-57654:

If both ($I$), (**U**), (**S'**), (**s**), (**S**), (**v**), (**F**), (**i**), (**N**), (**d**), (**K**), (**J**) and (**P**) are known, then its Tax Rate planned is:

$$t= 1-\{[N+K+J+P]/[1+I-U\}$$
$$/(\{S'[1+s]-Sv-F-Si\}[1-d])$$

Rule-57655:

If both ($I$), (**U**), (**S'**), (**s**), (**S**), (**v**), (**F**), (**i**), (**t**), (**N**), (**K**), (**J**) and (**P**) are known, then its Dividend Payout planned is:

$$d= 1-\{[N+K+J+P]/[1+I-U\}$$
$$/(\{S'[1+s]-Sv-F-Si\}[1-t])$$

Steve Asikin ISBN 13: **978-1564912361**, ISBN 10: **1564912363**

Rule-57656:

If both (**/**), (**U**), (**S'**), (**s**), (**S**), (**v**), (**F**), (**i**), (**t**), (**d**), (**N**), (**J**) and (**P**) are known, then its Kind of Cash planned is:

$$K= [1+/](U+\{S'[1+s]-Sv-F-Si\}[1-t][1-d])$$
$$-[N+J+P]$$

Rule-57657:

If both (**/**), (**U**), (**S'**), (**s**), (**S**), (**v**), (**F**), (**i**), (**t**), (**d**), (**K**), (**N**) and (**P**) are known, then its Job or Trade Account Receivable planned is:

$$J= [1+/](U+\{S'[1+s]-Sv-F-Si\}[1-t][1-d])$$
$$-[N+K+P]$$

Rule-57658:

If both (**/**), (**U**), (**S'**), (**s**), (**S**), (**v**), (**F**), (**i**), (**t**), (**d**), (**K**), (**J**) and (**N**) are known, then its Procured Inventory planned is:

$$P= [1+/](U+\{S'[1+s]-Sv-F-Si\}[1-t][1-d])$$
$$-[N+K+J]$$

Rule-57659:

If both (**/**), (**U**), (**S'**), (**s**), (**S**), (**v**), (**F**), (**i**), (**t**), (**d**), (**K**), (**J**), (**V**) and (**p**) are known, then its Non Current Asset planned is:

$$N= [1+/](U+\{S'[1+s]-Sv-F-Si\}[1-t][1-d])$$
$$-K-J-Vp/360$$

Steve Asikin ISBN 13: **978-1564912361**, ISBN 10: **1564912363**

Rule-57660:

If both (**N**), (**U**), (**S'**), (**s**), (**S**), (**v**), (**F**), (**i**), (**t**), (**d**), (**K**), (**J**), (**V**) and (**p**) are known, then its Leverage or Gearing Ratio planned is:

$$I= [\mathbf{N}+\mathbf{K}+\mathbf{J}+\mathbf{Vp}/360]$$
$$/(\mathbf{U}+\{\mathbf{S'}[1+\mathbf{s}]-\mathbf{Sv}-\mathbf{F}-\mathbf{Si}\}[1-\mathbf{t}][1-\mathbf{d}])-1$$

Rule-57661:

If both (**I**), (**N**), (**S'**), (**s**), (**S**), (**v**), (**F**), (**i**), (**t**), (**d**), (**K**), (**J**), (**V**) and (**p**) are known, then its Utilized or Starting Capital must be:

$$\mathbf{U}= ([\mathbf{N}+\mathbf{K}+\mathbf{J}+\mathbf{Vp}/360]$$
$$-[1+\mathbf{I}]\{\mathbf{S'}[1+\mathbf{s}]-\mathbf{Sv}-\mathbf{F}-\mathbf{Si}\}[1-\mathbf{t}][1-\mathbf{d}])/[1+\mathbf{I}]$$

Rule-57662:

If both (**I**), (**U**), (**N**), (**s**), (**S**), (**v**), (**F**), (**i**), (**t**), (**d**), (**K**), (**J**), (**V**) and (**p**) are known, then its Sales Past must be:

$$\mathbf{S'}= \{[\mathbf{N}+\mathbf{K}+\mathbf{J}+\mathbf{Vp}/360]/[1+\mathbf{I}]-\mathbf{U}+\mathbf{D}\}$$
$$/\{[1-\mathbf{t}][1-\mathbf{d}]\}+\mathbf{Sv}+\mathbf{F}+\mathbf{Si})/[1+\mathbf{s}]$$
$$\text{or}$$
$$\mathbf{S'}= \mathbf{S}/[1+\mathbf{s}]$$

430

Steve Asikin ISBN 13: **978-1564912361**, ISBN 10: **1564912363**

Rule-57663:

If both (**/**), (**U**), (**S'**), (**N**), (**S**), (**v**), (**F**), (**i**), (**t**), (**d**), (**K**), (**J**), (**V**) and (**p**) are known, then its Sales Growth planned is:

$$s= \{[N+K+J+Vp/360]/[1+/-U+D\}$$
$$/\{[1-t][1-d]\}+Sv+F+Si)/S'-1$$

or

$$s= S/S'-1$$


Rule-57664:

If both (**/**), (**U**), (**S'**), (**s**), (**N**), (**v**), (**F**), (**i**), (**t**), (**d**), (**K**), (**J**), (**V**) and (**p**) are known, then its Sales or Revenue planned is:

$$S= (S'[1+s]-F-\{[N+K+J+Vp/360]/[1+/-U\}$$
$$/\{[1-t][1-d]\})/[v+i]$$

or

$$S= S'[1+s]$$

or

$$S= V/v$$


Rule-57665:

If both (**/**), (**U**), (**S'**), (**s**), (**S**), (**N**), (**F**), (**i**), (**t**), (**d**), (**K**), (**J**), (**V**) and (**p**) are known, then its Variable Portion planned is:

$$v= [S'[1+s]-F-Si-\{[N+K+J+Vp/360]/[1+/-U\}$$
$$/\{[1-t][1-d]\}]/S$$

or

$$v= V/S$$

Steve Asikin ISBN 13: **978-1564912361**, ISBN 10: **1564912363**

Rule-57666:

If both (**/**), (**U**), (**S'**), (**s**), (**S**), (**v**), (**N**), (**i**), (**t**), (**d**), (**K**), (**J**), (**V**) and (**p**) are known, then its Fixed Cost planned is:

$$F= S'[1+s]-Sv-Si-\{[N+K+J+Vp/360]/[1+/-U\} /\{[1-t][1-d]\}$$

Rule-57667:

If both (**/**), (**U**), (**S'**), (**s**), (**S**), (**v**), (**F**), (**N**), (**t**), (**d**), (**K**), (**J**), (**V**) and (**p**) are known, then its Interest Portion planned is:

$$i= [S'[1+s]-Sv-F-\{[N+K+J+Vp/360]/[1+/-U\} /\{[1-t][1-d]\}]/S$$

Rule-57668:

If both (**/**), (**U**), (**S'**), (**s**), (**S**), (**v**), (**F**), (**i**), (**N**), (**d**), (**K**), (**J**), (**V**) and (**p**) are known, then its Tax Rate planned is:

$$t= 1-\{[N+K+J+Vp/360]/[1+/-U\} /(\{S'[1+s]-Sv-F-Si\}[1-d])$$

Rule-57669:

If both (**/**), (**U**), (**S'**), (**s**), (**S**), (**v**), (**F**), (**i**), (**t**), (**N**), (**K**), (**J**), (**V**) and (**p**) are known, then its Dividend Payout planned is:

$$d= 1-\{[N+K+J+Vp/360]/[1+/-U\} /(\{S'[1+s]-Sv-F-Si\}[1-t])$$

`

Steve Asikin ISBN 13: **978-1564912361**, ISBN 10: **1564912363**

Rule-57670:

If both (**/**), (**U**), (**S'**), (**s**), (**S**), (**v**), (**F**), (**i**), (**t**), (**d**), (**N**), (**J**), (**V**) and (**p**) are known, then its Kind of Cash planned is:

$$K= [1+/(U+\{S'[1+s]-Sv-F-Si\}[1-t][1-d])$$
$$-[N+J+Vp/360]$$

Rule-57671:

If both (**/**), (**U**), (**S'**), (**s**), (**S**), (**v**), (**F**), (**i**), (**t**), (**d**), (**K**), (**N**), (**V**) and (**p**) are known, then its Job or Trade Account Receivable planned is:

$$J= [1+/(U+\{S'[1+s]-Sv-F-Si\}[1-t][1-d])$$
$$-\{N+K+Vp/360\}$$

Rule-57672:

If both (**/**), (**U**), (**S'**), (**s**), (**S**), (**v**), (**F**), (**i**), (**t**), (**d**), (**K**), (**J**), (**N**) and (**p**) are known, then its Variable Cost planned is:

$$V= 360[[1+/(U+\{S'[1+s]-Sv-F-Si\}[1-t][1-d])$$
$$-[N+K+J]]/p$$

or

$$V= Sv$$

Rule-57673:

If both (**/**), (**U**), (**S'**), (**s**), (**S**), (**v**), (**F**), (**i**), (**t**), (**d**), (**K**), (**J**), (**V**) and (**N**) are known, then its Procured Inventory Days planned is:

$$p= 360[[1+/(U+\{S'[1+s]-Sv-F-Si\}[1-t][1-d])$$
$$-[N+K+J]]/V$$

Steve Asikin ISBN 13: **978-1564912361**, ISBN 10: **1564912363**

Rule-57674:

If both $(\mathit{l})$, $(\mathbf{U})$, $(\mathbf{S'})$, $(\mathbf{s})$, $(\mathbf{S})$, $(\mathbf{v})$, $(\mathbf{F})$, $(\mathbf{i})$, $(\mathbf{t})$, $(\mathbf{d})$, $(\mathbf{K})$, $(\mathbf{J})$ and $(\mathbf{p})$ are known, then its Non Current Asset planned is:

$$\mathbf{N} = [1 + \mathit{l}(\mathbf{U} + \{\mathbf{S'}[1+\mathbf{s}] - \mathbf{Sv} - \mathbf{F} - \mathbf{Si}\}[1-\mathbf{t}][1-\mathbf{d}])$$
$$- \mathbf{K} - \mathbf{J} - \mathbf{Svp}/360$$

Rule-57675:

If both $(\mathbf{N})$, $(\mathbf{U})$, $(\mathbf{S'})$, $(\mathbf{s})$, $(\mathbf{S})$, $(\mathbf{v})$, $(\mathbf{F})$, $(\mathbf{i})$, $(\mathbf{t})$, $(\mathbf{d})$, $(\mathbf{K})$, $(\mathbf{J})$ and $(\mathbf{p})$ are known, then its Leverage or Gearing Ratio planned is:

$$\mathit{l} = [\mathbf{N} + \mathbf{K} + \mathbf{J} + \mathbf{Svp}/360]/(\mathbf{U} + \{\mathbf{S'}[1+\mathbf{s}] - \mathbf{Sv} - \mathbf{F} - \mathbf{Si}\}$$
$$[1-\mathbf{t}][1-\mathbf{d}]) - 1$$

Rule-57676:

If both $(\mathit{l})$, $(\mathbf{N})$, $(\mathbf{S'})$, $(\mathbf{s})$, $(\mathbf{S})$, $(\mathbf{v})$, $(\mathbf{F})$, $(\mathbf{i})$, $(\mathbf{t})$, $(\mathbf{d})$, $(\mathbf{K})$, $(\mathbf{J})$ and $(\mathbf{p})$ are known, then its Utilized or Starting Capital must be:

$$\mathbf{U} = ([\mathbf{N} + \mathbf{K} + \mathbf{J} + \mathbf{Svp}/360]$$
$$- [1 + \mathit{l}]\{\mathbf{S'}[1+\mathbf{s}] - \mathbf{Sv} - \mathbf{F} - \mathbf{Si}\}[1-\mathbf{t}][1-\mathbf{d}])/[1 + \mathit{l}]$$

Rule-57677:

If both $(\mathit{l})$, $(\mathbf{U})$, $(\mathbf{N})$, $(\mathbf{s})$, $(\mathbf{S})$, $(\mathbf{v})$, $(\mathbf{F})$, $(\mathbf{i})$, $(\mathbf{t})$, $(\mathbf{d})$, $(\mathbf{K})$, $(\mathbf{J})$ and $(\mathbf{p})$ are known, then its Sales Past must be:

$$\mathbf{S'} = (\{[\mathbf{N} + \mathbf{K} + \mathbf{J} + \mathbf{Svp}/360]/[1 + \mathit{l}] - \mathbf{U}\}/\{[1-\mathbf{t}][1-\mathbf{d}]\}$$
$$+ \mathbf{Sv} + \mathbf{F} + \mathbf{Si})/[1+\mathbf{s}]$$

or

$$\mathbf{S'} = \mathbf{S}/[1+\mathbf{s}]$$

Steve Asikin ISBN 13: **978-1564912361**, ISBN 10: **1564912363**

Rule-57678:

If both (**/**), (**U**), (**S'**), (**N**), (**S**), (**v**), (**F**), (**i**), (**t**), (**d**), (**K**), (**J**) and (**p**) are known, then its Sales Growth planned is:

$$s= (\{[\mathbf{N}+\mathbf{K}+\mathbf{J}+\mathbf{Svp}/360]/[1+\mathbf{/}]-\mathbf{U}\}/\{[1-\mathbf{t}][1-\mathbf{d}]\}$$
$$+\mathbf{Sv}+\mathbf{F}+\mathbf{Si})/\mathbf{S'}-1$$

or

$$s= \mathbf{S}/\mathbf{S'}-1$$


Rule-57679:

If both (**/**), (**U**), (**S'**), (**s**), (**N**), (**v**), (**F**), (**i**), (**t**), (**d**), (**K**), (**J**) and (**p**) are known, then its Sales or Revenue planned is:

$$\mathbf{S}= [[1+\mathbf{/}](\mathbf{U}+\{\mathbf{S'}[1+\mathbf{s}]-\mathbf{F}\}[1-\mathbf{t}][1-\mathbf{d}])-[\mathbf{N}+\mathbf{K}+\mathbf{J}]]$$
$$/\{\mathbf{vp}/360+[1+\mathbf{/}][1-\mathbf{t}][1-\mathbf{d}][\mathbf{v}+\mathbf{i}]\}$$

or

$$\mathbf{S}= \mathbf{S'}[1+\mathbf{s}]$$


Rule-57680:

If both (**/**), (**U**), (**S'**), (**s**), (**S**), (**N**), (**F**), (**i**), (**t**), (**d**), (**K**), (**J**) and (**p**) are known, then its Variable Portion planned is:

$$\mathbf{v}= [[1+\mathbf{/}](\mathbf{U}+\{\mathbf{S'}[1+\mathbf{s}]-\mathbf{F}-\mathbf{Si}\}[1-\mathbf{t}][1-\mathbf{d}])-[\mathbf{N}+\mathbf{K}+\mathbf{J}]]$$
$$/(\mathbf{S}\{\mathbf{p}/360+[1+\mathbf{/}][1-\mathbf{t}][1-\mathbf{d}]\})$$

Steve Asikin ISBN 13: **978-1564912361**, ISBN 10: **1564912363**

Rule-57681:

If both ($I$), (**U**), (**S'**), (**s**), (**S**), (**v**), (**N**), (**i**), (**t**), (**d**), (**K**), (**J**) and (**p**) are known, then its Fixed Cost planned is:

$$F= S'[1+s]-Sv-Si-\{[N+K+J+Svp/360]/[1+I-U\}$$
$$/\{[1-t][1-d]\}$$

Rule-57682:

If both ($I$), (**U**), (**S'**), (**s**), (**S**), (**v**), (**F**), (**N**), (**t**), (**d**), (**K**), (**J**) and (**p**) are known, then its Interest Portion planned is:

$$i= [S'[1+s]-Sv-F-\{[N+K+J|Svp/360]/[1+I-U\}$$
$$/\{[1-t][1-d]\}]/S$$

Rule-57683:

If both ($I$), (**U**), (**S'**), (**s**), (**S**), (**v**), (**F**), (**i**), (**N**), (**d**), (**K**), (**J**) and (**p**) are known, then its Tax Rate planned is:

$$t= 1-\{[N+K+J+Svp/360]/[1+I-U\}$$
$$/(\{S'[1+s]-Sv-F-Si\}[1-d])$$

Rule-57684:

If both ($I$), (**U**), (**S'**), (**s**), (**S**), (**v**), (**F**), (**i**), (**t**), (**N**), (**K**), (**J**) and (**p**) are known, then its Dividend Payout planned is:

$$d= 1-\{[N+K+J+Svp/360]/[1+I-U\}$$
$$/(\{S'[1+s]-Sv-F-Si\}[1-t])$$

Steve Asikin ISBN 13: **978-1564912361**, ISBN 10: **1564912363**

<u>Rule-57685</u>:

If both (**/**), (**U**), (**S'**), (**s**), (**S**), (**v**), (**F**), (**i**), (**t**), (**d**), (**N**), (**J**) and (**p**) are known, then its Kind of Cash planned is:

$$K= [1+/](U+\{S'[1+s]-Sv-F-Si\}[1-t][1-d])$$
$$-[N+J+Svp/360]$$

<u>Rule-57686</u>:

If both (**/**), (**U**), (**S'**), (**s**), (**S**), (**v**), (**F**), (**i**), (**t**), (**d**), (**K**), (**N**) and (**p**) are known, then its Job or Trade Account Receivable planned is:

$$J= [1+/](U+\{S'[1+s]-Sv-F-Si\}[1-t][1-d])$$
$$-\{N+K+Svp/360\}$$

<u>Rule-57687</u>:

If both (**/**), (**U**), (**S'**), (**s**), (**S**), (**v**), (**F**), (**i**), (**t**), (**d**), (**K**), (**J**) and (**N**) are known, then its Procured Inventory Days planned is:

$$p= 360[[1+/](U+\{S'[1+s]-Sv-F-Si\}[1-t][1-d])$$
$$-[N+K+J]]/[Sv]$$

<u>Rule-57688</u>:

If both (**/**), (**U**), (**S'**), (**s**), (**S**), (**v**), (**F**), (**i**), (**t**), (**d**), (**K**), (**J**) and (**p**) are known, then its Non Current Asset planned is:

$$N= [1+/](U+\{S'[1+s]-Sv-F-Si\}[1-t][1-d])$$
$$-K-J-S'vp[1+s]/360$$

Steve Asikin ISBN 13: **978-1564912361**, ISBN 10: **1564912363**

Rule-57689:

If both (**N**), (**U**), (**S'**), (**s**), (**S**), (**v**), (**F**), (**i**), (**t**), (**d**), (**K**), (**J**) and (**p**) are known, then its Leverage or Gearing Ratio planned is:

$$I = \{N+K+J+S'vp[1+s]/360\}$$
$$/(U+\{S'[1+s]-Sv-F-Si\}[1-t][1-d])-1$$

Rule-57690:

If both (**I**), (**N**), (**S'**), (**s**), (**S**), (**v**), (**F**), (**i**), (**t**), (**d**), (**K**), (**J**) and (**p**) are known, then its Utilized or Starting Capital must be:

$$U = (\{N+K+J+S'vp[1+s]/360\}$$
$$-[1+I]\{S'[1+s]-Sv-F-Si\}[1-t][1-d])/[1+I]$$

Rule-57691:

If both (**I**), (**U**), (**S'**), (**s**), (**S**), (**v**), (**F**), (**i**), (**t**), (**d**), (**K**), (**J**) and (**p**) are known, then its Non Current Asset planned is:

$$S' = [[1+I](U-\{Sv+F+Si\}[1-t][1-d])-[N+K+J]]$$
$$/([1+s]\{vp/360+[1+I][1-t][1-d]\})$$

or

$$S' = S/[1+s]$$

Steve Asikin ISBN 13: **978-1564912361**, ISBN 10: **1564912363**

Rule-57692:

If both $(\textbf{\textit{I}})$, $(\textbf{U})$, $(\textbf{S'})$, $(\textbf{N})$, $(\textbf{S})$, $(\textbf{v})$, $(\textbf{F})$, $(\textbf{i})$, $(\textbf{t})$, $(\textbf{d})$, $(\textbf{K})$, $(\textbf{J})$ and $(\textbf{p})$ are known, then its Sales Growth planned is:

$$s = [[1+\textbf{\textit{I}}(\textbf{U}-\{\textbf{Sv}+\textbf{F}+\textbf{Si}\}[1-\textbf{t}][1-\textbf{d}])-[\textbf{N}+\textbf{K}+\textbf{J}]]$$
$$/(\textbf{S'}\{\textbf{vp}/360+[1+\textbf{\textit{I}}[1-\textbf{t}][1-\textbf{d}]\})-1$$

or

$$s = \textbf{S}/\textbf{S'}-1$$

Rule-57693:

If both $(\textbf{\textit{I}})$, $(\textbf{U})$, $(\textbf{S'})$, $(\textbf{s})$, $(\textbf{N})$, $(\textbf{v})$, $(\textbf{F})$, $(\textbf{i})$, $(\textbf{t})$, $(\textbf{d})$, $(\textbf{K})$, $(\textbf{J})$ and $(\textbf{p})$ are known, then its Sales or Revenue planned is:

$$S = [\textbf{S'}[1+s]-\textbf{F}-(\{\textbf{N}+\textbf{K}+\textbf{J}+\textbf{S'vp}[1+s]/360\}/[1+\textbf{\textit{I}}-\textbf{U})$$
$$/\{[1-\textbf{t}][1-\textbf{d}]\}]/[\textbf{v}+\textbf{i}]$$

or

$$S = \textbf{S'}[1+s]$$

Rule-57694:

If both $(\textbf{\textit{I}})$, $(\textbf{U})$, $(\textbf{S'})$, $(\textbf{s})$, $(\textbf{S})$, $(\textbf{N})$, $(\textbf{F})$, $(\textbf{i})$, $(\textbf{t})$, $(\textbf{d})$, $(\textbf{K})$, $(\textbf{J})$ and $(\textbf{p})$ are known, then its Variable Portion planned is:

$$v = [[1+\textbf{\textit{I}}(\textbf{U}+\{\textbf{S'}[1+s]-\textbf{F}-\textbf{Si}\}[1-\textbf{t}][1-\textbf{d}])-[\textbf{N}+\textbf{K}+\textbf{J}]]$$
$$/\{\textbf{S'p}[1+s]/360+\textbf{S}[1+\textbf{\textit{I}}[1-\textbf{t}][1-\textbf{d}]\}$$

Steve Asikin ISBN 13: **978-1564912361**, ISBN 10: **1564912363**

Rule-57695:

If both (**/**), (**U**), (**S'**), (**s**), (**S**), (**v**), (**N**), (**i**), (**t**), (**d**), (**K**), (**J**) and (**p**) are known, then its Fixed Cost planned is:

$$F= S'[1+s]-Sv-Si-(\{N+K+J+S'vp[1+s]/360\}$$
$$/[1+/-U)/\{[1-t][1-d]\}$$

Rule-57696:

If both (**/**), (**U**), (**S'**), (**s**), (**S**), (**v**), (**F**), (**N**), (**t**), (**d**), (**K**), (**J**) and (**p**) are known, then its Interest Portion planned is:

$$i= [S'[1+s]-Sv-F-(\{N+K+J+S'vp[1+s]/360\}$$
$$/[1+/-U)/\{[1-t][1-d]\}]/S$$

Rule-57697:

If both (**/**), (**U**), (**S'**), (**s**), (**S**), (**v**), (**F**), (**i**), (**N**), (**d**), (**K**), (**J**) and (**p**) are known, then its Tax Rate planned is:

$$t= 1-(\{N+K+J+S'vp[1+s]/360\}/[1+/-U)$$
$$/(\{S'[1+s]-Sv-F-Si\}[1-d])$$

Rule-57698:

If both (**/**), (**U**), (**S'**), (**s**), (**S**), (**v**), (**F**), (**i**), (**t**), (**N**), (**K**), (**J**) and (**p**) are known, then its Dividend Payout planned is:

$$d= 1-(\{N+K+J+S'vp[1+s]/360\}/[1+/-U)$$
$$/(\{S'[1+s]-Sv-F-Si\}[1-t])$$

Steve Asikin ISBN 13: **978-1564912361**, ISBN 10: **1564912363**

Rule-57699:

If both (**/**), (**U**), (**S'**), (**s**), (**S**), (**v**), (**F**), (**i**), (**t**), (**d**), (**N**), (**J**) and (**p**) are known, then its Kind of Cash planned is:

$$K = [1 + /(U + \{S'[1+s] - Sv - F - Si\}[1-t][1-d])$$
$$- \{N + J + S'vp[1+s]/360\}$$

Rule-57700:

If both (**/**), (**U**), (**S'**), (**s**), (**S**), (**v**), (**F**), (**i**), (**t**), (**d**), (**K**), (**N**) and (**p**) are known, then its Job or Trade Account Receivable planned is:

$$J = [[1 + /(U + \{S'[1+s] - Sv - F - Si\}[1-t][1-d])$$
$$- \{N + K + S'vp[1+s]/360\}$$

Rule-57701:

If both (**/**), (**U**), (**S'**), (**s**), (**S**), (**v**), (**F**), (**i**), (**t**), (**d**), (**K**), (**J**) and (**N**) are known, then its Procured Inventory Days planned is:

$$p = 360[[1 + /(U + \{S'[1+s] - Sv - F - Si\}[1-t][1-d])$$
$$- [N + K + J]]/\{S'v[1+s]\}$$

Rule-57702:

If both (**/**), (**U**), (**S'**), (**s**), (**S**), (**v**), (**F**), (**i**), (**t**), (**d**), (**K**), (**j**) and (**P**) are known, then its Non Current Asset planned is:

$$N = [1 + /(U + \{S'[1+s] - Sv - F - Si\}[1-t][1-d])$$
$$- K - Sj/360 - P$$

Steve Asikin ISBN 13: **978-1564912361**, ISBN 10: **1564912363**

Rule-57703:

If both (**/**), (**U**), (**S'**), (**s**), (**S**), (**v**), (**F**), (**i**), (**t**), (**d**), (**K**), (**j**) and (**P**) are known, then its Non Current Asset planned is:

$$/= \{\mathbf{N}+\mathbf{K}+\mathbf{Sj}/360+\mathbf{P}\}$$
$$/(\mathbf{U}+\{\mathbf{S'}[1+\mathbf{s}]-\mathbf{Sv}-\mathbf{F}-\mathbf{Si}\}[1-\mathbf{t}][1-\mathbf{d}])-1$$

Rule-57704:

If both (**/**), (**N**), (**S'**), (**s**), (**S**), (**v**), (**F**), (**i**), (**t**), (**d**), (**K**), (**j**) and (**P**) are known, then its Utilized or Starting Capital must be:

$$\mathbf{U}= (\{\mathbf{N}+\mathbf{K}+\mathbf{Sj}/360+\mathbf{P}\}$$
$$-[1+/]\{\mathbf{S'}[1+\mathbf{s}]-\mathbf{Sv}-\mathbf{F}-\mathbf{Si}\}[1-\mathbf{t}][1-\mathbf{d}])/[1+/]$$

Rule-57705:

If both (**/**), (**U**), (**N**), (**s**), (**S**), (**v**), (**F**), (**i**), (**t**), (**d**), (**K**), (**j**) and (**P**) are known, then its Sales Past must be:

$$\mathbf{S'}= (\{[\mathbf{N}+\mathbf{K}+\mathbf{Sj}/360+\mathbf{P}]/[1+/]-\mathbf{U}\}/\{[1-\mathbf{t}][1-\mathbf{d}]\}$$
$$+\mathbf{Sv}+\mathbf{F}+\mathbf{Si})/[1+\mathbf{s}]$$
$$\text{or}$$
$$\mathbf{S'}= \mathbf{S}/[1+\mathbf{s}]$$

Steve Asikin ISBN 13: **978-1564912361**, ISBN 10: **1564912363**

Rule-57706:

If both (**/**), (**U**), (**S'**), (**s**), (**S**), (**v**), (**F**), (**i**), (**t**), (**N**), (**K**), (**j**) and (**P**) are known, then its Dividend Payout planned is:

$$s= (\{[N+K+Sj/360+P]/[1+/\!-U\}$$
$$/\{[1-t][1-d]\}+Sv+F+Si)/S'-1$$

or

$$s= S/S'-1$$

Rule-57707:

If both (**/**), (**U**), (**S'**), (**s**), (**N**), (**v**), (**F**), (**i**), (**t**), (**d**), (**K**), (**j**) and (**P**) are known, then its Sales Past must be:

$$S= [[1+/\!(U+\{S'[1+s]-F\}[1-t][1-d])-[N+K+P]]$$
$$/\{j/360+[1+/\!][1-t][1-d] [v+i]\}$$

or

$$S= S'[1+s]$$

Rule-57708:

If both (**/**), (**U**), (**S'**), (**s**), (**S**), (**N**), (**F**), (**i**), (**t**), (**d**), (**K**), (**j**) and (**P**) are known, then its Variable Portion planned is:

$$v= [S'[1+s]-F-Si-\{[N+K+Sj/360+P]/[1+/\!-U\}$$
$$/\{[1-t][1-d]\}]/S$$

Rule-57709:

If both (**/**), (**U**), (**S'**), (**s**), (**S**), (**v**), (**N**), (**i**), (**t**), (**d**), (**K**), (**j**) and (**P**) are known, then its Fixed Cost planned is:

$$F= S'[1+s]-Sv-Si-\{[N+K+Sj/360+P]/[1+/\!-U\}$$
$$/\{[1-t][1-d]\}$$

Steve Asikin ISBN 13: **978-1564912361**, ISBN 10: **1564912363**

Rule-57710:

If both (**/**), (**U**), (**S'**), (**s**), (**S**), (**v**), (**F**), (**N**), (**t**), (**d**), (**K**), (**j**) and (**P**) are known, then its Interest Portion planned is:

$$i = [S'[1+s]-Sv-F-\{[N+K+Sj/360+P]/[1+I-U\}/\{[1-t][1-d]\}]/S$$

Rule-57711:

If both (**/**), (**U**), (**S'**), (**s**), (**S**), (**v**), (**F**), (**i**), (**N**), (**d**), (**K**), (**j**) and (**P**) are known, then its Tax Rate planned is:

$$t = 1-\{[N+K+Sj/360+P]/[1+I-U\}/(\{S'[1+s]-Sv-F-Si\}[1-d])$$

Rule-57712:

If both (**/**), (**U**), (**S'**), (**s**), (**S**), (**v**), (**F**), (**i**), (**S**), (**v**), (**N**), (**K**), (**j**) and (**P**) are known, then its Dividend Payout planned is:

$$d = 1-\{[N+K+Sj/360+P]/[1+I-U\}/(\{S'[1+s]-Sv-F-Si\}[1-t])$$

Rule-57713:

If both (**/**), (**U**), (**S'**), (**s**), (**S**), (**v**), (**F**), (**i**), (**t**), (**d**), (**N**), (**j**) and (**P**) are known, then its Kind of Cash planned is:

$$K = [1+I(U+\{S'[1+s]-Sv-F-Si\}[1-t][1-d])-[N+Sj/360+P]$$

Steve Asikin ISBN 13: **978-1564912361**, ISBN 10: **1564912363**

Rule-57714:

> If both $(I)$, $(U)$, $(S')$, $(s)$, $(S)$, $(v)$, $(F)$, $(i)$, $(t)$, $(d)$, $(K)$, $(N)$ and $(P)$ are known, then its Job or Trade Receivable Days planned is:
>
> $$j = 360[[1+I(U+\{S'[1+s]-Sv-F-Si\}[1-t][1-d]) -\{N+K+P\}]/S$$

Rule-57715:

> If both $(I)$, $(U)$, $(S')$, $(s)$, $(S)$, $(v)$, $(F)$, $(i)$, $(t)$, $(d)$, $(K)$, $(j)$ and $(N)$ are known, then its Procured Inventory planned is:
>
> $$P = [1+I(U+\{S'[1+s]-Sv-F-Si\}[1-t][1-d]) -\{N+K+Sj/360\}$$

Rule-57716:

> If both $(I)$, $(U)$, $(S')$, $(s)$, $(S)$, $(v)$, $(F)$, $(i)$, $(t)$, $(d)$, $(K)$, $(j)$, $(V)$ and $(p)$ are known, then its Non Current Asset planned is:
>
> $$N = [1+I(U+\{S'[1+s]-Sv-F-Si\}[1-t][1-d]) -K-Sj/360-Vp/360$$

Rule-57717:

> If both $(N)$, $(U)$, $(S')$, $(s)$, $(S)$, $(v)$, $(F)$, $(i)$, $(t)$, $(d)$, $(K)$, $(j)$, $(V)$ and $(p)$ are known, then its Leverage or Gearing Ratio planned is:
>
> $$I = [N+K+Sj/360+Vp/360] /(U+\{S'[1+s]-Sv-F-Si\}[1-t][1-d])-1$$

Steve Asikin ISBN 13: **978-1564912361**, ISBN 10: **1564912363**

Rule-57718:

If both (**∕**), (**N**), (**S'**), (**s**), (**$**), (**v**), (**F**), (**i**), (**t**), (**d**), (**K**), (**j**), (**V**) and (**p**) are known, then its Utilized or Starting Capital must be:

$$U= ([N+K+\$j/360+Vp/360]$$
$$-[1+\textit{∕}]\{S'[1+s]-\$v-F-\$i\}[1-t][1-d])/[1+\textit{∕}]$$

Rule-57719:

If both (**∕**), (**U**), (**N**), (**s**), (**$**), (**v**), (**F**), (**i**), (**t**), (**d**), (**K**), (**j**), (**V**) and (**p**) are known, then its Sales Past must be:

$$S'= (\{[N+K+\$j/360+Vp/360]/[1+\textit{∕}]-U\}$$
$$/\{[1-t][1-d]\}+\$v+F+\$i)/[1+s]$$
or
$$S'= \$[1+s]$$

Rule-57720:

If both (**∕**), (**U**), (**S'**), (**N**), (**$**), (**v**), (**F**), (**i**), (**t**), (**d**), (**K**), (**j**), (**V**) and (**p**) are known, then its Sales Growth planned is:

$$s= (\{[N+K+\$i/360+Vp/360]/[1+\textit{∕}]-U\}$$
$$/\{[1-t][1-d]\}+\$v+F+\$i)/S'-1$$
or
$$s= \$/S'-1$$

Steve Asikin ISBN 13: **978-1564912361**, ISBN 10: **1564912363**

Rule-57721:

If both $(\textbf{/})$, $(\textbf{U})$, $(\textbf{S'})$, $(\textbf{s})$, $(\textbf{N})$, $(\textbf{v})$, $(\textbf{F})$, $(\textbf{i})$, $(\textbf{t})$, $(\textbf{d})$, $(\textbf{K})$, $(\textbf{j})$, $(\textbf{V})$ and $(\textbf{p})$ are known, then its Sales or Revenue planned is:

$$S= [[1+\text{/}(U+\{S'[1+s]-F\}[1-t][1-d])$$
$$-[N+K+Vp/360]]$$
$$/\{j/360+[1+\text{/}][1-t][1-d]\ [v+i]\}$$

or

$$S= S'[1+s]$$

Rule-57722:

If both $(\textbf{/})$, $(\textbf{U})$, $(\textbf{S'})$, $(\textbf{s})$, $(\textbf{S})$, $(\textbf{N})$, $(\textbf{F})$, $(\textbf{i})$, $(\textbf{t})$, $(\textbf{d})$, $(\textbf{K})$, $(\textbf{j})$, $(\textbf{V})$ and $(\textbf{p})$ are known, then its Variable Portion planned is:

$$v= [S'[1+s]-F-Si-\{[N+K+Sj/360+Vp/360]$$
$$/[1+\text{/}-U\}/\{[1-t][1-d]\}]/S$$

or

$$v= V/S$$

Rule-57723:

If both $(\textbf{/})$, $(\textbf{U})$, $(\textbf{S'})$, $(\textbf{s})$, $(\textbf{S})$, $(\textbf{v})$, $(\textbf{N})$, $(\textbf{i})$, $(\textbf{t})$, $(\textbf{d})$, $(\textbf{K})$, $(\textbf{j})$, $(\textbf{V})$ and $(\textbf{p})$ are known, then its Fixed Cost planned is:

$$F= S'[1+s]-Sv-Si-\{[N+K+Sj/360+Vp/360]$$
$$/[1+\text{/}-U\}/\{[1-t][1-d]\}$$

Steve Asikin ISBN 13: **978-1564912361**, ISBN 10: **1564912363**

Rule-57724:

If both (**/**), (**U**), (**S'**), (**s**), (**S**), (**v**), (**F**), (**N**), (**t**), (**d**), (**K**), (**j**), (**V**) and (**p**) are known, then its Interest Portion planned is:

$$i= [S'[1+s]-Sv-F-\{[N+K+Sj/360+Vp/360]$$
$$/[1+/-U\}/\{[1-t][1-d]\}]/S$$

Rule-57725:

If both (**/**), (**U**), (**S'**), (**s**), (**S**), (**v**), (**F**), (**i**), (**N**), (**d**), (**K**), (**j**), (**V**) and (**p**) are known, then its Tax Rate planned is:

$$t= 1-\{[N+K+Sj/360+Vp/360]/[1+/-U\}$$
$$/(\{S'[1+s]-Sv-F-Si\}[1-d])$$

Rule-57726:

If both (**/**), (**U**), (**S'**), (**s**), (**S**), (**v**), (**F**), (**i**), (**t**), (**N**), (**K**), (**j**), (**V**) and (**p**) are known, then its Dividend Payout planned is:

$$d= 1-\{[N+K+Sj/360+Vp/360]/[1+/-U\}$$
$$/(\{S'[1+s]-Sv-F-Si\}[1-t])$$

Rule-57727:

If both (**/**), (**U**), (**S'**), (**s**), (**S**), (**v**), (**F**), (**i**), (**t**), (**d**), (**N**), (**j**), (**V**) and (**p**) are known, then its Kind of Cash planned is:

$$K= [1+/(U+\{S'[1+s]-Sv-F-Si\}[1-t][1-d])$$
$$-[N+Sj/360+Vp/360]$$

Steve Asikin ISBN 13: **978-1564912361**, ISBN 10: **1564912363**

Rule-57728:

If both (**/**), (**U**), (**S'**), (**s**), (**S**), (**v**), (**F**), (**i**), (**t**), (**d**), (**K**), (**N**), (**V**) and (**p**) are known, then its Job or Trade Receivable Days planned is:

$$j= 360[[1+/(U+\{S'[1+s]-Sv-F-Si\}[1-t][1-d]) -[N+K+Vp/360]]/S$$

Rule-57729:

If both (**/**), (**U**), (**S'**), (**s**), (**S**), (**v**), (**F**), (**i**), (**t**), (**d**), (**K**), (**j**), (**N**) and (**p**) are known, then its Variable Cost planned is:

$$V= 360[[1+/(U+\{S'[1+s]-Sv-F-Si\}[1-t][1-d]) -[N+K+Sj/360]]/p$$
$$\text{or}$$
$$V= Sv$$

Rule-57730:

If both (**/**), (**U**), (**S'**), (**s**), (**S**), (**v**), (**F**), (**i**), (**t**), (**d**), (**K**), (**j**), (**V**) and (**N**) are known, then its Procured Inventory Days planned is:

$$p= 360[[1+/(U+\{S'[1+s]-Sv-F-Si\}[1-t][1-d]) -[N+K+Sj/360]]/V$$

Rule-57731:

If both (**/**), (**U**), (**S'**), (**s**), (**S**), (**v**), (**F**), (**i**), (**t**), (**d**), (**K**), (**j**) and (**p**) are known, then its Non Current Asset planned is:

$$N= [1+/(U+\{S'[1+s]-Sv-F-Si\}[1-t][1-d]) -K-Sj/360-Svp/360$$

Steve Asikin ISBN 13: **978-1564912361**, ISBN 10: **1564912363**

<u>Rule-57732</u>:

If both (**N**), (**U**), (**S'**), (**s**), (**S**), (**v**), (**F**), (**i**), (**t**), (**d**), (**K**), (**j**) and (**p**) are known, then its Leverage or Gearing Ratio planned is:

$$I = [\mathbf{N} + \mathbf{K} + \mathbf{Sj}/360 + \mathbf{Svp}/360]$$
$$/(\mathbf{U} + \{\mathbf{S'}[1+\mathbf{s}] - \mathbf{Sv} - \mathbf{F} - \mathbf{Si}\}[1-\mathbf{t}][1-\mathbf{d}]) - 1$$

<u>Rule-57733</u>:

If both (**I**), (**N**), (**S'**), (**s**), (**S**), (**v**), (**F**), (**i**), (**t**), (**d**), (**K**), (**j**) and (**p**) are known, then its Utilized or Starting Capital must be:

$$\mathbf{U} = ([\mathbf{N} + \mathbf{K} + \mathbf{Sj}/360 + \mathbf{Svp}/360]$$
$$- [1+\mathbf{I}]\{\mathbf{S'}[1+\mathbf{s}] - \mathbf{Sv} - \mathbf{F} - \mathbf{Si}\}[1-\mathbf{t}][1-\mathbf{d}])/[1+\mathbf{I}]$$

<u>Rule-57734</u>:

If both (**I**), (**U**), (**N**), (**s**), (**S**), (**v**), (**F**), (**i**), (**t**), (**d**), (**K**), (**j**) and (**p**) are known, then its Sales Past must be:

$$\mathbf{S'} = (\{[\mathbf{N} + \mathbf{K} + \mathbf{Sj}/360 + \mathbf{Svp}/360]/[1+\mathbf{I}] - \mathbf{U}\}$$
$$/\{[1-\mathbf{t}][1-\mathbf{d}]\} + \mathbf{Sv} + \mathbf{F} + \mathbf{Si})/[1+\mathbf{s}]$$

or

$$\mathbf{S'} = \mathbf{S}/[1+\mathbf{s}]$$

Steve Asikin ISBN 13: **978-1564912361**, ISBN 10: **1564912363**

Rule-57735:

If both $(\textit{I})$, $(\textbf{U})$, $(\textbf{S'})$, $(\textbf{N})$, $(\textbf{S})$, $(\textbf{v})$, $(\textbf{F})$, $(\textbf{i})$, $(\textbf{t})$, $(\textbf{d})$, $(\textbf{K})$, $(\textbf{j})$ and $(\textbf{p})$ are known, then its Sales Growth planned is:

$$s = (\{[\textbf{N}+\textbf{K}+\textbf{Sj}/360+\textbf{Svp}/360]/[1+\textit{I}-\textbf{U}\}$$
$$/\{[1-\textbf{t}][1-\textbf{d}]\}+\textbf{Sv}+\textbf{F}+\textbf{Si})/\textbf{S'}-1$$
or
$$s = \textbf{S}/\textbf{S'}-1$$

Rule-57736:

If both $(\textit{I})$, $(\textbf{U})$, $(\textbf{S'})$, $(\textbf{s})$, $(\textbf{N})$, $(\textbf{v})$, $(\textbf{F})$, $(\textbf{i})$, $(\textbf{t})$, $(\textbf{d})$, $(\textbf{K})$, $(\textbf{j})$ and $(\textbf{p})$ are known, then its Sales or Revenue planned is:

$$\textbf{S} = [[1+\textit{I}(\textbf{U}+\{\textbf{S'}[1+\textbf{s}]-\textbf{F}\}[1-\textbf{t}][1-\textbf{d}])-[\textbf{N}+\textbf{K}]]$$
$$/\{\textbf{j}/360+\textbf{vp}/360+[1+\textit{I}[1-\textbf{t}][1-\textbf{d}][\textbf{v}+\textbf{i}]\}$$
or
$$\textbf{S} = \textbf{S'}[1+\textbf{s}]$$

Rule-57737:

If both $(\textit{I})$, $(\textbf{U})$, $(\textbf{S'})$, $(\textbf{s})$, $(\textbf{S})$, $(\textbf{N})$, $(\textbf{F})$, $(\textbf{i})$, $(\textbf{t})$, $(\textbf{d})$, $(\textbf{K})$, $(\textbf{j})$ and $(\textbf{p})$ are known, then its Variable Portion planned is:

$$\textbf{v} = [[1+\textit{I}(\textbf{U}+\{\textbf{S'}[1+\textbf{s}]-\textbf{F}-\textbf{Si}\}[1-\textbf{t}][1-\textbf{d}])$$
$$-[\textbf{N}+\textbf{K}+\textbf{Sj}/360]]$$
$$/(\textbf{S}\{\textbf{p}/360+[1+\textit{I}[1-\textbf{t}][1-\textbf{d}]\})$$

Steve Asikin ISBN 13: **978-1564912361**, ISBN 10: **1564912363**

Rule-57738:

If both $(\textbf{\textit{I}})$, $(\textbf{U})$, $(\textbf{S'})$, $(\textbf{s})$, $(\textbf{S})$, $(\textbf{v})$, $(\textbf{N})$, $(\textbf{i})$, $(\textbf{t})$, $(\textbf{d})$, $(\textbf{K})$, $(\textbf{j})$ and $(\textbf{p})$ are known, then its Fixed Cost planned is:

$$F= S'[1+s]-Sv-Si-\{[N+K+Sj/360+Svp/360]$$
$$/[1+I-U]/\{[1-t][1-d]\}$$

Rule-57739:

If both $(\textbf{\textit{I}})$, $(\textbf{U})$, $(\textbf{S'})$, $(\textbf{s})$, $(\textbf{S})$, $(\textbf{v})$, $(\textbf{F})$, $(\textbf{N})$, $(\textbf{t})$, $(\textbf{d})$, $(\textbf{K})$, $(\textbf{j})$ and $(\textbf{p})$ are known, then its Interest Portion planned is:

$$I= [S'[1+s]-Sv-F-\{[N \mid K \mid Sj/360+Svp/360]$$
$$/[1+I-U]/\{[1-t][1-d]\}]/S$$

Rule-57740:

If both $(\textbf{\textit{I}})$, $(\textbf{U})$, $(\textbf{S'})$, $(\textbf{s})$, $(\textbf{S})$, $(\textbf{v})$, $(\textbf{F})$, $(\textbf{i})$, $(\textbf{N})$, $(\textbf{d})$, $(\textbf{K})$, $(\textbf{j})$ and $(\textbf{p})$ are known, then its Tax Rate planned is:

$$t= 1-\{[N+K+Sj/360+Svp/360]/[1+I-U]$$
$$/(\{S'[1+s]-Sv-F-Si\}[1-d])$$

Rule-57741:

If both $(\textbf{\textit{I}})$, $(\textbf{U})$, $(\textbf{S'})$, $(\textbf{s})$, $(\textbf{S})$, $(\textbf{v})$, $(\textbf{F})$, $(\textbf{i})$, $(\textbf{t})$, $(\textbf{N})$, $(\textbf{K})$, $(\textbf{j})$ and $(\textbf{p})$ are known, then its Dividend Payout planned is:

$$d= 1-\{[N+K+Sj/360+Svp/360]/[1+I-U]$$
$$/(\{S'[1+s]-Sv-F-Si\}[1-t])$$

Steve Asikin ISBN 13: **978-1564912361**, ISBN 10: **1564912363**

Rule-57742:

If both (**/**), (**U**), (**S'**), (**s**), (**S**), (**v**), (**F**), (**i**), (**t**), (**d**), (**N**), (**j**) and (**p**) are known, then its Kind of Cash planned is:

$$K= [1+/](U+\{S'[1+s]-Sv-F-Si\}[1-t][1-d])$$
$$-[N+Sj/360+Svp/360]$$

Rule-57743:

If both (**/**), (**U**), (**S'**), (**s**), (**S**), (**v**), (**F**), (**N**), (**t**), (**d**), (**K**), (**j**) and (**p**) are known, then its Interest Portion planned is:

$$j= 360[[1+/](U+\{S'[1+s]-Sv-F-Si\}[1-t][1-d])$$
$$-\{N+K+Svp/360\}]/S$$

Rule-57744:

If both (**/**), (**U**), (**S'**), (**s**), (**S**), (**v**), (**F**), (**i**), (**t**), (**d**), (**K**), (**j**) and (**N**) are known, then its Procured Inventory Days planned is:

$$p= 360[[1+/](U+\{S'[1+s]-Sv-F-Si\}[1-t][1-d])$$
$$-[N+K+Sj/360]]/[Sv]$$

Rule-57745:

If both (**/**), (**U**), (**S'**), (**s**), (**S**), (**v**), (**F**), (**i**), (**t**), (**d**), (**K**), (**j**) and (**p**) are known, then its Non Current Asset planned is:

$$N= [1+/](U+\{S'[1+s]-Sv-F-Si\}[1-t][1-d])$$
$$-K-Sj/360-S'vp[1+s]/360$$

Steve Asikin ISBN 13: **978-1564912361**, ISBN 10: **1564912363**

<u>Rule-57746</u>:

If both (**N**), (**U**), (**S'**), (**s**), (**S**), (**v**), (**F**), (**i**), (**t**), (**d**), (**K**), (**j**) and (**p**) are known, then its Leverage or Gearing Ratio planned is:

$$I = \{\mathbf{N}+\mathbf{K}+\mathbf{S}j/360+\mathbf{S'vp}[1+\mathbf{s}]/360\}$$
$$/(\mathbf{U}+\{\mathbf{S'}[1+\mathbf{s}]-\mathbf{Sv}-\mathbf{F}-\mathbf{Si}\}[1-\mathbf{t}][1-\mathbf{d}])-1$$

<u>Rule-57747</u>:

If both (**I**), (**N**), (**S'**), (**s**), (**S**), (**v**), (**F**), (**i**), (**t**), (**d**), (**K**), (**j**) and (**p**) are known, then its Utilized or Starting Capital must be:

$$\mathbf{U} = (\{\mathbf{N}+\mathbf{K}+\mathbf{S}j/360+\mathbf{S'vp}[1+\mathbf{s}]/360\}$$
$$-[1+I]\{\mathbf{S'}[1+\mathbf{s}]-\mathbf{Sv}-\mathbf{F}-\mathbf{Si}\}[1-\mathbf{t}][1-\mathbf{d}])/\{[1+I]$$

<u>Rule-57748</u>:

If both (**I**), (**U**), (**N**), (**s**), (**S**), (**v**), (**F**), (**i**), (**t**), (**d**), (**K**), (**j**) and (**p**) are known, then its Sales Past must be:

$$\mathbf{S'} = [[1+I](\mathbf{U}+\{\mathbf{S'}[1+\mathbf{s}]-\mathbf{Sv}-\mathbf{F}-\mathbf{Si}\}[1-\mathbf{t}][1-\mathbf{d}])$$
$$-[\mathbf{N}+\mathbf{K}+\mathbf{S}j/360]]$$
$$/([1+\mathbf{s}]\{\mathbf{vp}/360-[1+I][1-\mathbf{t}][1-\mathbf{d}]\})$$

or

$$\mathbf{S'} = \mathbf{S}/[1+\mathbf{s}]$$

454

Rule-57749:

If both $(I)$, $(U)$, $(S')$, $(N)$, $(S)$, $(v)$, $(F)$, $(i)$, $(t)$, $(d)$, $(K)$, $(j)$ and $(p)$ are known, then its Sales Growth planned is:

$$s= [[1+I(U+\{S'[1+s]-Sv-F-Si\}[1-t][1-d])$$
$$-[N+K+Sj/360]]$$
$$/(S'\{vp/360-[1+I[1-t][1-d]\})-1$$

or

$$s= S/S'-1$$

Rule-57750:

If both $(I)$, $(U)$, $(S')$, $(s)$, $(N)$, $(v)$, $(F)$, $(i)$, $(t)$, $(d)$, $(K)$, $(j)$ and $(p)$ are known, then its Sales or Revenue planned is:

$$S= [[1+I(U+\{S'[1+s]-F\}[1-t][1-d])$$
$$-\{N+K+S'vp[1+s]/360\}]$$
$$/\{j/360+v[1+I[1-t][1-d]\}$$

or

$$S= S'[1+s]$$

Rule-57751:

If both $(I)$, $(U)$, $(S')$, $(s)$, $(S)$, $(N)$, $(F)$, $(i)$, $(t)$, $(d)$, $(K)$, $(j)$ and $(p)$ are known, then its Variable Portion planned is:

$$v= [S'[1+s]-F-Si-(\{N+K+Sj/360+S'vp[1+s]/360\}$$
$$/[1+I-U)/\{[1-t][1-d]\}]/S$$

Steve Asikin ISBN 13: **978-1564912361**, ISBN 10: **1564912363**

Rule-57752:

If both (**/**), (**U**), (**S'**), (**s**), (**S**), (**v**), (**N**), (**i**), (**t**), (**d**), (**K**), (**j**) and (**p**) are known, then its Fixed Cost planned is:

$$F= S'[1+s]-Sv-Si-(\{N+K+Sj/360+S'vp[1+s]/360\}$$
$$/[1+/-U)/\{[1-t][1-d]\}$$

Rule-57753:

If both (**/**), (**U**), (**S'**), (**s**), (**S**), (**v**), (**F**), (**N**), (**t**), (**d**), (**K**), (**j**) and (**p**) are known, then its Interest Portion planned is:

$$i= [S'[1+s]-Sv-F-(\{N+K+Sj/360+S'vp[1+s]/360\}$$
$$/[1+/-U)/\{[1-t][1-d]\}]/S$$

Rule-57754:

If both (**/**), (**U**), (**S'**), (**s**), (**S**), (**v**), (**F**), (**i**), (**N**), (**d**), (**K**), (**j**) and (**p**) are known, then its Tax Rate planned is:

$$t= 1-(\{N+K+Sj/360+S'vp[1+s]/360\}/[1+/-U)$$
$$/(\{S'[1+s]-Sv-F-Si\}[1-d])$$

Rule-57755:

If both (**/**), (**U**), (**S'**), (**s**), (**S**), (**v**), (**F**), (**i**), (**t**), (**N**), (**K**), (**j**) and (**p**) are known, then its Dividend Payout planned is:

$$d= 1-(\{N+K+Sj/360+S'vp[1+s]/360\}/[1+/-U)$$
$$/(\{S'[1+s]-Sv-F-Si\}[1-t])$$

Steve Asikin ISBN 13: **978-1564912361**, ISBN 10: **1564912363**

Rule-57756:

If both ($I$), (**U**), (**S'**), (**s**), (**S**), (**v**), (**F**), (**i**), (**t**), (**d**), (**N**), (**j**) and (**p**) are known, then its Kind of Cash planned is:

$$K= [1+I(U+\{S'[1+s]-Sv-F-Si\}[1-t][1-d])$$
$$-\{N+Sj/360+S'vp[1+s]/360\}$$

Rule-57757:

If both ($I$), (**U**), (**S'**), (**s**), (**S**), (**v**), (**F**), (**i**), (**t**), (**d**), (**K**), (**N**) and (**p**) are known, then its Job or Trade Receivable Days planned is:

$$j= 360[[1+I(U+\{S'[1+s]-Sv-F-Si\}[1-t][1-d])$$
$$-\{N+K+S'vp[1+s]/360\}]/S$$

Rule-57758:

If both ($I$), (**U**), (**S'**), (**s**), (**S**), (**v**), (**F**), (**i**), (**t**), (**d**), (**K**), (**j**) and (**N**) are known, then its Procured Inventory Days planned is:

$$p= 360[[1+I(U+\{S'[1+s]-Sv-F-Si\}[1-t][1-d])$$
$$-[N+K+Sj/360]]/\{S'v[1+s]\}$$

Rule-57759:

If both ($I$), (**U**), (**S'**), (**s**), (**S**), (**v**), (**F**), (**i**), (**t**), (**d**), (**K**), (**j**) and (**P**) are known, then its Non Current Asset planned is:

$$N= [1+I(U+\{S'[1+s]-Sv-F-Si\}[1-t][1-d])$$
$$-K-S'j[1+s]/360-P$$

Steve Asikin ISBN 13: **978-1564912361**, ISBN 10: **1564912363**

Rule-57760:

If both (**N**), (**U**), (**S'**), (**s**), (**S**), (**v**), (**F**), (**i**), (**t**), (**d**), (**K**), (**j**) and (**P**) are known, then its Leverage or Gearing Ratio planned is:

$$I = \{N+K+S'j[1+s]/360+P\}$$
$$/(U+\{S'[1+s]-Sv-F-Si\}[1-t][1-d])-1$$

Rule-57761:

If both (**I**), (**N**), (**S'**), (**s**), (**S**), (**v**), (**F**), (**i**), (**t**), (**d**), (**K**), (**j**) and (**P**) are known, then its Utilized or Starting Capital must be:

$$U = (\{N+K+S'j[1+s]/360+P\}$$
$$-[1+I]\{S'[1+s]-Sv-F-Si\}[1-t][1-d])/[1+I]$$

Rule-57762:

If both (**I**), (**U**), (**N**), (**s**), (**S**), (**v**), (**F**), (**i**), (**t**), (**d**), (**K**), (**j**) and (**P**) are known, then its Sales Past must be:

$$S' = [[1+I](U-\{Sv+F+Si\}[1-t][1-d])-[N+K+P]]$$
$$/([1+s]\{j/360-[1+I][1-t][1-d]\})$$
$$\text{or}$$
$$S' = S/[1+s]$$

Steve Asikin ISBN 13: **978-1564912361**, ISBN 10: **1564912363**

Rule-57763:

If both (**/**), (**U**), (**S'**), (**N**), (**S**), (**v**), (**F**), (**i**), (**t**), (**d**), (**K**), (**j**) and (**P**) are known, then its Sales Growth planned is:

$$s= [[1+\text{/}](\textbf{U}-\{\textbf{Sv}+\textbf{F}+\textbf{Si}\}[1-\textbf{t}][1-\textbf{d}])-[\textbf{N}+\textbf{K}+\textbf{P}]]$$
$$/(\textbf{S'}\{\textbf{j}/360-[1+\text{/}][1-\textbf{t}][1-\textbf{d}]\})-1$$

or

$$s= \textbf{S}/\textbf{S'}-1$$

Rule-57764:

If both (**/**), (**U**), (**S'**), (**s**), (**N**), (**v**), (**F**), (**i**), (**t**), (**d**), (**K**), (**j**) and (**P**) are known, then its Sales or Revenue planned is:

$$\textbf{S}= [\textbf{S'}[1+\textbf{s}]-\textbf{F}-(\{\textbf{N}+\textbf{K}+\textbf{S'j}[1+\textbf{s}]/360+\textbf{P}\}/[1+\text{/}]-\textbf{U})$$
$$/\{[1-\textbf{t}][1-\textbf{d}]\}]/[\textbf{v}+\textbf{i}]$$

or

$$\textbf{S}= \textbf{S'}[1+\textbf{s}]$$

Rule-57765:

If both (**/**), (**U**), (**S'**), (**s**), (**S**), (**N**), (**F**), (**i**), (**t**), (**d**), (**K**), (**j**) and (**P**) are known, then its Variable Portion planned is:

$$\textbf{v}= [\textbf{S'}[1+\textbf{s}]-\textbf{F}-\textbf{Si}-(\{\textbf{N}+\textbf{K}+\textbf{S'j}[1+\textbf{s}]/360+\textbf{P}\}$$
$$/[1+\text{/}]-\textbf{U})/\{[1-\textbf{t}][1-\textbf{d}]\}]/\textbf{S}$$

Steve Asikin ISBN 13: **978-1564912361**, ISBN 10: **1564912363**

Rule-57766:

If both (**I**), (**U**), (**S'**), (**s**), (**S**), (**v**), (**N**), (**i**), (**t**), (**d**), (**K**), (**j**) and (**P**) are known, then its Fixed Cost planned is:

$$F= S'[1+s]-Sv-Si-(\{N+K+S'j[1+s]/360+P\}$$
$$/[1+I-U]/\{[1-t][1-d]\}$$

Rule-57767:

If both (**I**), (**U**), (**S'**), (**s**), (**S**), (**v**), (**F**), (**N**), (**t**), (**d**), (**K**), (**j**) and (**P**) are known, then its Interest Portion planned is:

$$i= [S'[1+s] \; Sv \; F \; (\{N+K+S'j[1+s]/360+P\}$$
$$/[1+I-U]/\{[1-t][1-d]\}]/S$$

Rule-57768:

If both (**I**), (**U**), (**S'**), (**s**), (**S**), (**v**), (**F**), (**i**), (**N**), (**d**), (**K**), (**j**) and (**P**) are known, then its Tax Rate planned is:

$$t= 1-(\{N+K+S'j[1+s]/360+P\}/[1+I-U])$$
$$/(\{S'[1+s]-Sv-F-Si\}[1-d])$$

Rule-57769:

If both (**I**), (**U**), (**S'**), (**s**), (**S**), (**v**), (**F**), (**i**), (**t**), (**N**), (**K**), (**j**) and (**P**) are known, then its Dividend Payout planned is:

$$d= 1-(\{N+K+S'j[1+s]/360+P\}/[1+I-U])$$
$$/(\{S'[1+s]-Sv-F-Si\}[1-t])$$

`

Steve Asikin ISBN 13: **978-1564912361**, ISBN 10: **1564912363**

Rule-57770:

If both (**/**), (**U**), (**S'**), (**s**), (**S**), (**v**), (**F**), (**i**), (**t**), (**d**), (**N**), (**j**) and (**P**) are known, then its Kind of Cash planned is:

$$K= [1+/(U+\{S'[1+s]-Sv-F-Si\}[1-t][1-d])$$
$$-\{N+S'j[1+s]/360+P\}$$

Rule-57771:

If both (**/**), (**U**), (**S'**), (**s**), (**S**), (**v**), (**F**), (**i**), (**t**), (**d**), (**K**), (**N**) and (**P**) are known, then its Job or Trade Receivable Days planned is:

$$j= 360[[1+/(U+\{S'[1+s]-Sv-F-Si\}[1-t][1-d])$$
$$-\{N+K+P\}]/\{S'[1+s]\}$$

Rule-57772:

If both (**/**), (**U**), (**S'**), (**s**), (**S**), (**v**), (**F**), (**i**), (**t**), (**d**), (**K**), (**j**) and (**N**) are known, then its Procured Inventory planned is:

$$P= [1+/(U+\{S'[1+s]-Sv-F-Si\}[1-t][1-d])$$
$$-\{N+K+S'j[1+s]/360\}$$

Rule-57773:

If both (**/**), (**U**), (**S'**), (**s**), (**S**), (**v**), (**F**), (**i**), (**t**), (**d**), (**K**), (**j**), (**V**) and (**p**) are known, then its Non Current Asset planned is:

$$N= [1+/(U+\{S'[1+s]-Sv-F-Si\}[1-t][1-d])$$
$$-K-S'j[1+s]/360-Vp/360$$

Steve Asikin ISBN 13: **978-1564912361**, ISBN 10: **1564912363**

Rule-57774:

If both (**N**), (**U**), (**S'**), (**s**), (**S**), (**v**), (**F**), (**i**), (**t**), (**d**), (**K**), (**j**), (**V**) and (**p**) are known, then its Leverage or Gearing Ratio planned is:

$$I= \{N+K+S'j[1+s]/360+Vp/360\}$$
$$/(U+\{S'[1+s]-Sv-F-Si\}[1-t][1-d])-1$$

Rule-57775:

If both (**I**), (**N**), (**S'**), (**s**), (**S**), (**v**), (**F**), (**i**), (**t**), (**d**), (**K**), (**j**), (**V**) and (**p**) are known, then its Utilized or Starting Capital must be:

$$U= (\{N+K+S'j[1+s]/360+Vp/360\}$$
$$-[1+I]\{S'[1+s]-Sv-F-Si\}[1-t][1-d])/[1+I]$$

Rule-57776:

If both (**I**), (**U**), (**N**), (**s**), (**S**), (**v**), (**F**), (**i**), (**t**), (**d**), (**K**), (**j**), (**V**) and (**p**) are known, then its Sales Past must be:

$$S'= [[1+I](U-\{Sv+F+Si\}[1-t][1-d])$$
$$-[N+K+Vp/360]]$$
$$/([1+s]\{j/360-[1+I][1-t][1-d]\})$$

or

$$S'= S/[1+s]$$

Steve Asikin ISBN 13: **978-1564912361**, ISBN 10: **1564912363**

Rule-57777:

If both (**/**), (**U**), (**S'**), (**N**), (**$**), (**v**), (**F**), (**i**), (**t**), (**d**), (**K**), (**j**), (**V**) and (**p**) are known, then its Sales Growth planned is:

$$s= [[1+\text{/}(U-\{Sv+F+Si\}[1-t][1-d])$$
$$-[N+K+Vp/360]]$$
$$/(S'\{j/360-[1+\text{/}[1-t][1-d]\})-1$$

or

$$s= S/S'-1$$

Rule-57778:

If both (**/**), (**U**), (**S'**), (**s**), (**N**), (**v**), (**F**), (**i**), (**t**), (**d**), (**K**), (**j**), (**V**) and (**p**) are known, then its Sales or Revenue planned is:

$$S= [S'[1+s]-F-(\{N+K+S'j[1+s]/360+Vp/360\}$$
$$/[1+\text{/}-U)/\{[1-t][1-d]\}]/[v+i]$$

or

$$S= S'[1+s]$$

or

$$S= V/v$$

Rule-57779:

If both (**/**), (**U**), (**S'**), (**s**), (**$**), (**N**), (**F**), (**i**), (**t**), (**d**), (**K**), (**j**), (**V**) and (**p**) are known, then its Variable Portion planned is:

$$v= [S'[1+s]-F-Si-(\{N+K+S'j[1+s]/360+Vp/360\}$$
$$/[1+\text{/}-U)/\{[1-t][1-d]\}]/S$$

or

$$v= V/S$$

Steve Asikin ISBN 13: **978-1564912361**, ISBN 10: **1564912363**

Rule-57780:

If both (**/**), (**U**), (**S'**), (**s**), (**S**), (**v**), (**N**), (**i**), (**t**), (**d**), (**K**), (**j**), (**V**) and (**p**) are known, then its Fixed Cost planned is:

$$\mathbf{F} = \mathbf{S'}[1+\mathbf{s}]-\mathbf{Sv}-\mathbf{Si}-(\{\mathbf{N}+\mathbf{K}+\mathbf{S'j}[1+\mathbf{s}]/360+\mathbf{Vp}/360\}$$
$$/[1+\mathbf{/}-\mathbf{U})/\{[1-\mathbf{t}][1-\mathbf{d}]\}$$

Rule-57781:

If both (**/**), (**U**), (**S'**), (**s**), (**S**), (**v**), (**F**), (**N**), (**t**), (**d**), (**K**), (**j**), (**V**) and (**p**) are known, then its Interest Portion planned is:

$$\mathbf{i} = [\mathbf{S'}[1+\mathbf{s}]-\mathbf{Sv}-\mathbf{F}-(\{\mathbf{N}+\mathbf{K}+\mathbf{S'j}[1+\mathbf{s}]/360+\mathbf{Vp}/360\}$$
$$/[1+\mathbf{/}-\mathbf{U})/\{[1-\mathbf{t}][1-\mathbf{d}]\}]/\mathbf{S}$$

Rule-57782:

If both (**/**), (**U**), (**S'**), (**s**), (**S**), (**v**), (**F**), (**i**), (**N**), (**d**), (**K**), (**j**), (**V**) and (**p**) are known, then its Tax Rate planned is:

$$\mathbf{t} = 1-(\{\mathbf{N}+\mathbf{K}+\mathbf{S'j}[1+\mathbf{s}]/360+\mathbf{Vp}/360\}/[1+\mathbf{/}-\mathbf{U})$$
$$/(\{\mathbf{S'}[1+\mathbf{s}]-\mathbf{Sv}-\mathbf{F}-\mathbf{Si}\}[1-\mathbf{d}])$$

Rule-57783:

If both (**/**), (**U**), (**S'**), (**s**), (**S**), (**v**), (**F**), (**i**), (**t**), (**N**), (**K**), (**j**), (**V**) and (**p**) are known, then its Dividend Payout planned is:

$$\mathbf{d} = 1-(\{\mathbf{N}+\mathbf{K}+\mathbf{S'j}[1+\mathbf{s}]/360+\mathbf{Vp}/360\}/[1+\mathbf{/}-\mathbf{U})$$
$$/(\{\mathbf{S'}[1+\mathbf{s}]-\mathbf{Sv}-\mathbf{F}-\mathbf{Si}\}[1-\mathbf{t}])$$

Steve Asikin ISBN 13: **978-1564912361**, ISBN 10: **1564912363**

Rule-57784:

If both (**/**), (**U**), (**S'**), (**s**), (**S**), (**v**), (**F**), (**i**), (**t**), (**d**), (**N**), (**j**), (**V**) and (**p**) are known, then its Kind of Cash planned is:

$$K= [1+/](U+\{S'[1+s]-Sv-F-Si\}[1-t][1-d])$$
$$-\{N+S'j[1+s]/360+Vp/360\}$$

Rule-57785:

If both (**/**), (**U**), (**S'**), (**s**), (**S**), (**v**), (**F**), (**i**), (**t**), (**d**), (**K**), (**N**), (**V**) and (**p**) are known, then its Job or Trade Receivable Days planned is:

$$j= 360[[1+/](U+\{S'[1+s]-Sv-F-Si\}[1-t][1-d])$$
$$-\{N+K+Vp/360\}]/\{S'[1+s]\}$$

Rule-57786:

If both (**/**), (**U**), (**S'**), (**s**), (**S**), (**v**), (**F**), (**i**), (**t**), (**d**), (**K**), (**j**), (**N**) and (**p**) are known, then its Variable Cost planned is:

$$V= 360[[1+/](U+\{S'[1+s]-Sv-F-Si\}[1-t][1-d])$$
$$-\{N+K+S'j[1+s]/360\}]/p$$
or
$$V= Sv$$

Rule-57787:

If both (**/**), (**U**), (**S'**), (**s**), (**S**), (**v**), (**F**), (**i**), (**t**), (**d**), (**K**), (**j**), (**V**) and (**N**) are known, then its Procured Inventory Days planned is:

$$p= 360[[1+/](U+\{S'[1+s]-Sv-F-Si\}[1-t][1-d])$$
$$-\{N+K+S'j[1+s]/360\}]/V$$

Steve Asikin ISBN 13: **978-1564912361**, ISBN 10: **1564912363**

Rule-57788:

If both (**/**), (**U**), (**S'**), (**s**), (**S**), (**v**), (**F**), (**i**), (**t**), (**d**), (**K**), (**j**) and (**p**) are known, then its Non Current Asset planned is:

$$\mathbf{N}= [1+\mathbf{/}](\mathbf{U}+\{\mathbf{S'}[1+\mathbf{s}]-\mathbf{Sv}-\mathbf{F}-\mathbf{Si}\}[1-\mathbf{t}][1-\mathbf{d}])$$
$$-\mathbf{K}-\mathbf{S'j}[1+\mathbf{s}]/360-\mathbf{Svp}/360$$

Rule-57789:

If both (**N**), (**U**), (**S'**), (**s**), (**S**), (**v**), (**F**), (**i**), (**t**), (**d**), (**K**), (**j**) and (**p**) are known, then its Leverage or Gearing Ratio planned is:

$$\mathbf{/}= \{\mathbf{N}+\mathbf{K}+\mathbf{S'j}[1+\mathbf{s}]/360+\mathbf{Svp}/360\}$$
$$/(\mathbf{U}+\{\mathbf{S'}[1+\mathbf{s}]-\mathbf{Sv}-\mathbf{F}-\mathbf{Si}\}[1-\mathbf{t}][1-\mathbf{d}])-1$$

Rule-57790:

If both (**/**), (**N**), (**S'**), (**s**), (**S**), (**v**), (**F**), (**i**), (**t**), (**d**), (**K**), (**j**) and (**p**) are known, then its Utilized or Starting Capital must be:

$$\mathbf{U}= (\{\mathbf{N}+\mathbf{K}+\mathbf{S'j}[1+\mathbf{s}]/360+\mathbf{Svp}/360\}-[1+\mathbf{/}]$$
$$\{\mathbf{S'}[1+\mathbf{s}]-\mathbf{Sv}-\mathbf{F}-\mathbf{Si}\}[1-\mathbf{t}][1-\mathbf{d}])/\{[1+\mathbf{/}]\}$$

Rule-57791:

If both (**/**), (**U**), (**N**), (**s**), (**S**), (**v**), (**F**), (**i**), (**t**), (**d**), (**K**), (**j**) and (**p**) are known, then its Sales Past must be:

$$\mathbf{S'}= [[1+\mathbf{/}](\mathbf{U}-\{\mathbf{Sv}+\mathbf{F}+\mathbf{Si}\{[1-\mathbf{t}][1-\mathbf{d}])$$
$$-\{\mathbf{N}+\mathbf{K}+ \mathbf{Svp}/360\}]$$
$$/([1+\mathbf{s}]\{\mathbf{j}/360-[1+\mathbf{/}][1-\mathbf{t}][1-\mathbf{d}]\})$$

or

$$\mathbf{S'}= \mathbf{S}/[1+\mathbf{s}]$$

466

Steve Asikin ISBN 13: **978-1564912361**, ISBN 10: **1564912363**

Rule-57792:

If both (**/**), (**U**), (**S'**), (**s**), (**S**), (**v**), (**F**), (**i**), (**t**), (**d**), (**K**), (**j**) and (**p**) are known, then its Non Current Asset planned is:

$$s= [[1+/(U-\{Sv+F+Si\}[1-t][1-d])$$
$$-\{N+K+Svp/360\}]$$
$$/(S'\{j/360-[1+/[1-t]\})-1$$

or

$$s= S/S'-1$$

Rule-57793:

If both (**/**), (**U**), (**S'**), (**s**), (**N**), (**v**), (**F**), (**i**), (**t**), (**d**), (**K**), (**j**) and (**p**) are known, then its Sales or Revenue planned is:

$$S= [[1+/(U+\{S'[1+s]-F\}[1-t][1-d])$$
$$-\{N+K+S'j[1+s]/360\}]$$
$$/\{vp/360+[1+/[1-t][1-d]\,[v+i]\}$$

or

$$S= S'[1+s]$$

Rule-57794:

If both (**/**), (**U**), (**S'**), (**s**), (**S**), (**N**), (**F**), (**i**), (**t**), (**d**), (**K**), (**j**) and (**p**) are known, then its Variable Portion planned is:

$$v= [[1+/(U+\{S'[1+s]-F-Si\}[1-t][1-d])$$
$$-\{N+K+S'j[1+s]/360\}]$$
$$/\{Sp/360+S[1+/[1-t][1-d]\}$$

Steve Asikin ISBN 13: **978-1564912361**, ISBN 10: **1564912363**

<u>Rule-57795</u>:

If both (**/**), (**U**), (**S'**), (**s**), (**S**), (**v**), (**N**), (**i**), (**t**), (**d**), (**K**), (**j**) and (**p**) are known, then its Fixed Cost planned is:

$$F= S'[1+s]-Sv-Si-(\{N+K+S'j[1+s]/360+Svp/360\}$$
$$/[1+/-U)/\{[1-t][1-d]\}$$

<u>Rule-57796</u>:

If both (**/**), (**U**), (**S'**), (**s**), (**S**), (**v**), (**F**), (**N**), (**t**), (**d**), (**K**), (**j**) and (**p**) are known, then its Interest Portion planned is:

$$i= [S'[1+s]-Sv-F-(\{N+K+S'j[1+s]/360+Svp/360\}$$
$$/[1+/-U)/\{[1-t][1-d]\}]/S$$

<u>Rule-57797</u>:

If both (**/**), (**U**), (**S'**), (**s**), (**S**), (**v**), (**F**), (**i**), (**N**), (**d**), (**K**), (**j**) and (**p**) are known, then its Tax Rate planned is:

$$t= 1-(\{N+K+S'j[1+s]/360+Svp/360\}/[1+/-U)$$
$$/(\{S'[1+s]-Sv-F-Si\}[1-d])$$

<u>Rule-57798</u>:

If both (**/**), (**U**), (**S'**), (**s**), (**S**), (**v**), (**F**), (**i**), (**t**), (**N**), (**K**), (**j**) and (**p**) are known, then its Dividend Payout planned is:

$$d= 1-(\{N+K+S'j[1+s]/360+Svp/360\}/[1+/-U)$$
$$/(\{S'[1+s]-Sv-F-Si\}[1-t])$$

Steve Asikin ISBN 13: **978-1564912361**, ISBN 10: **1564912363**

Rule-57799:

If both (**/**), (**U**), (**$'**), (**s**), (**$**), (**v**), (**F**), (**i**), (**t**), (**d**), (**N**), (**j**) and (**p**) are known, then its Kind of Cash planned is:

$$K= [1+/(U+\{\$'[1+s]-\$v-F-\$i\}[1-t][1-d])$$
$$-\{N+\$'j[1+s]/360+\$vp/360\}$$

Rule-57800:

If both (**/**), (**U**), (**$'**), (**s**), (**$**), (**v**), (**F**), (**i**), (**t**), (**d**), (**K**), (**N**) and (**p**) are known, then its Job or Trade Receivable Days planned is:

$$j= 360[[1+/(U+\{\$'[1+s]-\$v-F-\$i\}[1-t][1-d])$$
$$-\{N+K+\$vp/360\}]/\{\$'[1+s]\}$$

Rule-57801:

If both (**/**), (**U**), (**$'**), (**s**), (**$**), (**v**), (**F**), (**i**), (**t**), (**d**), (**K**), (**j**) and (**N**) are known, then its Procured Inventory Days planned is:

$$p= 360[[1+/(U+\{\$'[1+s]-\$v-F-\$i\}[1-t][1-d])$$
$$-\{N+K+\$'j[1+s]/360\}]/[\$v]$$

Rule-57802:

If both (**/**), (**U**), (**$'**), (**s**), (**$**), (**v**), (**F**), (**i**), (**t**), (**d**), (**K**), (**j**) and (**p**) are known, then its Non Current Asset planned is:

$$N= [1+/(U+\{\$'[1+s]-\$v-F-\$i\}[1-t][1-d])$$
$$-K-\$'j[1+s]/360-\$'vp[1+s]/360$$

Steve Asikin ISBN 13: **978-1564912361**, ISBN 10: **1564912363**

Rule-57803:

If both (**N**), (**U**), (**S'**), (**s**), (**S**), (**v**), (**F**), (**i**), (**t**), (**d**), (**K**), (**j**) and (**p**) are known, then its Leverage or Gearing Ratio planned is:

$$/ = \{\mathbf{N+K+S'j}[1+\mathbf{s}]/360+\mathbf{S'vp}[1+\mathbf{s}]/360\}$$
$$/(\mathbf{U}+\{\mathbf{S'}[1+\mathbf{s}]-\mathbf{Sv-F-Si}\}[1-\mathbf{t}][1-\mathbf{d}])-1$$

Rule-57804:

If both (**/**), (**N**), (**S'**), (**s**), (**S**), (**v**), (**F**), (**i**), (**t**), (**d**), (**K**), (**j**) and (**p**) are known, then its Utilized or Starting Capital must be:

$$\mathbf{U} = (\{\mathbf{N+K+S'j}[1+\mathbf{s}]/360+\mathbf{S'vp}[1+\mathbf{s}]/360\}-[1+\mathbf{/}]$$
$$\{\mathbf{S'}[1+\mathbf{s}]-\mathbf{Sv-F-Si}\}[1-\mathbf{t}][1-\mathbf{d}])/[1+\mathbf{/}]$$

Rule-57805:

If both (**/**), (**U**), (**N**), (**s**), (**S**), (**v**), (**F**), (**i**), (**t**), (**d**), (**K**), (**j**) and (**p**) are known, then its Sales Past must be:

$$\mathbf{S'} = [[1+\mathbf{/}](\mathbf{U}-\{\mathbf{Sv+F+Si}\}[1-\mathbf{t}][1-\mathbf{d}])-[\mathbf{N+K}]]$$
$$/(\mathbf{S'}[1+\mathbf{s}]\{\mathbf{j}/360+\mathbf{vp}/360-[1+\mathbf{/}][1-\mathbf{t}][1-\mathbf{d}]\})$$
$$\text{or}$$
$$\mathbf{S'} = \mathbf{S}/[1+\mathbf{s}]$$

Steve Asikin ISBN 13: **978-1564912361**, ISBN 10: **1564912363**

Rule-57806:

If both (**/**), (**U**), (**S'**), (**N**), (**S**), (**v**), (**F**), (**i**), (**t**), (**d**), (**K**), (**j**) and (**p**) are known, then its Sales Growth planned is:

$$s= [[1+/](U-\{Sv+F+Si\}[1-t][1-d])-[N+K]]$$
$$/(S'\{j/360+vp/360-[1+/][1-t][1-d]\})-1$$

or

$$s= S/S'-1$$


Rule-57807:

If both (**/**), (**U**), (**S'**), (**s**), (**N**), (**v**), (**F**), (**i**), (**t**), (**d**), (**K**), (**j**) and (**p**) are known, then its Sales or Revenue planned is:

$$S= [S'[1+s]-F-(\{N+K+S'j[1+s]/360$$
$$+S'vp[1+s]/360\}/[1+/]-U)$$
$$/\{[1-t][1-d]\}]/[v+i]$$

or

$$S= S'[1+s]$$


Rule-57808:

If both (**/**), (**U**), (**S'**), (**s**), (**S**), (**N**), (**F**), (**i**), (**t**), (**d**), (**K**), (**j**) and (**p**) are known, then its Variable Portion planned is:

$$v= [[1+/](U+\{S'[1+s]-Sv-F-Si\}[1-t][1-d])$$
$$-\{N+S'j[1+s]/360+K\}]$$
$$/\{S'p[1+s]/360\}+S[1+/][1-t][1-d]\}$$

Steve Asikin ISBN 13: **978-1564912361**, ISBN 10: **1564912363**

Rule-57809:

If both $(\textbf{\textit{I}})$, $(\textbf{U})$, $(\textbf{S'})$, $(\textbf{s})$, $(\textbf{S})$, $(\textbf{v})$, $(\textbf{N})$, $(\textbf{i})$, $(\textbf{t})$, $(\textbf{d})$, $(\textbf{K})$, $(\textbf{j})$ and $(\textbf{p})$ are known, then its Fixed Portion planned is:

$$F = S'[1+s]-Sv-Si-(\{N+K+S'j[1+s]/360 \\ +S'vp[1+s]/360\}/[1+I-U)/\{[1-t][1-d]\}$$

Rule-57810:

If both $(\textbf{\textit{I}})$, $(\textbf{U})$, $(\textbf{S'})$, $(\textbf{s})$, $(\textbf{S})$, $(\textbf{v})$, $(\textbf{F})$, $(\textbf{N})$, $(\textbf{t})$, $(\textbf{d})$, $(\textbf{K})$, $(\textbf{j})$ and $(\textbf{p})$ are known, then its Interest Portion planned is:

$$i = [S'[1+s]-Sv-F-(\{N+K+S'j[1+s]/360 \\ +S'vp[1+s]/360\}/[1+I-U) \\ /\{[1-t][1-d]\}]/S$$

Rule-57811:

If both $(\textbf{\textit{I}})$, $(\textbf{U})$, $(\textbf{S'})$, $(\textbf{s})$, $(\textbf{S})$, $(\textbf{v})$, $(\textbf{F})$, $(\textbf{i})$, $(\textbf{N})$, $(\textbf{d})$, $(\textbf{K})$, $(\textbf{j})$ and $(\textbf{p})$ are known, then its Tax Rate planned is:

$$t = 1-(\{N+K+S'j[1+s]/360+S'vp[1+s]/360\} \\ /[1+I-U)/(\{S'[1+s]-Sv-F-Si\}[1-d])$$

Rule-57812:

If both $(\textbf{\textit{I}})$, $(\textbf{U})$, $(\textbf{S'})$, $(\textbf{s})$, $(\textbf{S})$, $(\textbf{v})$, $(\textbf{F})$, $(\textbf{i})$, $(\textbf{t})$, $(\textbf{N})$, $(\textbf{K})$, $(\textbf{j})$ and $(\textbf{p})$ are known, then its Dividend Payout planned is:

$$d = 1-(\{N+K+S'j[1+s]/360+S'vp[1+s]/360\} \\ /[1+I-U)/(\{S'[1+s]-Sv-F-Si\}[1-t])$$

Steve Asikin ISBN 13: **978-1564912361**, ISBN 10: **1564912363**

Rule-57813:

If both (**/**), (**U**), (**S'**), (**s**), (**$**), (**v**), (**F**), (**i**), (**t**), (**d**), (**N**), (**j**) and (**p**) are known, then its Kind of Cash planned is:

$$K= [1+\text{/}(U+\{S'[1+s]-Sv-F-Si\}[1-t][1-d])$$
$$-\{N+S'j[1+s]/360+S'vp[1+s]/360\}$$

Rule-57814:

If both (**/**), (**U**), (**S'**), (**s**), (**$**), (**v**), (**F**), (**i**), (**t**), (**d**), (**K**), (**N**) and (**p**) are known, then its Job or Trade Receivable Days planned is:

$$j= 360[[1+\text{/}(U+\{S'[1+s]-Sv-F-Si\}[1-t][1-d])$$
$$-\{N+K+S'vp[1+s]/360\}]/\{S'[1+s]\}$$

Rule-57815:

If both (**/**), (**U**), (**S'**), (**s**), (**$**), (**v**), (**F**), (**i**), (**t**), (**d**), (**K**), (**j**) and (**N**) are known, then its Procured Inventory Days planned is:

$$p= 360[[1+\text{/}(U+\{S'[1+s]-Sv-F-Si\}[1-t][1-d])$$
$$-\{N+K+S'j[1+s]/360\}]/\{S'v[1+s]\}$$

Rule-57816:

If both (**/**), (**U**), (**S'**), (**s**), (**$**), (**v**), (**F**), (**i**), (**T**), (**D**), (**K**), (**J**) and (**P**) are known, then its Non Current Asset planned is:

$$N= [1+\text{/}\{U+S'[1+s]-Sv-F-S'i[1+s]-T-D\}-K-J-P$$

Steve Asikin ISBN 13: **978-1564912361**, ISBN 10: **1564912363**

Rule-57817:

If both (**N**), (**U**), (**S'**), (**s**), (**S**), (**v**), (**F**), (**i**), (**T**), (**D**), (**K**), (**J**) and (**P**) are known, then its Leverage or Gearing Ratio planned is:

$$I = \{N+K+J+P\}$$
$$/(U+\{S'[1+s]-Sv-F-S'i[1+s]\}-T-D)-1$$

Rule-57818:

If both (**I**), (**N**), (**S'**), (**s**), (**S**), (**v**), (**F**), (**i**), (**T**), (**D**), (**K**), (**J**) and (**P**) are known, then its Utilized or starting Capital must be:

$$U = (\{N+K+J+P\}-[1+I]\{S'[1+s]-Sv-F-S'i[1+s]\}$$
$$-T-D)/[1+I]$$

Rule-57819:

If both (**I**), (**U**), (**N**), (**s**), (**S**), (**v**), (**F**), (**i**), (**T**), (**D**), (**K**), (**J**) and (**P**) are known, then its Sales Past must be:

$$S' = \{[N+K+J+P]/[1+I]-U+T+D\}+Sv+F)$$
$$/\{[1+s][1-i]\}$$

or

$$S' = S/[1+s]$$

Steve Asikin ISBN 13: **978-1564912361**, ISBN 10: **1564912363**

Rule-57820:

If both ($I$), (**U**), (**S'**), (**N**), (**S**), (**v**), (**F**), (**i**), (**T**), (**D**), (**K**), (**J**) and (**P**) are known, then its Sales Growth planned is:

$s= \{[\mathbf{N+K+J+P}]/[1+I\text{-}\mathbf{U+T+D}\}+\mathbf{Sv+F})$
$/\{\mathbf{S'}[1\text{-}\mathbf{i}]\}\text{-}1$

or

$s= \mathbf{S/S'}\text{-}1$


Rule-57821:

If both ($I$), (**U**), (**S'**), (**s**), (**N**), (**v**), (**F**), (**i**), (**T**), (**D**), (**K**), (**J**) and (**P**) are known, then its Sales or Revenue planned is:

$\mathbf{S}= (\{\mathbf{S'}[1+\mathbf{s}][1\text{-}\mathbf{i}]\}\text{-}\mathbf{F}\text{-}\{[\mathbf{N+K+J+P}]/[1+I$
$\text{-}\mathbf{U+T+D}\})/\mathbf{v}$

or

$\mathbf{S}= \mathbf{S'}[1+\mathbf{s}]$


Rule-57822:

If both ($I$), (**U**), (**S'**), (**s**), (**S**), (**N**), (**F**), (**i**), (**T**), (**D**), (**K**), (**J**) and (**P**) are known, then its Variable Portion planned is:

$\mathbf{v}= (\{\mathbf{S'}[1+\mathbf{s}][1\text{-}\mathbf{i}]\}\text{-}\mathbf{F}\text{-}\{[\mathbf{N+K+J+P}]/[1+I$
$\text{-}\mathbf{U+T+D}\})/\mathbf{S}$


Rule-57823:

If both ($I$), (**U**), (**S'**), (**s**), (**S**), (**v**), (**N**), (**i**), (**T**), (**D**), (**K**), (**J**) and (**P**) are known, then its Fixed Cost planned is:

$\mathbf{F}= \{\mathbf{S'}[1+\mathbf{s}][1\text{-}\mathbf{i}]\}\text{-}\mathbf{Sv}\text{-}\{[\mathbf{N+K+J+P}]/[1+I\text{-}\mathbf{U+T+D}\}$

Steve Asikin ISBN 13: **978-1564912361**, ISBN 10: **1564912363**

Rule-57824:

If both (**/**), (**U**), (**S'**), (**s**), (**S**), (**v**), (**F**), (**N**), (**T**), (**D**), (**K**), (**J**) and (**P**) are known, then its Interest Portion planned is:

$$i= 1-\{[N+K+J+P]/[1+/]-U +T+D\}+Sv+F)/\{S'[1+s]\}$$

Rule-57825:

If both (**/**), (**U**), (**S'**), (**s**), (**S**), (**v**), (**F**), (**i**), (**N**), (**D**), (**K**), (**J**) and (**P**) are known, then its Tax planned is:

$$T= U+\{S'[1+s]-Sv-F-S'i[1+s]\}-D -[N+K+J+P]/[1+/]$$

Rule-57826:

If both (**/**), (**U**), (**S'**), (**s**), (**S**), (**v**), (**F**), (**i**), (**T**), (**N**), (**K**), (**J**) and (**P**) are known, then its Dividend planned is:

$$D= U+\{S'[1+s]-Sv-F-S'i[1+s]\}-T -[N+K+J+P]/[1+/]$$

Rule-57827:

If both (**/**), (**U**), (**S'**), (**s**), (**S**), (**v**), (**F**), (**i**), (**T**), (**D**), (**N**), (**J**) and (**P**) are known, then its Kind of Cash planned is:

$$K= [1+/](U+\{S'[1+s]-Sv-F-S'i[1+s]\}-T-D) -[N+J+P]$$

Steve Asikin ISBN 13: **978-1564912361**, ISBN 10: **1564912363**

Rule-57828:

If both (**/**), (**U**), (**$'**), (**s**), (**$**), (**v**), (**F**), (**i**), (**T**), (**D**), (**K**), (**N**) and (**P**) are known, then its Job or Trade Account Receivable planned is:

$$J = [1+/](U+\{\$'[1+s]-\$v-F-\$'i[1+s]\}-T-D) -[N+K+P]$$

Rule-57829:

If both (**/**), (**U**), (**$'**), (**s**), (**$**), (**v**), (**F**), (**i**), (**T**), (**D**), (**K**), (**J**) and (**N**) are known, then its Procured Inventory planned is:

$$P = [1+/](U+\{\$'[1+s]-\$v-F-\$'i[1+s]\}-T-D) -[N+K+J]$$

Rule-57830:

If both (**/**), (**U**), (**$'**), (**s**), (**$**), (**v**), (**F**), (**i**), (**T**), (**D**), (**K**), (**J**), (**V**) and (**p**) are known, then its Non Current Asset planned is:

$$N = [1+/]\{U+\$'[1+s]-\$v-F-\$'i[1+s]-T-D\} -K-J-Vp/360$$

Rule-57831:

If both (**N**), (**U**), (**$'**), (**s**), (**$**), (**v**), (**F**), (**i**), (**T**), (**D**), (**K**), (**J**), (**V**) and (**p**) are known, then its Leverage or Gearing Ratio planned is:

$$/ = [N+K+J+Vp/360] /(U+\{\$'[1+s]-\$v-F-\$'i[1+s]\}-T-D)-1$$

Steve Asikin ISBN 13: **978-1564912361**, ISBN 10: **1564912363**

Rule-57832:

If both (**/**), (**N**), (**S'**), (**s**), (**S**), (**v**), (**F**), (**i**), (**T**), (**D**), (**K**), (**J**), (**V**) and (**p**) are known, then its Utilized or Starting Capital must be:

$$U= ([N+K+J+Vp/360]-[1+/]\{S'[1+s]-Sv-F$$
$$-S'i[1+s]\}-T-D)/[1+/]$$

Rule-57833:

If both (**/**), (**U**), (**N**), (**s**), (**S**), (**v**), (**F**), (**i**), (**T**), (**D**), (**K**), (**J**), (**V**) and (**p**) are known, then its Sales Past must be:

$$S'= (\{[N+K+J+Vp/360]/[1+/]-U+T+D\}+Sv-F)$$
$$/\{[1+s][1-i]\}$$
or
$$S'= S/[1+s]$$

Rule-57834:

If both (**/**), (**U**), (**S'**), (**N**), (**S**), (**v**), (**F**), (**i**), (**T**), (**D**), (**K**), (**J**), (**V**) and (**p**) are known, then its Sales Growth planned is:

$$s= (\{[N+K+J+Vp/360]/[1+/]-U+T+D\}+Sv-F)$$
$$/\{S'[1-i]\}-1$$
or
$$s= S/S'-1$$

Steve Asikin ISBN 13: **978-1564912361**, ISBN 10: **1564912363**

Rule-57835:

If both $(I)$, $(U)$, $(S')$, $(s)$, $(N)$, $(v)$, $(F)$, $(i)$, $(T)$, $(D)$, $(K)$, $(J)$, $(V)$ and $(p)$ are known, then its Sales or Revenue planned is:

$$S= (U-T-D+S'[1+s][1-i]-F$$
$$-\{[N+K+J+Vp/360]/[1+I])/v$$

or

$$S= S'[1+s]$$

or

$$S= V/v$$

Rule-57836:

If both $(I)$, $(U)$, $(S')$, $(s)$, $(S)$, $(N)$, $(F)$, $(i)$, $(T)$, $(D)$, $(K)$, $(J)$, $(V)$ and $(p)$ are known, then its Variable Portion planned is:

$$v= (U-T-D+S'[1+s][1-i]-F$$
$$-\{[N+K+J+Vp/360]/[1+I])/S$$

or

$$v= V/S$$

Rule-57837:

If both $(I)$, $(U)$, $(S')$, $(s)$, $(S)$, $(v)$, $(F)$, $(i)$, $(T)$, $(D)$, $(N)$, $(J)$, $(V)$ and $(p)$ are known, then its Fixed Cost planned is:

$$F= U-T-D+S'[1+s][1-i]-Sv$$
$$-\{[N+K+J+Vp/360]/[1+I]$$

Steve Asikin ISBN 13: **978-1564912361**, ISBN 10: **1564912363**

Rule-57838:

If both (**/**), (**U**), (**S'**), (**s**), (**S**), (**v**), (**F**), (**N**), (**T**), (**D**), (**K**), (**J**), (**V**) and (**p**) are known, then its Interest Portion planned is:

$$i= 1-(\{[N+K+J+Vp/360]/[1+/]-U+T+D\}+Sv-F)/\{S'[1+s]\}$$

Rule-57839:

If both (**/**), (**U**), (**S'**), (**s**), (**S**), (**v**), (**F**), (**i**), (**N**), (**D**), (**K**), (**J**), (**V**) and (**p**) are known, then its Tax planned is:

$$T= U+\{S'[1+s][1-i]-Sv-F\}-D -[N+K+J+Vp/360]/[1+/]$$

Rule-57840:

If both (**/**), (**U**), (**S'**), (**s**), (**S**), (**v**), (**F**), (**i**), (**T**), (**N**), (**K**), (**J**), (**V**) and (**p**) are known, then its Dividend planned is:

$$D= U+\{S'[1+s][1-i]-Sv-F\}-T -[N+K+J+Vp/360]/[1+/]$$

Rule-57841:

If both (**/**), (**U**), (**S'**), (**s**), (**S**), (**v**), (**F**), (**i**), (**T**), (**D**), (**N**), (**J**), (**V**) and (**p**) are known, then its Kind of Cash planned is:

$$K= [1+/](U+\{S'[1+s]-Sv-F-S'i[1+s]\}-T-D) -[N+J+Vp/360]$$

Steve Asikin ISBN 13: **978-1564912361**, ISBN 10: **1564912363**

Rule-57842:

> If both (**/**), (**U**), (**S'**), (**s**), (**S**), (**v**), (**F**), (**i**), (**T**), (**D**), (**K**), (**N**), (**V**) and (**p**) are known, then its Job or Trade Account Receivable planned is:
>
> $$J = [1+/(U+\{S'[1+s]-Sv-F-S'i[1+s]\}-T-D) -\{N+K+Vp/360\}$$

Rule-57843:

> If both (**/**), (**U**), (**S'**), (**s**), (**S**), (**v**), (**F**), (**i**), (**T**), (**D**), (**K**), (**J**), (**V**) and (**p**) are known, then its Non Current Asset planned is:
>
> $$V = 360[[1+/(U+\{S'[1+s]-Sv-F-S'i[1+s]\}-T-D) -[N+K+J]]/p$$
>
> or
>
> $$V = Sv$$

Rule-57844:

> If both (**/**), (**U**), (**S'**), (**s**), (**S**), (**v**), (**F**), (**i**), (**T**), (**D**), (**K**), (**J**), (**V**) and (**N**) are known, then its Procured Inventory Days planned is:
>
> $$p = 360[[1+/(U+\{S'[1+s]-Sv-F-S'i[1+s]\}-T-D) -[N+K+J]]/V$$

Rule-57845:

> If both (**/**), (**U**), (**S'**), (**s**), (**S**), (**v**), (**F**), (**i**), (**T**), (**D**), (**K**), (**J**) and (**p**) are known, then its Non Current Asset planned is:
>
> $$N = [1+/\{U+S'[1+s]-Sv-F-S'i[1+s]-T-D\} -K-J-Svp/360$$

Steve Asikin ISBN 13: **978-1564912361**, ISBN 10: **1564912363**

Rule-57846:

If both $(\textbf{\textit{I}}, (\textbf{U}), (\textbf{S'}), (\textbf{s}), (\textbf{S}), (\textbf{v}), (\textbf{F}), (\textbf{i}), (\textbf{T}), (\textbf{D}), (\textbf{K})$, $(\textbf{J})$ and $(\textbf{p})$ are known, then its Non Current Asset planned is:

$$I = [\textbf{N}+\textbf{K}+\textbf{J}+\textbf{Svp}/360]/(\textbf{U}+\{\textbf{S'}[1+\textbf{s}]-\textbf{Sv}-\textbf{F}$$
$$-\textbf{S'i}[1+\textbf{s}]\}-\textbf{T}-\textbf{D})-1$$

Rule-57847:

If both $(\textbf{\textit{I}}, (\textbf{N}), (\textbf{S'}), (\textbf{s}), (\textbf{S}), (\textbf{v}), (\textbf{F}), (\textbf{i}), (\textbf{T}), (\textbf{D}), (\textbf{K})$, $(\textbf{J})$ and $(\textbf{p})$ are known, then its Utilized or starting Capital must be:

$$\textbf{U} = ([\textbf{N}+\textbf{K}+\textbf{J}+\textbf{Svp}/360]-[1+\textbf{\textit{I}}]\{\textbf{S'}[1+\textbf{s}]-\textbf{Sv}-\textbf{F}$$
$$-\textbf{S'i}[1+\textbf{s}]\}-\textbf{T}-\textbf{D})/[1+\textbf{\textit{I}}]$$

Rule-57848:

If both $(\textbf{\textit{I}}), (\textbf{U}), (\textbf{S'}), (\textbf{s}), (\textbf{S}), (\textbf{v}), (\textbf{F}), (\textbf{i}), (\textbf{T}), (\textbf{D}), (\textbf{K})$, $(\textbf{J})$ and $(\textbf{p})$ are known, then its Sales Past must be:

$$\textbf{S'} = \{[\textbf{N}+\textbf{K}+\textbf{J}+\textbf{Svp}/360]/[1+\textbf{\textit{I}}]-\textbf{U}+\textbf{Sv}+\textbf{F}+\textbf{T}+\textbf{D}\}$$
$$/\{[1+\textbf{s}][1-\textbf{i}]\}$$
$$\text{or}$$
$$\textbf{S'} = \textbf{S}/[1+\textbf{s}]$$

Steve Asikin ISBN 13: **978-1564912361**, ISBN 10: **1564912363**

Rule-57849:

If both (**/**), (**U**), (**S'**), (**s**), (**S**), (**v**), (**F**), (**i**), (**T**), (**D**), (**K**), (**J**) and (**p**) are known, then its Sales Growth planned is:

$$s= \{[\mathbf{N+K+J+Sup}/360]/[1+\textit{/}]\mathbf{-U}$$
$$+\mathbf{Sv+F+T+D}\}/\{\mathbf{S'}[1\mathbf{-i}]\}-1$$

or

$$s= \mathbf{S/S'}-1$$

Rule-57850:

If both (**/**), (**U**), (**S'**), (**s**), (**N**), (**v**), (**F**), (**i**), (**T**), (**D**), (**K**), (**J**) and (**p**) are known, then its Sales or Revenue planned is:

$$\mathbf{S}= [[1+\textit{/}](\mathbf{U}+\{\mathbf{S'}[1\mathbf{+s}]\mathbf{-F-S'i}[1\mathbf{+s}]\}\mathbf{-T-D})$$
$$-[\mathbf{N+K+J}]]/(\mathbf{v}\{\mathbf{p}/360]+[1+\textit{/}]\})$$

or

$$\mathbf{S}= \mathbf{S'}[1\mathbf{+s}]$$

Rule-57851:

If both (**/**), (**U**), (**S'**), (**s**), (**S**), (**N**), (**F**), (**i**), (**T**), (**D**), (**K**), (**J**) and (**p**) are known, then its Variable Portion planned is:

$$\mathbf{v}= [[1+\textit{/}](\mathbf{U}+\{\mathbf{S'}[1\mathbf{+s}]\mathbf{-F-S'i}[1\mathbf{+s}]\}\mathbf{-T-D})$$
$$-[\mathbf{N+K+J}]]/(\mathbf{S}\{\mathbf{p}/360+[1+\textit{/}])$$

Steve Asikin ISBN 13: **978-1564912361**, ISBN 10: **1564912363**

Rule-57852:

If both $(I)$, $(U)$, $(S')$, $(s)$, $(S)$, $(v)$, $(N)$, $(i)$, $(T)$, $(D)$, $(K)$, $(J)$ and $(p)$ are known, then its Fixed Cost planned is:

$$F= U+\{S'[1+s][1-i]-Sv\}-T-D$$
$$-[N+K+J+Svp/360]/[1+I]$$

Rule-57853:

If both $(I)$, $(U)$, $(S')$, $(s)$, $(S)$, $(v)$, $(F)$, $(N)$, $(T)$, $(D)$, $(K)$, $(J)$ and $(p)$ are known, then its Interest Portion planned is:

$$i= 1-\{[N+K+J+Svp/360]/[1+I]-U$$
$$+Sv+F+T+D\}/\{S'[1+s]\}$$

Rule-57854:

If both $(I)$, $(U)$, $(S')$, $(s)$, $(S)$, $(v)$, $(F)$, $(i)$, $(N)$, $(D)$, $(K)$, $(J)$ and $(p)$ are known, then its Tax planned is:

$$T= U+\{S'[1+s][1-i]-Sv-F\}-D$$
$$-[N+K+J+Svp/360]/[1+I]$$

Rule-57855:

If both $(I)$, $(U)$, $(S')$, $(s)$, $(S)$, $(v)$, $(F)$, $(i)$, $(T)$, $(N)$, $(K)$, $(J)$ and $(p)$ are known, then its Dividend planned is:

$$D= U+\{S'[1+s][1-i]-Sv-F\}-T$$
$$-[N+K+J+Svp/360]/[1+I]$$

Steve Asikin ISBN 13: **978-1564912361**, ISBN 10: **1564912363**

Rule-57856:

If both (**/**), (**U**), (**S'**), (**s**), (**$**), (**v**), (**F**), (**i**), (**T**), (**D**), (**N**), (**J**) and (**p**) are known, then its Kind of Cash planned is:

$$K= [1+/(U+\{S'[1+s]-Sv-F-S'i[1+s]\}-T-D)$$
$$-[N+J+Svp/360]$$

Rule-57857:

If both (**/**), (**U**), (**S'**), (**s**), (**$**), (**v**), (**F**), (**i**), (**T**), (**D**), (**K**), (**N**) and (**p**) are known, then its Job or Trade Account Receivable planned is:

$$J= [1+/(U+\{S'[1+s]-Sv-F-S'i[1+s]\}-T-D)$$
$$-\{N+K+Svp/360\}$$

Rule-57858:

If both (**/**), (**U**), (**S'**), (**s**), (**$**), (**v**), (**F**), (**i**), (**T**), (**D**), (**K**), (**J**) and (**N**) are known, then its Procured Inventory Days planned is:

$$p= 360[[1+/(U+\{S'[1+s]-Sv-F-S'i[1+s]\}-T-D)$$
$$-[N+K+J]]/[Sv]$$

Rule-57859:

If both (**/**), (**U**), (**S'**), (**s**), (**$**), (**v**), (**F**), (**i**), (**T**), (**D**), (**K**), (**J**) and (**p**) are known, then its Non Current Asset planned is:

$$N= [1+/\{U+S'[1+s]-Sv-F-S'i[1+s]-T-D\}$$
$$-K-J-S'vp[1+s]/360$$

Steve Asikin ISBN 13: **978-1564912361**, ISBN 10: **1564912363**

Rule-57860:

If both (**N**), (**U**), (**S'**), (**s**), (**S**), (**v**), (**F**), (**i**), (**T**), (**D**), (**K**), (**J**) and (**p**) are known, then its Leverage or Gearing Ratio planned is:

$$\textit{l} = \{\textbf{N+K+J+S'vp}[1+\textbf{s}]/360\}$$
$$/(\textbf{U}+\{\textbf{S'}[1+\textbf{s}]-\textbf{Sv-F-S'i}[1+\textbf{s}]\}-\textbf{T-D})-1$$

Rule-57861:

If both (**l**), (**N**), (**S'**), (**s**), (**S**), (**v**), (**F**), (**i**), (**T**), (**D**), (**K**), (**J**) and (**p**) are known, then its Utilized or Starting Capital must be:

$$\textbf{U} = (\{\textbf{N+K+J+S'vp}[1+\textbf{s}]/360\}$$
$$-[1+\textit{l}\{\textbf{S'}[1+\textbf{s}]-\textbf{Sv-F-S'i}[1+\textbf{s}]\}-\textbf{T-D})/[1+\textit{l}$$

Rule-57862:

If both (**l**), (**U**), (**N**), (**s**), (**S**), (**v**), (**F**), (**i**), (**T**), (**D**), (**K**), (**J**) and (**p**) are known, then its Sales Past must be:

$$\textbf{S'} = [[1+\textit{l}(\textbf{U}+\{\textbf{Sv+F}\}-\textbf{T-D})-[\textbf{N+K+J}]]$$
$$/([1+\textbf{s}]\{\textbf{vp}/360\}-[1+\textit{l}][1-\textbf{i}]\})$$
$$\text{or}$$
$$\textbf{S'} = \textbf{S}/[1+\textbf{s}]$$

Steve Asikin ISBN 13: **978-1564912361**, ISBN 10: **1564912363**

Rule-57863:

If both (**ʃ**), (**U**), (**S'**), (**N**), (**S**), (**v**), (**F**), (**i**), (**T**), (**D**), (**K**), (**J**) and (**p**) are known, then its Sales Growth planned is:

$$s= [[1+ʃ(U+\{Sv+F\}-T-D)-[N+K+J]]$$
$$/(S'\{vp/360\}-[1+ʃ[1-i]\})-1$$

or

$$s= S/S'-1$$

Rule-57864:

If both (**ʃ**), (**U**), (**S'**), (**s**), (**N**), (**v**), (**F**), (**i**), (**T**), (**D**), (**K**), (**J**) and (**p**) are known, then its Sales or Revenue planned is:

$$S= [\{N+K+J+S'vp[1+s]/360\}/[1+ʃ-U+\{S'[1+s]$$
$$-F-S'i[1+s]\}-T-D]/v$$

or

$$S= S'[1+s]$$

Rule-57865:

If both (**ʃ**), (**U**), (**S'**), (**s**), (**S**), (**N**), (**F**), (**i**), (**T**), (**D**), (**K**), (**J**) and (**p**) are known, then its Variable Portion planned is:

$$v= [[1+ʃ(U+\{S'[1+s]-F-S'i[1+s]\}-T-D)$$
$$-[N+K+J]]/\{S'p[1+s]/360+S[1+ʃ]\}$$

Steve Asikin ISBN 13: **978-1564912361**, ISBN 10: **1564912363**

Rule-57866:

If both (**/**), (**U**), (**S'**), (**s**), (**S**), (**v**), (**N**), (**i**), (**T**), (**D**), (**K**), (**J**) and (**p**) are known, then its Fixed Cost planned is:

$$F= U+\{S'[1+s]-Sv-S'i[1+s]\}-T-D$$
$$-\{N+K+J+S'vp[1+s]/360\}/[1+/]$$

Rule-57867:

If both (**/**), (**U**), (**S'**), (**s**), (**S**), (**v**), (**F**), (**N**), (**T**), (**D**), (**K**), (**J**) and (**p**) are known, then its Interest Portion planned is:

$$i= (\{N+K+J+S'vp[1+s]/360\}$$
$$/[1+/]-U+\{S'[1+s]-Sv-F\}-T-D)/\{S'[1+s]\}$$

Rule-57868:

If both (**/**), (**U**), (**S'**), (**s**), (**S**), (**v**), (**F**), (**i**), (**N**), (**D**), (**K**), (**J**) and (**p**) are known, then its Tax planned is:

$$T= U+\{S'[1+s]-Sv-F-S'i[1+s]\}-D$$
$$-\{N+K+J+S'vp[1+s]/360\}/[1+/]$$

Rule-57869:

If both (**/**), (**U**), (**S'**), (**s**), (**S**), (**v**), (**F**), (**i**), (**T**), (**N**), (**K**), (**J**) and (**p**) are known, then its Dividend planned is:

$$D= U+\{S'[1+s]-Sv-F-S'i[1+s]\}-T$$
$$-\{N+K+J+S'vp[1+s]/360\}/[1+/]$$

Steve Asikin ISBN 13: **978-1564912361**, ISBN 10: **1564912363**

Rule-57870:

If both (**/**), (**U**), (**S'**), (**s**), (**S**), (**v**), (**F**), (**i**), (**T**), (**D**), (**N**), (**J**) and (**p**) are known, then its Kind of Cash planned is:

$$K= [1+/](U+\{S'[1+s]-Sv-F-S'i[1+s]\}-T-D)$$
$$-\{N+J+S'vp[1+s]/360\}$$

Rule-57871:

If both (**/**), (**U**), (**S'**), (**s**), (**S**), (**v**), (**F**), (**i**), (**T**), (**D**), (**K**), (**N**) and (**p**) are known, then its Job or Trade Account Receivable planned is:

$$J= [[1+/](U+\{S'[1+s]-Sv-F-S'i[1+s]\}-T-D)$$
$$-\{N+K+S'vp[1+s]/360\}$$

Rule-57872:

If both (**/**), (**U**), (**S'**), (**s**), (**S**), (**v**), (**F**), (**i**), (**T**), (**D**), (**K**), (**J**) and (**N**) are known, then its Procured Inventory Days planned is:

$$p= 360[[1+/](U+\{S'[1+s]-Sv-F-S'i[1+s]\}-T-D)$$
$$-[N+K+J]]/\{S'v[1+s]\}$$

Rule-57873:

If both (**/**), (**U**), (**S'**), (**s**), (**S**), (**v**), (**F**), (**i**), (**T**), (**D**), (**K**), (**j**) and (**P**) are known, then its Non Current Asset planned is:

$$N= [1+/]\{U+S'[1+s]-Sv-F-S'i[1+s]-T-D\}$$
$$-K-Sj/360-P$$

Steve Asikin ISBN 13: **978-1564912361**, ISBN 10: **1564912363**

Rule-57874:

If both (**N**), (**U**), (**S'**), (**s**), (**S**), (**v**), (**F**), (**i**), (**T**), (**D**), (**K**), (**j**) and (**P**) are known, then its Leverage or Gearing Ratio planned is:

$$I = \{N+K+Sj/360+P\}$$
$$/(U+\{S'[1+s]-Sv-F-S'i[1+s]\}-T-D)-1$$

Rule-57875:

If both (**I**), (**N**), (**S'**), (**s**), (**S**), (**v**), (**F**), (**i**), (**T**), (**D**), (**K**), (**j**) and (**P**) are known, then its Utilized or Starting Capital must be:

$$U = (\{N+K+Sj/360+P\}-[1+I]\{S'[1+s]-Sv-F$$
$$-S'i[1+s]\}-T-D)/[1+I]$$

Rule-57876:

If both (**I**), (**U**), (**N**), (**s**), (**S**), (**v**), (**F**), (**i**), (**T**), (**D**), (**K**), (**j**) and (**P**) are known, then its Sales Past must be:

$$S' = \{[N+K+Sj/360+P]/[1+I]-U+Sv+F-T-D\}$$
$$/\{[1+s][1-i]$$

or

$$S' = S/[1+s]$$

Steve Asikin ISBN 13: **978-1564912361**, ISBN 10: **1564912363**

Rule-57877:

If both (**ʃ**), (**U**), (**S'**), (**N**), (**S**), (**v**), (**F**), (**i**), (**T**), (**D**), (**K**), (**j**) and (**P**) are known, then its Sales Growth planned is:

$$s= \{[N+K+Sj/360+P]/[1+\textit{ʃ}-U+Sv+F-T-D\}$$
$$/\{S'[1-i]-1$$

or

$$s= S/S'-1$$

Rule-57878:

If both (**ʃ**), (**U**), (**S'**), (**s**), (**N**), (**v**), (**F**), (**i**), (**T**), (**D**), (**K**), (**j**) and (**P**) are known, then its Sales or Revenue planned is:

$$S= [[1+\textit{ʃ}(U+\{S'[1+s]-F-S'i[1+s]\}-T-D)$$
$$-[N+K+P]]/\{j/360+v[1+\textit{ʃ}]\}$$

or

$$S= S'[1+s]$$

Rule-57879:

If both (**ʃ**), (**U**), (**S'**), (**s**), (**S**), (**N**), (**F**), (**i**), (**T**), (**D**), (**K**), (**j**) and (**P**) are known, then its Variable Portion planned is:

$$v= (U+\{S'[1+s]-F-S'i[1+s]\}-T-D$$
$$-[N+K+Sj/360+P]/[1+\textit{ʃ}])/S$$

Steve Asikin ISBN 13: **978-1564912361**, ISBN 10: **1564912363**

Rule-57880:

If both ($I$), (**U**), (**$'**), (**s**), (**$**), (**v**), (**N**), (**i**), (**T**), (**D**), (**K**), (**j**) and (**P**) are known, then its Fixed Cost planned is:

$$F= U+\{\$'[1+s]-\$v-\$'i[1+s]\}-T-D$$
$$-[N+K+\$j/360+P]/[1+I]$$

Rule-57881:

If both ($I$), (**U**), (**$'**), (**s**), (**$**), (**v**), (**F**), (**N**), (**T**), (**D**), (**K**), (**j**) and (**P**) are known, then its Interest Portion planned is:

$$i= (U+\{\$'[1+s]-\$v-F\}-T-D$$
$$-[N+K+\$j/360+P]/[1+I])/\{\$'[1+s]\}$$

Rule-57882:

If both ($I$), (**U**), (**$'**), (**s**), (**$**), (**v**), (**F**), (**i**), (**N**), (**D**), (**K**), (**j**) and (**P**) are known, then its Tax planned is:

$$T= U+\{\$'[1+s]-\$v-F-\$'i[1+s]\}-D$$
$$-[N+K+\$j/360+P]/[1+I]$$

Rule-57883:

If both ($I$), (**U**), (**$'**), (**s**), (**$**), (**v**), (**F**), (**i**), (**T**), (**N**), (**K**), (**j**) and (**P**) are known, then its Dividend planned is:

$$D= U+\{\$'[1+s]-\$v-F-\$'i[1+s]\}-T$$
$$-[N+K+\$j/360+P]/[1+I]$$

Steve Asikin ISBN 13: **978-1564912361**, ISBN 10: **1564912363**

Rule-57884:

If both (**/**), (**U**), (**S'**), (**s**), (**S**), (**v**), (**F**), (**i**), (**T**), (**D**), (**K**), (**j**) and (**P**) are known, then its Non Current Asset planned is:

$$K= [1+/](U+\{S'[1+s]-Sv-F-S'i[1+s]\}-T-D)$$
$$-[N+Sj/360+P]$$

Rule-57885:

If both (**/**), (**U**), (**S'**), (**s**), (**S**), (**v**), (**F**), (**i**), (**T**), (**D**), (**K**), (**N**) and (**P**) are known, then its Job or Trade Receivable Days planned is:

$$j= 360[[1+/](U+\{S'[1+s]-Sv-F-S'i[1+s]\}-T-D)$$
$$-\{N+K+P\}]/S$$

Rule-57886:

If both (**/**), (**U**), (**S'**), (**s**), (**S**), (**v**), (**F**), (**i**), (**T**), (**D**), (**K**), (**j**) and (**N**) are known, then its Procured Inventory planned is:

$$P= [1+/](U+\{S'[1+s]-Sv-F-S'i[1+s]\}-T-D)$$
$$-\{N+K+Sj/360\}$$

Rule-57887:

If both (**/**), (**U**), (**S'**), (**s**), (**S**), (**v**), (**F**), (**i**), (**T**), (**D**), (**K**), (**j**), (**V**) and (**p**) are known, then its Non Current Asset planned is:

$$N= [1+/]\{U+S'[1+s]-Sv-F-S'i[1+s]-T-D\}$$
$$-K-Sj/360-Vp/360$$

Steve Asikin ISBN 13: **978-1564912361**, ISBN 10: **1564912363**

Rule-57888:

If both (**N**), (**U**), (**S'**), (**s**), (**S**), (**v**), (**F**), (**i**), (**T**), (**D**), (**K**), (**j**), (**V**) and (**p**) are known, then its Leverage or Gearing Ratio planned is:

$$I = [\mathbf{N+K+Sj}/360+\mathbf{Vp}/360]$$
$$/(\mathbf{U}+\{\mathbf{S'}[1+\mathbf{s}]-\mathbf{Sv-F-S'i}[1+\mathbf{s}]\}-\mathbf{T-D})-1$$

Rule-57889:

If both (**I**), (**N**), (**S'**), (**s**), (**S**), (**v**), (**F**), (**i**), (**T**), (**D**), (**K**), (**j**), (**V**) and (**p**) are known, then its Utilized or Starting Capital must be:

$$\mathbf{U} = ([\mathbf{N+K+Sj}/360+\mathbf{Vp}/360]$$
$$-[1+\mathbf{I}]\{\mathbf{S'}[1+\mathbf{s}]-\mathbf{Sv-F-S'i}[1+\mathbf{s}]\}-\mathbf{T-D})/[1+\mathbf{I}]$$

Rule-57890:

If both (**I**), (**U**), (**S'**), (**s**), (**S**), (**v**), (**F**), (**i**), (**T**), (**D**), (**K**), (**j**), (**V**) and (**p**) are known, then its Sales Past must be:

$$\mathbf{S'} = ([\mathbf{N+K+Sj}/360+\mathbf{Vp}/360]/[1+\mathbf{I}]-\mathbf{U}$$
$$+\mathbf{Sv+F+T+D})/\{[1+\mathbf{s}][1-\mathbf{i}]$$
or
$$\mathbf{S'} = \mathbf{S}[1+\mathbf{s}]$$

Steve Asikin ISBN 13: **978-1564912361**, ISBN 10: **1564912363**

Rule-57891:

If both (**/**), (**U**), (**S'**), (**N**), (**S**), (**v**), (**F**), (**i**), (**T**), (**D**), (**K**), (**j**), (**V**) and (**p**) are known, then its Sales Growth planned is:

$$s= ([\mathbf{N+K+Sj}/360+\mathbf{Vp}/360]/[1+\mathbf{/}]$$
$$-\mathbf{U+Sv+F+T+D})/\{\mathbf{S'}[1+\mathbf{s}]\}-1$$

or

$$s= \mathbf{S/S'}-1$$

Rule-57892:

If both (**/**), (**U**), (**S'**), (**s**), (**N**), (**v**), (**F**), (**i**), (**T**), (**D**), (**K**), (**j**), (**V**) and (**p**) are known, then its Sales or Revenue planned is:

$$\mathbf{S}= [[1+\mathbf{/}](\mathbf{U}+\{\mathbf{S'}[1+\mathbf{s}]-\mathbf{F-S'i}[1+\mathbf{s}]\}-\mathbf{T-D})$$
$$-[\mathbf{N+K+Vp}/360]]/\{\mathbf{j}/360+\mathbf{v}[1+\mathbf{/}]\}$$

or

$$\mathbf{S}= \mathbf{S'}[1+\mathbf{s}]$$

Rule-57893:

If both (**/**), (**U**), (**S'**), (**s**), (**S**), (**N**), (**F**), (**i**), (**T**), (**D**), (**K**), (**j**), (**V**) and (**p**) are known, then its Variable Portion planned is:

$$\mathbf{v}= (\mathbf{U}+\{\mathbf{S'}[1+\mathbf{s}]-\mathbf{F-S'i}[1+\mathbf{s}]\}-\mathbf{T-D}$$
$$-[\mathbf{N+K+Sj}/360+\mathbf{Vp}/360]/[1+\mathbf{/}])/\mathbf{S}$$

or

$$\mathbf{v}= \mathbf{V/S}$$

Steve Asikin ISBN 13: **978-1564912361**, ISBN 10: **1564912363**

Rule-57894:

If both ($I$), ($U$), ($S'$), ($s$), ($S$), ($v$), ($N$), ($i$), ($T$), ($D$), ($K$), ($j$), ($V$) and ($p$) are known, then its Fixed Cost planned is:

$$F= U+\{S'[1+s]-Sv-S'i[1+s]\}-T-D$$
$$-[N+K+Sj/360+Vp/360]/[1+I]$$

Rule-57895:

If both ($I$), ($U$), ($S'$), ($s$), ($S$), ($v$), ($F$), ($N$), ($T$), ($D$), ($K$), ($j$), ($V$) and ($p$) are known, then its Interest Portion planned is:

$$i= (U+\{S'[1+s]-Sv-F\}-T-D-[N+K+Sj/360$$
$$+Vp/360]/[1+I])/\{S'[1+s]\}$$

Rule-57896:

If both ($I$), ($U$), ($S'$), ($s$), ($S$), ($v$), ($F$), ($i$), ($N$), ($D$), ($K$), ($j$), ($V$) and ($p$) are known, then its Tax planned is:

$$T= U+\{S'[1+s]-Sv-F-S'i[1+s]\}-D$$
$$-[N+K+Sj/360+Vp/360]/[1+I]$$

Rule-57897:

If both ($I$), ($U$), ($S'$), ($s$), ($S$), ($v$), ($F$), ($i$), ($T$), ($N$), ($K$), ($j$), ($V$) and ($p$) are known, then its Dividend planned is:

$$D= U+\{S'[1+s]-Sv-F-S'i[1+s]\}-T$$
$$-[N+K+Sj/360+Vp/360]/[1+I]$$

Steve Asikin ISBN 13: **978-1564912361**, ISBN 10: **1564912363**

Rule-57898:

If both (**/**), (**U**), (**S'**), (**s**), (**S**), (**v**), (**F**), (**i**), (**T**), (**D**), (**n**), (**j**), (**V**) and (**p**) are known, then its Kind of Cash planned is:

$$\mathbf{K}= [1+\mathbf{/}(\mathbf{U}+\{\mathbf{S'}[1+\mathbf{s}]-\mathbf{Sv}-\mathbf{F}-\mathbf{S'i}[1+\mathbf{s}]\}-\mathbf{T}-\mathbf{D})$$
$$-[\mathbf{N}+\mathbf{Sj}/360+\mathbf{Vp}/360]$$

Rule-57899:

If both (**/**), (**U**), (**S'**), (**s**), (**S**), (**v**), (**F**), (**i**), (**T**), (**D**), (**K**), (**N**), (**V**) and (**p**) are known, then its Job or Trade Receivable Days planned is:

$$\mathbf{j}= 360[[1+\mathbf{/}(\mathbf{U}+\{\mathbf{S'}[1+\mathbf{s}]-\mathbf{Sv}-\mathbf{F}-\mathbf{S'i}[1+\mathbf{s}]\}-\mathbf{T}-\mathbf{D})$$
$$-[\mathbf{N}+\mathbf{K}+\mathbf{Vp}/360]]/\mathbf{S}$$

Rule-57900:

If both (**/**), (**U**), (**S'**), (**s**), (**S**), (**v**), (**F**), (**i**), (**T**), (**D**), (**K**), (**j**), (**N**) and (**p**) are known, then its Variable Cost planned is:

$$\mathbf{V}= 360[[1+\mathbf{/}(\mathbf{U}+\{\mathbf{S'}[1+\mathbf{s}]-\mathbf{Sv}-\mathbf{F}-\mathbf{S'i}[1+\mathbf{s}]\}-\mathbf{T}-\mathbf{D})$$
$$-[\mathbf{N}+\mathbf{K}+\mathbf{Sj}/360]]/\mathbf{p}$$
or
$$\mathbf{V}= \mathbf{Sv}$$

Rule-57901:

If both (**/**), (**U**), (**S'**), (**s**), (**S**), (**v**), (**F**), (**i**), (**T**), (**D**), (**K**), (**j**), (**V**) and (**N**) are known, then its Procured Inventory Days planned is:

$$\mathbf{p}= 360[[1+\mathbf{/}(\mathbf{U}+\{\mathbf{S'}[1+\mathbf{s}]-\mathbf{Sv}-\mathbf{F}-\mathbf{S'i}[1+\mathbf{s}]\}-\mathbf{T}-\mathbf{D})$$
$$-[\mathbf{N}+\mathbf{K}+\mathbf{Sj}/360]]/\mathbf{V}$$

Steve Asikin ISBN 13: **978-1564912361**, ISBN 10: **1564912363**

Rule-57902:

If both (**/**), (**U**), (**S'**), (**s**), (**S**), (**v**), (**F**), (**i**), (**T**), (**D**), (**K**), (**j**) and (**p**) are known, then its Non Current Asset planned is:

$$\mathbf{N} = [1 + \mathbf{/}]\{\mathbf{U} + \mathbf{S'}[1+\mathbf{s}] - \mathbf{Sv} - \mathbf{F} - \mathbf{S'i}[1+\mathbf{s}] - \mathbf{T} - \mathbf{D}\}$$
$$- \mathbf{K} - \mathbf{Sj}/360 - \mathbf{Svp}/360$$

Rule-57903:

If both (**N**), (**U**), (**S'**), (**s**), (**S**), (**v**), (**F**), (**i**), (**T**), (**D**), (**K**), (**j**) and (**p**) are known, then its Leverage or Gearing Ratio planned is:

$$\mathbf{/} = [\mathbf{N} + \mathbf{K} + \mathbf{Sj}/360 + \mathbf{Svp}/360]$$
$$/(\mathbf{U} + \{\mathbf{S'}[1+\mathbf{s}] - \mathbf{Sv} - \mathbf{F} - \mathbf{S'i}[1+\mathbf{s}]\} - \mathbf{T} - \mathbf{D}) - 1$$

Rule-57904:

If both (**/**), (**N**), (**S'**), (**s**), (**S**), (**v**), (**F**), (**i**), (**T**), (**D**), (**K**), (**j**) and (**p**) are known, then its Utilized or Starting Capital must be:

$$\mathbf{U} = ([\mathbf{N} + \mathbf{K} + \mathbf{Sj}/360 + \mathbf{Svp}/360] - [1 + \mathbf{/}]\{\mathbf{S'}[1+\mathbf{s}] - \mathbf{Sv} - \mathbf{F}$$
$$- \mathbf{S'i}[1+\mathbf{s}]\} - \mathbf{T} - \mathbf{D})/[1 + \mathbf{/}]$$

Rule-57905:

If both (**/**), (**U**), (**N**), (**s**), (**S**), (**v**), (**F**), (**i**), (**T**), (**D**), (**K**), (**j**) and (**p**) are known, then its Sales Past must be:

$$\mathbf{S'} = \{[\mathbf{N} + \mathbf{K} + \mathbf{Sj}/360 + \mathbf{Svp}/360]/[1 + \mathbf{/}] - \mathbf{U}$$
$$+ \mathbf{Sv} + \mathbf{F} - \mathbf{T} - \mathbf{D}\}/\{[1+\mathbf{s}][1-\mathbf{i}]\}$$
$$\text{or}$$
$$\mathbf{S'} = \mathbf{S}/[1+\mathbf{s}]$$

`

Steve Asikin ISBN 13: **978-1564912361**, ISBN 10: **1564912363**

Rule-57906:

If both (**/**), (**U**), (**S'**), (**N**), (**S**), (**v**), (**F**), (**i**), (**T**), (**D**), (**K**), (**j**) and (**p**) are known, then its Sales Growth planned is:

$$s= \{[N+K+Sj/360+Svp/360]/[1+/-U$$
$$+Sv+F-T-D\}/\{S'[1-i]\}-1$$

or

$$s= S/S'-1$$

Rule-57907:

If both (**/**), (**U**), (**S'**), (**s**), (**N**), (**v**), (**F**), (**i**), (**T**), (**D**), (**K**), (**j**) and (**p**) are known, then its Sales or Revenue planned is:

$$S= ([1+/](U+\{S'[1+s]-F-S'i[1+s]\}-T-D)$$
$$-[N+K])/(j/360+v\{p/360+[1+/]\})$$

or

$$S= S'[1+s]$$

Rule-57908:

If both (**/**), (**U**), (**S'**), (**s**), (**S**), (**N**), (**F**), (**i**), (**T**), (**D**), (**K**), (**j**) and (**p**) are known, then its Variable Portion planned is:

$$v= [[1+/](U+\{S'[1+s]-F-S'i[1+s]\}-T-D)$$
$$-[N+K+Sj/360]]/(S\{p/360+[1+/]\})$$

Steve Asikin ISBN 13: **978-1564912361**, ISBN 10: **1564912363**

Rule-57909:

If both (**/**), (**U**), (**S'**), (**s**), (**S**), (**v**), (**N**), (**i**), (**T**), (**D**), (**K**), (**j**) and (**p**) are known, then its Fixed Cost planned is:

$$F= U+\{S'[1+s]-Sv-S'i[1+s]\}-T-D$$
$$-[N+K+Sj/360+Svp/360]/[1+/]$$

Rule-57910:

If both (**/**), (**U**), (**S'**), (**s**), (**S**), (**v**), (**F**), (**N**), (**T**), (**D**), (**K**), (**j**) and (**p**) are known, then its Interest Portion planned is:

$$i= (U+\{S'[1+s]-Sv-F\}-T-D-[N+K+Sj/360$$
$$+Svp/360]/[1+/])/\{S'[1+s]\}$$

Rule-57911:

If both (**/**), (**U**), (**S'**), (**s**), (**S**), (**v**), (**F**), (**i**), (**N**), (**D**), (**K**), (**j**) and (**p**) are known, then its Tax planned is:

$$T= U+\{S'[1+s]-Sv-F-S'i[1+s]\}-D$$
$$-[N+K+Sj/360+Svp/360]/[1+/]$$

Rule-57912:

If both (**/**), (**U**), (**S'**), (**s**), (**S**), (**v**), (**F**), (**i**), (**T**), (**N**), (**K**), (**j**) and (**p**) are known, then its Dividend planned is:

$$D= U+\{S'[1+s]-Sv-F-S'i[1+s]\}-T$$
$$-[N+K+Sj/360+Svp/360]/[1+/]$$

Steve Asikin ISBN 13: **978-1564912361**, ISBN 10: **1564912363**

Rule-57913:

If both (**/**), (**U**), (**S'**), (**s**), (**S**), (**v**), (**F**), (**i**), (**T**), (**D**), (**N**), (**j**) and (**p**) are known, then its Kind of Cash planned is:

$$K= [1+/](U+\{S'[1+s]-Sv-F-S'i[1+s]\}-T-D)$$
$$-[N+Sj/360+Svp/360]$$

Rule-57914:

If both (**/**), (**U**), (**S'**), (**s**), (**S**), (**v**), (**F**), (**i**), (**T**), (**D**), (**K**), (**N**) and (**p**) are known, then its Job or Trade Receivable Days planned is:

$$j= 360[[1+/](U+\{S'[1+s]-Sv-F-S'i[1+s]\}-T-D)$$
$$-\{N+K+Svp/360\}]/S$$

Rule-57915:

If both (**/**), (**U**), (**S'**), (**s**), (**S**), (**v**), (**F**), (**i**), (**T**), (**D**), (**K**), (**j**) and (**N**) are known, then its Procured Inventory Days planned is:

$$p= 360[[1+/](U+\{S'[1+s]-Sv-F-S'i[1+s]\}-T-D)$$
$$-[N+K+Sj/360]]/[Sv]$$

Rule-57916:

If both (**/**), (**U**), (**S'**), (**s**), (**S**), (**v**), (**F**), (**i**), (**T**), (**D**), (**K**), (**j**) and (**p**) are known, then its Non Current Asset planned is:

$$N= [1+/]\{U+S'[1+s]-Sv-F-S'i[1+s]-T-D\}$$
$$-K-Sj/360-S'vp[1+s]/360$$

Steve Asikin ISBN 13: **978-1564912361**, ISBN 10: **1564912363**

<u>Rule-57917</u>:

If both (**N**), (**U**), (**S'**), (**s**), (**S**), (**v**), (**F**), (**i**), (**T**), (**D**), (**K**), (**j**) and (**p**) are known, then its Leverage or Gearing Ratio planned is:

$$I = \{N+K+Sj/360+S'vp[1+s]/360\}$$
$$/(U+\{S'[1+s]-Sv-F-S'i[1+s]\}-T-D)-1$$

<u>Rule-57918</u>:

If both (**I**), (**N**), (**S'**), (**s**), (**S**), (**v**), (**F**), (**i**), (**T**), (**D**), (**K**), (**j**) and (**p**) are known, then its Utilized or Starting Capital must be:

$$U = (\{N+K+Sj/360+S'vp[1+s]/360\}$$
$$-[1+I]\{S'[1+s]-Sv-F-S'i[1+s]\}-T-D)/[1+I]$$

<u>Rule-57919</u>:

If both (**I**), (**U**), (**N**), (**s**), (**S**), (**v**), (**F**), (**i**), (**T**), (**D**), (**K**), (**j**) and (**p**) are known, then its Sales Past must be:

$$S' = [[1+I](U+[Sv-F]-T-D)-[N+K+Sj/360]]$$
$$/([1+s]\{vp/360-[1+I][1-i]\})$$

or

$$S' = S/[1+s]$$

Steve Asikin ISBN 13: **978-1564912361**, ISBN 10: **1564912363**

Rule-57920:

If both (**/**), (**U**), (**$'**), (**N**), (**$**), (**v**), (**F**), (**i**), (**T**), (**D**), (**K**), (**j**) and (**p**) are known, then its Sales Growth planned is:

$= [[1+**/**(**U**+[**$v-F**]-**T-D**)-[**N**+**K**+**$j**/360]]
/($'{**vp**/360-[1+**/**[1-**i**]})-1

or

$= **$/$'**-1

Rule-57921:

If both (**/**), (**U**), (**$'**), (**$**), (**N**), (**v**), (**F**), (**i**), (**T**), (**D**), (**K**), (**j**) and (**p**) are known, then its Sales or Revenue planned is:

$= [[1+**/**(**U**+{**$'**[1+$]-**F-$'i**[1+$]}-**T-D**)
-{**N**+**K**+**$'vp**[1+$]/360}]/{**j**/360+**v**[1+**/**]}

or

$= **$'**[1+$]

Rule-57922:

If both (**/**), (**U**), (**$'**), (**$**), (**$**), (**N**), (**F**), (**i**), (**T**), (**D**), (**K**), (**j**) and (**p**) are known, then its Variable Portion planned is:

v= (**U**+{**$'**[1+$]-**F**}-**T-D**-{**N**+**K**+**$j**/360
+**$'i**[1+$]/360}/[1+**/**])/{**$'p**[1+$]+**$**[1+**/**]}

Steve Asikin ISBN 13: **978-1564912361**, ISBN 10: **1564912363**

<u>Rule-57923</u>:

If both (**/**), (**U**), (**S'**), (**s**), (**S**), (**v**), (**F**), (**N**), (**T**), (**D**), (**K**), (**j**) and (**p**) are known, then its Interest Portion planned is:

$$i = (U + \{S'[1+s] - Sv - F\} - T - D - \{N + K + Sj/360 + S'vp[1+s]/360\}/[1+/])/\{S'[1+s]\}$$

<u>Rule-57924</u>:

If both (**/**), (**U**), (**S'**), (**s**), (**S**), (**v**), (**N**), (**i**), (**T**), (**D**), (**K**), (**j**) and (**p**) are known, then its Fixed Cost planned is:

$$F = U + \{S'[1+s] - Sv - S'i[1+s]\} - T - D - \{N + K + Sj/360 + S'vp[1+s]/360\}/[1+/]$$

<u>Rule-57925</u>:

If both (**/**), (**U**), (**S'**), (**s**), (**S**), (**v**), (**F**), (**i**), (**N**), (**D**), (**K**), (**j**) and (**p**) are known, then its Tax planned is:

$$T = U + \{S'[1+s] - Sv - F - S'i[1+s]\} - D - \{N + K + Sj/360 + S'vp[1+s]/360\}/[1+/]$$

<u>Rule-57926</u>:

If both (**/**), (**U**), (**S'**), (**s**), (**S**), (**v**), (**F**), (**i**), (**T**), (**N**), (**K**), (**j**) and (**p**) are known, then its Dividend planned is:

$$D = U + \{S'[1+s] - Sv - F - S'i[1+s]\} - T - \{N + K + Sj/360 + S'vp[1+s]/360\}/[1+/]$$

Steve Asikin ISBN 13: **978-1564912361**, ISBN 10: **1564912363**

Rule-57927:

If both (**/**), (**U**), (**S'**), (**s**), (**S**), (**v**), (**F**), (**i**), (**T**), (**D**), (**N**), (**j**) and (**p**) are known, then its Kind of Cash planned is:

$$K= [1+/(U+\{S'[1+s]-Sv-F-S'i[1+s]\}-T-D) -\{N+Sj/360+S'vp[1+s]/360\}$$

Rule-57928:

If both (**/**), (**U**), (**S'**), (**s**), (**S**), (**v**), (**F**), (**i**), (**T**), (**D**), (**K**), (**N**) and (**p**) are known, then its Job or Trade Receivable Days planned is:

$$j= 360[[1+/(U+\{S'[1+s]-Sv-F-S'i[1+s]\}-T-D) -\{N+K+S'vp[1+s]/360\}]/S$$

Rule-57929:

If both (**/**), (**U**), (**S'**), (**s**), (**S**), (**v**), (**F**), (**i**), (**T**), (**D**), (**K**), (**j**) and (**N**) are known, then its Procured Inventory Days planned is:

$$p= 360[[1+/(U+\{S'[1+s]-Sv-F-S'i[1+s]\}-T-D) -[N+K+Sj/360]]/\{S'v[1+s]\}$$

Rule-57930:

If both (**/**), (**U**), (**S'**), (**s**), (**S**), (**v**), (**F**), (**i**), (**T**), (**D**), (**K**), (**j**) and (**P**) are known, then its Non Current Asset planned is:

$$N= [1+/\{U+S'[1+s]-Sv-F-S'i[1+s]-T-D\} -K-S'j[1+s]/360-P$$

Steve Asikin ISBN 13: **978-1564912361**, ISBN 10: **1564912363**

Rule-57931:

If both (**N**), (**U**), (**S'**), (**s**), (**S**), (**v**), (**F**), (**i**), (**T**), (**D**), (**K**), (**j**) and (**P**) are known, then its Leverage or Gearing Ratio planned is:

$$l= \{N+K+S'j[1+s]/360+P\}$$
$$/(U+\{S'[1+s]-Sv-F-S'i[1+s]\}-T-D)-1$$

Rule-57932:

If both (**l**), (**N**), (**S'**), (**s**), (**S**), (**v**), (**F**), (**i**), (**T**), (**D**), (**K**), (**j**) and (**P**) are known, then its Utilized or Starting Capital must be:

$$U= (\{N+K+S'j[1+s]/360+P\}-[1+l]\{S'[1+s]-Sv-F$$
$$-S'i[1+s]\}-T-D)/[1+l]$$

Rule-57933:

If both (**l**), (**U**), (**N**), (**s**), (**S**), (**v**), (**F**), (**i**), (**T**), (**D**), (**K**), (**j**) and (**P**) are known, then its Sales Past must be:

$$S'= [[1+l](U+\{Sv+F\}-T-D)-[N+K+P]]$$
$$/([1+s]\{j/360-[1+l][1-i]\})$$

or

$$S'= S/[1+s]$$

Steve Asikin ISBN 13: **978-1564912361**, ISBN 10: **1564912363**

Rule-57934:

If both (**/**), (**U**), (**S'**), (**N**), (**S**), (**v**), (**F**), (**i**), (**T**), (**D**), (**K**), (**j**) and (**P**) are known, then its Sales Growth planned is:

$$s= [[1+\textit{/}(U+\{Sv+F\}-T-D)-[N+K+P]]$$
$$/(S'\{j/360-[1+\textit{/}[1-i]\})-1$$

or

$$s= S/S'-1$$


Rule-57935:

If both (**/**), (**U**), (**S'**), (**s**), (**N**), (**v**), (**F**), (**i**), (**T**), (**D**), (**K**), (**j**) and (**P**) are known, then its Sales or Revenue planned is:

$$S= (U+\{S'[1+s]-F-S'i[1+s]\}-T-D$$
$$-\{N+K+S'j[1+s]/360+P\}/[1+\textit{/}])/v$$

or

$$S= S'[1+s]$$


Rule-57936:

If both (**/**), (**U**), (**S'**), (**s**), (**S**), (**N**), (**F**), (**i**), (**T**), (**D**), (**K**), (**j**) and (**P**) are known, then its Variable Portion planned is:

$$v= (U+\{S'[1+s]-F-S'i[1+s]\}-T-D$$
$$-\{N+K+S'j[1+s]/360+P\}/[1+\textit{/}])/S$$

Steve Asikin ISBN 13: **978-1564912361**, ISBN 10: **1564912363**

Rule-57937:

> If both (**ʃ**), (**U**), (**S'**), (**s**), (**S**), (**v**), (**N**), (**i**), (**T**), (**D**), (**K**), (**j**) and (**P**) are known, then its Fixed Cost planned is:
>
> $$F= U+\{S'[1+s]-Sv-S'i[1+s]\}-T-D$$
> $$-\{N+K+S'j[1+s]/360+P\}/[1+\textit{ʃ}]$$

Rule-57938:

> If both (**ʃ**), (**U**), (**S'**), (**s**), (**S**), (**v**), (**F**), (**N**), (**T**), (**D**), (**K**), (**j**) and (**P**) are known, then its Interest Portion planned is:
>
> $$i= (U+\{S'[1+s]-Sv-F\}-T-D$$
> $$-\{N+K+S'j[1+s]/360+P\}/[1+\textit{ʃ}])/\{S'[1+s]\}$$

Rule-57939:

> If both (**ʃ**), (**U**), (**S'**), (**s**), (**S**), (**v**), (**F**), (**i**), (**N**), (**D**), (**K**), (**j**) and (**P**) are known, then its Tax planned is:
>
> $$T= U+\{S'[1+s]-Sv-F-S'i[1+s]\}-D$$
> $$-\{N+K+S'j[1+s]/360+P\}/[1+\textit{ʃ}]$$

Rule-57940:

> If both (**ʃ**), (**U**), (**S'**), (**s**), (**S**), (**v**), (**F**), (**i**), (**T**), (**N**), (**K**), (**j**) and (**P**) are known, then its Dividend planned is:
>
> $$D= U+\{S'[1+s]-Sv-F-S'i[1+s]\}-T$$
> $$-\{N+K+S'j[1+s]/360+P\}/[1+\textit{ʃ}]$$

Steve Asikin ISBN 13: **978-1564912361**, ISBN 10: **1564912363**

Rule-57941:

If both (**/**), (**U**), (**S'**), (**s**), (**S**), (**v**), (**F**), (**i**), (**T**), (**D**), (**N**), (**j**) and (**P**) are known, then its Kind of Cash planned is:

$$K = [1 + \text{/}(U + \{S'[1+s] - Sv - F - S'i[1+s]\} - T - D) - \{N + S'j[1+s]/360 + P\}$$

Rule-57942:

If both (**/**), (**U**), (**S'**), (**s**), (**S**), (**v**), (**F**), (**i**), (**T**), (**D**), (**K**), (**N**) and (**P**) are known, then its Job or Trade Receivable Days planned is:

$$j = 360[[1 + \text{/}(U + \{S'[1+s] - Sv - F - S'i[1+s]\} - T - D) - \{N + K + P\}]/\{S'[1+s]\}$$

Rule-57943:

If both (**/**), (**U**), (**S'**), (**s**), (**S**), (**v**), (**F**), (**i**), (**T**), (**D**), (**K**), (**j**) and (**N**) are known, then its Procured Inventory planned is:

$$P = [1 + \text{/}(U + \{S'[1+s] - Sv - F - S'i[1+s]\} - T - D) - \{N + K + S'j[1+s]/360\}$$

Rule-57944:

If both (**/**), (**U**), (**S'**), (**s**), (**S**), (**v**), (**F**), (**i**), (**T**), (**D**), (**K**), (**j**), (**V**) and (**p**) are known, then its Non Current Asset planned is:

$$N = [1 + \text{/}\{U + S'[1+s] - Sv - F - S'i[1+s] - T - D\} - K - S'j[1+s]/360 - Vp/360$$

Steve Asikin ISBN 13: **978-1564912361**, ISBN 10: **1564912363**

Rule-57945:

If both (**N**), (**U**), (**S'**), (**s**), (**S**), (**v**), (**F**), (**i**), (**T**), (**D**), (**K**), (**j**), (**V**) and (**p**) are known, then its Leverage or Gearing Ratio planned is:

$$l = \{N+K+S'j[1+s]/360+Vp/360\}$$
$$/(U+\{S'[1+s]-Sv-F-S'i[1+s]\}-T-D)-1$$

Rule-57946:

If both (**l**), (**N**), (**S'**), (**s**), (**S**), (**v**), (**F**), (**i**), (**T**), (**D**), (**K**), (**j**), (**V**) and (**p**) are known, then its Utilized or Starting Capital must be:

$$U = (\{N+K+S'j[1+s]/360+Vp/360\}$$
$$-[1+l]\{S'[1+s]-Sv-F-S'i[1+s]\}-T-D)/[1+l]$$

Rule-57947:

If both (**l**), (**U**), (**N**), (**s**), (**S**), (**v**), (**F**), (**i**), (**T**), (**D**), (**K**), (**j**), (**V**) and (**p**) are known, then its Sales Past must be:

$$S' = [[1+l](U+\{Sv-F\}-T-D)-\{N+K+Vp/360\}]$$
$$/([1+s]\{j/360-[1+l][1-i]\})$$

or

$$S' = S/[1+s]$$

Steve Asikin ISBN 13: **978-1564912361**, ISBN 10: **1564912363**

Rule-57948:

If both ($I$), (**U**), (**S'**), (**N**), (**S**), (**v**), (**F**), (**i**), (**T**), (**D**), (**K**), (**j**), (**V**) and (**p**) are known, then its Sales Growth planned is:

$$s= [[1+I(U+\{Sv-F\}-T-D)-\{N+K+Vp/360\}]$$
$$/(S'\{j/360-[1+I[1-i]\})-1$$

or

$$s= S/S'-1$$

Rule-57949:

If both ($I$), (**U**), (**S'**), (**s**), (**N**), (**v**), (**F**), (**i**), (**T**), (**D**), (**K**), (**j**), (**V**) and (**p**) are known, then its Sales or Revenue planned is:

$$S= (U+\{S'[1+s]-F- S'i[1+s]\}-T-D$$
$$-\{N+K+S'j[1+s]/360+Vp/360\}/[1+I])/v$$

or

$$S= S'[1+s]$$

or

$$S= V/v$$

Rule-57950:

If both ($I$), (**U**), (**S'**), (**s**), (**S**), (**N**), (**F**), (**i**), (**T**), (**D**), (**K**), (**j**), (**V**) and (**p**) are known, then its Variable Cost planned is:

$$v= (U+\{S'[1+s]-F- S'i[1+s]\}-T-D$$
$$-\{N+K+S'j[1+s]/360+Vp/360\}/[1+I])/S$$

or

$$v= V/S$$

Steve Asikin ISBN 13: **978-1564912361**, ISBN 10: **1564912363**

Rule-57951:

If both (**/**), (**U**), (**S'**), (**s**), (**S**), (**v**), (**N**), (**i**), (**T**), (**D**), (**K**), (**j**), (**V**) and (**p**) are known, then its Fixed Cost planned is:

$$F= U+\{S'[1+s]-Sv- S'i[1+s]\}-T-D$$
$$-\{N+K+S'j[1+s]/360+Vp/360\}/[1+/]$$

Rule-57952:

If both (**/**), (**U**), (**S'**), (**s**), (**S**), (**v**), (**F**), (**N**), (**T**), (**D**), (**K**), (**j**), (**V**) and (**p**) are known, then its Interest Portion planned is:

$$i= (U+\{S'[1+s]-Sv-F-T-D-\{N+K+S'j[1+s]/360$$
$$+Vp/360\}/[1+/])/\{S'[1+s]\}$$

Rule-57953:

If both (**/**), (**U**), (**S'**), (**s**), (**S**), (**v**), (**F**), (**i**), (**N**), (**D**), (**K**), (**j**), (**V**) and (**p**) are known, then its Tax planned is:

$$T= U+\{S'[1+s]-Sv-F-S'i[1+s]\}-D$$
$$-\{N+K+S'j[1+s]/360+Vp/360\}/[1+/]$$

Rule-57954:

If both (**/**), (**U**), (**S'**), (**s**), (**S**), (**v**), (**F**), (**i**), (**T**), (**N**), (**K**), (**j**), (**V**) and (**p**) are known, then its Dividend planned is:

$$D= U+\{S'[1+s]-Sv-F-S'i[1+s]\}-T$$
$$-\{N+K+S'j[1+s]/360+Vp/360\}/[1+/]$$

Steve Asikin ISBN 13: **978-1564912361**, ISBN 10: **1564912363**

Rule-57955:

If both $(\textbf{\textit{I}})$, $(\textbf{U})$, $(\textbf{S'})$, $(\textbf{s})$, $(\textbf{S})$, $(\textbf{v})$, $(\textbf{F})$, $(\textbf{i})$, $(\textbf{T})$, $(\textbf{D})$, $(\textbf{N})$, $(\textbf{j})$, $(\textbf{V})$ and $(\textbf{p})$ are known, then its Kind of Cash planned is:

$$K = [1 + I(U + \{S'[1+s] - Sv - F - S'i[1+s]\} - T - D)$$
$$- \{N + S'j[1+s]/360 + Vp/360\}$$

Rule-57956:

If both $(\textbf{\textit{I}})$, $(\textbf{U})$, $(\textbf{S'})$, $(\textbf{s})$, $(\textbf{S})$, $(\textbf{v})$, $(\textbf{F})$, $(\textbf{i})$, $(\textbf{T})$, $(\textbf{D})$, $(\textbf{K})$, $(\textbf{N})$, $(\textbf{V})$ and $(\textbf{p})$ are known, then its Job or Trade Receivable Days planned is:

$$j = 360[[1 + I(U + \{S'[1+s] - Sv - F - S'i[1+s]\} - T - D)$$
$$- \{N + K + Vp/360\}]/\{S'[1+s]\}$$

Rule-57957:

If both $(\textbf{\textit{I}})$, $(\textbf{U})$, $(\textbf{S'})$, $(\textbf{s})$, $(\textbf{S})$, $(\textbf{v})$, $(\textbf{F})$, $(\textbf{i})$, $(\textbf{T})$, $(\textbf{D})$, $(\textbf{K})$, $(\textbf{j})$, $(\textbf{N})$ and $(\textbf{p})$ are known, then its Variable Cost planned is:

$$V = 360[[1 + I(U + \{S'[1+s] - Sv - F - S'i[1+s]\} - T - D)$$
$$- \{N + K + S'j[1+s]/360\}]/p$$
or
$$V = Sv$$

Rule-57958:

If both $(\textbf{\textit{I}})$, $(\textbf{U})$, $(\textbf{S'})$, $(\textbf{s})$, $(\textbf{S})$, $(\textbf{v})$, $(\textbf{F})$, $(\textbf{i})$, $(\textbf{T})$, $(\textbf{D})$, $(\textbf{K})$, $(\textbf{j})$, $(\textbf{V})$ and $(\textbf{N})$ are known, then its Procured Inventory Days planned is:

$$p = 360[[1 + I(U + \{S'[1+s] - Sv - F - S'i[1+s]\} - T - D)$$
$$- \{N + K + S'j[1+s]/360\}]/V$$

Steve Asikin ISBN 13: **978-1564912361**, ISBN 10: **1564912363**

Rule-57959:

If both (**/**), (**U**), (**S'**), (**s**), (**S**), (**v**), (**F**), (**i**), (**T**), (**D**), (**K**), (**j**) and (**p**) are known, then its Non Current Asset planned is:

$$N= [1+/]\{U+S'[1+s]-Sv-F-S'i[1+s]-T-D\}$$
$$-K-S'j[1+s]/360-Svp/360$$

Rule-57960:

If both (**N**), (**U**), (**S'**), (**s**), (**S**), (**v**), (**F**), (**i**), (**T**), (**D**), (**K**), (**j**) and (**p**) are known, then its Leverage or Gearing Ratio planned is:

$$/= \{N+K+S'j[1+s]/360+Svp/360\}$$
$$/(U+\{S'[1+s]-Sv-F-S'i[1+s]\}-T-D)-1$$

Rule-57961:

If both (**/**), (**N**), (**S'**), (**s**), (**S**), (**v**), (**F**), (**i**), (**T**), (**D**), (**K**), (**j**) and (**p**) are known, then its Utilized or Starting Capital must be:

$$U= (\{N+K+S'j[1+s]/360+Svp/360\}-[1+/]$$
$$\{S'[1+s]-Sv-F-S'i[1+s]\}-T-D)/[1+/]$$

Rule-57962:

If both (**/**), (**U**), (**N**), (**s**), (**S**), (**v**), (**F**), (**i**), (**T**), (**D**), (**K**), (**j**) and (**p**) are known, then its Sales Past must be:

$$S'= [[1+/](U+\{Sv-F\}-T-D)-[N+K+Svp/360]]$$
$$/([1+s]\{j/360-[1+/][1-i]\})$$

or

$$S'= S/[1+s]$$

Steve Asikin ISBN 13: **978-1564912361**, ISBN 10: **1564912363**

Rule-57963:

If both (**/**), (**U**), (**S'**), (**N**), (**S**), (**v**), (**F**), (**i**), (**T**), (**D**), (**K**), (**j**) and (**p**) are known, then its Sales Growth planned is:

$$s= [[1+/(U+\{Sv-F\}-T-D)-[N+K+Svp/360]]$$
$$/(S'\{j/360-[1+/[1-i]\})-1$$

or

$$s= S/S'-1$$

Rule-57964:

If both (**/**), (**U**), (**S'**), (**s**), (**N**), (**v**), (**F**), (**i**), (**T**), (**D**), (**K**), (**j**) and (**p**) are known, then its Sales or Revenue planned is:

$$S= [[1+/(U+\{S'[1+s]-F-S'i[1+s]\}-T-D)$$
$$-\{N+K+S'j[1+s]/360\}]/(v\{p/360+[1+/]\})$$

or

$$S= S'[1+s]$$

Rule-57965:

If both (**/**), (**U**), (**S'**), (**s**), (**S**), (**N**), (**F**), (**i**), (**T**), (**D**), (**K**), (**j**) and (**p**) are known, then its Variable Portion planned is:

$$v= [[1+/(U+\{S'[1+s]-F-S'i[1+s]\}-T-D)$$
$$-\{N+K+S'j[1+s]/360\}]/(S\{p/360+[1+/]\})$$

Steve Asikin ISBN 13: **978-1564912361**, ISBN 10: **1564912363**

Rule-57966:

If both (**/**), (**U**), (**S'**), (**s**), (**S**), (**v**), (**N**), (**i**), (**T**), (**D**), (**K**), (**j**) and (**p**) are known, then its Fixed Cost planned is:

$$F= U+\{S'[1+s]-Sv-S'i[1+s]\}-T-D$$
$$-\{N+K+S'j[1+s]/360+Svp/360\}/[1+/]$$

Rule-57967:

If both (**/**), (**U**), (**S'**), (**s**), (**S**), (**v**), (**F**), (**N**), (**T**), (**D**), (**K**), (**j**) and (**p**) are known, then its Interest Portion planned is:

$$i= (U+\{S'[1+s]-Sv-F\}- T-D-\{N \mid K \mid S'j[1+s]/360$$
$$+Svp/360\}/[1+/])/\{S'[1+s]\}$$

Rule-57968:

If both (**/**), (**U**), (**S'**), (**s**), (**S**), (**v**), (**F**), (**i**), (**N**), (**D**), (**K**), (**j**) and (**p**) are known, then its Tax planned is:

$$T= U+\{S'[1+s]-Sv-F-S'i[1+s]\}-D$$
$$-\{N+K+S'j[1+s]/360+Svp/360\}/[1+/]$$

Rule-57969:

If both (**/**), (**U**), (**S'**), (**s**), (**S**), (**v**), (**F**), (**i**), (**T**), (**N**), (**K**), (**j**) and (**p**) are known, then its Dividend planned is:

$$D= U+\{S'[1+s]-Sv-F-S'i[1+s]\}-T$$
$$-\{N+K+S'j[1+s]/360+Svp/360\}/[1+/]$$

Steve Asikin ISBN 13: **978-1564912361**, ISBN 10: **1564912363**

Rule-57970:

If both $(\textit{l})$, $(\textbf{U})$, $(\textbf{S'})$, $(\textbf{s})$, $(\textbf{S})$, $(\textbf{v})$, $(\textbf{F})$, $(\textbf{i})$, $(\textbf{T})$, $(\textbf{D})$, $(\textbf{N})$, $(\textbf{j})$ and $(\textbf{p})$ are known, then its Kind of Cash planned is:

$$K= [1+\textit{l}(U+\{S'[1+s]\text{-}Sv\text{-}F\text{-}S'i[1+s]\}\text{-}T\text{-}D)$$
$$-\{N+S'j[1+s]/360+Svp/360\}$$

Rule-57971:

If both $(\textit{l})$, $(\textbf{U})$, $(\textbf{S'})$, $(\textbf{s})$, $(\textbf{S})$, $(\textbf{v})$, $(\textbf{F})$, $(\textbf{i})$, $(\textbf{T})$, $(\textbf{D})$, $(\textbf{K})$, $(\textbf{N})$ and $(\textbf{p})$ are known, then its Job or Trade Receivable Days planned is:

$$j= 360[[1+\textit{l}(U+\{S'[1+s]\text{-}Sv\text{-}F\text{-}S'i[1+s]\}\text{-}T\text{-}D)$$
$$-\{N+K+Svp/360\}]/\{S'[1+s]\}$$

Rule-57972:

If both $(\textit{l})$, $(\textbf{U})$, $(\textbf{S'})$, $(\textbf{s})$, $(\textbf{S})$, $(\textbf{v})$, $(\textbf{F})$, $(\textbf{i})$, $(\textbf{T})$, $(\textbf{D})$, $(\textbf{K})$, $(\textbf{j})$ and $(\textbf{N})$ are known, then its Procured Inventory Days planned is:

$$p= 360[[1+\textit{l}(U+\{S'[1+s]\text{-}Sv\text{-}F\text{-}S'i[1+s]\}\text{-}T\text{-}D)$$
$$-\{N+K+S'j[1+s]/360\}]/[Sv]$$

Rule-57973:

If both $(\textit{l})$, $(\textbf{U})$, $(\textbf{S'})$, $(\textbf{s})$, $(\textbf{S})$, $(\textbf{v})$, $(\textbf{F})$, $(\textbf{i})$, $(\textbf{T})$, $(\textbf{D})$, $(\textbf{K})$, $(\textbf{j})$ and $(\textbf{p})$ are known, then its Non Current Asset planned is:

$$N= [1+\textit{l}\{U+S'[1+s]\text{-}Sv\text{-}F\text{-}S'i[1+s]\text{-}T\text{-}D\}$$
$$-K\text{-}S'j[1+s]/360\text{-}S'vp[1+s]/360$$

Steve Asikin ISBN 13: **978-1564912361**, ISBN 10: **1564912363**

Rule-57974:

If both $(\mathbf{N})$, $(\mathbf{U})$, $(\mathbf{S'})$, $(\mathbf{s})$, $(\mathbf{S})$, $(\mathbf{v})$, $(\mathbf{F})$, $(\mathbf{i})$, $(\mathbf{T})$, $(\mathbf{D})$, $(\mathbf{K})$, $(\mathbf{j})$ and $(\mathbf{p})$ are known, then its Leverage or Gearing Ratio planned is:

$$l = \{\mathbf{N} + \mathbf{K} + \mathbf{S'j}[1+\mathbf{s}]/360 + \mathbf{S'vp}[1+\mathbf{s}]/360\}$$
$$/(\mathbf{U} + \{\mathbf{S'}[1+\mathbf{s}] - \mathbf{Sv} - \mathbf{F} - \mathbf{S'i}[1+\mathbf{s}]\} - \mathbf{T} - \mathbf{D}) - 1$$

Rule-57975:

If both $(\mathit{l})$, $(\mathbf{N})$, $(\mathbf{S'})$, $(\mathbf{s})$, $(\mathbf{S})$, $(\mathbf{v})$, $(\mathbf{F})$, $(\mathbf{i})$, $(\mathbf{T})$, $(\mathbf{D})$, $(\mathbf{K})$, $(\mathbf{j})$ and $(\mathbf{p})$ are known, then its Utilized or Starting Capital must be:

$$\mathbf{U} = (\{\mathbf{N} + \mathbf{K} + \mathbf{S'j}[1+\mathbf{s}]/360 + \mathbf{S'vp}[1+\mathbf{s}]/360\}$$
$$-[1+\mathit{l}]\{\mathbf{S'}[1+\mathbf{s}] - \mathbf{Sv} - \mathbf{F} - \mathbf{S'i}[1+\mathbf{s}]\} - \mathbf{T} - \mathbf{D})/[1+\mathit{l}]$$

Rule-57976:

If both $(\mathit{l})$, $(\mathbf{U})$, $(\mathbf{N})$, $(\mathbf{s})$, $(\mathbf{S})$, $(\mathbf{v})$, $(\mathbf{F})$, $(\mathbf{i})$, $(\mathbf{T})$, $(\mathbf{D})$, $(\mathbf{K})$, $(\mathbf{j})$ and $(\mathbf{p})$ are known, then its Sales Past must be:

$$\mathbf{S'} = [[1+\mathit{l}](\mathbf{U} + \{\mathbf{Sv} + \mathbf{F}\} - \mathbf{T} - \mathbf{D}) - [\mathbf{N} + \mathbf{K}]]$$
$$/([1+\mathbf{s}]\{\mathbf{j}/360 + \mathbf{vp}/360 - [1+\mathit{l}][1-\mathbf{i}]\})$$
$$\text{or}$$
$$\mathbf{S'} = \mathbf{S}/[1+\mathbf{s}]$$

Steve Asikin ISBN 13: **978-1564912361**, ISBN 10: **1564912363**

Rule-57977:

If both **(/)**, **(U)**, **(S')**, **(N)**, **(S)**, **(v)**, **(F)**, **(i)**, **(T)**, **(D)**, **(K)**, **(j)** and **(p)** are known, then its Sales Growth planned is:

$$s= [[1+/(U+\{Sv+F\}-T-D)-[N+K]]$$
$$/(S'\{j/360+vp/360-[1+/][1-i]\})-1$$

or

$$s= S/S'-1$$

Rule-57978:

If both **(/)**, **(U)**, **(S')**, **(s)**, **(N)**, **(v)**, **(F)**, **(i)**, **(T)**, **(D)**, **(K)**, **(j)** and **(p)** are known, then its Sales or Revenue planned is:

$$S= (U+\{S'[1+s]-F-S'i[1+s]\}-T-D-\{N+K$$
$$+S'j[1+s]/360+S'vp[1+s]/360\}/[1+/])/v$$

or

$$S= S'[1+s]$$

Rule-57979:

If both **(/)**, **(U)**, **(S')**, **(s)**, **(S)**, **(N)**, **(F)**, **(i)**, **(T)**, **(D)**, **(K)**, **(j)** and **(p)** are known, then its Variable Portion planned is:

$$v= [[1+/(U+\{S'[1+s]-F-S'i[1+s]\}-T-D)-\{N+K$$
$$+S'j[1+s]/360\}]/\{S'p[1+s]/360+S[1+/]\}$$

Steve Asikin ISBN 13: **978-1564912361**, ISBN 10: **1564912363**

Rule-57980:

If both (**/**), (**U**), (**S'**), (**s**), (**S**), (**v**), (**N**), (**i**), (**T**), (**D**), (**K**), (**j**) and (**p**) are known, then its Fixed Cost planned is:

$$F= U+\{S'[1+s]-Sv-S'i[1+s]\}-T-D-\{N+K$$
$$+S'j[1+s]/360+S'vp[1+s]/360\}/[1+/]$$

Rule-57981:

If both (**/**), (**U**), (**S'**), (**s**), (**S**), (**v**), (**F**), (**N**), (**T**), (**D**), (**K**), (**j**) and (**p**) are known, then its Interest Portion planned is:

$$i= (U+\{S'[1+s]-Sv-F\}-T-D-\{N+K|S'j[1|s]/360$$
$$+S'vp[1+s]/360\}/[1+/])/\{S'[1+s]\}$$

Rule-57982:

If both (**/**), (**U**), (**S'**), (**s**), (**S**), (**v**), (**F**), (**i**), (**N**), (**D**), (**K**), (**j**) and (**p**) are known, then its Tax planned is:

$$T= U+\{S'[1+s]-Sv-F-S'i[1+s]\}-D-\{N+K$$
$$+S'j[1+s]/360+S'vp[1+s]/360\}/[1+/]$$

Rule-57983:

If both (**/**), (**U**), (**S'**), (**s**), (**S**), (**v**), (**F**), (**i**), (**T**), (**N**), (**K**), (**j**) and (**p**) are known, then its Dividend planned is:

$$D= U+\{S'[1+s]-Sv-F-S'i[1+s]\}-T-\{N+K$$
$$+S'j[1+s]/360+S'vp[1+s]/360\}/[1+/]$$

Steve Asikin ISBN 13: **978-1564912361**, ISBN 10: **1564912363**

Rule-57984:

If both ($I$), ($U$), ($S'$), ($s$), ($S$), ($v$), ($F$), ($i$), ($T$), ($D$), ($N$), ($j$) and ($p$) are known, then its Kind of Cash planned is:

$$K= [1+I(U+\{S'[1+s]-Sv-F-S'i[1+s]\}-T-D)$$
$$-\{N+S'j[1+s]/360+S'vp[1+s]/360\}$$

Rule-57985:

If both ($I$), ($U$), ($S'$), ($s$), ($S$), ($v$), ($F$), ($i$), ($T$), ($D$), ($K$), ($N$) and ($p$) are known, then its Job or Trade Receivable Days planned is:

$$j= 360[[1+I(U+\{S'[1+s]-Sv-F-S'i[1+s]\}-T-D)$$
$$-\{N+K+S'vp[1+s]/360\}]/\{S'[1+s]\}$$

Rule-57986:

If both ($I$), ($U$), ($S'$), ($s$), ($S$), ($v$), ($F$), ($i$), ($T$), ($D$), ($K$), ($j$) and ($N$) are known, then its Procured Inventory Days planned is:

$$p= 360[[1+I(U+\{S'[1+s]-Sv-F-S'i[1+s]\}-T-D)$$
$$-\{N+K+S'j[1+s]/360\}]/\{S'v[1+s]\}$$

Rule-57987:

If both ($I$), ($U$), ($S'$), ($s$), ($S$), ($v$), ($F$), ($i$), ($T$), ($d$), ($K$), ($J$) and ($P$) are known, then its Non Current Asset planned is:

$$N= [1+I(U+\{S'[1+s]-Sv-F-S'i[1+s]-T\}[1-d])$$
$$-K-J-P$$

Steve Asikin ISBN 13: **978-1564912361**, ISBN 10: **1564912363**

Rule-57988:

If both (**I**), (**U**), (**S'**), (**s**), (**S**), (**v**), (**F**), (**i**), (**T**), (**d**), (**K**), (**J**) and (**P**) are known, then its Non Current Asset planned is:

$$I = \{N+K+J+P\}/(U+\{S'[1+s]-Sv-F-S'i[1+s]-T\}[1-d])-1$$

Rule-57989:

If both (**I**), (**N**), (**S'**), (**s**), (**S**), (**v**), (**F**), (**i**), (**T**), (**d**), (**K**), (**J**) and (**P**) are known, then its Utilized or Starting Capital must be:

$$U = (\{N+K+J+P\}-[1+I]\{S'[1+s]-Sv-F-S'i[1+s]-T\}[1-d])/[1+I]$$

Rule-57990:

If both (**I**), (**U**), (**S'**), (**s**), (**S**), (**v**), (**F**), (**i**), (**T**), (**d**), (**K**), (**J**) and (**P**) are known, then its Sales Past must be:

$$S' = (\{[N+K+J+P]/[1+I]-U\}/[1-d]+Sv+F+T)/\{[1+s][1-i]\}$$

or

$$S' = S/[1+s]$$

`

Steve Asikin ISBN 13: **978-1564912361**, ISBN 10: **1564912363**

Rule-57991:

If both $(\textbf{\textit{I}})$, $(\textbf{U})$, $(\textbf{S'})$, $(\textbf{N})$, $(\textbf{S})$, $(\textbf{v})$, $(\textbf{F})$, $(\textbf{i})$, $(\textbf{T})$, $(\textbf{d})$, $(\textbf{K})$, $(\textbf{J})$ and $(\textbf{P})$ are known, then its Sales Growth planned is:

$$s = (\{[\textbf{N}+\textbf{K}+\textbf{J}+\textbf{P}]/[1+\textbf{\textit{I}}-\textbf{U}\}/[1-\textbf{d}]+\textbf{S}\textbf{v}+\textbf{F}+\textbf{T})$$
$$/\{\textbf{S'}[1-\textbf{i}]\}-1$$

or

$$s = \textbf{S}/\textbf{S'}-1$$


Rule-57992:

If both $(\textbf{\textit{I}})$, $(\textbf{U})$, $(\textbf{S'})$, $(\textbf{s})$, $(\textbf{N})$, $(\textbf{v})$, $(\textbf{F})$, $(\textbf{i})$, $(\textbf{T})$, $(\textbf{d})$, $(\textbf{K})$, $(\textbf{J})$ and $(\textbf{P})$ are known, then its Sales or Revenue planned is:

$$S = (\{\textbf{S'}[1+\textbf{s}]-\textbf{F}-\textbf{S'}\textbf{i}[1+\textbf{s}]\}-\textbf{T}-\{[\textbf{N}+\textbf{K}+\textbf{J}+\textbf{P}]$$
$$/[1+\textbf{\textit{I}}-\textbf{U}\}/[1-\textbf{d}])/\textbf{v}$$

or

$$S = \textbf{S'}[1+\textbf{s}]$$


Rule-57993:

If both $(\textbf{\textit{I}})$, $(\textbf{U})$, $(\textbf{S'})$, $(\textbf{s})$, $(\textbf{S})$, $(\textbf{N})$, $(\textbf{F})$, $(\textbf{i})$, $(\textbf{T})$, $(\textbf{d})$, $(\textbf{K})$, $(\textbf{J})$ and $(\textbf{P})$ are known, then its Variable Portion planned is:

$$v = (\{\textbf{S'}[1+\textbf{s}]-\textbf{F}-\textbf{S'}\textbf{i}[1+\textbf{s}]\}-\textbf{T}-\{[\textbf{N}+\textbf{K}+\textbf{J}+\textbf{P}]$$
$$/[1+\textbf{\textit{I}}-\textbf{U}\}/[1-\textbf{d}])/\textbf{S}$$

Steve Asikin ISBN 13: **978-1564912361**, ISBN 10: **1564912363**

Rule-57994:

If both (**/**), (**U**), (**S'**), (**s**), (**S**), (**v**), (**N**), (**i**), (**T**), (**d**), (**K**), (**J**) and (**P**) are known, then its Fixed Cost planned is:

$$F= \{S'[1+s]-Sv-S'i[1+s]\}-T-\{[N+K+J+P] /[1+/]-U\}/[1-d]$$

Rule-57995:

If both (**/**), (**U**), (**S'**), (**s**), (**S**), (**v**), (**F**), (**N**), (**T**), (**d**), (**K**), (**J**) and (**P**) are known, then its Interest Portion planned is:

$$i= (\{S'[1+s]-Sv-F\}-T-\{[N+K+J+P]/[1+/] U\} /[1-d])/\{S'[1+s]\}$$

Rule-57996:

If both (**/**), (**U**), (**S'**), (**s**), (**S**), (**v**), (**F**), (**i**), (**N**), (**d**), (**K**), (**J**) and (**P**) are known, then its Tax planned is:

$$T= S'[1+s]-Sv-F-S'i[1+s]-\{[N+K+J+P]/[1+/]-U\} /[1-d]$$

Rule-57997:

If both (**/**), (**U**), (**S'**), (**s**), (**S**), (**v**), (**F**), (**i**), (**T**), (**N**), (**K**), (**J**) and (**P**) are known, then its Dividend Payout planned is:

$$d= 1-\{[N+K+J+P]/[1+/]-U\} /\{S'[1+s]-Sv-F-S'i[1+s]-T\}$$

Steve Asikin ISBN 13: **978-1564912361**, ISBN 10: **1564912363**

Rule-57998:

> If both (**/**), (**U**), (**S'**), (**s**), (**S**), (**v**), (**F**), (**i**), (**T**), (**d**), (**K**), (**J**) and (**P**) are known, then its Non Current Asset planned is:
>
> $$K = [1+/](U+\{S'[1+s]-Sv-F-S'i[1+s]-T\}[1-d]) \\ -[N+J+P]$$

Rule-57999:

> If both (**/**), (**U**), (**S'**), (**s**), (**S**), (**v**), (**F**), (**i**), (**T**), (**d**), (**K**), (**J**) and (**P**) are known, then its Non Current Asset planned is:
>
> $$J = [1+/](U+\{S'[1+s]-Sv-F-S'i[1+s]-T\}[1-d]) \\ -[N+K+P]$$

Rule-58000:

> If both (**/**), (**U**), (**S'**), (**s**), (**S**), (**v**), (**F**), (**i**), (**T**), (**D**), (**K**), (**j**) and (**N**) are known, then its Non Current Asset planned is:
>
> $$P = [1+/](U+\{S'[1+s]-Sv-F-S'i[1+s]-T\}[1-d]) \\ -[N+K+J]$$

Rule-58001:

> If both (**/**), (**U**), (**S'**), (**s**), (**S**), (**v**), (**F**), (**i**), (**T**), (**d**), (**K**), (**J**), (**V**) and (**p**) are known, then its Non Current Asset planned is:
>
> $$N = [1+/](U+\{S'[1+s]-Sv-F-S'i[1+s]-T\}[1-d]) \\ -K-J-Vp/360$$

Steve Asikin ISBN 13: **978-1564912361**, ISBN 10: **1564912363**

Rule-58002:

If both (**N**), (**U**), (**S'**), (**s**), (**S**), (**v**), (**F**), (**i**), (**T**), (**d**), (**K**), (**J**), (**V**) and (**p**) are known, then its Leverage or Gearing Ratio planned is:

$$I = [N+K+J+Vp/360]$$
$$/(U+\{S'[1+s]-Sv-F-S'i[1+s]-T\}[1-d])-1$$

Rule-58003:

If both (**I**), (**N**), (**S'**), (**s**), (**S**), (**v**), (**F**), (**i**), (**T**), (**d**), (**K**), (**J**), (**V**) and (**p**) are known, then its Utilized or Starting Capital must be:

$$U = ([N+K+J+Vp/360]-[1+I]\{S'[1+s]-Sv-F$$
$$-S'i[1+s]-T\}[1-d])/[1+I]$$

Rule-58004:

If both (**I**), (**U**), (**N**), (**s**), (**S**), (**v**), (**F**), (**i**), (**T**), (**d**), (**K**), (**J**), (**V**) and (**p**) are known, then its Sales Past must be:

$$S' = (\{[N+K+J+Vp/360]/[1+I]-U\}/[1-d]+Sv+F+T)$$
$$/\{[1+s][1-i]\}$$
or
$$S' = S/[1+s]$$

Steve Asikin ISBN 13: **978-1564912361**, ISBN 10: **1564912363**

Rule-58005:

If both (**∫**), (**U**), (**S'**), (**N**), (**S**), (**v**), (**F**), (**i**), (**T**), (**d**), (**K**), (**J**), (**V**) and (**p**) are known, then its Sales Growth planned is:

$$s = (\{[\mathbf{N}+\mathbf{K}+\mathbf{J}+\mathbf{V}\mathbf{p}/360]/[1+\mathbf{\int}-\mathbf{U}\}/[1-\mathbf{d}]+\mathbf{S}\mathbf{v}+\mathbf{F}+\mathbf{T})$$
$$/\{\mathbf{S'}[1-\mathbf{i}]\}-1$$

or

$$s = \mathbf{S}/\mathbf{S'}-1$$

Rule-58006:

If both (**∫**), (**U**), (**S'**), (**s**), (**N**), (**v**), (**F**), (**i**), (**T**), (**d**), (**K**), (**J**), (**V**) and (**p**) are known, then its Sales or Revenue planned is:

$$\mathbf{S} = (\{\mathbf{S'}[1+\mathbf{s}]-\mathbf{F}-\mathbf{S'}\mathbf{i}[1+\mathbf{s}]-\mathbf{T}-\{[\mathbf{N}+\mathbf{K}+\mathbf{J}+\mathbf{V}\mathbf{p}/360]$$
$$/[1+\mathbf{\int}-\mathbf{U}\}/[1-\mathbf{d}])/\mathbf{v}$$

or

$$\mathbf{S} = \mathbf{S'}[1+\mathbf{s}]$$

or

$$\mathbf{S} = \mathbf{V}/\mathbf{v}$$

Rule-58007:

If both (**∫**), (**U**), (**S'**), (**s**), (**S**), (**N**), (**F**), (**i**), (**T**), (**d**), (**K**), (**J**), (**V**) and (**p**) are known, then its Variable Portion planned is:

$$\mathbf{v} = (\{\mathbf{S'}[1+\mathbf{s}]-\mathbf{F}-\mathbf{S'}\mathbf{i}[1+\mathbf{s}]-\mathbf{T}-\{[\mathbf{N}+\mathbf{K}+\mathbf{J}+\mathbf{V}\mathbf{p}/360]$$
$$/[1+\mathbf{\int}-\mathbf{U}\}/[1-\mathbf{d}])/\mathbf{S}$$

or

$$\mathbf{v} = \mathbf{V}/\mathbf{S}$$

Steve Asikin ISBN 13: **978-1564912361**, ISBN 10: **1564912363**

Rule-58008:

If both (**/**), (**U**), (**S'**), (**s**), (**S**), (**v**), (**N**), (**i**), (**T**), (**d**), (**K**), (**J**), (**V**) and (**p**) are known, then its Non Current Asset planned is:

$$F = S'[1+s] - Sv - S'i[1+s] - T - \{[N+K+J+Vp/360] / [1+/-U\} / [1-d]$$

Rule-58009:

If both (**/**), (**U**), (**S'**), (**s**), (**S**), (**v**), (**F**), (**N**), (**T**), (**d**), (**K**), (**J**), (**V**) and (**p**) are known, then its Interest Portion planned is:

$$i = (\{S'[1+s] - Sv - F - T - \{[N+K+J+Vp/360] / [1+/-U\} / [1-d]]) / \{S'[1+s]\}$$

Rule-58010:

If both (**/**), (**U**), (**S'**), (**s**), (**S**), (**v**), (**F**), (**i**), (**N**), (**d**), (**K**), (**J**), (**V**) and (**p**) are known, then its Tax planned is:

$$T = \{S'[1+s] - Sv - F - S'i[1+s] - \{[N+K+J+Vp/360] / [1+/-U\} / [1-d]$$

Rule-58011:

If both (**/**), (**U**), (**S'**), (**s**), (**S**), (**v**), (**F**), (**i**), (**T**), (**d**), (**K**), (**J**), (**V**) and (**p**) are known, then its Non Current Asset planned is:

$$d = 1 - \{[N+K+J+Vp/360] / [1+/-U\} / \{S'[1+s] - Sv - F - S'i[1+s] - T\}$$

Steve Asikin ISBN 13: **978-1564912361**, ISBN 10: **1564912363**

Rule-58012:

If both (**⁄**), (**U**), (**S'**), (**s**), (**S**), (**v**), (**F**), (**i**), (**T**), (**D**), (**N**), (**J**), (**V**) and (**p**) are known, then its Kind of Cash planned is:

$$K= [1+/(U+\{S'[1+s]-Sv-F-S'i[1+s]-T\}[1-d])$$
$$-[N+J+Vp/360]$$

Rule-58013:

If both (**⁄**), (**U**), (**S'**), (**s**), (**S**), (**v**), (**F**), (**i**), (**T**), (**D**), (**K**), (**N**), (**V**) and (**p**) are known, then its Job or Trade Account Receivable planned is:

$$J= [1+/(U+\{S'[1+s]-Sv-F-S'i[1+s]-T\}[1-d])$$
$$-\{N+K+Vp/360\}$$

Rule-58014:

If both (**⁄**), (**U**), (**S'**), (**s**), (**S**), (**v**), (**F**), (**i**), (**T**), (**D**), (**K**), (**J**), (**V**) and (**p**) are known, then its Variable Portion planned is:

$$V= 360[[1+/(U+\{S'[1+s]-Sv-F-S'i[1+s]-T\}[1-d])$$
$$-[N+K+J]]/p$$
or
$$V= Sv$$

Rule-58015:

If both (**⁄**), (**U**), (**S'**), (**s**), (**S**), (**v**), (**F**), (**i**), (**T**), (**D**), (**K**), (**J**), (**V**) and (**N**) are known, then its Procured Inventory Days planned is:

$$p= 360[[1+/(U+\{S'[1+s]-Sv-F-S'i[1+s]-T\}[1-d])$$
$$-[N+K+J]]/V$$

Steve Asikin ISBN 13: **978-1564912361**, ISBN 10: **1564912363**

Rule-58016:

If both (**/**), (**U**), (**S'**), (**s**), (**S**), (**v**), (**F**), (**i**), (**T**), (**d**), (**K**), (**J**) and (**p**) are known, then its Non Current Asset planned is:

$$N = [1+/](U+\{S'[1+s]-Sv-F-S'i[1+s]-T\}[1-d])$$
$$-K-J-Svp/360$$

Rule-58017:

If both (**N**), (**U**), (**S'**), (**s**), (**S**), (**v**), (**F**), (**i**), (**T**), (**d**), (**K**), (**J**) and (**p**) are known, then its Leverage or Gearing Ratio planned is:

$$/ = [N+K+J+Svp/360]$$
$$/(U+\{S'[1+s]-Sv-F-S'i[1+s]-T\}[1-d])-1$$

Rule-58018:

If both (**/**), (**N**), (**S'**), (**s**), (**S**), (**v**), (**F**), (**i**), (**T**), (**d**), (**K**), (**J**) and (**p**) are known, then its Utilized or Starting Capital must be:

$$U = ([N+K+J+Svp/360]-[1+/]\{S'[1+s]-Sv-F$$
$$-S'i[1+s]-T\}[1-d])/[1+/]$$

Rule-58019:

If both (**/**), (**U**), (**N**), (**s**), (**S**), (**v**), (**F**), (**i**), (**T**), (**d**), (**K**), (**J**) and (**p**) are known, then its Sales Past must be:

$$S' = (\{[N+K+J+Svp/360]/[1+/]-U\}/[1-d]$$
$$+Sv+F+T)/\{[1+s][1-i]\}$$

or

$$S' = S/[1+s]$$

Steve Asikin ISBN 13: **978-1564912361**, ISBN 10: **1564912363**

Rule-58020:

If both (**/**), (**U**), (**$'**), (**N**), (**$**), (**v**), (**F**), (**i**), (**T**), (**d**), (**K**), (**J**) and (**p**) are known, then its Sales Growth planned is:

$$s= (\{[N+K+J+\$vp/360]/[1+/-U\}/[1-d]+\$v+F+T)$$
$$/\{\$'[1-i]\}-1$$

or

$$s= \$/\$'-1$$

Rule-58021:

If both (**/**), (**U**), (**$'**), (**s**), (**N**), (**v**), (**F**), (**i**), (**T**), (**d**), (**K**), (**J**) and (**p**) are known, then its Sales or Revenue planned is:

$$\$= [[1+/(U+\{\$'[1+s]-F-\$'i[1+s]-T\}[1-d])$$
$$-[N+K+J]]/(v\{p/360]+[1+/[1-d]\})$$

or

$$\$= \$'[1+s]$$

Rule-58022:

If both (**/**), (**U**), (**$'**), (**s**), (**$**), (**N**), (**F**), (**i**), (**T**), (**d**), (**K**), (**J**) and (**p**) are known, then its Variable Portion planned is:

$$v= [[1+/(U+\{\$'[1+s]-F-\$'i[1+s]-T\}[1-d])$$
$$-[N+K+J]]/(\${p/360+[1+/[1-d]])$$

Steve Asikin ISBN 13: **978-1564912361**, ISBN 10: **1564912363**

Rule-58023:

If both (**/**), (**U**), (**S'**), (**s**), (**S**), (**v**), (**N**), (**i**), (**T**), (**d**), (**K**), (**J**) and (**p**) are known, then its Fixed Cost planned is:

$$F = S'[1+s][1-i]-Sv-T$$
$$-\{[N+K+J+Svp/360]/[1+/-U\}/[1-d]$$

Rule-58024:

If both (**/**), (**U**), (**S'**), (**s**), (**S**), (**v**), (**F**), (**N**), (**T**), (**d**), (**K**), (**J**) and (**p**) are known, then its Interest Portion planned is:

$$i = (S'[1+s]-Sv-F-T-\{[N+K+J+Svp/360]/[1+/-U\}$$
$$/[1-d])/\{S'[1+s]\}$$

Rule-58025:

If both (**/**), (**U**), (**S'**), (**s**), (**S**), (**v**), (**F**), (**i**), (**N**), (**d**), (**K**), (**J**) and (**p**) are known, then its Tax planned is:

$$T = S'[1+s]-Sv-F-S'i[1+s]$$
$$-\{[N+K+J+Svp/360]/[1+/-U\}/[1-d]$$

Rule-58026:

If both (**/**), (**U**), (**S'**), (**s**), (**S**), (**v**), (**F**), (**i**), (**T**), (**N**), (**K**), (**J**) and (**p**) are known, then its Dividend Payout planned is:

$$d = 1-\{[N+K+J+Svp/360]/[1+/-U\}$$
$$/\{S'[1+s]-Sv-F-S'i[1+s]-T\}$$

Steve Asikin ISBN 13: **978-1564912361**, ISBN 10: **1564912363**

Rule-58027:

If both (**/**), (**U**), (**S'**), (**s**), (**S**), (**v**), (**F**), (**i**), (**T**), (**d**), (**N**), (**J**) and (**p**) are known, then its Kind of Cash planned is:

$$K= [1+/(U+\{S'[1+s]-Sv-F-S'i[1+s]-T\}[1-d])$$
$$-[N+J+Svp/360]$$

Rule-58028:

If both (**/**), (**U**), (**S'**), (**s**), (**S**), (**v**), (**F**), (**i**), (**T**), (**d**), (**K**), (**N**) and (**p**) are known, then its Job or Trade Account Receivable planned is:

$$J= [1+/(U+\{S'[1+s]-Sv-F-S'i[1+s]-T\}[1-d])$$
$$-\{N+K+Svp/360\}$$

Rule-58029:

If both (**/**), (**U**), (**S'**), (**s**), (**S**), (**v**), (**F**), (**i**), (**T**), (**d**), (**K**), (**J**) and (**N**) are known, then its Procured Inventory Days planned is:

$$p= 360[[1+/(U+\{S'[1+s]-Sv-F-S'i[1+s]-T\}[1-d])$$
$$-[N+K+J]]/[Sv]$$

Rule-58030:

If both (**/**), (**U**), (**S'**), (**s**), (**S**), (**v**), (**F**), (**i**), (**T**), (**d**), (**K**), (**J**) and (**p**) are known, then its Non Current Asset planned is:

$$N= [1+/(U+\{S'[1+s]-Sv-F-S'i[1+s]-T\}[1-d])$$
$$-K-J-S'vp[1+s]/360$$

Steve Asikin ISBN 13: **978-1564912361**, ISBN 10: **1564912363**

Rule-58031:

If both (**N**), (**U**), (**S'**), (**s**), (**S**), (**v**), (**F**), (**i**), (**T**), (**d**), (**K**), (**J**) and (**p**) are known, then its Leverage or Gearing Ratio planned is:

$$I= \{\mathbf{N}+\mathbf{K}+\mathbf{J}+\mathbf{S'vp}[1+\mathbf{s}]/360\}$$
$$/(\mathbf{U}+\{\mathbf{S'}[1+\mathbf{s}]-\mathbf{Sv}-\mathbf{F}-\mathbf{S'i}[1+\mathbf{s}]-\mathbf{T}\}[1-\mathbf{d}])-1$$

Rule-58032:

If both (**I**), (**N**), (**S'**), (**s**), (**S**), (**v**), (**F**), (**i**), (**T**), (**d**), (**K**), (**J**) and (**p**) are known, then its Utilized or Starting Capital must be:

$$\mathbf{U}= (\{\mathbf{N}+\mathbf{K}+\mathbf{J}+\mathbf{S'vp}[1+\mathbf{s}]/360\}-[1+I]\{\mathbf{S'}[1+\mathbf{s}]-\mathbf{Sv}$$
$$-\mathbf{F}-\mathbf{S'i}[1+\mathbf{s}]-\mathbf{T}\}[1-\mathbf{d}])/[1+I]$$

Rule-58033:

If both (**I**), (**U**), (**N**), (**s**), (**S**), (**v**), (**F**), (**i**), (**T**), (**d**), (**K**), (**J**) and (**p**) are known, then its Sales Past must be:

$$\mathbf{S'}= [[1+I](\mathbf{U}+\{\mathbf{Sv}-\mathbf{F}-\mathbf{T}\}[1-\mathbf{d}])-[\mathbf{N}+\mathbf{K}+\mathbf{J}]]$$
$$/([1+\mathbf{s}]\{\mathbf{vp}/360-[1+I][1-\mathbf{d}][1-\mathbf{i}]\})$$

or

$$\mathbf{S'}= \mathbf{S}/[1+\mathbf{s}]$$

Steve Asikin ISBN 13: **978-1564912361**, ISBN 10: **1564912363**

Rule-58034:

If both $(\mathit{l})$, $(\mathbf{U})$, $(\mathbf{S'})$, $(\mathbf{N})$, $(\mathbf{S})$, $(\mathbf{v})$, $(\mathbf{F})$, $(\mathbf{i})$, $(\mathbf{T})$, $(\mathbf{d})$, $(\mathbf{K})$, $(\mathbf{J})$ and $(\mathbf{p})$ are known, then its Sales Growth planned is:

$$s= [[1+\mathit{l}(\mathbf{U}+\{\mathbf{Sv}\text{-}\mathbf{F}\text{-}\mathbf{T}\}[1\text{-}\mathbf{d}])\text{-}[\mathbf{N}+\mathbf{K}+\mathbf{J}]]$$
$$/(\mathbf{S'}\{\mathbf{vp}/360\text{-}[1+\mathit{l}[1\text{-}\mathbf{d}][1\text{-}\mathbf{i}]\})\text{-}1$$

or

$$s= \mathbf{S}/\mathbf{S'}\text{-}1$$


Rule-58035:

If both $(\mathit{l})$, $(\mathbf{U})$, $(\mathbf{S'})$, $(\mathbf{s})$, $(\mathbf{N})$, $(\mathbf{v})$, $(\mathbf{F})$, $(\mathbf{i})$, $(\mathbf{T})$, $(\mathbf{d})$, $(\mathbf{K})$, $(\mathbf{J})$ and $(\mathbf{p})$ are known, then its Sales or Revenue planned is:

$$S= [(\{\mathbf{S'}[1+\mathbf{s}]\text{-}\mathbf{F}\text{-}\mathbf{S'i}[1+\mathbf{s}]\text{-}\mathbf{T}\}\text{-}\{\mathbf{N}+\mathbf{K}+\mathbf{J}$$
$$+\mathbf{S'vp}[1+\mathbf{s}]/360\}/[1+\mathit{l}\text{-}\mathbf{U}]/[1\text{-}\mathbf{d}]]/\mathbf{v}$$

or

$$S= \mathbf{S'}[1+\mathbf{s}]$$


Rule-58036:

If both $(\mathit{l})$, $(\mathbf{U})$, $(\mathbf{S'})$, $(\mathbf{s})$, $(\mathbf{S})$, $(\mathbf{N})$, $(\mathbf{F})$, $(\mathbf{i})$, $(\mathbf{T})$, $(\mathbf{d})$, $(\mathbf{K})$, $(\mathbf{J})$ and $(\mathbf{p})$ are known, then its Variable Portion planned is:

$$v= [[1+\mathit{l}(\mathbf{U}+\{\mathbf{S'}[1+\mathbf{s}]\text{-}\mathbf{F}\text{-}\mathbf{S'i}[1+\mathbf{s}]\text{-}\mathbf{T}\}[1\text{-}\mathbf{d}])$$
$$\text{-}[\mathbf{N}+\mathbf{K}+\mathbf{J}]]/\{\mathbf{S'p}[1+\mathbf{s}]/360+\mathbf{S}[1+\mathit{l}[1\text{-}\mathbf{d}]\}$$

Steve Asikin ISBN 13: **978-1564912361**, ISBN 10: **1564912363**

Rule-58037:

If both (**/**), (**U**), (**S'**), (**s**), (**S**), (**v**), (**N**), (**i**), (**T**), (**d**), (**K**), (**J**) and (**p**) are known, then its Fixed Cost planned is:

$$F= S'[1+s]-Sv-S'i[1+s]-T-(\{N+K+J$$
$$+S'vp[1+s]/360\}/[1+/\,-U)/[1-d]\}$$

Rule-58038:

If both (**/**), (**U**), (**S'**), (**s**), (**S**), (**v**), (**F**), (**N**), (**T**), (**d**), (**K**), (**J**) and (**p**) are known, then its Interest Portion planned is:

$$i= (S'[1+s]-Sv-F-T-(\{N+K+J+S'vp[1\mid s]/360\}$$
$$/[1+/\,-U)/\{S'[1+s]\}$$

Rule-58039:

If both (**/**), (**U**), (**S'**), (**s**), (**S**), (**v**), (**F**), (**i**), (**N**), (**d**), (**K**), (**J**) and (**p**) are known, then its Tax planned is:

$$T= S'[1+s]-Sv-F-S'i[1+s]-(\{N+K+J$$
$$+S'vp[1+s]/360\}/[1+/\,-U)/[1-d]$$

Rule-58040:

If both (**/**), (**U**), (**S'**), (**s**), (**S**), (**v**), (**F**), (**i**), (**T**), (**N**), (**K**), (**J**) and (**p**) are known, then its Dividend Payout planned is:

$$d= 1-(\{N+K+J+S'vp[1+s]/360\}/[1+/\,-U)$$
$$/\{S'[1+s]-Sv-F-S'i[1+s]-T\}$$

Steve Asikin ISBN 13: **978-1564912361**, ISBN 10: **1564912363**

Rule-58041:

If both (**/**), (**U**), (**$'**), (**s**), (**$**), (**v**), (**F**), (**i**), (**T**), (**d**), (**N**), (**J**) and (**p**) are known, then its Kind of Cash planned is:

$$K= [1+/](U+\{\$'[1+s]-\$v-F-\$'i[1+s]-T\}[1-d])$$
$$-\{N+J+\$'vp[1+s]/360\}$$


Rule-58042:

If both (**/**), (**U**), (**$'**), (**s**), (**$**), (**v**), (**F**), (**i**), (**T**), (**d**), (**K**), (**N**) and (**p**) are known, then its Job or Trade Account Receivable planned is:

$$J= [[1+/](U+\{\$'[1+s]-\$v-F-\$'i[1+s]-T\}[1-d])$$
$$-\{N+K+\$'vp[1+s]/360\}$$


Rule-58043:

If both (**/**), (**U**), (**$'**), (**s**), (**$**), (**v**), (**F**), (**i**), (**T**), (**d**), (**K**), (**J**) and (**N**) are known, then its Procured Inventory Days planned is:

$$p= 360[[1+/](U+\{\$'[1+s]-\$v-F-\$'i[1+s]-T\}[1-d])$$
$$-[N+K+J]]/\{\$'v[1+s]\}$$


Rule-58044:

If both (**/**), (**U**), (**$'**), (**s**), (**$**), (**v**), (**F**), (**i**), (**T**), (**d**), (**K**), (**j**) and (**P**) are known, then its Non Current Asset planned is:

$$N= [1+/](U+\{\$'[1+s]-\$v-F-\$'i[1+s]-T\}[1-d])$$
$$-K-\$j/360-P$$

Steve Asikin ISBN 13: **978-1564912361**, ISBN 10: **1564912363**

Rule-58045:

If both (**N**), (**U**), (**S'**), (**s**), (**S**), (**v**), (**F**), (**i**), (**T**), (**d**), (**K**), (**j**) and (**P**) are known, then its Leverage or Gearing Ratio planned is:

$$I = \{N+K+Sj/360+P\}$$
$$/(U+\{S'[1+s]-Sv-F-S'i[1+s]-T\}[1-d])-1$$

Rule-58046:

If both (**I**), (**N**), (**S'**), (**s**), (**S**), (**v**), (**F**), (**i**), (**T**), (**d**), (**K**), (**j**) and (**P**) are known, then its Utilized or Starting Capital must be:

$$U = (\{N+K+Sj/360+P\}-[1+I]\{S'[1+s]-Sv-F$$
$$-S'i[1+s]-T\}[1-d])/[1+I]$$

Rule-58047:

If both (**I**), (**U**), (**N**), (**s**), (**S**), (**v**), (**F**), (**i**), (**T**), (**d**), (**K**), (**j**) and (**P**) are known, then its Sales Past must be:

$$S' = (\{[N+K+Sj/360+P]/[1+I]-U\}/[1-d]+Sv+F+T)$$
$$/\{[1+s][1-i]\}$$

or

$$S' = S/[1+s]$$

Steve Asikin ISBN 13: **978-1564912361**, ISBN 10: **1564912363**

Rule-58048:

If both $(\mathit{l})$, $(\mathbf{U})$, $(\mathbf{S'})$, $(\mathbf{N})$, $(\mathbf{S})$, $(\mathbf{v})$, $(\mathbf{F})$, $(\mathbf{i})$, $(\mathbf{T})$, $(\mathbf{d})$, $(\mathbf{K})$, $(\mathbf{j})$ and $(\mathbf{P})$ are known, then its Sales Growth planned is:

$$s = (\{[\mathbf{N}+\mathbf{K}+\mathbf{S}\mathbf{j}/360+\mathbf{P}]/[1+\mathit{l}-\mathbf{U}\}/[1-\mathbf{d}]+\mathbf{S}\mathbf{v}+\mathbf{F}+\mathbf{T})$$
$$/\{\mathbf{S'}[1-\mathbf{i}]\}-1$$

or

$$s = \mathbf{S}/\mathbf{S'}-1$$


Rule-58049:

If both $(\mathit{l})$, $(\mathbf{U})$, $(\mathbf{S'})$, $(\mathbf{s})$, $(\mathbf{N})$, $(\mathbf{v})$, $(\mathbf{F})$, $(\mathbf{i})$, $(\mathbf{T})$, $(\mathbf{d})$, $(\mathbf{K})$, $(\mathbf{j})$ and $(\mathbf{P})$ are known, then its Sales or Revenue planned is:

$$\mathbf{S} = [[1+\mathit{l}(\mathbf{U}+\{\mathbf{S'}[1+\mathbf{s}]-\mathbf{F}-\mathbf{S'}\mathbf{i}[1+\mathbf{s}]-\mathbf{T}\}[1-\mathbf{d}])$$
$$-[\mathbf{N}+\mathbf{K}+\mathbf{P}]]/\{\mathbf{j}/360+\mathbf{v}[1+\mathit{l}[1-\mathbf{d}]\}$$

or

$$\mathbf{S} = \mathbf{S'}[1+\mathbf{s}]$$


Rule-58050:

If both $(\mathit{l})$, $(\mathbf{U})$, $(\mathbf{S'})$, $(\mathbf{s})$, $(\mathbf{S})$, $(\mathbf{n})$, $(\mathbf{F})$, $(\mathbf{i})$, $(\mathbf{T})$, $(\mathbf{d})$, $(\mathbf{K})$, $(\mathbf{j})$ and $(\mathbf{P})$ are known, then its Variable Portion planned is:

$$\mathbf{v} = (\mathbf{U}+\{\mathbf{S'}[1+\mathbf{s}]-\mathbf{F}-\mathbf{S'}\mathbf{i}[1+\mathbf{s}]-\mathbf{T}\}[1-\mathbf{d}]$$
$$-[\mathbf{N}+\mathbf{K}+\mathbf{S}\mathbf{j}/360+\mathbf{P}]/[1+\mathit{l})/\{\mathbf{S}[1-\mathbf{d}]\}$$

Steve Asikin ISBN 13: **978-1564912361**, ISBN 10: **1564912363**

Rule-58051:

If both (**I**), (**U**), (**S'**), (**s**), (**S**), (**v**), (**N**), (**i**), (**T**), (**d**), (**K**), (**j**) and (**P**) are known, then its Fixed Cost planned is:

$$F = S'[1+s] - Sv - S'i[1+s] - T$$
$$- \{[N+K+Sj/360+P]/[1+I-U]\}/[1-d]$$

Rule-58052:

If both (**I**), (**U**), (**S'**), (**s**), (**S**), (**v**), (**F**), (**N**), (**T**), (**d**), (**K**), (**j**) and (**P**) are known, then its Interest Portion planned is:

$$i = (S'[1+s] - Sv - F - T - \{[N+K+Sj/360+P]/[1+I-U]\}$$
$$/[1-d])/\{S'[1+s]\}$$

Rule-58053:

If both (**I**), (**U**), (**S'**), (**s**), (**S**), (**v**), (**F**), (**i**), (**N**), (**d**), (**K**), (**j**) and (**P**) are known, then its Tax planned is:

$$T = S'[1+s] - Sv - F - S'i[1+s]$$
$$- \{[N+K+Sj/360+P]/[1+I-U]\}/[1-d]$$

Rule-58054:

If both (**I**), (**U**), (**S'**), (**s**), (**S**), (**v**), (**F**), (**i**), (**T**), (**N**), (**K**), (**j**) and (**P**) are known, then its Dividend Payout planned is:

$$d = 1 - \{[N+K+Sj/360+P]/[1+I-U]\}$$
$$/\{S'[1+s] - Sv - F - S'i[1+s] - T\}$$

Steve Asikin ISBN 13: **978-1564912361**, ISBN 10: **1564912363**

Rule-58055:

If both (**/**), (**U**), (**S'**), (**s**), (**S**), (**v**), (**F**), (**i**), (**T**), (**d**), (**N**), (**j**) and (**P**) are known, then its Kind of Cash planned is:

$$K= [1+/(U+\{S'[1+s]-Sv-F-S'i[1+s]-T\}[1-d])$$
$$-[N+Sj/360+P]$$

Rule-58056:

If both (**/**), (**U**), (**S'**), (**s**), (**S**), (**v**), (**F**), (**i**), (**T**), (**d**), (**K**), (**N**) and (**P**) are known, then its Job or Trade Receivable Days planned is:

$$j= 360[[1+/(U+\{S'[1+s]-Sv-F-S'i[1+s]-T\}[1-d])$$
$$-\{N+K+P\}]/S$$

Rule-58057:

If both (**/**), (**U**), (**S'**), (**s**), (**S**), (**v**), (**F**), (**i**), (**T**), (**d**), (**K**), (**j**) and (**N**) are known, then its Procured Inventory planned is:

$$P= [1+/(U+\{S'[1+s]-Sv-F-S'i[1+s]-T\}[1-d])$$
$$-\{N+K+Sj/360\}$$

Rule-58058:

If both (**/**), (**U**), (**S'**), (**s**), (**S**), (**v**), (**F**), (**i**), (**T**), (**d**), (**K**), (**j**), (**V**) and (**p**) are known, then its Non Current Asset planned is:

$$N= [1+/(U+\{S'[1+s]-Sv-F-S'i[1+s]-T\}[1-d])$$
$$-K-Sj/360-Vp/360$$

Steve Asikin ISBN 13: **978-1564912361**, ISBN 10: **1564912363**

Rule-58059:

If both (**N**), (**U**), (**S'**), (**s**), (**S**), (**v**), (**F**), (**i**), (**T**), (**d**), (**K**), (**j**), (**V**) and (**p**) are known, then its Leverage or Gearing Ratio planned is:

$$\textit{l} = [\mathbf{N}+\mathbf{K}+\mathbf{Sj}/360+\mathbf{Vp}/360]$$
$$/(\mathbf{U}+\{\mathbf{S'}[1+\mathbf{s}]-\mathbf{Sv}-\mathbf{F}-\mathbf{S'i}[1+\mathbf{s}]-\mathbf{T}\}[1-\mathbf{d}])-1$$

Rule-58060:

If both (**l**), (**N**), (**S'**), (**s**), (**S**), (**v**), (**F**), (**i**), (**T**), (**d**), (**K**), (**j**), (**V**) and (**p**) are known, then its Utilized or Starting Capital must be:

$$\mathbf{U} = ([\mathbf{N}+\mathbf{K}+\mathbf{Sj}/360+\mathbf{Vp}/360]-[1+\textit{l}]\{\mathbf{S'}[1+\mathbf{s}]-\mathbf{Sv}-\mathbf{F}$$
$$-\mathbf{S'i}[1+\mathbf{s}]-\mathbf{T}\}[1-\mathbf{d}])/[1+\textit{l}]$$

Rule-58061:

If both (**l**), (**U**), (**N**), (**s**), (**S**), (**v**), (**F**), (**i**), (**T**), (**d**), (**K**), (**j**), (**V**) and (**p**) are known, then its Sales Past must be:

$$\mathbf{S'} = (\{[\mathbf{N}+\mathbf{K}+\mathbf{Sj}/360+\mathbf{Vp}/360]/[1+\textit{l}]-\mathbf{U}\}/[1-\mathbf{d}]$$
$$+\mathbf{Sv}+\mathbf{F}+\mathbf{T}\})/\{\mathbf{S'}[1+\mathbf{s}][1-\mathbf{i}]\}$$
$$\text{or}$$
$$\mathbf{S'} = \mathbf{S}[1+\mathbf{s}]$$

Steve Asikin ISBN 13: **978-1564912361**, ISBN 10: **1564912363**

Rule-58062:

If both (**/**), (**U**), (**S'**), (**N**), (**S**), (**v**), (**F**), (**i**), (**T**), (**d**), (**K**), (**j**), (**V**) and (**p**) are known, then its Sales Growth planned is:

$$s= (\{[\mathbf{N}+\mathbf{K}+\mathbf{S}\mathbf{j}/360+\mathbf{V}\mathbf{p}/360]/[1+\mathbf{/}-\mathbf{U}]/[1-\mathbf{d}] +\mathbf{S}\mathbf{v}+\mathbf{F}+\mathbf{T}\})/\{\mathbf{S}'[1-\mathbf{i}]\}-1$$

or

$$s= \mathbf{S}/\mathbf{S}'-1$$

Rule-58063:

If both (**/**), (**U**), (**S'**), (**s**), (**N**), (**v**), (**F**), (**i**), (**T**), (**d**), (**K**), (**j**), (**V**) and (**p**) are known, then its Sales or Revenue planned is:

$$\mathbf{S}= (\mathbf{U}+\{\mathbf{S}'[1+\mathbf{s}]-\mathbf{F}-\mathbf{S}'\mathbf{i}[1+\mathbf{s}]-\mathbf{T}\}[1-\mathbf{d}]-[\mathbf{N}+\mathbf{K} +\mathbf{V}\mathbf{p}/360]/[1+\mathbf{/}])/\{\mathbf{j}/360+\mathbf{v}[1+\mathbf{/}][1-\mathbf{d}]\}$$

or

$$\mathbf{S}= \mathbf{S}'[1+\mathbf{s}]$$

Rule-58064:

If both (**/**), (**U**), (**S'**), (**s**), (**S**), (**N**), (**F**), (**i**), (**T**), (**d**), (**K**), (**j**), (**V**) and (**p**) are known, then its Variable Portion planned is:

$$\mathbf{v}= (\mathbf{U}+\{\mathbf{S}'[1+\mathbf{s}]-\mathbf{F}-\mathbf{S}'\mathbf{i}[1+\mathbf{s}]-\mathbf{T}\}[1-\mathbf{d}] -[\mathbf{N}+\mathbf{K}+\mathbf{S}\mathbf{j}/360+\mathbf{V}\mathbf{p}/360]/[1+\mathbf{/}])/\{\mathbf{S}[1-\mathbf{d}]\}$$

or

$$\mathbf{v}= \mathbf{V}/\mathbf{S}$$

Steve Asikin ISBN 13: **978-1564912361**, ISBN 10: **1564912363**

Rule-58065:

If both (**/**), (**U**), (**S'**), (**s**), (**S**), (**v**), (**N**), (**i**), (**T**), (**d**), (**K**), (**j**), (**V**) and (**p**) are known, then its Fixed Cost planned is:

$$F = S'[1+s]-Sv-S'i[1+s]-T$$
$$-\{[N+K+Sj/360+Vp/360]/[1+/-U\}/[1-d]$$

Rule-58066:

If both (**/**), (**U**), (**S'**), (**s**), (**S**), (**v**), (**F**), (**N**), (**T**), (**d**), (**K**), (**j**), (**V**) and (**p**) are known, then its Interest Portion planned is:

$$i = (S'[1+s]-Sv-F-T-\{[N+K+Sj/360+Vp/360]$$
$$/[1+/-U\}/[1-d])/\{S'[1+s]\}$$

Rule-58067:

If both (**/**), (**U**), (**S'**), (**s**), (**S**), (**v**), (**F**), (**i**), (**N**), (**d**), (**K**), (**j**), (**V**) and (**p**) are known, then its Tax planned is:

$$T = S'[1+s]-Sv-F-S'i[1+s]-\{[N+K+Sj/360+Vp/360]$$
$$/[1+/-U\}/[1-d]$$

Rule-58068:

If both (**/**), (**U**), (**S'**), (**s**), (**S**), (**v**), (**F**), (**i**), (**T**), (**N**), (**K**), (**j**), (**V**) and (**p**) are known, then its Dividend Payout planned is:

$$d = 1-\{[N+K+Sj/360+Vp/360]/[1+/-U\}$$
$$/\{S'[1+s]-Sv-F-S'i[1+s]-T\}$$

Steve Asikin ISBN 13: **978-1564912361**, ISBN 10: **1564912363**

Rule-58069:

If both (**∫**), (**U**), (**S'**), (**s**), (**\$**), (**v**), (**F**), (**i**), (**T**), (**d**), (**N**), (**j**), (**V**) and (**p**) are known, then its Kind of Capital must be planned is:

$$K= [1+\text{∫}(U+\{S'[1+s]\text{-}Sv\text{-}F\text{-}S'i[1+s]\text{-}T\}[1\text{-}d])$$
$$-[N+Sj/360+Vp/360]$$

Rule-58070:

If both (**∫**), (**U**), (**S'**), (**s**), (**\$**), (**v**), (**F**), (**i**), (**T**), (**d**), (**K**), (**N**), (**V**) and (**p**) are known, then its Job or Trade Receivable Days planned is:

$$j= 360[[1+\text{∫}(U+\{S'[1+s]\text{-}Sv\text{-}F\text{-}S'i[1+s]\text{-}T\}[1\text{-}d])$$
$$-[N+K+Vp/360]]/S$$

Rule-58071:

If both (**∫**), (**U**), (**S'**), (**s**), (**\$**), (**v**), (**F**), (**i**), (**T**), (**d**), (**K**), (**j**), (**N**) and (**p**) are known, then its Variable Cost planned is:

$$V= 360[[1+\text{∫}(U+\{S'[1+s]\text{-}Sv\text{-}F\text{-}S'i[1+s]\text{-}T\}[1\text{-}d])$$
$$-[N+K+Sj/360]]/p$$
$$\text{or}$$
$$V= Sv$$

Rule-58072:

If both (**∫**), (**U**), (**S'**), (**s**), (**\$**), (**v**), (**F**), (**i**), (**T**), (**d**), (**K**), (**j**), (**V**) and (**N**) are known, then its Procured Inventory planned is:

$$p= 360[[1+\text{∫}(U+\{S'[1+s]\text{-}Sv\text{-}F\text{-}S'i[1+s]\text{-}T\}[1\text{-}d])$$
$$-[N+K+Sj/360]]/V$$

Steve Asikin ISBN 13: **978-1564912361**, ISBN 10: **1564912363**

Rule-58073:

If both (**I**), (**U**), (**S'**), (**s**), (**S**), (**v**), (**F**), (**i**), (**T**), (**d**), (**K**), (**j**) and (**p**) are known, then its Non Current Asset planned is:

$$\mathbf{N} = [1+\mathbf{I}](\mathbf{U}+\{\mathbf{S'}[1+\mathbf{s}]-\mathbf{Sv}-\mathbf{F}-\mathbf{S'i}[1+\mathbf{s}]-\mathbf{T}\}[1-\mathbf{d}])$$
$$-\mathbf{K}-\mathbf{Sj}/360-\mathbf{Svp}/360$$

Rule-58074:

If both (**N**), (**U**), (**S'**), (**s**), (**S**), (**v**), (**F**), (**i**), (**T**), (**d**), (**K**), (**j**) and (**p**) are known, then its Leverage or Gearing Ratio planned is:

$$\mathbf{I} = [\mathbf{N}+\mathbf{K}+\mathbf{Sj}/360+\mathbf{Svp}/360]$$
$$/(\mathbf{U}+\{\mathbf{S'}[1+\mathbf{s}]-\mathbf{Sv}-\mathbf{F}-\mathbf{S'i}[1+\mathbf{s}]-\mathbf{T}\}[1-\mathbf{d}])-1$$

Rule-58075:

If both (**I**), (**N**), (**S'**), (**s**), (**S**), (**v**), (**F**), (**i**), (**T**), (**d**), (**K**), (**j**) and (**p**) are known, then its Utilized or Starting Capital must be:

$$\mathbf{U} = ([\mathbf{N}+\mathbf{K}+\mathbf{Sj}/360+\mathbf{Svp}/360]-[1+\mathbf{I}]\{\mathbf{S'}[1+\mathbf{s}]-\mathbf{Sv}-\mathbf{F}$$
$$-\mathbf{S'i}[1+\mathbf{s}]-\mathbf{T}\}[1-\mathbf{d}])/[1+\mathbf{I}]$$

Rule-58076:

If both (**I**), (**U**), (**N**), (**s**), (**S**), (**v**), (**F**), (**i**), (**T**), (**d**), (**K**), (**j**) and (**p**) are known, then its Sales Past must be:

$$\mathbf{S'} = (\{[\mathbf{N}+\mathbf{K}+\mathbf{Sj}/360+\mathbf{Svp}/360]/[1+\mathbf{I}]-\mathbf{U}$$
$$+\{\mathbf{Sv}+\mathbf{F}-\mathbf{T}\}[1-\mathbf{d}])/\{[1+\mathbf{s}][1-\mathbf{i}][1-\mathbf{d}]\}$$
$$\text{or}$$
$$\mathbf{S'} = \mathbf{S}/[1+\mathbf{s}]$$

`

Steve Asikin ISBN 13: **978-1564912361**, ISBN 10: **1564912363**

Rule-58077:

If both (**/**), (**U**), (**$'**), (**N**), (**$**), (**v**), (**F**), (**i**), (**T**), (**d**), (**K**), (**j**) and (**p**) are known, then its Sales Growth planned is:

$$s = (\{[\textbf{N}+\textbf{K}+\textbf{\$j}/360+\textbf{\$vp}/360]/[1+\textbf{/}]-\textbf{U}+\{\textbf{\$v}+\textbf{F-T}\}$$
$$[1-\textbf{d}])/\{\textbf{\$'}[1-\textbf{i}][1-\textbf{d}]\}-1$$

or

$$s = \textbf{\$}/\textbf{\$'}-1$$

Rule-58078:

If both (**/**), (**U**), (**$'**), (**s**), (**N**), (**v**), (**F**), (**i**), (**T**), (**d**), (**K**), (**j**) and (**p**) are known, then its Sales or Revenue planned is:

$$\textbf{\$} = ([1+\textbf{/}](\textbf{U}+\{\textbf{\$'}[1+\textbf{s}]-\textbf{F}-\textbf{\$'i}[1+\textbf{s}]-\textbf{T}\}[1-\textbf{d}])-[\textbf{N}+\textbf{K}])$$
$$/(\textbf{j}/360+\textbf{v}\{\textbf{p}/360+[1+\textbf{/}][1-\textbf{d}]\}$$

or

$$\textbf{\$} = \textbf{\$'}[1+\textbf{s}]$$

Rule-58079:

If both (**/**), (**U**), (**$'**), (**s**), (**$**), (**N**), (**F**), (**i**), (**T**), (**d**), (**K**), (**j**) and (**p**) are known, then its Variable Portion planned is:

$$\textbf{v} = [[1+\textbf{/}](\textbf{U}+\{\textbf{\$'}[1+\textbf{s}]-\textbf{F}-\textbf{\$'i}[1+\textbf{s}]-\textbf{T}\}[1-\textbf{d}])$$
$$-[\textbf{N}+\textbf{K}+\textbf{\$j}/360]]/(\textbf{\$}\{\textbf{p}/360+[1+\textbf{/}][1-\textbf{d}]\})$$

Steve Asikin ISBN 13: **978-1564912361**, ISBN 10: **1564912363**

Rule-58080:

If both ($I$), (**U**), (**S'**), (**s**), (**S**), (**v**), (**N**), (**i**), (**T**), (**d**), (**K**), (**j**) and (**p**) are known, then its Fixed Cost planned is:

$$F= S'[1+s]-Sv-S'i[1+s]-T$$
$$-\{[N+K+Sj/360+Svp/360]/[1+I-U\}/[1-d]$$

Rule-58081:

If both ($I$), (**U**), (**S'**), (**s**), (**S**), (**v**), (**F**), (**N**), (**T**), (**d**), (**K**), (**j**) and (**p**) are known, then its Interest Portion planned is:

$$i= (T= S'[1+s]-Sv-F-T-\{[N+K+Sj/360 + Svp/360]$$
$$/[1+I-U\}/[1-d])/\{S'[1+s]\}$$

Rule-58082.

If both ($I$), (**U**), (**S'**), (**s**), (**S**), (**v**), (**F**), (**i**), (**N**), (**d**), (**K**), (**j**) and (**p**) are known, then its Tax planned is:

$$T= S'[1+s]-Sv-F-S'i[1+s]$$
$$-\{[N+K+Sj/360+Svp/360]/[1+I-U\}/[1-d]$$

Rule-58083:

If both ($I$), (**U**), (**S'**), (**s**), (**S**), (**v**), (**F**), (**i**), (**T**), (**N**), (**K**), (**j**) and (**p**) are known, then its Dividend Payout planned is:

$$d= 1-\{[N+K+Sj/360+Svp/360]/[1+I-U\}$$
$$/\{S'[1+s]-Sv-F-S'i[1+s]-T\}$$

Steve Asikin ISBN 13: **978-1564912361**, ISBN 10: **1564912363**

Rule-58084:
If both (**/**), (**U**), (**S'**), (**s**), (**$**), (**v**), (**F**), (**i**), (**T**), (**d**), (**N**), (**j**) and (**p**) are known, then its Kind of Cash planned is:

$$K = [1+/](U+\{S'[1+s]-Sv-F-S'i[1+s]-T\}[1-d])$$
$$-[N+Sj/360+Svp/360]$$

Rule-58085:
If both (**/**), (**U**), (**S'**), (**s**), (**$**), (**v**), (**F**), (**i**), (**T**), (**d**), (**K**), (**N**) and (**p**) are known, then its Job or Trade Receivable Days planned is:

$$j = 360[[1+/](U+\{S'[1+s]-Sv-F-S'i[1+s]-T\}[1-d])$$
$$-\{N+K+Svp/360\}]/S$$

Rule-58086:
If both (**/**), (**U**), (**S'**), (**s**), (**$**), (**v**), (**F**), (**i**), (**T**), (**d**), (**K**), (**j**) and (**N**) are known, then its Procured Inventory Days planned is:

$$p = 360[[1+/](U+\{S'[1+s]-Sv-F-S'i[1+s]-T\}[1-d])$$
$$-[N+K+Sj/360]]/[Sv]$$

Rule-58087:
If both (**/**), (**U**), (**S'**), (**s**), (**$**), (**v**), (**F**), (**i**), (**T**), (**d**), (**K**), (**j**) and (**p**) are known, then its Non Current Asset planned is:

$$N = [1+/](U+\{S'[1+s]-Sv-F-S'i[1+s]-T\}[1-d])$$
$$-K-Sj/360-S'vp[1+s]/360$$

Steve Asikin ISBN 13: **978-1564912361**, ISBN 10: **1564912363**

Rule-58088:

If both (**N**), (**U**), (**S'**), (**s**), (**S**), (**v**), (**F**), (**i**), (**T**), (**d**), (**K**), (**j**) and (**p**) are known, then its Leverage or Gearing Ratio planned is:

$$I = \{N+K+Sj/360+S'vp[1+s]/360\}$$
$$/(U+\{S'[1+s]-Sv-F-S'i[1+s]-T\}[1-d])-1$$

Rule-58089:

If both (**I**), (**N**), (**S'**), (**s**), (**S**), (**v**), (**F**), (**i**), (**T**), (**d**), (**K**), (**j**) and (**p**) are known, then its Utilized or Starting Capital must be:

$$U = (\{N+K+Sj/360+S'vp[1+s]/360\}-[1+I]$$
$$\{S'[1+s]-Sv-F-S'i[1+s]-T\}[1-d])/[1+I]$$

Rule-58090:

If both (**I**), (**U**), (**N**), (**s**), (**S**), (**v**), (**F**), (**i**), (**T**), (**d**), (**K**), (**j**) and (**p**) are known, then its Sales Past must be:

$$S' = [[1+I](U+\{Sv+F+T\}[1-d])-[N+K+Sj/360]]$$
$$/([1+s]\{vp/360-[1+I][1-d][1-i]\})$$

or

$$S' = S/[1+s]$$

Steve Asikin ISBN 13: **978-1564912361**, ISBN 10: **1564912363**

Rule-58091:

If both (**/**), (**U**), (**S'**), (**N**), (**S**), (**v**), (**F**), (**i**), (**T**), (**d**), (**K**), (**j**) and (**p**) are known, then its Sales Growth planned is:

$$s= [[1+/(U+\{Sv+F+T\}[1-d])-[N+K+Sj/360]]$$
$$/(S'\{vp/360-[1+/[1-d][1-i]\})-1$$

or

$$s= S/S'-1$$

Rule-58092:

If both (**/**), (**U**), (**S'**), (**s**), (**N**), (**v**), (**F**), (**i**), (**T**), (**d**), (**K**), (**j**) and (**p**) are known, then its Sales or Revenue planned is:

$$S= [[1+/(U+\{S'[1+s]-F-S'i[1+s]-T\}[1-d])$$
$$-\{N+K+S'vp[1+s]/360\}]$$
$$/\{j/360+v[1+/[1-d]\}$$

or

$$S= S'[1+s]$$

Rule-58093:

If both (**/**), (**U**), (**S'**), (**s**), (**S**), (**N**), (**F**), (**i**), (**T**), (**d**), (**K**), (**j**) and (**p**) are known, then its Variable Portion planned is:

$$v= [S'[1+s]-F-S'i[1+s]-T-(\{N+K+Sj/360$$
$$+S'vp[1+s]/360\}/[1+/-U)/[1-d]]/S$$

Steve Asikin ISBN 13: **978-1564912361**, ISBN 10: **1564912363**

Rule-58094:

If both ($l$), (**U**), (**S'**), (**s**), (**S**), (**v**), (**F**), (**N**), (**T**), (**d**), (**K**), (**j**) and (**p**) are known, then its Interest Portion planned is:

$$i= [S'[1+s]-Sv-F-S'i[1+s]-T-(\{N+K+Sj/360 \\ +S'vp[1+s]/360\}/[1+l-U]/\{S'[1+s]\}$$

Rule-58095:

If both ($l$), (**U**), (**S'**), (**s**), (**S**), (**v**), (**N**), (**i**), (**T**), (**d**), (**K**), (**j**) and (**p**) are known, then its Fixed Cost planned is:

$$F= S'[1+s]-Sv-S'i[1+s]-T-(\{N+K+Sj/360 \\ +S'vp[1+s]/360\}/[1+l-U)/[1-d]$$

Rule-58096:

If both ($l$), (**U**), (**S'**), (**s**), (**S**), (**v**), (**F**), (**i**), (**N**), (**d**), (**K**), (**j**) and (**p**) are known, then its Tax planned is:

$$T= S'[1+s]-Sv-F-S'i[1+s]-(\{N+K+Sj/360 \\ +S'vp[1+s]/360\}/[1+l-U)/[1-d]$$

Rule-58097:

If both ($l$), (**U**), (**S'**), (**s**), (**S**), (**v**), (**F**), (**i**), (**T**), (**N**), (**K**), (**j**) and (**p**) are known, then its Dividend Payout planned is:

$$d= 1-(\{N+K+Sj/360+S'vp[1+s]/360\}/[1+l-U) \\ /\{S'[1+s]-Sv-F-S'i[1+s]-T\}$$

Steve Asikin ISBN 13: **978-1564912361**, ISBN 10: **1564912363**

Rule-58098:

If both $(f)$, $(U)$, $(S')$, $(s)$, $(S)$, $(v)$, $(F)$, $(i)$, $(T)$, $(d)$, $(N)$, $(j)$ and $(p)$ are known, then its Kind of Cash planned is:

$$K= [1+f(U+\{S'[1+s]-Sv-F-S'i[1+s]-T\}[1-d])$$
$$-\{N+Sj/360+S'vp[1+s]/360\}$$

Rule-58099:

If both $(f)$, $(U)$, $(S')$, $(s)$, $(S)$, $(v)$, $(F)$, $(i)$, $(T)$, $(d)$, $(K)$, $(N)$ and $(p)$ are known, then its Job or Trade Receivable Days planned is:

$$j= 360[[1+f(U+\{S'[1+s]-Sv-F-S'i[1+s]-T\}[1-d])$$
$$-\{N+K+S'vp[1+s]/360\}]/S$$

Rule-58100:

If both $(f)$, $(U)$, $(S')$, $(s)$, $(S)$, $(v)$, $(F)$, $(i)$, $(T)$, $(d)$, $(K)$, $(j)$ and $(N)$ are known, then its Procured Inventory Days planned is:

$$p= 360[[1+f(U+\{S'[1+s]-Sv-F-S'i[1+s]-T\}[1-d])$$
$$-[N+K+Sj/360]]/\{S'v[1+s]\}$$

Rule-58101:

If both $(f)$, $(U)$, $(S')$, $(s)$, $(S)$, $(v)$, $(F)$, $(i)$, $(T)$, $(d)$, $(K)$, $(j)$ and $(P)$ are known, then its Non Current Asset planned is:

$$N= [1+f(U+\{S'[1+s]-Sv-F-S'i[1+s]-T\}[1-d])$$
$$-K-S'j[1+s]/360-P$$

Steve Asikin ISBN 13: **978-1564912361**, ISBN 10: **1564912363**

Rule-58102:

If both (**N**), (**U**), (**S'**), (**s**), (**S**), (**v**), (**F**), (**i**), (**T**), (**d**), (**K**), (**j**) and (**P**) are known, then its Leverage oir Gearing Ratio planned is:

$$\textit{l} = \{\textbf{N}+\textbf{K}+\textbf{S'j}[1+\textbf{s}]/360+\textbf{P}\}/(\textbf{U}+\{\textbf{S'}[1+\textbf{s}]-\textbf{Sv}-\textbf{F}$$
$$-\textbf{S'i}[1+\textbf{s}]-\textbf{T}\}[1-\textbf{d}])-1$$

Rule-58103:

If both (**l**), (**N**), (**S'**), (**s**), (**S**), (**v**), (**F**), (**i**), (**T**), (**d**), (**K**), (**j**) and (**P**) are known, then its Utilized or Starting Capital must be:

$$\textbf{U} = (\{\textbf{N}+\textbf{K}+\textbf{S'j}[1+\textbf{s}]/360+\textbf{P}\}-[1+\textit{l}]\{\textbf{S'}[1+\textbf{s}]-\textbf{Sv}-\textbf{F}$$
$$-\textbf{S'i}[1+\textbf{s}]-\textbf{T}\}[1-\textbf{d}])/[1+\textit{l}]$$

Rule-58104:

If both (**l**), (**U**), (**N**), (**s**), (**S**), (**v**), (**F**), (**i**), (**T**), (**d**), (**K**), (**j**) and (**P**) are known, then its Sales Past must be:

$$\textbf{S'} = [[1+\textit{l}](\textbf{U}-\{\textbf{Sv}+\textbf{F}+\textbf{T}\}[1-\textbf{d}])-[\textbf{N}+\textbf{K}+\textbf{P}]]$$
$$/([1+\textbf{s}]\{\textbf{j}/360-[1+\textit{l}][1-\textbf{d}][1-\textbf{i}]\})$$

or

$$\textbf{S'} = \textbf{S}/[1+\textbf{s}]$$

`

Steve Asikin ISBN 13: **978-1564912361**, ISBN 10: **1564912363**

Rule-58105:

If both $(\textbf{\textit{l}})$, $(\textbf{U})$, $(\textbf{S'})$, $(\textbf{N})$, $(\textbf{S})$, $(\textbf{v})$, $(\textbf{F})$, $(\textbf{i})$, $(\textbf{T})$, $(\textbf{d})$, $(\textbf{K})$, $(\textbf{j})$ and $(\textbf{P})$ are known, then its Sales Growth planned is:

$$s= [[1+\textbf{\textit{l}}(\textbf{U}-\{\textbf{Sv}+\textbf{F}+\textbf{T}\}[1-\textbf{d}])-[\textbf{N}+\textbf{K}+\textbf{P}]]$$
$$/(\textbf{S'}\{\textbf{j}/360-[1+\textbf{\textit{l}}[1-\textbf{d}][1-\textbf{i}]\})-1$$

or

$$s= \textbf{S}/\textbf{S'}-1$$


Rule-58106:

If both $(\textbf{\textit{l}})$, $(\textbf{U})$, $(\textbf{S'})$, $(\textbf{s})$, $(\textbf{N})$, $(\textbf{v})$, $(\textbf{F})$, $(\textbf{i})$, $(\textbf{T})$, $(\textbf{d})$, $(\textbf{K})$, $(\textbf{j})$ and $(\textbf{P})$ are known, then its Sales or Revenue planned is:

$$S= [\textbf{S'}[1+\textbf{s}]-\textbf{F}-\textbf{S'i}[1+\textbf{s}]-\textbf{T}\}-(\{\textbf{N}+\textbf{K}+\textbf{S'j}[1+\textbf{s}]/360$$
$$+\textbf{P}\}/[1+\textbf{\textit{l}}-\textbf{U}]/[1-\textbf{d}]]/\textbf{v}$$

or

$$S= \textbf{S'}[1+\textbf{s}]$$


Rule-58107:

If both $(\textbf{\textit{l}})$, $(\textbf{U})$, $(\textbf{S'})$, $(\textbf{s})$, $(\textbf{S})$, $(\textbf{N})$, $(\textbf{F})$, $(\textbf{i})$, $(\textbf{T})$, $(\textbf{d})$, $(\textbf{K})$, $(\textbf{j})$ and $(\textbf{P})$ are known, then its Variable Portion planned is:

$$v= [\textbf{S'}[1+\textbf{s}]-\textbf{F}-\textbf{S'i}[1+\textbf{s}]-\textbf{T}\}-(\{\textbf{N}+\textbf{K}+\textbf{S'j}[1+\textbf{s}]/360$$
$$+\textbf{P}\}/[1+\textbf{\textit{l}}-\textbf{U}]/[1-\textbf{d}]]/\textbf{S}$$

Steve Asikin ISBN 13: **978-1564912361**, ISBN 10: **1564912363**

Rule-58108:

If both (**I**), (**U**), (**S'**), (**s**), (**S**), (**v**), (**N**), (**i**), (**T**), (**d**), (**K**), (**j**) and (**P**) are known, then its Fixed Cost planned is:

$$F= S'[1+s]-Sv-S'i[1+s]-T-(\{N+K+S'j[1+s]/360+P\}/[1+I-U)/[1-d]$$

Rule-58109:

If both (**I**), (**U**), (**S'**), (**s**), (**S**), (**v**), (**F**), (**N**), (**T**), (**d**), (**K**), (**j**) and (**P**) are known, then its Interest Portion planned is:

$$i= [S'[1+s]-Sv-F-T-(\{N+K+S'j[1+s]/360+P\}/[1+I-U)/[1-d]]/\{S'[1+s]\}$$

Rule-58110:

If both (**I**), (**U**), (**S'**), (**s**), (**S**), (**v**), (**F**), (**i**), (**N**), (**d**), (**K**), (**j**) and (**P**) are known, then its Tax planned is:

$$T= S'[1+s]-Sv-F-S'i[1+s]-(\{N+K+S'j[1+s]/360+P\}/[1+I-U)/[1-d]$$

Rule-58111:

If both (**I**), (**U**), (**S'**), (**s**), (**S**), (**v**), (**F**), (**i**), (**T**), (**N**), (**K**), (**j**) and (**P**) are known, then its Dividend Payout planned is:

$$d= 1-(\{N+K+S'j[1+s]/360+P\}/[1+I-U)/\{S'[1+s]-Sv-F-S'i[1+s]-T\}$$

Steve Asikin ISBN 13: **978-1564912361**, ISBN 10: **1564912363**

Rule-58112:

If both ($I$), ($U$), ($S'$), ($s$), ($S$), ($v$), ($F$), ($i$), ($T$), ($d$), ($N$), ($j$) and ($P$) are known, then its Kind of Cash planned is:

$$K= [1+I(U+\{S'[1+s]-Sv-F-S'i[1+s]-T\}[1-d])$$
$$-\{N+S'j[1+s]/360+P\}$$

Rule-58113:

If both ($I$), ($U$), ($S'$), ($s$), ($S$), ($v$), ($F$), ($i$), ($T$), ($d$), ($K$), ($N$) and ($P$) are known, then its Job or Trade Receivable Days planned is:

$$j= 360[[1+I(U+\{S'[1+s]-Sv-F-S'i[1+s]-T\}[1-d])$$
$$-\{N+K+P\}]/\{S'[1+s]\}$$

Rule-58114:

If both ($I$), ($U$), ($S'$), ($s$), ($S$), ($v$), ($F$), ($i$), ($T$), ($d$), ($K$), ($j$) and ($N$) are known, then its Procured Inventory planned is:

$$P= [1+I(U+\{S'[1+s]-Sv-F-S'i[1+s]-T\}[1-d])$$
$$-\{N+K+S'j[1+s]/360\}$$

Rule-58115:

If both ($I$), ($U$), ($S'$), ($s$), ($S$), ($v$), ($F$), ($i$), ($T$), ($d$), ($K$), ($j$), ($V$) and ($p$) are known, then its Non Current Asset planned is:

$$N= [1+I(U+\{S'[1+s]-Sv-F-S'i[1+s]-T\}[1-d])$$
$$-K-S'j[1+s]/360-Vp/360$$

557

Steve Asikin ISBN 13: **978-1564912361**, ISBN 10: **1564912363**

<u>Rule-58116</u>:

If both $(\mathbf{N})$, $(\mathbf{U})$, $(\mathbf{S'})$, $(\mathbf{s})$, $(\mathbf{S})$, $(\mathbf{v})$, $(\mathbf{F})$, $(\mathbf{i})$, $(\mathbf{T})$, $(\mathbf{d})$, $(\mathbf{K})$, $(\mathbf{j})$, $(\mathbf{V})$ and $(\mathbf{p})$ are known, then its Leverage or Gearing Ratio planned is:

$$I = \{\mathbf{N}+\mathbf{K}+\mathbf{S'j}[1+\mathbf{s}]/360+\mathbf{Vp}/360\}$$
$$/(\mathbf{U}+\{\mathbf{S'}[1+\mathbf{s}]-\mathbf{Sv}-\mathbf{F}-\mathbf{S'i}[1+\mathbf{s}]-\mathbf{T}\}[1-\mathbf{d}])-1$$

<u>Rule-58117</u>:

If both $(I)$, $(\mathbf{N})$, $(\mathbf{S'})$, $(\mathbf{s})$, $(\mathbf{S})$, $(\mathbf{v})$, $(\mathbf{F})$, $(\mathbf{i})$, $(\mathbf{T})$, $(\mathbf{d})$, $(\mathbf{K})$, $(\mathbf{j})$, $(\mathbf{V})$ and $(\mathbf{p})$ are known, then its Utilized or Starting Capital must be:

$$\mathbf{U} = (\{\mathbf{N}+\mathbf{K}+\mathbf{S'j}[1+\mathbf{s}]/360+\mathbf{Vp}/360\}-[1+I]\{\mathbf{S'}[1+\mathbf{s}]$$
$$-\mathbf{Sv}-\mathbf{F}-\mathbf{S'i}[1+\mathbf{s}]-\mathbf{T}\}[1-\mathbf{d}])/[1+I]$$

<u>Rule-58118</u>:

If both $(I)$, $(\mathbf{U})$, $(\mathbf{N})$, $(\mathbf{s})$, $(\mathbf{S})$, $(\mathbf{v})$, $(\mathbf{F})$, $(\mathbf{i})$, $(\mathbf{T})$, $(\mathbf{d})$, $(\mathbf{K})$, $(\mathbf{j})$, $(\mathbf{V})$ and $(\mathbf{p})$ are known, then its Sales Past must be:

$$\mathbf{S'} = [[1+I](\mathbf{U}-\{\mathbf{Sv}+\mathbf{F}+\mathbf{T}\}[1-\mathbf{d}])-[\mathbf{N}+\mathbf{K}+\mathbf{Vp}/360]]$$
$$/([1+\mathbf{s}]\{\mathbf{j}/360-[1+I][1-\mathbf{d}][1-\mathbf{i}]\})$$
$$\text{or}$$
$$\mathbf{S'} = \mathbf{S}/[1+\mathbf{s}]$$

Steve Asikin ISBN 13: **978-1564912361**, ISBN 10: **1564912363**

Rule-58119:

If both $(I)$, $(U)$, $(S')$, $(N)$, $(S)$, $(v)$, $(F)$, $(i)$, $(T)$, $(d)$, $(K)$, $(j)$, $(V)$ and $(p)$ are known, then its Sales Growth planned is:

$$s= [[1+I](U-\{Sv+F+T\}[1-d])-[N+K+Vp/360]]$$
$$/(S'\{j/360-[1+I][1-d][1-i]\})-1$$

or

$$s= S/S'-1$$

Rule-58120:

If both $(I)$, $(U)$, $(S')$, $(s)$, $(S)$, $(v)$, $(F)$, $(i)$, $(T)$, $(d)$, $(K)$, $(j)$, $(V)$ and $(p)$ are known, then its Non Current Asset planned is:

$$S= [S'[1+s]-F- S'i[1+s]-T\}[1-d]$$
$$-\{N+K+S'j[1+s]/360+Vp/360\}$$
$$/[1+I-U)/[1-d]]/v$$

or

$$S= S'[1+s]$$

or

$$S= V/v$$

Rule-58121:

If both $(I)$, $(U)$, $(S')$, $(s)$, $(S)$, $(N)$, $(F)$, $(i)$, $(T)$, $(d)$, $(K)$, $(j)$, $(V)$ and $(p)$ are known, then its Variable Portion planned is:

$$v= [S'[1+s]-F- S'i[1+s]-T\}[1-d]-\{N+K$$
$$+S'j[1+s]/360+Vp/360\}/[1+I-U)/[1-d]]/S$$

or

$$v= V/S$$

Steve Asikin ISBN 13: **978-1564912361**, ISBN 10: **1564912363**

Rule-58122:

If both ($\mathit{l}$), (**U**), (**S'**), (**s**), (**S**), (**v**), (**N**), (**i**), (**T**), (**d**), (**K**), (**j**), (**V**) and (**p**) are known, then its Fixed Cost planned is:

$$F= S'[1+s]-Sv- S'i[1+s]-T\}-(\{N+K+S'j[1+s]/360$$
$$+Vp/360\}/[1+\mathit{l}-U)/[1-d]$$

Rule-58123:

If both ($\mathit{l}$), (**U**), (**S'**), (**s**), (**S**), (**v**), (**F**), (**N**), (**T**), (**d**), (**K**), (**j**), (**V**) and (**p**) are known, then its Interest Portion planned is:

$$i= [S'[1+s]-Sv-F-T\}-(\{N+K+S'j[1+s]/360$$
$$+Vp/360\}/[1+\mathit{l}-U)/[1-d]]/\{S'[1+s]\}$$

Rule-58124:

If both ($\mathit{l}$), (**U**), (**S'**), (**s**), (**S**), (**v**), (**F**), (**i**), (**T**), (**d**), (**K**), (**j**), (**N**) and (**p**) are known, then its Tax planned is:

$$T= [S'[1+s]-Sv-F-S'i[1+s]\}[1-d]-\{N+K$$
$$+S'j[1+s]/360+Vp/360\}/[1+\mathit{l}-U)/[1-d]]/S$$

Rule-58125:

If both ($\mathit{l}$), (**U**), (**S'**), (**s**), (**S**), (**v**), (**F**), (**i**), (**T**), (**N**), (**K**), (**j**), (**V**) and (**p**) are known, then its Dividend Payout planned is:

$$d= 1-(\{N+K+S'j[1+s]/360+Vp/360\}/[1+\mathit{l}-U)$$
$$/\{S'[1+s]-Sv-F-S'i[1+s]-T\}$$

Steve Asikin ISBN 13: **978-1564912361**, ISBN 10: **1564912363**

Rule-58126:

If both (**/**), (**U**), (**S'**), (**s**), (**S**), (**v**), (**F**), (**i**), (**T**), (**d**), (**N**), (**j**), (**V**) and (**p**) are known, then its Kind of Cash planned is:

$$\mathbf{K} = [1+\mathbf{/}(\mathbf{U}+\{\mathbf{S'}[1+\mathbf{s}]-\mathbf{Sv}-\mathbf{F}-\mathbf{S'i}[1+\mathbf{s}]-\mathbf{T}\}[1-\mathbf{d}])$$
$$-\{\mathbf{N}+\mathbf{S'j}[1+\mathbf{s}]/360+\mathbf{Vp}/360\}$$

Rule-58127:

If both (**/**), (**U**), (**S'**), (**s**), (**S**), (**v**), (**F**), (**i**), (**T**), (**d**), (**K**), (**N**), (**V**) and (**p**) are known, then its Job or Trade Receivable Days planned is:

$$\mathbf{j} = 360[[1+\mathbf{/}(\mathbf{U}+\{\mathbf{S'}[1+\mathbf{s}]-\mathbf{Sv}-\mathbf{F}-\mathbf{S'i}[1+\mathbf{s}]-\mathbf{T}\}[1-\mathbf{d}])$$
$$-\{\mathbf{N}+\mathbf{K}+\mathbf{Vp}/360\}]/\{\mathbf{S'}[1+\mathbf{s}]\}$$

Rule-58128:

If both (**/**), (**U**), (**S'**), (**s**), (**S**), (**v**), (**F**), (**i**), (**T**), (**d**), (**K**), (**j**), (**N**) and (**p**) are known, then its Variable Cost planned is:

$$\mathbf{V} = 360[[1+\mathbf{/}(\mathbf{U}+\{\mathbf{S'}[1+\mathbf{s}]-\mathbf{Sv}-\mathbf{F}-\mathbf{S'i}[1+\mathbf{s}]-\mathbf{T}\}[1-\mathbf{d}])$$
$$-\{\mathbf{N}+\mathbf{K}+\mathbf{S'j}[1+\mathbf{s}]/360\}]/\mathbf{p}$$

or

$$\mathbf{V} = \mathbf{Sv}$$

Rule-58129:

If both (**/**), (**U**), (**S'**), (**s**), (**S**), (**v**), (**F**), (**i**), (**T**), (**d**), (**K**), (**j**), (**V**) and (**N**) are known, then its Procured Invenotry Days planned is:

$$\mathbf{p} = 360[[1+\mathbf{/}(\mathbf{U}+\{\mathbf{S'}[1+\mathbf{s}]-\mathbf{Sv}-\mathbf{F}-\mathbf{S'i}[1+\mathbf{s}]-\mathbf{T}\}[1-\mathbf{d}])$$
$$-\{\mathbf{N}+\mathbf{K}+\mathbf{S'j}[1+\mathbf{s}]/360\}]/\mathbf{V}$$

Steve Asikin ISBN 13: **978-1564912361**, ISBN 10: **1564912363**

Rule-58130:

If both (**/**), (**U**), (**S'**), (**s**), (**S**), (**v**), (**F**), (**i**), (**T**), (**d**), (**K**), (**j**) and (**p**) are known, then its Non Current Asset planned is:

$$\mathbf{N} = [1+/](\mathbf{U}+\{\mathbf{S'}[1+\mathbf{s}]-\mathbf{Sv}-\mathbf{F}-\mathbf{S'i}[1+\mathbf{s}]-\mathbf{T}\}[1-\mathbf{d}])$$
$$-\mathbf{K}-\mathbf{S'j}[1+\mathbf{s}]/360-\mathbf{Svp}/360$$

Rule-58131:

If both (**N**), (**U**), (**S'**), (**s**), (**S**), (**v**), (**F**), (**i**), (**T**), (**d**), (**K**), (**j**) and (**p**) are known, then its Leverage or Gearing Ratio planned is:

$$/ = \{\mathbf{N}+\mathbf{K}+\mathbf{S'j}[1+\mathbf{s}]/360+\mathbf{Svp}/360\}$$
$$/(\mathbf{U}+\{\mathbf{S'}[1+\mathbf{s}]-\mathbf{Sv}-\mathbf{F}-\mathbf{S'i}[1+\mathbf{s}]-\mathbf{T}\}[1-\mathbf{d}])-1$$

Rule-58132:

If both (**/**), (**N**), (**S'**), (**s**), (**S**), (**v**), (**F**), (**i**), (**T**), (**d**), (**K**), (**j**) and (**p**) are known, then its Utilized or Starting Capital must be:

$$\mathbf{U} = (\{\mathbf{N}+\mathbf{K}+\mathbf{S'j}[1+\mathbf{s}]/360+\mathbf{Svp}/360\}-[1+/]$$
$$\{\mathbf{S'}[1+\mathbf{s}]-\mathbf{Sv}-\mathbf{F}-\mathbf{S'i}[1+\mathbf{s}]-\mathbf{T}\}[1-\mathbf{d}])/[1+/]$$

Rule-58133:

If both (**/**), (**U**), (**N**), (**s**), (**S**), (**v**), (**F**), (**i**), (**T**), (**d**), (**K**), (**j**) and (**p**) are known, then its Sales Past must be:

$$\mathbf{S'} = [[1+/](\mathbf{U}+\{\mathbf{Sv}+\mathbf{F}+\mathbf{T}\}[1-\mathbf{d}])-[\mathbf{N}+\mathbf{K}+\mathbf{Svp}/360]]$$
$$/(\mathbf{S'}[1+\mathbf{s}]\{\mathbf{j}/360-[1+/][1-\mathbf{d}][1-\mathbf{i}]\})$$

or

$$\mathbf{S'} = \mathbf{S}/[1+\mathbf{s}]$$

Steve Asikin ISBN 13: **978-1564912361**, ISBN 10: **1564912363**

Rule-58134:

If both (**/**), (**U**), (**S'**), (**N**), (**S**), (**v**), (**F**), (**i**), (**T**), (**d**), (**K**), (**j**) and (**p**) are known, then its Sales Growth planned is:

$$s= [[1+/(U+\{Sv+F+T\}[1-d])-[N+K+Svp/360]]$$
$$/(S'\{j/360-[1+/[1-d][1-i]\})-1$$

or

$$s= S/S'-1$$

Rule-58135:

If both (**/**), (**U**), (**S'**), (**s**), (**N**), (**v**), (**F**), (**i**), (**T**), (**d**), (**K**), (**j**) and (**p**) are known, then its Sales or Revenue planned is:

$$S= [[1+/(U+\{S'[1+s]-F-S'i[1+s]-T\}[1-d])$$
$$-\{N+K+S'j[1+s]/360\}]$$
$$/(v\{p/360+[1+/[1-d]\})$$

or

$$S= S'[1+s]$$

Rule-58136:

If both (**/**), (**U**), (**S'**), (**s**), (**S**), (**N**), (**F**), (**i**), (**T**), (**d**), (**K**), (**j**) and (**p**) are known, then its Variable Portion planned is:

$$v= [[1+/(U+\{S'[1+s]-F-S'i[1+s]-T\}[1-d])$$
$$-\{N+K+S'j[1+s]/360\}]$$
$$/(S\{p/360+[1+/[1-d]\})$$

Steve Asikin ISBN 13: **978-1564912361**, ISBN 10: **1564912363**

Rule-58137:

If both (**/**), (**U**), (**S'**), (**s**), (**S**), (**v**), (**N**), (**i**), (**T**), (**d**), (**K**), (**j**) and (**p**) are known, then its Fixed Cost planned is:

$$\mathbf{F} = \mathbf{S'}[1+\mathbf{s}] - \mathbf{Sv} - \mathbf{S'i}[1+\mathbf{s}] - \mathbf{T} - (\{\mathbf{N} + \mathbf{K} + \mathbf{S'j}[1+\mathbf{s}]/360 + \mathbf{Svp}/360\}/[1+\mathbf{/} - \mathbf{U})/[1-\mathbf{d}]$$

Rule-58138:

If both (**/**), (**U**), (**S'**), (**s**), (**S**), (**v**), (**F**), (**N**), (**T**), (**d**), (**K**), (**j**) and (**p**) are known, then its Interest Portion planned is:

$$\mathbf{i} = (\mathbf{S'}[1+\mathbf{s}] - \mathbf{Sv} - \mathbf{F} - \mathbf{T} - (\{\mathbf{N} + \mathbf{K} + \mathbf{S'j}[1+\mathbf{s}]/360 + \mathbf{Svp}/360\}/[1+\mathbf{/} - \mathbf{U})/[1-\mathbf{d}])/\{\mathbf{S'}[1+\mathbf{s}]\}$$

Rule-58139:

If both (**/**), (**U**), (**S'**), (**s**), (**S**), (**v**), (**F**), (**i**), (**N**), (**d**), (**K**), (**j**) and (**p**) are known, then its Tax planned is:

$$\mathbf{T} = \mathbf{S'}[1+\mathbf{s}] - \mathbf{Sv} - \mathbf{F} - \mathbf{S'i}[1+\mathbf{s}] - (\{\mathbf{N} + \mathbf{K} + \mathbf{S'j}[1+\mathbf{s}]/360 + \mathbf{Svp}/360\}/[1+\mathbf{/} - \mathbf{U})/[1-\mathbf{d}]$$

Rule-58140:

If both (**/**), (**U**), (**S'**), (**s**), (**S**), (**v**), (**F**), (**i**), (**T**), (**N**), (**K**), (**j**) and (**p**) are known, then its Dividend Payout planned is:

$$\mathbf{d} = 1 - (\{\mathbf{N} + \mathbf{K} + \mathbf{S'j}[1+\mathbf{s}]/360 + \mathbf{Svp}/360\}/[1+\mathbf{/} - \mathbf{U}) / \{\mathbf{S'}[1+\mathbf{s}] - \mathbf{Sv} - \mathbf{F} - \mathbf{S'i}[1+\mathbf{s}] - \mathbf{T}\}$$

Steve Asikin ISBN 13: **978-1564912361**, ISBN 10: **1564912363**

Rule-58141:

If both $(\textit{I})$, $(\mathbf{U})$, $(\mathbf{S'})$, $(\mathbf{s})$, $(\mathbf{\$})$, $(\mathbf{v})$, $(\mathbf{F})$, $(\mathbf{i})$, $(\mathbf{T})$, $(\mathbf{d})$, $(\mathbf{N})$, $(\mathbf{j})$ and $(\mathbf{p})$ are known, then its Kind of Cash planned is:

$$\mathbf{K} = [1+\textit{I}(\mathbf{U}+\{\mathbf{S'}[1+\mathbf{s}]-\mathbf{\$v}-\mathbf{F}-\mathbf{S'i}[1+\mathbf{s}]-\mathbf{T}\}[1-\mathbf{d}])$$
$$-\{\mathbf{N}+\mathbf{S'j}[1+\mathbf{s}]/360+\mathbf{\$vp}/360\}$$

Rule-58142:

If both $(\textit{I})$, $(\mathbf{U})$, $(\mathbf{S'})$, $(\mathbf{s})$, $(\mathbf{\$})$, $(\mathbf{v})$, $(\mathbf{F})$, $(\mathbf{i})$, $(\mathbf{T})$, $(\mathbf{d})$, $(\mathbf{K})$, $(\mathbf{N})$ and $(\mathbf{p})$ are known, then its Job or Trade Receivable Days planned is:

$$\mathbf{j} = 360[[1+\textit{I}(\mathbf{U}+\{\mathbf{S'}[1+\mathbf{s}]-\mathbf{\$v}-\mathbf{F}-\mathbf{S'i}[1+\mathbf{s}]-\mathbf{T}\}[1-\mathbf{d}])$$
$$-\{\mathbf{N}+\mathbf{K}+\mathbf{\$vp}/360\}]/\{\mathbf{S'}[1+\mathbf{s}]\}$$

Rule-58143:

If both $(\textit{I})$, $(\mathbf{U})$, $(\mathbf{S'})$, $(\mathbf{s})$, $(\mathbf{\$})$, $(\mathbf{v})$, $(\mathbf{F})$, $(\mathbf{i})$, $(\mathbf{T})$, $(\mathbf{d})$, $(\mathbf{K})$, $(\mathbf{j})$ and $(\mathbf{N})$ are known, then its Procured Inventory Days planned is:

$$\mathbf{p} = 360[[1+\textit{I}(\mathbf{U}+\{\mathbf{S'}[1+\mathbf{s}]-\mathbf{\$v}-\mathbf{F}-\mathbf{S'i}[1+\mathbf{s}]-\mathbf{T}\}[1-\mathbf{d}])$$
$$-\{\mathbf{N}+\mathbf{K}+\mathbf{S'j}[1+\mathbf{s}]/360\}]/[\mathbf{\$v}]$$

Rule-58144:

If both $(\textit{I})$, $(\mathbf{U})$, $(\mathbf{S'})$, $(\mathbf{s})$, $(\mathbf{\$})$, $(\mathbf{v})$, $(\mathbf{F})$, $(\mathbf{i})$, $(\mathbf{T})$, $(\mathbf{d})$, $(\mathbf{K})$, $(\mathbf{j})$ and $(\mathbf{p})$ are known, then its Non Current Asset planned is:

$$\mathbf{N} = [1+\textit{I}(\mathbf{U}+\{\mathbf{S'}[1+\mathbf{s}]-\mathbf{\$v}-\mathbf{F}-\mathbf{S'i}[1+\mathbf{s}]-\mathbf{T}\}[1-\mathbf{d}])$$
$$-\mathbf{K}-\mathbf{S'j}[1+\mathbf{s}]/360-\mathbf{S'vp}[1+\mathbf{s}]/360$$

Steve Asikin ISBN 13: **978-1564912361**, ISBN 10: **1564912363**

Rule-58145:

If both (**N**), (**U**), (**S'**), (**s**), (**S**), (**v**), (**F**), (**i**), (**T**), (**d**), (**K**), (**j**) and (**p**) are known, then its Leverage or Gearing Ratio planned is:

$$l = \{N+K+S'j[1+s]/360+S'vp[1+s]/360\}$$
$$/(U+\{S'[1+s]-Sv-F-S'i[1+s]-T\}[1-d])-1$$

Rule-58146:

If both (**l**), (**N**), (**S'**), (**s**), (**S**), (**v**), (**F**), (**i**), (**T**), (**d**), (**K**), (**j**) and (**p**) are known, then its Utilized or Starting Capital must be:

$$U = (\{N+K+S'j[1+s]/360+S'vp[1+s]/360\}-[1+l]$$
$$\{S'[1+s]-Sv-F-S'i[1+s]-T\}[1-d])/[1+l]$$

Rule-58147:

If both (**l**), (**U**), (**N**), (**s**), (**S**), (**v**), (**F**), (**i**), (**T**), (**d**), (**K**), (**j**) and (**p**) are known, then its Sales Past must be:

$$S' = [[1+l](U-\{Sv+F+T\}[1-d])-[N+K]]$$
$$/([1+s]\{j/360+vp/360-[1+l][1-d][1-i]\})$$
$$\text{or}$$
$$S' = S/[1+s]$$

Steve Asikin ISBN 13: **978-1564912361**, ISBN 10: **1564912363**

Rule-58148:

If both (**/**), (**U**), (**S'**), (**N**), (**S**), (**v**), (**F**), (**i**), (**T**), (**d**), (**K**), (**j**) and (**p**) are known, then its Sales Growth planned is:

$$s= [[1+/](U-\{Sv+F+T\}[1-d])-[N+K]]$$
$$/(S'\{j/360+vp/360-[1+/][1-d][1-i]\})-1$$

or

$$s= S/S'-1$$


Rule-58149:

If both (**/**), (**U**), (**S'**), (**s**), (**S**), (**N**), (**F**), (**i**), (**T**), (**d**), (**K**), (**j**) and (**p**) are known, then its Variable Portion planned is:

$$S= [S'[1+s]-F-S'i[1+s]-T- (\{N+K+S'j[1+s]/360$$
$$+S'vp[1+s]/360\}/[1+/]-U)/[1-d]]/v$$

or

$$S= S'[1+s]$$


Rule-58150:

If both (**/**), (**U**), (**S'**), (**s**), (**S**), (**N**), (**F**), (**i**), (**T**), (**d**), (**K**), (**j**) and (**p**) are known, then its Variable Portion planned is:

$$v= ([1+/](U+\{S'[1+s]-F-S'i[1+s]\}[1-t]-D)$$
$$-\{N+K+S'j[1+s]/360\})$$
$$/\{S'p[1+s]/360+S[1+/][1-t]\}$$

Steve Asikin ISBN 13: **978-1564912361**, ISBN 10: **1564912363**

Rule-58151:

If both (**/**), (**U**), (**S'**), (**s**), (**S**), (**v**), (**N**), (**i**), (**T**), (**d**), (**K**), (**j**) and (**p**) are known, then its Fixed Cost planned is:

$$F= S'[1+s]-Sv-S'i[1+s]-T- (\{N+K+S'j[1+s]/360 \\ +S'vp[1+s]/360\}/[1+/-U)/[1-d]$$

Rule-58152:

If both (**/**), (**U**), (**S'**), (**s**), (**S**), (**v**), (**F**), (**N**), (**T**), (**d**), (**K**), (**j**) and (**p**) are known, then its Interest Portion planned is:

$$i= (S'[1+s]-Sv-F-T- (\{N+K+S'j[1+s]/360 \\ +S'vp[1+s]/360\}/[1+/-U]/[1-d]) \\ /\{S'[1+s]\}$$

Rule-58153:

If both (**/**), (**U**), (**S'**), (**s**), (**S**), (**v**), (**F**), (**i**), (**N**), (**d**), (**K**), (**j**) and (**p**) are known, then its Tax planned is:

$$T= S'[1+s]-Sv-F-S'i[1+s]- (\{N+K+S'j[1+s]/360 \\ +S'vp[1+s]/360\}/[1+/-U)/[1-d]$$

Rule-58154:

If both (**/**), (**U**), (**S'**), (**s**), (**S**), (**v**), (**F**), (**i**), (**T**), (**N**), (**K**), (**j**) and (**p**) are known, then its Dividend Payout planned is:

$$d= 1-(\{N+K+S'j[1+s]/360+S'vp[1+s]/360\} \\ /[1+/-U)/\{S'[1+s]-Sv-F-S'i[1+s]-T\}$$

Steve Asikin ISBN 13: **978-1564912361**, ISBN 10: **1564912363**

Rule-58155:

If both (**/**), (**U**), (**S'**), (**s**), (**$**), (**v**), (**F**), (**i**), (**T**), (**d**), (**N**), (**j**) and (**p**) are known, then its Kind of Cash planned is:

$$K= [1+/(U+\{S'[1+s]-Sv-F-S'i[1+s]-T\}[1-d])$$
$$-\{N+S'j[1+s]/360+S'vp[1+s]/360\}$$

Rule-58156:

If both (**/**), (**U**), (**S'**), (**s**), (**$**), (**v**), (**F**), (**i**), (**T**), (**d**), (**K**), (**N**) and (**p**) are known, then its Job or Trade Receivable Days planned is:

$$j= 360[[1+/(U+\{S'[1+s]-Sv-F-S'i[1+s]-T\}[1-d])$$
$$-\{N+K+S'vp[1+s]/360\}]/\{S'[1+s]\}$$

Rule-58157:

If both (**/**), (**U**), (**S'**), (**s**), (**$**), (**v**), (**F**), (**i**), (**T**), (**d**), (**K**), (**j**) and (**N**) are known, then its Procured Inventory Days planned is:

$$p= 360[[1+/(U+\{S'[1+s]-Sv-F-S'i[1+s]-T\}[1-d])$$
$$-\{N+K+S'j[1+s]/360\}]/\{S'v[1+s]\}$$

Rule-58158:

If both (**/**), (**U**), (**S'**), (**s**), (**$**), (**v**), (**F**), (**i**), (**t**), (**D**), (**K**), (**J**) and (**P**) are known, then its Non Current Asset planned is:

$$N= [1+/(U+\{S'[1+s]-Sv-F-S'i[1+s]\}[1-t]-D)$$
$$-K-J-P$$

Steve Asikin ISBN 13: **978-1564912361**, ISBN 10: **1564912363**

<u>Rule-58159</u>:

If both (**N**), (**U**), (**S'**), (**s**), (**S**), (**v**), (**F**), (**i**), (**t**), (**D**), (**K**), (**J**) and (**P**) are known, then its Leverage or Gearing Ratio planned is:

$$\text{/} = \{\mathbf{N+K+J+P}\}/(\mathbf{U}+\{\mathbf{S'}[1+\mathbf{s}]-\mathbf{Sv}-\mathbf{F}-\mathbf{S'i}[1+\mathbf{s}]\}[1-\mathbf{t}]-\mathbf{D})-1$$

<u>Rule-58160</u>:

If both (**/**), (**N**), (**S'**), (**s**), (**S**), (**v**), (**F**), (**i**), (**t**), (**D**), (**K**), (**J**) and (**P**) are known, then its Utilized or Starting Capital must be:

$$\mathbf{U} = (\{\mathbf{N+K+J+P}\}-[1+\text{/}]\{\mathbf{S'}[1+\mathbf{s}]-\mathbf{Sv}-\mathbf{F}-\mathbf{S'i}[1+\mathbf{s}]\}[1-\mathbf{t}]-\mathbf{D})/[1+\text{/}]$$

<u>Rule-58162</u>:

If both (**/**), (**U**), (**N**), (**s**), (**S**), (**v**), (**F**), (**i**), (**t**), (**D**), (**K**), (**J**) and (**P**) are known, then its Sales Past must be:

$$\mathbf{S'} = (\{[\mathbf{N+K+J+P}]/[1+\text{/}]-\mathbf{U}+\mathbf{D}\}/[1-\mathbf{t}]+\mathbf{Sv}+\mathbf{F})/\{[1+\mathbf{s}][1-\mathbf{i}]\}$$

or

$$\mathbf{S'} = \mathbf{S}/[1+\mathbf{s}]$$

Steve Asikin ISBN 13: **978-1564912361**, ISBN 10: **1564912363**

Rule-58163:

If both $(\textit{\textbf{I}})$, $(\textbf{U})$, $(\textbf{S'})$, $(\textbf{N})$, $(\textbf{S})$, $(\textbf{v})$, $(\textbf{F})$, $(\textbf{i})$, $(\textbf{t})$, $(\textbf{D})$, $(\textbf{K})$, $(\textbf{J})$ and $(\textbf{P})$ are known, then its Sales Growth planned is:

$$s = (\{[\textbf{N}+\textbf{K}+\textbf{J}+\textbf{P}]/[1+\textit{\textbf{I}}-\textbf{U}+\textbf{D}\}/[1-\textbf{t}]+\textbf{Sv}+\textbf{F})$$
$$/\{\textbf{S'}[1-\textbf{i}]\}-1$$

or

$$s = \textbf{S}/\textbf{S'}-1$$

Rule-58164:

If both $(\textit{\textbf{I}})$, $(\textbf{U})$, $(\textbf{S'})$, $(\textbf{s})$, $(\textbf{N})$, $(\textbf{v})$, $(\textbf{F})$, $(\textbf{i})$, $(\textbf{t})$, $(\textbf{D})$, $(\textbf{K})$, $(\textbf{J})$ and $(\textbf{P})$ are known, then its Sales or Revenue planned is:

$$S = (\textbf{S'}[1+\textbf{s}]-\textbf{F}-\textbf{S'i}[1+\textbf{s}]-\{[\textbf{N}+\textbf{K}+\textbf{J}+\textbf{P}]/[1+\textit{\textbf{I}}-\textbf{U}+\textbf{D}\}$$
$$/[1-\textbf{t}])/\textbf{v}$$

or

$$S = \textbf{S'}[1+\textbf{s}]$$

Rule-58165:

If both $(\textit{\textbf{I}})$, $(\textbf{U})$, $(\textbf{S'})$, $(\textbf{s})$, $(\textbf{S})$, $(\textbf{N})$, $(\textbf{F})$, $(\textbf{i})$, $(\textbf{t})$, $(\textbf{D})$, $(\textbf{K})$, $(\textbf{J})$ and $(\textbf{P})$ are known, then its Variable Portion planned is:

$$v = (\textbf{S'}[1+\textbf{s}]-\textbf{F}-\textbf{S'i}[1+\textbf{s}]-\{[\textbf{N}+\textbf{K}+\textbf{J}+\textbf{P}]/[1+\textit{\textbf{I}}-\textbf{U}+\textbf{D}\}$$
$$/[1-\textbf{t}])/\textbf{S}$$

Steve Asikin ISBN 13: **978-1564912361**, ISBN 10: **1564912363**

Rule-58166:

If both (**/**), (**U**), (**S'**), (**s**), (**S**), (**v**), (**N**), (**i**), (**t**), (**D**), (**K**), (**J**) and (**P**) are known, then its Fixed Cost planned is:

$$F= S'[1+s]-Sv-S'i[1+s]-\{[N+K+J+P]/[1+/]-U+D\}$$
$$/[1-t]$$

Rule-58167:

If both (**/**), (**U**), (**S'**), (**s**), (**S**), (**v**), (**F**), (**N**), (**t**), (**D**), (**K**), (**J**) and (**P**) are known, then its Interest Portion planned is:

$$i= (S'[1+s]-Sv-F-\{[N+K+J+P]/[1+/]-U+D\}/[1-t])$$
$$/\{S'[1+s]\}$$

Rule-58168:

If both (**/**), (**U**), (**S'**), (**s**), (**S**), (**v**), (**F**), (**i**), (**N**), (**D**), (**K**), (**J**) and (**P**) are known, then its Tax Rate planned is:

$$t= 1-\{[N+K+J+P]/[1+/]-U+D\}$$
$$/\{S'[1+s]-Sv-F-S'i[1+s]\}$$

Rule-58169:

If both (**/**), (**U**), (**S'**), (**s**), (**S**), (**v**), (**F**), (**i**), (**t**), (**N**), (**K**), (**J**) and (**P**) are known, then its Dividend planned is:

$$D= U+\{S'[1+s]-Sv-F-S'i[1+s]\}[1-t]$$
$$- [N+K+J+P]/[1+/]$$

Steve Asikin ISBN 13: **978-1564912361**, ISBN 10: **1564912363**

Rule-58170:

If both (**/**), (**U**), (**S'**), (**s**), (**S**), (**v**), (**F**), (**i**), (**t**), (**D**), (**N**), (**J**) and (**P**) are known, then its Kind of Cash planned is:

$$K= [1+/](U+\{S'[1+s]-Sv-F-S'i[1+s]\}[1-t]-D)$$
$$-[N+J+P]$$

Rule-58171:

If both (**/**), (**U**), (**S'**), (**s**), (**S**), (**v**), (**F**), (**i**), (**t**), (**D**), (**K**), (**N**) and (**P**) are known, then its Job or Trade Account Receivable planned is:

$$J= [1+/](U+\{S'[1+s]-Sv-F-S'i[1+s]\ ]\}[1-t]-D)$$
$$-[N+K+P]$$

Rule-58172:

If both (**/**), (**U**), (**S'**), (**s**), (**S**), (**v**), (**F**), (**i**), (**t**), (**D**), (**K**), (**J**) and (**N**) are known, then its Procured Inventory planned is:

$$P= [1+/](U+\{S'[1+s]-Sv-F-S'i[1+s]\}[1-t]-D)$$
$$-[N+K+J]$$

Rule-58173:

If both (**/**), (**U**), (**S'**), (**s**), (**S**), (**v**), (**F**), (**i**), (**t**), (**D**), (**K**), (**J**), (**V**) and (**p**) are known, then its Non Current Asset planned is:

$$N= [1+/](U+\{S'[1+s]-Sv-F-S'i[1+s]\}[1-t]-D)$$
$$-K-J-Vp/360$$

Steve Asikin ISBN 13: **978-1564912361**, ISBN 10: **1564912363**

Rule-58174:

If both (**N**), (**U**), (**S'**), (**s**), (**S**), (**v**), (**F**), (**i**), (**t**), (**D**), (**K**), (**J**), (**V**) and (**p**) are known, then its Leverage or Gearing Ratio planned is:

$$l = [N+K+J+Vp/360]/(U+\{S'[1+s]-Sv-F-S'i[1+s]\}[1-t]-D)-1$$

Rule-58175:

If both (**l**), (**N**), (**S'**), (**s**), (**S**), (**v**), (**F**), (**i**), (**t**), (**D**), (**K**), (**J**), (**V**) and (**p**) are known, then its Utilized or Starting Capital must be:

$$U = ([N+K+J+Vp/360]-[1+l]\{S'[1+s]-Sv-F-S'i[1+s]\}[1-t]-D)/[1+l]$$

Rule-58176:

If both (**l**), (**U**), (**N**), (**s**), (**S**), (**v**), (**F**), (**i**), (**t**), (**D**), (**K**), (**J**), (**V**) and (**p**) are known, then its Sales Past must be:

$$S' = (\{[N+K+J+Vp/360]/[1+l]-U+D\}/[1-t]+Sv+F)/\{[1+s][1-i]\}$$

or

$$S' = S/[1+s]$$

Steve Asikin ISBN 13: **978-1564912361**, ISBN 10: **1564912363**

Rule-58177:

If both (**/**), (**U**), (**S'**), (**N**), (**S**), (**v**), (**F**), (**i**), (**t**), (**D**), (**K**), (**J**), (**V**) and (**p**) are known, then its Sales Portion planned is:

$$s= (\{[N+K+J+Vp/360]/[1+\text{/}-U+D\}/[1-t]+Sv+F)$$
$$/\{S'[1-i]\}-1$$

or

$$s= S/S'-1$$

Rule-58178:

If both (**/**), (**U**), (**S'**), (**s**), (**N**), (**v**), (**F**), (**i**), (**t**), (**D**), (**K**), (**J**), (**V**) and (**p**) are known, then its Sales or Revenue planned is:

$$S= (S'[1+s]-F-S'i[1+s]-\{[N+K+J+Vp/360]$$
$$/[1+\text{/}-U+D\}/[1-t])/v$$

or

$$S= S'[1+s]$$

or

$$S= V/v$$

Rule-58179:

If both (**/**), (**U**), (**S'**), (**s**), (**S**), (**N**), (**F**), (**i**), (**t**), (**D**), (**K**), (**J**), (**V**) and (**p**) are known, then its Variable Portion planned is:

$$v= (S'[1+s]-F-S'i[1+s]-\{[N+K+J+Vp/360]$$
$$/[1+\text{/}-U+D\}/[1-t])/S$$

or

$$v= V/S$$

Steve Asikin ISBN 13: **978-1564912361**, ISBN 10: **1564912363**

Rule-58180:

If both (**/**), (**U**), (**S'**), (**s**), (**S**), (**v**), (**N**), (**i**), (**t**), (**D**), (**K**), (**J**), (**V**) and (**p**) are known, then its Fixed Cost planned is:

$$F = S'[1+s] - Sv - S'i[1+s] - \{[N+K+J+Vp/360]$$
$$/[1+/] - U + D\}/[1-t]$$

Rule-58181:

If both (**/**), (**U**), (**S'**), (**s**), (**S**), (**v**), (**F**), (**N**), (**t**), (**D**), (**K**), (**J**), (**V**) and (**p**) are known, then its Interest Portion planned is:

$$i = (S'[1+s] - Sv - F - \{[N+K+J+Vp/360]$$
$$/[1+/] - U + D\}/[1-t])/\{S'[1+s]\}$$

Rule-58182:

If both (**/**), (**U**), (**S'**), (**s**), (**S**), (**v**), (**F**), (**i**), (**N**), (**D**), (**K**), (**J**), (**V**) and (**p**) are known, then its Tax Rate planned is:

$$t = 1 - \{[N+K+J+Vp/360]/[1+/] - U + D\}$$
$$/\{S'[1+s] - Sv - F - S'i[1+s]\}$$

Rule-58183:

If both (**/**), (**U**), (**S'**), (**s**), (**S**), (**v**), (**F**), (**i**), (**t**), (**N**), (**K**), (**J**), (**V**) and (**p**) are known, then its Dividend planned is:

$$D = U + \{S'[1+s] - Sv - F - S'i[1+s]\}[1-t]$$
$$- [N+K+J+Vp/360]/[1+/]$$

Steve Asikin ISBN 13: **978-1564912361**, ISBN 10: **1564912363**

Rule-58184:

   If both (**/**), (**U**), (**S'**), (**s**), (**S**), (**v**), (**F**), (**i**), (**t**), (**D**), (**N**), (**J**), (**V**) and (**p**) are known, then its Kind of Cash planned is:

$$K= [1+/(U+\{S'[1+s]-Sv-F-S'i[1+s]\}[1-t]-D) -[N+J+Vp/360]$$

Rule-58185:

   If both (**/**), (**U**), (**S'**), (**s**), (**S**), (**v**), (**F**), (**i**), (**t**), (**D**), (**K**), (**N**), (**V**) and (**p**) are known, then its Job or Trade Account Receivable planned is:

$$J= [1+/(U+\{S'[1+s]-Sv-F-S'i[1+s]\}[1-t]-D) -\{N+K+Vp/360\}$$

Rule-58186:

   If both (**/**), (**U**), (**S'**), (**s**), (**S**), (**v**), (**F**), (**i**), (**t**), (**D**), (**K**), (**J**), (**N**) and (**p**) are known, then its Variable Cost planned is:

$$V= 360[[1+/(U+\{S'[1+s]-Sv-F-S'i[1+s]\}[1-t]-D) -[N+K+J]]/p$$

        or

$$V= Sv$$

Rule-58187:

   If both (**/**), (**U**), (**S'**), (**s**), (**S**), (**v**), (**F**), (**i**), (**t**), (**D**), (**K**), (**J**), (**V**) and (**N**) are known, then its Procured Inventory Days planned is:

$$p= 360[[1+/(U+\{S'[1+s]-Sv-F-S'i[1+s]\}[1-t]-D) -[N+K+J]]/V$$

577

Steve Asikin ISBN 13: **978-1564912361**, ISBN 10: **1564912363**

Rule-58188:

If both $(\mathit{l})$, $(\mathbf{U})$, $(\mathbf{S'})$, $(\mathbf{s})$, $(\mathbf{S})$, $(\mathbf{v})$, $(\mathbf{F})$, $(\mathbf{i})$, $(\mathbf{t})$, $(\mathbf{D})$, $(\mathbf{K})$, $(\mathbf{J})$ and $(\mathbf{p})$ are known, then its Non Current Asset planned is:

$$N = [1+\mathit{l}](U+\{S'[1+s]-Sv-F-S'i[1+s]\}[1-t]-D) -K-J-Svp/360$$

Rule-58189:

If both $(\mathbf{N})$, $(\mathbf{U})$, $(\mathbf{S'})$, $(\mathbf{s})$, $(\mathbf{S})$, $(\mathbf{v})$, $(\mathbf{F})$, $(\mathbf{i})$, $(\mathbf{t})$, $(\mathbf{D})$, $(\mathbf{K})$, $(\mathbf{J})$ and $(\mathbf{p})$ are known, then its Leverage or Gearing Ratio planned is:

$$\mathit{l} = [N+K+J+Svp/360]/(U+\{S'[1+s]-Sv-F -S'i[1+s]\}[1-t]-D)-1$$

Rule-58190:

If both $(\mathit{l})$, $(\mathbf{N})$, $(\mathbf{S'})$, $(\mathbf{s})$, $(\mathbf{S})$, $(\mathbf{v})$, $(\mathbf{F})$, $(\mathbf{i})$, $(\mathbf{t})$, $(\mathbf{D})$, $(\mathbf{K})$, $(\mathbf{J})$ and $(\mathbf{p})$ are known, then its Utilized or Starting Capital must be:

$$U = ([N+K+J+Svp/360]-[1+\mathit{l}]\{S'[1+s]-Sv-F -S'i[1+s]\}[1-t]-D)/[1+\mathit{l}]$$

Rule-58191:

If both $(\mathit{l})$, $(\mathbf{U})$, $(\mathbf{N})$, $(\mathbf{s})$, $(\mathbf{S})$, $(\mathbf{v})$, $(\mathbf{F})$, $(\mathbf{i})$, $(\mathbf{t})$, $(\mathbf{D})$, $(\mathbf{K})$, $(\mathbf{J})$ and $(\mathbf{p})$ are known, then its Sales Past must be:

$$S' = (\{[N+J+Svp/360]/[1+\mathit{l}]-U+D\}/[1-t] +Sv+F)/\{[1+s][1-i]\}$$

or

$$S' = S/[1+s]$$

Steve Asikin ISBN 13: **978-1564912361**, ISBN 10: **1564912363**

Rule-58192:

If both (**/**), (**U**), (**S'**), (**N**), (**S**), (**v**), (**F**), (**i**), (**t**), (**D**), (**K**), (**J**) and (**p**) are known, then its Sales Growth planned is:

$$s= (\{[N+J+Svp/360]/[1+/-U+D\}/[1-t]$$
$$+Sv+F)/\{S'[1-i]\}-1$$

or

$$s= S/S'-1$$

Rule-58193:

If both (**/**), (**U**), (**S'**), (**N**), (**S**), (**v**), (**F**), (**i**), (**t**), (**D**), (**K**), (**J**) and (**p**) are known, then its Sales or Revenue planned is:

$$S= [[1+/(U+\{S'[1+s]-F-S'i[1+s]\}[1-t]-D)$$
$$-[N+K+J]]/(v\{p/360+[1+/[1-t]\})$$

or

$$S= S'[1+s]$$

Rule-58194:

If both (**/**), (**U**), (**S'**), (**s**), (**S**), (**N**), (**F**), (**i**), (**t**), (**D**), (**K**), (**J**) and (**p**) are known, then its Variable Portion planned is:

$$v= [[1+/(U+\{S'[1+s]-F-S'i[1+s]\}[1-t]-D)$$
$$-[N+K+J]]/(S\{p/360+[1+/[1-t]\})$$

Steve Asikin ISBN 13: **978-1564912361**, ISBN 10: **1564912363**

Rule-58195:

If both (**/**), (**U**), (**S'**), (**s**), (**S**), (**v**), (**N**), (**i**), (**t**), (**D**), (**K**), (**J**) and (**p**) are known, then its Fixed Cost planned is:

$$\mathbf{F} = \{\mathbf{S'}[1+\mathbf{s}]-\mathbf{Sv}-\mathbf{S'i}[1+\mathbf{s}]\}-\{[\mathbf{N}+\mathbf{K}+\mathbf{J}+\mathbf{Svp}/360]$$
$$/[1+\mathbf{/}]-\mathbf{U}+\mathbf{D}\}/[1-\mathbf{t}]$$

Rule-58196:

If both (**/**), (**U**), (**S'**), (**s**), (**S**), (**v**), (**F**), (**N**), (**t**), (**D**), (**K**), (**J**) and (**p**) are known, then its Interest Portion planned is:

$$\mathbf{i} = \mathbf{S'}[1+\mathbf{s}]-\mathbf{Sv}-\mathbf{F}-\{[\mathbf{N}+\mathbf{K}+\mathbf{J}+\mathbf{Svp}/360]$$
$$/[1+\mathbf{/}]-\mathbf{U}+\mathbf{D}\}/[1-\mathbf{t}])/\{\mathbf{S'}[1+\mathbf{s}]\}$$

Rule-58197:

If both (**/**), (**U**), (**S'**), (**s**), (**S**), (**v**), (**F**), (**i**), (**N**), (**D**), (**K**), (**J**) and (**p**) are known, then its Tax Rate planned is:

$$\mathbf{t} = 1-\{[\mathbf{N}+\mathbf{K}+\mathbf{J}+\mathbf{Svp}/360]/[1+\mathbf{/}]-\mathbf{U}+\mathbf{D}\}$$
$$/\{\mathbf{S'}[1+\mathbf{s}]-\mathbf{Sv}-\mathbf{F}-\mathbf{S'i}[1+\mathbf{s}]\}$$

Rule-58198:

If both (**/**), (**U**), (**S'**), (**s**), (**S**), (**v**), (**F**), (**i**), (**t**), (**N**), (**K**), (**J**) and (**p**) are known, then its Dividend planned is:

$$\mathbf{D} = \mathbf{U}+\{\mathbf{S'}[1+\mathbf{s}]-\mathbf{Sv}-\mathbf{F}-\mathbf{S'i}[1+\mathbf{s}]\}[1-\mathbf{t}]$$
$$-[\mathbf{N}+\mathbf{K}+\mathbf{J}+\mathbf{Svp}/360]/[1+\mathbf{/}]$$

Steve Asikin ISBN 13: **978-1564912361**, ISBN 10: **1564912363**

Rule-58199:

If both (*l*), (**U**), (**S'**), (**s**), (**S**), (**v**), (**F**), (**i**), (**t**), (**D**), (**N**), (**J**) and (**p**) are known, then its Kind of Cash planned is:

$$K= [1+l(U+\{S'[1+s]-Sv-F-S'i[1+s]\}[1-t]-D)$$
$$-[N+J+Svp/360]$$

Rule-58200:

If both (*l*), (**U**), (**S'**), (**s**), (**S**), (**v**), (**F**), (**i**), (**t**), (**D**), (**K**), (**N**) and (**p**) are known, then its Job or Trade Account Receivable planned is:

$$J= [1+l(U+\{S'[1+s]-Sv-F-S'i[1+s]\}[1-t]-D)$$
$$-\{N+K+Svp/360\}$$

Rule-58201:

If both (*l*), (**U**), (**S'**), (**s**), (**S**), (**v**), (**F**), (**i**), (**t**), (**D**), (**K**), (**J**) and (**N**) are known, then its Procured Inventory Days planned is:

$$p= 360[[1+l(U+\{S'[1+s]-Sv-F-S'i[1+s]\}[1-t]-D)$$
$$-[N+K+J]]/[Sv]$$

Rule-58202:

If both (*l*), (**U**), (**S'**), (**s**), (**S**), (**v**), (**F**), (**i**), (**t**), (**D**), (**K**), (**J**) and (**p**) are known, then its Non Current Asset planned is:

$$N= [1+l(U+\{S'[1+s]-Sv-F-S'i[1+s]\}[1-t]-D)$$
$$-K-J-S'vp[1+s]/360$$

Steve Asikin ISBN 13: **978-1564912361**, ISBN 10: **1564912363**

Rule-58203:

If both (**N**), (**U**), (**S'**), (**s**), (**S**), (**v**), (**F**), (**i**), (**t**), (**D**), (**K**), (**J**) and (**p**) are known, then its Leverage or Gearing Ratio planned is:

$$I = \{N+K+J+S'vp[1+s]/360\}$$
$$/(U+\{S'[1+s]-Sv-F-S'i[1+s]\}[1-t]-D)-1$$

Rule-58204:

If both (**I**), (**N**), (**S'**), (**s**), (**S**), (**v**), (**F**), (**i**), (**t**), (**D**), (**K**), (**J**) and (**p**) are known, then its Utilized or Starting Capital must be:

$$U = (\{N+K+J+S'vp[1+s]/360\}-[1+I]\{S'[1+s]-Sv$$
$$-F-S'i[1+s]\}[1-t]-D)/[1+I]$$

Rule-58205:

If both (**I**), (**U**), (**N**), (**s**), (**S**), (**v**), (**F**), (**i**), (**t**), (**D**), (**K**), (**J**) and (**p**) are known, then its Sales Past must be:

$$S' = [[1+I](U-\{Sv+F\}[1-t]-D)-[N+K+J]]$$
$$/([1+s]\{vp/360-[1+I][1-t][1-i]\})$$
$$\text{or}$$
$$S' = S/[1+s]$$

Steve Asikin ISBN 13: **978-1564912361**, ISBN 10: **1564912363**

Rule-58206:

If both (**/**), (**U**), (**S'**), (**N**), (**S**), (**v**), (**F**), (**i**), (**t**), (**D**), (**K**), (**J**) and (**p**) are known, then its Sales Growth planned is:

$$s= [[1+/(U-\{Sv+F\}[1-t]-D)-[N+K+J]]$$
$$/(S'\{vp/360-[1+/[1-t][1-i]\})-1$$

or

$$s= S/S'-1$$


Rule-58207:

If both (**/**), (**U**), (**S'**), (**s**), (**N**), (**v**), (**F**), (**i**), (**t**), (**D**), (**K**), (**J**) and (**p**) are known, then its Sales or Revenue planned is:

$$S= [S'[1+s]-F-S'[1+s]-(\{N+K+J+S'vp[1+s]/360\}$$
$$/[1+/-U+D)/[1-t]]/v$$

or

$$S= S'[1+s]$$


Rule-58208:

If both (**/**), (**U**), (**S'**), (**s**), (**S**), (**N**), (**F**), (**i**), (**t**), (**D**), (**K**), (**J**) and (**p**) are known, then its Variable Portion planned is:

$$v= [S'[1+s]-F-S'[1+s]-(\{N+K+J+S'vp[1+s]/360\}$$
$$/[1+/-U+D)/[1-t]]/S$$

Steve Asikin ISBN 13: **978-1564912361**, ISBN 10: **1564912363**

Rule-58209:

If both (**/**), (**U**), (**S'**), (**s**), (**S**), (**v**), (**N**), (**i**), (**t**), (**D**), (**K**), (**J**) and (**p**) are known, then its Fixed Cost planned is:

$$F= S'[1+s]-Sv-S'[1+s]-(\{N+K+J+S'vp[1+s]/360\}$$
$$/[1+/-U+D)/[1-t]$$

Rule-58210:

If both (**/**), (**U**), (**S'**), (**s**), (**S**), (**v**), (**F**), (**N**), (**t**), (**D**), (**K**), (**J**) and (**p**) are known, then its Interest Portion planned is:

$$i= [S'[1+s]-Sv-F-(\{N+K+J+S'vp[1+s]/360\}$$
$$/[1+/-U+D)/[1-t]]/\{S'[1+s]\}$$

Rule-58211:

If both (**/**), (**U**), (**S'**), (**s**), (**S**), (**v**), (**F**), (**i**), (**N**), (**D**), (**K**), (**J**) and (**p**) are known, then its Tax Rate planned is:

$$t= 1-(\{N+K+J+S'vp[1+s]/360\}/[1+/-U+D)$$
$$/\{S'[1+s]-Sv-F-S'i[1+s]\}$$

Rule-58212:

If both (**/**), (**U**), (**S'**), (**s**), (**S**), (**v**), (**F**), (**i**), (**t**), (**N**), (**K**), (**J**) and (**p**) are known, then its Dividend planned is:

$$D= U+\{S'[1+s]-Sv-F-S'i[1+s]\}[1-t]$$
$$-\{N+K+J+S'vp[1+s]/360\}/[1+/$$

584

Steve Asikin ISBN 13: **978-1564912361**, ISBN 10: **1564912363**

Rule-58213:

If both (**/**), (**U**), (**S'**), (**s**), (**S**), (**v**), (**F**), (**i**), (**t**), (**D**), (**N**), (**J**) and (**p**) are known, then its Kind of Cash planned is:

$$K= [1+/](U+\{S'[1+s]-Sv-F-S'i[1+s]\}[1-t]-D)$$
$$-\{N+J+S'vp[1+s]/360\}$$

Rule-58214:

If both (**/**), (**U**), (**S'**), (**s**), (**S**), (**v**), (**F**), (**i**), (**t**), (**D**), (**K**), (**N**) and (**p**) are known, then its Job or Trade Account Receivable planned is:

$$J= [[1+/](U+\{S'[1+s]-Sv-F-S'i[1+s]\}[1-t]-D)$$
$$-\{N+K+S'vp[1+s]/360\}$$

Rule-58215:

If both (**/**), (**U**), (**S'**), (**s**), (**S**), (**v**), (**F**), (**i**), (**t**), (**D**), (**K**), (**J**) and (**N**) are known, then its Procured Inventory Days planned is:

$$p= 360[[1+/](U+\{S'[1+s]-Sv-F-S'i[1+s]\}[1-t]-D)$$
$$-[N+K+J]]/\{S'v[1+s]\}$$

Rule-58216:

If both (**/**), (**U**), (**S'**), (**s**), (**S**), (**v**), (**F**), (**i**), (**t**), (**D**), (**K**), (**j**) and (**P**) are known, then its Non Current Asset planned is:

$$N= [1+/](U+\{S'[1+s]-Sv-F-S'i[1+s]\}[1-t]-D)$$
$$-K-Sj/360-P$$

Steve Asikin ISBN 13: **978-1564912361**, ISBN 10: **1564912363**

Rule-58217:

If both ($I$), (**U**), (**S'**), (**s**), (**S**), (**v**), (**F**), (**i**), (**t**), (**D**), (**K**), (**j**) and (**P**) are known, then its Non Current Asset planned is:

$$I = \{N+K+Sj/360+P\}/(U+\{S'[1+s]-Sv-F-S'i[1+s]\}$$
$$[1-t]-D)-1$$

Rule-58218:

If both ($I$), (**N**), (**S'**), (**s**), (**S**), (**v**), (**F**), (**i**), (**t**), (**D**), (**K**), (**j**) and (**P**) are known, then its Utilized or Starting Capital must be:

$$U = (\{N+K+Sj/360+P\}-[1+I]\{S'[1+s]-Sv-F$$
$$-S'i[1+s]\}[1-t]-D)/[1+I]$$

Rule-58219:

If both ($I$), (**U**), (**N**), (**s**), (**S**), (**v**), (**F**), (**i**), (**t**), (**D**), (**K**), (**j**) and (**P**) are known, then its Starting Capital must be:

$$S' = (\{[N+K+Sj/360+P]/[1+I]-U+D\}/[1-t]$$
$$-\{Sv+F\})/\{S'[1+s][1-i]\}$$
$$\text{or}$$
$$S' = S/[1+s]$$

Steve Asikin ISBN 13: **978-1564912361**, ISBN 10: **1564912363**

Rule-58220:

If both (**/**), (**U**), (**S'**), (**N**), (**S**), (**v**), (**F**), (**i**), (**t**), (**D**), (**K**), (**j**) and (**P**) are known, then its Sales Growth planned is:

$$s = (\{[\mathbf{N+K+Sj}/360\mathbf{+P}]/[1\mathbf{+/}]\mathbf{-U+D}\}/[1\mathbf{-t}]\mathbf{-}\{\mathbf{Sv+F}\})$$
$$/\{\mathbf{S'}\{1\mathbf{-i}]\}\mathbf{-}1$$

or

$$s = \mathbf{S}/\mathbf{S'}\mathbf{-}1$$


Rule-58221:

If both (**/**), (**U**), (**S'**), (**s**), (**N**), (**v**), (**F**), (**i**), (**t**), (**D**), (**K**), (**j**) and (**P**) are known, then its Sales or Revenue planned is:

$$S = [[1\mathbf{+/}(\mathbf{U}\mathbf{-}\{\mathbf{Sv+F}\}[1\mathbf{-t}]\mathbf{-D})\mathbf{-}[\mathbf{N+K+P}]]$$
$$/\{\mathbf{j}/360\mathbf{+v}[1\mathbf{+/}][1\mathbf{-t}][1\mathbf{-i}]\}$$

or

$$S = \mathbf{S'}[1\mathbf{+s}]$$


Rule-58222:

If both (**/**), (**U**), (**S'**), (**s**), (**S**), (**N**), (**F**), (**i**), (**t**), (**D**), (**K**), (**j**) and (**P**) are known, then its Variable Portion planned is:

$$v = [\mathbf{S'}[1\mathbf{+s}]\mathbf{-F}\mathbf{-S'i}[1\mathbf{+s}]\mathbf{-}\{[\mathbf{N+K+Sj}/360\mathbf{+P}]/[1\mathbf{+/}]$$
$$\mathbf{-U+D}\}/[1\mathbf{-t}]]/\mathbf{S}$$

Steve Asikin ISBN 13: **978-1564912361**, ISBN 10: **1564912363**

## Rule-58223:

If both ($I$), (**U**), (**S'**), (**s**), (**S**), (**v**), (**N**), (**i**), (**t**), (**D**), (**K**), (**j**) and (**P**) are known, then its Fixed Cost planned is:

$$F= S'[1+s]-Sv-S'i[1+s]-\{[N+K+Sj/360+P]/[1+I]$$
$$-U+D\}/[1-t]$$

## Rule-58224:

If both ($I$), (**U**), (**S'**), (**s**), (**S**), (**v**), (**F**), (**N**), (**t**), (**D**), (**K**), (**j**) and (**P**) are known, then its Interest Portion planned is:

$$i= \{S'[1+s]-Sv-F-\{[N+K+Sj/360+P]/[1+I]-U+D\}$$
$$/[1-t]\}/\{S'[1+s]\}$$

## Rule-58225:

If both ($I$), (**U**), (**S'**), (**s**), (**S**), (**v**), (**F**), (**i**), (**N**), (**D**), (**K**), (**j**) and (**P**) are known, then its Tax Rate planned is:

$$t= 1-\{[N+K+Sj/360+P]/[1+I]-U+D\}$$
$$/\{S'[1+s]-Sv-F-S'i[1+s]\}$$

## Rule-58226:

If both ($I$), (**U**), (**S'**), (**s**), (**S**), (**v**), (**F**), (**i**), (**t**), (**N**), (**K**), (**j**) and (**P**) are known, then its Dividend planned is:

$$D= U+\{S'[1+s]-Sv-F-S'i[1+s]\}[1-t]$$
$$- [N+K+Sj/360+P]/[1+I]$$

`

Steve Asikin ISBN 13: **978-1564912361**, ISBN 10: **1564912363**

Rule-58227:

If both (**/**), (**U**), (**S'**), (**s**), (**S**), (**v**), (**F**), (**i**), (**t**), (**D**), (**N**), (**j**) and (**P**) are known, then its Kind of Cash planned is:

$$K= [1+/(U+\{S'[1+s]-Sv-F-S'i[1+s]\}[1-t]-D) -[N+Sj/360+P]$$

Rule-58228:

If both (**/**), (**U**), (**S'**), (**s**), (**S**), (**v**), (**F**), (**i**), (**t**), (**D**), (**K**), (**N**) and (**P**) are known, then its Job or Trade Receivable Days planned is:

$$j= 360[[1+/(U+\{S'[1+s]-Sv-F-S'i[1+s]\}[1-t]-D) -\{N+K+P\}]/S$$

Rule-58229:

If both (**/**), (**U**), (**S'**), (**s**), (**S**), (**v**), (**F**), (**i**), (**t**), (**D**), (**K**), (**j**) and (**N**) are known, then its Procured Inventory planned is:

$$P= [1+/(U+\{S'[1+s]-Sv-F-S'i[1+s]\}[1-t]-D) -\{N+K+Sj/360\}$$

Rule-58230:

If both (**/**), (**U**), (**S'**), (**s**), (**S**), (**v**), (**F**), (**i**), (**t**), (**D**), (**K**), (**j**), (**V**) and (**p**) are known, then its Non Current Asset planned is:

$$N= [1+/(U+\{S'[1+s]-Sv-F-S'i[1+s]\}[1-t]-D) -K-Sj/360-Vp/360$$

Steve Asikin ISBN 13: **978-1564912361**, ISBN 10: **1564912363**

Rule-58231:

If both (**N**), (**U**), (**S'**), (**s**), (**S**), (**v**), (**F**), (**i**), (**t**), (**D**), (**K**), (**j**), (**V**) and (**p**) are known, then its Leverage or Gearing Ratio planned is:

$$\textit{l}= [\mathbf{N+K+Sj}/360+\mathbf{Vp}/360]$$
$$/(\mathbf{U}+\{\mathbf{S'}[1+\mathbf{s}]-\mathbf{Sv-F-S'i}[1+\mathbf{s}]\}[1-\mathbf{t}]-\mathbf{D})-1$$

Rule-58232:

If both (**l**), (**N**), (**S'**), (**s**), (**S**), (**v**), (**F**), (**i**), (**t**), (**D**), (**K**), (**j**), (**V**) and (**p**) are known, then its Utilized or Starting Capital must be:

$$\mathbf{U}= ([\mathbf{N+K+Sj}/360+\mathbf{Vp}/360]-[1+\textit{l}]\{\mathbf{S'}[1+\mathbf{s}]-\mathbf{Sv-F}$$
$$-\mathbf{S'i}[1+\mathbf{s}]\}[1-\mathbf{t}]-\mathbf{D})/[1+\textit{l}]$$

Rule-58233:

If both (**l**), (**U**), (**N**), (**s**), (**S**), (**v**), (**F**), (**i**), (**t**), (**D**), (**K**), (**j**), (**V**) and (**p**) are known, then its Sales or Revenue planned is:

$$\mathbf{S'}= ([\mathbf{N+K+Sj}/360+\mathbf{Vp}/360]/[1+\textit{l}]-\mathbf{U+D}\}/[1-\mathbf{t}]\}$$
$$-[\mathbf{Sv+F}])/\{[1+\mathbf{s}][1-\mathbf{i}]\}$$
$$\text{or}$$
$$\mathbf{S'}= \mathbf{S}[1+\mathbf{s}]$$

Steve Asikin ISBN 13: **978-1564912361**, ISBN 10: **1564912363**

Rule-58234:

If both $(\textbf{\textit{I}})$, $(\textbf{U})$, $(\textbf{S'})$, $(\textbf{N})$, $(\textbf{S})$, $(\textbf{v})$, $(\textbf{F})$, $(\textbf{i})$, $(\textbf{t})$, $(\textbf{D})$, $(\textbf{K})$, $(\textbf{j})$, $(\textbf{V})$ and $(\textbf{p})$ are known, then its Sales Growth planned is:

$$s = ([\textbf{N}+\textbf{K}+\textbf{Sj}/360+\textbf{Vp}/360]/[1+\textbf{\textit{I}}-\textbf{U}+\textbf{D}\}/[1-\textbf{t}]\}$$
$$-[\textbf{Sv}+\textbf{F}])/\{\textbf{S'}[1-\textbf{i}]\}-1$$

or

$$s = \textbf{S}/\textbf{S'}-1$$


Rule-58235:

If both $(\textbf{\textit{I}})$, $(\textbf{U})$, $(\textbf{S'})$, $(\textbf{s})$, $(\textbf{N})$, $(\textbf{v})$, $(\textbf{F})$, $(\textbf{i})$, $(\textbf{t})$, $(\textbf{D})$, $(\textbf{K})$, $(\textbf{j})$, $(\textbf{V})$ and $(\textbf{p})$ are known, then its Sales or Revenue planned is:

$$\textbf{S} = [[1+\textbf{\textit{I}}(\textbf{U}+\{\textbf{S'}[1+\textbf{s}]-\textbf{F}-\textbf{S'i}[1+\textbf{s}]\}[1-\textbf{t}]-\textbf{D})$$
$$-[\textbf{N}+\textbf{K}+\textbf{Vp}/360]]/\{\textbf{j}/360+\textbf{v}[1+\textbf{\textit{I}}][1-\textbf{t}]\}$$

or

$$\textbf{S} = \textbf{S'}[1+\textbf{s}]$$


Rule-58236:

If both $(\textbf{\textit{I}})$, $(\textbf{U})$, $(\textbf{S'})$, $(\textbf{s})$, $(\textbf{S})$, $(\textbf{N})$, $(\textbf{F})$, $(\textbf{i})$, $(\textbf{t})$, $(\textbf{D})$, $(\textbf{K})$, $(\textbf{j})$, $(\textbf{V})$ and $(\textbf{p})$ are known, then its Variable Portion planned is:

$$v = (\{\textbf{S'}[1+\textbf{s}]-\textbf{S'i}[1+\textbf{s}]\}-\{[\textbf{N}+\textbf{K}+\textbf{Sj}/360+\textbf{Vp}/360]$$
$$/[1+\textbf{\textit{I}}-\textbf{U}+\textbf{D}\}/[1-\textbf{t}])/\textbf{S}$$

or

$$v = \textbf{V}/\textbf{S}$$

Steve Asikin ISBN 13: **978-1564912361**, ISBN 10: **1564912363**

Rule-58237:

If both (**I**), (**U**), (**S'**), (**s**), (**S**), (**v**), (**N**), (**i**), (**t**), (**D**), (**K**), (**j**), (**V**) and (**p**) are known, then its Fixed Cost planned is:

$$F= \{S'[1+s]-Sv-S'i[1+s]\}-\{[N+K+Sj/360+Vp/360]$$
$$/[1+I]-U+D\}/[1-t]$$

Rule-58238:

If both (**I**), (**U**), (**S'**), (**s**), (**S**), (**v**), (**F**), (**N**), (**t**), (**D**), (**K**), (**j**), (**V**) and (**p**) are known, then its Interest Portion planned is:

$$i= (\{S'[1+s]-Sv-F\}-\{[N+K+Sj/360+Vp/360]$$
$$/[1+I]-U+D\}/[1-t])/\{S'[1+s]\}$$

Rule-58239:

If both (**I**), (**U**), (**S'**), (**s**), (**S**), (**v**), (**F**), (**i**), (**N**), (**D**), (**K**), (**j**), (**V**) and (**p**) are known, then its Tax Rate planned is:

$$t= 1-\{[N+K+Sj/360+Vp/360]/[1+I]-U+D\}$$
$$/\{S'[1+s]-Sv-F-S'i[1+s]\}$$

Rule-58240:

If both (**I**), (**U**), (**S'**), (**s**), (**S**), (**v**), (**F**), (**i**), (**t**), (**N**), (**K**), (**j**), (**V**) and (**p**) are known, then its Dividend planned is:

$$D= U+\{S'[1+s]-Sv-F-S'i[1+s]\}[1-t]$$
$$-[N+K+Sj/360+Vp/360]/[1+I]$$

`

Steve Asikin ISBN 13: **978-1564912361**, ISBN 10: **1564912363**

Rule-58241:
>  If both (**/**), (**U**), (**S'**), (**s**), (**S**), (**v**), (**F**), (**i**), (**t**), (**D**), (**N**),
>  (**j**), (**V**) and (**p**) are known, then its Kind of Cash
>  planned is:
>
>  $$K = [1 + /(U + \{S'[1+s] - Sv - F - S'i[1+s]\}[1-t] - D) - [N + Sj/360 + Vp/360]$$

Rule-58242:
>  If both (**/**), (**U**), (**S'**), (**s**), (**S**), (**v**), (**F**), (**i**), (**t**), (**D**), (**K**),
>  (**N**), (**V**) and (**p**) are known, then its Job or Trade
>  Receivable Days planned is:
>
>  $$j = 360[[1 + /(U + \{S'[1+s] - Sv - F - S'i[1+s]\}[1-t] - D) - [N + K + Vp/360]]/S$$

Rule-58243:
>  If both (**/**), (**U**), (**S'**), (**s**), (**S**), (**v**), (**F**), (**i**), (**t**), (**D**), (**K**),
>  (**j**), (**N**) and (**p**) are known, then its Variable Cost
>  planned is:
>
>  $$V = 360[[1 + /(U + \{S'[1+s] - Sv - F - S'i[1+s]\}[1-t] - D) - [N + K + Sj/360]]/p$$
>
>  or
>
>  $$V = Sv$$

Rule-58244:
>  If both (**/**), (**U**), (**S'**), (**s**), (**S**), (**v**), (**F**), (**i**), (**t**), (**D**), (**K**),
>  (**j**), (**V**) and (**N**) are known, then its Procured
>  Inventory Days planned is:
>
>  $$p = 360[[1 + /(U + \{S'[1+s] - Sv - F - S'i[1+s]\}[1-t] - D) - [N + K + Sj/360]]/V$$

Steve Asikin ISBN 13: **978-1564912361**, ISBN 10: **1564912363**

Rule-58245:

If both ($I$), ($U$), ($S'$), ($s$), ($S$), ($v$), ($F$), ($i$), ($t$), ($D$), ($K$), ($j$) and ($p$) are known, then its Non Current Asset planned is:

$$N= [1+I(U+\{S'[1+s]-Sv-F-S'i[1+s]\}[1-t]-D)$$
$$-K-Sj/360-Svp/360$$

Rule-58246:

If both ($N$), ($U$), ($S'$), ($s$), ($S$), ($v$), ($F$), ($i$), ($t$), ($D$), ($K$), ($j$) and ($p$) are known, then its Leverage or Gearing Ratio planned is:

$$I= [N+K+Sj/360+Svp/360]$$
$$/(U+\{S'[1+s]-Sv-F-S'i[1+s]\}[1-t]-D)-1$$

Rule-58247:

If both ($I$), ($N$), ($S'$), ($s$), ($S$), ($v$), ($F$), ($i$), ($t$), ($D$), ($K$), ($j$) and ($p$) are known, then its Utilized or Starting Capital must be:

$$U= ([N+K+Sj/360+Svp/360]-[1+I]\{S'[1+s]-Sv-F$$
$$-S'i[1+s]\}[1-t]-D)/[1+I]$$

Rule-58248:

If both ($I$), ($U$), ($N$), ($s$), ($S$), ($v$), ($F$), ($i$), ($t$), ($D$), ($K$), ($j$) and ($p$) are known, then its Sales Past must be:

$$S'= (\{[N+K+Sj/360+Svp/360]/[1+I]-U$$
$$+\{Sv+F\}[1-t]-D)/\{[1+s][1-i][1-t]\}$$
$$\text{or}$$
$$S'= S/[1+s]$$

Steve Asikin ISBN 13: **978-1564912361**, ISBN 10: **1564912363**

Rule-58249:

If both (**/**), (**U**), (**$'**), (**N**), (**$**), (**v**), (**F**), (**i**), (**t**), (**D**), (**K**), (**j**) and (**p**) are known, then its Sales Growth planned is:

$$s = (\{[N+K+\$j/360+\$vp/360]/[1+/]-U$$
$$+\{\$v+F\}[1-t]-D)/\{\$'[1-i][1-t]\}-1$$

or

$$s = \$/\$'-1$$

Rule-58250:

If both (**/**), (**U**), (**$'**), (**s**), (**N**), (**v**), (**F**), (**i**), (**t**), (**D**), (**K**), (**j**) and (**p**) are known, then its Sales or Revenue planned is:

$$\$ = [[1+/](U+\{\$'[1+s]-F-\$'i[1+s]\}[1-t]-D)-[N+K]]$$
$$/(j/360+v\{p/360+[1+/][1-t]\})$$

or

$$\$ = \$'[1+s]$$

Rule-58251:

If both (**/**), (**U**), (**$'**), (**s**), (**$**), (**N**), (**F**), (**i**), (**t**), (**D**), (**K**), (**j**) and (**p**) are known, then its Variable Portion planned is:

$$v = [[1+/](U+\{\$'[1+s]-F-\$'i[1+s]\}[1-t]-D)$$
$$-[N+K+\$j/360]]/(\${p/360+[1+/][1-t]\})$$

Steve Asikin ISBN 13: **978-1564912361**, ISBN 10: **1564912363**

Rule-58252:

If both $(I)$, $(U)$, $(S')$, $(s)$, $(S)$, $(v)$, $(N)$, $(i)$, $(t)$, $(D)$, $(K)$, $(j)$ and $(p)$ are known, then its Fixed Cost planned is:

$$F= S'[1+s]-Sv-S'i[1+s]-\{[N+K+Sj/360+Svp/360]/[1+I-U+D\}/[1-t]$$

Rule-58253:

If both $(I)$, $(U)$, $(S')$, $(s)$, $(S)$, $(v)$, $(F)$, $(N)$, $(t)$, $(D)$, $(K)$, $(j)$ and $(p)$ are known, then its Interest Portion planned is:

$$i= (S'[1+s]-Sv-F-\{[N+K+Sj/360+Svp/360]/[1+I-U+D\}/[1-t])/\{S'[1+s]\}$$

Rule-58254:

If both $(I)$, $(U)$, $(S')$, $(s)$, $(S)$, $(v)$, $(F)$, $(i)$, $(N)$, $(D)$, $(K)$, $(j)$ and $(p)$ are known, then its Tax Rate planned is:

$$t= 1-\{[N+K+Sj/360+Svp/360]/[1+I-U+D\}/\{S'[1+s]-Sv-F-S'i[1+s]\}$$

Rule-58255:

If both $(I)$, $(U)$, $(S')$, $(s)$, $(S)$, $(v)$, $(F)$, $(i)$, $(t)$, $(N)$, $(K)$, $(j)$ and $(p)$ are known, then its Dividend planned is:

$$D= U+\{S'[1+s]-Sv-F-S'i[1+s]\}[1-t]-[N+K+Sj/360+Svp/360]/[1+I]$$

Steve Asikin ISBN 13: **978-1564912361**, ISBN 10: **1564912363**

Rule-58256:

If both (**/**), (**U**), (**S'**), (**s**), (**S**), (**v**), (**F**), (**i**), (**t**), (**D**), (**N**), (**j**) and (**p**) are known, then its Kind of Cash planned is:

$$\mathbf{K} = [1+\mathbf{/}(\mathbf{U}+\{\mathbf{S'}[1+\mathbf{s}]-\mathbf{Sv}-\mathbf{F}-\mathbf{S'i}[1+\mathbf{s}]\}[1-\mathbf{t}]-\mathbf{D})$$
$$-[\mathbf{N}+\mathbf{Sj}/360+\mathbf{Svp}/360]$$

Rule-58257:

If both (**/**), (**U**), (**S'**), (**s**), (**S**), (**v**), (**F**), (**i**), (**t**), (**D**), (**K**), (**N**) and (**p**) are known, then its Job or Trade Receivable Days planned is:

$$\mathbf{j} = 360[[1+\mathbf{/}(\mathbf{U}+\{\mathbf{S'}[1+\mathbf{s}]-\mathbf{Sv}-\mathbf{F}-\mathbf{S'i}[1+\mathbf{s}]\}[1-\mathbf{t}]-\mathbf{D})$$
$$-\{\mathbf{N}+\mathbf{K}+\mathbf{Svp}/360\}]/\mathbf{S}$$

Rule-58258:

If both (**/**), (**U**), (**S'**), (**s**), (**S**), (**v**), (**F**), (**i**), (**t**), (**D**), (**K**), (**j**) and (**N**) are known, then its Procured Inventory Days planned is:

$$\mathbf{p} = 360[[1+\mathbf{/}(\mathbf{U}+\{\mathbf{S'}[1+\mathbf{s}]-\mathbf{Sv}-\mathbf{F}-\mathbf{S'i}[1+\mathbf{s}]\}[1-\mathbf{t}]-\mathbf{D})$$
$$-[\mathbf{N}+\mathbf{K}+\mathbf{Sj}/360]]/[\mathbf{Sv}]$$

Rule-58259:

If both (**/**), (**U**), (**S'**), (**s**), (**S**), (**v**), (**F**), (**i**), (**t**), (**D**), (**K**), (**j**) and (**p**) are known, then its Non Current Asset planned is:

$$\mathbf{N} = [1+\mathbf{/}(\mathbf{U}+\{\mathbf{S'}[1+\mathbf{s}]-\mathbf{Sv}-\mathbf{F}-\mathbf{S'i}[1+\mathbf{s}]\}[1-\mathbf{t}]-\mathbf{D})$$
$$-\mathbf{K}-\mathbf{Sj}/360-\mathbf{S'vp}[1+\mathbf{s}]/360$$

Steve Asikin ISBN 13: **978-1564912361**, ISBN 10: **1564912363**

Rule-58260:

If both (**N**), (**U**), (**S'**), (**s**), (**S**), (**v**), (**F**), (**i**), (**t**), (**D**), (**K**), (**j**) and (**p**) are known, then its Leverage or Gearing Ratio planned is:

$$\begin{aligned} l = \{&\mathbf{N}+\mathbf{K}+\mathbf{S}\mathbf{j}/360+\mathbf{S'vp}[1+\mathbf{s}]/360\} \\ &/(\mathbf{U}+\{\mathbf{S'}[1+\mathbf{s}]-\mathbf{S}\mathbf{v}-\mathbf{F}-\mathbf{S'}\mathbf{i}[1+\mathbf{s}]\}[1-\mathbf{t}]-\mathbf{D})-1 \end{aligned}$$

Rule-58261:

If both (**l**), (**N**), (**S'**), (**s**), (**S**), (**v**), (**F**), (**i**), (**t**), (**D**), (**K**), (**j**) and (**p**) are known, then its Utilized or Starting Capital must be:

$$\begin{aligned} \mathbf{U}= (\{&\mathbf{N}+\mathbf{K}+\mathbf{S}\mathbf{j}/360+\mathbf{S'vp}[1+\mathbf{s}]/360\}-[1+l] \\ &\{\mathbf{S'}[1+\mathbf{s}]-\mathbf{S}\mathbf{v}-\mathbf{F}-\mathbf{S'}\mathbf{i}[1+\mathbf{s}]\}[1-\mathbf{t}]-\mathbf{D})/[1+l] \end{aligned}$$

Rule-58262:

If both (**l**), (**U**), (**N**), (**s**), (**S**), (**v**), (**F**), (**i**), (**t**), (**D**), (**K**), (**j**) and (**p**) are known, then its Sales Past must be:

$$\begin{aligned} \mathbf{S'}= [[&1+l](\mathbf{U}+\{\mathbf{S'}[1+\mathbf{s}]-\mathbf{S}\mathbf{v}-\mathbf{F}-\mathbf{S'}\mathbf{i}[1+\mathbf{s}]\}[1-\mathbf{t}]-\mathbf{D}) \\ &-[\mathbf{N}+\mathbf{K}+\mathbf{S}\mathbf{j}/360]] \\ &\qquad/([1+\mathbf{s}]\{\mathbf{vp}/360-[1+l][1-\mathbf{t}][1-\mathbf{i}]\}) \end{aligned}$$

or

$$\mathbf{S'}= \mathbf{S}/[1+\mathbf{s}]$$

Steve Asikin ISBN 13: **978-1564912361**, ISBN 10: **1564912363**

Rule-58263:

If both $(I)$, $(U)$, $(S')$, $(N)$, $(S)$, $(v)$, $(F)$, $(i)$, $(t)$, $(D)$, $(K)$, $(j)$ and $(p)$ are known, then its Sales Growth planned is:

$$s = [[1+I(U+\{S'[1+s]-Sv-F-S'i[1+s]\}[1-t]-D)$$
$$-[N+K+Sj/360]]$$
$$/(S'\{vp/360-[1+I][1-t][1-i]\})-1$$

or

$$s = S/S'-1$$

Rule-58264:

If both $(I)$, $(U)$, $(S')$, $(s)$, $(N)$, $(v)$, $(F)$, $(i)$, $(t)$, $(D)$, $(K)$, $(j)$ and $(p)$ are known, then its Sales or Revenue planned is:

$$S = [[1+I(U+\{S'[1+s]-Sv-F-S'i[1+s]\}[1-t]-D)$$
$$-\{N+K+S'vp[1+s]/360\}]$$
$$/\{j/360+v[1+I][1-t]\}$$

or

$$S = S'[1+s]$$

Rule-58265:

If both $(I)$, $(U)$, $(S')$, $(s)$, $(S)$, $(N)$, $(F)$, $(i)$, $(t)$, $(D)$, $(K)$, $(j)$ and $(p)$ are known, then its Variable Portion planned is:

$$v = [[1+I(U+\{S'[1+s]-F-S'i[1+s]\}[1-t]-D)$$
$$-[N+K+Sj/360]]$$
$$/\{S'p[1+s]/360+S[1+I][1-t]\}$$

Steve Asikin ISBN 13: **978-1564912361**, ISBN 10: **1564912363**

Rule-58266:

If both (**/**), (**U**), (**S'**), (**s**), (**S**), (**v**), (**N**), (**i**), (**t**), (**D**), (**K**), (**j**) and (**p**) are known, then its Fixed Cost planned is:

$$F= S'[1+s]-Sv-S'i[1+s]-(\{N+K+Sj/360$$
$$+S'vp[1+s]/360\}/[1+/-U+D)/[1-t]$$

Rule-58267:

If both (**/**), (**U**), (**S'**), (**s**), (**S**), (**v**), (**F**), (**N**), (**t**), (**D**), (**K**), (**j**) and (**p**) are known, then its Interest Portion planned is:

$$i= [S'[1+s]-Sv-F-(\{N+K+Sj/360+S'vp[1+s]/360\}$$
$$/[1+/-U+D)/[1-t]]/\{S'[1+s]\}$$

Rule-58268:

If both (**/**), (**U**), (**S'**), (**s**), (**S**), (**v**), (**F**), (**i**), (**N**), (**D**), (**K**), (**j**) and (**p**) are known, then its Tax Rate planned is:

$$t= 1-(\{N+K+Sj/360+S'vp[1+s]/360\}/[1+/-U+D)$$
$$/\{S'[1+s]-Sv-F-S'i[1+s]\}$$

Rule-58269:

If both (**/**), (**U**), (**S'**), (**s**), (**S**), (**v**), (**F**), (**i**), (**t**), (**N**), (**K**), (**j**) and (**p**) are known, then its Dividend planned is:

$$D= U+\{S'[1+s]-Sv-F-S'i[1+s]\}[1-t]$$
$$-\{N+K+Sj/360+S'vp[1+s]/360\}/[1+/$$

Steve Asikin ISBN 13: **978-1564912361**, ISBN 10: **1564912363**

Rule-58270:

If both (**/**), (**U**), (**S'**), (**s**), (**S**), (**v**), (**F**), (**i**), (**t**), (**D**), (**N**), (**j**) and (**p**) are known, then its Kind of Cash planned is:

$$K= [1+/(U+\{S'[1+s]-Sv-F-S'i[1+s]\}[1-t]-D) -\{N+Sj/360+S'vp[1+s]/360\}$$

Rule-58271:

If both (**/**), (**U**), (**S'**), (**s**), (**S**), (**v**), (**F**), (**i**), (**t**), (**D**), (**K**), (**N**) and (**p**) are known, then its Job or Trade Receivable Days planned is:

$$j= 360[[1+/(U+\{S'[1+s]-Sv-F-S'i[1+s]\}[1-t]-D) -\{N+K+S'vp[1+s]/360\}]/S$$

Rule-58272:

If both (**/**), (**U**), (**S'**), (**s**), (**S**), (**v**), (**F**), (**i**), (**t**), (**D**), (**K**), (**j**) and (**N**) are known, then its Procured Inventory Days planned is:

$$p= 360[[1+/(U+\{S'[1+s]-Sv-F-S'i[1+s]\}[1-t]-D) -[N+K+Sj/360]]/\{S'v[1+s]\}$$

Rule-58273:

If both (**/**), (**U**), (**S'**), (**s**), (**S**), (**v**), (**F**), (**i**), (**t**), (**D**), (**K**), (**j**) and (**P**) are known, then its Non Current Asset planned is:

$$N= [1+/(U+\{S'[1+s]-Sv-F-S'i[1+s]\}[1-t]-D) -K-S'j[1+s]/360-P$$

601

Steve Asikin ISBN 13: **978-1564912361**, ISBN 10: **1564912363**

Rule-58274:

If both (**N**), (**U**), (**S'**), (**s**), (**S**), (**v**), (**F**), (**i**), (**t**), (**D**), (**K**), (**j**) and (**P**) are known, then its Leverage or Gearing Ratio planned is:

$$\textit{l}= \{\mathbf{N+K+S'j}[1+\mathbf{s}]/360+\mathbf{P}\}$$
$$/(\mathbf{U}+\{\mathbf{S'}[1+\mathbf{s}]-\mathbf{Sv-F-S'i}[1+\mathbf{s}]\}[1-\mathbf{t}]-\mathbf{D})-1$$

Rule-58275:

If both (**l**), (**N**), (**S'**), (**s**), (**S**), (**v**), (**F**), (**i**), (**t**), (**D**), (**K**), (**j**) and (**P**) are known, then its Utilized or Starting Capital must be:

$$\mathbf{U}= (\{\mathbf{N+K+S'j}[1+\mathbf{s}]/360+\mathbf{P}\}-[1+\textit{l}]\{\mathbf{S'}[1+\mathbf{s}]-\mathbf{Sv-F}$$
$$-\mathbf{S'i}[1+\mathbf{s}]\}[1-\mathbf{t}]-\mathbf{D})/[1+\textit{l}]$$

Rule-58276:

If both (**l**), (**U**), (**N**), (**s**), (**S**), (**v**), (**F**), (**i**), (**t**), (**D**), (**K**), (**j**) and (**P**) are known, then its Sales Past must be:

$$\mathbf{S'}= \{[\mathbf{N+K+P}]/[1+\textit{l}]-\mathbf{U+D}\}/[1-\mathbf{t}]+\mathbf{Sv+F})$$
$$/\{[1+\mathbf{s}][1-\mathbf{i}]\}$$
$$\text{or}$$
$$\mathbf{S'}= \mathbf{S}/[1+\mathbf{s}]$$

Steve Asikin ISBN 13: **978-1564912361**, ISBN 10: **1564912363**

Rule-58277:

If both (**/**), (**U**), (**S'**), (**N**), (**S**), (**v**), (**F**), (**i**), (**t**), (**D**), (**K**), (**j**) and (**P**) are known, then its Sales or Revenue planned is:

$$s = \{[\mathbf{N+K+P}]/[1+\mathbf{/}-\mathbf{U}+\mathbf{D}]/[1-\mathbf{t}]+\mathbf{Sv}+\mathbf{F})$$
$$/\{\mathbf{S'}[1-\mathbf{i}]\}-1$$

or

$$s = \mathbf{S}/\mathbf{S'}-1$$


Rule-58278:

If both (**/**), (**U**), (**S'**), (**s**), (**N**), (**v**), (**F**), (**i**), (**t**), (**D**), (**K**), (**j**) and (**P**) are known, then its Sales or Revenue planned is:

$$S = [\mathbf{S'}[1+\mathbf{s}]-\mathbf{F}-\mathbf{S'i}[1+\mathbf{s}]-(\{\mathbf{N+K+S'j}[1+\mathbf{s}]/360+\mathbf{P}\}$$
$$/[1+\mathbf{/}-\mathbf{U}+\mathbf{D})/[1-\mathbf{t}]]/\mathbf{v}$$

or

$$S = \mathbf{S'}[1+\mathbf{s}]$$


Rule-58279:

If both (**/**), (**U**), (**S'**), (**s**), (**S**), (**N**), (**F**), (**i**), (**t**), (**D**), (**K**), (**j**) and (**P**) are known, then its Variable Portion planned is:

$$v = [\mathbf{S'}[1+\mathbf{s}]-\mathbf{F}-\mathbf{S'i}[1+\mathbf{s}]-(\{\mathbf{N+K+S'j}[1+\mathbf{s}]/360+\mathbf{P}\}$$
$$/[1+\mathbf{/}-\mathbf{U}+\mathbf{D})/[1-\mathbf{t}]]/\mathbf{S}$$

Steve Asikin ISBN 13: **978-1564912361**, ISBN 10: **1564912363**

Rule-58280:

If both (**/**), (**U**), (**S'**), (**s**), (**S**), (**v**), (**N**), (**i**), (**t**), (**D**), (**K**),
(**j**) and (**P**) are known, then its Fixed Cost planned is:

$$F= S'[1+s]-Sv-S'i[1+s]-(\{N+K+S'j[1+s]/360+P\}$$
$$/[1+/]-U+D)/[1-t]$$

Rule-58281:

If both (**/**), (**U**), (**S'**), (**s**), (**S**), (**v**), (**F**), (**N**), (**t**), (**D**), (**K**),
(**j**) and (**P**) are known, then its Interest Portion
planned is:

$$i= [S'[1+s]-Sv-F-(\{N+K+S'j[1+s]/360+P\}$$
$$/[1+/]-U+D)/[1-t]]/\{S'[1+s]\}$$

Rule-58282:

If both (**/**), (**U**), (**S'**), (**s**), (**S**), (**v**), (**F**), (**i**), (**N**), (**D**), (**K**),
(**j**) and (**P**) are known, then its Tax Rate planned is:

$$t= 1-(\{N+K+S'j[1+s]/360+P\}/[1+/]-U+D)$$
$$/\{S'[1+s]-Sv-F-S'i[1+s]\}$$

Rule-58283:

If both (**/**), (**U**), (**S'**), (**s**), (**S**), (**v**), (**F**), (**i**), (**t**), (**N**), (**K**),
(**j**) and (**P**) are known, then its Dividend planned is:

$$D= U+\{S'[1+s]-Sv-F-S'i[1+s]\}[1-t]$$
$$-\{N+K+S'j[1+s]/360+P\}/[1+/]$$

Steve Asikin ISBN 13: **978-1564912361**, ISBN 10: **1564912363**

Rule-58284:

If both (**/**), (**U**), (**S'**), (**s**), (**S**), (**v**), (**F**), (**i**), (**t**), (**D**), (**N**), (**j**) and (**P**) are known, then its Kind of Cash planned is:

$$K= [1+/(U+\{S'[1+s]-Sv-F-S'i[1+s]\}[1-t]-D)$$
$$-\{N+S'j[1+s]/360+P\}$$

Rule-58285:

If both (**/**), (**U**), (**S'**), (**s**), (**S**), (**v**), (**F**), (**i**), (**t**), (**D**), (**K**), (**N**) and (**P**) are known, then its Job or Trade Receivable Days planned is:

$$j= 360[[1+/(U+\{S'[1+s]-Sv-F-S'i[1+s]\}[1-t]-D)$$
$$-\{N+K+P\}]/\{S'[1+s]\}$$

Rule-58286:

If both (**/**), (**U**), (**S'**), (**s**), (**S**), (**v**), (**F**), (**i**), (**t**), (**D**), (**K**), (**j**) and (**N**) are known, then its Procured Inventory planned is:

$$P= [1+/(U+\{S'[1+s]-Sv-F-S'i[1+s]\}[1-t]-D)$$
$$-\{N+K+S'j[1+s]/360\}$$

Rule-58287:

If both (**/**), (**U**), (**S'**), (**s**), (**S**), (**v**), (**F**), (**i**), (**t**), (**D**), (**K**), (**j**), (**V**) and (**p**) are known, then its Non Current Asset planned is:

$$N= [1+/(U+\{S'[1+s]-Sv-F-S'i[1+s]\}[1-t]-D)$$
$$-K-S'j[1+s]/360-Vp/360$$

Steve Asikin ISBN 13: **978-1564912361**, ISBN 10: **1564912363**

<u>Rule-58288</u>:

If both (**N**), (**U**), (**S'**), (**s**), (**S**), (**v**), (**F**), (**i**), (**t**), (**D**), (**K**), (**j**), (**V**) and (**p**) are known, then its Leverage or Gearing Ratio planned is:

$$I= \{N+K+S'j[1+s]/360+Vp/360\}$$
$$/(U+\{S'[1+s]-Sv-F-S'i[1+s]\}[1-t]-D)-1$$

<u>Rule-58289</u>:

If both (**I**), (**N**), (**S'**), (**s**), (**S**), (**v**), (**F**), (**i**), (**t**), (**D**), (**K**), (**j**), (**V**) and (**p**) are known, then its Utilized or Starting Capital must be:

$$U= (\{N+K+S'j[1+s]/360+Vp/360\}-[1+I]\{S'[1+s]$$
$$-Sv-F-S'i[1+s]\}[1-t]-D)/[1+I]$$

<u>Rule-58290</u>:

If both (**I**), (**U**), (**N**), (**s**), (**S**), (**v**), (**F**), (**i**), (**t**), (**D**), (**K**), (**j**), (**V**) and (**p**) are known, then its Sales Past must be:

$$S'= [[1+I](U-\{Sv+F\}[1-t]-D)-[N+K+Vp/360]]$$
$$/([1+s]\{j/360-[1+I][1-t][1-i]\})$$

or

$$S'= S/[1+s]$$

Steve Asikin ISBN 13: **978-1564912361**, ISBN 10: **1564912363**

Rule-58291:

If both (**/**), (**U**), (**S'**), (**N**), (**S**), (**v**), (**F**), (**i**), (**t**), (**D**), (**K**), (**j**), (**V**) and (**p**) are known, then its Sales Growth planned is:

$$s= [[1+/](U-\{Sv+F\}[1-t]-D)-[N+K+Vp/360]] /(S'\{j/360-[1+/][1-t][1-i]\})-1$$

or

$$s= S/S'-1$$


Rule-58292:

If both (**/**), (**U**), (**S'**), (**s**), (**N**), (**v**), (**F**), (**i**), (**t**), (**D**), (**K**), (**j**), (**V**) and (**p**) are known, then its Sales or Revenue planned is:

$$S= [S'[1+s]-F-S'i[1+s]- (\{N+K+S'j[1+s]/360 +Vp/360\}/[1+/]-U+D)/[1-t]]/v$$

or

$$S= S'[1+s]$$

or

$$S= V/v$$


Rule-58293:

If both (**/**), (**U**), (**S'**), (**s**), (**S**), (**N**), (**F**), (**i**), (**t**), (**D**), (**K**), (**j**), (**V**) and (**p**) are known, then its Variable Portion planned is:

$$v= [S'[1+s]-F-S'i[1+s]- (\{N+K+S'j[1+s]/360 +Vp/360\}/[1+/]-U+D)/[1-t]]/S$$

or

$$v= V/S$$

Steve Asikin ISBN 13: **978-1564912361**, ISBN 10: **1564912363**

Rule-58294:

If both (**/**), (**U**), (**S'**), (**s**), (**S**), (**v**), (**N**), (**i**), (**t**), (**D**), (**K**), (**j**), (**V**) and (**p**) are known, then its Fixed Cost planned is:

$$\textbf{F} = \textbf{S'}[1+\textbf{s}]-\textbf{Sv}-\textbf{S'i}[1+\textbf{s}]-(\{\textbf{N}+\textbf{K}+\textbf{S'j}[1+\textbf{s}]/360 +\textbf{Vp}/360\}/[1+\textbf{/}-\textbf{U}+\textbf{D})/[1-\textbf{t}]$$

Rule-58295:

If both (**/**), (**U**), (**S'**), (**s**), (**S**), (**v**), (**F**), (**N**), (**t**), (**D**), (**K**), (**j**), (**V**) and (**p**) are known, then its Interest Portion planned is:

$$\textbf{i} = [\textbf{S'}[1+\textbf{s}]-\textbf{Sv}-\textbf{F}-(\{\textbf{N}+\textbf{K}+\textbf{S'j}[1+\textbf{s}]/360+\textbf{Vp}/360\} /[1+\textbf{/}-\textbf{U}+\textbf{D})/[1-\textbf{t}]]/\{\textbf{S'}[1+\textbf{s}]\}$$

Rule-58296:

If both (**/**), (**U**), (**S'**), (**s**), (**S**), (**v**), (**F**), (**i**), (**N**), (**D**), (**K**), (**j**), (**V**) and (**p**) are known, then its Tax Rate planned is:

$$\textbf{t} = 1-(\{\textbf{N}+\textbf{K}+\textbf{S'j}[1+\textbf{s}]/360+\textbf{Vp}/360\}/[1+\textbf{/}-\textbf{U}+\textbf{D}) /\{\textbf{S'}[1+\textbf{s}]-\textbf{Sv}-\textbf{F}-\textbf{S'i}[1+\textbf{s}]\}$$

Rule-58297:

If both (**/**), (**U**), (**S'**), (**s**), (**S**), (**v**), (**F**), (**i**), (**t**), (**N**), (**K**), (**j**), (**V**) and (**p**) are known, then its Dividend planned is:

$$\textbf{D} = \textbf{U}+\{\textbf{S'}[1+\textbf{s}]-\textbf{Sv}-\textbf{F}-\textbf{S'i}[1+\textbf{s}]\}[1-\textbf{t}] -\{\textbf{N}+\textbf{K}+\textbf{S'j}[1+\textbf{s}]/360+\textbf{Vp}/360\}/[1+\textbf{/}]$$

Steve Asikin ISBN 13: **978-1564912361**, ISBN 10: **1564912363**

Rule-58298:

If both (**/**), (**U**), (**S'**), (**s**), (**S**), (**v**), (**F**), (**i**), (**t**), (**D**), (**N**), (**j**), (**V**) and (**p**) are known, then its Kind of Cash planned is:

$$\mathbf{K}= [1+\mathbf{/}(\mathbf{U}+\{\mathbf{S'}[1+\mathbf{s}]-\mathbf{Sv}-\mathbf{F}-\mathbf{S'i}[1+\mathbf{s}]\}[1-\mathbf{t}]-\mathbf{D})$$
$$-\{\mathbf{N}+\mathbf{S'j}[1+\mathbf{s}]/360+\mathbf{Vp}/360\}$$

Rule-58299:

If both (**/**), (**U**), (**S'**), (**s**), (**S**), (**v**), (**F**), (**i**), (**t**), (**D**), (**K**), (**N**), (**V**) and (**p**) are known, then its Job or Trade Receivable Days planned is:

$$\mathbf{j}= 360[[1+\mathbf{/}(\mathbf{U}+\{\mathbf{S'}[1+\mathbf{s}]-\mathbf{Sv}-\mathbf{F}-\mathbf{S'i}[1+\mathbf{s}]\}[1-\mathbf{t}]-\mathbf{D})$$
$$-\{\mathbf{N}+\mathbf{K}+\mathbf{Vp}/360\}]/\{\mathbf{S'}[1+\mathbf{s}]\}$$

Rule-58300:

If both (**/**), (**U**), (**S'**), (**s**), (**S**), (**v**), (**F**), (**i**), (**t**), (**D**), (**K**), (**j**), (**N**) and (**p**) are known, then its Variable Cost planned is:

$$\mathbf{V}= 360[[1+\mathbf{/}(\mathbf{U}+\{\mathbf{S'}[1+\mathbf{s}]-\mathbf{Sv}-\mathbf{F}-\mathbf{S'i}[1+\mathbf{s}]\}[1-\mathbf{t}]-\mathbf{D})$$
$$-\{\mathbf{N}+\mathbf{K}+\mathbf{S'j}[1+\mathbf{s}]/360\}]/\mathbf{p}$$

or

$$\mathbf{V}= \mathbf{Sv}$$

Rule-58301:

If both (**/**), (**U**), (**S'**), (**s**), (**S**), (**v**), (**F**), (**i**), (**t**), (**D**), (**K**), (**j**), (**V**) and (**N**) are known, then its Procured Inventory Days planned is:

$$\mathbf{p}= 360[[1+\mathbf{/}(\mathbf{U}+\{\mathbf{S'}[1+\mathbf{s}]-\mathbf{Sv}-\mathbf{F}-\mathbf{S'i}[1+\mathbf{s}]\}[1-\mathbf{t}]-\mathbf{D})$$
$$-\{\mathbf{N}+\mathbf{K}+\mathbf{S'j}[1+\mathbf{s}]/360\}]/\mathbf{V}$$

Steve Asikin ISBN 13: **978-1564912361**, ISBN 10: **1564912363**

<u>Rule-58302</u>:

If both (**/**), (**U**), (**S'**), (**s**), (**S**), (**v**), (**F**), (**i**), (**t**), (**D**), (**K**), (**j**) and (**p**) are known, then its Non Current Asset planned is:

$$\mathbf{N} = [1 + \mathbf{/}(\mathbf{U} + \{\mathbf{S'}[1+\mathbf{s}] - \mathbf{Sv} - \mathbf{F} - \mathbf{S'i}[1+\mathbf{s}]\}[1-\mathbf{t}] - \mathbf{D})$$
$$-\mathbf{K} - \mathbf{S'j}[1+\mathbf{s}]/360 - \mathbf{Svp}/360$$

<u>Rule-58303</u>:

If both (**N**), (**U**), (**S'**), (**s**), (**S**), (**v**), (**F**), (**i**), (**t**), (**D**), (**K**), (**j**) and (**p**) are known, then its Leverage or Gearing Ratio planned is:

$$\mathbf{/} = \{\mathbf{N} + \mathbf{K} + \mathbf{S'j}[1+\mathbf{s}]/360 + \mathbf{Svp}/360\}$$
$$/(\mathbf{U} + \{\mathbf{S'}[1+\mathbf{s}] - \mathbf{Sv} - \mathbf{F} - \mathbf{S'i}[1+\mathbf{s}]\}[1-\mathbf{t}] - \mathbf{D}) - 1$$

<u>Rule-58304</u>:

If both (**/**), (**N**), (**S'**), (**s**), (**S**), (**v**), (**F**), (**i**), (**t**), (**D**), (**K**), (**j**) and (**p**) are known, then its Utilized or Starting Capital must be:

$$\mathbf{U} = (\{\mathbf{N} + \mathbf{K} + \mathbf{S'j}[1+\mathbf{s}]/360 + \mathbf{Svp}/360\} - [1 + \mathbf{/}$$
$$\{\mathbf{S'}[1+\mathbf{s}] - \mathbf{Sv} - \mathbf{F} - \mathbf{S'i}[1+\mathbf{s}]\}[1-\mathbf{t}] - \mathbf{D})/[1 + \mathbf{/}$$

<u>Rule-58305</u>:

If both (**/**), (**U**), (**N**), (**s**), (**S**), (**v**), (**F**), (**i**), (**t**), (**D**), (**K**), (**j**) and (**p**) are known, then its Sales Past must be:

$$\mathbf{S'} = [[1 + \mathbf{/}(\mathbf{U} - \{\mathbf{Sv} + \mathbf{F}\}[1-\mathbf{t}] - \mathbf{D}) - [\mathbf{N} + \mathbf{K} + \mathbf{Svp}/360]]$$
$$/([1+\mathbf{s}]\{\mathbf{j}/360 - [1 + \mathbf{/}\ [1-\mathbf{t}][1-\mathbf{i}]\})$$

or

$$\mathbf{S'} = \mathbf{S}/[1+\mathbf{s}]$$

`

610

Rule-58306:

If both (**/**), (**U**), (**$'**), (**N**), (**$**), (**v**), (**F**), (**i**), (**t**), (**D**), (**K**), (**j**) and (**p**) are known, then its Sales Growth planned is:

$$s= [[1+/(U-\{Sv+F\}[1-t]-D)-[N+K+Svp/360]]$$
$$/(S'\{j/360-[1+/\ [1-t][1-i]\})-1$$

or

$$s= S/S'-1$$

Rule-58307:

If both (**/**), (**U**), (**$'**), (**s**), (**N**), (**v**), (**F**), (**i**), (**t**), (**D**), (**K**), (**j**) and (**p**) are known, then its Sales or Revenue planned is:

$$S= [[1+/(U+\{S'[1+s]-F-S'i[1+s]\}[1-t]-D)-\{N+K$$
$$+S'j[1+s]/360\}]/(v\{p/360+[1+/[1-t]\})$$

or

$$S= S'[1+s]$$

Rule-58308:

If both (**/**), (**U**), (**$'**), (**s**), (**$**), (**N**), (**F**), (**i**), (**t**), (**D**), (**K**), (**j**) and (**p**) are known, then its Variable Portion planned is:

$$v= [[1+/(U+\{S'[1+s]-F-S'i[1+s]\}[1-t]-D)-\{N+K$$
$$+S'j[1+s]/360\}]/(S\{p/360+[1+/[1-t]\})$$

Steve Asikin ISBN 13: **978-1564912361**, ISBN 10: **1564912363**

Rule-58309:

If both (**I**), (**U**), (**S'**), (**s**), (**S**), (**v**), (**N**), (**i**), (**t**), (**D**), (**K**), (**j**) and (**p**) are known, then its Fixed Cost planned is:

$$\textbf{F} = \textbf{S'}[1+\textbf{s}] - \textbf{Sv} - \textbf{S'i}[1+\textbf{s}] - (\{\textbf{N} + \textbf{K} + \textbf{S'j}[1+\textbf{s}]/360 + \textbf{Svp}/360\}/[1+\textbf{I} - \textbf{U} + \textbf{D})/[1-\textbf{t}]$$

Rule-58310:

If both (**I**), (**U**), (**S'**), (**s**), (**S**), (**v**), (**F**), (**N**), (**t**), (**D**), (**K**), (**j**) and (**p**) are known, then its Interest Portion planned is:

$$\textbf{i} = [\textbf{S'}[1+\textbf{s}] - \textbf{Sv} - \textbf{F} - (\{\textbf{N} + \textbf{K} + \textbf{S'j}[1+\textbf{s}]/360 + \textbf{Svp}/360\}/[1+\textbf{I} - \textbf{U} + \textbf{D})/[1-\textbf{t}]]/\{\textbf{S'}[1+\textbf{s}]\}$$

Rule-58311:

If both (**I**), (**U**), (**S'**), (**s**), (**S**), (**v**), (**F**), (**i**), (**N**), (**D**), (**K**), (**j**) and (**p**) are known, then its Tax Rate planned is:

$$\textbf{t} = 1 - (\{\textbf{N} + \textbf{K} + \textbf{S'j}[1+\textbf{s}]/360 + \textbf{Svp}/360\}/[1+\textbf{I} - \textbf{U} + \textbf{D})/\{\textbf{S'}[1+\textbf{s}] - \textbf{Sv} - \textbf{F} - \textbf{S'i}[1+\textbf{s}]\}$$

Rule-58312:

If both (**I**), (**U**), (**S'**), (**s**), (**S**), (**v**), (**F**), (**i**), (**t**), (**N**), (**K**), (**j**) and (**p**) are known, then its Dividend planned is:

$$\textbf{D} = \textbf{U} + \{\textbf{S'}[1+\textbf{s}] - \textbf{Sv} - \textbf{F} - \textbf{S'i}[1+\textbf{s}]\}[1-\textbf{t}] - \{\textbf{N} + \textbf{K} + \textbf{S'j}[1+\textbf{s}]/360 + \textbf{Svp}/360\}/[1+\textbf{I}]$$

Steve Asikin ISBN 13: **978-1564912361**, ISBN 10: **1564912363**

Rule-58313:

If both (**/**), (**U**), (**S'**), (**s**), (**S**), (**v**), (**F**), (**i**), (**t**), (**D**), (**N**), (**j**) and (**p**) are known, then its Kind of Cash planned is:

$$K = [1+/(U+\{S'[1+s]-Sv-F-S'i[1+s]\}[1-t]-D) \\ -\{N+S'j[1+s]/360+Svp/360\}$$

Rule-58314:

If both (**/**), (**U**), (**S'**), (**s**), (**S**), (**v**), (**F**), (**i**), (**t**), (**D**), (**K**), (**N**) and (**p**) are known, then its Job or Trade Receivable Days planned is:

$$j = 360[[1+/(U+\{S'[1+s]-Sv-F-S'i[1+s]\}[1-t]-D) \\ -\{N+K+Svp/360\}]/\{S'[1+s]\}$$

Rule-58315:

If both (**/**), (**U**), (**S'**), (**s**), (**S**), (**v**), (**F**), (**i**), (**t**), (**D**), (**K**), (**j**) and (**N**) are known, then its Procured Inventory Days planned is:

$$p = 360[[1+/(U+\{S'[1+s]-Sv-F-S'i[1+s]\}[1-t]-D) \\ -\{N+K+S'j[1+s]/360\}]/[Sv]$$

Rule-58316:

If both (**/**), (**U**), (**S'**), (**s**), (**S**), (**v**), (**F**), (**i**), (**t**), (**D**), (**K**), (**j**) and (**p**) are known, then its Non Current Asset planned is:

$$N = [1+/(U+\{S'[1+s]-Sv-F-S'i[1+s]\}[1-t]-D) \\ -K-S'j[1+s]/360-S'vp[1+s]/360$$

Steve Asikin ISBN 13: **978-1564912361**, ISBN 10: **1564912363**

Rule-58317:

If both (**N**), (**U**), (**S'**), (**s**), (**S**), (**v**), (**F**), (**i**), (**t**), (**D**), (**K**), (**j**) and (**p**) are known, then its Leverage or Gearing Ratio planned is:

$$l = \{N+K+S'j[1+s]/360+S'vp[1+s]/360\}$$
$$/(U+\{S'[1+s]-Sv-F-S'i[1+s]\}[1-t]-D)-1$$

Rule-58318:

If both (**l**), (**N**), (**S'**), (**s**), (**S**), (**v**), (**F**), (**i**), (**t**), (**D**), (**K**), (**j**) and (**p**) are known, then its Utilized or Starting Capital must be:

$$U = (\{N+K+S'j[1+s]/360+S'vp[1+s]/360\}-[1+l$$
$$\{S'[1+s]-Sv-F-S'i[1+s]\}[1-t]-D)/[1+l$$

Rule-58319:

If both (**l**), (**U**), (**N**), (**s**), (**S**), (**v**), (**F**), (**i**), (**t**), (**D**), (**K**), (**j**) and (**p**) are known, then its Sales Past must be:

$$S' = [[1+l(U+\{S'[1+s]-Sv-F-S'i[1+s]\}[1-t]-D)$$
$$-[N+K]]/([1+s]\{j/360+vp/360-[1+l][1-t])$$

or

$$S' = S/[1+s]$$

Steve Asikin ISBN 13: **978-1564912361**, ISBN 10: **1564912363**

Rule-58320:

If both (**/**), (**U**), (**S'**), (**N**), (**S**), (**v**), (**F**), (**i**), (**t**), (**D**), (**K**), (**j**) and (**p**) are known, then its Sales Growth planned is:

$$s= [[1+/](U+\{S'[1+s]-Sv-F-S'i[1+s]\}[1-t]-D)$$
$$-[N+K]]/(S'\{j/360+vp/360-[1+/][1-t])-1$$

or

$$s= S/S'-1$$


Rule-58321:

If both (**/**), (**U**), (**S'**), (**s**), (**N**), (**v**), (**F**), (**i**), (**t**), (**D**), (**K**), (**j**) and (**p**) are known, then its Sales or Revenue planned is:

$$S= [S'[1+s]-F-S'i[1+s]-(\{N+K+S'j[1+s]/360$$
$$+S'vp[1+s]/360\}/[1+/]-U+D)/[1-t]]/v$$

or

$$S= S'[1+s]$$


Rule-58322:

If both (**/**), (**U**), (**S'**), (**s**), (**S**), (**N**), (**F**), (**i**), (**t**), (**D**), (**K**), (**j**) and (**p**) are known, then its Variable Portion planned is:

$$v= [[1+/](U+\{S'[1+s]-F-S'i[1+s]\}[1-t]-D)$$
$$-\{N+K+S'j[1+s]/360\}]$$
$$/\{S'p[1+s]/360+S[1+/][1-t]\}$$

Steve Asikin ISBN 13: **978-1564912361**, ISBN 10: **1564912363**

Rule-58323:

If both $(I)$, $(U)$, $(S')$, $(s)$, $(S)$, $(v)$, $(N)$, $(i)$, $(t)$, $(D)$, $(K)$, $(j)$ and $(p)$ are known, then its Fixed Cost planned is:

$$F= S'[1+s]-Sv-S'i[1+s]-(\{N+K+S'j[1+s]/360$$
$$+S'vp[1+s]/360\}/[1+I-U+D)/[1-t]$$

Rule-58324:

If both $(I)$, $(U)$, $(S')$, $(s)$, $(S)$, $(v)$, $(F)$, $(N)$, $(t)$, $(D)$, $(K)$, $(j)$ and $(p)$ are known, then its Interest Portion planned is:

$$i= [S'[1+s]-Sv-F-(\{N+K+S'j[1+s]/360$$
$$+S'vp[1+s]/360\}/[1+I-U+D)/[1-t]]$$
$$/\{S'[1+s]\}$$

Rule-58325:

If both $(I)$, $(U)$, $(S')$, $(s)$, $(S)$, $(v)$, $(F)$, $(i)$, $(N)$, $(D)$, $(K)$, $(j)$ and $(p)$ are known, then its Tax Rate planned is:

$$t= 1-(\{N+K+S'j[1+s]/360+S'vp[1+s]/360\}/[1+I$$
$$-U+D)/\{S'[1+s]-Sv-F-S'i[1+s]\}$$

Rule-58326:

If both $(I)$, $(U)$, $(S')$, $(s)$, $(S)$, $(v)$, $(F)$, $(i)$, $(t)$, $(N)$, $(K)$, $(j)$ and $(p)$ are known, then its Dividend planned is:

$$D= U+\{S'[1+s]-Sv-F-S'i[1+s]\}[1-t]- \{N+K$$
$$+S'j[1+s]/360+S'vp[1+s]/360\}/[1+I$$

Steve Asikin ISBN 13: **978-1564912361**, ISBN 10: **1564912363**

Rule-58327:

If both ($I$), (**U**), (**S'**), (**s**), (**S**), (**v**), (**F**), (**i**), (**t**), (**D**), (**N**), (**j**) and (**p**) are known, then its Kind of Cash planned is:

$$K= [1+I(U+\{S'[1+s]-Sv-F-S'i[1+s]\}[1-t]-D)$$
$$-\{N+S'j[1+s]/360+S'vp[1+s]/360\}$$


Rule-58328:

If both ($I$), (**U**), (**S'**), (**s**), (**S**), (**v**), (**F**), (**i**), (**t**), (**D**), (**K**), (**N**) and (**p**) are known, then its Job or Trade Receivable Days planned is:

$$j= 360[[1+I(U+\{S'[1+s]-Sv-F-S'i[1+s]\}[1-t]-D)$$
$$-\{N+K+S'vp[1+s]/360\}]/\{S'[1+s]\}$$


Rule-58329:

If both ($I$), (**U**), (**S'**), (**s**), (**S**), (**v**), (**F**), (**i**), (**t**), (**D**), (**K**), (**j**) and (**N**) are known, then its Procured Inventory Days planned is:

$$p= 360[[1+I(U+\{S'[1+s]-Sv-F-S'i[1+s]\}[1-t]-D)$$
$$-\{N+K+S'j[1+s]/360\}]/\{S'v[1+s]\}$$


Rule-58330:

If both ($I$), (**U**), (**S'**), (**s**), (**S**), (**v**), (**F**), (**i**), (**t**), (**d**), (**K**), (**J**) and (**P**) are known, then its Non Current Asset planned is:

$$N= [1+I(U+\{S'[1+s]-Sv-F-S'i[1+s]\}[1-t][1-d])$$
$$-K-J-P$$

Steve Asikin ISBN 13: **978-1564912361**, ISBN 10: **1564912363**

Rule-58331:

If both (**N**), (**U**), (**S'**), (**s**), (**S**), (**v**), (**F**), (**i**), (**t**), (**d**), (**K**), (**J**) and (**P**) are known, then its Leverage or Gearing Ratio planned is:

$$I = \{N+K+J+P\}/(U+\{S'[1+s]-Sv-F-S'i[1+s]\}$$
$$[1-t][1-d])-1$$

Rule-58332:

If both (**I**), (**N**), (**S'**), (**s**), (**S**), (**v**), (**F**), (**i**), (**t**), (**d**), (**K**), (**J**) and (**P**) are known, then its Utilized or Starting Capital must be:

$$U = (\{N+K+J+P\}-[1+I]\{S'[1+s]-Sv-F-S'i[1+s]\}$$
$$[1-t][1-d])/[1+I]$$

Rule-58333:

If both (**I**), (**U**), (**N**), (**s**), (**S**), (**v**), (**F**), (**i**), (**t**), (**d**), (**K**), (**J**) and (**P**) are known, then its Sales Past must be:

$$S' = (\{[N+K+J+P]/[1+I]-U\}/\{[1-t][1-d]\}$$
$$+Sv+F)/\{[1+s][1-i]\}$$

or

$$S' = S/[1+s]$$

Rule-58334:

If both (**I**), (**U**), (**S'**), (**N**), (**S**), (**v**), (**F**), (**i**), (**t**), (**d**), (**K**), (**J**) and (**P**) are known, then its Sales Past must be:

$$s = (\{[N+K+J+P]/[1+I]-U\}/\{[1-t][1-d]\}$$
$$+Sv+F)/\{S'[1-i]\}-1$$

or

$$s = S/S'-1$$

Steve Asikin ISBN 13: **978-1564912361**, ISBN 10: **1564912363**

Rule-58335:

If both (**/**), (**U**), (**S'**), (**s**), (**N**), (**v**), (**F**), (**i**), (**t**), (**d**), (**K**), (**J**) and (**P**) are known, then its Sales or Revenue planned is:

$$S= [S'[1+s]-F-S'i[1+s]-[N+K+J+P]/[1+/-U}$$
$$/\{[1-t][1-d]]/v$$

or

$$S= S'[1+s]$$

Rule-58336:

If both (**/**), (**U**), (**S'**), (**s**), (**S**), (**N**), (**F**), (**i**), (**t**), (**d**), (**K**), (**J**) and (**P**) are known, then its Variable Portion planned is:

$$v= [S'[1+s]-F-S'i[1+s]-[N+K+J+P]/[1+/-U}$$
$$/\{[1-t][1-d]]/S$$

Rule-58337:

If both (**/**), (**U**), (**S'**), (**s**), (**S**), (**v**), (**N**), (**i**), (**t**), (**d**), (**K**), (**J**) and (**P**) are known, then its Fixed Cost planned is:

$$F= S'[1+s]-Sv-S'i[1+s]-[N+K+J+P]/[1+/-U}$$
$$/\{[1-t][1-d]$$

Rule-58338:

If both (**/**), (**U**), (**S'**), (**s**), (**S**), (**v**), (**F**), (**N**), (**t**), (**d**), (**K**), (**J**) and (**P**) are known, then its Interest Portion planned is:

$$i= (S'[1+s]-Sv-F-[N+K+J+P]/[1+/-U}$$
$$/\{[1-t][1-d]\})/\{S'[1+s]\}$$

Steve Asikin ISBN 13: **978-1564912361**, ISBN 10: **1564912363**

Rule-58339:

If both (**/**), (**U**), (**S'**), (**s**), (**S**), (**v**), (**F**), (**i**), (**N**), (**d**), (**K**), (**J**) and (**P**) are known, then its Tax Rate planned is:

$$t = 1 - \{[N+K+J+P]/[1+/-U\}$$
$$/(\{S'[1+s]-Sv-F-S'i[1+s]\}[1-d])$$

Rule-58340:

If both (**/**), (**U**), (**S'**), (**s**), (**S**), (**v**), (**F**), (**i**), (**t**), (**N**), (**K**), (**J**) and (**P**) are known, then its Dividend Payout planned is:

$$d = 1 - \{[N+K+J+P]/[1+/-U\}$$
$$/(\{S'[1+s]-Sv-F-S'i[1+s]\}[1-t])$$

Rule-58341:

If both (**/**), (**U**), (**S'**), (**s**), (**S**), (**v**), (**F**), (**i**), (**t**), (**d**), (**N**), (**J**) and (**P**) are known, then its Kind of Cash planned is:

$$K = [1+/](U+\{S'[1+s]-Sv-F-S'i[1+s]\}[1-t][1-d])$$
$$-[N+J+P]$$

Rule-58342:

If both (**/**), (**U**), (**S'**), (**s**), (**S**), (**v**), (**F**), (**i**), (**t**), (**d**), (**K**), (**N**) and (**P**) are known, then its Job or Trade Receivable Days planned is:

$$J = [1+/](U+\{S'[1+s]-Sv-F-S'i[1+s]\}[1-t][1-d])$$
$$-[N+K+P]$$

Steve Asikin ISBN 13: **978-1564912361**, ISBN 10: **1564912363**

Rule-58343:
   If both (**/**), (**U**), (**S'**), (**s**), (**S**), (**v**), (**F**), (**i**), (**t**), (**d**), (**K**),
   (**J**) and (**N**) are known, then its Procured Inventory
   planned is:
   $$P= [1+/](U+\{S'[1+s]-Sv-F-S'i[1+s]\}[1-t][1-d])$$
   $$-[N+K+J]$$

Rule-58344:
   If both (**/**), (**U**), (**S'**), (**s**), (**S**), (**v**), (**F**), (**i**), (**t**), (**d**), (**K**),
   (**J**), (**V**) and (**p**) are known, then its Non Current
   Asset planned is:
   $$N= [1+/](U+\{S'[1+s]-Sv-F-S'i[1+s]\}[1-t][1-d])$$
   $$-K-J-Vp/360$$

Rule-58345:
   If both (**N**), (**U**), (**S'**), (**s**), (**S**), (**v**), (**F**), (**i**), (**t**), (**d**), (**K**),
   (**J**), (**V**) and (**p**) are known, then its Leverage or
   Gearing Ratio planned is:
   $$/= [N+K+J+Vp/360]/(U+\{S'[1+s]-Sv-F$$
   $$-S'i[1+s]\}[1-t][1-d])-1$$

Rule-58346:
   If both (**/**), (**N**), (**S'**), (**s**), (**S**), (**v**), (**F**), (**i**), (**t**), (**d**), (**K**),
   (**J**), (**V**) and (**p**) are known, then its Utilized or
   Starting Capital must be:
   $$U= ([N+K+J+Vp/360]-[1+/]\{S'[1+s]-Sv-F$$
   $$-S'i[1+s]\}[1-t][1-d])/[1+/]$$

Steve Asikin ISBN 13: **978-1564912361**, ISBN 10: **1564912363**

Rule-58347:

If both $(I)$, $(U)$, $(N)$, $(s)$, $(S)$, $(v)$, $(F)$, $(i)$, $(t)$, $(d)$, $(K)$, $(J)$, $(V)$ and $(p)$ are known, then its Sales Past must be:

$$S' = (\{[N+K+J+Vp/360]/[1+I-U\}/\{[1-t][1-d]\}$$
$$+ Sv+F)/\{[1+s][1-i]\}$$

or

$$S' = S/[1+s]$$

Rule-58348:

If both $(I)$, $(U)$, $(S')$, $(N)$, $(S)$, $(v)$, $(F)$, $(I)$, $(t)$, $(d)$, $(K)$, $(J)$, $(V)$ and $(p)$ are known, then its Sales Growth planned is:

$$s = (\{[N+K+J+Vp/360]/[1+I-U\}/\{[1-t][1-d]\}$$
$$+ Sv+F)/\{S'[1-i]\} - 1$$

or

$$s = S/S' - 1$$

Rule-58349:

If both $(I)$, $(U)$, $(S')$, $(s)$, $(N)$, $(v)$, $(F)$, $(i)$, $(t)$, $(d)$, $(K)$, $(J)$, $(V)$ and $(p)$ are known, then its Sales or Revenue planned is:

$$S = [S'[1+s]-F-S'i[1+s]-\{[N+K+J+Vp/360]$$
$$/[1+I-U\}/\{[1-t][1-d]\}]/v$$

or

$$S = S'[1+s]$$

or

$$S = V/v$$

Steve Asikin ISBN 13: **978-1564912361**, ISBN 10: **1564912363**

Rule-58350:

If both (**/**), (**U**), (**S'**), (**s**), (**$**), (**υ**), (**F**), (**i**), (**t**), (**d**), (**K**), (**J**), (**V**) and (**p**) are known, then its Non Current Asset planned is:

$$υ= [\textbf{S'}[1+\textbf{s}]-\textbf{F}-\textbf{S'i}[1+\textbf{s}]-\{[\textbf{N}+\textbf{K}+\textbf{J}+\textbf{Vp}/360]$$
$$/[1+\textbf{/}-\textbf{U}\}/\{[1-\textbf{t}][1-\textbf{d}]\}\,]/\textbf{S}$$

or

$$υ= \textbf{V}/\textbf{S}$$

Rule-58351:

If both (**/**), (**U**), (**S'**), (**s**), (**$**), (**υ**), (**N**), (**i**), (**t**), (**d**), (**K**), (**J**), (**V**) and (**p**) are known, then its Fixed Cost planned is:

$$\textbf{F}= \textbf{S'}[1+\textbf{s}]-\textbf{Sυ}-\textbf{S'i}[1+\textbf{s}]-\{[\textbf{N}+\textbf{K}+\textbf{J}+\textbf{Vp}/360]$$
$$/[1+\textbf{/}-\textbf{U}\}/\{[1-\textbf{t}][1-\textbf{d}]\}$$

Rule-58352:

If both (**/**), (**U**), (**S'**), (**s**), (**$**), (**υ**), (**F**), (**N**), (**t**), (**d**), (**K**), (**J**), (**V**) and (**p**) are known, then its Interest Portion planned is:

$$\textbf{i}= [\textbf{S'}[1+\textbf{s}]-\textbf{Sυ}-\textbf{F}-\{[\textbf{N}+\textbf{K}+\textbf{J}+\textbf{Vp}/360]/[1+\textbf{/}-\textbf{U}\}$$
$$/\{[1-\textbf{t}][1-\textbf{d}]\}\,]/\{\textbf{S'}[1+\textbf{s}]\}$$

Rule-58353:

If both (**/**), (**U**), (**S'**), (**s**), (**$**), (**υ**), (**F**), (**i**), (**N**), (**d**), (**K**), (**J**), (**V**) and (**p**) are known, then its Tax Rate planned is:

$$\textbf{t}= 1-\{[\textbf{N}+\textbf{K}+\textbf{J}+\textbf{Vp}/360]/[1+\textbf{/}-\textbf{U}\}$$
$$/(\{\textbf{S'}[1+\textbf{s}]-\textbf{Sυ}-\textbf{F}-\textbf{S'i}[1+\textbf{s}]\}[1-\textbf{d}])$$

Steve Asikin ISBN 13: **978-1564912361**, ISBN 10: **1564912363**

Rule-58354:

If both ($\textit{I}$), ($\textbf{U}$), ($\textbf{S'}$), ($\textbf{s}$), ($\textbf{\$}$), ($\textbf{v}$), ($\textbf{F}$), ($\textbf{i}$), ($\textbf{t}$), ($\textbf{N}$), ($\textbf{K}$), ($\textbf{J}$), ($\textbf{V}$) and ($\textbf{p}$) are known, then its Dividend Payout planned is:

$$d = 1 - \{[N+K+J+Vp/360]/[1+\textit{I}-U\} \\ /(\{S'[1+s]-\$v-F-S'i[1+s]\}[1-t])$$

Rule-58355:

If both ($\textit{I}$), ($\textbf{U}$), ($\textbf{S'}$), ($\textbf{s}$), ($\textbf{\$}$), ($\textbf{v}$), ($\textbf{F}$), ($\textbf{i}$), ($\textbf{t}$), ($\textbf{d}$), ($\textbf{N}$), ($\textbf{J}$), ($\textbf{V}$) and ($\textbf{p}$) are known, then its Kind of Cash planned is:

$$K = [1+\textit{I}(U+\{S'[1+s]-\$v-F-S'i[1+s]\}[1-t][1-d]) \\ -[N+J+Vp/360]$$

Rule-58356:

If both ($\textit{I}$), ($\textbf{U}$), ($\textbf{S'}$), ($\textbf{s}$), ($\textbf{\$}$), ($\textbf{v}$), ($\textbf{F}$), ($\textbf{i}$), ($\textbf{t}$), ($\textbf{d}$), ($\textbf{K}$), ($\textbf{N}$), ($\textbf{V}$) and ($\textbf{p}$) are known, then its Job or Trade Receivable Days planned is:

$$J = [1+\textit{I}(U+\{S'[1+s]-\$v-F-S'i[1+s]\}[1-t][1-d]) \\ -\{N+K+Vp/360\}$$

Rule-58357:

If both ($\textit{I}$), ($\textbf{U}$), ($\textbf{S'}$), ($\textbf{s}$), ($\textbf{\$}$), ($\textbf{v}$), ($\textbf{F}$), ($\textbf{i}$), ($\textbf{t}$), ($\textbf{d}$), ($\textbf{K}$), ($\textbf{J}$), ($\textbf{N}$) and ($\textbf{p}$) are known, then its Variable Cost planned is:

$$V = 360[[1+\textit{I}(U+\{S'[1+s]-\$v-F-S'i[1+s]\} \\ [1-t][1-d])-[N+K+J]]/p$$

or

$$V = \$v$$

Steve Asikin ISBN 13: **978-1564912361**, ISBN 10: **1564912363**

Rule-58358:

If both (**/**), (**U**), (**$'**), (**s**), (**$**), (**v**), (**F**), (**i**), (**t**), (**d**), (**K**), (**J**), (**V**) and (**N**) are known, then its Procured Inventory Days planned is:

$$\mathbf{p}= 360[[1+/(\mathbf{U}+\{\$'[1+s]-\$v-F-\$'i[1+s]\}[1-t][1-d])-[\mathbf{N+K+J}]]/\mathbf{V}$$

Rule-58359:

If both (**/**), (**U**), (**$'**), (**s**), (**$**), (**v**), (**F**), (**i**), (**t**), (**d**), (**K**), (**J**), (**V**) and (**p**) are known, then its Non Current Asset planned is:

$$\mathbf{N}= [1+/(\mathbf{U}+\{\$'[1+s]-\$v-F-\$'i[1+s]\}[1-t][1-d])-\mathbf{K-J-\$vp}/360$$

Rule-58360:

If both (**N**), (**U**), (**$'**), (**s**), (**$**), (**v**), (**F**), (**i**), (**t**), (**d**), (**K**), (**J**), (**V**) and (**p**) are known, then its Leverage or Gearing Ratio planned is:

$$\mathbf{/}= [\mathbf{N+K+J+\$vp}/360]/(\mathbf{U}+\{\$'[1+s]-\$v-F-\$'i[1+s]\}[1-t][1-d])-1$$

Rule-58361:

If both (**/**), (**N**), (**$'**), (**s**), (**$**), (**v**), (**F**), (**i**), (**t**), (**d**), (**K**), (**J**), (**V**) and (**p**) are known, then its Utilized or Starting Capital must be:

$$\mathbf{U}= ([\mathbf{N+K+J+\$vp}/360]-[1+/\{\$'[1+s]-\$v-F-\$'i[1+s]\}[1-t][1-d])/[1+/]$$

Steve Asikin ISBN 13: **978-1564912361**, ISBN 10: **1564912363**

Rule-58362:

If both (**I**), (**U**), (**N**), (**s**), (**S**), (**v**), (**F**), (**i**), (**t**), (**d**), (**K**), (**J**), (**V**) and (**p**) are known, then its Sales Past must be:

$$S' = \{[N+K+J+Svp/360]/[1+I]-U\}$$
$$/\{[1-t][1-d]\}+Sv+F)/\{[1+s][1-i]\}$$

or

$$S' = S/[1+s]$$


Rule-58363:

If both (**I**), (**U**), (**S'**), (**N**), (**S**), (**v**), (**F**), (**i**), (**t**), (**d**), (**K**), (**J**), (**V**) and (**p**) are known, then its Sales Growth planned is:

$$s = \{[N+K+J+Svp/360]/[1+I]-U\}/\{[1-t][1-d]\}$$
$$+Sv+F)/\{S'[1-i]\}-1$$

or

$$s = S/S'-1$$


Rule-58364:

If both (**I**), (**U**), (**S'**), (**s**), (**N**), (**v**), (**F**), (**i**), (**t**), (**d**), (**K**), (**J**), (**V**) and (**p**) are known, then its Sales or Revenue planned is:

$$S = [[1+I](U+\{S'[1+s]-F-S'i[1+s]\}[1-t][1-d])$$
$$-[N+K+J]]/(v\{p/360+[1+I][1-t][1-d]\})$$

or

$$S = S'[1+s]$$

`

Steve Asikin ISBN 13: **978-1564912361**, ISBN 10: **1564912363**

Rule-58365:

If both (*I*), (**U**), (**S'**), (**s**), (**S**), (**N**), (**F**), (**i**), (**t**), (**d**), (**K**), (**J**), (**V**) and (**p**) are known, then its Variable Portion planned is:

$$v = [[1+I(U+\{S'[1+s]-F-S'i[1+s]\}[1-t][1-d]) -[N+K+J]]/(S\{p/360+[1+I[1-t][1-d]\})$$

Rule-58366:

If both (*I*), (**U**), (**S'**), (**s**), (**S**), (**v**), (**N**), (**i**), (**t**), (**d**), (**K**), (**J**), (**V**) and (**p**) are known, then its Fixed Cost planned is:

$$F = S'[1+s]-Sv-S'i[1+s]-\{[N+K+J+Svp/360] /[1+I-U\}/\{[1-t][1-d]\}$$

Rule-58367:

If both (*I*), (**U**), (**S'**), (**s**), (**S**), (**v**), (**F**), (**N**), (**t**), (**d**), (**K**), (**J**), (**V**) and (**p**) are known, then its Interest Portion planned is:

$$i = (S'[1+s]-Sv-F-\{[N+K+J+Svp/360]/[1+I-U\} /\{[1-t][1-d]\})/\{S'[1+s]\}$$

Rule-58368:

If both (*I*), (**U**), (**S'**), (**s**), (**S**), (**v**), (**F**), (**i**), (**N**), (**d**), (**K**), (**J**), (**V**) and (**p**) are known, then its Tax Rate planned is:

$$t = 1-\{[N+K+J+Svp/360]/[1+I-U\} /(\{S'[1+s]-Sv-F-S'i[1+s]\}[1-d])$$

Steve Asikin ISBN 13: **978-1564912361**, ISBN 10: **1564912363**

Rule-58369:

If both (**/**), (**U**), (**S'**), (**s**), (**S**), (**v**), (**F**), (**i**), (**t**), (**N**), (**K**), (**J**), (**V**) and (**p**) are known, then its Dividend Payout planned is:

$$\mathbf{d} = 1 - \{[\mathbf{N}+\mathbf{K}+\mathbf{J}+\mathbf{Svp}/360]/[1+\mathbf{/}]-\mathbf{U}\}$$
$$/(\{\mathbf{S'}[1+\mathbf{s}]-\mathbf{Sv}-\mathbf{F}-\mathbf{S'i}[1+\mathbf{s}]\}[1-\mathbf{t}])$$

Rule-58370:

If both (**/**), (**U**), (**S'**), (**s**), (**S**), (**v**), (**F**), (**i**), (**t**), (**d**), (**N**), (**J**), (**V**) and (**p**) are known, then its Kind of Cash planned is:

$$\mathbf{K} = [1+\mathbf{/}(\mathbf{U}+\{\mathbf{S'}[1+\mathbf{s}]-\mathbf{Sv}-\mathbf{F}-\mathbf{S'i}[1+\mathbf{s}]\}[1-\mathbf{t}][1-\mathbf{d}])$$
$$-[\mathbf{N}+\mathbf{J}+\mathbf{Svp}/360]$$

Rule-58371:

If both (**/**), (**U**), (**S'**), (**s**), (**S**), (**v**), (**F**), (**i**), (**t**), (**d**), (**K**), (**N**), (**V**) and (**p**) are known, then its Job or Trade Account Receivable planned is:

$$\mathbf{J} = [1+\mathbf{/}(\mathbf{U}+\{\mathbf{S'}[1+\mathbf{s}]-\mathbf{Sv}-\mathbf{F}-\mathbf{S'i}[1+\mathbf{s}]\}[1-\mathbf{t}][1-\mathbf{d}])$$
$$-\{\mathbf{N}+\mathbf{K}+\mathbf{Svp}/360\}$$

Rule-58372:

If both (**/**), (**U**), (**S'**), (**s**), (**S**), (**v**), (**F**), (**i**), (**t**), (**d**), (**K**), (**J**), (**V**) and (**N**) are known, then its Procured Inventory Days planned is:

$$\mathbf{p} = 360[[1+\mathbf{/}(\mathbf{U}+\{\mathbf{S'}[1+\mathbf{s}]-\mathbf{Sv}-\mathbf{F}-\mathbf{S'i}[1+\mathbf{s}]\}[1-\mathbf{t}]$$
$$[1-\mathbf{d}])-[\mathbf{N}+\mathbf{K}+\mathbf{J}]]/[\mathbf{Sv}]$$

Steve Asikin ISBN 13: **978-1564912361**, ISBN 10: **1564912363**

Rule-58373:

If both (**/**), (**U**), (**S'**), (**s**), (**S**), (**v**), (**F**), (**i**), (**t**), (**d**), (**K**), (**J**), (**V**) and (**p**) are known, then its Non Current Asset planned is:

$$\mathbf{N}= [1+\text{/}](\mathbf{U}+\{\mathbf{S'}[1+s]\text{-}\mathbf{Sv}\text{-}\mathbf{F}\text{-}\mathbf{S'i}[1+s]\}[1\text{-}\mathbf{t}][1\text{-}\mathbf{d}])$$
$$\text{-}\mathbf{K}\text{-}\mathbf{J}\text{-}\mathbf{S'vp}[1+s]/360$$

Rule-58374:

If both (**N**), (**U**), (**S'**), (**s**), (**S**), (**v**), (**F**), (**i**), (**t**), (**d**), (**K**), (**J**), (**V**) and (**p**) are known, then its Leverage or Gearing Ratio planned is:

$$\mathbf{/}= \{\mathbf{N}+\mathbf{K}+\mathbf{J}+\mathbf{S'vp}[1+s]/360\}/(\mathbf{U}+\{\mathbf{S'}[1+s]\text{-}\mathbf{Sv}\text{-}\mathbf{F}$$
$$\text{-}\mathbf{S'i}[1+s]\}[1\text{-}\mathbf{t}][1\text{-}\mathbf{d}])\text{-}1$$

Rule-58375:

If both (**/**), (**N**), (**S'**), (**s**), (**S**), (**v**), (**F**), (**i**), (**t**), (**d**), (**K**), (**J**), (**V**) and (**p**) are known, then its Utilized or Starting Capital must be:

$$\mathbf{U}= (\{\mathbf{N}+\mathbf{K}+\mathbf{J}+\mathbf{S'vp}[1+s]/360\}\text{-}[1+\text{/}]\{\mathbf{S'}[1+s]\text{-}\mathbf{Sv}$$
$$\text{-}\mathbf{F}\text{-}\mathbf{S'i}[1+s]\}[1\text{-}\mathbf{t}][1\text{-}\mathbf{d}])/[1+\text{/}]$$

Rule-58376:

If both (**/**), (**U**), (**N**), (**s**), (**S**), (**v**), (**F**), (**i**), (**t**), (**d**), (**K**), (**J**), (**V**) and (**p**) are known, then its Sales Past must be:

$$\mathbf{S'}= [[1+\text{/}](\mathbf{U}\text{-}\{\mathbf{Sv}+\mathbf{F}\}[1\text{-}\mathbf{t}][1\text{-}\mathbf{d}])\text{-}[\mathbf{N}+\mathbf{K}+\mathbf{J}]]$$
$$/([1+s]\{\mathbf{vp}/360\text{-}[1+\text{/}][1\text{-}\mathbf{t}][1\text{-}\mathbf{d}][1\text{-}\mathbf{i}]\})$$
$$\text{or}$$
$$\mathbf{S'}= \mathbf{S}/[1+s]$$

Steve Asikin ISBN 13: **978-1564912361**, ISBN 10: **1564912363**

Rule-58377:

If both (**/**), (**U**), (**S'**), (**N**), (**S**), (**v**), (**F**), (**i**), (**t**), (**d**), (**K**), (**J**), (**V**) and (**p**) are known, then its Sales Growth planned is:

$$s= [[1+\text{/}(\text{U}-\{\text{S}\text{v}+\text{F}\}[1-\text{t}][1-\text{d}])-[\text{N}+\text{K}+\text{J}]]$$
$$/(\text{S}'\{\text{vp}/360-[1+\text{/}][1-\text{t}][1-\text{d}][1-\text{i}]\})-1$$

or

$$s= \text{S}/\text{S}'-1$$

Rule-58378:

If both (**/**), (**U**), (**S'**), (**s**), (**N**), (**v**), (**F**), (**i**), (**t**), (**d**), (**K**), (**J**), (**V**) and (**p**) are known, then its Sales or Revenue planned is:

$$S= (\{\text{N}+\text{K}+\text{J}+\text{S}'\text{vp}[1+\text{s}]/360\}/[1+\text{/}]-\text{U})$$
$$/\{[1-\text{t}][1-\text{d}]\}-\text{S}'[1+\text{s}]+\text{F}+\text{S}'\text{i}[1+\text{s}]\})/\text{v}$$

or

$$S= \text{S}'[1+\text{s}]$$

Rule-58379:

If both (**/**), (**U**), (**S'**), (**s**), (**S**), (**N**), (**F**), (**i**), (**t**), (**d**), (**K**), (**J**), (**V**) and (**p**) are known, then its Variable Portion planned is:

$$v= [[1+\text{/}(\text{U}+\{\text{S}'[1+\text{s}]-\text{F}-\text{S}'\text{i}[1+\text{s}]\}[1-\text{t}][1-\text{d}])$$
$$-[\text{N}+\text{K}+\text{J}]]/\{\text{S}'\text{p}[1+\text{s}]/360$$
$$+\text{S}[1+\text{/}][1-\text{t}][1-\text{d}]\}$$

Steve Asikin ISBN 13: **978-1564912361**, ISBN 10: **1564912363**

Rule-58380:

If both $(I)$, $(U)$, $(S')$, $(s)$, $(S)$, $(v)$, $(N)$, $(i)$, $(t)$, $(d)$, $(K)$, $(J)$, $(V)$ and $(p)$ are known, then its Fixed Cost planned is:

$$F= S'[1+s]-Sv-S'i[1+s]-(\{N+K+J+S'vp[1+s]/360\}$$
$$/[1+I-U)/\{[1-t][1-d]\}$$

Rule-58381:

If both $(I)$, $(U)$, $(S')$, $(s)$, $(S)$, $(v)$, $(F)$, $(N)$, $(t)$, $(d)$, $(K)$, $(J)$, $(V)$ and $(p)$ are known, then its Interest Portion planned is:

$$i= [S'[1+s]-Sv-F-(\{N+K+J+S'vp[1+s]/360\}$$
$$/[1+I-U)/\{[1-t][1-d]\}]/\{S'[1+s]\}$$

Rule-58382:

If both $(I)$, $(U)$, $(S')$, $(s)$, $(S)$, $(v)$, $(F)$, $(i)$, $(N)$, $(d)$, $(K)$, $(J)$, $(V)$ and $(p)$ are known, then its Tax Rate planned is:

$$t= 1-\{\{N+K+J+S'vp[1+s]/360\}/[1+I-U\}$$
$$/(\{S'[1+s]-Sv-F-S'i[1+s]\}[1-d])$$

Rule-58383:

If both $(I)$, $(U)$, $(S')$, $(s)$, $(S)$, $(v)$, $(F)$, $(i)$, $(t)$, $(N)$, $(K)$, $(J)$, $(V)$ and $(p)$ are known, then its Dividend Payout planned is:

$$d= 1-\{\{N+K+J+S'vp[1+s]/360\}/[1+I-U\}$$
$$/(\{S'[1+s]-Sv-F-S'i[1+s]\}[1-t])$$

Steve Asikin ISBN 13: **978-1564912361**, ISBN 10: **1564912363**

Rule-58384:

If both (**/**), (**U**), (**$'**), (**s**), (**$**), (**v**), (**F**), (**i**), (**t**), (**d**), (**N**), (**J**), (**V**) and (**p**) are known, then its Kind of Cash planned is:

$$K= [1+\textbf{/}(\textbf{U}+\{\textbf{\$'}[1+\textbf{s}]-\textbf{\$v}-\textbf{F}-\textbf{\$'i}[1+\textbf{s}]\}[1-\textbf{t}][1-\textbf{d}])$$
$$-\{\textbf{N}+\textbf{J}+\textbf{\$'vp}[1+\textbf{s}]/360\}$$

Rule-58385:

If both (**/**), (**U**), (**$'**), (**s**), (**$**), (**v**), (**F**), (**i**), (**t**), (**d**), (**K**), (**N**), (**V**) and (**p**) are known, then its Job or Trade Account Receicable planned is:

$$J= [[1+\textbf{/}(\textbf{U}+\{\textbf{\$'}[1+\textbf{s}]-\textbf{\$v}-\textbf{F}-\textbf{\$'i}[1+\textbf{s}]\}[1-\textbf{t}][1-\textbf{d}])$$
$$-\{\textbf{N}+\textbf{K}+\textbf{\$'vp}[1+\textbf{s}]/360\}$$

Rule-58386:

If both (**/**), (**U**), (**$'**), (**s**), (**$**), (**v**), (**F**), (**i**), (**t**), (**d**), (**K**), (**J**), (**V**) and (**N**) are known, then its Procured Inventory Days planned is:

$$p= 360[[1+\textbf{/}(\textbf{U}+\{\textbf{\$'}[1+\textbf{s}]-\textbf{\$v}-\textbf{F}-\textbf{\$'i}[1+\textbf{s}]\}[1-\textbf{t}]$$
$$[1-\textbf{d}])-[\textbf{N}+\textbf{K}+\textbf{J}]]/\{\textbf{\$'v}[1+\textbf{s}]\}$$

Rule-58387:

If both (**/**), (**U**), (**$'**), (**s**), (**$**), (**v**), (**F**), (**i**), (**t**), (**d**), (**K**), (**j**) and (**P**) are known, then its Non Current Asset planned is:

$$N= [1+\textbf{/}(\textbf{U}+\{\textbf{\$'}[1+\textbf{s}]-\textbf{\$v}-\textbf{F}-\textbf{\$'i}[1+\textbf{s}]\}[1-\textbf{t}][1-\textbf{d}])$$
$$-\textbf{K}-\textbf{\$j}/360-\textbf{P}$$

Steve Asikin ISBN 13: **978-1564912361**, ISBN 10: **1564912363**

Rule-58388:

If both **(N)**, **(U)**, **(S')**, **(s)**, **(S)**, **(v)**, **(F)**, **(i)**, **(t)**, **(d)**, **(K)**, **(j)** and **(P)** are known, then its Leverage or Gearing Ratio planned is:

$$\mathcal{F} = \{N+K+Sj/360+P\}/(U+\{S'[1+s]-Sv-F$$
$$-S'i[1+s]\}[1-t][1-d])-1$$

Rule-58389:

If both **(∕)**, **(N)**, **(S')**, **(s)**, **(S)**, **(v)**, **(F)**, **(i)**, **(t)**, **(d)**, **(K)**, **(j)** and **(P)** are known, then its Utilized or Starting Capital must be:

$$U = (\{N+K+Sj/360+P\}-[1+\mathcal{I}]\{S'[1+s]-Sv-F$$
$$-S'i[1+s]\}[1-t][1-d])/[1+\mathcal{I}]$$

Rule-58390:

If both **(∕)**, **(U)**, **(N)**, **(s)**, **(S)**, **(v)**, **(F)**, **(i)**, **(t)**, **(d)**, **(K)**, **(j)** and **(P)** are known, then its Sales Past must be:

$$S' = (\{[N+K+Sj/360+P]/[1+\mathcal{I}]-U\}/\{[1-t][1-d]\}$$
$$+Sv+F)/\{[1+s][1-i]\}$$

$$\text{or}$$

$$S' = S/[1+s]$$

Steve Asikin ISBN 13: **978-1564912361**, ISBN 10: **1564912363**

Rule-58391:

If both (**/**), (**U**), (**$'**), (**N**), (**$**), (**v**), (**F**), (**i**), (**t**), (**d**), (**K**), (**j**) and (**P**) are known, then its Sales Growth planned is:

$$s= (\{[\mathbf{N+K+\$j}/360+\mathbf{P}]/[1+\textbf{/}-\mathbf{U}\}/\{[1-\mathbf{t}][1-\mathbf{d}]\}$$
$$+\mathbf{\$v+F})/\{\mathbf{\$'}[1-\mathbf{i}]\}-1$$

or

$$s= \mathbf{\$}/\mathbf{\$'}-1$$


Rule-58392:

If both (**/**), (**U**), (**$'**), (**s**), (**$**), (**v**), (**F**), (**i**), (**t**), (**d**), (**K**), (**j**) and (**P**) are known, then its Sales or Revenue planned is:

$$\mathbf{\$}= [[1+\textbf{/}(\mathbf{U}-\{\mathbf{\$v+F}\}[1-\mathbf{t}][1-\mathbf{d}])-[\mathbf{N+K+P}]]$$
$$/\{\mathbf{j}/360+\mathbf{v}[1+\textbf{/}[1-\mathbf{t}][1-\mathbf{d}][1-\mathbf{i}]\}$$

or

$$\mathbf{\$}= \mathbf{\$'}[1+\mathbf{s}]$$


Rule-58393:

If both (**/**), (**U**), (**$'**), (**s**), (**$**), (**N**), (**F**), (**i**), (**t**), (**d**), (**K**), (**j**) and (**P**) are known, then its Variable Portion planned is:

$$\mathbf{v}= (\mathbf{\$'}[1+\mathbf{s}]-\mathbf{F}-\mathbf{\$'i}[1+\mathbf{s}]-\{[\mathbf{N+K+\$j}/360+\mathbf{P}]$$
$$/[1+\textbf{/}-\mathbf{U}\}/\{[1-\mathbf{t}][1-\mathbf{d}]\})/\mathbf{\$}$$

Steve Asikin ISBN 13: **978-1564912361**, ISBN 10: **1564912363**

Rule-58394:

If both ($I$), (**U**), (**S'**), (**s**), (**S**), (**v**), (**N**), (**i**), (**t**), (**d**), (**K**), (**j**) and (**P**) are known, then its Fixed Cost planned is:

$$F= S'[1+s]-Sv-S'i[1+s]-\{[N+K+Sj/360+P]$$
$$/[1+I-U\}/\{[1-t][1-d]\}$$

Rule-58395:

If both ($I$), (**U**), (**S'**), (**s**), (**S**), (**v**), (**F**), (**N**), (**t**), (**d**), (**K**), (**j**) and (**P**) are known, then its Interest Portion planned is:

$$i= (S'[1+s]-Sv-F-\{[N+K+Sj/360+P]/[1+I-U\}$$
$$/\{[1-t][1-d]\})/\{S'[1+s]\}$$

Rule-58396:

If both ($I$), (**U**), (**S'**), (**s**), (**S**), (**v**), (**F**), (**i**), (**N**), (**d**), (**K**), (**j**) and (**P**) are known, then its Tax Rate planned is:

$$t= 1-\{[N+K+Sj/360+P]/[1+I-U\}/(\{S'[1+s]-Sv-F$$
$$-S'i[1+s]\}[1-d])$$

Rule-58397:

If both ($I$), (**U**), (**S'**), (**s**), (**S**), (**v**), (**F**), (**i**), (**t**), (**N**), (**K**), (**j**) and (**P**) are known, then its Dividend Payout planned is:

$$d= 1-\{[N+K+Sj/360+P]/[1+I-U\}/(\{S'[1+s]-Sv-F$$
$$-S'i[1+s]\}[1-t])$$

Steve Asikin ISBN 13: **978-1564912361**, ISBN 10: **1564912363**

Rule-58398:

If both (**/**), (**U**), (**$'**), (**s**), (**$**), (**v**), (**F**), (**i**), (**t**), (**d**), (**N**), (**j**) and (**P**) are known, then its Kind of Cash planned is:

$$K = [1+/](U+\{\$'[1+s]-\$v-F-\$'i[1+s]\}[1-t][1-d])$$
$$-[N+\$j/360+P]$$

Rule-58399:

If both (**/**), (**U**), (**$'**), (**s**), (**$**), (**v**), (**F**), (**i**), (**t**), (**d**), (**K**), (**N**) and (**P**) are known, then its Job or Trade Receivable Days planned is:

$$j = 360[[1+/](U+\{\$'[1+s]-\$v-F-\$'i[1+s]\}[1-t]$$
$$[1-d])-\{N+K+P\}]/\$$$

Rule-58400:

If both (**/**), (**U**), (**$'**), (**s**), (**$**), (**v**), (**F**), (**i**), (**t**), (**d**), (**K**), (**j**) and (**N**) are known, then its Procured Inventory planned is:

$$P = [1+/](U+\{\$'[1+s]-\$v-F-\$'i[1+s]\}[1-t][1-d])$$
$$-\{N+K+\$j/360\}$$

Rule-58401:

If both (**/**), (**U**), (**$'**), (**s**), (**$**), (**v**), (**F**), (**i**), (**t**), (**d**), (**K**), (**j**), (**V**) and (**p**) are known, then its Non Current Asset planned is:

$$N = [1+/](U+\{\$'[1+s]-\$v-F-\$'i[1+s]\}[1-t][1-d])$$
$$-K-\$j/360-Vp/360$$

Steve Asikin ISBN 13: **978-1564912361**, ISBN 10: **1564912363**

Rule-58402:

If both **(N)**, **(U)**, **($')**, **(s)**, **($)**, **(v)**, **(F)**, **(i)**, **(t)**, **(d)**, **(K)**, **(j)**, **(V)** and **(p)** are known, then its Leverage or Gearing Ratio planned is:

$$\textit{I} = [N+K+\$j/360+Vp/360]/(U+\{\$'[1+s]-\$v-F$$
$$-\$'i[1+s]\}[1-t][1-d])-1$$


Rule-58403:

If both **(I)**, **(N)**, **($')**, **(s)**, **($)**, **(v)**, **(F)**, **(i)**, **(t)**, **(d)**, **(K)**, **(j)**, **(V)** and **(p)** are known, then its Utilized or Starting Capital must be:

$$U = ([N+K+\$j/360+Vp/360]-[1+\textit{I}]\{\$'[1+s]-\$v-F$$
$$-\$'i[1+s]\}[1-t][1-d])/[1+\textit{I}]$$


Rule-58404:

If both **(I)**, **(U)**, **(N)**, **(s)**, **($)**, **(v)**, **(F)**, **(i)**, **(t)**, **(d)**, **(K)**, **(j)**, **(V)** and **(p)** are known, then its Sales Past must be:

$$\$' = (\{[N+K+\$j/360+Vp/360]/[1+\textit{I}]-U\}$$
$$/\{[1-t][1-d]\}+\$v+F)/\{[1+s][1-i]\}$$
$$\text{or}$$
$$\$' = \$[1+s]$$

Steve Asikin ISBN 13: **978-1564912361**, ISBN 10: **1564912363**

Rule-58405:

If both (**/**), (**U**), (**S'**), (**N**), (**$**), (**v**), (**F**), (**i**), (**t**), (**d**), (**K**), (**j**), (**V**) and (**p**) are known, then its Sales Growth planned is:

$$s= (\{[\mathbf{N+K+Sj}/360+\mathbf{Vp}/360]/[1+\mathbf{/-U}\}$$
$$/\{[1-\mathbf{t}][1-\mathbf{d}]\}+\mathbf{Sv+F})/\{\mathbf{S'}[1-\mathbf{i}]\}-1$$

or

$$s= \mathbf{S/S'}-1$$


Rule-58406:

If both (**/**), (**U**), (**S'**), (**s**), (**N**), (**v**), (**F**), (**i**), (**t**), (**d**), (**K**), (**j**), (**V**) and (**p**) are known, then its Sales or Revenue planned is:

$$S= \lfloor[1+\mathbf{/}](\mathbf{U}+\{\mathbf{S'}[1+\mathbf{s}]-\mathbf{F}-\mathbf{S'i}[1+\mathbf{s}]\}[1-\mathbf{t}][1-\mathbf{d}])$$
$$-[\mathbf{N+K+Vp}/360]\rfloor$$
$$/\{\mathbf{j}/360+\mathbf{v}[1+\mathbf{/}][1-\mathbf{t}][1-\mathbf{d}]\}$$

or

$$S= \mathbf{S'}[1+\mathbf{s}]$$


Rule-58407:

If both (**/**), (**U**), (**S'**), (**s**), (**$**), (**N**), (**F**), (**i**), (**t**), (**d**), (**K**), (**j**), (**V**) and (**p**) are known, then its Variable Portion planned is:

$$v= (\mathbf{S'}[1+\mathbf{s}]-\mathbf{F}-\mathbf{S'i}[1+\mathbf{s}]-\{[\mathbf{N+K+Sj}/360+\mathbf{Vp}/360]$$
$$/[1+\mathbf{/-U}\}/\{[1-\mathbf{t}][1-\mathbf{d}]\})/\mathbf{S}$$

or

$$v= \mathbf{V/S}$$

`

Steve Asikin ISBN 13: **978-1564912361**, ISBN 10: **1564912363**

Rule-58408:

If both (**/**), (**U**), (**S'**), (**s**), (**S**), (**v**), (**N**), (**i**), (**t**), (**d**), (**K**), (**j**), (**V**) and (**p**) are known, then its Fixed Cost planned is:

$$\mathbf{F} = \mathbf{S'}[1+\mathbf{s}]-\mathbf{F}-\mathbf{S'i}[1+\mathbf{s}]-\{[\mathbf{N}+\mathbf{K}+\mathbf{Sj}/360+\mathbf{Vp}/360]$$
$$/[1+\mathbf{/}-\mathbf{U}\}/\{[1-\mathbf{t}][1-\mathbf{d}]\}$$

Rule-58409:

If both (**/**), (**U**), (**S'**), (**s**), (**S**), (**v**), (**F**), (**N**), (**t**), (**d**), (**K**), (**j**), (**V**) and (**p**) are known, then its Interest Portion planned is:

$$\mathbf{i} = (\mathbf{S'}[1+\mathbf{s}]-\mathbf{Sv}-\mathbf{F}-\{[\mathbf{N}+\mathbf{K}+\mathbf{Sj}/360+\mathbf{Vp}/360]$$
$$/[1+\mathbf{/}-\mathbf{U}\}/\{[1-\mathbf{t}][1-\mathbf{d}]\})/\{\mathbf{S'}[1+\mathbf{s}]\}$$

Rule-58410:

If both (**/**), (**U**), (**S'**), (**s**), (**S**), (**v**), (**F**), (**i**), (**N**), (**d**), (**K**), (**j**), (**V**) and (**p**) are known, then its Tax Rate planned is:

$$\mathbf{t} = 1-\{[\mathbf{N}+\mathbf{K}+\mathbf{Sj}/360+\mathbf{Vp}/360]/[1+\mathbf{/}-\mathbf{U}\}$$
$$/(\{\mathbf{S'}[1+\mathbf{s}]-\mathbf{Sv}-\mathbf{F}-\mathbf{S'i}[1+\mathbf{s}]\}[1-\mathbf{d}])$$

Rule-58411:

If both (**/**), (**U**), (**S'**), (**s**), (**S**), (**v**), (**F**), (**i**), (**t**), (**N**), (**K**), (**j**), (**V**) and (**p**) are known, then its Dividend Payout planned is:

$$\mathbf{d} = 1-\{[\mathbf{N}+\mathbf{K}+\mathbf{Sj}/360+\mathbf{Vp}/360]/[1+\mathbf{/}-\mathbf{U}\}$$
$$/(\{\mathbf{S'}[1+\mathbf{s}]-\mathbf{Sv}-\mathbf{F}-\mathbf{S'i}[1+\mathbf{s}]\}[1-\mathbf{t}])$$

Steve Asikin ISBN 13: **978-1564912361**, ISBN 10: **1564912363**

Rule-58412:

If both (**/**), (**U**), (**S'**), (**s**), (**S**), (**v**), (**F**), (**i**), (**t**), (**d**), (**N**), (**j**), (**V**) and (**p**) are known, then its Kind of Cash planned is:

$$K= [1+/(U+\{S'[1+s]-Sv-F-S'i[1+s]\}[1-t][1-d])$$
$$-[N+Sj/360+Vp/360]$$

Rule-58413:

If both (**/**), (**U**), (**S'**), (**s**), (**S**), (**v**), (**F**), (**i**), (**t**), (**d**), (**K**), (**N**), (**V**) and (**p**) are known, then its Job or Trade Receivable Days planned is:

$$j= 360[[1+/(U+\{S'[1+s]-Sv-F-S'i[1+s]\}[1-t]$$
$$[1-d])-[N+K+Vp/360]]/S$$

Rule-58414:

If both (**/**), (**U**), (**S'**), (**s**), (**S**), (**v**), (**F**), (**i**), (**t**), (**d**), (**K**), (**j**), (**N**) and (**p**) are known, then its Variable Cost planned is:

$$V= 360[[1+/(U+\{S'[1+s]-Sv-F-S'i[1+s]\}[1-t]$$
$$[1-d])-[N+K+Sj/360]]/p$$
$$or$$
$$V= Sv$$

Rule-58415:

If both (**/**), (**U**), (**S'**), (**s**), (**S**), (**v**), (**F**), (**i**), (**t**), (**d**), (**K**), (**j**), (**V**) and (**N**) are known, then its Procured Inventory Days planned is:

$$p= 360[[1+/(U+\{S'[1+s]-Sv-F-S'i[1+s]\}[1-t]$$
$$[1-d])-[N+K+Sj/360]]/V$$

Steve Asikin ISBN 13: **978-1564912361**, ISBN 10: **1564912363**

## Rule-58416:

If both $(I)$, $(U)$, $(S')$, $(s)$, $(S)$, $(v)$, $(F)$, $(i)$, $(t)$, $(d)$, $(K)$, $(j)$ and $(p)$ are known, then its Non Current Asset planned is:

$$N= [1+I](U+\{S'[1+s]-Sv-F-S'i[1+s]\}[1-t][1-d])$$
$$-K-Sj/360-Svp/360$$

## Rule-58417:

If both $(N)$, $(U)$, $(S')$, $(s)$, $(S)$, $(v)$, $(F)$, $(i)$, $(t)$, $(d)$, $(K)$, $(j)$ and $(p)$ are known, then its Leverage or Gearing Ratio planned is:

$$I= [N+K+Sj/360+Svp/360]/(U+\{S'[1+s]-Sv-F$$
$$-S'i[1+s]\}[1-t][1-d])-1$$

## Rule-58418:

If both $(I)$, $(N)$, $(S')$, $(s)$, $(S)$, $(v)$, $(F)$, $(i)$, $(t)$, $(d)$, $(K)$, $(j)$ and $(p)$ are known, then its Utilized or Starting Capital must be:

$$U= ([N+K+Sj/360+Svp/360]-[1+I]\{S'[1+s]-Sv-F$$
$$-S'i[1+s]\}[1-t][1-d])/[1+I]$$

## Rule-58419:

If both $(I)$, $(U)$, $(S')$, $(s)$, $(S)$, $(v)$, $(F)$, $(i)$, $(t)$, $(d)$, $(K)$, $(j)$ and $(p)$ are known, then its Sales Past must be:

$$S'= \{[N+K+Sj/360+Svp/360]/[1+I]-U\}/\{[1-t]$$
$$[1-d]\}+Sv+F)/\{S'[1+s][1-i]\}$$

or

$$S'= S/[1+s]$$

Steve Asikin ISBN 13: **978-1564912361**, ISBN 10: **1564912363**

Rule-58420:

If both (**/**), (**U**), (**S'**), (**N**), (**S**), (**v**), (**F**), (**i**), (**t**), (**d**), (**K**), (**j**) and (**p**) are known, then its Sales Growth planned is:

$$s = \{[N+K+Sj/360+Svp/360]/[1+/]-U\}/\{[1-t]$$
$$[1-d]\}+Sv+F)/\{S'[1-i]\}-1$$

or

$$s = S/S'-1$$


Rule-58421:

If both (**/**), (**U**), (**S'**), (**s**), (**S**), (**v**), (**F**), (**i**), (**t**), (**d**), (**K**), (**j**) and (**p**) are known, then its Sales or Revenue planned is:

$$S = [[1+/](U+\{S'[1+s]-F-S'i[1+s]\}[1-t][1-d])$$
$$-[N+K]]/(j/360+v\{p/360$$
$$+[1+/][1-t][1-d]\})$$

or

$$S = S'[1+s]$$


Rule-58422:

If both (**/**), (**U**), (**S'**), (**s**), (**S**), (**N**), (**F**), (**i**), (**t**), (**d**), (**K**), (**j**) and (**p**) are known, then its Variable Portion planned is:

$$v = [[1+/](U+\{S'[1+s]-F-S'i[1+s]\}[1-t][1-d])$$
$$-[N+K+Sj/360]]$$
$$/(S\{p/360+[1+/][1-t][1-d]\})$$

Steve Asikin ISBN 13: **978-1564912361**, ISBN 10: **1564912363**

Rule-58423:

If both $(\textbf{\textit{I}})$, $(\textbf{U})$, $(\textbf{S'})$, $(\textbf{s})$, $(\textbf{\$})$, $(\textbf{v})$, $(\textbf{N})$, $(\textbf{i})$, $(\textbf{t})$, $(\textbf{d})$, $(\textbf{K})$, $(\textbf{j})$ and $(\textbf{p})$ are known, then its Fixed Cost planned is:

$$F = S'[1+s] - \$v - S'i[1+s] - \{[N+K+\$j/360+\$vp/360]$$
$$/[1+\textbf{\textit{I}}-U\} / \{[1-t][1-d]\}$$


Rule-58424:

If both $(\textbf{\textit{I}})$, $(\textbf{U})$, $(\textbf{S'})$, $(\textbf{s})$, $(\textbf{\$})$, $(\textbf{v})$, $(\textbf{F})$, $(\textbf{N})$, $(\textbf{t})$, $(\textbf{d})$, $(\textbf{K})$, $(\textbf{j})$ and $(\textbf{p})$ are known, then its Interest Portion planned is:

$$i = (S'[1+s] - \$v - F - \{[N+K+\$j/360+\$vp/360]$$
$$/[1+\textbf{\textit{I}}-U\} / \{[1-t][1-d]\}) / \{S'[1+s]\}$$


Rule-58425:

If both $(\textbf{\textit{I}})$, $(\textbf{U})$, $(\textbf{S'})$, $(\textbf{s})$, $(\textbf{\$})$, $(\textbf{v})$, $(\textbf{F})$, $(\textbf{i})$, $(\textbf{N})$, $(\textbf{d})$, $(\textbf{K})$, $(\textbf{j})$ and $(\textbf{p})$ are known, then its Tax Rate planned is:

$$t = 1 - \{[N+K+\$j/360+\$vp/360]/[1+\textbf{\textit{I}}-U\}$$
$$/(\{S'[1+s] - \$v - F - S'i[1+s]\}[1-d])$$


Rule-58426:

If both $(\textbf{\textit{I}})$, $(\textbf{U})$, $(\textbf{S'})$, $(\textbf{s})$, $(\textbf{\$})$, $(\textbf{v})$, $(\textbf{F})$, $(\textbf{i})$, $(\textbf{t})$, $(\textbf{N})$, $(\textbf{K})$, $(\textbf{j})$ and $(\textbf{p})$ are known, then its Dividend Payout planned is:

$$d = 1 - \{[N+K+\$j/360+\$vp/360]/[1+\textbf{\textit{I}}-U\}$$
$$/(\{S'[1+s] - \$v - F - S'i[1+s]\}[1-t])$$

Steve Asikin ISBN 13: **978-1564912361**, ISBN 10: **1564912363**

Rule-58427:

If both ($\mathbf{\Lambda}$), ($\mathbf{U}$), ($\mathbf{S'}$), ($\mathbf{s}$), ($\mathbf{S}$), ($\mathbf{v}$), ($\mathbf{F}$), ($\mathbf{i}$), ($\mathbf{t}$), ($\mathbf{d}$), ($\mathbf{N}$), ($\mathbf{j}$) and ($\mathbf{p}$) are known, then its Kind of Cash planned is:

$$\mathbf{K} = [1+\mathbf{\Lambda}(\mathbf{U}+\{\mathbf{S'}[1+\mathbf{s}]-\mathbf{Sv}-\mathbf{F}-\mathbf{S'i}[1+\mathbf{s}]\}[1-\mathbf{t}][1-\mathbf{d}])$$
$$-[\mathbf{N}+\mathbf{Sj}/360+\mathbf{Svp}/360]$$

Rule-58428:

If both ($\mathbf{\Lambda}$), ($\mathbf{U}$), ($\mathbf{S'}$), ($\mathbf{s}$), ($\mathbf{S}$), ($\mathbf{v}$), ($\mathbf{F}$), ($\mathbf{i}$), ($\mathbf{t}$), ($\mathbf{d}$), ($\mathbf{K}$), ($\mathbf{N}$) and ($\mathbf{p}$) are known, then its Job or Trade Rceivable Days planned is:

$$\mathbf{j} = 360[[1+\mathbf{\Lambda}(\mathbf{U}+\{\mathbf{S'}[1+\mathbf{s}]-\mathbf{Sv}-\mathbf{F}-\mathbf{S'i}[1+\mathbf{s}]\}[1-\mathbf{t}]$$
$$[1-\mathbf{d}])-\{\mathbf{N}+\mathbf{K}+\mathbf{Svp}/360\}]/\mathbf{S}$$

Rule-58429:

If both ($\mathbf{\Lambda}$), ($\mathbf{U}$), ($\mathbf{S'}$), ($\mathbf{s}$), ($\mathbf{S}$), ($\mathbf{v}$), ($\mathbf{F}$), ($\mathbf{i}$), ($\mathbf{t}$), ($\mathbf{d}$), ($\mathbf{K}$), ($\mathbf{j}$) and ($\mathbf{N}$) are known, then its Procured Inventory Days planned is:

$$\mathbf{p} = 360[[1+\mathbf{\Lambda}(\mathbf{U}+\{\mathbf{S'}[1+\mathbf{s}]-\mathbf{Sv}-\mathbf{F}-\mathbf{S'i}[1+\mathbf{s}]\}[1-\mathbf{t}]$$
$$[1-\mathbf{d}])-[\mathbf{N}+\mathbf{K}+\mathbf{Sj}/360]]/[\mathbf{Sv}]$$

Rule-58430:

If both ($\mathbf{\Lambda}$), ($\mathbf{U}$), ($\mathbf{S'}$), ($\mathbf{s}$), ($\mathbf{S}$), ($\mathbf{v}$), ($\mathbf{F}$), ($\mathbf{i}$), ($\mathbf{t}$), ($\mathbf{d}$), ($\mathbf{K}$), ($\mathbf{j}$) and ($\mathbf{p}$) are known, then its Non Current Asset planned is:

$$\mathbf{N} = [1+\mathbf{\Lambda}(\mathbf{U}+\{\mathbf{S'}[1+\mathbf{s}]-\mathbf{Sv}-\mathbf{F}-\mathbf{S'i}[1+\mathbf{s}]\}[1-\mathbf{t}][1-\mathbf{d}])$$
$$-\mathbf{K}-\mathbf{Sj}/360-\mathbf{S'vp}[1+\mathbf{s}]/360$$

Steve Asikin ISBN 13: **978-1564912361**, ISBN 10: **1564912363**

Rule-58431:

If both **(N)**, **(U)**, **(S')**, **(s)**, **(S)**, **(v)**, **(F)**, **(i)**, **(t)**, **(d)**, **(K)**, **(j)** and **(p)** are known, then its Leverage or Gearing Ratio planned is:

$$\text{┣} = \{N+K+Sj/360+S'vp[1+s]/360\}/(U+\{S'[1+s]$$
$$-Sv-F-S'i[1+s]\}[1-t][1-d])-1$$

Rule-58432:

If both **(┃)**, **(N)**, **(S')**, **(s)**, **(S)**, **(v)**, **(F)**, **(i)**, **(t)**, **(d)**, **(K)**, **(j)** and **(p)** are known, then its Utilized or Starting Capital must be:

$$U = (\{N+K+Sj/360+S'vp[1+s]/360\}-[1+\text{┃}]\{S'[1+s]$$
$$-Sv-F-S'i[1+s]\}[1-t][1-d])/[1+\text{┃}]$$

Rule-58433:

If both **(┃)**, **(U)**, **(N)**, **(s)**, **(S)**, **(v)**, **(F)**, **(i)**, **(t)**, **(d)**, **(K)**, **(j)** and **(p)** are known, then its Sales Past must be:

$$S' = [[1+\text{┃}](U+\{S'[1+s]-Sv-F-S'i[1+s]\}[1-t][1-d])$$
$$-[N+K+Sj/360]]/([1+s]\{vp/360$$
$$-[1+\text{┃}][1-t][1-d][1-i]\})$$

or

$$S' = S/[1+s]$$

Steve Asikin ISBN 13: **978-1564912361**, ISBN 10: **1564912363**

Rule-58434:

If both (**/**), (**U**), (**S'**), (**N**), (**S**), (**v**), (**F**), (**i**), (**t**), (**d**), (**K**), (**j**) and (**p**) are known, then its Sales Growth planned is:

$$s= [[1+/(U+\{S'[1+s]-Sv-F-S'i[1+s]\}[1-t][1-d])$$
$$-[N+K+Sj/360]]/(S'\{vp/360$$
$$-[1+/[1-t][1-d][1-i]\})-1$$

or

$$s= S/S'-1$$


Rule-58435:

If both (**/**), (**U**), (**S'**), (**s**), (**N**), (**v**), (**F**), (**i**), (**t**), (**d**), (**K**), (**j**) and (**p**) are known, then its Sales or Revenue planned is:

$$S= [[1+/(U+\{S'[1+s]-F-S'i[1+s]\}[1-t][1-d])$$
$$-\{N+K+S'vp[1+s]/360\}]$$
$$/\{j/360+v[1+/[1-t][1-d]\}$$

or

$$S= S'[1+s]$$


Rule-58436:

If both (**/**), (**U**), (**S'**), (**s**), (**S**), (**N**), (**F**), (**i**), (**t**), (**d**), (**K**), (**j**) and (**p**) are known, then its Variable Portion planned is:

$$v= ([1+/(U+\{S'[1+s]-Sv-F-S'i[1+s]\}[1-t][1-d])$$
$$-[N+K+Sj/360])/\{S'p[1+s]/360$$
$$+S[1+/[1-t][1-d]\}$$

Steve Asikin ISBN 13: **978-1564912361**, ISBN 10: **1564912363**

Rule-58437:

If both ($\pmb{I}$), ($\pmb{U}$), ($\pmb{S'}$), ($\pmb{s}$), ($\pmb{S}$), ($\pmb{v}$), ($\pmb{N}$), ($\pmb{i}$), ($\pmb{t}$), ($\pmb{d}$), ($\pmb{K}$), ($\pmb{j}$) and ($\pmb{p}$) are known, then its Fixed Cost planned is:

$$F = (S'[1+s]-Sv-S'i[1+s]-\{N+K+Sj/360$$
$$+S'vp[1+s]/360\}/[1+I-U)/\{[1-t][1-d]\})$$

Rule-58438:

If both ($\pmb{I}$), ($\pmb{U}$), ($\pmb{S'}$), ($\pmb{s}$), ($\pmb{S}$), ($\pmb{v}$), ($\pmb{F}$), ($\pmb{N}$), ($\pmb{t}$), ($\pmb{d}$), ($\pmb{K}$), ($\pmb{j}$) and ($\pmb{p}$) are known, then its Interest Portion planned is:

$$i = (S'[1+s]-Sv-F-\{N+K+Sj/360+S'vp[1+s]/360\}$$
$$/[1+I-U)/\{[1-t][1-d]\})/\{S'[1+s]\}$$

Rule-58439:

If both ($\pmb{I}$), ($\pmb{U}$), ($\pmb{S'}$), ($\pmb{s}$), ($\pmb{S}$), ($\pmb{v}$), ($\pmb{F}$), ($\pmb{i}$), ($\pmb{N}$), ($\pmb{d}$), ($\pmb{K}$), ($\pmb{j}$) and ($\pmb{p}$) are known, then its Tax Rate planned is:

$$t = 1-(\{N+K+Sj/360+S'vp[1+s]/360\}/[1+I-U)$$
$$/(\{S'[1+s]-Sv-F-S'i[1+s]\}[1-d])$$

Rule-58440:

If both ($\pmb{I}$), ($\pmb{U}$), ($\pmb{S'}$), ($\pmb{s}$), ($\pmb{S}$), ($\pmb{v}$), ($\pmb{F}$), ($\pmb{i}$), ($\pmb{t}$), ($\pmb{N}$), ($\pmb{K}$), ($\pmb{j}$) and ($\pmb{p}$) are known, then its Dividend Payout planned is:

$$d = 1-(\{N+K+Sj/360+S'vp[1+s]/360\}/[1+I-U)$$
$$/(\{S'[1+s]-Sv-F-S'i[1+s]\}[1-t])$$

Steve Asikin ISBN 13: **978-1564912361**, ISBN 10: **1564912363**

Rule-58441:

If both (**/**), (**U**), (**S'**), (**s**), (**S**), (**v**), (**F**), (**i**), (**t**), (**d**), (**N**), (**j**) and (**p**) are known, then its Kind of Cash planned is:

$$K= [1+/](U+\{S'[1+s]-Sv-F-S'i[1+s]\}[1-t][1-d])$$
$$-\{N+Sj/360+S'vp[1+s]/360\}$$

Rule-58442:

If both (**/**), (**U**), (**S'**), (**s**), (**S**), (**v**), (**F**), (**i**), (**t**), (**d**), (**K**), (**N**) and (**p**) are known, then its Job or Trade Receivable Days planned is:

$$j= 360[[1+/](U+\{S'[1+s]-Sv-F-S'i[1+s]\}[1-t]$$
$$[1-d])-\{N+K+S'vp[1+s]/360\}]/S$$

Rule-58443:

If both (**/**), (**U**), (**S'**), (**s**), (**S**), (**v**), (**F**), (**i**), (**t**), (**d**), (**K**), (**j**) and (**N**) are known, then its Procured Inventory Days planned is:

$$p= 360[[1+/](U+\{S'[1+s]-Sv-F-S'i[1+s]\}[1-t]$$
$$[1-d])-[N+K+Sj/360]]/\{S'v[1+s]\}$$

Rule-58444:

If both (**/**), (**U**), (**S'**), (**s**), (**S**), (**v**), (**F**), (**i**), (**t**), (**d**), (**K**), (**j**) and (**P**) are known, then its Non Current Asset planned is:

$$N= [1+/](U+\{S'[1+s]-Sv-F-S'i[1+s]\}[1-t][1-d])$$
$$-K-S'j[1+s]/360-P$$

Steve Asikin ISBN 13: **978-1564912361**, ISBN 10: **1564912363**

Rule-58445:

If both **(N)**, **(U)**, **(S')**, **(s)**, **(S)**, **(v)**, **(F)**, **(i)**, **(t)**, **(d)**, **(K)**, **(j)** and **(P)** are known, then its Leverage or Gearing Ratio planned is:

$$I = \{N+K+S'j[1+s]/360+P\}/(U+\{S'[1+s]-Sv-F$$
$$-S'i[1+s]\}[1-t][1-d])-1$$

Rule-58446:

If both **(I)**, **(N)**, **(S')**, **(s)**, **(S)**, **(v)**, **(F)**, **(i)**, **(t)**, **(d)**, **(K)**, **(j)** and **(P)** are known, then its Utilized or Starting Capital must be:

$$U = (\{N+K+S'j[1+s]/360+P\}-[1+I]\{S'[1+s]-Sv-F$$
$$-S'i[1+s]\}[1-t][1-d])/[1+I]$$

Rule-58447:

If both **(I)**, **(U)**, **(N)**, **(s)**, **(S)**, **(v)**, **(F)**, **(i)**, **(t)**, **(d)**, **(K)**, **(j)** and **(P)** are known, then its Sales Past must be:

$$S' = [[1+I](U-\{Sv-F\}[1-t][1-d])-[N+K+P]]$$
$$/([1+s]\{j/360-[1+I][1-t][1-d]\})$$

or

$$S' = S/[1+s]$$

Steve Asikin ISBN 13: **978-1564912361**, ISBN 10: **1564912363**

Rule-58448:

If both (**/**), (**U**), (**S'**), (**N**), (**S**), (**v**), (**F**), (**i**), (**t**), (**d**), (**K**), (**j**) and (**P**) are known, then its Sales Growth planned is:

$$s= [[1+/](U-\{Sv-F\}[1-t][1-d])-[N+K+P]]$$
$$/(S'\{j/360-[1+/][1-t][1-d]\})-1$$
or
$$s= S/S'-1$$

Rule-58449:

If both (**/**), (**U**), (**S'**), (**s**), (**N**), (**v**), (**F**), (**i**), (**t**), (**d**), (**K**), (**j**) and (**P**) are known, then its Sales or Revenue planned is:

$$S= [S'[1+s]-F-S'i[1+s]-(\{N+K+S'j[1+s]/360+P\}$$
$$/[1+/]-U)/\{[1-t][1-d]\}]/v$$
or
$$S= S'[1+s]$$

Rule-58450:

If both (**/**), (**U**), (**S'**), (**s**), (**S**), (**N**), (**F**), (**i**), (**t**), (**d**), (**K**), (**j**) and (**P**) are known, then its Variable Portion planned is:

$$v= [S'[1+s]-F-S'i[1+s]-(\{N+K+S'j[1+s]/360+P\}$$
$$/[1+/]-U)/\{[1-t][1-d]\}]/S$$

Steve Asikin ISBN 13: **978-1564912361**, ISBN 10: **1564912363**

Rule-58451:

If both (**/**), (**U**), (**S'**), (**s**), (**S**), (**v**), (**N**), (**i**), (**t**), (**d**), (**K**), (**j**) and (**P**) are known, then its Fixed Cost planned is:

$$F= S'[1+s]-Sv-S'i[1+s]-(\{N+K+S'j[1+s]/360+P\}$$
$$/[1+\text{/}-U)/\{[1-t][1-d]\}$$

Rule-58452:

If both (**/**), (**U**), (**S'**), (**s**), (**S**), (**v**), (**F**), (**N**), (**t**), (**d**), (**K**), (**j**) and (**P**) are known, then its Interest Portion planned is:

$$i= [S'[1+s]-Sv-F-(\{N+K+S'j[1+s]/360+P\}$$
$$/[1+\text{/}-U)/\{[1-t][1-d]\}]/\{S'[1+s]\}$$

Rule-58453:

If both (**/**), (**U**), (**S'**), (**s**), (**S**), (**v**), (**F**), (**i**), (**N**), (**d**), (**K**), (**j**) and (**P**) are known, then its Tax Rate planned is:

$$t= 1-(\{N+K+S'j[1+s]/360+P\}/[1+\text{/}-U)$$
$$/(\{S'[1+s]-Sv-F-S'i[1+s]\}[1-d])$$

Rule-58454:

If both (**/**), (**U**), (**S'**), (**s**), (**S**), (**v**), (**F**), (**i**), (**t**), (**N**), (**K**), (**j**) and (**P**) are known, then its Dividend Payout planned is:

$$d= 1-(\{N+K+S'j[1+s]/360+P\}/[1+\text{/}-U)$$
$$/(\{S'[1+s]-Sv-F-S'i[1+s]\}[1-t])$$

Steve Asikin ISBN 13: **978-1564912361**, ISBN 10: **1564912363**

Rule-58455:

If both (**/**), (**U**), (**S'**), (**s**), (**S**), (**v**), (**F**), (**i**), (**t**), (**d**), (**N**), (**j**) and (**P**) are known, then its Kind of Cash planned is:

$$\mathbf{K} = [1 + \mathbf{/}(\mathbf{U} + \{\mathbf{S'}[1+\mathbf{s}] - \mathbf{Sv} - \mathbf{F} - \mathbf{S'i}[1+\mathbf{s}]\}[1-\mathbf{t}][1-\mathbf{d}])$$
$$- \{\mathbf{N} + \mathbf{S'j}[1+\mathbf{s}]/360 + \mathbf{P}\}$$

Rule-58456:

If both (**/**), (**U**), (**S'**), (**s**), (**S**), (**v**), (**F**), (**i**), (**t**), (**d**), (**K**), (**N**) and (**P**) are known, then its Job or Trade Receivable Days planned is:

$$\mathbf{j} = 360[[1 + \mathbf{/}(\mathbf{U} + \{\mathbf{S'}[1+\mathbf{s}] - \mathbf{Sv} - \mathbf{F} - \mathbf{S'i}[1+\mathbf{s}]\}[1-\mathbf{t}]$$
$$[1-\mathbf{d}]) - \{\mathbf{N} + \mathbf{K} + \mathbf{P}\}] / \{\mathbf{S'}[1+\mathbf{s}]\}$$

Rule-58457:

If both (**/**), (**U**), (**S'**), (**s**), (**S**), (**v**), (**F**), (**i**), (**t**), (**d**), (**K**), (**j**) and (**N**) are known, then its Procured Inventory planned is:

$$\mathbf{P} = [1 + \mathbf{/}(\mathbf{U} + \{\mathbf{S'}[1+\mathbf{s}] - \mathbf{Sv} - \mathbf{F} - \mathbf{S'i}[1+\mathbf{s}]\}[1-\mathbf{t}][1-\mathbf{d}])$$
$$- \{\mathbf{N} + \mathbf{K} + \mathbf{S'j}[1+\mathbf{s}]/360\}$$

Rule-58458:

If both (**/**), (**U**), (**S'**), (**s**), (**S**), (**v**), (**F**), (**i**), (**t**), (**d**), (**K**), (**j**), (**V**) and (**p**) are known, then its Non Current Asset planned is:

$$\mathbf{N} = [1 + \mathbf{/}(\mathbf{U} + \{\mathbf{S'}[1+\mathbf{s}] - \mathbf{Sv} - \mathbf{F} - \mathbf{S'i}[1+\mathbf{s}]\}[1-\mathbf{t}][1-\mathbf{d}])$$
$$- \mathbf{K} - \mathbf{S'j}[1+\mathbf{s}]/360 - \mathbf{Vp}/360$$

Steve Asikin ISBN 13: **978-1564912361**, ISBN 10: **1564912363**

Rule-58459:

If both (**N**), (**U**), (**S'**), (**s**), (**S**), (**v**), (**F**), (**i**), (**t**), (**d**), (**K**), (**j**), (**V**) and (**p**) are known, then its Leverage or Gearing Ratio planned is:

$$I= \{N+K+S'j[1+s]/360+Vp/360\}/(U+\{S'[1+s]-Sv$$
$$-F-S'i[1+s]\}[1-t][1-d])-1$$

Rule-58460:

If both (**I**), (**N**), (**S'**), (**s**), (**S**), (**v**), (**F**), (**i**), (**t**), (**d**), (**K**), (**j**), (**V**) and (**p**) are known, then its Utilized or Starting Capital must be:

$$U= (\{N+K+S'j[1+s]/360+Vp/360\}-[1+I]\{S'[1+s]$$
$$-Sv-F-S'i[1+s]\}[1-t][1-d])/[1+I]$$

Rule-58461:

If both (**I**), (**U**), (**N**), (**s**), (**S**), (**v**), (**F**), (**i**), (**t**), (**d**), (**K**), (**j**), (**V**) and (**p**) are known, then its Sales Past must be:

$$S'= [[1+I](U-\{Sv+F\}[1-t][1-d])-[N+K+Vp/360]]$$
$$/([1+s]\{j/360-[1+I][1-t][1-d][1-i]\})$$

or

$$S'= S/[1+s]$$

Steve Asikin ISBN 13: **978-1564912361**, ISBN 10: **1564912363**

Rule-58462:

If both (**/**), (**U**), (**\$'**), (**N**), (**\$**), (**v**), (**F**), (**i**), (**t**), (**d**), (**K**), (**j**), (**V**) and (**p**) are known, then its Sales Growth planned is:

$$s= [[1+/(U-\{\$v+F\}[1-t][1-d])-[N+K+Vp/360]]$$
$$/(\$'\{j/360-[1+/][1-t][1-d][1-i]\})-1$$

or

$$s= \$/\$'-1$$

Rule-58463:

If both (**/**), (**U**), (**\$'**), (**s**), (**N**), (**v**), (**F**), (**i**), (**t**), (**d**), (**K**), (**j**), (**V**) and (**p**) are known, then its Sales or Revenue planned is:

$$\$= [\$'[1+s]-\$v-\$'i[1+s]-(\{N+K+\$'j[1+s]/360$$
$$+Vp/360\}/[1+/-U)/\{[1-t][1-d]\}]/v$$

or

$$\$= \$'[1+s]$$

or

$$\$= V/v$$

Rule-58464:

If both (**/**), (**U**), (**\$'**), (**s**), (**\$**), (**N**), (**F**), (**i**), (**t**), (**d**), (**K**), (**j**), (**V**) and (**p**) are known, then its Variable Portion planned is:

$$v= [\$'[1+s]-\$v-\$'i[1+s]-(\{N+K+\$'j[1+s]/360$$
$$+Vp/360\}/[1+/-U)/\{[1-t][1-d]\}]/\$$$

or

$$v= V/\$$$

Steve Asikin ISBN 13: **978-1564912361**, ISBN 10: **1564912363**

Rule-58465:

If both $(I)$, $(U)$, $(S')$, $(s)$, $(S)$, $(v)$, $(N)$, $(i)$, $(t)$, $(d)$, $(K)$, $(j)$, $(V)$ and $(p)$ are known, then its Fixed Cost planned is:

$$F= S'[1+s]-Sv-S'i[1+s]-(\{N+K+S'j[1+s]/360$$
$$+Vp/360\}/[1+I-U)/\{[1-t][1-d]\}$$

Rule-58466:

If both $(I)$, $(U)$, $(S')$, $(s)$, $(S)$, $(v)$, $(F)$, $(N)$, $(t)$, $(d)$, $(K)$, $(j)$, $(V)$ and $(p)$ are known, then its Interest Portion planned is:

$$i= [S'[1+s]-Sv-F-(\{N+K+S'j[1+s]/360+Vp/360\}$$
$$/[1+I-U)/\{[1-t][1-d]\}]/\{S'[1+s]\}$$

Rule-58467:

If both $(I)$, $(U)$, $(S')$, $(s)$, $(S)$, $(v)$, $(F)$, $(i)$, $(N)$, $(d)$, $(K)$, $(j)$, $(V)$ and $(p)$ are known, then its Tax Rate planned is:

$$t= 1-(\{N+K+S'j[1+s]/360+Vp/360\}/[1+I-U)$$
$$/(\{S'[1+s]-Sv-F-S'i[1+s]\}[1-d])$$

Rule-58468:

If both $(I)$, $(U)$, $(S')$, $(s)$, $(S)$, $(v)$, $(F)$, $(i)$, $(t)$, $(N)$, $(K)$, $(j)$, $(V)$ and $(p)$ are known, then its Dividend Payout planned is:

$$d= 1-(\{N+K+S'j[1+s]/360+Vp/360\}/[1+I-U)$$
$$/(\{S'[1+s]-Sv-F-S'i[1+s]\}[1-t])$$

Steve Asikin ISBN 13: **978-1564912361**, ISBN 10: **1564912363**

Rule-58469:

If both (**/**), (**U**), (**$'**), (**s**), (**$**), (**v**), (**F**), (**i**), (**t**), (**d**), (**N**), (**j**), (**V**) and (**p**) are known, then its Kind of Cash planned is:

$$K= [1+/(U+\{\$'[1+s]-\$v-F-\$'i[1+s]\}[1-t][1-d])$$
$$-\{N+\$'j[1+s]/360+Vp/360\}$$

Rule-58470:

If both (**/**), (**U**), (**$'**), (**s**), (**$**), (**v**), (**F**), (**i**), (**t**), (**d**), (**K**), (**N**), (**V**) and (**p**) are known, then its Job or Trade Receivable Days planned is:

$$j= 360[[1+/(U+\{\$'[1+s]-\$v-F-\$'i[1+s]\}[1-t]$$
$$[1-d])-\{N+K+Vp/360\}]/\{\$'[1+s]\}$$

Rule-58471:

If both (**/**), (**U**), (**$'**), (**s**), (**$**), (**v**), (**F**), (**i**), (**t**), (**d**), (**K**), (**j**), (**N**) and (**p**) are known, then its Variable Cost planned is:

$$V= 360[[1+/(U+\{\$'[1+s]-\$v-F-\$'i[1+s]\}[1-t]$$
$$[1-d])-\{N+K+\$'j[1+s]/360\}]/p$$

or

$$V= \$v$$

Rule-58472:

If both (**/**), (**U**), (**$'**), (**s**), (**$**), (**v**), (**F**), (**i**), (**t**), (**d**), (**K**), (**j**), (**V**) and (**N**) are known, then its Procured Inventpry Days planned is:

$$p= 360[[1+/(U+\{\$'[1+s]-\$v-F-\$'i[1+s]\}[1-t]$$
$$[1-d])-\{N+K+\$'j[1+s]/360\}]/V$$

`

656

Rule-58473:
>  If both (**/**), (**U**), (**S'**), (**s**), (**$**), (**v**), (**F**), (**i**), (**t**), (**d**), (**K**),
>  (**j**) and (**p**) are known, then its Non Current Asset
>  planned is:
>  $$N = [1+/](U+\{S'[1+s]-Sv-F-S'i[1+s]\}[1-t][1-d])$$
>  $$-K-S'j[1+s]/360-Svp/360$$

Rule-58474:
>  If both (**N**), (**U**), (**S'**), (**s**), (**$**), (**v**), (**F**), (**i**), (**t**), (**d**), (**K**),
>  (**j**) and (**p**) are known, then its Leverage or Gearing
>  Ratio planned is:
>  $$/= \{N+K+S'j[1+s]/360+Svp/360\}/(U+\{S'[1+s]$$
>  $$-Sv-F-S'i[1+s]\}[1-t][1-d])-1$$

Rule-58475:
>  If both (**/**), (**N**), (**S'**), (**s**), (**$**), (**v**), (**F**), (**i**), (**t**), (**d**), (**K**),
>  (**j**) and (**p**) are known, then its Utilized or Starting
>  Capital must be:
>  $$U = (\{N+K+S'j[1+s]/360+Svp/360\}-[1+/]\{S'[1+s]$$
>  $$-Sv-F-S'i[1+s]\}[1-t][1-d])/[1+/]$$

Rule-58476:
>  If both (**/**), (**U**), (**N**), (**s**), (**$**), (**v**), (**F**), (**i**), (**t**), (**d**), (**K**),
>  (**j**) and (**p**) are known, then its Sales Past must be:
>  $$S' = [[1+/](U-\{Sv+F\}[1-t][1-d])-[N+K+Svp/360]]$$
>  $$/([1+s]\{j/360-[1+/][1-t][1-d]\})$$
>  or
>  $$S' = S/[1+s]$$

Steve Asikin ISBN 13: **978-1564912361**, ISBN 10: **1564912363**

Rule-58477:

If both (**/**), (**U**), (**S'**), (**N**), (**S**), (**v**), (**F**), (**i**), (**t**), (**d**), (**K**), (**j**) and (**p**) are known, then its Sales Growth planned is:

$$s= [[1+/(U-\{Sv+F\}[1-t][1-d])-[N+K+Svp/360]]$$
$$/(S'\{j/360-[1+/][1-t][1-d]\})-1$$

or

$$s= S/S'-1$$

Rule-58478:

If both (**/**), (**U**), (**S'**), (**s**), (**N**), (**v**), (**F**), (**I**), (**t**), (**d**), (**K**), (**j**) and (**p**) are known, then its Sales or Revenue planned is:

$$S= [[1+/(U+\{S'[1+s]-F-S'i[1+s]\}[1-t][1-d])$$
$$-\{N+K+S'j[1+s]/360\}]$$
$$/\{vp/360+[1+/][1-t][1-d]\}$$

or

$$S= S'[1+s]$$

Rule-58479:

If both (**/**), (**U**), (**S'**), (**s**), (**S**), (**N**), (**F**), (**i**), (**t**), (**d**), (**K**), (**j**) and (**p**) are known, then its Variable Portion planned is:

$$v= [[1+/(U+\{S'[1+s]-F-S'i[1+s]\}[1-t][1-d])$$
$$-\{N+K+S'j[1+s]/360\}]$$
$$/(S\{p/360+[1+/][1-t][1-d]\})$$

Steve Asikin ISBN 13: **978-1564912361**, ISBN 10: **1564912363**

Rule-58480:

If both (**/**), (**U**), (**S'**), (**s**), (**S**), (**v**), (**N**), (**i**), (**t**), (**d**), (**K**), (**j**) and (**p**) are known, then its Fixed Cost planned is:

$$F= S'[1+s]-Sv-S'i[1+s]-(\{N+K+S'j[1+s]/360$$
$$+Svp/360\}/[1+/-U)/\{[1-t][1-d]\}$$

Rule-58481:

If both (**/**), (**U**), (**S'**), (**s**), (**S**), (**v**), (**F**), (**N**), (**t**), (**d**), (**K**), (**j**) and (**p**) are known, then its Interest Portion planned is:

$$i= [S'[1+s]-Sv-F-(\{N+K+S'j[1+s]/360+Svp/360\}$$
$$/[1+/-U)/\{[1-t][1-d]\}]/\{S'[1+s]\}$$

Rule-58482:

If both (**/**), (**U**), (**S'**), (**s**), (**S**), (**v**), (**F**), (**i**), (**N**), (**d**), (**K**), (**j**) and (**p**) are known, then its Tax Rate planned is:

$$t= 1-(\{N+K+S'j[1+s]/360+Svp/360\}/[1+/-U)$$
$$/(\{S'[1+s]-Sv-F-S'i[1+s]\}[1-d])$$

Rule-58483:

If both (**/**), (**U**), (**S'**), (**s**), (**S**), (**v**), (**F**), (**i**), (**t**), (**N**), (**K**), (**j**) and (**p**) are known, then its Dividend Payout planned is:

$$d= 1-(\{N+K+S'j[1+s]/360+Svp/360\}/[1+/-U)$$
$$/(\{S'[1+s]-Sv-F-S'i[1+s]\}[1-t])$$

Steve Asikin ISBN 13: **978-1564912361**, ISBN 10: **1564912363**

## Rule-58484:

If both (**/**), (**U**), (**S'**), (**s**), (**S**), (**v**), (**F**), (**i**), (**t**), (**d**), (**N**), (**j**) and (**p**) are known, then its Kind of Cash planned is:

$$K= [1+/](U+\{S'[1+s]-Sv-F-S'i[1+s]\}[1-t][1-d])$$
$$-\{N+S'j[1+s]/360+Svp/360\}$$

## Rule-58485:

If both (**/**), (**U**), (**S'**), (**s**), (**S**), (**v**), (**F**), (**i**), (**t**), (**d**), (**K**), (**N**) and (**p**) are known, then its Job or Trade Receivable Days planned is:

$$j= 360[[1+/](U+\{S'[1+s]-Sv-F-S'i[1+s]\}[1-t]$$
$$[1-d])-\{N+K+Svp/360\}]/\{S'[1+s]\}$$

## Rule-58486:

If both (**/**), (**U**), (**S'**), (**s**), (**S**), (**v**), (**F**), (**i**), (**t**), (**d**), (**K**), (**j**) and (**N**) are known, then its Procured Inventory planned is:

$$p= 360[[1+/](U+\{S'[1+s]-Sv-F-S'i[1+s]\}[1-t]$$
$$[1-d])-\{N+K+S'j[1+s]/360\}]/[Sv]$$

## Rule-58487:

If both (**/**), (**U**), (**S'**), (**s**), (**S**), (**v**), (**F**), (**i**), (**t**), (**d**), (**K**), (**j**) and (**p**) are known, then its Non Current Asset planned is:

$$N= [1+/](U+\{S'[1+s]-Sv-F-S'i[1+s]\}[1-t][1-d])$$
$$-K-S'j[1+s]/360-S'vp[1+s]/360$$

Steve Asikin ISBN 13: **978-1564912361**, ISBN 10: **1564912363**

Rule-58488:

If both (**N**), (**U**), (**S'**), (**s**), (**S**), (**v**), (**F**), (**i**), (**t**), (**d**), (**K**), (**j**) and (**p**) are known, then its Leverage or Gearing Ratio planned is:

$$\textit{l}= \{\textbf{N}+\textbf{K}+\textbf{S'j}[1+\textbf{s}]/360+\textbf{S'vp}[1+\textbf{s}]/360\}$$
$$/(\textbf{U}+\{\textbf{S'}[1+\textbf{s}]-\textbf{Sv}-\textbf{F}-\textbf{S'i}[1+\textbf{s}]\}[1-\textbf{t}][1-\textbf{d}])-1$$

Rule-58489:

If both (**l**), (**N**), (**S'**), (**s**), (**S**), (**v**), (**F**), (**i**), (**t**), (**d**), (**K**), (**j**) and (**p**) are known, then its Utilized or Starting Capital must be:

$$\textbf{U}= (\{\textbf{N}+\textbf{K}+\textbf{S'j}[1+\textbf{s}]/360+\textbf{S'vp}[1+\textbf{s}]/360\}-[1+\textit{l}]$$
$$\{\textbf{S'}[1+\textbf{s}]-\textbf{Sv}-\textbf{F}-\textbf{S'i}[1+\textbf{s}]\}[1-\textbf{t}][1-\textbf{d}])/[1+\textit{l}]$$

Rule-58490:

If both (**l**), (**U**), (**N**), (**s**), (**S**), (**v**), (**F**), (**i**), (**t**), (**d**), (**K**), (**j**) and (**p**) are known, then its Sales Past must be:

$$\textbf{S'}= [[1+\textit{l}](\textbf{U}+\{\textbf{S'}[1+\textbf{s}]-\textbf{Sv}-\textbf{F}-\textbf{S'i}[1+\textbf{s}]\}[1-\textbf{t}][1-\textbf{d}])$$
$$-[\textbf{N}+\textbf{K}]]/([1+\textbf{s}]\{\textbf{j}/360+\textbf{vp}/360$$
$$-[1+\textit{l}][1-\textbf{t}][1-\textbf{d}])$$

or

$$\textbf{S'}= \textbf{S}/[1+\textbf{s}]$$

Steve Asikin ISBN 13: **978-1564912361**, ISBN 10: **1564912363**

## Rule-58491:

If both $(\textbf{\textit{I}})$, $(\textbf{U})$, $(\textbf{S'})$, $(\textbf{N})$, $(\textbf{S})$, $(\textbf{v})$, $(\textbf{F})$, $(\textbf{i})$, $(\textbf{t})$, $(\textbf{d})$, $(\textbf{K})$, $(\textbf{j})$ and $(\textbf{p})$ are known, then its Sales Growth planned is:

$$s= [[1+I(\textbf{U}+\{\textbf{S'}[1+s]-\textbf{Sv}-\textbf{F}-\textbf{S'i}[1+s]\}[1-\textbf{t}][1-\textbf{d}])$$
$$-[\textbf{N}+\textbf{K}]]/(\textbf{S'}\{\textbf{j}/360+\textbf{vp}/360$$
$$-[1+I][1-\textbf{t}][1-\textbf{d}])-1$$

or

$$s= \textbf{S}/\textbf{S'}-1$$

## Rule-58492:

If both $(\textbf{\textit{I}})$, $(\textbf{U})$, $(\textbf{S'})$, $(\textbf{s})$, $(\textbf{N})$, $(\textbf{v})$, $(\textbf{F})$, $(\textbf{i})$, $(\textbf{t})$, $(\textbf{d})$, $(\textbf{K})$, $(\textbf{j})$ and $(\textbf{p})$ are known, then its Sales or Revenue planned is:

$$S= [\textbf{S'}[1+s]-\textbf{F}-\textbf{S'i}[1+s]-(\{\textbf{N}+\textbf{K}+\textbf{S'j}[1+s]/360$$
$$+\textbf{S'vp}[1+s]/360\}/[1+I]-\textbf{U})$$
$$/\{[1-\textbf{t}][1-\textbf{d}]\}]/\textbf{v}$$

or

$$S= \textbf{S'}[1+s]$$

## Rule-58493:

If both $(\textbf{\textit{I}})$, $(\textbf{U})$, $(\textbf{S'})$, $(\textbf{s})$, $(\textbf{S})$, $(\textbf{N})$, $(\textbf{F})$, $(\textbf{i})$, $(\textbf{t})$, $(\textbf{d})$, $(\textbf{K})$, $(\textbf{j})$ and $(\textbf{p})$ are known, then its Variable Portion planned is:

$$v= [[1+I(\textbf{U}+\{\textbf{S'}[1+s]-\textbf{F}-\textbf{S'i}[1+s]\}[1-\textbf{t}][1-\textbf{d}])$$
$$-\{\textbf{N}+\textbf{K}+\textbf{S'j}[1+s]/360\}]/\{\textbf{S'p}[1+s]/360$$
$$+[1+I][1-\textbf{t}][1-\textbf{d}]\}$$

`

Steve Asikin ISBN 13: **978-1564912361**, ISBN 10: **1564912363**

Rule-58494:

If both $(\textbf{\textit{I}})$, $(\textbf{U})$, $(\textbf{S'})$, $(\textbf{s})$, $(\textbf{S})$, $(\textbf{v})$, $(\textbf{N})$, $(\textbf{i})$, $(\textbf{t})$, $(\textbf{d})$, $(\textbf{K})$, $(\textbf{j})$ and $(\textbf{p})$ are known, then its Fixed Cost planned is:

$$\textbf{F} = \textbf{S'}[1+\textbf{s}] - \textbf{Sv} - \textbf{S'i}[1+\textbf{s}] - (\{\textbf{N}+\textbf{K}+\textbf{S'j}[1+\textbf{s}]/360$$
$$+\textbf{S'vp}[1+\textbf{s}]/360\}/[1+\textbf{\textit{I}}-\textbf{U})/\{[1-\textbf{t}][1-\textbf{d}]\}$$

Rule-58495:

If both $(\textbf{\textit{I}})$, $(\textbf{U})$, $(\textbf{S'})$, $(\textbf{s})$, $(\textbf{S})$, $(\textbf{v})$, $(\textbf{F})$, $(\textbf{N})$, $(\textbf{t})$, $(\textbf{d})$, $(\textbf{K})$, $(\textbf{j})$ and $(\textbf{p})$ are known, then its Interest Portion planned is:

$$\textbf{i} = [\textbf{S'}[1+\textbf{s}] - \textbf{Sv} - \textbf{F} - (\{\textbf{N}+\textbf{K}+\textbf{S'j}[1+\textbf{s}]/360$$
$$+\textbf{S'vp}[1+\textbf{s}]/360\}/[1+\textbf{\textit{I}}-\textbf{U})$$
$$/\{[1-\textbf{t}][1-\textbf{d}]\}]/\{\textbf{S'}[1+\textbf{s}]\}$$

Rule-58496:

If both $(\textbf{\textit{I}})$, $(\textbf{U})$, $(\textbf{S'})$, $(\textbf{s})$, $(\textbf{S})$, $(\textbf{v})$, $(\textbf{F})$, $(\textbf{i})$, $(\textbf{N})$, $(\textbf{d})$, $(\textbf{K})$, $(\textbf{j})$ and $(\textbf{p})$ are known, then its Tax Rate planned is:

$$\textbf{t} = 1 - (\{\textbf{N}+\textbf{K}+\textbf{S'j}[1+\textbf{s}]/360 + \textbf{S'vp}[1+\textbf{s}]/360\}$$
$$/[1+\textbf{\textit{I}}-\textbf{U})/(\{\textbf{S'}[1+\textbf{s}] - \textbf{Sv} - \textbf{F} - \textbf{S'i}[1+\textbf{s}]\}[1-\textbf{d}])$$

Rule-58497:

If both $(\textbf{\textit{I}})$, $(\textbf{U})$, $(\textbf{S'})$, $(\textbf{s})$, $(\textbf{S})$, $(\textbf{v})$, $(\textbf{F})$, $(\textbf{i})$, $(\textbf{t})$, $(\textbf{N})$, $(\textbf{K})$, $(\textbf{j})$ and $(\textbf{p})$ are known, then its Dividend Payout planned is:

$$\textbf{d} = 1 - (\{\textbf{N}+\textbf{K}+\textbf{S'j}[1+\textbf{s}]/360 + \textbf{S'vp}[1+\textbf{s}]/360\}$$
$$/[1+\textbf{\textit{I}}-\textbf{U})/(\{\textbf{S'}[1+\textbf{s}] - \textbf{Sv} - \textbf{F} - \textbf{S'i}[1+\textbf{s}]\}[1-\textbf{t}])$$

Steve Asikin ISBN 13: **978-1564912361**, ISBN 10: **1564912363**

<u>Rule-58498</u>:

If both (**/**), (**U**), (**$'**), (**s**), (**$**), (**v**), (**F**), (**i**), (**t**), (**d**), (**N**), (**j**) and (**p**) are known, then its Kind of Cash planned is:

$$K= [1+/(U+\{\$'[1+s]-\$v-F-\$'i[1+s]\}[1-t][1-d])$$
$$-\{N+\$'j[1+s]/360+\$'vp[1+s]/360\}$$

<u>Rule-58499</u>:

If both (**/**), (**U**), (**$'**), (**s**), (**$**), (**v**), (**F**), (**i**), (**t**), (**d**), (**K**), (**N**) and (**p**) are known, then its Job or Trade Receivable Days planned is:

$$j= 360[[1+/(U+\{\$'[1+s]-\$v-F-\$'i[1+s]\}[1-t]$$
$$[1-d])-\{N+K+\$'vp[1+s]/360\}]/\{\$'[1+s]\}$$

<u>Rule-58500</u>:

If both (**/**), (**U**), (**$'**), (**s**), (**$**), (**v**), (**F**), (**i**), (**t**), (**d**), (**K**), (**j**) and (**N**) are known, then its Procured Inventory Days planned is:

$$p= 360[[1+/(U+\{\$'[1+s]-\$v-F-\$'i[1+s]\}[1-t]$$
$$[1-d])-\{N+K+\$'j[1+s]/360\}]/\{\$'v[1+s]\}$$

<u>Rule-58501</u>:

If both (**/**), (**U**), (**$'**), (**s**), (**$**), (**v**), (**f**), (**I**), (**T**), (**D**), (**K**), (**J**) and (**P**) are known, then its Non Current Asset planned is:

$$N= [1+/\{U+\$'[1+s]-\$v-\$f-I-T-D\}-K-J-P$$

Steve Asikin ISBN 13: **978-1564912361**, ISBN 10: **1564912363**

Rule-58502:

If both **(N)**, **(U)**, **(S')**, **(s)**, **(S)**, **(v)**, **(f)**, **(I)**, **(T)**, **(D)**, **(K)**, **(J)** and **(P)** are known, then its Leverage or Gearing Ratio planned is:

$$I = \{N+K+J+P\}/(U+\{S'[1+s]-Sv-Sf-I\}-T-D)-1$$

Rule-58503:

If both **(I)**, **(N)**, **(S')**, **(s)**, **(S)**, **(v)**, **(f)**, **(I)**, **(T)**, **(D)**, **(K)**, **(J)** and **(P)** are known, then its Utilized or Starting Capital must be:

$$U = (\{N+K+J+P\}-[1+I]\{S'[1+s]-Sv-Sf-I\}-T-D)$$
$$/[1+I]$$

Rule-58504:

If both **(I)**, **(U)**, **(N)**, **(s)**, **(S)**, **(v)**, **(f)**, **(I)**, **(T)**, **(D)**, **(K)**, **(J)** and **(P)** are known, then its Sales Past must be:

$$S' = [N+K+J+P]/[1+I]-U+T+D+Sv+Sf+I)/[1+s]$$
$$\text{or}$$
$$S' = S/[1+s]$$

Rule-58505:

If both **(I)**, **(U)**, **(S')**, **(N)**, **(S)**, **(v)**, **(f)**, **(I)**, **(T)**, **(D)**, **(K)**, **(J)** and **(P)** are known, then its Sales Growth planned is:

$$s = [N+K+J+P]/[1+I]-U+T+D+Sv+Sf+I)/S'-1$$
$$\text{or}$$
$$s = S/S'-1$$

Steve Asikin ISBN 13: **978-1564912361**, ISBN 10: **1564912363**

Rule-58506:

If both (**/**), (**U**), (**S'**), (**s**), (**N**), (**v**), (**f**), (**I**), (**T**), (**D**), (**K**), (**J**) and (**P**) are known, then its Sales or Revenue planned is:

$$S= (U+\{S'[1+s]-I\}-T-D-[N+K+J+P]/[1+/])/[v+f]$$
or
$$S= S'[1+s]$$

Rule-58507:

If both (**/**), (**U**), (**S'**), (**s**), (**S**), (**N**), (**f**), (**I**), (**T**), (**D**), (**K**), (**J**) and (**P**) are known, then its Variable Portion planned is:

$$v= (U+\{S'[1+s]-Sf-I\}-T-D-[N+K+J+P]/[1+/])/S$$

Rule-58508:

If both (**/**), (**U**), (**S'**), (**s**), (**S**), (**v**), (**N**), (**I**), (**T**), (**D**), (**K**), (**J**) and (**P**) are known, then its Fixed Portion planned is:

$$f= (U+\{S'[1+s]-Sv-I\}-T-D-[N+K+J+P]/[1+/])/S$$

Rule-58509:

If both (**/**), (**U**), (**S'**), (**s**), (**S**), (**v**), (**f**), (**N**), (**T**), (**D**), (**K**), (**J**) and (**P**) are known, then its Interest Expense planned is:

$$I= U+\{S'[1+s]-Sv-Sf\}-T-D-[N+K+J+P]/[1+/]$$

Steve Asikin ISBN 13: **978-1564912361**, ISBN 10: **1564912363**

Rule-58510:

If both (**∫**), (**U**), (**S'**), (**s**), (**S**), (**v**), (**f**), (**I**), (**N**), (**D**), (**K**), (**J**) and (**P**) are known, then its Tax planned is:

$$T= U+\{S'[1+s]-Sv-Sf-I\}-D-[N+K+J+P]/[1+\int]$$

Rule-58511:

If both (**∫**), (**U**), (**S'**), (**s**), (**S**), (**v**), (**f**), (**I**), (**T**), (**N**), (**K**), (**J**) and (**P**) are known, then its Dividend planned is:

$$D= U+\{S'[1+s]-Sv-Sf-I\}-T-[N+K+J+P]/[1+\int]$$

Rule-58512:

If both (**∫**), (**U**), (**S'**), (**s**), (**S**), (**v**), (**f**), (**I**), (**T**), (**D**), (**N**), (**J**) and (**P**) are known, then its Kind of Cash planned is:

$$K= [1+\int(U+\{S'[1+s]-Sv-Sf-I\}-T-D)-[N+J+P]$$

Rule-58513:

If both (**∫**), (**U**), (**S'**), (**s**), (**S**), (**v**), (**f**), (**I**), (**T**), (**D**), (**K**), (**N**) and (**P**) are known, then its Job or Trade Account Receivable planned is:

$$J= [1+\int(U+\{S'[1+s]-Sv-Sf-I\}-T-D)-[N+K+P]$$

Rule-58514:

If both (**∫**), (**U**), (**S'**), (**s**), (**S**), (**v**), (**f**), (**I**), (**T**), (**D**), (**K**), (**J**) and (**N**) are known, then its Procured Inventory planned is:

$$P= [1+\int(U+\{S'[1+s]-Sv-Sf-I\}-T-D)-[N+K+J]$$

Steve Asikin ISBN 13: **978-1564912361**, ISBN 10: **1564912363**

Rule-58501:

If both (**∫**), (**U**), (**S'**), (**s**), (**S**), (**v**), (**f**), (**I**), (**T**), (**D**), (**K**), (**J**), (**V**) and (**p**) are known, then its Non Current Asset planned is:

$$\textbf{N} = [1+\textbf{∫}]\{\textbf{U}+\textbf{S'}[1+\textbf{s}]-\textbf{Sv}-\textbf{Sf}-\textbf{I}-\textbf{T}-\textbf{D}\}-\textbf{K}-\textbf{J}-\textbf{Vp}/360$$

Rule-58502:

If both (**N**), (**U**), (**S'**), (**s**), (**S**), (**v**), (**f**), (**I**), (**T**), (**D**), (**K**), (**J**), (**V**) and (**p**) are known, then its Leverage or Gearing Ratio planned is:

$$\textbf{∫} = [\textbf{N}+\textbf{K}+\textbf{J}+\textbf{Vp}/360]/(\textbf{U}+\{\textbf{S'}[1+\textbf{s}]-\textbf{Sv}-\textbf{Sf}-\textbf{I}\}-\textbf{T}-\textbf{D})-1$$

Rule-58503:

If both (**∫**), (**N**), (**S'**), (**s**), (**S**), (**v**), (**f**), (**I**), (**T**), (**D**), (**K**), (**J**), (**V**) and (**p**) are known, then its Utilized or Starting Capital must be:

$$\textbf{U} = ([\textbf{N}+\textbf{K}+\textbf{J}+\textbf{Vp}/360]-[1+\textbf{∫}]\{\textbf{S'}[1+\textbf{s}]-\textbf{Sv}-\textbf{Sf}-\textbf{I}\}$$
$$-\textbf{T}-\textbf{D})/[1+\textbf{∫}]$$

Rule-58504:

If both (**∫**), (**U**), (**N**), (**s**), (**S**), (**v**), (**f**), (**I**), (**T**), (**D**), (**K**), (**J**), (**V**) and (**p**) are known, then its Sales Past must be:

$$\textbf{S'} = [\textbf{N}+\textbf{K}+\textbf{J}+\textbf{Vp}/360]/[1+\textbf{∫}-\textbf{U}+\textbf{Sv}+\textbf{Sf}+\textbf{I}+\textbf{T}+\textbf{D})$$
$$/[1+\textbf{s}]$$
$$\text{or}$$
$$\textbf{S'} = \textbf{S}/[1+\textbf{s}]$$

Steve Asikin ISBN 13: **978-1564912361**, ISBN 10: **1564912363**

Rule-58505:

If both (**/**), (**U**), (**S'**), (**N**), (**S**), (**v**), (**f**), (**I**), (**T**), (**D**), (**K**), (**J**), (**V**) and (**p**) are known, then its Sales Growth planned is:

$$s= [N+K+J+Vp/360]/[1+/-U+Sv+Sf+I+T+D)/S'-1$$

or

$$s= S/S'-1$$

Rule-58506:

If both (**/**), (**U**), (**S'**), (**s**), (**N**), (**v**), (**f**), (**I**), (**T**), (**D**), (**K**), (**J**), (**V**) and (**p**) are known, then its Sales or Revenue planned is:

$$S= (U+\{S'[1+s]-I\}-T-D-[N+K+J+Vp/360]$$
$$/[1+/])/[v+f]$$

or

$$S= S'[1+s]$$

or

$$S= V/v$$

Rule-58507:

If both (**/**), (**U**), (**S'**), (**s**), (**S**), (**N**), (**f**), (**I**), (**T**), (**D**), (**K**), (**J**), (**V**) and (**p**) are known, then its Variable Portion planned is:

$$v= (U+\{S'[1+s]-Sf-I\}-T-D-[N+K+J+Vp/360]$$
$$/[1+/])/S$$

or

$$v= V/S$$

Steve Asikin ISBN 13: **978-1564912361**, ISBN 10: **1564912363**

## Rule-58508:

If both (**/**), (**U**), (**$'**), (**s**), (**$**), (**v**), (**N**), (**I**), (**T**), (**D**), (**K**), (**J**), (**V**) and (**p**) are known, then its Fixed Portion planned is:

$$f = (U + \{\$'[1+s] - \$v - I\} - T - D - [N+K+J+Vp/360]\ /[1+/])/\$$$

## Rule-58509:

If both (**/**), (**U**), (**$'**), (**s**), (**$**), (**v**), (**f**), (**N**), (**T**), (**D**), (**K**), (**J**), (**V**) and (**p**) are known, then its Interest Expense planned is:

$$I = U + \{\$'[1+s] - \$v - \$f\} - T - D - [N+K+J+Vp/360]\ /[1+/]$$

## Rule-58510:

If both (**/**), (**U**), (**$'**), (**s**), (**$**), (**v**), (**f**), (**I**), (**N**), (**D**), (**K**), (**J**), (**V**) and (**p**) are known, then its Tax planned is:

$$T = U + \{\$'[1+s] - \$v - \$f - I\} - D - [N+K+J+Vp/360]\ /[1+/]$$

## Rule-58511:

If both (**/**), (**U**), (**$'**), (**s**), (**$**), (**v**), (**f**), (**I**), (**T**), (**N**), (**K**), (**J**), (**V**) and (**p**) are known, then its Dividend planned is:

$$D = U + \{\$'[1+s] - \$v - \$f - I\} - T - [N+K+J+Vp/360]\ /[1+/]$$

Steve Asikin ISBN 13: **978-1564912361**, ISBN 10: **1564912363**

Rule-58512:

If both $(\textit{I})$, $(\textbf{U})$, $(\textbf{S'})$, $(\textbf{s})$, $(\textbf{\$})$, $(\textbf{v})$, $(\textbf{f})$, $(\textbf{I})$, $(\textbf{T})$, $(\textbf{D})$, $(\textbf{N})$, $(\textbf{J})$, $(\textbf{V})$ and $(\textbf{p})$ are known, then its Kind of Cash planned is:

$$\textbf{K} = [1 + \textit{I}(\textbf{U} + \{\textbf{S'}[1+\textbf{s}] - \textbf{Sv} - \textbf{Sf} - \textbf{I}\} - \textbf{T} - \textbf{D})$$
$$- [\textbf{N} + \textbf{J} + \textbf{Vp}/360]$$

Rule-58513:

If both $(\textit{I})$, $(\textbf{U})$, $(\textbf{S'})$, $(\textbf{s})$, $(\textbf{\$})$, $(\textbf{v})$, $(\textbf{f})$, $(\textbf{I})$, $(\textbf{T})$, $(\textbf{D})$, $(\textbf{K})$, $(\textbf{N})$, $(\textbf{V})$ and $(\textbf{p})$ are known, then its Job or Trade Account Receivable planned is:

$$\textbf{J} = [1 + \textit{I}(\textbf{U} + \{\textbf{S'}[1+\textbf{s}] - \textbf{Sv} - \textbf{Sf} - \textbf{I}\} - \textbf{T} - \textbf{D})$$
$$- \{\textbf{N} + \textbf{K} + \textbf{Vp}/360\}$$

Rule-58514:

If both $(\textit{I})$, $(\textbf{U})$, $(\textbf{S'})$, $(\textbf{s})$, $(\textbf{\$})$, $(\textbf{v})$, $(\textbf{f})$, $(\textbf{I})$, $(\textbf{T})$, $(\textbf{D})$, $(\textbf{K})$, $(\textbf{J})$, $(\textbf{N})$ and $(\textbf{p})$ are known, then its Variable Cost planned is:

$$\textbf{V} = 360[[1 + \textit{I}(\textbf{U} + \{\textbf{S'}[1+\textbf{s}] - \textbf{Sv} - \textbf{Sf} - \textbf{I}\} - \textbf{T} - \textbf{D})$$
$$- [\textbf{N} + \textbf{K} + \textbf{J}]]/\textbf{p}$$
$$\text{or}$$
$$\textbf{V} = \textbf{Sv}$$

Rule-58515:

If both $(\textit{I})$, $(\textbf{U})$, $(\textbf{S'})$, $(\textbf{s})$, $(\textbf{\$})$, $(\textbf{v})$, $(\textbf{f})$, $(\textbf{I})$, $(\textbf{T})$, $(\textbf{D})$, $(\textbf{K})$, $(\textbf{J})$, $(\textbf{V})$ and $(\textbf{N})$ are known, then its Procured Inventory Days planned is:

$$\textbf{p} = 360[[1 + \textit{I}(\textbf{U} + \{\textbf{S'}[1+\textbf{s}] - \textbf{Sv} - \textbf{Sf} - \textbf{I}\} - \textbf{T} - \textbf{D})$$
$$- [\textbf{N} + \textbf{K} + \textbf{J}]]/\textbf{V}$$

Steve Asikin ISBN 13: **978-1564912361**, ISBN 10: **1564912363**

Rule-58516:

If both (**/**), (**U**), (**S'**), (**s**), (**S**), (**v**), (**f**), (**I**), (**T**), (**D**), (**K**), (**J**), (**V**) and (**p**) are known, then its Non Current Asset planned is:

$$\mathbf{N} = [1+\mathbf{/}]\{\mathbf{U}+\mathbf{S'}[1+\mathbf{s}]-\mathbf{Sv}-\mathbf{Sf}-\mathbf{I}-\mathbf{T}-\mathbf{D}\}-\mathbf{K}-\mathbf{J}-\mathbf{Sup}/360$$

Rule-58517:

If both (**N**), (**U**), (**S'**), (**s**), (**S**), (**v**), (**f**), (**I**), (**T**), (**D**), (**K**), (**J**), (**V**) and (**p**) are known, then its Leverage or Gearing Ratio planned is:

$$\mathbf{/} = [\mathbf{N}+\mathbf{K}+\mathbf{J}+\mathbf{Sup}/360]$$
$$/(\mathbf{U}+\{\mathbf{S'}[1+\mathbf{s}]-\mathbf{Sv}-\mathbf{Sf}-\mathbf{I}\}-\mathbf{T}-\mathbf{D})-1$$

Rule-58518:

If both (**/**), (**N**), (**S'**), (**s**), (**S**), (**v**), (**f**), (**I**), (**T**), (**D**), (**K**), (**J**), (**V**) and (**p**) are known, then its Utilized or Starting Capital must be:

$$\mathbf{U} = ([\mathbf{N}+\mathbf{K}+\mathbf{J}+\mathbf{Sup}/360]-[1+\mathbf{/}]\{\mathbf{S'}[1+\mathbf{s}]-\mathbf{Sv}-\mathbf{Sf}-\mathbf{I}\}$$
$$-\mathbf{T}-\mathbf{D})/[1+\mathbf{/}]$$

Rule-58519:

If both (**/**), (**U**), (**N**), (**s**), (**S**), (**v**), (**f**), (**I**), (**T**), (**D**), (**K**), (**J**), (**V**) and (**p**) are known, then its Sales Past must be:

$$\mathbf{S'} = ([\mathbf{N}+\mathbf{K}+\mathbf{J}+\mathbf{Sup}/360]/[1+\mathbf{/}]-\mathbf{U}+\mathbf{Sv}+\mathbf{Sf}+\mathbf{I}+\mathbf{T}+\mathbf{D})$$
$$/[1+\mathbf{s}]$$
$$\text{or}$$
$$\mathbf{S'} = \mathbf{S}/[1+\mathbf{s}]$$

`

Steve Asikin ISBN 13: **978-1564912361**, ISBN 10: **1564912363**

Rule-58520:

If both (**/**), (**U**), (**S'**), (**N**), (**S**), (**v**), (**f**), (**I**), (**T**), (**D**), (**K**), (**J**), (**V**) and (**p**) are known, then its Sales Growth planned is:

$$s = ([N+K+J+Svp/360]/[1+/-U +Sv+Sf+I+T+D)/S'-1$$

or

$$s = S/S'-1$$

Rule-58521:

If both (**/**), (**U**), (**S'**), (**s**), (**N**), (**v**), (**f**), (**I**), (**T**), (**D**), (**K**), (**J**), (**V**) and (**p**) are known, then its Sales or Revenue planned is:

$$S = [[1+/(U+\{S'[1+s]-I\}-T-D) -[N+K+J]]/\{vp/360+[1+/[v+f]\}$$

or

$$S = S'[1+s]$$

Rule-58522:

If both (**/**), (**U**), (**S'**), (**s**), (**S**), (**N**), (**f**), (**I**), (**T**), (**D**), (**K**), (**J**), (**V**) and (**p**) are known, then its Variable Portion planned is:

$$v = [[1+/(U+\{S'[1+s]-Sf-I\}-T-D)-[N+K+J]] /(S\{p/360+[1+/]\})$$

Steve Asikin ISBN 13: **978-1564912361**, ISBN 10: **1564912363**

## Rule-58523:

If both (**/**), (**U**), (**S'**), (**s**), (**S**), (**v**), (**N**), (**I**), (**T**), (**D**), (**K**), (**J**), (**V**) and (**p**) are known, then its Fixed Portion planned is:

$$f = (U + \{S'[1+s] - Sv - I\} - T - D$$
$$- [N + K + J + Svp/360]/[1 + I])/S$$

## Rule-58524:

If both (**/**), (**U**), (**S'**), (**s**), (**S**), (**v**), (**f**), (**N**), (**T**), (**D**), (**K**), (**J**), (**V**) and (**p**) are known, then its Interest Expense planned is:

$$I = U + \{S'[1+s] - Sv - Sf\} - T - D$$
$$- [N + K + J + Svp/360]/[1 + I]$$

## Rule-58525:

If both (**/**), (**U**), (**S'**), (**s**), (**S**), (**v**), (**f**), (**I**), (**N**), (**D**), (**K**), (**J**), (**V**) and (**p**) are known, then its Tax Rate planned is:

$$T = U + \{S'[1+s] - Sv - Sf - I\} - D$$
$$- [N + K + J + Svp/360]/[1 + I]$$

## Rule-58526:

If both (**/**), (**U**), (**S'**), (**s**), (**S**), (**v**), (**f**), (**I**), (**T**), (**N**), (**K**), (**J**), (**V**) and (**p**) are known, then its Dividend planned is:

$$D = U + \{S'[1+s] - Sv - Sf - I\} - T$$
$$- [N + K + J + Svp/360]/[1 + I]$$

`

Steve Asikin ISBN 13: **978-1564912361**, ISBN 10: **1564912363**

Rule-58527:
   If both (**/**), (**U**), (**S'**), (**s**), (**S**), (**v**), (**f**), (**I**), (**T**), (**D**), (**N**), (**J**), (**V**) and (**p**) are known, then its Kind of Cash planned is:
   $$K= [1+/](U+\{S'[1+s]-Sv-Sf-I\}-T-D)$$
   $$-[N+J+Svp/360]$$


Rule-58528:
   If both (**/**), (**U**), (**S'**), (**s**), (**S**), (**v**), (**f**), (**I**), (**T**), (**D**), (**K**), (**N**), (**V**) and (**p**) are known, then its Job or Trade Account Receivable planned is:
   $$J= [1+/](U+\{S'[1+s]-Sv-Sf-I\}-T-D)$$
   $$-\{N+K+Svp/360\}$$


Rule-58529:
   If both (**/**), (**U**), (**S'**), (**s**), (**S**), (**v**), (**f**), (**I**), (**T**), (**D**), (**K**), (**J**), (**V**) and (**N**) are known, then its Procured Inventory Days planned is:
   $$p= 360[[1+/](U+\{S'[1+s]-Sv-Sf-I\}-T-D)$$
   $$-[N+K+J]]/[Sv]$$


Rule-58530:
   If both (**/**), (**U**), (**S'**), (**s**), (**S**), (**v**), (**f**), (**I**), (**T**), (**D**), (**K**), (**J**) and (**p**) are known, then its Non Current Asset planned is:
   $$N= [1+/]\{U+S'[1+s]-Sv-Sf-I-T-D\}-K-J$$
   $$-S'vp[1+s]/360$$

Steve Asikin ISBN 13: **978-1564912361**, ISBN 10: **1564912363**

## Rule-58531:

If both $(\mathbf{N})$, $(\mathbf{U})$, $(\mathbf{S'})$, $(\mathbf{s})$, $(\mathbf{\$})$, $(\mathbf{v})$, $(\mathbf{f})$, $(\mathbf{I})$, $(\mathbf{T})$, $(\mathbf{D})$, $(\mathbf{K})$, $(\mathbf{J})$ and $(\mathbf{p})$ are known, then its Leverage or Gearing Ratio planned is:

$$I = \{\mathbf{N+K+J+S'vp}[1+s]/360\}$$
$$/(\mathbf{U}+\{\mathbf{S'}[1+s]-\mathbf{Sv-Sf-I}\}-\mathbf{T-D})-1$$

## Rule-58532:

If both $(\mathbf{I})$, $(\mathbf{N})$, $(\mathbf{S'})$, $(\mathbf{s})$, $(\mathbf{\$})$, $(\mathbf{v})$, $(\mathbf{f})$, $(\mathbf{I})$, $(\mathbf{T})$, $(\mathbf{D})$, $(\mathbf{K})$, $(\mathbf{J})$ and $(\mathbf{p})$ are known, then its Utilized or Starting Capital must be:

$$\mathbf{U} = (\{\mathbf{N+K+J+S'vp}[1+s]/360\}-[1+I]\{\mathbf{S'}[1+s]$$
$$-\mathbf{Sv-Sf-I}\}-\mathbf{T-D})/[1+I]$$

## Rule-58533:

If both $(\mathbf{I})$, $(\mathbf{U})$, $(\mathbf{N})$, $(\mathbf{s})$, $(\mathbf{\$})$, $(\mathbf{v})$, $(\mathbf{f})$, $(\mathbf{I})$, $(\mathbf{T})$, $(\mathbf{D})$, $(\mathbf{K})$, $(\mathbf{J})$ and $(\mathbf{p})$ are known, then its Sales Past must be:

$$\mathbf{S'} = [[1+I](\mathbf{U}-\{\mathbf{Sv+Sf+I}\}+\mathbf{T+D})-[\mathbf{N+K+J}]]$$
$$/([1+s]\{\mathbf{vp}/360-[1+I]\})$$

or

$$\mathbf{S'} = \mathbf{\$}/[1+s]$$

Steve Asikin ISBN 13: **978-1564912361**, ISBN 10: **1564912363**

Rule-58534:

If both (**/**), (**U**), (**S'**), (**N**), (**S**), (**v**), (**f**), (**I**), (**T**), (**D**), (**K**), (**J**) and (**p**) are known, then its Sales Growth planned is:

$$s= [[1+\text{/}](U-\{Sv+Sf+I\}+T+D)-[N+K+J]]$$
$$/(S'\{vp/360-[1+\text{/}]\})-1$$

or

$$s= S/S'-1$$


Rule-58535:

If both (**/**), (**U**), (**S'**), (**s**), (**N**), (**v**), (**f**), (**I**), (**T**), (**D**), (**K**), (**J**) and (**p**) are known, then its Sales or Revenue planned is:

$$S= (U+\{S'[1+s]-I\}-T-D-\{N+K+J+S'vp[1+s]/360\}$$
$$/[1+\text{/}])/[v+f]$$

or

$$S= S'[1+s]$$


Rule-58536:

If both (**/**), (**U**), (**S'**), (**s**), (**S**), (**N**), (**f**), (**I**), (**T**), (**D**), (**K**), (**J**) and (**p**) are known, then its Variable Portion planned is:

$$v= [[1+\text{/}](U+\{S'[1+s]-Sf-I\}-T-D)-[N+K+J]]$$
$$/\{S'p[1+s]/360\}+S[1+\text{/}]\}$$

Steve Asikin ISBN 13: **978-1564912361**, ISBN 10: **1564912363**

Rule-58537:

If both $(I)$, $(U)$, $(S')$, $(s)$, $(S)$, $(v)$, $(N)$, $(I)$, $(T)$, $(D)$, $(K)$, $(J)$ and $(p)$ are known, then its Fixed Portion planned is:

$$f = (U + \{S'[1+s] - Sv - I\} - T - D - \{N + K + J + S'vp[1+s]/360\}/[1+I])/S$$

Rule-58538:

If both $(I)$, $(U)$, $(S')$, $(s)$, $(S)$, $(v)$, $(f)$, $(N)$, $(T)$, $(D)$, $(K)$, $(J)$ and $(p)$ are known, then its Interest Portion planned is:

$$I = U + \{S'[1+s] - Sv - Sf\} - T - D - \{N + K + J + S'vp[1+s]/360\}/[1+I]$$

Rule-58539:

If both $(I)$, $(U)$, $(S')$, $(s)$, $(S)$, $(v)$, $(f)$, $(I)$, $(N)$, $(D)$, $(K)$, $(J)$ and $(p)$ are known, then its Tax planned is:

$$T = U + \{S'[1+s] - Sv - Sf - I\} - D - \{N + K + J + S'vp[1+s]/360\}/[1+I]$$

Rule-58540:

If both $(I)$, $(U)$, $(S')$, $(s)$, $(S)$, $(v)$, $(f)$, $(I)$, $(T)$, $(N)$, $(K)$, $(J)$ and $(p)$ are known, then its Dividend planned is:

$$D = U + \{S'[1+s] - Sv - Sf - I\} - T - \{N + K + J + S'vp[1+s]/360\}/[1+I]$$

Steve Asikin ISBN 13: **978-1564912361**, ISBN 10: **1564912363**

Rule-58541:

If both (**/**), (**U**), (**S'**), (**s**), (**S**), (**v**), (**f**), (**I**), (**T**), (**D**), (**N**), (**J**) and (**p**) are known, then its Kind of Cash planned is:

$$K= [1+/](U+\{S'[1+s]-Sv-Sf-I\}-T-D)$$
$$-\{N+J+S'vp[1+s]/360\}$$

Rule-58542:

If both (**/**), (**U**), (**S'**), (**s**), (**S**), (**v**), (**f**), (**I**), (**T**), (**D**), (**K**), (**N**) and (**p**) are known, then its Job ot Trade Account Receivable planned is:

$$J= [[1+/](U+\{S'[1+s]-Sv-Sf-I\}-T-D)$$
$$-\{N+K+S'vp[1+s]/360\}$$

Rule-58543:

If both (**/**), (**U**), (**S'**), (**s**), (**S**), (**v**), (**f**), (**I**), (**T**), (**D**), (**K**), (**J**) and (**N**) are known, then its Procured Inventory Days planned is:

$$p= 360[[1+/](U+\{S'[1+s]-Sv-Sf-I\}-T-D)$$
$$-[N+K+J]]/\{S'v[1+s]\}$$

Rule-58544:

If both (**/**), (**U**), (**S'**), (**s**), (**S**), (**v**), (**f**), (**I**), (**T**), (**D**), (**K**), (**j**) and (**P**) are known, then its Non Current Asset planned is:

$$N= [1+/]\{U+S'[1+s]-Sv-Sf-I-T-D\}-K-Sj/360-P$$

Steve Asikin ISBN 13: **978-1564912361**, ISBN 10: **1564912363**

Rule-58545:

If both (**N**), (**U**), (**S'**), (**s**), (**S**), (**v**), (**f**), (**I**), (**T**), (**D**), (**K**), (**j**) and (**P**) are known, then its Leverage or Gearing Ratio planned is:

$$I= \{N+K+Sj/360+P\}$$
$$/(U+\{S'[1+s]-Sv-Sf-I\}-T-D)-1$$

Rule-58546:

If both (**I**), (**N**), (**S'**), (**s**), (**S**), (**v**), (**f**), (**I**), (**T**), (**D**), (**K**), (**j**) and (**P**) are known, then its Utilized or Starting Capital must be:

$$U= (\{N+K+Sj/360+P\}-[1+I]\{S'[1+s]-Sv-Sf-I\}$$
$$-T-D)/[1+I]$$

Rule-58547:

If both (**I**), (**U**), (**N**), (**s**), (**S**), (**v**), (**f**), (**I**), (**T**), (**D**), (**K**), (**j**) and (**P**) are known, then its Sales Past must be:

$$S'= ([N+K+Sj/360+P]/[1+I-U+Sv+Sf+I+T+D)$$
$$/[1+s]$$
or
$$S'= S/[1+s]$$

Rule-58548:

If both (**I**), (**U**), (**S'**), (**N**), (**S**), (**v**), (**f**), (**I**), (**T**), (**D**), (**K**), (**j**) and (**P**) are known, then its Sales Growth planned is:

$$s= ([N+K+Sj/360+P]/[1+I-U+Sv+Sf+I+T+D)/S'-1$$
or
$$s= S/S'-1$$

Steve Asikin ISBN 13: **978-1564912361**, ISBN 10: **1564912363**

Rule-58549:

If both (**/**), (**U**), (**S'**), (**s**), (**N**), (**v**), (**f**), (**I**), (**T**), (**D**), (**K**), (**j**) and (**P**) are known, then its Sales or Revenue planned is:

$$S= [[1+/](U+\{S'[1+s]-I\}-T-D)-[N+K+P]]$$
$$/\{j/360+[1+/][v+f]\}$$

or

$$S= S'[1+s]$$

Rule-58550:

If both (**/**), (**U**), (**S'**), (**s**), (**S**), (**N**), (**f**), (**I**), (**T**), (**D**), (**K**), (**j**) and (**P**) are known, then its Variable Portion planned is:

$$v= (U+\{S'[1+s]-Sf-I\}-T-D$$
$$-[N+K+Sj/360+P]/[1+/])/S$$

Rule-58551:

If both (**/**), (**U**), (**S'**), (**s**), (**S**), (**v**), (**N**), (**I**), (**T**), (**D**), (**K**), (**j**) and (**P**) are known, then its Fixed Portion planned is:

$$f= (U+\{S'[1+s]-Sv-I\}-T-D$$
$$-[N+K+Sj/360+P]/[1+/])/S$$

Rule-58552:

If both (**/**), (**U**), (**S'**), (**s**), (**S**), (**v**), (**f**), (**N**), (**T**), (**D**), (**K**), (**j**) and (**P**) are known, then its Interest Expense planned is:

$$I= U+\{S'[1+s]-Sv-Sf\}-T-D-[N+K+Sj/360+P]/[1+/]$$

Steve Asikin ISBN 13: **978-1564912361**, ISBN 10: **1564912363**

Rule-58553:

If both (**/**), (**U**), (**S'**), (**s**), (**$**), (**v**), (**f**), (**I**), (**N**), (**D**), (**K**), (**j**) and (**P**) are known, then its Tax planned is:

$$T= U+\{S'[1+s]-Sv-Sf-I\}-D-[N+K+Sj/360+P]/[1+/\!\!/$$

Rule-58554:

If both (**/**), (**U**), (**S'**), (**s**), (**$**), (**v**), (**f**), (**I**), (**T**), (**N**), (**K**), (**j**) and (**P**) are known, then its Dividend planned is:

$$D= U+\{S'[1+s]-Sv-Sf-I\}-T-[N+K+Sj/360+P]/[1+/\!\!/$$

Rule-58555:

If both (**/**), (**U**), (**S'**), (**s**), (**$**), (**v**), (**f**), (**I**), (**T**), (**D**), (**N**), (**j**) and (**P**) are known, then its Kind of Cash planned is:

$$K= [1+/\!\!/(U+\{S'[1+s]-Sv-Sf-I\}-T-D)$$
$$-[N+Sj/360+P]$$

Rule-58556:

If both (**/**), (**U**), (**S'**), (**s**), (**$**), (**v**), (**f**), (**I**), (**T**), (**D**), (**K**), (**N**) and (**P**) are known, then its Job or Trade Receivable Days planned is:

$$j= 360[[1+/\!\!/(U+\{S'[1+s]-Sv-Sf-I\}-T-D)$$
$$-\{N+K+P\}]/S$$

Steve Asikin ISBN 13: **978-1564912361**, ISBN 10: **1564912363**

Rule-58557:

If both ($I$), (**U**), (**S'**), (**s**), (**S**), (**v**), (**f**), (**I**), (**T**), (**D**), (**K**), (**j**) and (**N**) are known, then its Procured Inventory planned is:

$$P= [1+I(U+\{S'[1+s]-Sv-Sf-I\}-T-D)$$
$$-\{N+K+Sj/360\}$$

Rule-58558:

If both ($I$), (**U**), (**S'**), (**s**), (**S**), (**v**), (**f**), (**I**), (**T**), (**D**), (**K**), (**j**), (**V**) and (**p**) are known, then its Non Current Asset planned is:

$$N= [1+I\{U+S'[1+s]-Sv-Sf-I-T-D\}$$
$$-K-Sj/360-Vp/360$$

Rule-58559:

If both (**N**), (**U**), (**S'**), (**s**), (**S**), (**v**), (**f**), (**I**), (**T**), (**D**), (**K**), (**j**), (**V**) and (**p**) are known, then its Leverage or Gearing Ratio planned is:

$$I= [N+K+Sj/360+Vp/360]$$
$$/(U+\{S'[1+s]-Sv-Sf-I\}-T-D)-1$$

Rule-58560:

If both ($I$), (**N**), (**S'**), (**s**), (**S**), (**v**), (**f**), (**I**), (**T**), (**D**), (**K**), (**j**), (**V**) and (**p**) are known, then its Utilized or Starting Capital must be:

$$U= ([N+K+Sj/360+Vp/360]-[1+I\{S'[1+s]$$
$$-Sv-Sf-I\}-T-D)/[1+I]$$

Steve Asikin ISBN 13: **978-1564912361**, ISBN 10: **1564912363**

Rule-58561:

If both (**/**), (**U**), (**N**), (**s**), (**\$**), (**v**), (**f**), (**I**), (**T**), (**D**), (**K**), (**j**), (**V**) and (**p**) are known, then its Sales Past mustbe:

$$\$' = ([N+K+\$j/360+Vp/360]/[1+/]-U$$
$$+\$v+\$f+I+T+D)/[1+s]$$

or

$$\$' = \$[1+s]$$

Rule-58562:

If both (**/**), (**U**), (**\$'**), (**N**), (**\$**), (**v**), (**f**), (**I**), (**T**), (**D**), (**K**), (**j**), (**V**) and (**p**) are known, then its Sales Growth planned is:

$$s = ([N+K+\$j/360+Vp/360]/[1+/]-U$$
$$+\$v+\$f+I+T+D)/\$'-1$$

or

$$s = \$/\$'-1$$

Rule-58563:

If both (**/**), (**U**), (**\$'**), (**s**), (**N**), (**v**), (**f**), (**I**), (**T**), (**D**), (**K**), (**j**), (**V**) and (**p**) are known, then its Sales or Revenue planned is:

$$\$ = [[1+/](U+\{\$'[1+s]-I\}-T-D)-[N+K+Vp/360]]$$
$$/\{j/360+[1+/][v+f]\}$$

or

$$\$ = \$'[1+s]$$

`

Steve Asikin ISBN 13: **978-1564912361**, ISBN 10: **1564912363**

Rule-58564:

If both (**/**), (**U**), (**S'**), (**s**), (**S**), (**v**), (**f**), (**I**), (**T**), (**D**), (**K**), (**j**), (**V**) and (**p**) are known, then its Non Current Asset planned is:

$$v= (U+\{S'[1+s]-I-Sf\}-T-D-[N+K+Sj/360 \\ +Vp/360]/[1+/])/S$$

or

$$v= V/S$$

Rule-58565:

If both (**/**), (**U**), (**S'**), (**s**), (**S**), (**v**), (**N**), (**I**), (**T**), (**D**), (**K**), (**j**), (**V**) and (**p**) are known, then its Fixed Cost planned is:

$$f= (U+\{S'[1+s]-Sv-I\}-T-D-[N+K+Sj/360+Vp/360] \\ /[1+/])/S$$

Rule-58566:

If both (**/**), (**U**), (**S'**), (**s**), (**S**), (**v**), (**f**), (**N**), (**T**), (**D**), (**K**), (**j**), (**V**) and (**p**) are known, then its Interest Expense planned is:

$$I= U+\{S'[1+s]-Sv-Sf\}-T-D-[N+K+Sj/360 \\ +Vp/360]/[1+/]$$

Rule-58567:

If both (**/**), (**U**), (**S'**), (**s**), (**S**), (**v**), (**f**), (**I**), (**N**), (**D**), (**K**), (**j**), (**V**) and (**p**) are known, then its Tax planned is:

$$T= U+\{S'[1+s]-Sv-Sf-I\}-D-[N+K+Sj/360 \\ +Vp/360]/[1+/]$$

Steve Asikin ISBN 13: **978-1564912361**, ISBN 10: **1564912363**

Rule-58568:

If both (**/**), (**U**), (**S'**), (**s**), (**S**), (**v**), (**f**), (**I**), (**T**), (**N**), (**K**), (**j**), (**V**) and (**p**) are known, then its Dividend planned is:

$$D= U+\{S'[1+s]-Sv-Sf-I\}-T-[N+K+Sj/360 \\ +Vp/360]/[1+/]$$

Rule-58569:

If both (**/**), (**U**), (**S'**), (**s**), (**S**), (**v**), (**f**), (**I**), (**T**), (**D**), (**N**), (**j**), (**V**) and (**p**) are known, then its Kind of Cash planned is:

$$K= [1+/](U+\{S'[1+s]-Sv-Sf-I\}-T-D) \\ -[N+Sj/360+Vp/360]$$

Rule-58570:

If both (**/**), (**U**), (**S'**), (**s**), (**S**), (**v**), (**f**), (**I**), (**T**), (**D**), (**K**), (**N**), (**V**) and (**p**) are known, then its Job or Trade Receivable Days planned is:

$$j= 360[[1+/](U+\{S'[1+s]-Sv-Sf-I\}-T-D) \\ -[N+K+Vp/360]]/S$$

Rule-58571:

If both (**/**), (**U**), (**S'**), (**s**), (**S**), (**v**), (**f**), (**I**), (**T**), (**D**), (**K**), (**j**), (**N**) and (**p**) are known, then its Variable Cost planned is:

$$V= 360[[1+/](U+\{S'[1+s]-Sv-Sf-I\}-T-D) \\ -[N+K+Sj/360]]/p$$

or

$$V= Sv$$

Steve Asikin ISBN 13: **978-1564912361**, ISBN 10: **1564912363**

Rule-58572:

If both (**/**), (**U**), (**S'**), (**s**), (**S**), (**v**), (**f**), (**I**), (**T**), (**D**), (**K**), (**j**), (**V**) and (**N**) are known, then its Procured Inventory Days planned is:

$$\mathbf{p} = 360[[1+\mathbf{/}](\mathbf{U}+\{\mathbf{S'}[1+\mathbf{s}]-\mathbf{Sv\text{-}Sf\text{-}I}\}\text{-}\mathbf{T}\text{-}\mathbf{D})$$
$$-[\mathbf{N}+\mathbf{K}+\mathbf{Sj}/360]]/\mathbf{V}$$

Rule-58573:

If both (**/**), (**U**), (**S'**), (**s**), (**S**), (**v**), (**f**), (**I**), (**T**), (**D**), (**K**), (**j**) and (**p**) are known, then its Non Current Asset planned is:

$$\mathbf{N} = [1+\mathbf{/}]\{\mathbf{U}+\mathbf{S'}[1+\mathbf{s}]-\mathbf{Sv\text{-}Sf\text{-}I\text{-}T\text{-}D}\}$$
$$-\mathbf{K}\text{-}\mathbf{Sj}/360\text{-}\mathbf{Svp}/360$$

Rule-58574:

If both (**N**), (**U**), (**S'**), (**s**), (**S**), (**v**), (**f**), (**I**), (**T**), (**D**), (**K**), (**j**) and (**p**) are known, then its Leverage or Gearing Ratio planned is:

$$\mathbf{/} = [\mathbf{N}+\mathbf{K}+\mathbf{Sj}/360+\mathbf{Svp}/360]$$
$$/(\mathbf{U}+\{\mathbf{S'}[1+\mathbf{s}]-\mathbf{Sv\text{-}Sf\text{-}I}\}\text{-}\mathbf{T}\text{-}\mathbf{D})\text{-}1$$

Rule-58575:

If both (**/**), (**N**), (**S'**), (**s**), (**S**), (**v**), (**f**), (**I**), (**T**), (**D**), (**K**), (**j**) and (**p**) are known, then its Utilized or Starting Capital must be:

$$\mathbf{U} = ([\mathbf{N}+\mathbf{K}+\mathbf{Sj}/360+\mathbf{Svp}/360]\text{-}[1+\mathbf{/}]\{\mathbf{S'}[1+\mathbf{s}]$$
$$-\mathbf{Sv\text{-}Sf\text{-}I}\}\text{-}\mathbf{T}\text{-}\mathbf{D})/[1+\mathbf{/}]$$

Steve Asikin ISBN 13: **978-1564912361**, ISBN 10: **1564912363**

Rule-58576:

If both ($I$), (**U**), (**N**), ($s$), ($\$$), (**v**), (**f**), (**I**), (**T**), (**D**), (**K**), (**j**) and (**p**) are known, then its Sales Past must be:

$$\$' = ([\mathbf{N}+\mathbf{K}+\$\mathbf{j}/360+\$\mathbf{vp}/360]/[1+I-\mathbf{U}+\$\mathbf{v}+\$\mathbf{f}+\mathbf{I}+\mathbf{T}+\mathbf{D})/[1+s]$$

or

$$\$' = \$/[1+s]$$

Rule-58577:

If both ($I$), (**U**), ($\$'$), ($s$), ($\$$), (**v**), (**f**), (**I**), (**T**), (**D**), (**K**), (**j**) and (**p**) are known, then its Non Current Asset planned is:

$$s = ([\mathbf{N}+\mathbf{K}+\$\mathbf{j}/360+\$\mathbf{vp}/360]/[1+I-\mathbf{U}+\$\mathbf{v}+\$\mathbf{f}+\mathbf{I}+\mathbf{T}+\mathbf{D})/\$'-1$$

or

$$s = \$/\$'-1$$

Rule-58578:

If both ($I$), (**U**), ($\$'$), ($s$), (**N**), (**v**), (**f**), (**I**), (**T**), (**D**), (**K**), (**j**) and (**p**) are known, then its Sales or Revenue planned is:

$$\$ = [[1+I(\mathbf{U}+\{\$'[1+s]-\mathbf{I}\}-\mathbf{T}-\mathbf{D})-[\mathbf{N}+\mathbf{K}]] /\{\mathbf{j}/360+\mathbf{vp}/360+[1+I[\mathbf{v}+\mathbf{p}]\}$$

or

$$\$ = \$'[1+s]$$

Steve Asikin ISBN 13: **978-1564912361**, ISBN 10: **1564912363**

Rule-58579:

If both (**/**), (**U**), (**S'**), (**s**), (**S**), (**N**), (**f**), (**I**), (**T**), (**D**), (**K**), (**j**) and (**p**) are known, then its Variable Portion planned is:

$$v= [[1+/](U+\{S'[1+s]-Sv-Sf-I\}-T-D)$$
$$-[N+K+Sj/360]]/(S\{p/360+[1+/]\})$$

Rule-58580:

If both (**/**), (**U**), (**S'**), (**s**), (**S**), (**v**), (**N**), (**I**), (**T**), (**D**), (**K**), (**j**) and (**p**) are known, then its Fixed Portion planned is:

$$f= (U+\{S'[1+s]-Sv-I\}-T-D$$
$$-[N+K+Sj/360+Svp/360]/[1+/])/S$$

Rule-58581:

If both (**/**), (**U**), (**S'**), (**s**), (**S**), (**v**), (**f**), (**N**), (**T**), (**D**), (**K**), (**j**) and (**p**) are known, then its Interest Expense planned is:

$$I= U+\{S'[1+s]-Sv-Sf\}-T-D$$
$$-[N+K+Sj/360+Svp/360]/[1+/]$$

Rule-58582:

If both (**/**), (**U**), (**S'**), (**s**), (**S**), (**v**), (**f**), (**I**), (**N**), (**D**), (**K**), (**j**) and (**p**) are known, then its Tax planned is:

$$T= U+\{S'[1+s]-Sv-Sf-I\}-D$$
$$-[N+K+Sj/360+Svp/360]/[1+/]$$

Steve Asikin ISBN 13: **978-1564912361**, ISBN 10: **1564912363**

Rule-58583:
> If both (**/**), (**U**), (**S'**), (**s**), (**S**), (**v**), (**f**), (**I**), (**T**), (**N**), (**K**),
> (**j**) and (**p**) are known, then its Dividend planned is:
>
> $$D= U+\{S'[1+s]-Sv-Sf-I\}-T$$
> $$-[N+K+Sj/360+Svp/360]/[1+/]$$

Rule-58584:
> If both (**/**), (**U**), (**S'**), (**s**), (**S**), (**v**), (**f**), (**I**), (**T**), (**D**), (**N**),
> (**j**) and (**p**) are known, then its Kind of Cash planned
> is:
>
> $$K= [1+/](U+\{S'[1+s]-Sv-Sf-I\}-T-D)$$
> $$-[N+Sj/360+Svp/360]$$

Rule-58585:
> If both (**/**), (**U**), (**S'**), (**s**), (**S**), (**v**), (**f**), (**I**), (**T**), (**D**), (**K**),
> (**N**) and (**p**) are known, then its Job or Trade
> Receivable Days planned is:
>
> $$j= 360[[1+/](U+\{S'[1+s]-Sv-Sf-I\}-T-D)$$
> $$-\{N+K+Svp/360\}]/S$$

Rule-58586:
> If both (**/**), (**U**), (**S'**), (**s**), (**S**), (**v**), (**f**), (**I**), (**T**), (**D**), (**K**),
> (**j**) and (**N**) are known, then its Procured Inventory
> Days planned is:
>
> $$p= 360[[1+/](U+\{S'[1+s]-Sv-Sf-I\}-T-D)$$
> $$-[N+K+Sj/360]]/[Sv]$$

Steve Asikin ISBN 13: **978-1564912361**, ISBN 10: **1564912363**

Rule-58587:

If both $(I)$, $(U)$, $(S')$, $(s)$, $(S)$, $(v)$, $(f)$, $(I)$, $(T)$, $(D)$, $(K)$, $(j)$ and $(p)$ are known, then its Non Current Asset planned is:

$$N= [1+I]\{U+S'[1+s]-Sv-Sf-I-T-D\}-K-Sj/360$$
$$-S'vp[1+s]/360$$

Rule-58588:

If both $(N)$, $(U)$, $(S')$, $(s)$, $(S)$, $(v)$, $(f)$, $(I)$, $(T)$, $(D)$, $(K)$, $(j)$ and $(p)$ are known, then its Leverage or Gearing Ratio planned is:

$$I= \{N+K+Sj/360+S'vp[1+s]/360\}/(U+\{S'[1+s]$$
$$-Sv-Sf-I-T-D)-1$$

Rule-58589:

If both $(I)$, $(N)$, $(S')$, $(s)$, $(S)$, $(v)$, $(f)$, $(I)$, $(T)$, $(D)$, $(K)$, $(j)$ and $(p)$ are known, then its Utilized or Starting Capital must be:

$$U= (\{N+K+Sj/360+S'vp[1+s]/360\}$$
$$-[1+I]\{S'[1+s]-Sv-Sf-I\}-T-D)/[1+I]$$

Rule-58590:

If both $(I)$, $(U)$, $(N)$, $(s)$, $(S)$, $(v)$, $(f)$, $(I)$, $(T)$, $(D)$, $(K)$, $(j)$ and $(p)$ are known, then its Sales Past must be:

$$S'= [[1+I](U-\{Sv+Sf+I\}-T-D)-\{N+K+Sj/360\}]$$
$$/([1+s]\{vp/360-[1+I]\})$$
$$\text{or}$$
$$S'= S/[1+s]$$

Steve Asikin ISBN 13: **978-1564912361**, ISBN 10: **1564912363**

Rule-58591:

If both (**/**), (**U**), (**S'**), (**N**), (**S**), (**v**), (**f**), (**I**), (**T**), (**D**), (**K**), (**j**) and (**p**) are known, then its Sales Growth planned is:

$$s= [[1+/](U-\{Sv+Sf+I\}-T-D)-\{N+K+Sj/360\}]$$
$$/(S'\{vp/360-[1+/]\})-1$$

or

$$s= S/S'-1$$


Rule-58592:

If both (**/**), (**U**), (**S'**), (**s**), (**N**), (**v**), (**f**), (**I**), (**T**), (**D**), (**K**), (**j**) and (**p**) are known, then its Sales or Revenue planned is:

$$S= [[1+/](U+\{S'[1+s]-Sv-Sf-I\}-T-D)-\{N+K$$
$$+S'vp[1+s]/360\}]/\{j/360+[1+/][v+f]\}$$

or

$$S= S'[1+s]$$


Rule-58593:

If both (**/**), (**U**), (**S'**), (**s**), (**S**), (**N**), (**f**), (**I**), (**T**), (**D**), (**K**), (**j**) and (**p**) are known, then its Variable Portion planned is:

$$v= [[1+/](U+\{S'[1+s]-Sf-I\}-T-D)$$
$$-[N+K+Sj/360]]/\{S'p[1+s]/360+S[1+/]\}$$

Steve Asikin ISBN 13: **978-1564912361**, ISBN 10: **1564912363**

Rule-58594:

If both (**/**), (**U**), (**S'**), (**s**), (**S**), (**v**), (**N**), (**I**), (**T**), (**D**), (**K**), (**j**) and (**p**) are known, then its Fixed Portion planned is:

$$f= (U+\{S'[1+s]-Sv-I\}-T-D$$
$$-\{N+K+Sj/360+S'vp[1+s]/360\}/[1+I])/S$$

Rule-58595:

If both (**/**), (**U**), (**S'**), (**s**), (**S**), (**v**), (**f**), (**N**), (**T**), (**D**), (**K**), (**j**) and (**p**) are known, then its Interest Expense planned is:

$$I= U+\{S'[1+s]-Sv-Sf\}-T-D-\{N+K+Sj/360$$
$$+S'vp[1+s]/360\}/[1+I]$$

Rule-58596:

If both (**/**), (**U**), (**S'**), (**s**), (**S**), (**v**), (**f**), (**I**), (**N**), (**D**), (**K**), (**j**) and (**p**) are known, then its Tax planned is:

$$T= U+\{S'[1+s]-Sv-Sf-I\}-D-\{N+K+Sj/360$$
$$+S'vp[1+s]/360\}/[1+I]$$

Rule-58597:

If both (**/**), (**U**), (**S'**), (**s**), (**S**), (**v**), (**f**), (**I**), (**T**), (**N**), (**K**), (**j**) and (**p**) are known, then its Dividend planned is:

$$D= U+\{S'[1+s]-Sv-Sf-I\}-T-\{N+K+Sj/360$$
$$+S'vp[1+s]/360\}/[1+I]$$

Steve Asikin ISBN 13: **978-1564912361**, ISBN 10: **1564912363**

Rule-58598:

If both ($\Lambda$, (**U**), (**S'**), (**s**), (**S**), (**v**), (**f**), (**I**), (**T**), (**D**), (**N**), (**j**) and (**p**) are known, then its Kind of Cash planned is:

$$K= [1+\Lambda(U+\{S'[1+s]-Sv-Sf-I\}-T-D)-\{N+Sj/360 +S'vp[1+s]/360\}$$

Rule-58599:

If both ($\Lambda$, (**U**), (**S'**), (**s**), (**S**), (**v**), (**f**), (**I**), (**T**), (**D**), (**K**), (**N**) and (**p**) are known, then its Job or Trade Receivable Days planned is:

$$j= 360[[1+\Lambda(U+\{S'[1+s]-Sv-Sf-I\}-T-D) -\{N+K+S'vp[1+s]/360\}]/S$$

Rule-58600:

If both ($\Lambda$, (**U**), (**S'**), (**s**), (**S**), (**v**), (**f**), (**I**), (**T**), (**D**), (**K**), (**j**) and (**N**) are known, then its Procured Inventory Days planned is:

$$p= 360[[1+\Lambda(U+\{S'[1+s]-Sv-Sf-I\}-T-D) -[N+K+Sj/360]]/\{S'v[1+s]\}$$

Rule-58601:

If both ($\Lambda$, (**U**), (**S'**), (**s**), (**S**), (**v**), (**f**), (**I**), (**T**), (**D**), (**K**), (**j**) and (**P**) are known, then its Non Current Asset planned is:

$$N= [1+\Lambda\{U+S'[1+s]-Sv-Sf-I-T-D\} -K-S'j[1+s]/360-P$$

694

Steve Asikin ISBN 13: **978-1564912361**, ISBN 10: **1564912363**

Rule-58602:

If both (**N**), (**U**), (**S'**), (**s**), (**S**), (**v**), (**f**), (**I**), (**T**), (**D**), (**K**), (**j**) and (**P**) are known, then its Leverage or Gearing Ratio planned is:

$$I = \{N+K+S'j[1+s]/360+P\}/(U+\{S'[1+s]-Sv-Sf-I\} -T-D)-1$$

Rule-58603:

If both (**I**), (**N**), (**S'**), (**s**), (**S**), (**v**), (**f**), (**I**), (**T**), (**D**), (**K**), (**j**) and (**P**) are known, then its Utilized or Starting Capital must be:

$$U = (\{N+K+S'j[1+s]/360+P\} -[1+I]\{S'[1+s]-Sv-Sf-I\}-T-D)/[1+I]$$

Rule-58604:

If both (**I**), (**U**), (**N**), (**s**), (**S**), (**v**), (**f**), (**I**), (**T**), (**D**), (**K**), (**j**) and (**P**) are known, then its Sales Past must be:

$$S' = [[1+I](U-\{Sv-Sf-I\}-T-D)-[N+K+P]] /([1+s]\{j/360-[1+I]\})$$

or

$$S' = S/[1+s]$$

Steve Asikin ISBN 13: **978-1564912361**, ISBN 10: **1564912363**

Rule-58605:

If both (**/**), (**U**), (**S'**), (**N**), (**S**), (**v**), (**f**), (**I**), (**T**), (**D**), (**K**), (**j**) and (**P**) are known, then its Sales Growth planned is:

$$s = [[1+/](U-\{Sv-Sf-I\}-T-D)-[N+K+P]]$$
$$/(S'\{j/360-[1+/]\}-1)$$

or

$$s = S/S'-1$$

Rule-58606:

If both (**/**), (**U**), (**S'**), (**s**), (**N**), (**v**), (**f**), (**I**), (**T**), (**D**), (**K**), (**j**) and (**P**) are known, then its Sales or Revenue planned is:

$$S = (U+\{S'[1+s]-I\}-T-D-\{N+K+S'j[1+s]/360+P\}$$
$$/[1+/])/[v+f]$$

or

$$S = S'[1+s]$$

Rule-58607:

If both (**/**), (**U**), (**S'**), (**s**), (**S**), (**N**), (**f**), (**I**), (**T**), (**D**), (**K**), (**j**) and (**P**) are known, then its Variable Portion planned is:

$$v = (U+\{S'[1+s]-Sf-I\}-T-D-\{N+K$$
$$+S'j[1+s]/360+P\}/[1+/])/S$$

Steve Asikin ISBN 13: **978-1564912361**, ISBN 10: **1564912363**

Rule-58608:

If both (**/**), (**U**), (**S'**), (**s**), (**S**), (**v**), (**N**), (**I**), (**T**), (**D**), (**K**), (**j**) and (**P**) are known, then its Fixed Portion planned is:

$$f= (U+\{S'[1+s]-Sv-I\}-T-D-\{N+K$$
$$+S'j[1+s]/360+P\}/[1+/])/S$$

Rule-58609:

If both (**/**), (**U**), (**S'**), (**s**), (**S**), (**v**), (**f**), (**N**), (**T**), (**D**), (**K**), (**j**) and (**P**) are known, then its Interest Expense planned is:

$$I= U+\{S'[1+s]-Sv-Sf\}-T-D-\{N+K$$
$$+S'j[1+s]/360+P\}/[1+/]$$

Rule-58610:

If both (**/**), (**U**), (**S'**), (**s**), (**S**), (**v**), (**f**), (**I**), (**N**), (**D**), (**K**), (**j**) and (**P**) are known, then its Tax planned is:

$$T= U+\{S'[1+s]-Sv-Sf-I\}-D-\{N+K$$
$$+S'j[1+s]/360+P\}/[1+/]$$

Rule-58611:

If both (**/**), (**U**), (**S'**), (**s**), (**S**), (**v**), (**f**), (**I**), (**T**), (**N**), (**K**), (**j**) and (**P**) are known, then its Dividend planned is:

$$D= U+\{S'[1+s]-Sv-Sf-I\}-T-\{N+K$$
$$+S'j[1+s]/360+P\}/[1+/]$$

Steve Asikin ISBN 13: **978-1564912361**, ISBN 10: **1564912363**

Rule-58612:

If both (**/**), (**U**), (**S'**), (**s**), (**S**), (**v**), (**f**), (**I**), (**T**), (**D**), (**N**), (**j**) and (**P**) are known, then its Kind of Cash planned is:

$$K= [1+/](U+\{S'[1+s]-Sv-Sf-I\}-T-D)$$
$$-\{N+S'j[1+s]/360+P\}$$

Rule-58613:

If both (**/**), (**U**), (**S'**), (**s**), (**S**), (**v**), (**f**), (**I**), (**T**), (**D**), (**K**), (**N**) and (**P**) are known, then its Job or Trade Receivable Days planned is:

$$j= 360[[1+/](U+\{S'[1+s]-Sv-Sf-I\}-T-D)$$
$$-\{N+K+P\}]/\{S'[1+s]\}$$

Rule-58614:

If both (**/**), (**U**), (**S'**), (**s**), (**S**), (**v**), (**f**), (**I**), (**T**), (**D**), (**K**), (**j**) and (**N**) are known, then its Procured Inventory planned is:

$$P= [1+/](U+\{S'[1+s]-Sv-Sf-I\}-T-D)$$
$$-\{N+K+S'j[1+s]/360\}$$

Rule-58615:

If both (**/**), (**U**), (**S'**), (**s**), (**S**), (**v**), (**f**), (**I**), (**T**), (**D**), (**K**), (**j**), (**V**) and (**p**) are known, then its Non Current Asset planned is:

$$N= [1+/]\{U+S'[1+s]-Sv-Sf-I-T-D\}$$
$$-K-S'j[1+s]/360-Vp/360$$

Steve Asikin ISBN 13: **978-1564912361**, ISBN 10: **1564912363**

Rule-58616:

If both (**N**), (**U**), (**S'**), (**s**), (**S**), (**v**), (**f**), (**I**), (**T**), (**D**), (**K**), (**j**), (**V**) and (**p**) are known, then its Leverage or Gearing Ratio planned is:

$$I = \{N+K+S'j[1+s]/360+Vp/360\}$$
$$/(U+\{S'[1+s]-Sv-Sf-I\}-T-D)-1$$

Rule-58617:

If both (**I**), (**N**), (**S'**), (**s**), (**S**), (**v**), (**f**), (**I**), (**T**), (**D**), (**K**), (**j**), (**V**) and (**p**) are known, then its Utilized or Starting Capital must be:

$$U = (\{N+K+S'j[1+s]/360+Vp/360\}-[1+I]\{S'[1+s]$$
$$-Sv-Sf-I\}-T-D)/[1+I]$$

Rule-58618:

If both (**I**), (**U**), (**N**), (**s**), (**S**), (**v**), (**f**), (**I**), (**T**), (**D**), (**K**), (**j**), (**V**) and (**p**) are known, then its Sales Past must be:

$$S' = [[1+I](U-\{Sv+Sf+I\}-T-D)-[N+K+Vp/360]]$$
$$/([1+s]\{j/360-[1+I]\})$$

or

$$S' = S/[1+s]$$

Steve Asikin ISBN 13: **978-1564912361**, ISBN 10: **1564912363**

<u>Rule-58619:</u>

If both (**/**), (**U**), (**S'**), (**N**), (**S**), (**v**), (**f**), (**I**), (**T**), (**D**), (**K**), (**j**), (**V**) and (**p**) are known, then its Sales Growth planned is:

$$s = [[1+/](U-\{Sv+Sf+I\}-T-D)-[N+K+Vp/360]] / ([1+s]\{j/360-[1+/]\})-1$$

or

$$s = S/S'-1$$

<u>Rule-58620:</u>

If both (**/**), (**U**), (**S'**), (**s**), (**N**), (**v**), (**f**), (**I**), (**T**), (**D**), (**K**), (**j**), (**V**) and (**p**) are known, then its Sales or Revenue planned is:

$$S = (U+\{S'[1+s]\}-T-D-\{N+K+S'j[1+s]/360 +Vp/360\}/[1+/])/[f+v]$$

or

$$S = S'[1+s]$$

or

$$S = V/v$$

<u>Rule-58621:</u>

If both (**/**), (**U**), (**S'**), (**s**), (**S**), (**N**), (**f**), (**I**), (**T**), (**D**), (**K**), (**j**), (**V**) and (**p**) are known, then its Variable Portion planned is:

$$v = (U+\{S'[1+s]-Sv-I\}-T-D-\{N+K+S'j[1+s]/360 +Vp/360\}/[1+/])/S$$

or

$$v = V/S$$

Steve Asikin ISBN 13: **978-1564912361**, ISBN 10: **1564912363**

Rule-58622:

If both (**/**), (**U**), (**S'**), (**s**), (**$**), (**v**), (**N**), (**I**), (**T**), (**D**), (**K**), (**j**), (**V**) and (**p**) are known, then its Fixed Portion planned is:

$$f= (U+\{S'[1+s]-Sv-I\}-T-D-\{N+K+S'j[1+s]/360$$
$$+Vp/360\}/[1+/])/S$$

Rule-58623:

If both (**/**), (**U**), (**S'**), (**s**), (**$**), (**v**), (**f**), (**N**), (**T**), (**D**), (**K**), (**j**), (**V**) and (**p**) are known, then its Interest Expense planned is:

$$I= U+\{S'[1+s]-Sv-Sf\}-T-D-\{N+K+S'j[1+s]/360$$
$$+Vp/360\}/[1+/]$$

Rule-58624:

If both (**/**), (**U**), (**S'**), (**s**), (**$**), (**v**), (**f**), (**I**), (**N**), (**D**), (**K**), (**j**), (**V**) and (**p**) are known, then its Tax planned is:

$$T= U+\{S'[1+s]-Sv-Sf-I\}-D-\{N+K+S'j[1+s]/360$$
$$+Vp/360\}/[1+/]$$

Rule-58625:

If both (**/**), (**U**), (**S'**), (**s**), (**$**), (**v**), (**f**), (**I**), (**T**), (**N**), (**K**), (**j**), (**V**) and (**p**) are known, then its Dividend planned is:

$$D= U+\{S'[1+s]-Sv-Sf-I\}-T-\{N+K+S'j[1+s]/360$$
$$+Vp/360\}/[1+/]$$

Steve Asikin ISBN 13: **978-1564912361**, ISBN 10: **1564912363**

Rule-58626:

If both (**/**), (**U**), (**S'**), (**s**), (**S**), (**v**), (**f**), (**I**), (**T**), (**D**), (**N**), (**j**), (**V**) and (**p**) are known, then its Kind of Cash planned is:

$$K = [1 + /(U + \{S'[1+s] - Sv - Sf - I\} - T - D) - \{N + S'j[1+s]/360 + Vp/360\}$$

Rule-58627:

If both (**/**), (**U**), (**S'**), (**s**), (**S**), (**v**), (**f**), (**I**), (**T**), (**D**), (**K**), (**N**), (**V**) and (**p**) are known, then its Job or Trade Receivable Days planned is:

$$j = 360[[1 + /(U + \{S'[1+s] - Sv - Sf - I\} - T - D) - \{N + K + Vp/360\}]/\{S'[1+s]\}$$

Rule-58628:

If both (**/**), (**U**), (**S'**), (**s**), (**S**), (**v**), (**f**), (**I**), (**T**), (**D**), (**K**), (**j**), (**N**) and (**p**) are known, then its Variable Cost planned is:

$$V = 360[[1 + /(U + \{S'[1+s] - Sv - Sf - I\} - T - D) - \{N + K + S'j[1+s]/360\}]/p$$

or

$$V = Sv$$

Rule-58629:

If both (**/**), (**U**), (**S'**), (**s**), (**S**), (**v**), (**f**), (**I**), (**T**), (**D**), (**K**), (**j**), (**V**) and (**N**) are known, then its Procured Inventory Days planned is:

$$p = 360[[1 + /(U + \{S'[1+s] - Sv - Sf - I\} - T - D) - \{N + K + S'j[1+s]/360\}]/V$$

`

702

Steve Asikin ISBN 13: **978-1564912361**, ISBN 10: **1564912363**

Rule-58630:

If both (*I*), (**U**), (**S'**), (**s**), (**S**), (**v**), (**f**), (**I**), (**T**), (**D**), (**K**), (**j**) and (**p**) are known, then its Non Current Asset planned is:

$$N= [1+I]\{U+S'[1+s]-Sv-Sf-I-T-D\}-K$$
$$-S'j[1+s]/360-Svp/360$$

Rule-58631:

If both (**N**), (**U**), (**S'**), (**s**), (**S**), (**v**), (**f**), (**I**), (**T**), (**D**), (**K**), (**j**) and (**p**) are known, then its Leverage or Gearing Ratio planned is:

$$I= \{N+K+S'j[1+s]/360+Svp/360\}$$
$$/(U+\{S'[1+s]-Sv-Sf-I\}-T-D)-1$$

Rule-58632:

If both (*I*), (**N**), (**S'**), (**s**), (**S**), (**v**), (**f**), (**I**), (**T**), (**D**), (**K**), (**j**) and (**p**) are known, then its Utilized or Starting Cap[ital must be:

$$U= (\{N+K+S'j[1+s]/360+Svp/360\}$$
$$-[1+I]\{S'[1+s]-Sv-Sf-I\}-T-D)/[1+I]$$

Rule-58633:

If both (*I*), (**U**), (**N**), (**s**), (**S**), (**v**), (**f**), (**I**), (**T**), (**D**), (**K**), (**j**) and (**p**) are known, then its Sales Past must be:

$$S'= [[1+I](U-\{Sv+Sf+I\}-T-D)-[N+K+Svp/360]]$$
$$/([1+s]\{j/360-[1+I]\})$$

or

$$S'= S/[1+s]$$

Steve Asikin ISBN 13: **978-1564912361**, ISBN 10: **1564912363**

Rule-58634:

If both (**/**), (**U**), (**S'**), (**N**), (**S**), (**v**), (**f**), (**I**), (**T**), (**D**), (**K**), (**j**) and (**p**) are known, then its Sales Growth planned is:

$$s = [[1+/](U-\{Sv+Sf+I\}-T-D)-[N+K+Svp/360]]$$
$$/(S'\{j/360-[1+/]\})-1$$

or

$$s = S/S'-1$$

Rule-58635:

If both (**/**), (**U**), (**S'**), (**s**), (**N**), (**v**), (**f**), (**I**), (**T**), (**D**), (**K**), (**j**) and (**p**) are known, then its Sales or Revenue planned is:

$$S = [[1+/](U+\{S'[1+s]-I\}-T-D)-\{N+K$$
$$+S'j[1+s]/360\}]/(v\{p/360+[1+/]\}+f[1+/])$$

or

$$S = S'[1+s]$$

Rule-58636:

If both (**/**), (**U**), (**S'**), (**s**), (**S**), (**N**), (**f**), (**I**), (**T**), (**D**), (**K**), (**j**) and (**p**) are known, then its Variable Portion planned is:

$$v = [[1+/](U+\{S'[1+s]-Sf-I\}-T-D)$$
$$-\{N+K+S'j[1+s]/360\}]/(S\{p/360+[1+/]\})$$

Steve Asikin ISBN 13: **978-1564912361**, ISBN 10: **1564912363**

Rule-58637:

If both (**/**), (**U**), (**S'**), (**s**), (**S**), (**v**), (**N**), (**I**), (**T**), (**D**), (**K**), (**j**) and (**p**) are known, then its Fixed Portion planned is:

$$f= (U+\{S'[1+s]-Sv-I\}-T-D-\{N+K+S'j[1+s]/360$$
$$+Svp/360\}/[1+/])/S$$

Rule-58638:

If both (**/**), (**U**), (**S'**), (**s**), (**S**), (**v**), (**f**), (**N**), (**T**), (**D**), (**K**), (**j**) and (**p**) are known, then its Interest Expense planned is:

$$I= U+\{S'[1+s]-Sv-Sf\}-T-D-\{N+K+S'j[1+s]/360$$
$$+Svp/360\}/[1+/]$$

Rule-58639:

If both (**/**), (**U**), (**S'**), (**s**), (**S**), (**v**), (**f**), (**I**), (**N**), (**D**), (**K**), (**j**) and (**p**) are known, then its Tax planned is:

$$T= U+\{S'[1+s]-Sv-Sf-I\}-D$$
$$-\{N+K+S'j[1+s]/360+Svp/360\}/[1+/]$$

Rule-58640:

If both (**/**), (**U**), (**S'**), (**s**), (**S**), (**v**), (**f**), (**I**), (**T**), (**N**), (**K**), (**j**) and (**p**) are known, then its Dividend planned is:

$$D= U+\{S'[1+s]-Sv-Sf-I\}-T$$
$$-\{N+K+S'j[1+s]/360+Svp/360\}/[1+/]$$

Steve Asikin ISBN 13: **978-1564912361**, ISBN 10: **1564912363**

## Rule-58641:

If both (**I**), (**U**), (**S'**), (**s**), (**S**), (**v**), (**f**), (**I**), (**T**), (**D**), (**N**), (**j**) and (**p**) are known, then its Kind of Cash planned is:

$$K = [1+I(U+\{S'[1+s]-Sv-Sf-I\}-T-D) \\ -\{N+S'j[1+s]/360+Svp/360\}$$

## Rule-58642:

If both (**I**), (**U**), (**S'**), (**s**), (**S**), (**v**), (**f**), (**I**), (**T**), (**D**), (**K**), (**N**) and (**p**) are known, then its Job or Trade Receivable Days planned is:

$$j = 360[[1+I(U+\{S'[1+s]-Sv-Sf-I\}-T-D) \\ -\{N+K+Svp/360\}]/\{S'[1+s]\}$$

## Rule-58643:

If both (**I**), (**U**), (**S'**), (**s**), (**S**), (**v**), (**f**), (**I**), (**T**), (**D**), (**K**), (**j**) and (**N**) are known, then its Procured Inventory Days planned is:

$$p = 360[[1+I(U+\{S'[1+s]-Sv-Sf-I\}-T-D) \\ -\{N+K+S'j[1+s]/360\}]/[Sv]$$

## Rule-58644:

If both (**I**), (**U**), (**S'**), (**s**), (**S**), (**v**), (**f**), (**I**), (**T**), (**D**), (**K**), (**j**) and (**p**) are known, then its Non Current Asset planned is:

$$N = [1+I\{U+S'[1+s]-Sv-Sf-I-T-D\}-K \\ -S'j[1+s]/360-S'vp[1+s]/360$$

Steve Asikin ISBN 13: **978-1564912361**, ISBN 10: **1564912363**

Rule-58645:

If both (**N**), (**U**), (**S'**), (**s**), (**S**), (**v**), (**f**), (**I**), (**T**), (**D**), (**K**), (**j**) and (**p**) are known, then its Leverage or Gearing Ratio planned is:

$$I = \{N+K+S'j[1+s]/360+S'vp[1+s]/360\}$$
$$/(U+\{S'[1+s]-Sv-Sf-I\}-T-D)-1$$

Rule-58646:

If both (**I**), (**N**), (**S'**), (**s**), (**S**), (**v**), (**f**), (**I**), (**T**), (**D**), (**K**), (**j**) and (**p**) are known, then its Utilized or Starting Capital must be:

$$U = (\{N+K+S'j[1+s]/360+S'vp[1+s]/360\}$$
$$-[1+I]\{S'[1+s]-Sv-Sf-I\}-T-D)/[1+I]$$

Rule-58647:

If both (**I**), (**U**), (**N**), (**s**), (**S**), (**v**), (**f**), (**I**), (**T**), (**D**), (**K**), (**j**) and (**p**) are known, then its Sales Past must be:

$$S' = [(U-\{Sv+Sf+I\}-T-D)-[N+K]]$$
$$/([1+s]\{j/360+vp/360-[1+I]\})$$
$$\text{or}$$
$$S' = S/[1+s]$$

Steve Asikin ISBN 13: **978-1564912361**, ISBN 10: **1564912363**

Rule-58648:

If both (**/**), (**U**), (**S'**), (**N**), (**S**), (**v**), (**f**), (**I**), (**T**), (**D**), (**K**), (**j**) and (**p**) are known, then its Sales Growth planned is:

$$s= [(U-\{Sv+Sf+I\}-T-D)-[N+K]]$$
$$/(S'\{j/360+vp/360-[1+/]\})-1$$

or

$$s= S/S'-1$$


Rule-58649:

If both (**/**), (**U**), (**S'**), (**s**), (**N**), (**v**), (**f**), (**I**), (**T**), (**D**), (**K**), (**j**) and (**p**) are known, then its Sales or Revenue planned is:

$$S= (U+\{S'[1+s]-I\}-T-D-\{N+K+S'j[1+s]/360$$
$$+S'vp[1+s]/360\}/[1+/])/[f+v]$$

or

$$S= S'[1+s]$$


Rule-58650:

If both (**/**), (**U**), (**S'**), (**s**), (**S**), (**N**), (**f**), (**I**), (**T**), (**D**), (**K**), (**j**) and (**p**) are known, then its Variable Portion planned is:

$$v= [[1+/](U+\{S'[1+s]-Sf-I\}-T-D)-\{N+K$$
$$+S'j[1+s]/360\}]/\{S'p[1+s]/360+S[1+/]\}$$

Steve Asikin ISBN 13: **978-1564912361**, ISBN 10: **1564912363**

Rule-58651:

If both (**I**), (**U**), (**S'**), (**s**), (**S**), (**v**), (**N**), (**I**), (**T**), (**D**), (**K**), (**j**) and (**p**) are known, then its Fixed Portion planned is:

$$f = (U + \{S'[1+s] - Sv - I\} - T - D - \{N + K + S'j[1+s]/360 + S'vp[1+s]/360\}/[1+I])/S$$

Rule-58652:

If both (**I**), (**U**), (**S'**), (**s**), (**S**), (**v**), (**f**), (**N**), (**T**), (**D**), (**K**), (**j**) and (**p**) are known, then its Interest Expense planned is:

$$I = U + \{S'[1+s] - Sv - Sf\} - T - D - \{N + K + S'j[1+s]/360 + S'vp[1+s]/360\}/[1+I]$$

Rule-58653:

If both (**I**), (**U**), (**S'**), (**s**), (**S**), (**v**), (**f**), (**I**), (**N**), (**D**), (**K**), (**j**) and (**p**) are known, then its Tax planned is:

$$T = U + \{S'[1+s] - Sv - Sf - I\} - D - \{N + K + S'j[1+s]/360 + S'vp[1+s]/360\}/[1+I]$$

Rule-58654:

If both (**I**), (**U**), (**S'**), (**s**), (**S**), (**v**), (**f**), (**I**), (**T**), (**N**), (**K**), (**j**) and (**p**) are known, then its Dividend planned is:

$$D = U + \{S'[1+s] - Sv - Sf - I\} - T - \{N + K + S'j[1+s]/360 + S'vp[1+s]/360\}/[1+I]$$

Steve Asikin ISBN 13: **978-1564912361**, ISBN 10: **1564912363**

Rule-58655:

If both (**/**), (**U**), (**S'**), (**s**), (**S**), (**v**), (**f**), (**I**), (**T**), (**D**), (**N**), (**j**) and (**p**) are known, then its Kind of Cash planned is:

$$K= [1+/(U+\{S'[1+s]-Sv-Sf-I\}-T-D)$$
$$-\{N+S'j[1+s]/360+S'vp[1+s]/360\}$$

Rule-58656:

If both (**/**), (**U**), (**S'**), (**s**), (**S**), (**v**), (**f**), (**I**), (**T**), (**D**), (**K**), (**N**) and (**p**) are known, then its Job or Trade Receivable Days planned is:

$$j= 360[[1+/(U+\{S'[1+s]-Sv-Sf-I\}-T-D)$$
$$-\{N+K+S'vp[1+s]/360\}]/\{S'[1+s]\}$$

Rule-58657:

If both (**/**), (**U**), (**S'**), (**s**), (**S**), (**v**), (**f**), (**I**), (**T**), (**D**), (**K**), (**j**) and (**N**) are known, then its Procured Inventory Days planned is:

$$p= 360[[1+/(U+\{S'[1+s]-Sv-Sf-I\}-T-D)$$
$$-\{N+K+S'j[1+s]/360\}]/\{S'v[1+s]\}$$

Rule-58658:

If both (**/**), (**U**), (**S'**), (**s**), (**S**), (**v**), (**f**), (**I**), (**T**), (**d**), (**K**), (**J**) and (**P**) are known, then its Non Current Asset planned is:

$$N= [1+/(U+\{S'[1+s]-Sv-Sf-I-T\}[1-d])-K-J-P$$

Steve Asikin ISBN 13: **978-1564912361**, ISBN 10: **1564912363**

Rule-58659:

If both **(N)**, **(U)**, **(S')**, **(s)**, **(S)**, **(v)**, **(f)**, **(I)**, **(T)**, **(d)**, **(K)**, **(J)** and **(P)** are known, then its Leverage or Gearing Ratio planned is:

$$l = \{N+K+J+P\}/(U+\{S'[1+s]-Sv-Sf-I-T\}[1-d])-1$$

Rule-58660:

If both **(l)**, **(N)**, **(S')**, **(s)**, **(S)**, **(v)**, **(f)**, **(I)**, **(T)**, **(d)**, **(K)**, **(J)** and **(P)** are known, then its Utilized or Starting Capital must be:

$$U = (\{N+K+J+P\}-[1+l]\{S'[1+s]-Sv-Sf-I-T\}[1-d])$$
$$/[1+l]$$

Rule-58661:

If both **(l)**, **(U)**, **(N)**, **(s)**, **(S)**, **(v)**, **(f)**, **(I)**, **(T)**, **(d)**, **(K)**, **(J)** and **(P)** are known, then its Sales Past must be:

$$S' = (\{[N+K+J+P]/[1+l]-U\}/[1-d]+Sv+Sf+I+T)$$
$$/[1+s]$$
or
$$S' = S/[1+s]$$

Rule-58662:

If both **(l)**, **(U)**, **(S')**, **(N)**, **(S)**, **(v)**, **(f)**, **(I)**, **(T)**, **(d)**, **(K)**, **(J)** and **(P)** are known, then its Sales Growth planned is:

$$s = (\{[N+K+J+P]/[1+l]-U\}/[1-d]+Sv+Sf+I+T)/S'-1$$
or
$$s = S/S'-1$$

Steve Asikin ISBN 13: **978-1564912361**, ISBN 10: **1564912363**

Rule-58663:

If both (**/**), (**U**), (**S'**), (**s**), (**N**), (**v**), (**f**), (**I**), (**T**), (**d**), (**K**), (**J**) and (**P**) are known, then its Sales or Revenue planned is:

$$S = (U + \{S'[1+s]-I-T\}[1-d]-[N+K+J+P]/[1+/])$$
$$/\{[v+f][1-d]\}$$

or

$$S = S'[1+s]$$

Rule-58664:

If both (**/**), (**U**), (**S'**), (**s**), (**S**), (**N**), (**f**), (**I**), (**T**), (**d**), (**K**), (**J**) and (**P**) are known, then its Variable Portion planned is:

$$v = (U + \{S'[1+s]-Sf-I-T\}[1-d]-[N+K+J+P]/[1+/])$$
$$/\{S[1-d]\}$$

Rule-58665:

If both (**/**), (**U**), (**S'**), (**s**), (**S**), (**v**), (**N**), (**I**), (**T**), (**d**), (**K**), (**J**) and (**P**) are known, then its Fixed Portion planned is:

$$f = (U + \{S'[1+s]-Sv-I-T\}[1-d]-[N+K+J+P]/[1+/])$$
$$/\{S[1-d]\}$$

Rule-58666:

If both (**/**), (**U**), (**S'**), (**s**), (**S**), (**v**), (**f**), (**N**), (**T**), (**d**), (**K**), (**J**) and (**P**) are known, then its Interest Expense planned is:

$$I = S'[1+s]-Sv-Sf-T-\{[N+K+J+P]/[1+/-U\}/[1-d]$$

Steve Asikin ISBN 13: **978-1564912361**, ISBN 10: **1564912363**

Rule-58667:

If both (**/**), (**U**), (**S'**), (**s**), (**S**), (**v**), (**f**), (**I**), (**N**), (**d**), (**K**), (**J**) and (**P**) are known, then its Tax planned is:

$$T= S'[1+s]-Sv-Sf-I-\{[N+K+J+P]/[1+/]-U\}/[1-d]$$

Rule-58668:

If both (**/**), (**U**), (**S'**), (**s**), (**S**), (**v**), (**f**), (**I**), (**T**), (**N**), (**K**), (**J**) and (**P**) are known, then its Dividend Payout planned is:

$$d= 1-\{[N+K+J+P]/[1+/]-U\}/\{S'[1+s]-Sv-Sf-I-T\}$$

Rule-58669:

If both (**/**), (**U**), (**S'**), (**s**), (**S**), (**v**), (**f**), (**I**), (**T**), (**d**), (**N**), (**J**) and (**P**) are known, then its Kind of Cash planned is:

$$K= [1+/](U+\{S'[1+s]-Sv-Sf-I-T\}[1-d])-[N+J+P]$$

Rule-58670:

If both (**/**), (**U**), (**S'**), (**s**), (**S**), (**v**), (**f**), (**I**), (**T**), (**d**), (**K**), (**N**) and (**P**) are known, then its Job or Trade Account Receivable planned is:

$$J= [1+/](U+\{S'[1+s]-Sv-Sf-I-T\}[1-d])-[N+K+P]$$

Rule-58671:

If both (**/**), (**U**), (**S'**), (**s**), (**S**), (**v**), (**f**), (**I**), (**T**), (**d**), (**K**), (**J**) and (**N**) are known, then its Procured Inventory planned is:

$$P= [1+/](U+\{S'[1+s]-Sv-Sf-I-T\}[1-d])-[N+K+J]$$

Steve Asikin ISBN 13: **978-1564912361**, ISBN 10: **1564912363**

## Rule-58672:

If both (**I**), (**U**), (**S'**), (**s**), (**S**), (**v**), (**f**), (**I**), (**T**), (**d**), (**K**), (**J**), (**V**) and (**P**) are known, then its Non Current Asset planned is:

$$\mathbf{N} = [1 + \mathbf{I}](\mathbf{U} + \{\mathbf{S'}[1+\mathbf{s}] - \mathbf{Sv} - \mathbf{Sf} - \mathbf{I} - \mathbf{T}\}[1-\mathbf{d}])$$
$$-\mathbf{K} - \mathbf{J} - \mathbf{Vp}/360$$

## Rule-58673:

If both (**N**), (**U**), (**S'**), (**s**), (**S**), (**v**), (**f**), (**I**), (**T**), (**d**), (**K**), (**J**), (**V**) and (**P**) are known, then its Leverage or Gearing Ratio planned is:

$$\mathbf{I} = [\mathbf{N} + \mathbf{K} + \mathbf{J} + \mathbf{Vp}/360]$$
$$/(\mathbf{U} + \{\mathbf{S'}[1+\mathbf{s}] - \mathbf{Sv} - \mathbf{Sf} - \mathbf{I} - \mathbf{T}\}[1-\mathbf{d}]) - 1$$

## Rule-58674:

If both (**I**), (**N**), (**S'**), (**s**), (**S**), (**v**), (**f**), (**I**), (**T**), (**d**), (**K**), (**J**), (**V**) and (**P**) are known, then its Utilized or Starting Capital must be:

$$\mathbf{U} = ([\mathbf{N} + \mathbf{K} + \mathbf{J} + \mathbf{Vp}/360]$$
$$- [1 + \mathbf{I}]\{\mathbf{S'}[1+\mathbf{s}] - \mathbf{Sv} - \mathbf{Sf} - \mathbf{I} - \mathbf{T}\}[1-\mathbf{d}])/[1+\mathbf{I}]$$

## Rule-58675:

If both (**I**), (**U**), (**N**), (**s**), (**S**), (**v**), (**f**), (**I**), (**T**), (**d**), (**K**), (**J**), (**V**) and (**P**) are known, then its Sales Past must be:

$$\mathbf{S'} = (\{[\mathbf{N} + \mathbf{K} + \mathbf{J} + \mathbf{Vp}/360]/[1+\mathbf{I}] - \mathbf{U}\}/[1-\mathbf{d}]$$
$$+ \mathbf{Sv} + \mathbf{Sf} + \mathbf{I} + \mathbf{T})/[1+\mathbf{s}]$$

or

$$\mathbf{S'} = \mathbf{S}/[1+\mathbf{s}]$$

Steve Asikin ISBN 13: **978-1564912361**, ISBN 10: **1564912363**

Rule-58676:

If both (**/**), (**U**), (**S'**), (**N**), (**S**), (**v**), (**f**), (**I**), (**T**), (**d**), (**K**), (**J**), (**V**) and (**P**) are known, then its Sales Growth planned is:

$$s= (\{[\mathbf{N}+\mathbf{K}+\mathbf{J}+\mathbf{Vp}/360]/[1+\mathbf{/}\text{-}\mathbf{U}\}/[1\text{-}\mathbf{d}]$$
$$+\mathbf{Sv}+\mathbf{Sf}+\mathbf{I}+\mathbf{T})/\mathbf{S'}\text{-}1$$

or

$$s= \mathbf{S}/\mathbf{S'}\text{-}1$$

Rule-58677:

If both (**/**), (**U**), (**S'**), (**s**), (**N**), (**v**), (**f**), (**I**), (**T**), (**d**), (**K**), (**J**), (**V**) and (**P**) are known, then its Sales or Revenue planned is:

$$\mathbf{S}= (\mathbf{U}+\{\mathbf{S'}[1+\mathbf{s}]\text{-}\mathbf{I}\text{-}\mathbf{T}\}[1\text{-}\mathbf{d}]\text{-}[\mathbf{N}+\mathbf{K}+\mathbf{J}+\mathbf{Vp}/360]$$
$$/[1+\mathbf{/}])/\{[1\text{-}\mathbf{d}][\mathbf{v}+\mathbf{f}]\}$$

or

$$\mathbf{S}= \mathbf{S'}[1+\mathbf{s}]$$

or

$$\mathbf{S}= \mathbf{V}/\mathbf{v}$$

Rule-58678:

If both (**/**), (**U**), (**S'**), (**s**), (**S**), (**N**), (**f**), (**I**), (**T**), (**d**), (**K**), (**J**), (**V**) and (**P**) are known, then its Variable Portion planned is:

$$\mathbf{v}= (\mathbf{U}+\{\mathbf{S'}[1+\mathbf{s}]\text{-}\mathbf{Sf}\text{-}\mathbf{I}\text{-}\mathbf{T}\}[1\text{-}\mathbf{d}]\text{-}[\mathbf{N}+\mathbf{K}+\mathbf{J}+\mathbf{Vp}/360]$$
$$/[1+\mathbf{/}])/\{\mathbf{S}[1\text{-}\mathbf{d}]\}$$

or

$$\mathbf{v}= \mathbf{V}/\mathbf{S}$$

Steve Asikin ISBN 13: **978-1564912361**, ISBN 10: **1564912363**

Rule-58679:

If both (**/**), (**U**), (**S'**), (**s**), (**S**), (**v**), (**N**), (**I**), (**T**), (**d**), (**K**), (**J**), (**V**) and (**P**) are known, then its Fixed Portion planned is:

$$f = (U + \{S'[1+s] - Sv - I - T\}[1-d] - [N + K + J + Vp/360] / [1+/])/\{S[1-d]\}$$

Rule-58680:

If both (**/**), (**U**), (**S'**), (**s**), (**S**), (**v**), (**f**), (**N**), (**T**), (**d**), (**K**), (**J**), (**V**) and (**P**) are known, then its Interest Expense planned is:

$$I = S'[1+s] - Sv - Sf - T - \{[N + K + J + Vp/360]/[1+/] - U\} / [1-d]$$

Rule-58681:

If both (**/**), (**U**), (**S'**), (**s**), (**S**), (**v**), (**f**), (**I**), (**N**), (**d**), (**K**), (**J**), (**V**) and (**P**) are known, then its Tax planned is:

$$T = S'[1+s] - Sv - Sf - I - \{[N + K + J + Vp/360]/[1+/] - U\} / [1-d]$$

Rule-58682:

If both (**/**), (**U**), (**S'**), (**s**), (**S**), (**v**), (**f**), (**I**), (**T**), (**N**), (**K**), (**J**), (**V**) and (**P**) are known, then its Dividend Payout planned is:

$$d = 1 - \{[N + K + J + Vp/360]/[1+/] - U\} / \{S'[1+s] - Sv - Sf - I - T\}$$

Steve Asikin ISBN 13: **978-1564912361**, ISBN 10: **1564912363**

Rule-58683:

If both ($I$), (**U**), (**S'**), (**s**), (**S**), (**v**), (**f**), (**I**), (**T**), (**d**), (**N**), (**J**), (**V**) and (**P**) are known, then its Kind of Cash planned is:

$$K= [1+I(U+\{S'[1+s]-Sv-Sf-I-T\}[1-d])$$
$$-[N+J+Vp/360]$$

Rule-58684:

If both ($I$), (**U**), (**S'**), (**s**), (**S**), (**v**), (**f**), (**I**), (**T**), (**d**), (**K**), (**N**), (**V**) and (**P**) are known, then its Job or Trade Account Receivable planned is:

$$J= [1+I(U+\{S'[1+s]-Sv-Sf-I-T\}[1-d])$$
$$-\{N+K+Vp/360\}$$

Rule-58685:

If both ($I$), (**U**), (**S'**), (**s**), (**S**), (**v**), (**f**), (**I**), (**T**), (**d**), (**K**), (**J**), (**N**) and (**p**) are known, then its Variable Cost planned is:

$$V= 360[[1+I(U+\{S'[1+s]-Sv-Sf-I-T\}[1-d])$$
$$-[N+K+J]]/p$$
or
$$V= Sv$$

Rule-58686:

If both ($I$), (**U**), (**S'**), (**s**), (**S**), (**v**), (**f**), (**I**), (**T**), (**d**), (**K**), (**J**), (**V**) and (**N**) are known, then its Procured Inventory planned is:

$$p= 360[[1+I(U+\{S'[1+s]-Sv-Sf-I-T\}[1-d])$$
$$-[N+K+J]]/V$$

Steve Asikin ISBN 13: **978-1564912361**, ISBN 10: **1564912363**

Rule-58687:

If both (*I*), (**U**), (**S'**), (**s**), (**S**), (**v**), (**f**), (**I**), (**T**), (**d**), (**K**), (**J**) and (**p**) are known, then its Non Current Asset planned is:

$$\mathbf{N} = [1+I(\mathbf{U}+\{\mathbf{S'}[1+\mathbf{s}]-\mathbf{Sv}-\mathbf{Sf}-\mathbf{I}-\mathbf{T}\}[1-\mathbf{d}]) -\mathbf{K}-\mathbf{J}-\mathbf{Svp}/360$$

Rule-58688:

If both (**N**), (**U**), (**S'**), (**s**), (**S**), (**v**), (**f**), (**I**), (**T**), (**d**), (**K**), (**J**) and (**p**) are known, then its Leverage or Gearing Ratio planned is:

$$I = [\mathbf{N}+\mathbf{K}+\mathbf{J}+\mathbf{Svp}/360]/(\mathbf{U}+\{\mathbf{S'}[1+\mathbf{s}]-\mathbf{Sv}-\mathbf{Sf}-\mathbf{I}-\mathbf{T}\}[1-\mathbf{d}])-1$$

Rule-58689:

If both (*I*), (**N**), (**S'**), (**s**), (**S**), (**v**), (**f**), (**I**), (**T**), (**d**), (**K**), (**J**) and (**p**) are known, then its Utilized or Starting Capital must be:

$$\mathbf{U} = ([\mathbf{N}+\mathbf{K}+\mathbf{J}+\mathbf{Svp}/360]-[1+I\{\mathbf{S'}[1+\mathbf{s}]-\mathbf{Sv}-\mathbf{Sf}-\mathbf{I}-\mathbf{T}\}[1-\mathbf{d}])/[1+I]$$

Rule-58690:

If both (*I*), (**U**), (**N**), (**s**), (**S**), (**v**), (**f**), (**I**), (**T**), (**d**), (**K**), (**J**) and (**p**) are known, then its Sales Past must be:

$$\mathbf{S'} = (\{[\mathbf{N}+\mathbf{K}+\mathbf{J}+\mathbf{Svp}/360]/[1+I]-\mathbf{U}\}/[1-\mathbf{d}] +\mathbf{Sv}+\mathbf{Sf}+\mathbf{I}+\mathbf{T})/[1+\mathbf{s}]$$

or

$$\mathbf{S'} = \mathbf{S}/[1+\mathbf{s}]$$

Steve Asikin ISBN 13: **978-1564912361**, ISBN 10: **1564912363**

Rule-58691:

If both (**/**), (**U**), (**S'**), (**N**), (**S**), (**v**), (**f**), (**I**), (**T**), (**d**), (**K**), (**J**) and (**p**) are known, then its Sales Growth planned is:

$$s = (\{[\mathbf{N}+\mathbf{K}+\mathbf{J}+\mathbf{Svp}/360]/[1+\mathbf{/}-\mathbf{U}\}/[1-\mathbf{d}]$$
$$+\mathbf{Sv}+\mathbf{Sf}+\mathbf{I}+\mathbf{T})/\mathbf{S'}-1$$

or

$$s = \mathbf{S}/\mathbf{S'}-1$$


Rule-58692:

If both (**/**), (**U**), (**S'**), (**s**), (**N**), (**v**), (**f**), (**I**), (**T**), (**d**), (**K**), (**J**) and (**p**) are known, then its Sales or Revenue planned is:

$$\mathbf{S} = [[1+\mathbf{/}(\mathbf{U}+\{\mathbf{S'}[1+\mathbf{s}]-\mathbf{I}-\mathbf{T}\}[1-\mathbf{d}])-[\mathbf{N}+\mathbf{K}+\mathbf{J}]]$$
$$/\{\mathbf{vp}/360+[1+\mathbf{/}][1-\mathbf{d}][\mathbf{f}+\mathbf{v}]\}$$

or

$$\mathbf{S} = \mathbf{S'}[1+\mathbf{s}]$$


Rule-58693:

If both (**/**), (**U**), (**S'**), (**s**), (**S**), (**N**), (**f**), (**I**), (**T**), (**d**), (**K**), (**J**) and (**p**) are known, then its Variable Portion planned is:

$$\mathbf{v} = [[1+\mathbf{/}(\mathbf{U}+\{\mathbf{S'}[1+\mathbf{s}]-\mathbf{Sf}-\mathbf{I}-\mathbf{T}\}[1-\mathbf{d}])-[\mathbf{N}+\mathbf{K}+\mathbf{J}]]$$
$$/(\mathbf{S}\{\mathbf{p}/360+[1+\mathbf{/}][1-\mathbf{d}]\})$$

Steve Asikin ISBN 13: **978-1564912361**, ISBN 10: **1564912363**

Rule-58694:

If both $(I)$, $(U)$, $(S')$, $(s)$, $(S)$, $(v)$, $(N)$, $(I)$, $(T)$, $(d)$, $(K)$, $(J)$ and $(p)$ are known, then its Fixed Portion planned is:

$$f = (S'[1+s]-Sv-I-\{[N+K+J+Svp/360]/[1+I-U\}$$
$$/[1-d])/S$$

Rule-58695:

If both $(I)$, $(U)$, $(S')$, $(s)$, $(S)$, $(v)$, $(f)$, $(N)$, $(T)$, $(d)$, $(K)$, $(J)$ and $(p)$ are known, then its Interest Expense planned is:

$$I = S'[1+s]-Sv-Sf-T-\{[N+K+J+Svp/360]/[1+I-U\}$$
$$/[1-d]$$

Rule-58696:

If both $(I)$, $(U)$, $(S')$, $(s)$, $(S)$, $(v)$, $(f)$, $(I)$, $(N)$, $(d)$, $(K)$, $(J)$ and $(p)$ are known, then its Tax planned is:

$$T = S'[1+s]-Sv-Sf-I-\{[N+K+J+Svp/360]/[1+I-U\}$$
$$/[1-d]$$

Rule-58697:

If both $(I)$, $(U)$, $(S')$, $(s)$, $(S)$, $(v)$, $(f)$, $(I)$, $(T)$, $(N)$, $(K)$, $(J)$ and $(p)$ are known, then its Dividend Payout planned is:

$$d = 1-\{[N+K+J+Svp/360]/[1+I-U\}$$
$$/\{S'[1+s]-Sv-Sf-I-T\}$$

Steve Asikin ISBN 13: **978-1564912361**, ISBN 10: **1564912363**

Rule-58698:
> If both (**Ɩ**), (**U**), (**S'**), (**s**), (**S**), (**v**), (**f**), (**I**), (**T**), (**d**), (**N**),
> (**J**) and (**p**) are known, then its Kind of Cash planned
> is:
> $$K= [1+Ɩ(U+\{S'[1+s]-Sv-Sf-I-T\}[1-d])$$
> $$-[N+J+Svp/360]$$

Rule-58699:
> If both (**Ɩ**), (**U**), (**S'**), (**s**), (**S**), (**v**), (**f**), (**I**), (**T**), (**d**), (**K**),
> (**N**) and (**p**) are known, then its Job or Trade
> Receivable Days planned is:
> $$J= [1+Ɩ(U+\{S'[1+s]-Sv-Sf-I-T\}[1-d])$$
> $$-\{N+K+Svp/360\}$$

Rule-58700:
> If both (**Ɩ**), (**U**), (**S'**), (**s**), (**S**), (**v**), (**f**), (**I**), (**T**), (**d**), (**K**),
> (**J**) and (**N**) are known, then its Procured Inventory
> Days planned is:
> $$p= 360[[1+Ɩ(U+\{S'[1+s]-Sv-Sf-I-T\}[1-d])$$
> $$-[N+K+J]]/[Sv]$$

Rule-58701:
> If both (**Ɩ**), (**U**), (**S'**), (**s**), (**S**), (**v**), (**f**), (**I**), (**T**), (**d**), (**K**),
> (**J**) and (**p**) are known, then its Non Current Asset
> planned is:
> $$N= [1+Ɩ(U+\{S'[1+s]-Sv-Sf-I-T\}[1-d])$$
> $$-K-J-S'vp[1+s]/360$$

Steve Asikin ISBN 13: **978-1564912361**, ISBN 10: **1564912363**

## Rule-58702:

If both $(\textit{I})$, $(\mathbf{U})$, $(\mathbf{S'})$, $(\mathbf{s})$, $(\mathbf{S})$, $(\mathbf{v})$, $(\mathbf{f})$, $(\mathbf{I})$, $(\mathbf{T})$, $(\mathbf{d})$, $(\mathbf{K})$, $(\mathbf{J})$ and $(\mathbf{p})$ are known, then its Non Current Asset planned is:

$$\textit{I} = \{\mathbf{N}+\mathbf{K}+\mathbf{J}+\mathbf{S'vp}[1+\mathbf{s}]/360\}$$
$$/(\mathbf{U}+\{\mathbf{S'}[1+\mathbf{s}]-\mathbf{Sv}-\mathbf{Sf}-\mathbf{I}-\mathbf{T}\}[1-\mathbf{d}])-1$$

## Rule-58703:

If both $(\textit{I})$, $(\mathbf{N})$, $(\mathbf{S'})$, $(\mathbf{s})$, $(\mathbf{S})$, $(\mathbf{v})$, $(\mathbf{f})$, $(\mathbf{I})$, $(\mathbf{T})$, $(\mathbf{d})$, $(\mathbf{K})$, $(\mathbf{J})$ and $(\mathbf{p})$ are known, then its Utilized or Starting Capital must be:

$$\mathbf{U} = (\{\mathbf{N}+\mathbf{K}+\mathbf{J}+\mathbf{S'vp}[1+\mathbf{s}]/360\}$$
$$-[1+\textit{I}]\{\mathbf{S'}[1+\mathbf{s}]-\mathbf{Sv}-\mathbf{Sf}-\mathbf{I}-\mathbf{T}\}[1-\mathbf{d}])/[1+\textit{I}]$$

## Rule-58704:

If both $(\textit{I})$, $(\mathbf{U})$, $(\mathbf{N})$, $(\mathbf{s})$, $(\mathbf{S})$, $(\mathbf{v})$, $(\mathbf{f})$, $(\mathbf{I})$, $(\mathbf{T})$, $(\mathbf{d})$, $(\mathbf{K})$, $(\mathbf{J})$ and $(\mathbf{p})$ are known, then its Sales Past must be:

$$\mathbf{S'} = [[1+\textit{I}](\mathbf{U}-\{\mathbf{Sv}+\mathbf{Sf}+\mathbf{I}+\mathbf{T}\}[1-\mathbf{d}])-[\mathbf{N}+\mathbf{K}+\mathbf{J}]]$$
$$/([1+\mathbf{s}]\{\mathbf{vp}/360-[1+\textit{I}][1-\mathbf{d}]\})$$
$$\text{or}$$
$$\mathbf{S'} = \mathbf{S}/[1+\mathbf{s}]$$

Steve Asikin ISBN 13: **978-1564912361**, ISBN 10: **1564912363**

Rule-58705:

If both (**/**), (**U**), (**S'**), (**N**), (**S**), (**v**), (**f**), (**I**), (**T**), (**d**), (**K**), (**J**) and (**p**) are known, then its Sales Growth planned is:

$$s= [[1+\text{/}](U-\{Sv+Sf+I+T\}[1-d])-[N+K+J]]$$
$$/(S'\{vp/360-[1+\text{/}][1-d]\})-1$$

or

$$s= S/S'-1$$


Rule-58706:

If both (**/**), (**U**), (**S'**), (**s**), (**N**), (**v**), (**f**), (**I**), (**T**), (**d**), (**K**), (**J**) and (**p**) are known, then its Sales or Revenue planned is:

$$S= (U+\{S'[1+s]-I-T\}[1-d]-\{N+K+J$$
$$+S'vp[1+s]/360\}/[1+\text{/}])/\{[v+f][1-d]\}$$

or

$$S= S'[1+s]$$


Rule-58707:

If both (**/**), (**U**), (**S'**), (**s**), (**S**), (**N**), (**f**), (**I**), (**T**), (**d**), (**K**), (**J**) and (**p**) are known, then its Variable Portion planned is:

$$v= [[1+\text{/}](U+\{S'[1+s]-Sf-I-T\}[1-d])-[N+K+J]]$$
$$/\{S'p[1+s]/360+S[1+\text{/}][1-d]\}$$

Steve Asikin ISBN 13: **978-1564912361**, ISBN 10: **1564912363**

Rule-58708:

If both (**/**), (**U**), (**S'**), (**s**), (**S**), (**v**), (**N**), (**I**), (**T**), (**d**), (**K**), (**J**) and (**p**) are known, then its Fixed Portion planned is:

$$f= [S'[1+s]-Sv-I-T-(\{N+K+J+S'vp[1+s]/360\}/[1+/-U)/[1-d]]/S$$

Rule-58709:

If both (**/**), (**U**), (**S'**), (**s**), (**S**), (**v**), (**f**), (**N**), (**T**), (**d**), (**K**), (**J**) and (**p**) are known, then its Interest Expense planned is:

$$I= S'[1+s]-Sv-Sf-T-(\{N+K+J+S'vp[1+s]/360\}/[1+/-U)/[1-d]$$

Rule-58710:

If both (**/**), (**U**), (**S'**), (**s**), (**S**), (**v**), (**f**), (**I**), (**N**), (**d**), (**K**), (**J**) and (**p**) are known, then its Tax planned is:

$$T= S'[1+s]-Sv-Sf-I-(\{N+K+J+S'vp[1+s]/360\}/[1+/-U)/[1-d]$$

Rule-58711:

If both (**/**), (**U**), (**S'**), (**s**), (**S**), (**v**), (**f**), (**I**), (**T**), (**N**), (**K**), (**J**) and (**p**) are known, then its Dividend Payout planned is:

$$d= 1-(\{N+K+J+S'vp[1+s]/360\}/[1+/-U)/\{S'[1+s]-Sv-Sf-I-T\}$$

Steve Asikin ISBN 13: **978-1564912361**, ISBN 10: **1564912363**

Rule-58712:

If both (**/**), (**U**), (**S'**), (**s**), (**S**), (**v**), (**f**), (**I**), (**T**), (**d**), (**N**), (**J**) and (**p**) are known, then its Kind of Cash planned is:

$$K= [1+/(U+\{S'[1+s]-Sv-Sf-I-T\}[1-d])$$
$$-\{N+J+S'vp[1+s]/360\}$$

Rule-58713:

If both (**/**), (**U**), (**S'**), (**s**), (**S**), (**v**), (**f**), (**I**), (**T**), (**d**), (**K**), (**N**) and (**p**) are known, then its Job or Trade Receivable Days planned is:

$$J= [[1+/(U+\{S'[1+s]-Sv-Sf-I-T\}[1-d])$$
$$-\{N+K+S'vp[1+s]/360\}$$

Rule-58714:

If both (**/**), (**U**), (**S'**), (**s**), (**S**), (**v**), (**f**), (**I**), (**T**), (**d**), (**K**), (**J**) and (**N**) are known, then its Procured Inventory Days planned is:

$$p= 360[[1+/(U+\{S'[1+s]-Sv-Sf-I-T\}[1-d])$$
$$-[N+K+J]]/\{S'v[1+s]\}$$

Rule-58715:

If both (**/**), (**U**), (**S'**), (**s**), (**S**), (**v**), (**f**), (**I**), (**T**), (**d**), (**K**), (**j**) and (**p**) are known, then its Non Current Asset planned is:

$$N= [1+/(U+\{S'[1+s]-Sv-Sf-I-T\}[1-d])$$
$$-K-Sj/360-P$$

Steve Asikin ISBN 13: **978-1564912361**, ISBN 10: **1564912363**

Rule-58716:

If both (**N**), (**U**), (**S'**), (**s**), (**S**), (**v**), (**f**), (**I**), (**T**), (**d**), (**K**), (**j**) and (**p**) are known, then its Leverage or Gearing Ratio planned is:

$$I= \{N+K+Sj/360+P\}$$
$$/(U+\{S'[1+s]-Sv-Sf-I-T\}[1-d])-1$$

Rule-58717:

If both (**I**), (**N**), (**S'**), (**s**), (**S**), (**v**), (**f**), (**I**), (**T**), (**d**), (**K**), (**j**) and (**p**) are known, then its Utilized or Starting Capital must be:

$$U= (\{N+K+Sj/360+P\}-[1+I]\{S'[1+s]-Sv-Sf-I-T\}$$
$$[1-d])/[1+I]$$

Rule-58718:

If both (**I**), (**U**), (**N**), (**s**), (**S**), (**v**), (**f**), (**I**), (**T**), (**d**), (**K**), (**j**) and (**p**) are known, then its Sales Past must be:

$$S'= (\{[N+K+Sj/360+P]-[1+I]-U\}/[1-d]$$
$$+Sv+Sf+I+T\})/[1+s]$$

or

$$S'= S/[1+s]$$

Steve Asikin ISBN 13: **978-1564912361**, ISBN 10: **1564912363**

Rule-58719:

If both $(\textit{\textbf{l}})$, $(\textbf{U})$, $(\textbf{S'})$, $(\textbf{N})$, $(\textbf{S})$, $(\textbf{v})$, $(\textbf{f})$, $(\textbf{I})$, $(\textbf{T})$, $(\textbf{d})$, $(\textbf{K})$, $(\textbf{j})$ and $(\textbf{p})$ are known, then its Sales Growth planned is:

$$s = (\{[N+K+Sj/360+P]-[1+\textit{l}-U\}/[1-d] + Sv+Sf+I+T\})/S'-1$$

or

$$s = S/S'-1$$

Rule-58720:

If both $(\textit{\textbf{l}})$, $(\textbf{U})$, $(\textbf{S'})$, $(\textbf{s})$, $(\textbf{N})$, $(\textbf{v})$, $(\textbf{f})$, $(\textbf{I})$, $(\textbf{T})$, $(\textbf{d})$, $(\textbf{K})$, $(\textbf{j})$ and $(\textbf{p})$ are known, then its Sales or Revenue planned is:

$$S = [[1+\textit{l}(U+\{S'[1+s]-I-T\}[1-d])-[N+K+P]] / \{j/360+[1+\textit{l}[1-d][v+f]\}$$

or

$$S = S'[1+s]$$

Rule-58721:

If both $(\textit{\textbf{l}})$, $(\textbf{U})$, $(\textbf{S'})$, $(\textbf{s})$, $(\textbf{S})$, $(\textbf{N})$, $(\textbf{f})$, $(\textbf{I})$, $(\textbf{T})$, $(\textbf{d})$, $(\textbf{K})$, $(\textbf{j})$ and $(\textbf{p})$ are known, then its Variable Portion planned is:

$$v = (S'[1+s]-Sf-I-T-\{[N+K+Sj/360+P]-[1+\textit{l}-U\} /[1-d])/S$$

Steve Asikin ISBN 13: **978-1564912361**, ISBN 10: **1564912363**

Rule-58722:

If both (**/**), (**U**), (**S'**), (**s**), (**S**), (**v**), (**N**), (**I**), (**T**), (**d**), (**K**), (**j**) and (**p**) are known, then its Fixed Portion planned is:

$$f = (S'[1+s]-Sv-I-T-\{[N+K+Sj/360+P]-[1+/]-U\}/[1-d])/S$$

Rule-58723:

If both (**/**), (**U**), (**S'**), (**s**), (**S**), (**v**), (**f**), (**N**), (**T**), (**d**), (**K**), (**j**) and (**p**) are known, then its Interest Expense planned is:

$$I = S'[1+s]-Sv-Sf-T-\{[N+K+Sj/360+P]-[1+/]-U\}/[1-d]$$

Rule-58724:

If both (**/**), (**U**), (**S'**), (**s**), (**S**), (**v**), (**f**), (**I**), (**N**), (**d**), (**K**), (**j**) and (**p**) are known, then its Tax planned is:

$$T = S'[1+s]-Sv-Sf-I-\{[N+K+Sj/360+P]-[1+/]-U\}/[1-d]$$

Rule-58725:

If both (**/**), (**U**), (**S'**), (**s**), (**S**), (**v**), (**f**), (**I**), (**T**), (**N**), (**K**), (**j**) and (**p**) are known, then its Dividend Payout planned is:

$$d = 1-\{[N+K+Sj/360+P]-[1+/]-U\}/\{S'[1+s]-Sv-Sf-I-T\}$$

Steve Asikin ISBN 13: **978-1564912361**, ISBN 10: **1564912363**

## Rule-58726:

If both $(I)$, $(U)$, $(S')$, $(s)$, $(S)$, $(v)$, $(f)$, $(I)$, $(T)$, $(d)$, $(N)$, $(j)$ and $(p)$ are known, then its Kind of Cash planned is:

$$K = [1+I](U+\{S'[1+s]-Sv-Sf-I-T\}[1-d])$$
$$-[N+Sj/360+P]$$

## Rule-58727:

If both $(I)$, $(U)$, $(S')$, $(s)$, $(S)$, $(v)$, $(f)$, $(I)$, $(T)$, $(d)$, $(K)$, $(N)$ and $(p)$ are known, then its Job or Trade Receivable Days planned is:

$$j = 360[[1+I](U+\{S'[1+s]-Sv-Sf-I-T\}[1-d])$$
$$-\{N+K+P\}]/S$$

## Rule-58728:

If both $(I)$, $(U)$, $(S')$, $(s)$, $(S)$, $(v)$, $(f)$, $(I)$, $(T)$, $(d)$, $(K)$, $(j)$ and $(P)$ are known, then its Non Current Asset planned is:

$$P = [1+I](U+\{S'[1+s]-Sv-Sf-I-T\}[1-d])$$
$$-\{N+K+Sj/360\}$$

## Rule-58729:

If both $(I)$, $(U)$, $(S')$, $(s)$, $(S)$, $(v)$, $(f)$, $(I)$, $(T)$, $(d)$, $(K)$, $(j)$, $(V)$ and $(p)$ are known, then its Non Current Asset planned is:

$$N = [1+I](U+\{S'[1+s]-Sv-Sf-I-T\}[1-d])$$
$$-K-Sj/360-Vp/360$$

Steve Asikin ISBN 13: **978-1564912361**, ISBN 10: **1564912363**

Rule-58730:

If both (**N**), (**U**), (**S'**), (**s**), (**S**), (**v**), (**f**), (**I**), (**T**), (**d**), (**K**), (**j**), (**V**) and (**p**) are known, then its Leverage or Gearing Ratio planned is:

$$\textit{l} = [N+K+Sj/360+Vp/360]$$
$$/(U+\{S'[1+s]-Sv-Sf-I-T\}[1-d])-1$$

Rule-58731:

If both (**l**), (**N**), (**S'**), (**s**), (**S**), (**v**), (**f**), (**I**), (**T**), (**d**), (**K**), (**j**), (**V**) and (**p**) are known, then its Utilized or starting Capital must be:

$$U = ([N+K+Sj/360+Vp/360]$$
$$-\lceil 1+\textit{l}\rceil\{S'[1+s]-Sv-Sf-I-T\}[1-d])/[1+\textit{l}]$$

Rule-58732:

If both (**l**), (**U**), (**N**), (**s**), (**S**), (**v**), (**f**), (**I**), (**T**), (**d**), (**K**), (**j**), (**V**) and (**p**) are known, then its Sales Past must be:

$$S' = (\{[N+K+Sj/360+Vp/360]/[1+\textit{l}]-U\}/[1-d]$$
$$+Sv+Sf+I+T)/[1+s]$$

or

$$S' = S[1+s]$$

Steve Asikin ISBN 13: **978-1564912361**, ISBN 10: **1564912363**

Rule-58733:

If both (**/**), (**U**), (**S'**), (**N**), (**S**), (**v**), (**f**), (**I**), (**T**), (**d**), (**K**), (**j**), (**V**) and (**p**) are known, then its Sales Growth planned is:

$$s= (\{[\mathbf{N+K+Sj}/360+\mathbf{Vp}/360]/[1+\mathbf{/}-\mathbf{U}\}/[1-\mathbf{d}]$$
$$+\mathbf{Sv+Sf+I+T})/\mathbf{S'}-1$$

or

$$s= \mathbf{S}/\mathbf{S'}-1$$


Rule-58734:

If both (**/**), (**U**), (**S'**), (**s**), (**N**), (**v**), (**f**), (**I**), (**T**), (**d**), (**K**), (**j**), (**V**) and (**p**) are known, then its Sales or Revenue planned is:

$$S= [[1+\mathbf{/}(\mathbf{U}+\{\mathbf{S'}[1+\mathbf{s}]-\mathbf{I}-\mathbf{T}\}[1-\mathbf{d}])-[\mathbf{N+K+Vp}/360]]$$
$$/\{\mathbf{j}/360+[1+\mathbf{/}][1-\mathbf{d}][\mathbf{v+f}]\}$$

or

$$S= \mathbf{S'}[1+\mathbf{s}]$$


Rule-58735:

If both (**/**), (**U**), (**S'**), (**s**), (**S**), (**N**), (**f**), (**I**), (**T**), (**d**), (**K**), (**j**), (**V**) and (**p**) are known, then its Variable Portion planned is:

$$v= (\mathbf{S'}[1+\mathbf{s}]-\mathbf{Sf}-\mathbf{I}-\mathbf{T}-\{[\mathbf{N+K+Sj}/360+\mathbf{Vp}/360]$$
$$/[1+\mathbf{/}-\mathbf{U}\}/[1-\mathbf{d}])/\mathbf{S}$$

or

$$v= \mathbf{V}/\mathbf{S}$$

Steve Asikin ISBN 13: **978-1564912361**, ISBN 10: **1564912363**

## Rule-58736:

If both (**/**), (**U**), (**S'**), (**s**), (**S**), (**v**), (**N**), (**I**), (**T**), (**d**), (**K**), (**j**), (**V**) and (**p**) are known, then its Fixed Portion planned is:

$$f = (S'[1+s]-Sv-I-T-\{[N+K+Sj/360+Vp/360] /[1+/-U\}/[1-d])/S$$

## Rule-58737:

If both (**/**), (**U**), (**S'**), (**s**), (**S**), (**v**), (**f**), (**N**), (**T**), (**d**), (**K**), (**j**), (**V**) and (**p**) are known, then its Interest Expense planned is:

$$I = S'[1+s]-Sv-Sf-T-\{[N+K+Sj/360+Vp/360] /[1+/-U\}/[1-d]$$

## Rule-58738:

If both (**/**), (**U**), (**S'**), (**s**), (**S**), (**v**), (**f**), (**I**), (**N**), (**d**), (**K**), (**j**), (**V**) and (**p**) are known, then its Tax planned is:

$$T = S'[1+s]-Sv-Sf-I-\{[N+K+Sj/360+Vp/360] /[1+/-U\}/[1-d]$$

## Rule-58739:

If both (**/**), (**U**), (**S'**), (**s**), (**S**), (**v**), (**f**), (**I**), (**T**), (**N**), (**K**), (**j**), (**V**) and (**p**) are known, then its Dividend Payout planned is:

$$d = 1-\{[N+K+Sj/360+Vp/360]/[1+/-U\} /\{S'[1+s]-Sv-Sf-I-T\}$$

Steve Asikin ISBN 13: **978-1564912361**, ISBN 10: **1564912363**

Rule-58740:
>  If both (**/**), (**U**), (**S'**), (**s**), (**S**), (**v**), (**f**), (**I**), (**T**), (**d**), (**N**), (**j**), (**V**) and (**p**) are known, then its Kind of Cash planned is:
>> $K= [1+/(U+\{S'[1+s]-Sv-Sf-I-T\}[1-d])$
>> $-[N+Sj/360+Vp/360]$

Rule-58741:
>  If both (**/**), (**U**), (**S'**), (**s**), (**S**), (**v**), (**f**), (**I**), (**T**), (**d**), (**K**), (**N**), (**V**) and (**p**) are known, then its Job or Trade Receivable Days planned is:
>> $j= 360[[1+/(U+\{S'[1+s]-Sv-Sf-I-T\}[1-d])$
>> $-[N+K+Vp/360]]/S$

Rule-58742:
>  If both (**/**), (**U**), (**S'**), (**s**), (**S**), (**v**), (**f**), (**I**), (**T**), (**d**), (**K**), (**j**), (**V**) and (**p**) are known, then its Non Current Asset planned is:
>> $V= 360[[1+/(U+\{S'[1+s]-Sv-Sf-I-T\}[1-d])$
>> $-[N+K+Sj/360]]/p$
>>> or
>> $V= Sv$

Rule-58743:
>  If both (**/**), (**U**), (**S'**), (**s**), (**S**), (**v**), (**f**), (**I**), (**T**), (**d**), (**K**), (**j**), (**V**) and (**N**) are known, then its Procured Inventory Days planned is:
>> $p= 360[[1+/(U+\{S'[1+s]-Sv-Sf-I-T\}[1-d])$
>> $-[N+K+Sj/360]]/V$

Steve Asikin ISBN 13: **978-1564912361**, ISBN 10: **1564912363**

## Rule-58744:

If both ($I$), (**U**), (**S'**), (**s**), (**S**), (**v**), (**f**), (**I**), (**T**), (**d**), (**K**), (**j**) and (**p**) are known, then its Non Current Asset planned is:

$$\mathbf{N} = [1+I](\mathbf{U}+\{\mathbf{S'}[1+\mathbf{s}]-\mathbf{Sv}-\mathbf{Sf}-\mathbf{I}-\mathbf{T}\}[1-\mathbf{d}])$$
$$-\mathbf{K}-\mathbf{Sj}/360-\mathbf{Svp}/360$$

## Rule-58745:

If both (**N**), (**U**), (**S'**), (**s**), (**S**), (**v**), (**f**), (**I**), (**T**), (**d**), (**K**), (**j**) and (**p**) are known, then its Leverage or Gearing Ratio planned is:

$$I = [\mathbf{N}+\mathbf{K}+\mathbf{Sj}/360+\mathbf{Svp}/360]$$
$$/(\mathbf{U}+\{\mathbf{S'}[1+\mathbf{s}]-\mathbf{Sv}-\mathbf{Sf}-\mathbf{I}-\mathbf{T}\}[1-\mathbf{d}])-1$$

## Rule-58746:

If both ($I$), (**N**), (**S'**), (**s**), (**S**), (**v**), (**f**), (**I**), (**T**), (**d**), (**K**), (**j**) and (**p**) are known, then its Utilized or Starting Capital must be:

$$\mathbf{U} = ([\mathbf{N}+\mathbf{K}+\mathbf{Sj}/360+\mathbf{Svp}/360]$$
$$-[1+I]\{\mathbf{S'}[1+\mathbf{s}]-\mathbf{Sv}-\mathbf{Sf}-\mathbf{I}-\mathbf{T}\}[1-\mathbf{d}])/[1+I]$$

## Rule-58747:

If both ($I$), (**U**), (**N**), (**s**), (**S**), (**v**), (**f**), (**I**), (**T**), (**d**), (**K**), (**j**) and (**p**) are known, then its Sales Past must be:

$$\mathbf{S'} = (\{[\mathbf{N}+\mathbf{K}+\mathbf{Sj}/360+\mathbf{Svp}/360]/[1+I]-\mathbf{U}\}/[1-\mathbf{d}]$$
$$+\mathbf{Sv}+\mathbf{Sf}+\mathbf{I}+\mathbf{T})/[1+\mathbf{s}]$$

or

$$\mathbf{S'} = \mathbf{S}/[1+\mathbf{s}]$$

Steve Asikin ISBN 13: **978-1564912361**, ISBN 10: **1564912363**

Rule-58748:

If both (**/**), (**U**), (**S'**), (**N**), (**S**), (**v**), (**f**), (**I**), (**T**), (**d**), (**K**), (**j**) and (**p**) are known, then its Sales Growth planned is:

$$s= (\{[N+K+Sj/360+Svp/360]/[1+/]-U\}/[1-d]$$
$$+Sv+Sf+I+T)/S'-1$$

or

$$s= S/S'-1$$

Rule-58749:

If both (**/**), (**U**), (**S'**), (**s**), (**N**), (**v**), (**f**), (**I**), (**T**), (**d**), (**K**), (**j**) and (**p**) are known, then its Sales or Revenue planned is:

$$S= [[1+/](U+\{S'[1+s]-Sv-Sf-I-T\}[1-d])-[N+K]]$$
$$/(j/360+v\{p/360+[1+/][1-d]\})$$

or

$$S= S'[1+s]$$

Rule-58750:

If both (**/**), (**U**), (**S'**), (**s**), (**S**), (**N**), (**f**), (**I**), (**T**), (**d**), (**K**), (**j**) and (**p**) are known, then its Variable Portion planned is:

$$v= [[1+/](U+\{S'[1+s]-Sf-I-T\}[1-d])$$
$$-[N+K+Sj/360]]/\{Sp/360+[1+/][1-d]\}$$

Steve Asikin ISBN 13: **978-1564912361**, ISBN 10: **1564912363**

## Rule-58751:

If both ($\mathit{l}$), (**U**), (**S'**), (**s**), (**S**), (**v**), (**N**), (**I**), (**T**), (**d**), (**K**), (**j**) and (**p**) are known, then its Fixed Portion planned is:

$$f = (S'[1+s]-Sv-I-T-\{[N+K+Sj/360+Svp/360]\\ /[1+\mathit{l}-U\}/[1-d])/S$$

## Rule-58752:

If both ($\mathit{l}$), (**U**), (**S'**), (**s**), (**S**), (**v**), (**f**), (**N**), (**T**), (**d**), (**K**), (**j**) and (**p**) are known, then its Interest Expense planned is:

$$I = S'[1+s]-Sv-Sf-T-\{[N+K+Sj/360+Svp/360]\\ /[1+\mathit{l}-U\}/[1-d]$$

## Rule-58753:

If both ($\mathit{l}$), (**U**), (**S'**), (**s**), (**S**), (**v**), (**f**), (**I**), (**N**), (**d**), (**K**), (**j**) and (**p**) are known, then its Tax planned is:

$$T = S'[1+s]-Sv-Sf-I-\{[N+K+Sj/360+Svp/360]\\ /[1+\mathit{l}-U\}/[1-d]$$

## Rule-58754:

If both ($\mathit{l}$), (**U**), (**S'**), (**s**), (**S**), (**v**), (**f**), (**I**), (**T**), (**D**), (**K**), (**j**) and (**p**) are known, then its Dividend Payout planned is:

$$d = 1-\{[N+K+Sj/360+Svp/360]/[1+\mathit{l}-U\}\\ /\{S'[1+s]-Sv-Sf-I-T\}$$

Steve Asikin ISBN 13: **978-1564912361**, ISBN 10: **1564912363**

Rule-58755:

If both (**/**), (**U**), (**S'**), (**s**), (**S**), (**v**), (**f**), (**I**), (**T**), (**d**), (**N**), (**j**) and (**p**) are known, then its Kind of Cash planned is:

$$\mathbf{K} = [1 + \mathbf{/}(\mathbf{U} + \{\mathbf{S'}[1+\mathbf{s}] - \mathbf{Sv} - \mathbf{Sf} - \mathbf{I} - \mathbf{T}\}[1-\mathbf{d}])$$
$$-[\mathbf{N} + \mathbf{Sj}/360 + \mathbf{Svp}/360]$$

Rule-58756:

If both (**/**), (**U**), (**S'**), (**s**), (**S**), (**v**), (**f**), (**I**), (**T**), (**d**), (**K**), (**N**) and (**p**) are known, then its Job or Trade Receivable Days planned is:

$$\mathbf{j} = 360[[1 + \mathbf{/}(\mathbf{U} + \{\mathbf{S'}[1+\mathbf{s}] - \mathbf{Sv} - \mathbf{Sf} - \mathbf{I} - \mathbf{T}\}[1-\mathbf{d}])$$
$$-\{\mathbf{N} + \mathbf{K} + \mathbf{Svp}/360\}]/\mathbf{S}$$

Rule-58757:

If both (**/**), (**U**), (**S'**), (**s**), (**S**), (**v**), (**f**), (**I**), (**T**), (**d**), (**K**), (**j**) and (**N**) are known, then its Procured Inventory Days planned is:

$$\mathbf{p} = 360[[1 + \mathbf{/}(\mathbf{U} + \{\mathbf{S'}[1+\mathbf{s}] - \mathbf{Sv} - \mathbf{Sf} - \mathbf{I} - \mathbf{T}\}[1-\mathbf{d}])$$
$$-[\mathbf{N} + \mathbf{K} + \mathbf{Sj}/360]]/[\mathbf{Sv}]$$

Rule-58758:

If both (**/**), (**U**), (**S'**), (**s**), (**S**), (**v**), (**f**), (**I**), (**T**), (**d**), (**K**), (**j**) and (**p**) are known, then its Non Current Asset planned is:

$$\mathbf{N} = [1 + \mathbf{/}(\mathbf{U} + \{\mathbf{S'}[1+\mathbf{s}] - \mathbf{Sv} - \mathbf{Sf} - \mathbf{I} - \mathbf{T}\}[1-\mathbf{d}])$$
$$-\mathbf{K} - \mathbf{Sj}/360 - \mathbf{S'vp}[1+\mathbf{s}]/360$$

Steve Asikin ISBN 13: **978-1564912361**, ISBN 10: **1564912363**

## Rule-58759:

If both (**N**), (**U**), (**S'**), (**s**), (**S**), (**v**), (**f**), (**I**), (**T**), (**d**), (**K**), (**j**) and (**p**) are known, then its Leverage or Gearing Ratio planned is:

$$I = \{N+K+Sj/360+S'vp[1+s]/360\} / (U+\{S'[1+s]-Sv-Sf-I-T\}[1-d])-1$$

## Rule-58760:

If both (**I**), (**N**), (**S'**), (**s**), (**S**), (**v**), (**f**), (**I**), (**T**), (**d**), (**K**), (**j**) and (**p**) are known, then its Utilized or Starting Capital must be:

$$U = (\{N+K+Sj/360+S'vp[1+s]/360\}-[1+I] \{S'[1+s]-Sv-Sf-I-T\}[1-d])/[1+I]$$

## Rule-58761:

If both (**I**), (**U**), (**S'**), (**s**), (**S**), (**v**), (**f**), (**I**), (**T**), (**d**), (**K**), (**j**) and (**p**) are known, then its Sales Past must be:

$$S' = [[1+I](U-\{Sv+Sf+I+T\}[1-d])-[N+K+Sj/360]] /([1+s]\{vp/360-[1+I][1-d]\})$$

or

$$S' - S/[1+s]$$

Steve Asikin ISBN 13: **978-1564912361**, ISBN 10: **1564912363**

Rule-58762:

If both $(\textit{\textbf{I}})$, $(\textbf{U})$, $(\textbf{S'})$, $(\textbf{N})$, $(\textbf{S})$, $(\textbf{v})$, $(\textbf{f})$, $(\textbf{I})$, $(\textbf{T})$, $(\textbf{d})$, $(\textbf{K})$, $(\textbf{j})$ and $(\textbf{p})$ are known, then its Sales Growth planned is:

$$s= [[1+\textit{I}(U-\{Sv+Sf+I+T\}[1-d])-[N+K+Sj/360]]$$
$$/(S'\{vp/360-[1+\textit{I}[1-d]\})-1$$

or

$$s= S/S'-1$$

Rule-58763:

If both $(\textit{\textbf{I}})$, $(\textbf{U})$, $(\textbf{S'})$, $(\textbf{s})$, $(\textbf{N})$, $(\textbf{v})$, $(\textbf{f})$, $(\textbf{I})$, $(\textbf{T})$, $(\textbf{d})$, $(\textbf{K})$, $(\textbf{j})$ and $(\textbf{p})$ are known, then its Sales or Revenue planned is:

$$S= [[1+\textit{I}(U+\{S'[1+s]-I-T\}[1-d])$$
$$-\{N+K+S'vp[1+s]/360\}]$$
$$/\{j/360+[1+\textit{I}[1-d][v+f]\}$$

or

$$S= S'[1+s]$$

Rule-58764:

If both $(\textit{\textbf{I}})$, $(\textbf{U})$, $(\textbf{S'})$, $(\textbf{s})$, $(\textbf{S})$, $(\textbf{N})$, $(\textbf{f})$, $(\textbf{I})$, $(\textbf{T})$, $(\textbf{d})$, $(\textbf{K})$, $(\textbf{j})$ and $(\textbf{p})$ are known, then its Variable Portion planned is:

$$v= [[1+\textit{I}(U+\{S'[1+s]-Sf-I-T\}[1-d])-[N+K$$
$$+Sj/360]]/\{S'p[1+s]/360+S[1+\textit{I}[1-d]\}$$

Steve Asikin ISBN 13: **978-1564912361**, ISBN 10: **1564912363**

Rule-58765:

If both $(I)$, $(U)$, $(S')$, $(s)$, $(S)$, $(v)$, $(N)$, $(I)$, $(T)$, $(d)$, $(K)$, $(j)$ and $(p)$ are known, then its Fixed Portion planned is:

$$f = [S'[1+s]-Sv-I-T-(\{N+K+Sj/360 \\ +S'vp[1+s]/360\}/[1+I]-U)/[1-d]]/S$$

Rule-58766:

If both $(I)$, $(U)$, $(S')$, $(s)$, $(S)$, $(v)$, $(f)$, $(N)$, $(T)$, $(d)$, $(K)$, $(j)$ and $(p)$ are known, then its Interest Expense planned is:

$$I = S'[1+s]-Sv-Sf-T-(\{N+K+Sj/360 \\ +S'vp[1+s]/360\}/[1+I]-U)/[1-d]$$

Rule-58767:

If both $(I)$, $(U)$, $(S')$, $(s)$, $(S)$, $(v)$, $(f)$, $(I)$, $(N)$, $(d)$, $(K)$, $(j)$ and $(p)$ are known, then its Tax planned is:

$$T = S'[1+s]-Sv-Sf-I-(\{N+K+Sj/360 \\ +S'vp[1+s]/360\}/[1+I]-U)/[1-d]$$

Rule-58768:

If both $(I)$, $(U)$, $(S')$, $(s)$, $(S)$, $(v)$, $(f)$, $(I)$, $(T)$, $(N)$, $(K)$, $(j)$ and $(p)$ are known, then its Dividend Payout planned is:

$$d = 1-(\{N+K+Sj/360+S'vp[1+s]/360\}/[1+I]-U) \\ /\{S'[1+s]-Sv-Sf-I-T\}$$

`

Steve Asikin ISBN 13: **978-1564912361**, ISBN 10: **1564912363**

Rule-58769:

If both (**/**), (**U**), (**S'**), (**s**), (**S**), (**v**), (**f**), (**I**), (**T**), (**d**), (**N**), (**j**) and (**p**) are known, then its Kind of Cash planned is:

$$K= [1+/](U+\{S'[1+s]-Sv-Sf-I-T\}[1-d])$$
$$-\{N+Sj/360+S'vp[1+s]/360\}$$

Rule-58770:

If both (**/**), (**U**), (**S'**), (**s**), (**S**), (**v**), (**f**), (**I**), (**T**), (**d**), (**K**), (**N**) and (**p**) are known, then its Job or Trade Receivable Days planned is:

$$j= 360[[1+/](U+\{S'[1+s]-Sv-Sf-I-T\}[1-d])$$
$$-\{N+K+S'vp[1+s]/360\}]/S$$

Rule-58771:

If both (**/**), (**U**), (**S'**), (**s**), (**S**), (**v**), (**f**), (**I**), (**T**), (**d**), (**K**), (**j**) and (**N**) are known, then its Procured Inventory Days planned is:

$$p= 360[[1+/](U+\{S'[1+s]-Sv-Sf-I-T\}[1-d])$$
$$-[N+K+Sj/360]]/\{S'v[1+s]\}$$

Rule-58772:

If both (**/**), (**U**), (**S'**), (**s**), (**S**), (**v**), (**f**), (**I**), (**T**), (**d**), (**K**), (**j**) and (**P**) are known, then its Non Current Asset planned is:

$$N= [1+/](U+\{S'[1+s]-Sv-Sf-I-T\}[1-d])$$
$$-K-S'j[1+s]/360-P$$

Steve Asikin ISBN 13: **978-1564912361**, ISBN 10: **1564912363**

Rule-58773:

If both $(N)$, $(U)$, $(S')$, $(s)$, $(S)$, $(v)$, $(f)$, $(I)$, $(T)$, $(d)$, $(K)$, $(j)$ and $(P)$ are known, then its Leverage or Gearing Ratio planned is:

$$I = \{N+K+S'j[1+s]/360+P\}$$
$$/(U+\{S'[1+s]-Sv-Sf-I-T\}[1-d])-1$$

Rule-58774:

If both $(I)$, $(N)$, $(S')$, $(s)$, $(S)$, $(v)$, $(f)$, $(I)$, $(T)$, $(d)$, $(K)$, $(j)$ and $(P)$ are known, then its Utilized or Starting Capital must be:

$$U = (\{N+K+S'j[1+s]/360+P\}$$
$$-[1+I]\{S'[1+s]-Sv-Sf-I-T\}[1-d])/[1+I]$$

Rule-58775:

If both $(I)$, $(U)$, $(N)$, $(s)$, $(S)$, $(v)$, $(f)$, $(I)$, $(T)$, $(d)$, $(K)$, $(j)$ and $(P)$ are known, then its Sales past must be:

$$S' = [[1+I](U-\{Sv+Sf+I+T\}[1-d])-[N+K+P]]$$
$$/([1+s]\{j/360-[1+I][1-d]\})$$

or

$$S' = S/[1+s]$$

Steve Asikin ISBN 13: **978-1564912361**, ISBN 10: **1564912363**

Rule-58776:

If both (**/**), (**U**), (**S'**), (**N**), (**S**), (**v**), (**f**), (**I**), (**T**), (**d**), (**K**), (**j**) and (**P**) are known, then its Sales Growth planned is:

$$s= [[1+\text{/}(U-\{Sv+Sf+I+T\}[1-d])-[N+K+P]]$$
$$/(S'\{j/360-[1+\text{/}[1-d]\})-1$$

or

$$s= S/S'-1$$


Rule-58777:

If both (**/**), (**U**), (**S'**), (**s**), (**N**), (**v**), (**f**), (**I**), (**T**), (**d**), (**K**), (**j**) and (**P**) are known, then its Sales or Revenue planned is:

$$S= (S'[1+s]-I-T-(\{N+K+S'j[1+s]/360+P\}$$
$$/[1+\text{/}-U)/[1-d])/ [v+f]$$

or

$$S= S'[1+s]$$


Rule-58778:

If both (**/**), (**U**), (**S'**), (**s**), (**S**), (**N**), (**f**), (**I**), (**T**), (**d**), (**K**), (**j**) and (**P**) are known, then its Variable Portion planned is:

$$v= (S'[1+s]-Sf-I-T-(\{N+K+S'j[1+s]/360+P\}$$
$$/[1+\text{/}-U)/[1-d])/S$$

Steve Asikin ISBN 13: **978-1564912361**, ISBN 10: **1564912363**

Rule-58779:

If both (**/**), (**U**), (**S'**), (**s**), (**S**), (**v**), (**N**), (**I**), (**T**), (**d**), (**K**), (**j**) and (**P**) are known, then its Fixed Portion planned is:

$$f= [S'[1+s]-Sv-I-T-(\{N+K+S'j[1+s]/360+P\}/[1+/-U)/[1-d]]/S$$

Rule-58780:

If both (**/**), (**U**), (**S'**), (**s**), (**S**), (**v**), (**f**), (**N**), (**T**), (**d**), (**K**), (**j**) and (**P**) are known, then its Interest Expense planned is:

$$I= S'[1+s]-Sv-Sf-T-(\{N+K+S'j[1+s]/360+P\}/[1+/-U)/[1-d]$$

Rule-58781:

If both (**/**), (**U**), (**S'**), (**s**), (**S**), (**v**), (**f**), (**I**), (**N**), (**d**), (**K**), (**j**) and (**P**) are known, then its Tax planned is:

$$T= S'[1+s]-Sv-Sf-I-(\{N+K+S'j[1+s]/360+P\}/[1+/-U)/[1-d]$$

Rule-58782:

If both (**/**), (**U**), (**S'**), (**s**), (**S**), (**v**), (**f**), (**I**), (**T**), (**N**), (**K**), (**j**) and (**P**) are known, then its Dividend Payout planned is:

$$d= 1-(\{N+K+S'j[1+s]/360+P\}/[1+/-U)/\{S'[1+s]-Sv-Sf-I-T\}$$

Steve Asikin ISBN 13: **978-1564912361**, ISBN 10: **1564912363**

Rule-58783:

If both $(\textit{I})$, $(\textbf{U})$, $(\textbf{S'})$, $(\textbf{s})$, $(\textbf{S})$, $(\textbf{v})$, $(\textbf{f})$, $(\textbf{I})$, $(\textbf{T})$, $(\textbf{d})$, $(\textbf{N})$, $(\textbf{j})$ and $(\textbf{P})$ are known, then its Kind of Cash planned is:

$$K = [1+\textit{I}(U+\{S'[1+s]\text{-}Sv\text{-}Sf\text{-}I\text{-}T\}[1\text{-}d])$$
$$-\{N+S'j[1+s]/360+P\}$$

Rule-58784:

If both $(\textit{I})$, $(\textbf{U})$, $(\textbf{S'})$, $(\textbf{s})$, $(\textbf{S})$, $(\textbf{v})$, $(\textbf{f})$, $(\textbf{I})$, $(\textbf{T})$, $(\textbf{d})$, $(\textbf{K})$, $(\textbf{N})$ and $(\textbf{P})$ are known, then its Job or Trade Receivable Days planned is:

$$j = 360[[1+\textit{I}(U+\{S'[1+s]\text{-}Sv\text{-}Sf\text{-}I\text{-}T\}[1\text{-}d])$$
$$-\{N+K+P\}]/\{S'[1+s]\}$$

Rule-58785:

If both $(\textit{I})$, $(\textbf{U})$, $(\textbf{S'})$, $(\textbf{s})$, $(\textbf{S})$, $(\textbf{v})$, $(\textbf{f})$, $(\textbf{I})$, $(\textbf{T})$, $(\textbf{d})$, $(\textbf{K})$, $(\textbf{j})$ and $(\textbf{N})$ are known, then its Procured Inventory planned is:

$$P = [1+\textit{I}(U+\{S'[1+s]\text{-}Sv\text{-}Sf\text{-}I\text{-}T\}[1\text{-}d])$$
$$-\{N+K+S'j[1+s]/360\}$$

Rule-58786:

If both $(\textit{I})$, $(\textbf{U})$, $(\textbf{S'})$, $(\textbf{s})$, $(\textbf{S})$, $(\textbf{v})$, $(\textbf{f})$, $(\textbf{I})$, $(\textbf{T})$, $(\textbf{d})$, $(\textbf{K})$, $(\textbf{j})$, $(\textbf{V})$ and $(\textbf{p})$ are known, then its Non Current Asset planned is:

$$N = [1+\textit{I}(U+\{S'[1+s]\text{-}Sv\text{-}Sf\text{-}I\text{-}T\}[1\text{-}d])$$
$$-K\text{-}S'j[1+s]/360\text{-}Vp/360$$

Steve Asikin ISBN 13: **978-1564912361**, ISBN 10: **1564912363**

## Rule-58787:

If both $(\mathbf{N})$, $(\mathbf{U})$, $(\mathbf{S'})$, $(\mathbf{s})$, $(\mathbf{S})$, $(\mathbf{v})$, $(\mathbf{f})$, $(\mathbf{I})$, $(\mathbf{T})$, $(\mathbf{d})$, $(\mathbf{K})$, $(\mathbf{j})$, $(\mathbf{V})$ and $(\mathbf{p})$ are known, then its Leverage or Gearing Ratio planned is:

$$I = \{\mathbf{N}+\mathbf{K}+\mathbf{S'j}[1+\mathbf{s}]/360+\mathbf{Vp}/360\}$$
$$/(\mathbf{U}+\{\mathbf{S'}[1+\mathbf{s}]-\mathbf{Sv}-\mathbf{Sf}-\mathbf{I}-\mathbf{T}\}[1-\mathbf{d}])-1$$

## Rule-58788:

If both $(\mathbf{I})$, $(\mathbf{N})$, $(\mathbf{S'})$, $(\mathbf{s})$, $(\mathbf{S})$, $(\mathbf{v})$, $(\mathbf{f})$, $(\mathbf{I})$, $(\mathbf{T})$, $(\mathbf{d})$, $(\mathbf{K})$, $(\mathbf{j})$, $(\mathbf{V})$ and $(\mathbf{p})$ are known, then its Utilized or Starting Capital must be:

$$\mathbf{U} = (\{\mathbf{N}+\mathbf{K}+\mathbf{S'j}[1+\mathbf{s}]/360+\mathbf{Vp}/360\}$$
$$-[1+\mathbf{I}]\{\mathbf{S'}[1+\mathbf{s}]-\mathbf{Sv}-\mathbf{Sf}-\mathbf{I}-\mathbf{T}\}[1-\mathbf{d}])/[1+\mathbf{I}]$$

## Rule-58789:

If both $(\mathbf{I})$, $(\mathbf{U})$, $(\mathbf{N})$, $(\mathbf{s})$, $(\mathbf{S})$, $(\mathbf{v})$, $(\mathbf{f})$, $(\mathbf{I})$, $(\mathbf{T})$, $(\mathbf{d})$, $(\mathbf{K})$, $(\mathbf{j})$, $(\mathbf{V})$ and $(\mathbf{p})$ are known, then its Sales Past must be:

$$\mathbf{S'} = [[1+\mathbf{I}](\mathbf{U}-\{\mathbf{Sv}+\mathbf{Sf}+\mathbf{I}+\mathbf{T}\}[1-\mathbf{d}])-[\mathbf{N}+\mathbf{K}+\mathbf{Vp}/360]]$$
$$/([1+\mathbf{s}]\{\mathbf{j}/360-[1+\mathbf{I}][1-\mathbf{d}]\})$$

or

$$\mathbf{S'} = \mathbf{S}/[1+\mathbf{s}]$$

Steve Asikin ISBN 13: **978-1564912361**, ISBN 10: **1564912363**

Rule-58790:

If both (**/**), (**U**), (**S'**), (**N**), (**S**), (**v**), (**f**), (**I**), (**T**), (**d**), (**K**), (**j**), (**V**) and (**p**) are known, then its Sales Growth planned is:

$$s= [[1+/(U-\{Sv+Sf+I+T\}[1-d])-[N+K+Vp/360]]$$
$$/(S'\{j/360-[1+/][1-d]\})-1$$

or

$$s= S/S'-1$$

Rule-58791:

If both (**/**), (**U**), (**S'**), (**s**), (**N**), (**v**), (**f**), (**I**), (**T**), (**d**), (**K**), (**j**), (**V**) and (**p**) are known, then its Sales or Revenue planned is:

$$S= [S'[1+s]-I-T-(\{N+K+S'j[1+s]/360+Vp/360\}$$
$$/[1+/-U)/[1-d]]/[v+f]$$

or

$$S= S'[1+s]$$

or

$$S= V/v$$

Rule-58792:

If both (**/**), (**U**), (**S'**), (**s**), (**S**), (**N**), (**f**), (**I**), (**T**), (**d**), (**K**), (**j**), (**V**) and (**p**) are known, then its Variable Portion planned is:

$$v= [S'[1+s]-Sf-I-T-(\{N+K+S'j[1+s]/360+Vp/360\}$$
$$/[1+/-U)/[1-d]]/S$$

or

$$v= V/S$$

Steve Asikin ISBN 13: **978-1564912361**, ISBN 10: **1564912363**

Rule-58793:

If both (**/**), (**U**), (**S'**), (**s**), (**S**), (**v**), (**N**), (**I**), (**T**), (**d**), (**K**), (**j**), (**V**) and (**p**) are known, then its Fixed Portion planned is:

$$f= [S'[1+s]-Sv-I-T-(\{N+K+S'j[1+s]/360+Vp/360\}/[1+/-U)/[1-d]]/S$$

Rule-58794:

If both (**/**), (**U**), (**S'**), (**s**), (**S**), (**v**), (**f**), (**N**), (**T**), (**d**), (**K**), (**j**), (**V**) and (**p**) are known, then its Interest Portion planned is:

$$I= S'[1+s]-Sv-Sf-T-(\{N+K+S'j[1+s]/360+Vp/360\}/[1+/-U)/[1-d]$$

Rule-58795:

If both (**/**), (**U**), (**S'**), (**s**), (**S**), (**v**), (**f**), (**I**), (**N**), (**d**), (**K**), (**j**), (**V**) and (**p**) are known, then its Tax planned is:

$$T= S'[1+s]-Sv-Sf-I-(\{N+K+S'j[1+s]/360+Vp/360\}/[1+/-U)/[1-d]$$

Rule-58796:

If both (**/**), (**U**), (**S'**), (**s**), (**S**), (**v**), (**f**), (**I**), (**T**), (**N**), (**K**), (**j**), (**V**) and (**p**) are known, then its Dividend Payout planned is:

$$d= 1-(\{N+K+S'j[1+s]/360+Vp/360\}/[1+/-U)/\{S'[1+s]-Sv-Sf-I-T\}$$

Steve Asikin ISBN 13: **978-1564912361**, ISBN 10: **1564912363**

Rule-58797:

If both (**/**), (**U**), (**S'**), (**s**), (**S**), (**v**), (**f**), (**I**), (**T**), (**d**), (**N**), (**j**), (**V**) and (**p**) are known, then its Kind of Cash planned is:

$$K= [1+/(U+\{S'[1+s]-Sv-Sf-I-T\}[1-d])$$
$$-\{N+S'j[1+s]/360+Vp/360\}$$

Rule-58798:

If both (**/**), (**U**), (**S'**), (**s**), (**S**), (**v**), (**f**), (**I**), (**T**), (**d**), (**K**), (**N**), (**V**) and (**p**) are known, then its Job or Trade Receivable Days planned is:

$$j= 360[[1+/(U+\{S'[1+s]-Sv-Sf-I-T\}[1-d])$$
$$-\{N+K+Vp/360\}]/\{S'[1+s]\}$$

Rule-58799:

If both (**/**), (**U**), (**S'**), (**s**), (**S**), (**v**), (**f**), (**I**), (**T**), (**d**), (**K**), (**j**), (**N**) and (**p**) are known, then its Variable Cost planned is:

$$V= 360[[1+/(U+\{S'[1+s]-Sv-Sf-I-T\}[1-d])$$
$$-\{N+K+S'j[1+s]/360\}]/p$$

or

$$V= Sv$$

Rule-58800:

If both (**/**), (**U**), (**S'**), (**s**), (**S**), (**v**), (**f**), (**I**), (**T**), (**d**), (**K**), (**j**), (**V**) and (**N**) are known, then its Procured Inventory Days planned is:

$$p= 360[[1+/(U+\{S'[1+s]-Sv-Sf-I-T\}[1-d])$$
$$-\{N+K+S'j[1+s]/360\}]/V$$

Steve Asikin ISBN 13: **978-1564912361**, ISBN 10: **1564912363**

## Rule-58801:

If both $(\textbf{\textit{l}})$, $(\textbf{U})$, $(\textbf{S'})$, $(\textbf{s})$, $(\textbf{S})$, $(\textbf{v})$, $(\textbf{f})$, $(\textbf{I})$, $(\textbf{T})$, $(\textbf{d})$, $(\textbf{K})$, $(\textbf{j})$ and $(\textbf{p})$ are known, then its Non Current Asset planned is:

$$\textbf{N} = [1 + \textbf{\textit{l}}(\textbf{U} + \{\textbf{S'}[1+\textbf{s}] - \textbf{Sv} - \textbf{Sf} - \textbf{I} - \textbf{T}\}[1-\textbf{d}])$$
$$- \textbf{K} - \textbf{S'j}[1+\textbf{s}]/360 - \textbf{Svp}/360$$

## Rule-58802:

If both $(\textbf{N})$, $(\textbf{U})$, $(\textbf{S'})$, $(\textbf{s})$, $(\textbf{S})$, $(\textbf{v})$, $(\textbf{f})$, $(\textbf{I})$, $(\textbf{T})$, $(\textbf{d})$, $(\textbf{K})$, $(\textbf{j})$ and $(\textbf{p})$ are known, then its Leverage or Gearing Ratio planned is:

$$\textbf{\textit{l}} = \{\textbf{N} + \textbf{K} + \textbf{S'j}[1+\textbf{s}]/360 + \textbf{Svp}/360\}$$
$$/(\textbf{U} + \{\textbf{S'}[1+\textbf{s}] - \textbf{Sv} - \textbf{Sf} - \textbf{I} - \textbf{T}\}[1-\textbf{d}]) - 1$$

## Rule-58803:

If both $(\textbf{\textit{l}})$, $(\textbf{N})$, $(\textbf{S'})$, $(\textbf{s})$, $(\textbf{S})$, $(\textbf{v})$, $(\textbf{f})$, $(\textbf{I})$, $(\textbf{T})$, $(\textbf{d})$, $(\textbf{K})$, $(\textbf{j})$ and $(\textbf{p})$ are known, then its Utilized or Starting Capital must be:

$$\textbf{U} = (\{\textbf{N} + \textbf{K} + \textbf{S'j}[1+\textbf{s}]/360 + \textbf{Svp}/360\}$$
$$- [1+\textbf{\textit{l}}]\{\textbf{S'}[1+\textbf{s}] - \textbf{Sv} - \textbf{Sf} - \textbf{I} - \textbf{T}\}[1-\textbf{d}])/[1+\textbf{\textit{l}}$$

## Rule-58804:

If both $(\textbf{\textit{l}})$, $(\textbf{U})$, $(\textbf{N})$, $(\textbf{s})$, $(\textbf{S})$, $(\textbf{v})$, $(\textbf{f})$, $(\textbf{I})$, $(\textbf{T})$, $(\textbf{d})$, $(\textbf{K})$, $(\textbf{j})$ and $(\textbf{p})$ are known, then its Sales Past must be:

$$\textbf{S'} = [[1 + \textbf{\textit{l}}(\textbf{U} - \{\textbf{Sv} + \textbf{Sf} + \textbf{I} + \textbf{T}\}[1-\textbf{d}]) - [\textbf{N} + \textbf{K}$$
$$+ \textbf{Svp}/360]]/([1+\textbf{s}]\{\textbf{j}/360 - [1+\textbf{\textit{l}}][1-\textbf{d}]\})$$

or

$$\textbf{S'} = \textbf{S}/[1+\textbf{s}]$$

Steve Asikin ISBN 13: **978-1564912361**, ISBN 10: **1564912363**

Rule-58805:

If both (**/**), (**U**), (**S'**), (**N**), (**S**), (**v**), (**f**), (**I**), (**T**), (**d**), (**K**), (**j**) and (**p**) are known, then its Sales Growth planned is:

$$s= [[1+/(U-\{Sv+Sf+I+T\}[1-d])-[N+K+Svp/360]]$$
$$/(S'\{j/360-[1+/ [1-d]\})-1$$

or

$$s= S/S'-1$$


Rule-58806:

If both (**/**), (**U**), (**S'**), (**s**), (**N**), (**v**), (**f**), (**I**), (**T**), (**d**), (**K**), (**j**) and (**p**) are known, then its Sales or Revenue planned is:

$$S= [[1+/(U+\{S'[1+s]-I-T\}[1-d])$$
$$-\{N+K+S'j[1+s]/360\}]$$
$$/\{vp/360+[1+/][1-d][v+f]\}$$

or

$$S= S'[1+s]$$


Rule-58807:

If both (**/**), (**U**), (**S'**), (**s**), (**S**), (**N**), (**f**), (**I**), (**T**), (**d**), (**K**), (**j**) and (**p**) are known, then its Variable Portion planned is:

$$v= [[1+/(U+\{S'[1+s]-Sf-I-T\}[1-d])-\{N+K$$
$$+S'j[1+s]/360\}]/(S\{p/360+[1+/][1-d]\})$$

Steve Asikin ISBN 13: **978-1564912361**, ISBN 10: **1564912363**

Rule-58808:

If both ($\mathit{f}$), (**U**), (**$'**), (**s**), (**$**), (**v**), (**N**), (**I**), (**T**), (**d**), (**K**), (**j**) and (**p**) are known, then its Fixed Portion planned is:

$$f= [\$'[1+s]-\$v-\$f-I-(\{N+K+\$'j[1+s]/360 \\ +\$vp/360\}/[1+\mathit{f}-U)/[1-d]]/\$$$

Rule-58809:

If both ($\mathit{f}$), (**U**), (**$'**), (**s**), (**$**), (**v**), (**f**), (**N**), (**T**), (**d**), (**K**), (**j**) and (**p**) are known, then its Interest Expense planned is:

$$I= \$'[1+s]-\$v-\$f-T-(\{N+K+\$'j[1+s]/360 \\ +\$vp/360\}/[1+\mathit{f}-U)/[1-d]$$

Rule-58810:

If both ($\mathit{f}$), (**U**), (**$'**), (**s**), (**$**), (**v**), (**f**), (**I**), (**N**), (**d**), (**K**), (**j**) and (**p**) are known, then its Tax planned is:

$$T= \$'[1+s]-\$v-\$f-I-(\{N+K+\$'j[1+s]/360 \\ +\$vp/360\}/[1+\mathit{f}-U)/[1-d]$$

Rule-58811:

If both ($\mathit{f}$), (**U**), (**$'**), (**s**), (**$**), (**v**), (**f**), (**I**), (**T**), (**N**), (**K**), (**j**) and (**p**) are known, then its Dividend Payout planned is:

$$d= 1-(\{N+K+\$'j[1+s]/360+\$vp/360\}/[1+\mathit{f}-U) \\ /\{\$'[1+s]-\$v-\$f-I-T\}$$

Steve Asikin ISBN 13: **978-1564912361**, ISBN 10: **1564912363**

Rule-58812:

If both (**/**), (**U**), (**$'**), (**s**), (**$**), (**v**), (**f**), (**I**), (**T**), (**d**), (**N**), (**j**) and (**p**) are known, then its Kind of Cash planned is:

$$\mathbf{K}= [1+/](\mathbf{U}+\{\mathbf{\$'}[1+s]-\mathbf{\$v}-\mathbf{\$f}-\mathbf{I}-\mathbf{T}\}[1-\mathbf{d}])$$
$$-\{\mathbf{N}+\mathbf{\$'j}[1+s]/360+\mathbf{\$vp}/360\}$$

Rule-58813:

If both (**/**), (**U**), (**$'**), (**s**), (**$**), (**v**), (**f**), (**I**), (**T**), (**d**), (**K**), (**N**) and (**p**) are known, then its Job or Trade Receivable Days planned is:

$$\mathbf{j}= 360[[1+/](\mathbf{U}+\{\mathbf{\$'}[1+s]-\mathbf{\$v}-\mathbf{\$f}-\mathbf{I}-\mathbf{T}\}[1-\mathbf{d}])$$
$$-\{\mathbf{N}+\mathbf{K}+\mathbf{\$vp}/360\}]/\{\mathbf{\$'}[1+s]\}$$

Rule-58814:

If both (**/**), (**U**), (**$'**), (**s**), (**$**), (**v**), (**f**), (**I**), (**T**), (**d**), (**K**), (**j**) and (**N**) are known, then its Procured Inventory Days planned is:

$$\mathbf{p}= 360[[1+/](\mathbf{U}+\{\mathbf{\$'}[1+s]-\mathbf{\$v}-\mathbf{\$f}-\mathbf{I}-\mathbf{T}\}[1-\mathbf{d}])$$
$$-\{\mathbf{N}+\mathbf{K}+\mathbf{\$'j}[1+s]/360\}]/[\mathbf{\$v}]$$

Rule-58815:

If both (**/**), (**U**), (**$'**), (**s**), (**$**), (**v**), (**f**), (**I**), (**T**), (**d**), (**K**), (**j**) and (**p**) are known, then its Non Current Asset planned is:

$$\mathbf{N}= [1+/](\mathbf{U}+\{\mathbf{\$'}[1+s]-\mathbf{\$v}-\mathbf{\$f}-\mathbf{I}-\mathbf{T}\}[1-\mathbf{d}])$$
$$-\mathbf{K}-\mathbf{\$'j}[1+s]/360-\mathbf{\$'vp}[1+s]/360$$

Steve Asikin ISBN 13: **978-1564912361**, ISBN 10: **1564912363**

Rule-58816:

If both $(\mathbf{N})$, $(\mathbf{U})$, $(\mathbf{S'})$, $(\mathbf{s})$, $(\mathbf{S})$, $(\mathbf{v})$, $(\mathbf{f})$, $(\mathbf{I})$, $(\mathbf{T})$, $(\mathbf{d})$, $(\mathbf{K})$, $(\mathbf{j})$ and $(\mathbf{p})$ are known, then its Leverage or Gearing Ratio planned is:

$$I = \{\mathbf{N+K+S'j}[1+\mathbf{s}]/360+\mathbf{S'vp}[1+\mathbf{s}]/360\}$$
$$/(\mathbf{U}+\{\mathbf{S'}[1+\mathbf{s}]-\mathbf{Sv-Sf-I-T}\}[1-\mathbf{d}])-1$$

Rule-58817:

If both $(I)$, $(\mathbf{N})$, $(\mathbf{S'})$, $(\mathbf{s})$, $(\mathbf{S})$, $(\mathbf{v})$, $(\mathbf{f})$, $(\mathbf{I})$, $(\mathbf{T})$, $(\mathbf{d})$, $(\mathbf{K})$, $(\mathbf{j})$ and $(\mathbf{p})$ are known, then its Utilized or Starting Capital must be:

$$\mathbf{U}= (\{\mathbf{N+K+S'j}[1+\mathbf{s}]/360+\mathbf{S'vp}[1+\mathbf{s}]/360\}$$
$$-\lceil1+I\rceil\{\mathbf{S'}[1+\mathbf{s}]-\mathbf{Sv-Sf-I-T}\}[1-\mathbf{d}])/[1+I]$$

Rule-58818:

If both $(I)$, $(\mathbf{U})$, $(\mathbf{N})$, $(\mathbf{s})$, $(\mathbf{S})$, $(\mathbf{v})$, $(\mathbf{f})$, $(\mathbf{I})$, $(\mathbf{T})$, $(\mathbf{d})$, $(\mathbf{K})$, $(\mathbf{j})$ and $(\mathbf{p})$ are known, then its Sales Past must be:

$$\mathbf{S'}= [[1+I](\mathbf{U}-\{\mathbf{Sv+Sf+I+T}\}[1-\mathbf{d}])-[\mathbf{N+K}]]$$
$$/([1+\mathbf{s}]\{\mathbf{j}/360+\mathbf{vp}/360-[1+I][1-\mathbf{d}]\})$$

or

$$\mathbf{S'}=\mathbf{S}/[1+\mathbf{s}]$$

Steve Asikin ISBN 13: **978-1564912361**, ISBN 10: **1564912363**

Rule-58819:

If both ($I$), (**U**), (**S'**), (**N**), (**S**), (**v**), (**f**), (**I**), (**T**), (**d**), (**K**), (**j**) and (**p**) are known, then its Sales Growth planned is:

$$s= [[1+I(U-\{Sv+Sf+I+T\}[1-d])-[N+K]]$$
$$/(S'\{j/360+vp/360-[1+I[1-d]\})-1$$

or

$$s= S/S'-1$$

Rule-58820:

If both ($I$), (**U**), (**S'**), (**s**), (**N**), (**v**), (**f**), (**I**), (**T**), (**d**), (**K**), (**j**) and (**p**) are known, then its Sales or Revenue planned is:

$$S= (U+\{S'[1+s]-I-T\}[1-d]-\{N+K+S'j[1+s]/360$$
$$+S'vp[1+s]/360\}/[1+I])/[f+v]$$

or

$$S= S'[1+s]$$

Rule-58821:

If both ($I$), (**U**), (**S'**), (**s**), (**S**), (**N**), (**f**), (**I**), (**T**), (**d**), (**K**), (**j**) and (**p**) are known, then its Variable Portion planned is:

$$v= [S'[1+s]-Sf-I-T-\{N+K+S'j[1+s]/360\}/[1+I]-U\}$$
$$/[1-d]]/\{S'p[1+s]/360+S[1+I]\}$$

Steve Asikin ISBN 13: **978-1564912361**, ISBN 10: **1564912363**

Rule-58822:

If both ($I$), ($U$), ($S'$), ($s$), ($S$), ($v$), ($N$), ($I$), ($T$), ($d$), ($K$), ($j$) and ($p$) are known, then its Fixed Portion planned is:

$$f = [S'[1+s]-Sv-I-T-\{N+K+S'j[1+s]/360$$
$$+S'vp[1+s]/360\}/[1+I-U\}/[1-d]]/S$$

Rule-58823:

If both ($I$), ($U$), ($S'$), ($s$), ($S$), ($v$), ($f$), ($N$), ($T$), ($d$), ($K$), ($j$) and ($p$) are known, then its Interest Expense planned is:

$$I = S'[1+s]-Sv-Sf-T-\{N+K+S'j[1+s]/360$$
$$+S'vp[1+s]/360\}/[1+I-U\}/[1-d]$$

Rule-58824:

If both ($I$), ($U$), ($S'$), ($s$), ($S$), ($v$), ($f$), ($I$), ($N$), ($d$), ($K$), ($j$) and ($p$) are known, then its Tax planned is:

$$T = S'[1+s]-Sv-Sf-I-\{N+K+S'j[1+s]/360$$
$$+S'vp[1+s]/360\}/[1+I-U\}/[1-d]$$

Rule-58825:

If both ($I$), ($U$), ($S'$), ($s$), ($S$), ($v$), ($f$), ($I$), ($T$), ($N$), ($K$), ($j$) and ($p$) are known, then its Dividend Payout planned is:

$$d = 1-\{N+K+S'j[1+s]/360+S'vp[1+s]/360\}$$
$$/[1+I-U\}/\{S'[1+s]-Sv-Sf-I-T\}$$

Steve Asikin ISBN 13: 978-1564912361, ISBN 10: 1564912363

Rule-58826:

If both (**/**), (**U**), (**S'**), (**s**), (**S**), (**v**), (**f**), (**I**), (**T**), (**d**), (**N**), (**j**) and (**p**) are known, then its Kind of Cash planned is:

$$K= [1+/(U+\{S'[1+s]-Sv-Sf-I-T\}[1-d])$$
$$-\{N+S'j[1+s]/360+S'vp[1+s]/360\}$$

Rule-58827:

If both (**/**), (**U**), (**S'**), (**s**), (**S**), (**v**), (**f**), (**I**), (**T**), (**d**), (**K**), (**N**) and (**p**) are known, then its Job or Trade Receivable Days planned is:

$$j= 360[[1+/(U+\{S'[1+s]-Sv-Sf-I-T\}[1-d])$$
$$-\{N+K+S'vp[1+s]/360\}]/\{S'[1+s]\}$$

Rule-58828:

If both (**/**), (**U**), (**S'**), (**s**), (**S**), (**v**), (**f**), (**I**), (**T**), (**d**), (**K**), (**j**) and (**N**) are known, then its Procured Inventory Days planned is:

$$p= 360[[1+/(U+\{S'[1+s]-Sv-Sf-I-T\}[1-d])$$
$$-\{N+K+S'j[1+s]/360\}]/\{S'v[1+s]\}$$

Rule-58829:

If both (**/**), (**U**), (**S'**), (**s**), (**S**), (**v**), (**f**), (**I**), (**t**), (**D**), (**K**), (**J**) and (**P**) are known, then its Non Current Asset planned is:

$$N= [1+/(U+\{S'[1+s]-Sv-Sf-I\}[1-t]-D)-K-J-P$$

Steve Asikin ISBN 13: **978-1564912361**, ISBN 10: **1564912363**

### Rule-58830:

If both $(\mathbf{N})$, $(\mathbf{U})$, $(\mathbf{S'})$, $(\mathbf{s})$, $(\mathbf{S})$, $(\mathbf{v})$, $(\mathbf{f})$, $(\mathbf{I})$, $(\mathbf{t})$, $(\mathbf{D})$, $(\mathbf{K})$, $(\mathbf{J})$ and $(\mathbf{P})$ are known, then its Leverage or Gearing Ratio planned is:

$$\mathbf{/}= \{\mathbf{N}+\mathbf{K}+\mathbf{J}+\mathbf{P}\}/(\mathbf{U}+\{\mathbf{S'}[1+\mathbf{s}]-\mathbf{Sv}-\mathbf{Sf}-\mathbf{I}\}[1-\mathbf{t}]-\mathbf{D})-1$$

### Rule-58831:

If both $(\mathbf{/})$, $(\mathbf{N})$, $(\mathbf{S'})$, $(\mathbf{s})$, $(\mathbf{S})$, $(\mathbf{v})$, $(\mathbf{f})$, $(\mathbf{I})$, $(\mathbf{t})$, $(\mathbf{D})$, $(\mathbf{K})$, $(\mathbf{J})$ and $(\mathbf{P})$ are known, then its Utilized or Starting Capital must be:

$$\mathbf{U}= (\{\mathbf{N}+\mathbf{K}+\mathbf{J}+\mathbf{P}\}-[1+\mathbf{/}]\{\mathbf{S'}[1+\mathbf{s}]-\mathbf{Sv}-\mathbf{Sf}-\mathbf{I}\}[1-\mathbf{t}]-\mathbf{D})$$
$$/[1+\mathbf{/}]$$

### Rule-58832:

If both $(\mathbf{/})$, $(\mathbf{U})$, $(\mathbf{N})$, $(\mathbf{s})$, $(\mathbf{S})$, $(\mathbf{v})$, $(\mathbf{f})$, $(\mathbf{I})$, $(\mathbf{t})$, $(\mathbf{D})$, $(\mathbf{K})$, $(\mathbf{J})$ and $(\mathbf{P})$ are known, then its Sales Past must be:

$$\mathbf{S'}= ([\mathbf{N}+\mathbf{K}+\mathbf{J}+\mathbf{P}]/[1+\mathbf{/}]-\mathbf{U}+\{\mathbf{Sv}+\mathbf{Sf}+\mathbf{I}\}[1-\mathbf{t}]+\mathbf{D})$$
$$/[[1+\mathbf{s}][1-\mathbf{t}]\}$$

or

$$\mathbf{S'}= \mathbf{S}/[1+\mathbf{s}]$$

### Rule-58833:

If both $(\mathbf{/})$, $(\mathbf{U})$, $(\mathbf{S'})$, $(\mathbf{N})$, $(\mathbf{S})$, $(\mathbf{v})$, $(\mathbf{f})$, $(\mathbf{I})$, $(\mathbf{t})$, $(\mathbf{D})$, $(\mathbf{K})$, $(\mathbf{J})$ and $(\mathbf{P})$ are known, then its Sales Growth planned is:

$$\mathbf{s}= ([\mathbf{N}+\mathbf{K}+\mathbf{J}+\mathbf{P}]/[1+\mathbf{/}]-\mathbf{U}+\{\mathbf{Sv}+\mathbf{Sf}+\mathbf{I}\}[1-\mathbf{t}]+\mathbf{D})$$
$$/\{\mathbf{S'}[1-\mathbf{t}]\}-1$$

or

$$\mathbf{s}= \mathbf{S}/\mathbf{S'}-1$$

Steve Asikin ISBN 13: **978-1564912361**, ISBN 10: **1564912363**

Rule-58834:

> If both ($I$), (**U**), (**S'**), (**s**), (**N**), (**v**), (**f**), (**I**), (**t**), (**D**), (**K**), (**J**) and (**P**) are known, then its Sales or Revenue planned is:
>
> $$S= (S'[1+s]-I-\{[N+K+J+P]/[1+I]-U+D\}/[1-t])/[v+f]$$
>
> or
>
> $$S= S'[1+s]$$

Rule-58835:

> If both ($I$), (**U**), (**S'**), (**s**), (**S**), (**N**), (**f**), (**I**), (**t**), (**D**), (**K**), (**J**) and (**P**) are known, then its Variable Portion planned is:
>
> $$v= (S'[1+s]-Sf-I-\{[N+K+J+P]/[1+I]-U+D\}/[1-t])/S$$

Rule-58836:

> If both ($I$), (**U**), (**S'**), (**s**), (**S**), (**v**), (**N**), (**I**), (**t**), (**D**), (**K**), (**J**) and (**P**) are known, then its Fixed Portion planned is:
>
> $$f= (S'[1+s]-Sv-I-\{[N+K+J+P]/[1+I]-U+D\}/[1-t])/S$$

Rule-58837:

> If both ($I$), (**U**), (**S'**), (**s**), (**S**), (**v**), (**f**), (**N**), (**t**), (**D**), (**K**), (**J**) and (**P**) are known, then its Interest Expense planned is:
>
> $$I= S'[1+s]-Sv-Sf-\{[N+K+J+P]/[1+I]-U+D\}/[1-t]$$

Steve Asikin ISBN 13: **978-1564912361**, ISBN 10: **1564912363**

Rule-58838:
If both (**/**), (**U**), (**S'**), (**s**), (**S**), (**v**), (**f**), (**I**), (**N**), (**D**), (**K**), (**J**) and (**P**) are known, then its Tax Rate planned is:
$$t= 1-\{[\mathbf{N+K+J+P}]/[1+\mathbf{/}]-\mathbf{U+D}\}/\{\mathbf{S'}[1+\mathbf{s}]-\mathbf{Sv-Sf-I}\}$$

Rule-58839:
If both (**/**), (**U**), (**S'**), (**s**), (**S**), (**v**), (**f**), (**I**), (**t**), (**N**), (**K**), (**J**) and (**P**) are known, then its Dividend planned is:
$$D= U+\{\mathbf{S'}[1+\mathbf{s}]-\mathbf{Sv-Sf-I}\}[1-\mathbf{t}]-[\mathbf{N+K+J+P}]/[1+\mathbf{/}]$$

Rule-58840:
If both (**/**), (**U**), (**S'**), (**s**), (**S**), (**v**), (**f**), (**I**), (**t**), (**D**), (**N**), (**J**) and (**P**) are known, then its Kind of Cash planned is:
$$K= [1+\mathbf{/}](\mathbf{U}+\{\mathbf{S'}[1+\mathbf{s}]-\mathbf{Sv-Sf-I}\}[1-\mathbf{t}]-\mathbf{D})-[\mathbf{N+J+P}]$$

Rule-58841:
If both (**/**), (**U**), (**S'**), (**s**), (**S**), (**v**), (**f**), (**I**), (**t**), (**D**), (**K**), (**N**) and (**P**) are known, then its Job or Trade Account Receivable planned is:
$$J= [1+\mathbf{/}](\mathbf{U}+\{\mathbf{S'}[1+\mathbf{s}]-\mathbf{Sv-Sf-I}\}[1-\mathbf{t}]-\mathbf{D})-[\mathbf{N+K+P}]$$

Rule-58842:
If both (**/**), (**U**), (**S'**), (**s**), (**S**), (**v**), (**f**), (**I**), (**t**), (**D**), (**K**), (**J**) and (**N**) are known, then its Procured Inventory planned is:
$$P= [1+\mathbf{/}](\mathbf{U}+\{\mathbf{S'}[1+\mathbf{s}]-\mathbf{Sv-Sf-I}\}[1-\mathbf{t}]-\mathbf{D})-[\mathbf{N+K+J}]$$

Steve Asikin ISBN 13: **978-1564912361**, ISBN 10: **1564912363**

Rule-58843:

If both $(I)$, $(U)$, $(S')$, $(s)$, $(S)$, $(v)$, $(f)$, $(I)$, $(t)$, $(D)$, $(K)$, $(J)$, $(V)$ and $(p)$ are known, then its Non Current Asset planned is:

$$N = [1+I(U+\{S'[1+s]-Sv-Sf-I\}[1-t]-D)$$
$$-K-J-Vp/360$$

Rule-58844:

If both $(N)$, $(U)$, $(S')$, $(s)$, $(S)$, $(v)$, $(f)$, $(I)$, $(t)$, $(D)$, $(K)$, $(J)$, $(V)$ and $(p)$ are known, then its Leverage or Gearing Ratio planned is:

$$I = [N+K+J+Vp/360]/(U+\{S'[1+s]-Sv-Sf-I\}$$
$$[1-t]-D)-1$$

Rule-58845:

If both $(I)$, $(N)$, $(S')$, $(s)$, $(S)$, $(v)$, $(f)$, $(I)$, $(t)$, $(D)$, $(K)$, $(J)$, $(V)$ and $(p)$ are known, then its Utilized or Starting Capital must be:

$$U = ([N+K+J+Vp/360]-[1+I]\{S'[1+s]-Sv-Sf-I\}$$
$$[1-t]-D)/[1+I]$$

Rule-58846:

If both $(I)$, $(U)$, $(N)$, $(s)$, $(S)$, $(v)$, $(f)$, $(I)$, $(t)$, $(D)$, $(K)$, $(J)$, $(V)$ and $(p)$ are known, then its Sales Past must be:

$$S' = (\{[N+K+J+Vp/360]/[1+I]-U+D\}/[1-t]$$
$$+Sv+Sf+I)/[1+s]$$
$$\text{or}$$
$$S' = S/[1+s]$$

Steve Asikin ISBN 13: **978-1564912361**, ISBN 10: **1564912363**

## Rule-58847:

If both (**/**), (**U**), (**S'**), (**N**), (**S**), (**v**), (**f**), (**I**), (**t**), (**D**), (**K**), (**J**), (**V**) and (**p**) are known, then its Sales Growth planned is:

$$s = (\{[N+K+J+Vp/360]/[1+/]-U+D\}/[1-t]$$
$$+Sv+Sf+I)/S-1$$

or

$$s = S/S'-1$$

## Rule-58848:

If both (**/**), (**U**), (**S'**), (**s**), (**N**), (**v**), (**f**), (**I**), (**t**), (**D**), (**K**), (**J**), (**V**) and (**p**) are known, then its Sales or Revenue planned is:

$$S = ((S'[1+s]-I-\{[N+K+J+Vp/360]/[1+/]-U+D\}$$
$$/[1-t])/[v+f]$$

or

$$S = S'[1+s]$$

or

$$S = V/v$$

## Rule-58849:

If both (**/**), (**U**), (**S'**), (**s**), (**S**), (**N**), (**f**), (**I**), (**t**), (**D**), (**K**), (**J**), (**V**) and (**p**) are known, then its Variable Portion planned is:

$$v = (S'[1+s]-I-Sf-\{[N+K+J+Vp/360]/[1+/]$$
$$-U+D\}/[1-t])/S$$

or

$$v = V/S$$

`

Steve Asikin ISBN 13: **978-1564912361**, ISBN 10: **1564912363**

Rule-58850:

If both (**/**), (**U**), (**S'**), (**s**), (**S**), (**v**), (**N**), (**I**), (**t**), (**D**), (**K**), (**J**), (**V**) and (**p**) are known, then its Fixed Portion planned is:

$$f= (S'[1+s]-Sv-I-\{[N+K+J+Vp/360]/[1+/]$$
$$-U+D\}/[1-t])/S$$

Rule-58851:

If both (**/**), (**U**), (**S'**), (**s**), (**S**), (**v**), (**f**), (**N**), (**t**), (**D**), (**K**), (**J**), (**V**) and (**p**) are known, then its Interest Expense planned is:

$$I= S'[1+s]-Sv-Sf-\{[N+K+J+Vp/360]/[1+/]-U+D\}$$
$$/[1-t]$$

Rule-58852:

If both (**/**), (**U**), (**S'**), (**s**), (**S**), (**v**), (**f**), (**I**), (**N**), (**D**), (**K**), (**J**), (**V**) and (**p**) are known, then its Tax Rate planned is:

$$t= 1-\{[N+K+J+Vp/360]/[1+/]-U+D\}$$
$$/\{S'[1+s]-Sv-Sf-I\}$$

Rule-58853:

If both (**/**), (**U**), (**S'**), (**s**), (**S**), (**v**), (**f**), (**I**), (**t**), (**N**), (**K**), (**J**), (**V**) and (**p**) are known, then its Dividend planned is:

$$D= U+\{S'[1+s]-Sv-Sf-I\}[1-t]$$
$$-[N+K+J+Vp/360]/[1+/]$$

Steve Asikin ISBN 13: **978-1564912361**, ISBN 10: **1564912363**

Rule-58854:

If both (**/**), (**U**), (**S'**), (**s**), (**S**), (**v**), (**f**), (**I**), (**t**), (**D**), (**N**), (**J**), (**V**) and (**p**) are known, then its Kind of Cash planned is:

$$K = [1 + /(U + \{S'[1+s] - Sv - Sf - I\}[1-t] - D) - [N + J + Vp/360]$$

Rule-58855:

If both (**/**), (**U**), (**S'**), (**s**), (**S**), (**v**), (**f**), (**I**), (**t**), (**D**), (**K**), (**N**), (**V**) and (**p**) are known, then its Job or Trade Account Receivable planned is:

$$J = [1 + /(U + \{S'[1+s] - Sv - Sf - I\}[1-t] - D) - \{N + K + Vp/360\}$$

Rule-58856:

If both (**/**), (**U**), (**S'**), (**s**), (**S**), (**v**), (**f**), (**I**), (**t**), (**D**), (**K**), (**J**), (**N**) and (**p**) are known, then its Variable Cost planned is:

$$V = 360[[1 + /(U + \{S'[1+s] - Sv - Sf - I\}[1-t] - D) - [N + K + J]]/p$$

or

$$V = Sv$$

Rule-58857:

If both (**/**), (**U**), (**S'**), (**s**), (**S**), (**v**), (**f**), (**I**), (**t**), (**D**), (**K**), (**J**), (**V**) and (**N**) are known, then its Procured Inventory Days planned is:

$$p = 360[[1 + /(U + \{S'[1+s] - Sv - Sf - I\}[1-t] - D) - [N + K + J]]/V$$

Steve Asikin ISBN 13: **978-1564912361**, ISBN 10: **1564912363**

Rule-58858:

If both (**I**), (**U**), (**S'**), (**s**), (**S**), (**v**), (**f**), (**I**), (**t**), (**D**), (**K**), (**J**), (**V**) and (**p**) are known, then its Non Current Asset planned is:

$$N = [1+I](U+\{S'[1+s]-Sv-Sf-I\}[1-t]-D)$$
$$-K-J-Svp/360$$

Rule-58859:

If both (**N**), (**U**), (**S'**), (**s**), (**S**), (**v**), (**f**), (**I**), (**t**), (**D**), (**K**), (**J**), (**V**) and (**p**) are known, then its Leverage or Gearing Ratio planned is:

$$I = [N+K+J+Svp/360]/(U+\{S'[1+s]-Sv-Sf-I\}$$
$$[1-t]-D)-1$$

Rule-58860:

If both (**I**), (**N**), (**S'**), (**s**), (**S**), (**v**), (**f**), (**I**), (**t**), (**D**), (**K**), (**J**), (**V**) and (**p**) are known, then its Utilized or Starting Capital must be:

$$U = ([N+K+J+Svp/360]-[1+I]\{S'[1+s]-Sv-Sf-I\}$$
$$[1-t]-D)/[1+I]$$

Rule-58861:

If both (**I**), (**U**), (**N**), (**s**), (**S**), (**v**), (**f**), (**I**), (**t**), (**D**), (**K**), (**J**), (**V**) and (**p**) are known, then its Sales Past must be:

$$S' = (\{[N+K+J+Svp/360]/[1+I]-U+D\}/[1-t]$$
$$+Sv+Sf+I)/[1+s]$$

or

$$S' = S/[1+s]$$

Steve Asikin ISBN 13: **978-1564912361**, ISBN 10: **1564912363**

Rule-58862:

If both (**⁄**), (**U**), (**S'**), (**N**), (**S**), (**v**), (**f**), (**I**), (**t**), (**D**), (**K**), (**J**), (**V**) and (**p**) are known, then its Sales Growth planned is:

$$s = (\{[N+K+J+Svp/360]/[1+\text{⁄}-U+D]\}/[1-t] + Sv+Sf+I)/S'-1$$

or

$$s = S/S'-1$$

Rule-58863:

If both (**⁄**), (**U**), (**S'**), (**s**), (**N**), (**v**), (**f**), (**I**), (**t**), (**D**), (**K**), (**J**), (**V**) and (**p**) are known, then its Sales or Revenue planned is:

$$S = [[1+\text{⁄}(U+\{S'[1+s]-I\}[1-t]-D)-[N+K+J]] / \{vp/360]+[1+\text{⁄}[1-t][v+f]\}$$

or

$$S = S'[1+s]$$

Rule-58864:

If both (**⁄**), (**U**), (**S'**), (**s**), (**S**), (**N**), (**f**), (**I**), (**t**), (**D**), (**K**), (**J**), (**V**) and (**p**) are known, then its Variable Portion planned is:

$$v = [[1+\text{⁄}(U+\{S'[1+s]-Sf-I\}[1-t]-D)-[N+K+J]] / (S\{p/360]+[1+\text{⁄}[1-t]\})$$

Steve Asikin ISBN 13: **978-1564912361**, ISBN 10: **1564912363**

Rule-58865:

If both (**/**), (**U**), (**S'**), (**s**), (**S**), (**v**), (**N**), (**I**), (**t**), (**D**), (**K**), (**J**), (**V**) and (**p**) are known, then its Fixed Portion planned is:

$$f= (S'[1+s]-Sf-I-\{[N+K+J+Svp/360]/[1+/]-U+D\}/[1-t])/S$$

Rule-58866:

If both (**/**), (**U**), (**S'**), (**s**), (**S**), (**v**), (**f**), (**N**), (**t**), (**D**), (**K**), (**J**), (**V**) and (**p**) are known, then its Interest Expense planned is:

$$I= S'[1+s]-Sv-Sf-\{[N+K+J+Svp/360]/[1+/]-U+D\}/[1-t]$$

Rule-58867:

If both (**/**), (**U**), (**S'**), (**s**), (**S**), (**v**), (**f**), (**I**), (**N**), (**D**), (**K**), (**J**), (**V**) and (**p**) are known, then its Tax Rate planned is:

$$t= 1-\{[N+K+J+Svp/360]/[1+/]-U+D\}/\{S'[1+s]-Sv-Sf-I\}$$

Rule-58868:

If both (**/**), (**U**), (**S'**), (**s**), (**S**), (**v**), (**f**), (**I**), (**t**), (**N**), (**K**), (**J**), (**V**) and (**p**) are known, then its Dividend planned is:

$$D= U+\{S'[1+s]-Sv-Sf-I\}[1-t]-[N+K+J+Svp/360]/[1+/]$$

Steve Asikin ISBN 13: **978-1564912361**, ISBN 10: **1564912363**

Rule-58869:

If both (**/**), (**U**), (**S'**), (**s**), (**S**), (**v**), (**f**), (**I**), (**t**), (**D**), (**N**), (**J**), (**V**) and (**p**) are known, then its Kind of Cash planned is:

$$K = [1 + /](U + \{S'[1+s] - Sv - Sf - I\}[1-t] - D) - [N + J + Svp/360]$$

Rule-58870:

If both (**/**), (**U**), (**S'**), (**s**), (**S**), (**v**), (**f**), (**I**), (**t**), (**D**), (**K**), (**N**), (**V**) and (**p**) are known, then its Job or Trade Account Receivable planned is:

$$J = [1 + /](U + \{S'[1+s] - Sv - Sf - I\}[1-t] - D) - \{N + K + Svp/360\}$$

Rule-58871:

If both (**/**), (**U**), (**S'**), (**s**), (**S**), (**v**), (**f**), (**I**), (**t**), (**D**), (**K**), (**J**), (**V**) and (**N**) are known, then its Procured Inventory Days planned is:

$$p = 360[[1 + /](U + \{S'[1+s] - Sv - Sf - I\}[1-t] - D) - [N + K + J]]/[Sv]$$

Rule-58872:

If both (**/**), (**U**), (**S'**), (**s**), (**S**), (**v**), (**f**), (**I**), (**t**), (**D**), (**K**), (**J**) and (**p**) are known, then its Non Current Asset planned is:

$$N = [1 + /](U + \{S'[1+s] - Sv - Sf - I\}[1-t] - D) - K - J - S'vp[1+s]/360$$

Steve Asikin ISBN 13: **978-1564912361**, ISBN 10: **1564912363**

Rule-58873:

If both $(\mathbf{N})$, $(\mathbf{U})$, $(\mathbf{S'})$, $(\mathbf{s})$, $(\mathbf{S})$, $(\mathbf{v})$, $(\mathbf{f})$, $(\mathbf{I})$, $(\mathbf{t})$, $(\mathbf{D})$, $(\mathbf{K})$, $(\mathbf{J})$ and $(\mathbf{p})$ are known, then its Leverage or Gearing Ratio planned is:

$$\mathbf{I} = \{\mathbf{N}+\mathbf{K}+\mathbf{J}+\mathbf{S'vp}[1+\mathbf{s}]/360\}/(\mathbf{U}+\{\mathbf{S'}[1+\mathbf{s}]-\mathbf{Sv}-\mathbf{Sf}-\mathbf{I}\}[1-\mathbf{t}]-\mathbf{D})-1$$

Rule-58874:

If both $(\mathbf{I})$, $(\mathbf{N})$, $(\mathbf{S'})$, $(\mathbf{s})$, $(\mathbf{S})$, $(\mathbf{v})$, $(\mathbf{f})$, $(\mathbf{I})$, $(\mathbf{t})$, $(\mathbf{D})$, $(\mathbf{K})$, $(\mathbf{J})$ and $(\mathbf{p})$ are known, then its Utilized or Starting Capital must be:

$$\mathbf{U} = (\{\mathbf{N}+\mathbf{K}+\mathbf{J}+\mathbf{S'vp}[1+\mathbf{s}]/360\} -[1+\mathbf{I}]\{\mathbf{S'}[1+\mathbf{s}]-\mathbf{Sv}-\mathbf{Sf}-\mathbf{I}\}[1-\mathbf{t}]-\mathbf{D})/[1+\mathbf{I}]$$

Rule-58875:

If both $(\mathbf{I})$, $(\mathbf{U})$, $(\mathbf{N})$, $(\mathbf{s})$, $(\mathbf{S})$, $(\mathbf{v})$, $(\mathbf{f})$, $(\mathbf{I})$, $(\mathbf{t})$, $(\mathbf{D})$, $(\mathbf{K})$, $(\mathbf{J})$ and $(\mathbf{p})$ are known, then its Sales Past must be:

$$\mathbf{S'} = [[1+\mathbf{I}](\mathbf{U}-\{\mathbf{Sv}+\mathbf{Sf}+\mathbf{I}\}[1-\mathbf{t}]-\mathbf{D})-[\mathbf{N}+\mathbf{K}+\mathbf{J}]] /([1+\mathbf{s}]\{\mathbf{vp}/360-[1+\mathbf{I}][1-\mathbf{t}]\})$$

or

$$\mathbf{S'} = \mathbf{S}/[1+\mathbf{s}]$$

Steve Asikin ISBN 13: **978-1564912361**, ISBN 10: **1564912363**

Rule-58876:

If both $(I)$, $(U)$, $(S')$, $(N)$, $(S)$, $(v)$, $(f)$, $(I)$, $(t)$, $(D)$, $(K)$, $(J)$ and $(p)$ are known, then its Sales Growth planned is:

$$s= [[1+I(U-\{Sv+Sf+I\}[1-t]-D)-[N+K+J]]$$
$$/(S'\{vp/360-[1+I[1-t]\})-1$$

or

$$s= S/S'-1$$

Rule-58877:

If both $(I)$, $(U)$, $(S')$, $(s)$, $(N)$, $(v)$, $(f)$, $(I)$, $(t)$, $(D)$, $(K)$, $(J)$ and $(p)$ are known, then its Sales or Revenue planned is:

$$S= [S'[1+s]-Sf-I-(\{N+K+J+S'vp[1+s]/360\}/[1+I$$
$$-U+D)/[1-t]]/v$$

or

$$S= S'[1+s]$$

Rule-58878:

If both $(I)$, $(U)$, $(S')$, $(s)$, $(S)$, $(N)$, $(f)$, $(I)$, $(t)$, $(D)$, $(K)$, $(J)$ and $(p)$ are known, then its Variable Portion planned is:

$$v= [[1+I(U+\{S'[1+s]-Sf-I\}[1-t]-D)-[N+K+J]]$$
$$/\{S'p[1+s]/360+S[1+I][1-t]\}$$

Steve Asikin ISBN 13: **978-1564912361**, ISBN 10: **1564912363**

Rule-58879:

If both (**/**), (**U**), (**S'**), (**s**), (**S**), (**v**), (**N**), (**I**), (**t**), (**D**), (**K**), (**J**) and (**p**) are known, then its Fixed Portion planned is:

$$f= [S'[1+s]-Sv-I-(\{N+K+J+S'vp[1+s]/360\}/[1+/$$
$$-U+D)/[1-t]]/S$$

Rule-58880:

If both (**/**), (**U**), (**S'**), (**s**), (**S**), (**v**), (**f**), (**N**), (**t**), (**D**), (**K**), (**J**) and (**p**) are known, then its Interest Expense planned is:

$$I= S'[1+s]-Sv-Sf-(\{N+K+J+S'vp[1+s]/360\}/[1+/$$
$$-U+D)/[1-t]$$

Rule-58881:

If both (**/**), (**U**), (**S'**), (**s**), (**S**), (**v**), (**f**), (**I**), (**N**), (**D**), (**K**), (**J**) and (**p**) are known, then its Tax Rate planned is:

$$t= 1-(\{N+K+J+S'vp[1+s]/360\}/[1+/]-U+D)$$
$$/\{S'[1+s]-Sv-Sf-I\}$$

Rule-58882:

If both (**/**), (**U**), (**S'**), (**s**), (**S**), (**v**), (**f**), (**I**), (**t**), (**N**), (**K**), (**J**) and (**p**) are known, then its Dividend planned is:

$$D= U+\{S'[1+s]-Sv-Sf-I\}[1-t]$$
$$-\{N+K+J+S'vp[1+s]/360\}/[1+/$$

Steve Asikin ISBN 13: **978-1564912361**, ISBN 10: **1564912363**

Rule-58883:

If both (**/**), (**U**), (**S'**), (**s**), (**S**), (**v**), (**f**), (**I**), (**t**), (**D**), (**N**), (**J**) and (**p**) are known, then its Kind of Cash planned is:

$$K = [1+\text{/}(U+\{S'[1+s]-Sv-Sf-I\}[1-t]-D) -\{N+J+S'vp[1+s]/360\}$$

Rule-58884:

If both (**/**), (**U**), (**S'**), (**s**), (**S**), (**v**), (**f**), (**I**), (**t**), (**D**), (**K**), (**N**) and (**p**) are known, then its Job or Trade Account Receivable planned is:

$$J = [[1+\text{/}(U+\{S'[1+s]-Sv-Sf-I\}[1-t]-D) -\{N+K+S'vp[1+s]/360\}$$

Rule-58885:

If both (**/**), (**U**), (**S'**), (**s**), (**S**), (**v**), (**f**), (**I**), (**t**), (**D**), (**K**), (**J**) and (**N**) are known, then its Procured Inventory Days planned is:

$$p = 360[[1+\text{/}(U+\{S'[1+s]-Sv-Sf-I\}[1-t]-D) -[N+K+J]]/\{S'v[1+s]\}$$

Rule-58886:

If both (**/**), (**U**), (**S'**), (**s**), (**S**), (**v**), (**f**), (**I**), (**t**), (**D**), (**K**), (**j**) and (**P**) are known, then its Non Current Asset planned is:

$$N = [1+\text{/}(U+\{S'[1+s]-Sv-Sf-I\}[1-t]-D)-K-Sj/360-P$$

Steve Asikin ISBN 13: **978-1564912361**, ISBN 10: **1564912363**

Rule-58887:

If both (**N**), (**U**), (**S'**), (**s**), (**S**), (**v**), (**f**), (**I**), (**t**), (**D**), (**K**), (**j**) and (**P**) are known, then its Leverage or Gearing Ratio planned is:

$$I = \{N+K+Sj/360+P\}/(U+\{S'[1+s]-Sv-Sf-I\}$$
$$[1-t]-D)-1$$

Rule-58888:

If both (**I**), (**N**), (**S'**), (**s**), (**S**), (**v**), (**f**), (**I**), (**t**), (**D**), (**K**), (**j**) and (**P**) are known, then its Utilized or Starting Capital must be:

$$U = (\{N+K+Sj/360+P\}-[1+I]\{S'[1+s]-Sv-Sf-I\}$$
$$[1-t]-D)/[1+I]$$

Rule-58889:

If both (**I**), (**U**), (**N**), (**s**), (**S**), (**v**), (**f**), (**I**), (**t**), (**D**), (**K**), (**j**) and (**P**) are known, then its Sales Past must be:

$$S' = (\{[N+K+Sj/360+P]/[1+I]-U+D\}/[1-t]$$
$$+Sv+Sf+I)/[1+s]$$

or

$$S' = S/[1+s]$$

Steve Asikin ISBN 13: **978-1564912361**, ISBN 10: **1564912363**

Rule-58890:

If both (**/**), (**U**), (**S'**), (**N**), (**S**), (**v**), (**f**), (**I**), (**t**), (**D**), (**K**), (**j**) and (**P**) are known, then its Sales Growth planned is:

$$s = (\{[N+K+Sj/360+P]/[1+/\,-U+D\}/[1-t]$$
$$+Sv+Sf+I)/S'-1$$

or

$$s = S/S'-1$$

Rule-58891:

If both (**/**), (**U**), (**S'**), (**s**), (**N**), (**v**), (**f**), (**I**), (**t**), (**D**), (**K**), (**j**) and (**P**) are known, then its Sales or Revenue planned is:

$$S = [[1+/\,](U+\{S'[1+s]-I\}[1-t]-D)-[N+K+P]]$$
$$/\{j/360+[1+/\,][1-t][v+f]\}$$

or

$$S = S'[1+s]$$

Rule-58892:

If both (**/**), (**U**), (**S'**), (**s**), (**S**), (**N**), (**f**), (**I**), (**t**), (**D**), (**K**), (**j**) and (**P**) are known, then its Variable Portion planned is:

$$v = [S'[1+s]-Sf-I-([N+K+Sj/360+P]/[1+/\,-U+D)$$
$$/[1-t]]/S$$

Steve Asikin ISBN 13: **978-1564912361**, ISBN 10: **1564912363**

Rule-58893:

If both ($I$), (**U**), (**S'**), (**s**), (**$**), (**v**), (**N**), (**I**), (**t**), (**D**), (**K**), (**j**) and (**P**) are known, then its Fixed Portion planned is:

$$f= [S'[1+s]-Sv-I-([N+K+Sj/360+P]/[1+I-U+D)/[1-t]]/S$$

Rule-58894:

If both ($I$), (**U**), (**S'**), (**s**), (**$**), (**v**), (**f**), (**N**), (**t**), (**D**), (**K**), (**j**) and (**P**) are known, then its Interest Expense planned is:

$$I= \{S'[1+s]-Sv-Sf-([N+K+Sj/360+P]/[1+I-U+D)/[1-t]$$

Rule-58895:

If both ($I$), (**U**), (**S'**), (**s**), (**$**), (**v**), (**f**), (**I**), (**N**), (**D**), (**K**), (**j**) and (**P**) are known, then its Tax Rate planned is:

$$t= 1-([N+K+Sj/360+P]/[1+I-U+D)/\{S'[1+s]-Sv-Sf-I\}$$

Rule-58896:

If both ($I$), (**U**), (**S'**), (**s**), (**$**), (**v**), (**f**), (**I**), (**t**), (**N**), (**K**), (**j**) and (**P**) are known, then its Dividend planned is:

$$D= U+\{S'[1+s]-Sv-Sf-I\}[1-t]-[N+K+Sj/360+P]/[1+I$$

Steve Asikin ISBN 13: **978-1564912361**, ISBN 10: **1564912363**

Rule-58897:

If both $(I)$, $(U)$, $(S')$, $(s)$, $(S)$, $(v)$, $(f)$, $(I)$, $(t)$, $(D)$, $(N)$, $(j)$ and $(P)$ are known, then its Kind of Cash planned is:

$$K = [1+I(U+\{S'[1+s]-Sv-Sf-I\}[1-t]-D) \\ -[N+Sj/360+P]$$

Rule-58898:

If both $(I)$, $(U)$, $(S')$, $(s)$, $(S)$, $(v)$, $(f)$, $(I)$, $(t)$, $(D)$, $(K)$, $(N)$ and $(P)$ are known, then its Job or Trade Receivable Days planned is:

$$j = 360[[1+I(U+\{S'[1+s]-Sv-Sf-I\}[1-t]-D) \\ -\{N+K+P\}]/S$$

Rule-58899:

If both $(I)$, $(U)$, $(S')$, $(s)$, $(S)$, $(v)$, $(f)$, $(I)$, $(t)$, $(D)$, $(K)$, $(j)$ and $(N)$ are known, then its Procured Inventory planned is:

$$P = [1+I(U+\{S'[1+s]-Sv-Sf-I\}[1-t]-D) \\ -\{N+K+Sj/360\}$$

Rule-58900:

If both $(I)$, $(U)$, $(S')$, $(s)$, $(S)$, $(v)$, $(f)$, $(I)$, $(t)$, $(D)$, $(K)$, $(j)$, $(V)$ and $(p)$ are known, then its Non Current Asset planned is:

$$N = [1+I(U+\{S'[1+s]-Sv-Sf-I\}[1-t]-D) \\ -K-Sj/360-Vp/360$$

Steve Asikin ISBN 13: **978-1564912361**, ISBN 10: **1564912363**

Rule-58901:

If both (**N**), (**U**), (**S'**), (**s**), (**S**), (**v**), (**f**), (**I**), (**t**), (**D**), (**K**), (**j**), (**V**) and (**p**) are known, then its Leverage or Gearing Ratio planned is:

$$I = [N+K+Sj/360+Vp/360]$$
$$/(U+\{S'[1+s]-Sv-Sf-I\}[1-t]-D)-1$$


Rule-58902:

If both (**I**), (**N**), (**S'**), (**s**), (**S**), (**v**), (**f**), (**I**), (**t**), (**D**), (**K**), (**j**), (**V**) and (**p**) are known, then its Utilized or Starting Capital must be:

$$U = ([N+K+Sj/360+Vp/360]$$
$$-[1+I]\{S'[1+s]-Sv-Sf-I\}[1-t]-D)/[1+I]$$


Rule-58903:

If both (**I**), (**U**), (**N**), (**s**), (**S**), (**v**), (**f**), (**I**), (**t**), (**D**), (**K**), (**j**), (**V**) and (**p**) are known, then its Sales Past must be:

$$S' = (\{[N+K+Sj/360+Vp/360]/[1+I]-U+D\}/[1-t]$$
$$+Sv+Sf+I)/[1+s]$$

or

$$S' = S[1+s]$$

Steve Asikin ISBN 13: **978-1564912361**, ISBN 10: **1564912363**

Rule-58904:

If both (**/**), (**U**), (**$'**), (**N**), (**$**), (**v**), (**f**), (**I**), (**t**), (**D**), (**K**), (**j**), (**V**) and (**p**) are known, then its Sales Growth planned is:

$$s = (\{[N+K+\$j/360+Vp/360]/[1+/-U+D]\}/[1-t]$$
$$+\$v+\$f+I)/\$'-1$$

or

$$s = \$/\$'-1$$

Rule-58905:

If both (**/**), (**U**), (**$'**), (**s**), (**N**), (**v**), (**f**), (**I**), (**t**), (**D**), (**K**), (**j**), (**V**) and (**p**) are known, then its Sales or Revenue planned is:

$$\$ = [[1+/(U+\{\$'[1+s]-I\}[1-t]-D)-[N+K+Vp/360]]$$
$$/\{j/360+[1+/][1-t][v+f]\}$$

or

$$\$ = \$'[1+s]$$

Rule-58906:

If both (**/**), (**U**), (**$'**), (**s**), (**$**), (**N**), (**f**), (**I**), (**t**), (**D**), (**K**), (**j**), (**V**) and (**p**) are known, then its Variable Portion planned is:

$$v = (\$'[1+s]-\$f-I-\{[N+K+\$j/360+Vp/360]$$
$$/[1+/-U+D]\}/[1-t])/\$$$

or

$$v = V/\$$$

Steve Asikin ISBN 13: **978-1564912361**, ISBN 10: **1564912363**

Rule-58907:

If both (**/**), (**U**), (**S'**), (**s**), (**S**), (**v**), (**N**), (**I**), (**t**), (**D**), (**K**), (**j**), (**V**) and (**p**) are known, then its Fixed Portion planned is:

$$f= (S'[1+s]-Sv-I-\{[N+K+Sj/360+Vp/360]$$
$$/[1+/]-U+D\}/[1-t])/S$$

Rule-58908:

If both (**/**), (**U**), (**S'**), (**s**), (**S**), (**v**), (**f**), (**N**), (**t**), (**D**), (**K**), (**j**), (**V**) and (**p**) are known, then its Interest Expense planned is:

$$I= S'[1+s]-Sv-Sf-\{[N+K+Sj/360+Vp/360]$$
$$/[1+/]-U+D\}/[1-t]$$

Rule-58909:

If both (**/**), (**U**), (**S'**), (**s**), (**S**), (**v**), (**f**), (**I**), (**N**), (**D**), (**K**), (**j**), (**V**) and (**p**) are known, then its Tax Rate planned is:

$$t= 1-\{[N+K+Sj/360+Vp/360]/[1+/]-U+D\}$$
$$/\{S'[1+s]-Sv-Sf-I\}$$

Rule-58910:

If both (**/**), (**U**), (**S'**), (**s**), (**S**), (**v**), (**f**), (**I**), (**t**), (**N**), (**K**), (**j**), (**V**) and (**p**) are known, then its Dividend planned is:

$$D= U+\{S'[1+s]-Sv-Sf-I\}[1-t]$$
$$- [N+K+Sj/360+Vp/360]/[1+/]$$

Steve Asikin ISBN 13: **978-1564912361**, ISBN 10: **1564912363**

Rule-58911:
   If both (**/**), (**U**), (**S'**), (**s**), (**$**), (**v**), (**f**), (**I**), (**t**), (**D**), (**N**), (**j**), (**V**) and (**p**) are known, then its Kind of Cash planned is:
   $$K= [1+/(U+\{S'[1+s]-Sv-Sf-I\}[1-t]-D)$$
   $$-[N+Sj/360+Vp/360]$$

Rule-58912:
   If both (**/**), (**U**), (**S'**), (**s**), (**$**), (**v**), (**f**), (**I**), (**t**), (**D**), (**K**), (**N**), (**V**) and (**p**) are known, then its Job or Trade Receivable Days planned is:
   $$j= 360[[1+/(U+\{S'[1+s]-Sv-Sf-I\}[1-t]-D)$$
   $$-[N+K+Vp/360]]/S$$

Rule-58913:
   If both (**/**), (**U**), (**S'**), (**s**), (**$**), (**v**), (**f**), (**I**), (**t**), (**D**), (**K**), (**j**), (**N**) and (**p**) are known, then its Variable Cost planned is:
   $$V= 360[[1+/(U+\{S'[1+s]-Sv-Sf-I\}[1-t]-D)$$
   $$-[N+K+Sj/360]]/p$$
   or
   $$V= Sv$$

Rule-58914:
   If both (**/**), (**U**), (**S'**), (**s**), (**$**), (**v**), (**f**), (**I**), (**t**), (**D**), (**K**), (**j**), (**V**) and (**N**) are known, then its Procured Inventory Days planned is:
   $$p= 360[[1+/(U+\{S'[1+s]-Sv-Sf-I\}[1-t]-D)$$
   $$-[N+K+Sj/360]]/V$$

Steve Asikin ISBN 13: **978-1564912361**, ISBN 10: **1564912363**

Rule-58915:

If both $(I)$, $(U)$, $(S')$, $(s)$, $(S)$, $(v)$, $(f)$, $(I)$, $(t)$, $(D)$, $(K)$, $(j)$ and $(p)$ are known, then its Non Current Asset planned is:

$$N= [1+I(U+\{S'[1+s]-Sv-Sf-I\}[1-t]-D)$$
$$-K-Sj/360-Svp/360$$

Rule-58916:

If both $(N)$, $(U)$, $(S')$, $(s)$, $(S)$, $(v)$, $(f)$, $(I)$, $(t)$, $(D)$, $(K)$, $(j)$ and $(p)$ are known, then its Leverage or Gearing Ratio planned is:

$$I= [N+K+Sj/360+Svp/360]$$
$$/(U+\{S'[1+s]-Sv-Sf-I\}[1-t]-D)-1$$

Rule-58917:

If both $(I)$, $(N)$, $(S')$, $(s)$, $(S)$, $(v)$, $(f)$, $(I)$, $(t)$, $(D)$, $(K)$, $(j)$ and $(p)$ are known, then its Utilized or Starting Capital must be:

$$U= ([N+K+Sj/360+Svp/360]$$
$$-[1+I]\{S'[1+s]-Sv-Sf-I\}[1-t]-D)/[1+I]$$

Rule-58918:

If both $(I)$, $(U)$, $(N)$, $(s)$, $(S)$, $(v)$, $(f)$, $(I)$, $(t)$, $(D)$, $(K)$, $(j)$ and $(p)$ are known, then its Sales Past must be:

$$S'= (\{[N+K+Sj/360+Svp/360]/[1+I]-U+D\}$$
$$/[1-t]+Sv+Sf+I)/[1+s]$$

or

$$S'= S/[1+s]$$

Steve Asikin ISBN 13: **978-1564912361**, ISBN 10: **1564912363**

## Rule-58919:

If both (**/**), (**U**), (**S'**), (**N**), (**S**), (**v**), (**f**), (**I**), (**t**), (**D**), (**K**), (**j**) and (**p**) are known, then its Sales Growth planned is:

$$s = (\{[N+K+Sj/360+Svp/360]/[1+/-U+D\}$$
$$/[1-t]+Sv+Sf+I)/S'-1$$

or

$$s = S/S'-1$$

## Rule-58920:

If both (**/**), (**U**), (**S'**), (**s**), (**N**), (**v**), (**f**), (**I**), (**t**), (**D**), (**K**), (**j**) and (**p**) are known, then its Sales or Revenue planned is:

$$S = [[I+/(U+\{S'[1+s]-I\}[1-t] \; D) \; [N+K]]$$
$$/\{j/360+vp/360+[1+/][1-t][v+f]\}$$

or

$$S = S'[1+s]$$

## Rule-58921:

If both (**/**), (**U**), (**S'**), (**s**), (**S**), (**N**), (**f**), (**I**), (**t**), (**D**), (**K**), (**j**) and (**p**) are known, then its Variable Portion planned is:

$$v = [[1+/(U+\{S'[1+s]-Sf-I\}[1-t]-D)$$
$$-[N+K+Sj/360]]/(S\{p/360+[1+/][1-t]\})$$

Steve Asikin ISBN 13: **978-1564912361**, ISBN 10: **1564912363**

Rule-58922:

If both $(\textbf{\textit{I}})$, $(\textbf{U})$, $(\textbf{S'})$, $(\textbf{s})$, $(\textbf{S})$, $(\textbf{v})$, $(\textbf{N})$, $(\textbf{I})$, $(\textbf{t})$, $(\textbf{D})$, $(\textbf{K})$, $(\textbf{j})$ and $(\textbf{p})$ are known, then its Fixed Portion planned is:

$$f = (S'[1+s]-Sv-I-\{[N+K+Sj/360+Svp/360]$$
$$/[1+I-U+D]/[1-t])/S$$

Rule-58923:

If both $(\textbf{\textit{I}})$, $(\textbf{U})$, $(\textbf{S'})$, $(\textbf{s})$, $(\textbf{S})$, $(\textbf{v})$, $(\textbf{f})$, $(\textbf{N})$, $(\textbf{t})$, $(\textbf{D})$, $(\textbf{K})$, $(\textbf{j})$ and $(\textbf{p})$ are known, then its Interest Expense planned is:

$$I = S'[1+s]-Sv-Sf-\{[N+K+Sj/360+Svp/360]$$
$$/[1+I-U+D]/[1-t]$$

Rule-58924:

If both $(\textbf{\textit{I}})$, $(\textbf{U})$, $(\textbf{S'})$, $(\textbf{s})$, $(\textbf{S})$, $(\textbf{v})$, $(\textbf{f})$, $(\textbf{I})$, $(\textbf{N})$, $(\textbf{D})$, $(\textbf{K})$, $(\textbf{j})$ and $(\textbf{p})$ are known, then its Tax Rate planned is:

$$t = 1-\{[N+K+Sj/360+Svp/360]/[1+I-U+D\}$$
$$/\{S'[1+s]-Sv-Sf-I\}$$

Rule-58925:

If both $(\textbf{\textit{I}})$, $(\textbf{U})$, $(\textbf{S'})$, $(\textbf{s})$, $(\textbf{S})$, $(\textbf{v})$, $(\textbf{f})$, $(\textbf{I})$, $(\textbf{t})$, $(\textbf{N})$, $(\textbf{K})$, $(\textbf{j})$ and $(\textbf{p})$ are known, then its Dividend planned is:

$$D = U+\{S'[1+s]-Sv-Sf-I\}[1-t]$$
$$- [N+K+Sj/360+Svp/360]/[1+I]$$

Steve Asikin ISBN 13: **978-1564912361**, ISBN 10: **1564912363**

Rule-58926:

If both (**/**), (**U**), (**S'**), (**s**), (**S**), (**v**), (**f**), (**I**), (**t**), (**D**), (**N**), (**j**) and (**p**) are known, then its Kind of Cash planned is:

$$K= [1+/(U+\{S'[1+s]-Sv-Sf-I\}[1-t]-D)$$
$$-[N+Sj/360+Svp/360]$$

Rule-58927:

If both (**/**), (**U**), (**S'**), (**s**), (**S**), (**v**), (**f**), (**I**), (**t**), (**D**), (**K**), (**N**) and (**p**) are known, then its Job or Trade Receivable Days planned is:

$$j= 360[[1+/(U+\{S'[1+s]-Sv-Sf-I\}[1-t]-D)$$
$$-\{N+K+Svp/360\}]/S$$

Rule-58928:

If both (**/**), (**U**), (**S'**), (**s**), (**S**), (**v**), (**f**), (**I**), (**t**), (**D**), (**K**), (**j**) and (**N**) are known, then its Procured Inventory Days planned is:

$$p= 360[[1+/(U+\{S'[1+s]-Sv-Sf-I\}[1-t]-D)$$
$$-[N+K+Sj/360]]/[Sv]$$

Rule-58929:

If both (**/**), (**U**), (**S'**), (**s**), (**S**), (**v**), (**f**), (**I**), (**t**), (**D**), (**K**), (**j**) and (**p**) are known, then its Non Current Asset planned is:

$$N= [1+/(U+\{S'[1+s]-Sv-Sf-I\}[1-t]-D)-K-Sj/360-$$
$$S'vp[1+s]/360$$

Steve Asikin ISBN 13: **978-1564912361**, ISBN 10: **1564912363**

## Rule-58930:

If both **(N)**, **(U)**, **(S')**, **(s)**, **(S)**, **(v)**, **(f)**, **(I)**, **(t)**, **(D)**, **(K)**, **(j)** and **(p)** are known, then its Leverage or Gearing Ratio planned is:

$$\textit{l} = \{N+K+Sj/360+S'vp[1+s]/360\}$$
$$/(U+\{S'[1+s]-Sv-Sf-I\}[1-t]-D)-1$$

## Rule-58931:

If both **(l)**, **(N)**, **(S')**, **(s)**, **(S)**, **(v)**, **(f)**, **(I)**, **(t)**, **(D)**, **(K)**, **(j)** and **(p)** are known, then its Utilized or Starting Capital must be:

$$U = (\{N+K+Sj/360+S'vp[1+s]/360\}$$
$$-[1+\textit{l}]\{S'[1+s]-Sv-Sf-I\}[1-t]-D)/[1+\textit{l}]$$

## Rule-58932:

If both **(l)**, **(U)**, **(N)**, **(s)**, **(S)**, **(v)**, **(f)**, **(I)**, **(t)**, **(D)**, **(K)**, **(j)** and **(p)** are known, then its Sales Past must be:

$$S' = [[1+\textit{l}](U-\{Sv+Sf+I\}[1-t]-D)-[N+K+Sj/360]]$$
$$/([1+s]\{vp/360-[1+\textit{l}][1-t]\})$$

$$\text{or}$$

$$S' = S/[1+s]$$

Steve Asikin ISBN 13: **978-1564912361**, ISBN 10: **1564912363**

## Rule-58933:

If both (**/**), (**U**), (**S'**), (**N**), (**S**), (**v**), (**f**), (**I**), (**t**), (**D**), (**K**), (**j**) and (**p**) are known, then its Sales Growth planned is:

$$s= [[1+/(U-\{Sv+Sf+I\}[1-t]-D)-[N+K+Sj/360]]$$
$$/(S'\{vp/360-[1+/][1-t]\})-1$$

or

$$s= S/S'-1$$

## Rule-58934:

If both (**/**), (**U**), (**S'**), (**s**), (**N**), (**v**), (**f**), (**I**), (**t**), (**D**), (**K**), (**j**) and (**p**) are known, then its Sales or Revenue planned is:

$$S= [[1+/(U+\{S'[1+s]-I\}[1-t]-D)$$
$$-\{N+K+S'vp[1+s]/360\}]$$
$$/\{j/360+[1+/][1-t][v+f]\}$$

or

$$S= S'[1+s]$$

## Rule-58935:

If both (**/**), (**U**), (**S'**), (**s**), (**S**), (**N**), (**f**), (**I**), (**t**), (**D**), (**K**), (**j**) and (**p**) are known, then its Variable Portion planned is:

$$v= [[1+/(U+\{S'[1+s]-Sf-I\}[1-t]-D)-[N+K$$
$$+Sj/360]]/\{S'p[1+s]/360+S[1+/][1-t]\}$$

Steve Asikin ISBN 13: **978-1564912361**, ISBN 10: **1564912363**

Rule-58936:

If both (**/**), (**U**), (**S'**), (**s**), (**S**), (**v**), (**N**), (**I**), (**t**), (**D**), (**K**), (**j**) and (**p**) are known, then its Fixed Portion planned is:

$$f= [S'[1+s]-Sv-I-(\{N+K+Sj/360+S'vp[1+s]/360\}$$
$$/[1+/-U+D)/[1-t]]/S$$

Rule-58937:

If both (**/**), (**U**), (**S'**), (**s**), (**S**), (**v**), (**f**), (**N**), (**t**), (**D**), (**K**), (**j**) and (**p**) are known, then its Interest Expense planned is:

$$I= S'[1+s]-Sv-Sf-(\{N+K+Sj/360+S'vp[1+s]/360\}$$
$$/[1+/-U+D)/[1-t]$$

Rule-58938:

If both (**/**), (**U**), (**S'**), (**s**), (**S**), (**v**), (**f**), (**I**), (**N**), (**D**), (**K**), (**j**) and (**p**) are known, then its Tax Rate planned is:

$$t= 1-(\{N+K+Sj/360+S'vp[1+s]/360\}/[1+/-U+D)$$
$$/\{S'[1+s]-Sv-Sf-I\}$$

Rule-58939:

If both (**/**), (**U**), (**S'**), (**s**), (**S**), (**v**), (**f**), (**I**), (**t**), (**N**), (**K**), (**j**) and (**p**) are known, then its Dividend planned is:

$$D= U+\{S'[1+s]-Sv-Sf-I\}[1-t]- \{N+K+Sj/360$$
$$+S'vp[1+s]/360\}/[1+/]$$

Steve Asikin ISBN 13: **978-1564912361**, ISBN 10: **1564912363**

Rule-58940:

If both (**/**), (**U**), (**$'**), (**s**), (**$**), (**v**), (**f**), (**I**), (**t**), (**D**), (**N**), (**j**) and (**p**) are known, then its Kind of Cash planned is:

$$K= [1+/\!(U+\{\$'[1+s]-\$v-\$f-I\}[1-t]-D)$$
$$-\{N+\$j/360+\$'vp[1+s]/360\}$$

Rule-58941:

If both (**/**), (**U**), (**$'**), (**s**), (**$**), (**v**), (**f**), (**I**), (**t**), (**D**), (**K**), (**N**) and (**p**) are known, then its Job or Trade Receivable Days planned is:

$$j= 360[[1+/\!(U+\{\$'[1+s]-\$v-\$f-I\}[1-t]-D)$$
$$-\{N+K+\$'vp[1+s]/360\}]/\$$$

Rule-58942:

If both (**/**), (**U**), (**$'**), (**s**), (**$**), (**v**), (**f**), (**I**), (**t**), (**D**), (**K**), (**j**) and (**N**) are known, then its Procured Inventory Days planned is:

$$p= 360[[1+/\!(U+\{\$'[1+s]-\$v-\$f-I\}[1-t]-D)$$
$$-[N+K+\$j/360]]/\{\$'v[1+s]\}$$

Steve Asikin ISBN 13: **978-1564912361**, ISBN 10: **1564912363**

The Rest will be Continued in
The Series of Book

## MATH FIN LAW 19
Mathematical Financial Law

At
**http: //www.Amazon.Com**

Steve Asikin ISBN 13: **978-1564912361**, ISBN 10: **1564912363**

# REFERENCES:

## A. Books:

1) **Amin, A.R. & Asikin, S.**(2011a- 110 pages), *CEO Time Matrix: Productivity Management Skill that Increase CEO Effectiveness by 12400%*, Seattle (US): Amazon Createspace. http://www.amazon.com/CEO-Time-Matrix-Productivity-Effectiveness/dp/1461066069/ref=sr_1_1?s=books&ie=UTF8&qid=1387091880&sr=1-1&keywords=ceo+time+matrix. ISBN-13: 978-1461066064, ISBN-10: 1461066069

2) **Amin, A.R. & Asikin, S.** (2014- 76 pages), *No Urgency CEO: A High-Tech Book in Modern Management*, Seattle (US): Amazon Createspace. http://www.amazon.com/No-Urgency-CEO-Hi-Tech-Management/dp/1489509356/ref=sr_1_1?s=books&ie=UTF8&qid=1396440688&sr=1-1&keywords=no+urgency. ISBN-13: 978-1489509352, ISBN-10: 1489509356

3} **Amin, A.R. & Asikin, S.** (2011b- 292 pages), *The Celestial Management: Spiritual Wisdom for All Human Beings*, Seattle (US): Amazon Createspace. http://www.amazon.com/Celestial-Management-Islamic-Wisdom-Beings/dp/1456450700/ref=sr_1_1?s=books&ie=UTF8&qid=1387092036&sr=1-1&keywords=celestial+management. ISDN-13: 978-1456450700, ISBN 10: 1456450700

4) **Asikin, S.** (2009- 548 pages), *Arigato, Obrigado... (Thanks): Portuguese-Japan 1550-1647 Historic Novel*, Seattle (US): Amazon Createspace. http://www.amazon.com/Arigato-Obrigado-Thanks-Portuguese-Japan-1550-1647/dp/1453786368/ref=sr_1_1?s=books&ie=UTF8&qid=1396442162&sr=1-1&keywords=arigato. ISBN-13: 978-1453786369, ISBN-10: 1453786368

5) **Asikin, S.** (2013a- 824 pages), *Econometric Rex: 888 Marketing Promo 6666 Six Sigma Parametric Theories (of 10 Data)*, Seattle (US): Amazon Createspace. http://www.amazon.co.uk/Econometric-Rex-Marketing-Parametric-Theories/dp/1482745259, ISBN-13: 978-1482745252, ISBN-10: 1482745259

6) **Asikin, S.** (2010b- 224 pages), *FINANCE Architecture: VISUAL art, for Corporate Finance's Beauty, Safety and Simplicity*, Seattle (US): Amazon Createspace. http://www.amazon.co.uk/FINANCE-Architecture-Corporate-Finances-Simplicity/dp/1453829547/ref=sr_1_14?s=books&ie=UTF8&qid=1385371842&sr=1-14&keywords=finance+architecture. ISBN-13: 978-1453829547, ISBN-10: 1453829547

7) **Asikin, S.** (2010c- 150 pages), *Geometric FINANCE: Useful concepts that work better than Economic Noble Laureates'* Seattle (US): Amazon Createspace. http://www.amazon.co.uk/Magic-Geometric-FINANCE-concepts-Laureates/dp/1453790284/ref=tmm_pap_title_0, ISBN-13: 978-1453790281, ISBN-10: 1453790284

Steve Asikin ISBN 13: **978-1564912361**, ISBN 10: **1564912363**

8) **Asikin, S.** (2010a- 290 pages), *Investment ALGEBRA: Strategic Profitable Comprehensive Business Finance Engineering Technology,* Seattle (US): Amazon Createspace. http://www.amazon.co.uk/Investment-ALGEBRA-Profitable-Comprehensive-Engineering/dp/1453856811/ref=tmm_pap_title_0?ie=UTF8&qid=1385371309&sr=1-1, ISBN-13: 978-1453856819, ISBN-10: 1453856811

9) **Asikin, S.** (2011a- 352 pages), *State Economics: Comprehensive Macro-Micro Economics' Simple Fiscal-Monetary Export-Import Accouting, Integrated Supply-Demand Managerial ... Mathematical Engineering Visual .* Seattle (US): Amazon Createspace. http://www.amazon.co.uk/STATE-ECONOMICS-Steve-Asikin-ebook/dp/B00G8905SS, http://www.amazon.co.uk State-Economics-Steve-Asikin-ebook/dp/B00BBPLQDI, ISBN-13: 978-1461066095, ISBN-10: 1461066093

10) **Asikin, S.** (2013b- 260 pages), *Technoeconomistat: Economic &Statistical Technology for Financial Planning,* Seattle (US): Amazon Createspace. http://www.amazon.co.uk/Technoeconomistat-Economic-Statistical-Technology-Financial/dp/1482304244, ISBN-13: 978-1482304244, ISBN-10: 1482304244

11) **Asikin, S.** (2011b- 540 pages), *Terima Kasih: Novel Internasional Sejarah Jepang-Portugis 1550-1647,* Seattle (US): Amazon Createspace. http://www.amazon.com/Terima-Kasih-Internasional-Jepang-Portugis-1550-1647/dp/146351106X/ref=sr_1_1?s=books&ie=UTF8&qid=1396442361&sr=1-1&keywords=terima, ISBN-13: 978-1463511067, ISBN-10: 146351106X

12) **Asikin, S.** (2010d- 294 pages), *TREASURY Planology: Strategic Profitable Comprehensive Business Finance Engineering Technology,* Seattle (US): Amazon Createspace. http://www.amazon.co.uk/TREASURY-PLANOLOGY-Steve-Asikin-ebook/dp/B00G88MTFW/ref=sr_1_1?s=books&ie=UTF8&qid=1385403690&sr=1-1&keywords=Treasury+Planology. ISBN-13: 978-1453891476, ISBN-10: 1453891471

13) **Asikin, S.** (2014a- 190 pages), *Calculus Accountancy: Leibnitz Newton Pacioli's Polynomial Quantitative Finance Risk Modelling Optimization,* Seattle (US): Amazon Createspace. http://www.amazon.com/Calculus-Accountancy-Polynomial-Quantitative-Optimization/dp/1495463028/ref=sr_1_1?s=books&ie=UTF8&qid=1396441338&sr=1-1&keywords=calculus+accountancy, ISBN-13: 978-1495463020, ISBN-10: 1495463028

14) **Asikin, S.**(2014b- 318 pages), *No Parametric: No Parametric: Hyperbolic Octahedron Central 3-Dimensional Orthogonal Risk Distribution Topology,* Seattle (US): Amazon Createspace. http://www.amazon.com/Parametric-Hyperbolic-Octahedron-3-Dimensional-Distribution/dp/149608973l/ref=sr_1_1?s=books&ie=UTF8&qid=1396441543&sr=1-1&keywords=No+Parametric, ISBN-13: 978-1496089731, ISBN-10: 1496089731

Steve Asikin ISBN 13: **978-1564912361**, ISBN 10: **1564912363**

15) **Asikin, S.** (2014c- 138 pages), *Quick Help: Religion vs Real-Legion Philosophy*, Seattle (US): Amazon Createspace.
http:/www.amazon.com/Quick-Help-Religion-Real-Legion-Philosophy/dp/1497466520/ref=sr_1_2?s=books&ie=UTF8&qid=1396441740&sr=1-2&keywords=quick+help, ISBN-13: 978-1497466524, ISBN-10: 1497466520

16) **Asikin, S.** (2014d- 824 pages), *Anne of Denmark: King James I the Bible, Historic Plays Drama*, Seattle (US): Amazon Createspace.
http:/www.amazon.com/Anne-Denmark-James-Bible-Drama/dp/149299734X/ref=sr_1_1?s=books&ie=UTF8&qid=1396441857&sr=1-1&keywords=anne+asikin, ISBN-13: 978-1492997344, ISBN-10: 149299734X

17) **Asikin, S.** (2014e- 184 pages), *Constructive Spells: Positive Spiritual Mentality Personal Development*: Seattle (US): Amazon Createspace.
http:/www.amazon.com/Constructive-Spells-Spiritual-Mentality-Development/dp/1497498899/ref=sr_1_1?s=books&ie=UTF8&qid=1397664512&sr=1-1&keywords=constructive+spells; ISBN-13: 978-1497498891, ISBN-10: 1497498899

18) **Asikin, S.** (2014f- 466 pages), *Elizabeth of Russia: Another Virgin Queen Gloariana: A Romanov Plays Drama*, Seattle (US): Amazon Createspace.
http:/www.amazon.com/Elizabeth-RUSSIA-Another-Gloriana-Romanov/dp/1505542480/ref=sr_1_4?s=books&ie=UTF8&qid=1423888482&sr=1-4&keywords=elizabeth+russia; ISBN-13: 978-1505542486, ISBN-10: 1505542480

19) **Asikin, S.** (2014f- 708 pages), *Shinmen Munisai:Miyamoto Musashi's Father, A Sword Samurai Plays Drama*, Seattle (US): Amazon Createspace.
http:/www.amazon.com/Shinmen-Munisai-Miyamoto-Musashis-Samurai/dp/1507857799/ref=sr_1_1?s=books&ie=UTF8&qid=1428094280&sr=1-1&keywords=munisai; ISBN-13: 978-1507857793, ISBN-10: 1507857799

20) **Asikin, S.** (2015a-767 pages), *Math Fin Law-1: Mathematical Financial Law for Public Listed Firms Rule 1-4441*, Seattle (US): Amazon Createspace.
http://www.amazon.com/Math-Fin-Law-Mathematical-Financial-ebook/dp/B00WLNX3Y4/ref=asap_bc?ie=UTF8; ISBN-13: 978-1511792219, ISBN-10: 1511792213

21) **Asikin, S.** (2015b-813 pages), *Math Fin Law-2: Mathematical Financial Law for Public Listed Firms Rule 4442-8712*, Seattle (US): Amazon Createspace.
http://www.amazon.com/Math-Fin-Law-Mathematical-No-4442-8712/dp/1512285080/ref=sr_1_1?s=books&ie=UTF8&qid=1432732199&sr=1-1&keywords=math+fin+law+2; ISBN-13: 978-1512285086, ISBN-10: 1512285080

22) **Asikin, S.** (2015c-813 pages), *Math Fin Law-3: Mathematical Financial Law for Public Listed Firms Rule 8713-12575*, Seattle (US): Amazon Createspace. http://www.amazon.com/Math-Fin-Law-Mathematical-8712-12575/dp/1514276763/ref=sr_1_1?s=books&ie=UTF8&qid=1433982149&sr=1-1&keywords=math+fin+law+3, ISBN-13: 978-1514276761. ISBN-10: 1514276763

Steve Asikin ISBN 13: **978-1564912361**, ISBN 10: **1564912363**

23) **Asikin, S.** (2015d-813 pages), _Math Fin Law-4: Mathematical Financial Law for Public Listed Firms Rule 12576-16333,_ Seattle (US): Amazon Createspace. http://www.amazon.com/Math-Fin-Law-Mathematical-12576-16333/dp/1514685132/ref=sr_1_1?s=books&ie=UTF8&qid=1435393628&sr=1-1&keywords=math+fin+law+4, ISBN-13: 978-1514685136, ISBN-10: 1514685132

24) **Asikin, S.** (2015d-816 pages), _Math Fin Law-5: Mathematical Financial Law for Public Listed Firms Rule 16334-19904,_ Seattle (US): Amazon Createspace. http://www.amazon.com/Math-Fin-Law-Mathematical-No-16334-19904/dp/1515073009/ref=sr_1_1?s=books&ie=UTF8&qid=1437224693&sr=1-1&keywords=math+fin+law+5. ISBN-13: 978-1515073000, ISBN-10: 1515073009

25) **Asikin, S.** (2015d-728 pages), _Math Fin Law-6: Mathematical Financial Law for Public Listed Firms Rule 19905-23237,_ Seattle (US): Amazon Createspace. https://www.amazon.com/Math-Fin-Law-Mathematical-No-19905-23238-ebook/dp/B0129ILVNK/ref=sr_1_2?ie=UTF8&qid=1478017624&sr=8-2&keywords=Math+Fin+Law+Asikin. ISBN-13: 978-1515158691, ISBN-10: 1515158691

26) **Asikin, S.** (2016a-728 pages), _Math Fin Law-7: Mathematical Financial Law for Public Listed Firms Rule 23238-27115,_ Seattle (US): Amazon Createspace. https://www.amazon.com/Math-Fin-Law-Mathematical-No-23238-27115/dp/1508518521/ref=la_B00428V6JU_1_27?s=books. ISBN-13: 978-1508518525, ISBN-10: 1508518521

27) **Asikin, S.** (2016b-728 pages), _Math Fin Law-8: Mathematical Financial Law for Public Listed Firms Rule 27116-30330,_ Seattle (US): Amazon Createspace. https://www.amazon.com/dp/1541091884, https://www.amazon.com/Math-Fin-Law-Mathematical-No-27116-30330/dp/1541091884/ref=la_B00428V6JU_1_27?s=books&ie=UTF8&qid=1479398538& sr=1-27&refinements=p_82%3AB00428V6JU, ISBN-13: 978-1508518525, ISBN-10: 1508518521

28) **Asikin, S.** (2017a-828 pages), _Math Fin Law-9: Mathematical Financial Law for Public Listed Firms Rule 30330-33373,_ Seattle (US): Amazon Createspace. https://www.amazon.com/Math-Fin-Law-Mathematical-30331-33373-ebook/dp/B01MRA8HZQ/ref=la_B00428V6JU_1_4?s=books&ie=UTF8&qid=1485921040&sr=1-4, ISBN-13: 978-1541092433, ISBN-10: 1541092430

29) **Asikin, S.** (2017b-823 pages), _Math Fin Law-10: Mathematical Financial Law for Public Listed Firms Rule 33374-36242,_ Seattle (US): Amazon Createspace. https://www.amazon.com/s/ref=nb_sb_noss?url=search-alias%3Dstripbooks&field-keywords=math+fin+law+10, https://www.amazon.com/s/ref=nb_sb_noss?url=search-alias%3Dstripbooks&field-keywords=math+fin+law+10, ISBN-13: 978-1541092761, ISBN-10: 1541092767

Steve Asikin ISBN 13: **978-1564912361**, ISBN 10: **1564912363**

30) **Asikin, S**. (2017c-823 pages), *Math Finance Law-11: Mathematical Financial Law for Public Listed Firms Rule 36243-39158,* Seattle (US): Amazon Createspace. https://www.amazon.com/s/ref=nb_sb_noss?url=search-alias%3Dstripbooks&field-keywords=math+finance+law+11. https://www.amazon.com/dp/1541215265/ref=sr_1_1?s=books&ie=UTF8&qid=1488029978&sr =1-1&keywords=math+finance+law. ISBN-13: 978-1541215269. ISBN-10: 1541215265

31) **Asikin, S**. (2017d-823 pages), *Math Finance Law-12: Mathematical Financial Law for Public Listed Firms Rule 30150-42152,* Seattle (US): Amazon Createspace. https://www.amazon.com/s/ref=nb_sb_noss?url=search-alias%3Dstripbooks&field-keywords=math+finance+law+12. https://www.amazon.com/Math-Finance-Law-Mathematical-39159-42152/dp/1541215516/ref=sr_1_2?s=books&ie=UTF8&qid=1490990342&sr=1-2&keywords=math+finance+law+12. ISBN-13: 978-1541215511, ISBN-10: 1541215516

32) **Asikin, S**. (2017e-824 pages), *Math Finance Law-13: Mathematical Financial Law for Public Listed Firms Rule 42153-44938,* Seattle (US): Amazon Createspace. https://www.amazon.com/dp/1542675219/ref=sr_1_1?s=books&ie=UTF8&qid=1491146227&sr=1 -1&keywords=math+finance+law+13 . ISBN-13: 978-1542675215, ISBN-10: 1542675219

33) **Asikin, S**. (2017f-824 pages), *Math Finance Law-14: Mathematical Financial Law for Public Listed Firms Rule 44939-47745,* Seattle (US): Amazon Createspacc. https://www.amazon.com/dp/1542675219/ref=sr_1_1?s=books&ie=UTF8&qid=1491146227&sr=1 -1&keywords=math+finance+law+14 . ISBN-13: 978-1544628967, ISBN-10: 154462896X

34) **Asikin, S**. (2017g-810 pages), *Math Finance Law-15: Mathematical Financial Law for Public Listed Firms Rule 47746-50465,* Seattle (US): Amazon Createspace. https://www.amazon.com/dp/1545118647/ref=sr_1_1?s=books&ie=UTF8&qid=1493626425&sr= 1-1&keywords=math+finance+law+15. . ISBN-13: 978-1545118641, ISBN-10: 1545118647

35) **Asikin, S**. (2017h-816 pages), *Math Finance Law 16: Mathematical Financial Law Public Listed Firm Rule No.50,466-53,320,* Seattle (US), Amazon CreateSpace, https://www.amazon.com/dp/1545320179/ref=sr_1_1?s=books&ie=UTF8&qid=1495692695&sr =1-1&keywords=math+finance+law+16. ISBN-13: 978-1545320174, ISBN-10: 1545320179

36) **Asikin, S**. (2017i-809 pages), *Math Finance Law 17: Mathematical Financial Law Public Listed Firm Rule No.53,231-56,124,* Seattle (US), Amazon CreateSpace, https://www.amazon.com/dp/1545320179/ref=sr_1_1?s=books&ie=UTF8&qid=1495692695&sr =1-1&keywords=math+finance+law+17. ISBN-13: 978-1545486474, ISBN-10: 1545486476

Steve Asikin ISBN 13: **978-1564912361**, ISBN 10: **1564912363**

37) **Asikin, S**. (2014-47 pages), *Doing Business in Comprehensive Interactive Interdisciplinary Micro-Macro Eco nomic International Accounting Architecture of Serial Well Known General Great Theories*, Proceedings in Finance and Risk Perspective 13, ACRN Oxford Publishing House, UK, p.197-243, https://books.google.com/books?id=4e4KAwAAQBAJ&pg=PA2198lpg=PA219&dq=steve+asikin+ oxford&source=bl&ots=qVq9KyAmem&sig=ABdBliwuzUwRigIsuLimuJ5ZdW4&hl=en&sa=X&sqi=2 &ved=0ahUKEwjZkMrF7KjTAhUDSiYKHREDB- q0GAEILjAC#v=onepage&q=steve%20asikin%20oxford&f=false, ISBN 978-3-9503518-1-1.

38) **Asikin, T.& Asikin, S.** (2013- 828 pages), *448 Poems: KARAOKE Song Book: 24 Hours Story of 30 Languages*, Seattle (US): Amazon Createspace. http://www.amazon.com/448-Poems-KARAOKE-Song- Book/dp/1480034363/ref=sr_1_1?s=books&ie=UTF8&qid=1396442034&sr=1- 1&keywords=448+poems, ISBN-13: 978-1480034365, ISBN-10: 1480034363

39) **Asikin, T, Asikin, S, Senihardja, I.** (2017- 508 pages), Finance Construction 1: *Corporate IFRS-GAAP (B/S-I/S) Engineering Technologies No. 1-1,000 of 111,111 Laws*, Seattle (US): Amazon Createspace. https://www.amazon.com/s/ref=nb_sb_noss?url=search- alias%3Dstripbooks&field-keywords=finance+construction+asikin. ISBN-13: 978- 1544059525, ISBN-10: 1544059523

40) **Brigham, EF & Gapenski, LC, .** (1988- Textbook), *Financial Management: Theory and Practice (Study Guide)*, NY (US): Dryden Press. http://www.amazon.com/Financial-Management-Theory-Practice- Study/dp/0030125391/ref=sr_1_1?s=books&ie=UTF8&qid=1429380835&sr=1-1 ;

41) **Chandonet, RE, .** (2011-32 pages), *Balancesheet Management, Trends & Strategies*, NY (US): Sandler O'Neil. https://www.ncbankers.org/uploads/File/Meetings/Presentations/2011CONV-Chandonnet.pdf;

42) **Cornuejols G & Tutuncu, R, .** (2006-358 pages), *Optimization Methods in Finance*, Cambridge (UK): Cambridge University Press. http://www.math.ku.dk/~rolf/CT_FinOpt.pdf;

43) **Crook, M.** (2015-6 pages), *Optimization Methods in Finance*, UBS FS (Swiss): CIO WM Research. https://www.ubs.com/content/dam/WealthManagementAmericas/documents/investment- strategy-insights-2015-01-09-balance-sheet-optimization.pdf;

44) **Fee, Greg.** (2012-127 pages), *Catalan's Constant: (Ramanujan's Formula}: Catalan Constant to 300,000 digits*, Seattle (US): Amazon Kindle. http://www.Amazon.com/gp/aw/d/B0082Z5IEW/ref=mp_sa_1_4?ie=UTF8%Qqid=148 4367818&sr=8- 4&pi=AC_SX236_SY340_QL65&keywords=ramanujan&dpPl=1&dplD=5IFqlwDZpRL&ref= plSrch

45) **Gardenfors, Peter.** (2014) *The Geometry of Meaning: Semantics Based on Conceptual Spaces.* Cambridge, Mass.: MIT Press. ISBN 0- 262-02678-3.

795

Steve Asikin ISBN 13: **978-1564912361**, ISBN 10: **1564912363**

46) **Gardenfors, Peter.** (2000), *Conceptual Spaces: The Geometry of Thought*. Cambridge, Mass.: MIT Press. ISBN 0-262-07199-1.

47) **Kanigel, Robert.** (2013-465 pages), *The Man Who Knew Infinity:A Life of the Genius Ramanujan,* Washington (US): Washington Square Press&Amazon Kindle.
http://www.Amazon.com/gp/aw/d/B008BW4VEGM/ref=pd_aw_sim35l_1?ie=UTF8&psc=1&refRID=VYN345GVDBJ6ATDQ35W2&dpID=91Zgpc4ETdL

48) **Limlingan, Victor S.** (1993-1 page), *The Limlingan Financial Model,* Manila: Asian Manager. http://limlingan.com/images/article.gif

49) **Newton, Isaac.** (2015-631 pages), *Principia:The Mathematical Principles of Natural Philosophy (Active Content),* London (UK): RSM&Amazon Kindle.
http://www.Amazon.com/gp/aw/sitb/B011SFJSJ0?ref=sib_dp_aw_kb_udp

50) **Nikisa, Evets.** (2014g- 828 pages), *Wonderful Live Spells: 144-Steps Recreative Nice Children Story, Succesful Development:* Seattle (US): Amazon Createspace. http://www.amazon.com/WONDERFUL-Live-Spells-Recreative-Development/dp/1499248423/ref=sr_1_1?s=books&ie=UTF8&qid=1418698379&sr=1-1&keywords=wonderful asikin, ISBN-13: 978-1499248425, ISBN-10: 1499248423

51) **Nielson, SS.** (2007- 430 pages), *Practical Financial Optimization*: *Decision Making for Financial Engineers:* NY (US): Wiley.
https://books.google.com/books/about/Practical_Financial_Optimization.html?id=_XkwLAAACAAJ,

52) **Niswonger CR, Fess PE & Warren CE.** (1993-Textbook), *Accounting Principles:* Seattle (US): South Western Publishing Co.
http://www.amazon.com/Accounting-Principles-Chapters-C-Rollin-Niswonger/dp/0538818603

53) **Omerod, P.** (1993-Textbook), (1994), *The Death of Ecoomics,* St Martin's Press, London, . http://www.amazon.com/Accounting-Principles-Chapters-C-Rollin-Niswonger/dp/0538818603, ISBN-13: 978-0312134648, ISBN-10: 0312134649

54) **Paccioli, L. (1989-Textbook),** *Summa de Arithmetica Geometria Proportioni et Proportionalita,* Venice 1494. http://www.amazon.com/Arithmetica-Geometria-Proportioni-Proportionalita-Venice/dp/B00T4HXT1W/ref=sr_1_fkmr0_2?s=books&ie=UTF8&qid=1429597966&sr=1-2-fkmr0&keywords=summa+mhematica+arithmetica

55) **Persson, Torsten & Tabellini, Guido** (2000), *Political Economics – Explaining Economic Policy,* MIT Press. https://mitpress.mit.edu/authors/torsten-persson

56) **Persson, Torsten & Tabellini, Guido** (2003), *The Economic Effects of Constitutions,* MIT Press. https://mitpress.mit.edu/authors/torsten-persson

57) **Persson, Torsten & Tabellini, Guido** (2000), *Monetary and Fiscal Policy –1 Credibility,* MIT Press. https://mitpress.mit.edu/authors/torsten-persson

Steve Asikin ISBN 13: **978-1564912361**, ISBN 10: **1564912363**

58) **Persson, Torsten & Tabellini, Guido** (2003), *Monetary and Fiscal Policy –2 Politics*, MIT Press. https://mitpress.mit.edu/authors/torsten-persson

59) **Puts, J** (2012-102 pages), *Balancesheet Optimization Under Basel III*, Amsterdam (Netherlads): Vrije Universiteit.
http://www.math.vu.nl/~sbhulai/papers/thesis-puts.pdf

60) **Timmermans, K** (2012-30 pages), *Balancesheet Optimization Under Basel III*, Amsterdam (Netherlads): ING Investor Day.
https://www.ing.com/web/file?uuid=09c803d9-79a1-47ec-bfe5...owner...

61 **Weston JF, Copeland TE & Shastri, K.** (2004-Textbook), *Financial Theory and Corporate Policy:* NY (US): Addison Wesley.
http://www.amazon.com/Financial-Corporate-Copeland-Kuldeep-Paperback/dp/B008YSUE4M/ref=sr_1_1?s=books&ie=UTF8&qid=1429381700&sr=1-1&keywords=weston+copeland

62) **Zenios SA.** (2002-350 pages), *Financial Optimization:* Cambridge (UK): Cambridge University Press.
https://books.google.com/books/about/Financial_Optimization.html?id=-FJdqjZOqSAC, .

Steve Asikin ISBN 13: **978-1564912361**, ISBN 10: **1564912363**

# B. Intellectual Properties:

1) **Indonesian Banker Certification 1993**, *Sertifikasi Bankir Indonesia 1993*;
2) **MATHFINPLAN** (Mathematical Financial Planning);
   *Perencanaan Keuangan Matematis*
3) **BANK MATHPLAN** (Bank's Mathematical Planning);
   *Perencanaan Matematis Bank*
4) **TECHNO-ECONOMISTAT** (Technology for Economical Statistics);
   *Teknologi Statistik untuk Ekonomi*
5) **FINANCIAL GEOMETRY** (IT Program); *Ilmu Geometri Keuangan*
6) **BAK DOBEL** (Car Gasoline Tank: Left and Right Side Fuel Fill-in).
   *Tangki Mobil yang bisa diisi dari Kiri atau Kanan*
7) **Mister GO-CHENG** (Cheap Food Franchising). *Waralaba Pangan Murah*

Steve Asikin ISBN 13: **978-1564912361**, ISBN 10: **1564912363**

## C. SCIENTIFIC PAPERS:

1) **Eichhornia Crassipes**, as Anti Polutant and Fertile Sediment, *Pemanfaatan Eceng Gondok untuk Menetralkan Polusi dan Pengurukan Subur pada Genangan Air*, LKIR, Lomba Karya Ilmiah Remaja, LIPI (1979)

2) **Archimedes Catesian Buoyancy Law**, *Penyimpangan Hukum Archimedes pada Model Pelampung Cartesian*, Lomba Karya Ilmiah Remaja, Depdikbud-RI (1980)

3) **SURAMADU**, Surabaya-Madura Inter Island Bridge, Design Supervisor, *Sumarsono Sumali Engineering Contractor*, Tambak-V, Tandes, Surabaya (1982),

4) **GAZOLINE Taxi-Pool Depot**, Jemur Handayani, Surabaya, *Sumarsono Sumali Engineering Contractor*, Tambak-V, Tandes, Surabaya (1982),

5) **SOEKARNO-HATTA**, Jakarta International Airport, Garuda Maintenance Facilities, *Junior Architect*, As Bulit Drawings, Coinstruction Management, Aerospatiale-Encona Engineering (1983-1985),

6) **ADISUCIPTO**, Yogya International Airport, *Junior Architect*, Preliminary Technical Drawings, Budiono Surasno-Encona Engineering (1986),

7) **BLOK-M PLAZA**, "New Garden Hall", Kebayoran Baru Supermall, *Preliminary Design Architect*, Danisworo-James Ferry- Jasa Ferry-Encona (1986),

8) **NORTH JAKARTA PUBLIC LIBRARY**, Semper, Tanjung Priok, *Chief Architect*, Agung Darma Sarana-Asep Hermawan (1987),

9) **The Banker's Trainer**: A Manual for The Training Head in Banking. *Profesi Bankir Pelatih, Tuntunan bagi Kepala Diklat Perbankan*, (1200 Pages, 1993),

10) **Indonesian Bankers Certification**, a Strategic Move for the National Banking Improvement. *Sertifikasi Bankir Indonesia, Langkah Strategis Pembenahan Perbankan Nasional*, (KOMPAS, 20 Juli 1993).

11) **Indonesian Rupiah's Exchange Rate**, on a Quiet but Steady Movement. *Nilai Tukar Rupiah, Diam-diam Meningkat Pesat*, (MANAJEMEN, Jan-Feb 1994).

12) **The Bank's Credit**, for the Best Practices. *Kredit Praktis* (200 pages, 1994),

13) **Down to Earth Credit Meeting Credo**, *Kiat Rapat Pemberian Kredit yang Realistik*. (MANAJEMEN, Sep-Oct 1994).

14) **SCHIPHOL**, Amsterdam International Airport, 1st Anthony Fokker Weg, Sloten, *Airport Engineer*, Samsung- Fokker-Beurlin, BV (1996),

15) **Ultrapeneurship: The Secret of Starting Business Without Capital**. Ultrapreneur: *Kiat Usaha Tanpa Modal*, (MANAJEMEN, Nov-Dec 1996),

16) **Recipe from the Corporate Raider's Kitchen**, *Menguak Dapur Corporate Raiders*, (MANAJEMEN, Jan-Feb 1997).

17) **The 10-Tips and Tricks of Engineered Conglomeration**. *Sepuluh Rekayasa Keuangan Konglomerat*, (MANAJEMEN, Mar-Apr 1997),

18) **The Management of Succession in Five Parts of the World**. *Manajemen Suksesi di Lima bagian Dunia*, (MANAJEMEN, Mei-Jun 1997),

19) **Auditing**: The Global Language for the 2003's Human Resources. *Audit : Bahasa Global untuk SDM 2003*, (MANAJEMEN, Jul-Aug 1997),

Steve Asikin ISBN 13: **978-1564912361**, ISBN 10: **1564912363**

20)<u>The Elected Parliament 1997-2002</u>, and the Hope of Professionals. *DPR 1997–2002 dan Harapan Profesional,* (MANAJEMEN, Sep-Okt 1997),

21)<u>Disappearance of World's Soft Currencies</u>, and Proposition for Global Economic Thinking. *Lenyapnya Mata Uang Lunak Dunia dan Usulan Pemikiran Ekonomi Global,* (MANAJEMEN, Nov-Dec 1997),

22)<u>The Human Resources and Its Accountability</u>, in Critical Situation. *Akuntabilitas Sumberdaya Manusia di Masa Krisis,* (MANAJEMEN, Jan-Feb 1998), '

23)<u>The Investment Economic Model</u>, and the Failure of Growth Economic Theories. *Model Ekonomi Investasi dan Kegagalan Teori Ekonomi Pertumbuhan,* (MANAJEMEN, Nov 1998),

24)<u>The Exact Equilibrium of Capital Budgeting</u>, *Persamaan-persamaan akurat dalam Penganggaran Modal Perusahaan.* (MANAGEMENT EXPOSE, November 1998),

25)<u>The Corruptive Economic Model</u>, and Its Solution for Long Term Debt's Problem. *Model Ekonomi Korupsi dan Penyelesaian Utang Jangka Panjang,* (MANAJEMEN, Dec 1998),

26)<u>The Self Equity Budgeting</u>, and Operations Research in Investments. *Penganggaran Modal Sendiri dan Riset Operasi dalam Investasi,* (MANAJEMEN, Jan 1999),

27)<u>Actual Impact of Inflation</u>, and the Need of Geometric Approach in Statistic. *Pengaruh Riil Inflasi dan Perlunya Statistika Geometri,* (MANAJEMEN, Feb 1999),

28)<u>Multiplication of Money</u>, to Overcome the Crisis. *Melipat Gandakan Uang untuk Mengusir Krisis,* (MANAJEMEN, Apr 1999),

29)<u>The Interactive Mathematical Model</u>, for Multi Constraints Financial Budgeting, *Model Matematika Ekonomi Interaktif bagi Penganggaran Keuangan Multi Kendala.* (SCRIPTA ECONOMICA, Vol. 2, April 1999)

30)<u>Application of Ethics</u>, for the Management Accounting Practices. *Aplikasi Etika dalam Praktek Akuntansi Manajemen,* (MANAJEMEN, May 1999),

31)<u>The Curriculum of Magistrate of Management in Finance</u>: Its Needs and Practices. *Kurikulum Magister Manajemen Keuangan 1988-1998: Antara Kebutuhan dan Praktek,* (MANAJEMEN, Jun 1999),

32)<u>The Attractiveness</u>, Personal Power and Management. *Pesona Pribadi: "Personal Power" dan Manajemen,* (MANAJEMEN, Jul 1999),

33)<u>The Market Research</u>, for Magistrate Degree in Management. *Soal Jawab Riset Pasar untuk Pascasarjana Manajemen* (126 Pages, 1999),

34)<u>The Thesis Manual</u>, Question and Answer for Magistrate Degree in Management. *Soal Jawab Tugas Akhir untuk Pascasarjana Manajemen* (126 Pages, 1999),

35)<u>The Risk Management</u>, Question and Answers for Magistrate Degree in Management. *Soal Jawab Manajemen Resiko untuk Pascasarjana Manajemen* (126 Pages, 1999),

Steve Asikin ISBN 13: **978-1564912361**, ISBN 10: **1564912363**

36)**The Credit Manual**, Question and Answers for Magistrate Degree in Management. *Soal Jawab Manajemen Kredit untuk Pascasarjana Manajemen* (126 Pages, 1999),

37)**Enhancing the Morality Structure**. *Solidkan Struktur Moralitas,* (MANAJEMEN, Oct 1999)

38)**Performance of PT. Astra Otoparts**, Plc, in its Press Releases and Public Financial Reports. *Kinerja PT. Astra Otoparts dalam Siaran Pers dan Laporan Keuangan Publik,* (MANAJEMEN, Oct 1999)

39)**The Management of Conflicts**, for the New Indonesian's "Rainbow Colored" Cabinet. *Manajemen Konflik dan Kabinet Pelangi,* (MANAJEMEN, Dec 1999),

40)**Technoeconometric**: An Engineering Elaboration to Matrix Multi Variate Econometrics, *Tekno Ekonometrik sebagai sumbangan teknologi untuk Ekonometrika Matrik Multi Variat.* (MANAGEMENT EXPOSE, November 1999),

41)**Technoeconometric** : A Mathematical Shortcut for Non Linear Bivariate Econometrics, *Tekno Ekonometrik sebagai Usulan Jalan Pintas Matematik untuk Ekonometrika non Linear Bivariat.* (SCRIPTA ECONOMICA, Vol. 2, No. 3, Desember 1999),

42)**Developing "New Image"**, for the New Indonesia. *Membangun Citra Indonesia Baru,* (MANAJEMEN, Jan 2000),

43)**Segregations of Duties**, the Needs of Both Indonesian Governments and Private Sectors. *Pemerintah dan Swasta Perlu Berbagi Tugas,* MANAJEMEN, Mar 2000),

44)**The 812.500 Financial Models**, for the Autonomous Country Governments. *Tentang 812.500 Model Keuangan Otonomi Daerah,* (MANAJEMEN, Mar 2000),

45)**The Forecast of 10 Most Attractive Share Values**, in the Jakarta Stock Exchange, April 2000. *Prakiraan Harga Sepuluh Saham Unggulan di Bursa Efek Jakarta, April 2000,* (MANAJEMEN, Mar 2000),

46)**Let Us Merchandise the Intelectual Properties**!. *Jual "Intelectual Property" Saja!* (MANAJEMEN, Apr 2000),

47)**The Forecast of ten Most Attractive Share Values**, in the Jakarta Stock Exchange May 2000. *Prakiraan Harga Sepuluh Saham Unggulan di Bursa Efek Jakarta Mei 2000,* (MANAJEMEN, Apr 2000),

48)**HEATHROW**, London International Airport, Second Runway, *Q-Basic Algorithmic Engineer,* Beasley-Krishnamoorthy, Imperial College, Cambridge, Post Doctoral Project (2000),

49)**Why the Banking Sector Should be Healthy**, Before Others?. *Mengapa Perbankan Harus Sehat?* (MANAJEMEN, Mei 2000),

50)**Recycling of Wastes as a Negative Costing**. *Memanfaatkan Sampah atau Negative Costing,* (MANAJEMEN, Agt 2000),

51)**The Guidance for Indonesia Corporate Restructuring**. *Kiat Restrukturisasi Perusahaan Indonesia,* (MANAJEMEN, Agt 2000),

52)**Redefinition of the Ethical Comprehension**. *Pemaknaan Kembali atas Pemahaman Etika,* (MANAJEMEN, Sep 2000),

Steve Asikin ISBN 13: **978-1564912361**, ISBN 10: **1564912363**

53)**The Critical View the 1999's Indonesians Central Bank Financial Statement**. *Analisa Laporan Keuangan Bank Indonesia 1999*, (MANAJEMEN, Oct 2000),

54)**Becoming an "Achiever"**, not Just Common "Worker". *Bukan Sekadar "Kerja", Tapi Menghasilkan Karya!* (MANAJEMEN, Feb 2001),

55)**Ten Steps to Real Wealth!** *Sepuluh Langkah Menjadi Kaya,* (MANAJEMEN, Jun 2001),

56)**Creative Financial Breakthrough**, *Seminar Sehari tentang Terobosan Keuangan yang Kreatif,* (Gramedia-PPM dan MANAJEMEN, Jun 2001),

57)**The Marketing Practices of Financial Softwares**, for Publics Listed Corporation. *Praktek Pemasaran Program Perangkat Lunak untuk Keuangan Perusahaan Publik* (Univ Satyagama, Jakarta Juni 24, 2001).

58)**The Mathematical Financial Planning**, *Perancangan Keuangan Secara Eksak dengan Matematik.* (MANAJEMEN, Jul 2001),

59)**The Real Research and Development, not Rework and Destruction**. *Litbang yang Bukan "Sulit Berkembang",* (MANAJEMEN, Jul 2001),

60)**The Treatise on the Advantages and Disadvantages**, of Revolutionary Learning Methods. *Baik Buruknya: Revolusi Cara Belajar,* (MANAJEMEN, Jul 2001),

61)**Better Make an Indonesian "Corruption Prevention" Body**, not Just a "Corruption Watch". *Buat Indonesian Corruption Prevention (ICP), Bukan Indonesian Corruption Watch, ICW!* (MANAJEMEN, Okt 2001),

62)**Preparation of the Worlds Business Leaders**, for our Century and the Century After. *Mempersiapkan Manusia-manusia yang akan menjadi Khafilah Bisnis Dunia Abad ini dan Abad Mendatang.* (Univ Satyagama – Gregorio Araneta Univ, Jakarta, Oktober 27, 2001).

63)**The Good Corporate Governance**, is not just a Morale Condition. *Kepemerintahan Perusahaan yang Baik, Bukan Sekedar Moralitas,* (MANAJEMEN, Nov 2001),

64)**Corporate Healthiness and Cleanliness as a Challenge**. *Bersih, Sehat Perusahaan, sebuah Tantangan* (MANAJEMEN, Jan 2002),

65)**Developing the Cultural Base for Future Competitions**. *Budaya untuk Bersaing di Masa Depan* (MANAJEMEN, Jan 2002),

66)**"Catapult", "Bullet" or "Ballistic Missiles" Approac**h, for Corporate Healthiness and Cleanliness. *Pola "Ketapel", "Peluru Biasa" atau "Rudal" untuk Keuangan Perusahaan yang Bersih Sehat.* (MANAJEMEN, Feb 2002),

67)**The Mathematical Invesment Solution Mode**l, for Indonesian Foreign Debt and Its Corruption, *Solusi Matematika Investasi bagi Penyelesaian Masalah Utang Luar Negeri dan Korupsi di Indonesia.* (MANAGEMENT EXPOSE, Mar 2002),

68)**The Attitude, Skills and Knowledge (ASK)**, in Educational System Approach. *Sikap, Keterampilan dan Pengetahuan Menurut Pendekatan Sistemik Pendidikan,* (MANAJEMEN, Apr 2002),

Steve Asikin ISBN 13: **978-1564912361**, ISBN 10: **1564912363**

69)**Strategic Treatise on Regional Autonomy**, in Futurization, Digitalization, Securitization and Globalization Era in Accordance to the Awakening of the Mercantile Nations. *Kajian Strategik Otonomi Daerah : Era Futurisasi, Digitalisasi, Strukturisasi, dan Globalisasi dalam Kebangkitan Negara-Negara Dagang.* (Training for Provincial Government Regional Management, Indonesian Association of Provincial Governments, *Pelatihan Pengembangan Manajemen Pemerintah Daerah bagi Perangkat Pemerintah Propinsi, Angkatan II,* (APPSI), June 27, 2002.

70)**The Suggestion for the Doctoral Course**, for Islamic Financial Management System. *Usulan-usulan Perbaikan dalam Kurikulum Doktor Manajemen Keuangan Islam.* (December, 17-19, 2002, Canadian International Development Agency - CIDA, ADSGM, Universiti Utara Malaysia, Kedah, Da'arul Sintok).

71)**The Defensive Strategy for the Market Leader**, a Nokia Cellular Telephone's Case. *Strategi Bertahan Sang Pemimpin Pasar dengan Menggunakan Multi Merk, Kasus Telepon Seluler Nokia.* (March, 12, 2004, Univ Tarumanegara, Jakarta).

Steve Asikin ISBN 13: **978-1564912361**, ISBN 10: **1564912363**

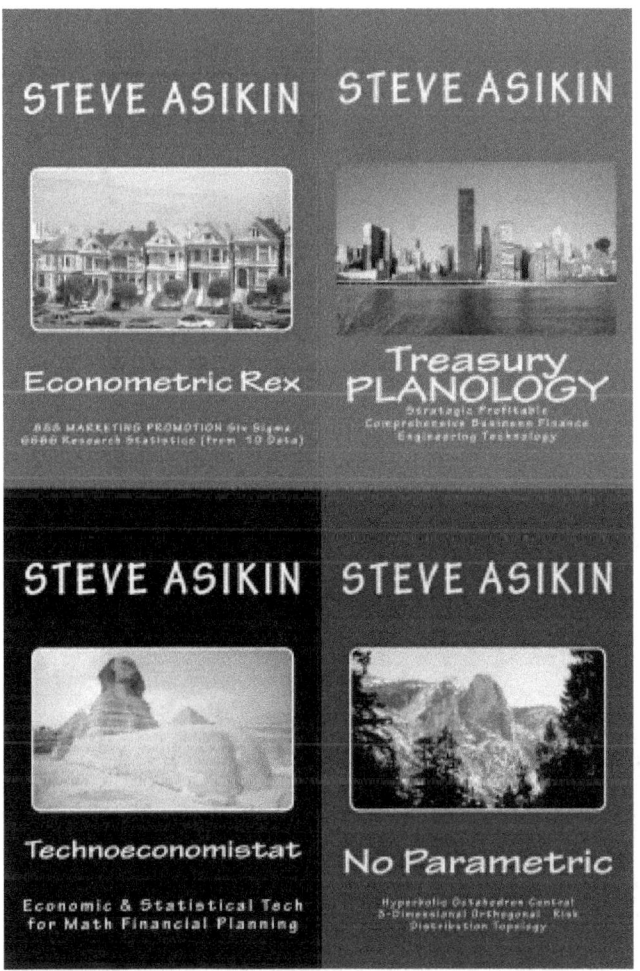

Steve Asikin ISBN 13: **978-1564912361**, ISBN 10: **1564912363**

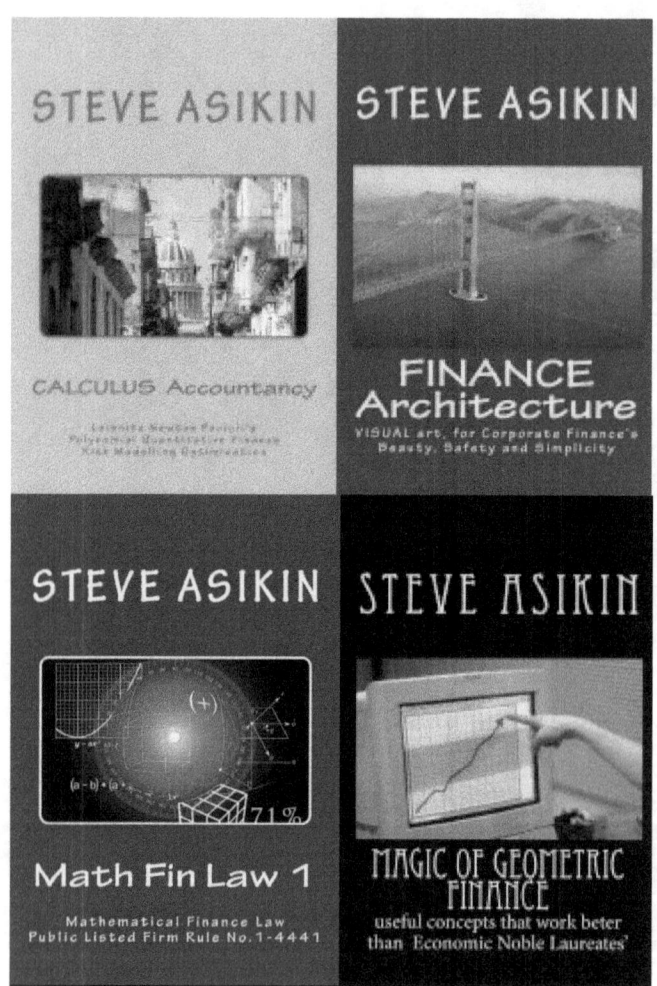

Steve Asikin ISBN 13: **978-1564912361**, ISBN 10: **1564912363**

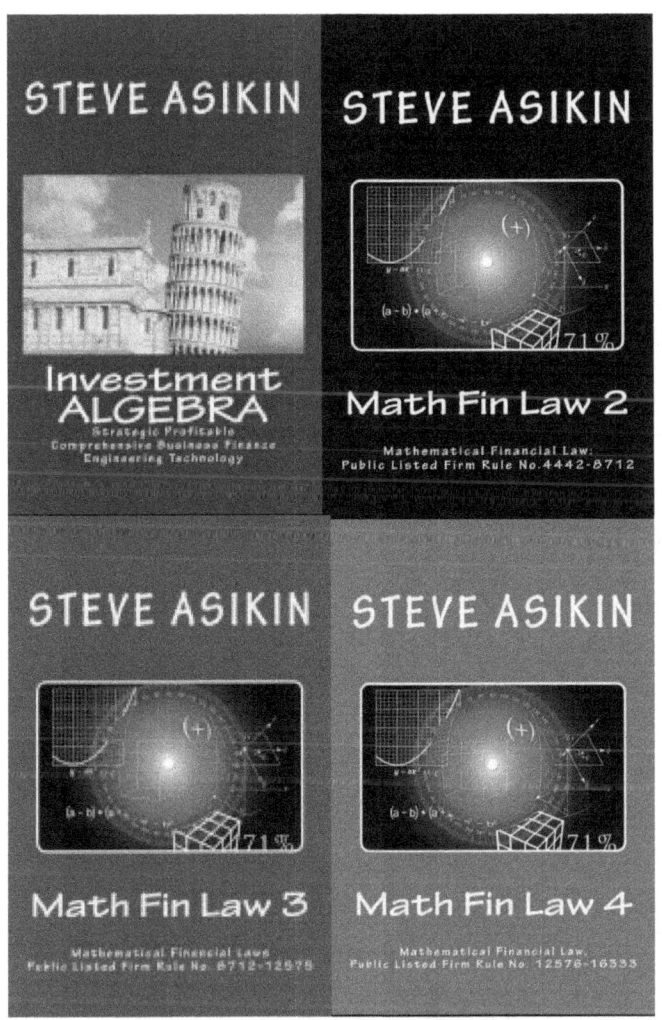

Steve Asikin ISBN 13: **978-1564912361**, ISBN 10: **1564912363**

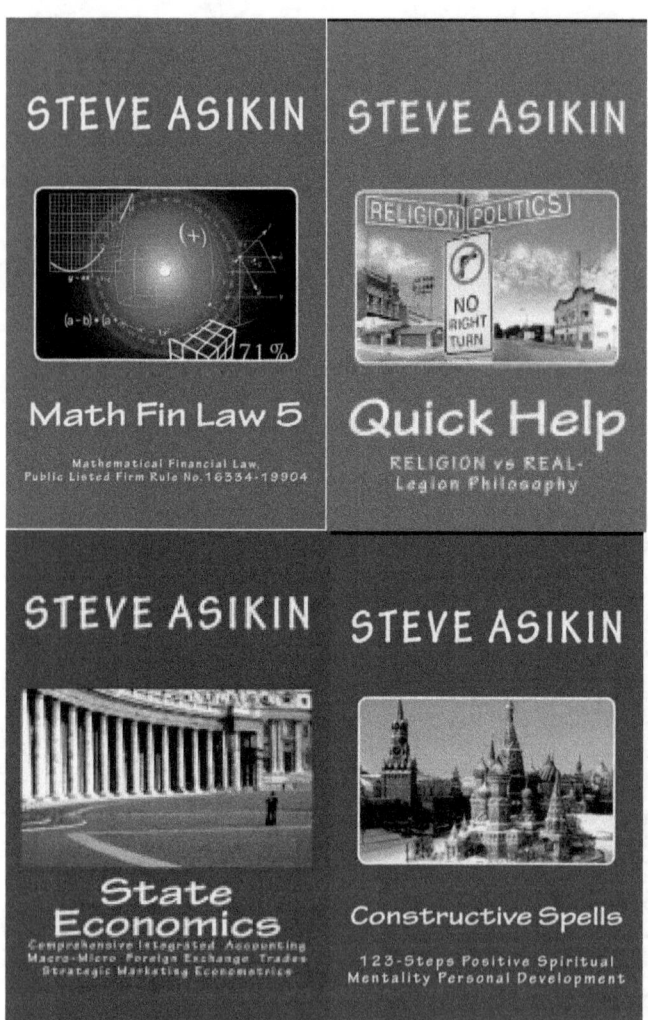

Steve Asikin ISBN 13: **978-1564912361**, ISBN 10: **1564912363**

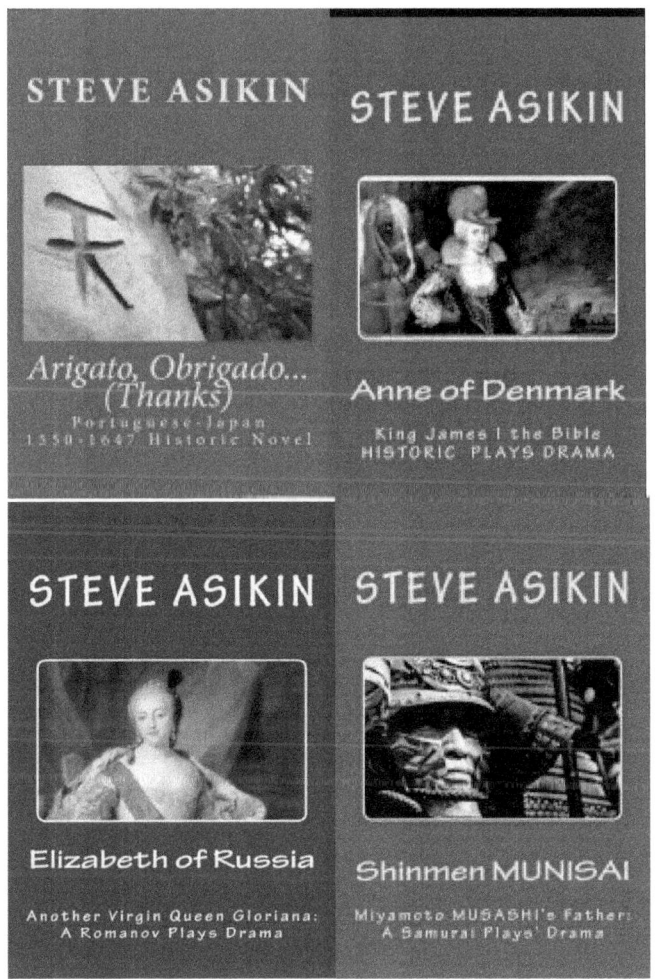

STEVE ASIKIN

*Arigato, Obrigado...*
*(Thanks)*
Portuguese-Japan
1550-1647 Historic Novel

STEVE ASIKIN

Anne of Denmark
King James I the Bible
HISTORIC PLAYS DRAMA

STEVE ASIKIN

Elizabeth of Russia
Another Virgin Queen Gloriana:
A Romanov Plays Drama

STEVE ASIKIN

Shinmen MUNISAI
Miyamoto MUSASHI's Father:
A Samurai Plays' Drama

Steve Asikin ISBN 13: **978-1564912361**, ISBN 10: **1564912363**

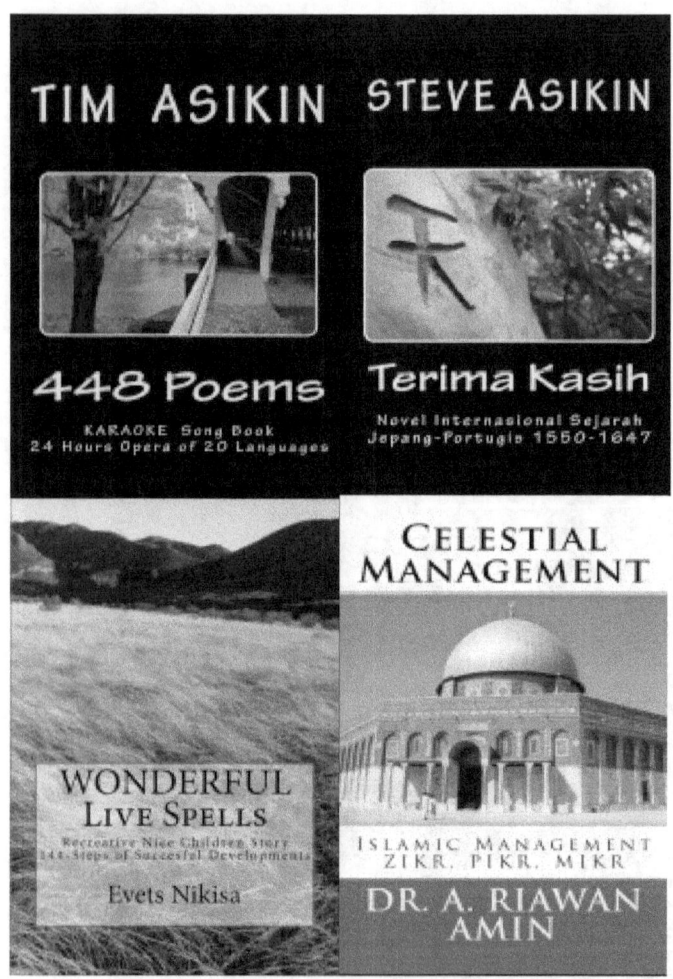

Steve Asikin ISBN 13: **978-1564912361**, ISBN 10: **1564912363**

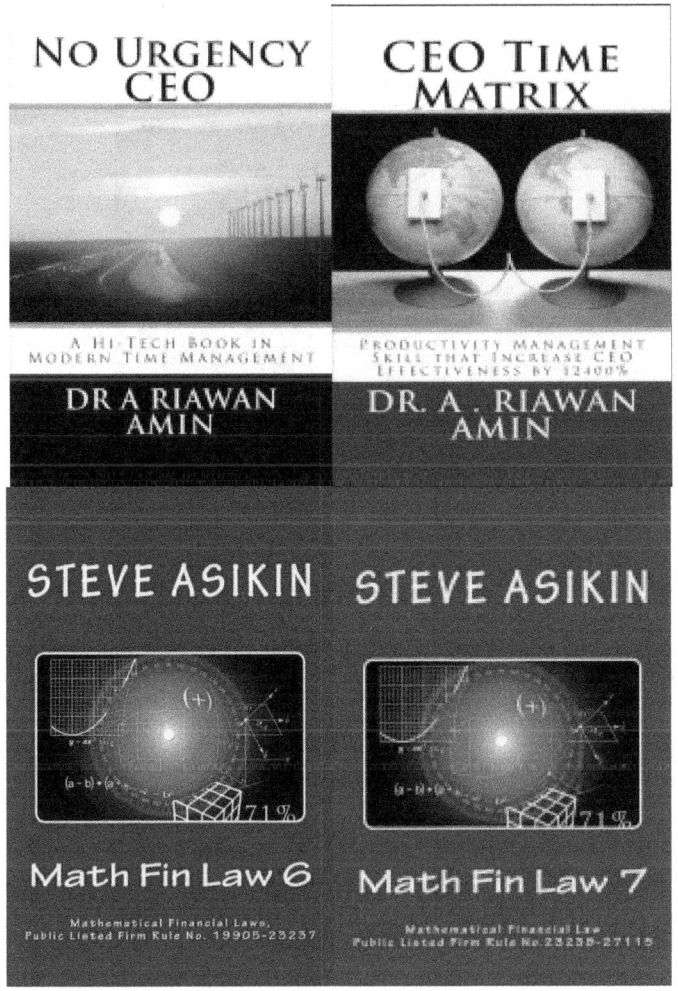

Steve Asikin ISBN 13: **978-1564912361**, ISBN 10: **1564912363**

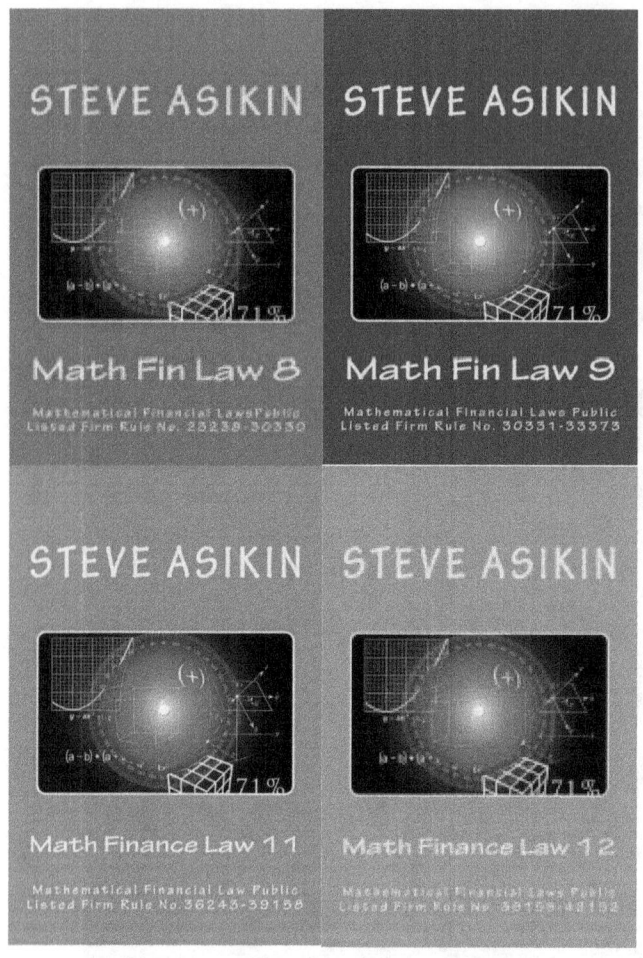

Steve Asikin ISBN 13: **978-1564912361**, ISBN 10: **1564912363**

**Math Finance Law 18**, *(Math Fin Law 18)*, Public Listed Firm Rule No.56125-58942

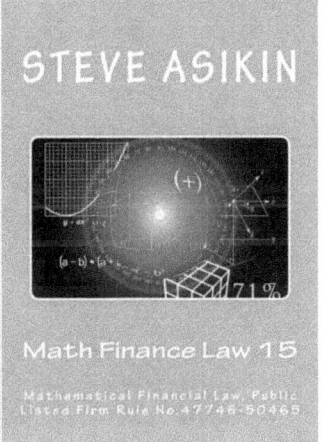

Steve Asikin ISBN 13: **978-1564912361**, ISBN 10: **1564912363**

**Math Finance Law 18**, *(Math Fin Law 18)*, Public Listed Firm Rule No.56125-58942

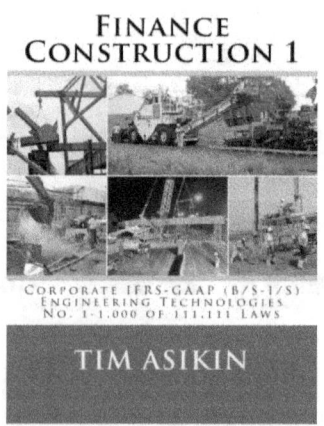

Steve Asikin ISBN 13: **978-1564912361**, ISBN 10: **1564912363**